Joseph Ames, William Herbert

Typographical antiquities

the history of printing in England, Scotland and Ireland: containing memoirs of our ancient printers, and a register of the books printed by them

Joseph Ames, William Herbert

Typographical antiquities
the history of printing in England, Scotland and Ireland: containing memoirs of our ancient printers, and a register of the books printed by them

ISBN/EAN: 9783742823908

Manufactured in Europe, USA, Canada, Australia, Japa

Cover: Foto ©Thomas Meinert / pixelio.de

Manufactured and distributed by brebook publishing software (www.brebook.com)

Joseph Ames, William Herbert

Typographical antiquities

Typographical Antiquities:

OR AN

HISTORICAL ACCOUNT

OF THE

ORIGIN AND PROGRESS

OF

PRINTING

IN

GREAT BRITAIN and *IRELAND*:

CONTAINING

MEMOIRS OF OUR ANCIENT PRINTERS,

AND A

REGISTER of BOOKS printed by them,

From the Year MCCCCLXXI to the Year MDC.

Begun by the late JOSEPH AMES, F.R. & A.S.S.
And SECRETARY to the SOCIETY of ANTIQUARIES.

Confiderably augmented, both in the MEMOIRS and NUMBER of BOOKS,
By WILLIAM HERBERT, of CHESHUNT, HERTS.

IN THREE VOLUMES.
VOL. II.

LONDON:

PRINTED FOR THE EDITOR, AND SOLD BY MR. T. PAYNE AND SON, MR. BENJAMIN WHITE, MR. L. DAVIS, MR. DODSLEY, MR. ROBSON, MR. NICHOL, MESS. LEIGH AND SOTHEBY, MR. COLLINS, AND THE REST OF THE BOOKSELLERS OF GREAT BRITAIN AND IRELAND.

MDCCLXXXVI.

HENRY DYSZELL, DISLEY, or DISLE,

KEPT a shop at the South-west door of St. Paul's, in St. Paul's church-yard. He was a Stationer by company, having been bound apprentice to W. Jones for 13 years, from St. John Baptist, 1563. He seems to have taken up his freedom soon after he was out of his time, for he had licence from the master and wardens of the company, 30 Dec. 1576, to print " An epitaphe vppon the death of Syr Edw. Saunders knight, late Chief baron of Thexchequer." June 20, 1577, he was fined " for printinge a booke vnlawfullie, and vnallowed, xx. s." The book is not mentioned. ." The Paradyse of daynty deuises" was printed by him this year; and is probably the book intended, as i do not find any licence he had for it. January 26, 1579-80, he had licence from the bishop of London, as well as from the wardens, to print " The englishe skoolmaster, set forth by James Bellot for teaching of straungers to pronounce Englishe." To all appearance he died a young man; for July 26, 1582, the court of assistants granted to Timothy Rider " A copie which perteined to Henry Disley deceased: Intituled, A Paradyce of Daintie deuices." This is the only book of his printing which has escaped the devastations of time.

" THE PARADYSE of daynty deuises. Conteyning sundry pithy pre- 1577. cepts, learned Counsels, and excellent inuentions, right pleasant and profitable for all estates. Deuised and written for the most part by M. Edwards[a], sometimes of her Majesties Chappell: the rest, by sundry learned gentlemen, both of honor and woorship, whose names hereafter folowe." The device, an angel crowned, holding in his right hand a flaming heart, and in his left a crosse. The motto about it, EGO SUM VIA ET VERITAS. " ¶ Jmprinted at London, by Henry Disle, dwellyng in Paules Churchyard, at the Southwest doore of Saint Paules Churche, and are there to be solde. 1577." On the next leaf is the coat armour in 12 escutcheons, with crest, supporters and motto of the right honourable Sir Henry Compton, knight, Lord Compton of Compton, to whom the book is dedicated by H. D. the printer. Under this coat of arms are the names referred to in the title-page, viz. Saint Barnard; E. O; Lord Vaux, the elder; W. Hunis; Jasper Heywood; F. Kindlemarsh; D. Sand; M. Yloop; but there are several other names and signatures under the several poems, as W. R; R. Hill; R. D; M. T; D. S; T. H. C. H; M. D; F. M; M. S; F. G; Lodowick Lloyd; E. S; M. K; M. Thorn; L. V; E. Oxf; W. H; R. L; T. Marshall; M. Edwardes; and one who subscribes with this anagram, " My lucke is losie." In the collection of Sir John Hawkins. Quarto.

[a] His Christian name was Richard, || c. 151. but all his poems in this book are according to Wood, Ath. Oxon. Vol. 1. || subscribed M. Edwards.

1578. The same mentioned by Dr. Percy[b]; and Mr. Warton[c], who mentions also an edition in 1573[d]; a misprint, probably, for 1578, seeing the dedication intimates its being first printed by this printer, who had not served out his apprenticeship 'till Midsummer, 1576.

ROBERT REDBORNE

PROBABLY the person inserted in the company's charter, by the name of Rob. Badborne; perhaps a mistake for Radborne. One John Radborne was apprentice to Reginalde Wolfe in 1559; but i do not find either of these names in the company's register afterwards.

"Arthur of Brytayn." On a ribbon; under which, "¶ The hystory of the moost noble and valyaunt knyght Arthur of lytell brytayne, translated out of frenshe in to englyshe by the noble Johan bourghcher knyght lorde Barners, newly Jmprynted." Over a cut of the knight and his squire, inclosed in a border of four odd pieces. On the back is the translator's prologue. On the next leaf begins "The table of thys present hystorie." Ten pages, double columns. Contains 174 leaves, with cuts, though numbered only Fol. lxix. "Here endeth the hystory of Arthur of lytell Brytayne. ¶ Jmprynted at London in Powles churche yearde at the sygne of the Cock by Robert Redborne." John Baynes Esq. Folio.

WILLIAM SERES

WAS at first concerned in printing with John Day, but probably not so early as Mr. Ames supposed[f]. Though they printed many books together, they do not appear to have formed a strict partnership; for Seres was occasionally concerned with Anthony Scoloker, Richard

[b] Reliques of ancient English Poetry, Vol. 11. p. 138.
[c] Hist. of Eng. Poetry, Vol. 111. p. 44, and 285, note i.
[d] Ibid p. 388.
[e] It is very probable this may be the same edition as that by one of the Coplands, but the printer's name changed in the colophon, after a certain number of copies had been worked off, in like manner as Chaucer's works, &c. See p. 351, and p. 557.
[f] See p. 615, and p. 622.

Kele,

WILLIAM SERES.

Kele, William Hill, and perhaps others; Day alſo printed ſome books on his ſole account, within the time of their connection, which ſeems to have been from their commencement in buſineſs, until ſome time in the year 1550; when they appear to have had ſeparate intereſts, as i find no book with both their names dated after that time. Their earlieſt books intimate their living together on Snow-hill for ſome time. In 1548 he appears to have had connections with Scoloker, and lived without Alderſgate. In 1549 he dwelt in Peter College.[r] Afterwards he ſettled at the weſt end of St. Paul's Church-yard, at the ſign of the hedge-hog, near Peter College, his former dwelling, which at this time was converted into a hall, for the worſhipful company of Stationers.

[r] The original foundation and preciſe ſite of this college is not now certainly known. Biſhop Tanner conjectured it might be the ſame with Holmes College, of which ſee Dugdale's Hiſt. of Paul's, p. 35, 36, edit. 1658; or Lancaſter College, ſee p. 37, 38; alſo, Notitia Monaſtica, p. 323. But, with great deference to this learned author's opinion, it may be queried, as to either of them, why, and when was the name changed: beſides at their ſuppreſſion in the reign of K. Edw. vi, each of them appear to have been granted to different proprietors. Peter College thus appearing to have been a ſeparate building from either Holmes or Lancaſter Colleges: i ſhall endeavour to ſhew to what uſe it was appropriated. There were ſeveral other chanteries in St. Paul's cathedral, beſides thoſe of Holmes and Lancaſter, enumerated in Dugdale to the number of 44. Theſe, "upon a viſitation made by Robert de Braybroke, biſhop of London, in 14 Rich. 2. were found ſo ſlenderly endowed, that divine ſervice could not be maintained thereby, according to the tenor of their foundations; the ſaid biſhop, therefore, for the better performance of thoſe duties,—by ſpecial licenſe from the king, did incorporate them." Theſe ſeem to be the chantry prieſts or chaplains mentioned in M.S. penes Will. Pierpont inſerted in the appendix to the Hiſt. of St. Paul's Cathedral, p. 253. " Omnes unà in Collegio S. Petri maneant & commeant, & Cameram & menſam illic teneant;" and again, "—in domo Collegii S. Petri, in menſa cum aliis commeant, & ſuis illic Cameris condormiant." To find out the ſite of this college, as near as may be, at this day; Stow, in the firſt edition of his ſurvey, p. 303, after pointing out the Dean's houſe, &c. as " directly againſt the (Biſhop's) palace," which was on the N. W. ſide of St. Paul's church-yard; and therefore, conſequently the deanery was ſituated on the S. W. ſide of it, ſays " Then is Stationers Hall on the ſame ſide, lately builded for them;" and in his ſecond edition is added, " in Place of Peters College." The wardens of the company in their account of expences from July, 1582, to July 1583, charge for " reparations, 1l. 5s. 6d. and in the next article " for a labourer cleaning the Dean's yard, 4d." This is a plain indication that their hall then adjoined to the Dean's yard. They had another hall after this, at the N.W. corner of Ave Maria lane, before called Abergavenny houſe. The ground on which their preſent hall is built, was probably part of the ſame eſtate. In Strype's edition of Stow, 1720, towards the concluſion of the then preſent ſtate of Caſtle Baynard ward, after the mention of Deans Court, by many called Prerogative Court, " More towards Ludgate Hill, and almoſt in the S. W. corner is a new court, called Paul's College, made uſe of for the ſinging men belonging to St. Paul's Cathedral" This account has in the margin R. B. (Richard Blome) who probably wrote his remarks ſoon after the rebuilding of London, after the great conflagration, & then the ſaid Paul's College might well be called a new court. This place ſeems to me to be nearly, if not preciſely, the ſame ſpot on which Peter College formerly ſtood.

WILLIAM SERES.

It is obfervable that Day's name is always fet firſt in thofe books they printed together, and yet Seres appears to have been his fenior in the Stationers' company, of which they both were members before the company's charter from Phil. and Mary, in 1556; and Seres was one of the old livery, whereas Day was only the firſt of the new. Seres accordingly took the lead of him in ferving the offices of warden and mafter; to the latter he had the honour of being chofen five times[h] in feven years.

Mr. Bagford fuppofed him to have been a fervant to Sir Henry Sidney, from his fetting up the fign of the hedgehog, which was Sir Henry Sidney's badge, or creſt. Mr. Strype fpeaks of him thus: "Sir William Cecil, principal fecretary of ſtate to king Edward, procured for him, being his fervant, a licence[i] to print all manner of private prayers called primers, as ſhould be agreeable to the common-prayer, eſtabliſhed in the court of parliament; and that none elſe ſhould print the fame. Provided, that before the faid Seres, or his aſſigns, did begin to print off the fame, he or they ſhould prefent a copy thereof, to be allowed by the lords of the privy-council, or by the lord chancellor for the time being, or by the king's four ordinary chaplains, or two of them. And when the fame was or ſhould be from time to time printed, that by the faid lords, and other of the faid privy-council, or by the lord-chancellor, or with the advice of the wardens of the occupation, the reafonable price thereof be fett, as well in the leaves, as being bound in paſt or board, in like manner as was expreſſed in the end of the book of common prayer." Strype alfo fays "Seres had a privilege for the printing of all pfalters; all manner of primers, Engliſh and Latin, and all manner of prayer books; that as this privilege was taken away by queen Mary, fo it was reſtored again by queen Elizabeth by means of lord Cecil, with the addition of the grant to him and to his fon[k] during the life of the longeſt liver, and that this gave occafion to a great cafe: for Seres, the father, in his latter years not being well able to follow his bufinefs, aſſigned his privilege, with all his preſſes, letters, ſtocks, and copies to one Henry Denham, for a yearly rent. Denham took feven young men of the company of Stationers to joyn him in the fame; but certain inferior perfons of the company, fetting up preſſes, more than England might bear, did print other mens copies forbidden to them, and privileged to others by the queen's letters patents.' Thefe endeavoured for their own gain to have the privilege taken away, preferring a petition to the privy-council, wherein they pretended that in juſtice it ſtood with the beſt policy of this realm, that the printing of all good and ufeful books ſhould be at the liberty for every man to do, without granting or allowing any privilege by the prince to the contrary. And they faid it was againſt law, and

---

[h] In 1570, 1571, 1575, 1576, and 1577. | Edw. vi. Eccl. Memor. Vol. 1, p. 378, 504.
[i] This licence bore date, 4 March, 7, | [k] His name was William, alfo.

that

that the queen ought not to grant any such. Seres upon this, in a petition to the lord-treasurer, urged against these men, that privilege for special books was ever granted by the prince; for that for the most part in all antient books we read these words, Cum privilegio ad imprimendum solum; and that many records might be found of the same, whereby it appeared, that the prince or magistrate had ever care to commit the printing of all good books, especially of the best sort, to some special men well known and tried for their fidelity, skill, and ability. Examples whereof might be shewed as well in England, as other christian countries. And that the reason hereof was, that printing of itself was most dangerous and pernicious, if it were not straitned and restrained by politic order of the prince or magistrate. This affair at last was made up by a friendly agreement. The expedient was this, that those who had privileges were to grant some allowances unto the company of stationers for the maintenance of their charge and their poor. This was about the year 1583."

He had liccence to print the following books, viz. From July, 1558, to July 1559. " The Image of idelnes. The Psalmes in mytre noted. Proverbis in mytre. A songe Exortinge to the laude of God. A ballet called Mercyes fortte." March 4, 1559-60 " ij Sermons of Caluyns." 1562—1563. " The perfett newes out of Fraunce. Dialogus cōtra papistem. A dyalogue in englesshe agaynste papystrye. The burnynge of Powlis, made by the besshop of Duram. Certayn noble storyes cōtaynynge Rare worthy matter. Onosandro platonico, Of the generall captayne and of his Office." 1564—1565; " Julius Cefears Cōmentaryes. A Catechisme of the County Palyntyne. The dyall of Agnes, cōtaynynge ȳ names in Greke, Laten and Englesshe." 1565—1566. " A Pasquin. A prodegious history." 1566—1567. " Peter Marters prayers vpon the hole psalter. Christian prayers and godly meditations." 1567—1568. The prodegious historye, by Geo. Grafton. Serten versis in laten by Hypocrits. Mr. doctor Haddons workes. A proclamation, Articles with ȳ edict in Fraunce, 4. Marche. 1567. The natures and properties of wynes." 1568—1569. " The shippe of saulf yarde. Arthographe, cōtaynynge the due order of Reason." 1569—1570, " The answere to the proclamation of the Rebelles in the Northe partes. Johānes *Caius de canibus, de animal um historia, de libris suis.* A very breafe & profitable treatise of the state of counsellers."

Besides the books printed with John Day, which may be found under his name, he printed these following:

" A Boke made by Johan Fryth, prysonner in the Tower of London/ 1548. answering vnto. M. Mores letter/ which He wrote against the fyrst lytle treatyse that Johan Fryth made concerning the sacrament of the body and bloud of Christ: vnto which boke are added in the ende the artycles of his Examination before the Bysshoppes of London/ Winchester and Lincolne/ in Paules Churche at London/ for whych John Frith was condempned and after brente in Smythfelde without Newgate/ the forth
day

day of July Anno 1533. Now newely reuifed corrected and printed in the Yeare of our Lord. 1548. the laft day of June. Deade men fhall ryfe agayne." The whole contains P 6, in eights. Colophon, " Jmprinted at London, by Anthony Scoloker. And Wyllyā Seres Dwelling wythout Alderfgate. Cum Gratia et priuilegio ad Jmprimendū' folum." On the laft leaf is Scoloker's device. W.H. Octavo.

1548. " A Notable collection of diuers and fōdry places of the facred fcriptures/ which make to the declaratyon of the Lordes prayer/ Comenly called the Pater nofter. Gathered by the famous Clerke Mafter Peter Viret/ Frenchman. And tranflated out of Frenche into Inglyfh by Anthony Scoloker. The. viii. Daye of June. Anno. 1548. Imprynted ———by Anthony Scoloker, Dwelling wythout Alderfgate. And Wyllyā Seres Dwellyng Jn the Elye rentes in holborne. Cum Gratia & Priuilegio—folum/ Per Septennium." With cuts. C 4, in eights. W.H. Octavo.

1548. " A new Dialoge called the endightment agaynfte mother Meffe. ———.Jmprinted———by William Hill & and William Seres,—in Paules Church yeard, At the figne of the grene Hyll. Anno M.D.XLVIII. 17 December. Cum Gratia" &c. Octavo.

1550. " The ymage of bothe paftoures, fette forthe by that moofte famoufe clerck, Huldrych Zwinglius, and now tranflated out of Latin into Englifhe, by John Vernon (*Veron*) Sinonoys. A moft fruitefull and neceffary boke, to be had and redde in all churches, therwyth to enarme all fymple and ignorant folkes, agaynft the raueninge wolues and falfe prophetes." At the end " ¶ Of the metynge of Mayfter John Hooper, byfhop of Gloceter, and of maifter doctoure Cole, quondam chaunceler of London, and now wardeyn of the new college in Oxforde." In 8 leaves. Cum priv. folum. Printed with Kele. Octavo.

1552. Fuller mentions " A Primer or Booke of priuate Prayer, &c. Ex officina Wilhelmi Seres, 1552." Church Hift. Book vii. p,1,385.

1553. " The Actes of the Apoftles, tranflated into Englyfl.e ᵞMetre, and dedicated to the Kynges mofte excellent Maieftye, by Chriftofer Tye, Doctor in Mufyke, and one of the Gentylmen of his graces mofte honourable Chappell, wyth notes to eche Chapter, to fynge and alfo to play upon the Lute, very neceffarye for ftudentes after theyr ftudye, to fyle theyr wyttes, and alfo for all Chriftians that cannot fynge, to reade the good and Godlye ftoryes of the lyues of Chryft hys Apoftles." 1553." Here are only the firft 14 chapters. N 4, in eights. At the end,

---

¹ Printed the fame year alfo by Richard Jugge, Octavo.
⁂ For further particulars concerning this book and it's author, fee Warton's Hift. of Eng. Poetry, Vol. III, p. 190, &c.

" Jmprynted

Jmprynted at London by Nycolas Hyll, for *Wyllyam Seres. Cum priuilegio ad imprimendum folum." W.H. Octavo.

"Certayne Pfalmes felect out of the Pfalter of Dauid, and drawen into Englyfhe Metre, wyth Notes to euery Pfalme in iiij. parts to Synge by F. ☞ S. ¶ Imprinted at London by vvyllyam Seres, at the fygne of the Hedge Hogge." In an architective compartment ufed by Nic. Hyll, on the fell of which is the date " 1553." It is dedicated, in Sternhold's Stanza " To the ryght honorable lorde Ruffell," and fubfcribed " Your lordefhyps humble órator Francys Seager." There are 19 pfalms, at the end of which is " A difcription of the lyfe of man, the worlde, and vanities therof," in the fame metre, which concludes with " Deo foli honor et gloria. q. F. S. Cum priuilegio——folum." W.H. Octavo. 1553.

" The pādectes of the Euangelycall Lawe. Comprifyng the whole Hiftorye of Chriftes Gofpell. Set forth by Thomas Paynell. Anno Dominj. 1553". In the fame compartment as the laft article ; but with N. H. in the place where the date was inferted in that. It is dedicated " To the ryghte worfhipful fyr John Baker knight, Chauncelour of the kinges Maiefties courte of firfte frutes and tenthes, & under Trefurour of Englande." This is a kind of harmony of the Gofpels, reciting the hiftories of the Evangelifts in chronological order. It is divided into three parts ; the firft containing the hiftory of Chrift's eternal generation, and of his coming into the world ; the 2d treats of what Jefus did and faid from his baptifm unto Maundy Thurfday; the 3d contains the hiftory of his going out of the world. Contains Fol. 196, befides the Dedication, and two tables at the end ; one of the titles of the feveral chapters, or fections in their order, the other alphabetical. "Jmprynted—— by Nycolas Hyll, for Wyllyam Seres, and Abraham Vele. Cum priuilegio——folum. Anno Domini. 1553." W.H. Octavo. 1553.

" THE DECADES of the newe worlde or Weft-India," &c. as p. 587. Colophon, ¶ "Jmprynted—— in Paules church-yarde at the figne of the Hedge hogge, by Wyllyam° Seres Anno DnI. M.D.LV." Quarto. 1555.

" The Jmage of gouernance——Anno M. D. LVI." In a compartment with W. P. on a fhield fupported by Cupids, at bottom. See it printed by T. Berthelet, p. 438. Contains 157 leaves befides the preface and table of contents prefixed, and an alphabetical table at the end. "¶ Jmprynted—in Powles Churchyarde at the figne of the Hedgehogge by Wyllyam Seres." W.H. Octavo. 1556.

---

* Printed alfo by him, without date.
° This book was printed by Will. Powell, and publifhed in fhares, in like manner as the Bible, Chaucer's Works, &c. See p. 515,

and p. 557. Some copies having their colophon with the name, &c. of Robert Toy, others that of Edward Sutton, in the place of Will. Seres.

" 1557

1557. "The Schoole of Vertue and booke of Good Nurture for Children, and Youth to learne theyre dutie by. Newly perufed, corrected, and augmented by the fyrſt auctour. F. S. (*Francis Seager*) wrote. With a declaration of the dutie of eche degree. Anno 1557." In verſe. Octavo.

1559. "AN EPITOME OF CRONICLES. Conteyninge the whole difcourfe of hiftories as well of this realme of England as al other coũtreys, with the fucceffion of their kinges, the time of their reigne, and what notable actes they did: much profitable to be redde, namelye of Magiſtrates, and fuch as haue auctoritee in commõ weales, gathered out of moſt probable auctours. Firſte by Thomas Lanquet, from the beginning of the worlde to the incarnation of Chriſte, Secondely to the reigne of our foueraigne lord king Edward the fixt by Thomas Cooper, and thirdly to the reigne of our foueraigne Ladye Quene Elizabeth, by Robert Crowley. Anno 1559. Londini, In ædibus Thomæ Marſhe." Mr Cooper proteſts againſt the "vnhoneſt dealynge" of this book in his next edition printed "in the houſe late Thomas Berthelettes," 1560. See p. 465, &c. The firſt edition was printed in 1549. See p. 453. Colophon, "Imprinted at London by William Seres at the Weſte Ende of Poules towarde Ludgate at the ſigne of the Hedgehogge and are there to be folde. 1559. the v. day of Apryll." W.H. Quarto.

"Pfalmes or prayers taken out of holy Scripture." &c. as page 357. Sixteens.

1560. "The Pfalter or Pfalmes of Dauid. Corrected and pointed as they ſhal be fong in Churches, after the tranſlation of the great Byble: With certayne addicions of Collectes and other ordinary feruice, gathered out of the booke of Common prayer, confirmed by Acte of Parliament, in the fyrſt yere of the raigne of our foueraigne Lady Quene Elizabeth. Anno Chriſti. 1560. Cum priuilegio." This title printed in red and black is incloſed in a border of four pieces: in the bottom one, W. S. in a circle with this motto, "DE SVPER OMNIA." On the back, "An Almanacke for xxii. yeres," beginning M.D.lix. On the next leaf " The order how the Pfalter is appointed to be read." On the back thereof " A Table for the order of the Pfalmes to be fayde at Mornyng and Euenyng prayer." Then a "Kalender" for the 12 months with the leffons at morning and evening prayer. The feruice is printed in like manner as in modern uſe to bind up fometimes with the Bible; but the collects have no directions annexed to them where to find the epiftles and gofpels. Contains, befides the prefixes, J 3, in fours. My copy wants the laſt leaf, which perhaps had a colophon, &c. W.H. Quarto.

The Pfalter was ufually printed feparately; that bound up with my copy of the feruice was printed by Jugge and Cawood, without date; but doubtleſs was printed alfo by Seres, whoſe privilege it was.

"¶ Aggeus the Prophte declared by a large commentary. The earneſt loue that I beare to thy houfe hath eaten me. Pfal. 69, John 2. Phinees hath

WILLIAM SERES.

hath turned awaye my anger becaufe he was moued with loue of me. Numer. 25. J. P. L. C. Jmprinted at London by Wyllyam Seres the 2 of September. 1560." In a compartment with two demi-cupids fupporting a blank fhield at the bottom. It is introduced with " A preface to all that loue the earneft promotynge of Gods glorye in hys Churche by true religion," which concludes thus : " Among many other thinges that J a poore workeman in gods houfe, wold haue fayde to encourage other workemen, and fpecially thofe that fhoulde be the chiefe buylders and pyllers of his church, thefe few things at this time fhal ferue, becaufe the printer makes haft, and J haue not leafure. James Pylkynton Maifter of S. Johns College in Cambridge," &c. Ff 6, in eights ; but Z is omitted. ¶ Jmprinted——by Wylliam Seres dwelling at the Weft ende of Poules church, at the figneof the Hedgehogge. Cum priuilegio——folum". W.H.
                       Octavo.

¶ ORARIVM SEV LIBELLVS Precationum per Regiam maieftatem, Latiné 1560. æditus. 1560. Cum priuilegio——folum." Red and black, in a compartment with a cherubic head on each jamb. L 4. in eights ; Roman letter. " Londini ex officina Wilhelmi Seres typographi. Cum priuilegio——folum. 1560." W.H.   Octavo.

" A primer or book of priuate praier, needful to be vfed of all faithfull 1560. chriftians.—Which book is to be vfed of all our louyng fubiects. Nou. xii."   Octavo.

" A Godly Boke wherein is contayned certayne fruitefull, godlye, and 1561 neceffarye Rules, to bee exercifed & put in practife by all Chriftes Souldiers lyuynge in the campe of this worlde. 1561. Cum priuilegio——folum." In the fame compartment as the Orarium—libellus. The prologue is thus addreffed, " John Govvghe Paftor to the Paryfhe of Saynete Peters in Cornehull of London, to the Chriftian Reader." At the end of it " God preferue the Queene. Fare well in the Lorde. J. G." Q 6 in eights. " Jmprynted—by Wyllyam Seres,—at the vvefte eude of Paules Churche, at the fygne of the Hedgehogge. Anno 1561. Cum priuilegio—folum." W.H.   Octavo.

" ¶ Three ... ᵖTreatifes, no leffe pleafant than neceffary for all men to 1561. read wherof the one is called the Learned Prince, the other the Fruites of Foes, the thyrde the Porte of reft. ¶ Jmprynted—by Wyllyam Seres,—at the weft ende of Poules at the—Hedghogge. Cum priuilegio—folum." The firft is written in four-lined ftanzas, and dedicated " To the Queenes highnes" by " Thomas Blondeville." C 3, in fours.

The fecond tract has a feparate title-page : " ¶ The fruytes of Foes. Newly corrected and cleanfed of many faultes efcaped in the former printing. Anno domini. M.D. LXJ. Cum priuilegio——folum.'" This

ᵖ The word torn off in my copy.

is recommended in 3 four-lined ſtanzas by " Roger Aſcham Secretory to the Queenes maieſtie, for the latin tongue." As we have but few ſpecimens of this learned man's poetry, and the book being very ſcarce, i have tranſcribed them below.¹ The treatiſe is in the ſame ſtanza, and dedicated " To the Queenes highnes" alſo by " Thomas Blundeuille." E 2, in fours.

The third tract has alſo a ſeparate title-page, which contains only " ¶ The porte of reſte." On the back,

" Lyke as the mightie Oke whoſe rootes, Jn thearth are fixed faſt:
Js hable to withſtande eache winde, That blowes moſt boyſtrous blaſt.
Euen ſo eache frowarde Fortunes happe That euer maie beryde:
The conſtant minde with vertue fraught is hable to abyde."

It is dedicated " To the true louers of wiſedome, John Aſteley, mayſter of the Queenes Maieſties Jewell houſe, and John Harington Eſquier" by " Thomas Blundeuille," This dedication is in the ſame ſtanza as the former treatiſes: but this tract is in proſe. They are all taken from Plutarch. K, in fours. The colophon, " Jmprinted—at the Weſt end of Poules at the—Hedgehogge, the vii. daye of June. An. domini. 1561. Cum priuilegio—ſolum." W.H.   Octavo.

1561. " THE COVRTYER OF COVNT BALDESSAR caſtilio diuided into foure bookes. Very neceſſary and profitable for yonge Gentilmen and Gentilwomen abiding in Court, Palaice or Place, done into Englyſhe by Thomas Hoby. Imprinted——by wyllyam Seres—at the——Hedghogge. 1561." In a compartment with Jugge's cypher at top, terminuſſes on the ſides, and 2 lions couchant at the bottom. On the back, " The contentes of the booke. The firſt booke, entreateth of the perfect qualities of a Courtier. The ſecond, of the vſe of them, and of merie Jeſtes and Pranckes. The thirde, of the condicions and qualities of a vvaytinge Gentillvvoman. The fourth, of the end of a Courtier, and of honeſt loue." On the next leaf is the printer's addreſs to the reader, acquainting him that the reaſon why it was not printed ſooner was becauſe " there were certain places in it whiche of late yearas beeing miſliked of ſome, that had the peruſing of it—the Author thought it much better to keepe it in darknes a while, then to put it in light vnperfect and in peecemeale to ſerue the time." On the back are ſome verſes by " Thomas Sackevylle in commendation of the vvorke." It is dedicated by the tranſlator " To the right honorable the Lord Henry

¹ " Of Engliſh bokes, as J could find,
  J haue peruſed many one:
  Yet ſo wel done vnto my mind,
  As this is, yet haue J found none.

  " The woordes of matter here doe riſe,
  So fitly and ſo naturally,

As heart can wiſhe or witte deuiſe
Jn my conceit and fantaſie:

" The woordes well choſen & well ſette,
Doe bryng ſuche light vnto the ſenſe:
As if J lackt J woulde not lette
To bye this booke for forty penſe."
                                                    Haſtinges

Haſtinges, ſonne and heire apparant to the noble Erle of Huntyngton.
—1556." Then, the dedication by the author " Vnto the reverend and
honorable Lorde Mychaell de Sylua Biſhop of Viſeo." Contains Z z 5, beſides the prefixes. On the laſt leaf is " A letter of ſyr I. Cheekes."
This letter, being omitted in the ſubſequent editions, for the ſingularity
of it's orthography, and the learned writer's opinion concerning our language, i have given entire below.' " Jmprinted——at the Weſt end of
Poules, at the——Hedghog." W.H. Quarto.

" A dialogue of the tiranie of the papiſtes, tranſlated out of Latine 1562.
into Englyſhe by E. C. 17. Septembris. 1562. W.S." William Bayntun Eſq. Octavo.

" Aggeus and Abdias Prophetes, the one corrected, the other newly 1562.
added, and both at large declared." The mottos the ſame as to Aggeus.
P. 692. " Jmprinted—by Willyam Seres. 1562." In a compartment
with the queen's arms at top, his mark on each ſide, and the Stationers'
arms at bottom. At the end of the preface, " I. P. L. C. D." Aggeus
contains Z, in eights.

Before Abdias is " A preface to all the Enemies of God, his woorde,
people and Religion, to leaue their wickednes: and to comforte the
good manfully too beare their madnes, and patiently to loke for Gods

' " To his louing frind M. Thomas Hoby. For your opinion of my gud will vnto you as you write, you can not be de-
ceiued: for ſubmitting your doinges to mi judgement. I thanke you: for taking this pain of your tranſlation, you worthilie deſeru great thankes of all ſortes. I haue taken ſum pain at your requeſt cheflie in your preface, not in the reading of it for that was pleaſaunt vnto me boath for the roundnes of your ſaienges and welſpeakinges of the ſaam, but in changing certein wordes which might verie well be let aloan, but that I am verie curious in mi freendes matters, not to determijn, but to debaat what is beſt. Whearin, I ſeek not the beſtnes haplie bi truth, but bi mijn own phanſie, and ſhew of goodnes.

I am of this opinion that our own tung ſhold be written cleane and pure, vnmixt and vnmangeled with borowing of other tunges, wherin if we take not heed bi tijm, euer borowing and neuer payeng, ſhe ſhall be fain to keep her houſe as bankrupt. For then doth our tung naturallie and praiſa-
blie vtter her meaning, when ſhe bourow-
eth no counterfeitnes of other tunges to

attire herſelf withall, but vſeth plainlie her own with ſuch ſhift, as nature, craft, experiens and folowing of other excellent doth lead her vnto, and if ſhe want at ani tijm (as being vnperſight ſhe muſt) yet let her borow with ſuche baſhfulnes, that it mai appeer, that if either the mould of our own tung could ſerue vs to faſcion a woord of our own, or if the old deniſoned wordes could content and eaſe this neede, we wold not boldly venture of vnknowen wordes. This I ſay not for reproof of you, who haue ſcarſlie and neceſſarily vſed whear occaſion ſerueth a ſtrange word ſo, as it ſeemeth to grow out of the matter and not to be fought for: but for mijn own de-
fens, who might be counted ouerſtraight a deemer of thinges, if I gaue not thys ac-
compt to you, mi freend and wijf, of mi marring this your handiwork. But I am called awai, I prai you pardon my ſhortnes, the reſt of mi ſaienges ſhould be but praiſe and exhortacion in this your doinges, which at moar leiſor I ſhould do better. From my houſe in Woodſtreete the 16. of July, 1557. Yours aſſured Joan Cheek." See Strype's Life of Sir John Cheke, p. 211.

goodneſs.

WILLIAM SERES.

goodnes. At the end of the book, " Iaco. P. Ep. D." Abdias contains J 3, in eights. " ¶ Jmprynted—at the weft ende of Paules Churche, at the ——Hedgehogge Anno 1562. Cum priuilegio——folum." W.H.
Octavo.

1562. All the Statutes of the Stannary. Menfe Maij. 32 leaves. Quarto.

1563. " The burnynge of Paules church in London in the yeare of oure Lord 1561. and the iiii. day of June by lyghtnynge, at three of the clocke, at after noone, which continued terrible and helpleffe vnto nyght.' VVere thefe greater finners, than the reft? No: I faye vnto you except ye repent, ye fhall all lykevvyfe perythe. Luc. 13. ¶ Jmprinted—at the weft ende of Powles, at the—Hedgehog." Pilkington then bifhop of Durham, foon after this conflagration, preaching at Paul's crofs, took an occafion in his fermon to point out feveral abufes by which the church had been profaned as the caufes of the Divine difpleafure, fuch as idle talking, buying, felling, fighting, brawling, &c. In order to refute this, a libel was printed with this title: " An Addicion vvith an Appologie to the caufes of brinnynge of Paules Church, the which caufes were vttred at Pauls Croffe by the reuerend Bysfhop of Durefme the viii. of June. 1561." This is reprinted and fet before the book, bearing the following head-title, which is the propereft: " A Confutation of An Addicion. vvith an Appologye written and caft in the ftretes of VVeft Chefter, agaynft the caufes of burnyng Paules Church in London : whych caufes, the reuerend Bysfhop of Durefme declared at Paules Croffe 8 Iunij. 1561." After this confutation, " ¶ Here folowe alfo certaine queftions propounded by him,' whiche are fullye althoughe fhortly aunfwered." Thefe are 13 in number. The whole contains S 4, in eights. " Jmprynted" &c. as in the title-page, " ¶ The tenth of March. Anno. 1563;" W.H. Octavo.

1563. " Onofandro Platonico of the generall captaine, and of his office, tranflated out of Greke into Italian, by Fabio Cotta, a Romayne, and out of Italian into Englyfh, by Peter Whytehorne." Dedicated to Thomas duke of Norfolk. Octavo.

1564. Preces priuatæ. Again 1568, 1573, &c. Sixteens.

1565. The Book of Common Prayer. Octavo.

1565. " THE FYRST FOVVER BOOKES OF P. Ouidius Nafos worke, intitled Metamorphofis, tranflated oute of Latin into Englifhe meter by Arthur Golding Gent. A woorke very pleafaunt and delectable.
" With fkill, heede, and iudgement, thys woorke muft bee red,
For els too the reader it ftands in fmall ftead.

' Confeffed by the Sexton not to be || p. 296.
fo before he died. See Heylin's Britan. || ' The author of the libel.

" Imprinted

"Imprinted at London by VVillyam Seres. Anno. 1565." In a compartment fomewhat like that of Aggeus and Abdias, but enlarged. On the back is a dedication " Too the right honorable—Lorde, Robert Erle of Leycefter, Baron of Denbygh, Knyght of the mofte noble order of the Garter, &c.—At Cecill houfe, the xxiij. of December. Anno 1564." Then a preface " Too the Reader" in verfe. Each book has a frefh fet of fignatures, with a colophon, fo as to fell feparate occafionally. The firft book contains Fol. 12; the fecond 14; the third 12; and the fourth 12; exclufive of the dedication and preface. " Imprinted ——at the vveft ende of Paules churche—at the hedgehogge. Cum priuilegio——folum." W.H. Quarto.

The Pfalter* or Pfalmes of Dauid, &c. as p. 692. Octavo. 1565

" Regina literata, five de fereniffimae dominae Elizabethae Angl. &c. narratio Abr. Hartuelii, Cantab." Octavo. 155.

" The eyght bookes of Caius Iulius Cæfar conteyning his martiall exployptes in the Realme of Gallia and the Countries bordering vppon the fame, tranflated oute of latin into Englifh by Arthur Goldinge G. ¶ Imprinted——Anno 1565." In the fame compartment as Aggeus and Abdias, 1562. It is dedicated " To the ryghte honorable Syr Willyam Cecill knight, principal Secretorye to the Queenes maieftie, and maifter of her highnes Courtes of wardes and liueries.——At Powles Belchamp the xii. of October. Anno. 1565.——Arthur Golding." After this dedication is a preface, defcribing the boundaries of Gaul in Cæfar's time; then a table of " Errours efcaped in printing." The commentaries contain 272 leaves befides the prefixes, and at the end " An expofition of the old names of the Countries, Cities, Townes, Hilles, and Riuers mencioned in this booke, together with a defcription of certaine Engines of warre vfed in thofe dayes by the Romanes. Jmprinted——at the weft ende of Paules Church, at the——Hedgehogge." W.H. Octavo. 1565.

" The iudgement of the godly and learned father, M. Henry Bullinger &c. declaring it to be lawful for the minifters of the church of Englande to weare the apparel prefcribed by the lawes and orders of the fame realme." In Englifh and Latin. Octavo. 1566.

Jo. Jones Doctor in Phificke, " His Diall of Agues, wherein may be feene, the diuerfitie of them, with their names, the Definitions, fimple and compound, proper and accidentall, diuifions, caufes and fignes." Maunfell; part 2, p. 14. Octavo. 1566.

* As this was a privileged book, and ferting it any more under this printer. very frequently printed, i fhall omit in-

The

1566. "The Primer and Catechifme fet out at large wyth many godly praires neceffarie for all faithfull Chriftians to reade. Imprinted" &c. William Bayntun, Efq. Octavo.

1567. "G. HADDONI, LEGUM DOCTORIS, S. REGINÆ Elizabethæ à fupplicum libellis, lucubrationes pafsim collectae & editæ. Studio & labore Thomæ Hatcheri, Cantabrigienfis. Ad Lectorem.
" Ne quæras lector, cur quæ iuueniliter olim
Scripferat Haddonus, publica facta legas.
Omnia coniunxi, quia mel, quia fuccus in illis,
Et condita fuo quæq; lepôre iuuant.
" Londini, Apud Gulielmum Serefium. 1567." On the back, " In D. Gualteri Haddoni præclariffimas lucubrationes, per Thomam Hatcherum collectas, epigramma Ioannis Freri medicinæ doctoris.—In eafdem, Abrahamus Hartwellus." It is dedicated " Clariffimo viro D. Gulielmo Cecilio, equiti, fereniffimæ reginæ Elifabethæ, cum à fecretis, tum à confilijs, & academiæ Cantabrigienfis cancellario dignifsimo.—Vale. Londini. 20. Nouembris. 1567. Tuæ dignitatis ftudiofifsimus Thomas Hatcherus." Next, " Thomas Wilfonus Legum doctor, Thomæ Hatchero fuo. S.D.P.——Vale Xen dochio apud fanctam Katherinam juxta turrim Londinenfem, Calendis Julij. 1567." Then an index of the orations and epiftles of which this book confifts. Contains befides thefe prefixes 350 pages, in double pica roman. W.H. Quarto.

1567. "D. GVALTERI HADDONI, LEGVM DOCTORIS, ferenifsimæ reginæ Elizabethæ à fupplicum libellis Poëmata. Studio & labore Thomæ Hatcheri, Cantabrigienfis fparfim collecta & edita. Londini, Apud Gulielmum Serefium. 1567." On the back, " Gualterus Haddonus pio Lectori." Contains 140 pages, in double-pica Italian, and a table to correct the errors of both books. W.H. Quarto

1567. John Calvin his " Treatife concerning offences, whereby at this day diuers are feared, & many alfo are quite withdrawen from the pure doctrine of the Gofpell : a worke very needfull and profitable. tranfl. out of Latine by Arthur Golding." Maunfell, p. 27. Octavo.

1567. " The. xv. Bookes of P. Ouidius Nafo, entytuled Metamorphofis, tranflated oute of Latin into Englifh meeter, by Arthur Golding Gentleman, A worke very pleafaunt and delectable. With fkill, &c. as p. 696. The earl of Leicefter's creft, encircled with the garter. Under it " 1567. Imprynted at London by Willyam Seres." It is infcribed " To the ryght Honourable—Robert Erle of Leycefter," &c. in a long epiftle, in Alexandrine verfe, declaring therein the morals of feveral of the fables. "——At Barwicke the. xx. of Aprill. 1567.——Arthur Golding." The preface " Too the Reader" is the fame as to The firft four books in 1565. Contains 200 leaves, befides the prefixes. At the end " Laus & honor foli Deo. Imprinted——at the Hedge-hogge." W.H. Quarto

"GODL-

"Godly Meditations vppon the ten Commaundementes, the Ar- 1567.
ticles of the fayth, and the Lords prayer. Whervnto is ioyned a treatife againſt the feare of death: Alſo a compariſon betweene the old
man and the new: the lawe and the goſple. &c. Made by Maiſter John
Bradford. Seene and allowed according to the Queenes Iniunctions.
1567. Imprinted at London by William Seres." It has prefixed "A
Preface ſhewing the true vnderſtanding of Gods word, and the right vſe
of Gods workes and benefites," by Tho. Leuer. Contains beſides,
96 leaves. W.H. Octavo.

"Three pleaſant treatiſes; the learned prince, &c. as p. 693. Quarto. 1568.

"A new boke of the natures and proporties of all Wines that are 1568
commonlye vſed here in England, with 'a confutation of an errour of
ſome men, that holde that Rhenniſh and other ſmall white wines ought
not to be drunken of them that either haue, or are in daunger of the
ſtone, the reume, and diuers other diſeaſes. Made by William Turner,
doctor of Phiſicke. Whervnto is annexed the Booke of the natures and
vertues of Triacles, newly corrected and ſet foorth againe by the ſayde
Wm. Turner. Imprinted——Anno 1568." The epiſtle dedicatory is
thus inſcribed: " To the right honorable Sir William Cecill Knight,
chiefe Secretarie vnto the Queenes Maieſtye, & maiſter of hir Highneſſe
Courts of Wardes and Liueries. &c. and ſomtime his Conſtudent in the
Vniverſitie of Cambridge, Wm. Turner wiſheth all proſperitie," &c.
The treatiſe of Wines ends on ſignature E 4; then follows that of the
virtues of treacles, with a ſeparate title-page. " This Booke ſheweth
at large the powers, commodities, vertues and properties of the three
moſt renouned & famous Preſeruatiues or Triacles: to weete, of the
great Triacle called in Latine Theriaca Andromachi: of the Triacle
Salt: and of it that is called by the name of the firſt finder out & maker,
Mithriadatium: Gathered out of Galen and Aëtius, by the labours &
paines of Wm. Turner, Doctor of Phiſicke. Newly corrected and
amended. Mellis ſi nimia eſt copia, bilis erit." The ſignatures proceed
to G 4, in eights. " ¶ Imprinted——at the Weſt ende of Paules, at
the——Hedgehogge." W.H. Octavo.

"Robert Filles Tranſlation of Praiers and meditations on all the 1568.
pſalmes of* Dauid." Maunſell, p. 50. Sixteens.

"Oratio pia, religiosa, et solatii plena, de vera animi tran- 1568.
quilitate. Authore Joanne Bernardo, Cantabrigienſi. Londini. Apud
Gulielmum Sereſium. 1568." Dedicated " Præclaro viro Petro Oſburno
theſaurarij Scaccarij illuſtriſsimæ reginæ Eliſabethæ, rememoratori fide-
liſſimo," by Thomas Bernard, the author's brother. Contains 118 leaves

---

* This book was reprinted by Richard || anatomie of the ſoule, &c."
Watkins, in 1590, and entitled " The ||

beſides

besides prefixes commendatory, neatly printed in double-pica Roman." "Excusum Londini apud Gulielmum Sereſium." W.H. Quarto.

1568. "Preces priuatæ, in ſtudioſorum gratiā collectæ, & Regia authoritate approbatæ. Math. 26. Vigilate & orate ne intretis in tentationem. ¶ Londini, excudebat Gulielmus Sereſius. Anno. 1568. Cum priuilegio." In a border of metal flowers. On the back, "Typographus Lectori." Giving him to underſtand that theſe were deſigned only for thoſe who underſtood the Latin tongue, &c. Then, "Index Contentorum in hoc libello. Calendarium vna cum varijs regulis. Catechiſmus puerilis. Precis Matutinæ. Litania. Preces Veſpertinæ. Preces ſelectæ: De Natiuitate Chriſti: De Paſſione: De Reſurrectione: De Aſcentione: De miſsione Spiritus ſancti: De ſancta Trinitate. Pſalmi (quos vocant) Pœnitentiales. Alij Pſalmi ſelecti. Selecti Pſalmi pro Regina. Flores Pſalmorum. Piæ Meditationes de vitæ fragilitate & ſpe reſurrectionis, &c. Preces Biblicæ. Preces ſeu Eiaculationes ſanctæ. Aliæ preces Piæ. Benedictiones menſæ, &c. Alia Miſcellanea in calce adiecta." Qq, in eights, half ſheets, beſides the prefixes. "Londini. Per Gulielmum Sereſium, ſub ſigno Herinacei in cæmiterio Paulino. Anno. 1568." W.H. Sixteens.

1569. The hurt of Sedition, how grieuous it is to a common welth. Set out by Sir John Cheeke Knight. 1549. And now newly pervſed and printed the 14. of December. 1569. ¶ Imprinted at London by Willyam Seres." In the ſame compartment as Aggeus &c. p. 695. On the back is the queen's arms crowned and gartered. M, in fours, half ſheets. At the end, "God ſaue the Queene. ¶ Imprinted—at the vveſt ende of Paules Church, at the—Hedgehogge. Cum priuilegio——ſolum." W.H. Octavo.

1569. "An Abridgemēt of all the Canonical books of the olde Teſtament, written in Sternholds metre by VV. Samuel miniſter. The names of the books are in the next leaf following. Imprinted——Anno Domini. 1569. Cum priuilegio——ſolum." In the ſame compartment as the laſt article. On the back of the table of contents, referred to in the title page, begins "The Preface;" at the end of which is a ſcheme to aſſiſt the memory in finding the ſtaff, or ſtanza where any particular ſubject is mentioned as taught in the ſaid preface, and ſome verſes following. Then a table of "Faults eſcaped in the Printing." The whole contains Z, in eights. At the end,

"The Prophets thus are finiſhed, and books Canonicall: Apocripha ye ſhall haue next, if death doo not me call.'

ʸ I do not find any more of this intended work printed, but with my copy is bound up much blank paper, and the whole goſ- pel of St. Mathew written in a ſimilar manner, though ſeemingly by another author.

Then

Then fome I hope wil finifhe that, which my good wil began:
All fhall be one in me or him, fo that you haue it than."
W.H. Octavo.
" An Awnfwere to the proclamation of the rebels in the North." 1569.
In verfe.

" An orthographie, conteyning the due order and reafon how to write or 1569.
paint thimage of mannes voice mofte like to the life or nature, compofed
by I. H. (*John Hart*) Chefter herault." Octavo.

" Moft Godly prayers, compiled out of Dauids Pfalmes, by D. Peter 1569.
Martyr. Tranflated out of Latin into Englifh by Charles Glemham, G.
Seene and allowed according to the order appointed. Imprinted 1569.
Cum priuilegio." In the fame compartment as Aggeus, &c. p. 695. On
the back are directions to correct the book in the arguments of the
14. and 15. prayers. The epiftle dedicatory is thus infcribed, " Iofias
Simler fendeth hartie greeting to the right worfhipfull and his finguler
good friend Doctor Herman Folkerfheimer Frifelander." Dated, " Zurike
the xx. day of June———1564." Ll 4. in eights. W.H. Octavo.

" JOANNIS CAII BRITANNI DE CANIBVS BRITANNICIS, LIBER VNVS. DE 1570.
RARIORVM ANIMALIVM ET STIRPIVM HISTORIA, LIBER VNVS. DE LIBRIS
PROPRIIS, LIBER VNUS. Iam primum excufi. Londini per Gulielmum
Serefium typographum. Anno. 1750." Thefe three books are neatly printed
in fmall pica Roman, and have diftinct title-pages and indexes, fo as to
fell feparate, occafionally. W.H Octavo.

" A very briefe and profitable Treatife declaring hovve many Coun- 1570.
fels, and what manner of Counfelers a Prince that will gouerne well
ought to haue. The book fpeaketh.
  " All you that Honours woulde atcheeue,
  And Counflers eke defire to bee,
  Of felfe loue flee the falfe beleeue,
  And learne my lore that you may fee
  What worthyneffe in you doth reygne,
  Such worthy ftate thereby tattayne.
" Imprinted at London by William Seres." On the back, " Firmo Ap-
poggio," over the earl of Leicefter's creft encircled with the garter, under
which is " 15 70." It is dedicated " To the ryght Noble Erle of Ley-
cefter, one of hir highneffe moft Honorable, wife, and graue Counfelers.
———From Newton Flotman, the firft of Aprill. 1570. Moft bound to
your Honour, Thomas Blundeuill." Contains Q, in fours, half Sheets. the
type, Roman; it's fize, Englifh. W.H. Octavo.

" Of Phifiognomie, a finguler difcourfe on all the members and parts of 1571.
man : alfo a treatife of the fignification of fundrie Lines feene in moft
mens

mens fore-heads, with diuers examples of his own knowledge, and a little treatife of Moles, feene in any part eyther of man or woman, fette forth by Tho. Hill." Maunfell, part, 2, p. 20. Octavo.

1571. "Articles of enquiry in the vifitation of the diocefs of London, by Edwyn bifhop of London. an. 1571." Tho. Baker's Maunfell, p. 5. Quarto.

1572. "The Key of knowledge, containing fundry praiers and meditations, to occupie the minde of well difpofed perfons, by Tho. Achelley." Maunfell, p. 86. Sixteens.

1573. "Preces privatae," &c. as p. 700. "Nouiter impreffæ & quibufdam in locis etiam auctæ.——Anno Domini 1573. Cum priuilegio Reginæ." H h, in eights, half fheets. A and B twice. "Londini. Per Gulielmum Seres, fub figno Erinacei in cœmeterio Paulino prope Ludgate. Anno Domini. 1573. & fælicifsimi Regni Reginæ noftræ Elizabethæ. 15. Cum priuilegio——folum." Neatly printed in fmall pica Roman; red and black. W.H. Sixteens.

1574. Printed again in 1574, and perhaps often.

1574. "¶ The true order and Methode of wryting and reading Hyftories, according to the precepts of Francifco Patricio, and Accontio Tridentino, two Italian writers, no leffe plainly than briefly, fet forth in our vulgar fpeach, to the great profite and commoditye of all thofe that delight in Hyftories. By Thomas Blundeuill of Newton Flotman in Norfolke. Anno. 1574. ¶ Imprinted——by VVillyam Seres." On the back, the earl of Leicefter's motto over his creft, encircled with the garter, under which is " 15 74." Dedicated "To the moft Noble Erle of Leycefter." H, in fours, half fheets, Roman letter. W.H. Octavo.

1575. "The xv. bookes of P. Ouidius Nafo," &c. as p. 698. This edition contains, befides the prefixes, 192 leaves, numbered only 191; but No. 113 is omitted. W.H. Quarto.

1576. Articles of enquiry by Edmonde, Archbifhop of Canterbury. Anno. 18. Eliz. Quarto.

1576. The hurt of Sedition,ª &c. as p. 700. newly perufed.ᵇ Octavo.

—— "Two godly and notable fermons, preached by the excellent and famous clarke, mafter John Caluine, in the yere 1555. The one concernynge pacience in aduerfitie, the other touchyng the moft comfortable affurance of oure faluation in Chryfte Jefu. Tranflated out of Frenche into Englyfhe." Text, 2 Tim: 1; 8, 9, 10. Octavo.

ª. This book was printed again in Quarto, 1641, with a preface, applicable to thofe times, and the life of Sir John Cheek, by ‖ Dr. Gerrard Langbane. See Ath. Oxon. Vol. 11, c. 220.

" Three

"Three moral treatifes," &c. as p. 693. Octavo.

"Pafquine in a Traunce. A Chriftian and learned Dialogue (contayning wonderfull and moft ftrange newes out of Heauen, Purgatorie, and Hell) Wherein befydes Chriftes truth playnely fet forth, ye fhall alfo finde a numbre of pleafaunt hyftories, difcouering all the crafty conueyaunces of Antechrift. Wherunto are added certayne Queftions then put forth by Pafquine, to haue bene difputed in the Councell of Trent. Turned but lately out of the Italian into this tongue by W.P. Seene and allowed according to the order appointed in the Queenes Maiefties Iniunctions. Luke 19. Verily I tell you, that if thefe fhould holde their peace, the ftones would cry. Imprinted——at the Weaft ende of Paules, at the——Hedgehogge." The preface informs us, that Celius Secundus Curio, an Italian, was the author; but i do not find the tranflator's name. After the preface are fome verfes " To the Reader," figned " Ber. Gar." Contains 112 leaves, befides prefixes. ," Imprinted——Cum priuilegio——folum." W.H. Quarto.

"¶ A newe booke containing the arte of ryding and breakinge greate Horfes, together with the fhapes and Figures, of many and diuers kyndes of Byttes, mete to ferue diuers mouthes. Very neceffary for all Gentlemen Souldyours, Seruingmen, and for any man that delighteth in a horfe." The preface is addreffed " To the right honorable——the L. Roberte Dudley, knight of the honorable order of the garter, and mafter of the Queenes highnes horfes, and one of her Maiefties mofte honorable pryuy counfayl, by Thomas Blundeuill." To which is annexed " A chapter to the reader touching the order obferued in thys booke." Next is " A Table declaring the contentes of thefe three bookes," into which it is divided. Then, " Examples recyted by Gryfon ——, commending the aptnes of a horfe to learne, &c. Whereunto is adioyned, who firfte founde out riding, & who inuented the bittes, & who firft vfed the feruice of horfes in the fielde." Contains, befides thefe prefixes, F 4, in eights. " ¶ Jmprinted——at the Weft ende of Poules, at the——Hedgehog." Afterwards are 50 wood-cuts of the halter and various forts of Bitts. W.H. Octavo.

" The fower chiefyft offices* belonging to horfemanfhippe, that is to fay, the office of the breeder, of the rider, of the keper, and of the ferrer" &c. By Thomas Blundevill, of Newton Flotman in Norfolk. About 1566. Quarto.

" The Practyfe of prelates. Compyled by the Faythfull & Godlye learned man, Wyllyam Tyndale. Imprinted——by Anthony Scoloker. And Wyllyam Seres. Dwellynge in the Sauoy rentes Wythoute Temple barre. Cum Priuilegio——folum." On the back begins W. Tyndale's preface. Contains J 4, in eights, half fheets. W.H. Sixteens.

* Each of thefe offices are treated of in Denham, Seres's Affignee, in 1580. feparate books. See them printed by Hen.

" A

"A Notable collection of diuers and fōdry places of the facred fcriptures, which make to the declaratyon of the Lordes prayer, Comenly called the Pater nofter. Gathered by the famous Clerke Mafter Peter viret, Frenchman. And tranflated oute of Frenche into Jnglyſh, by Anthony Scoloker, The. viij. Daye of June. Anno 1548. Imprinted —/ by Anthony Scoloker, Dwelling wythout Algerfgate. And Wyllyā Seres Dwellyng. Jn Elye rentes in holborne. Cum gratia et priuilegio—folum, Per Septennium." C 4, in eights; with cuts. On the laft leaf is Scoloker's device, Verbum Dei. W.H.   Sixteens.

——— "The image of idlenefs, contayning certain matters moued betwene Walter Wedlock and Bawdyn Bachelor. Tranflated out of the Troiane or Cornifh tongue into Englifh by Olyuer Oldwanton, and dedicated to the lady Luft."   Octavo.

——— "A fight of the Portugall Pearle, that is, THE AWNSVVERE OF D. Haddon Maifter of the requefts vnto our foueraigne Lady Elizabeth. by the grace of God quene of England, Fraunce, and Irelande, defendour of the faith, &c. againft the epiftle of Hieronimus Oforius a Portugall, entitled a Pearle for a Prince. Tranflated out of lattyn into englifhe by Abraham Hartwell, Student in the kynges colledge in Cambridge." On the back are nine diftichs of Latin verfes "Ad amicum P. Sextum. A.H." Next, an epiftle "To Mayfter[b] Shacklock." Then, "The preface," in which is introduced "Certaine fentences collected out of Oforius in Englyfhe"; and concludes "Dat. at Cambridge the. xxvij. of May. 1565. Abraham Hartwell." Contains befides, F 6, in eights. "¶ Jmprinted——at the weft ende of Paules Church, at the—Hedgehogge." W.H.   Octavo.

——— "A Goodly Dyalogue betwene Knowledge and Symplicitie. Imprinted—, by Anthony Scoloker And Willyam Seres. Dwellyng wythout Alderfgate. Cum gratia et priuilegio——folum." Begins on the back: in the octave ftanza. Eight leaves. W.H.   Octavo.

"A brefe Chronycle concerning the Examinacion and death of the Bleffed martir of Chrift, Sir Johan Oldecaftell the Lord Cobham, collected together by Johan[c] Bale." This title is over a wood-cut, reprefenting fir John in a warlike pofture, in a Roman habit, with a drawn fword, flamed at the point, in his right hand, and a fhield in his left, on which is depicted a crucifix with the V. Mary and St. John ftanding by it; about the whole is this infcription: "☞ Sir. Johan. Oldcaftel. ỹ. worthy. ✠ Lord. Cobham. and. moofte valyaunt. Warryoure. of. JESV. Chryfte. ✠ fuffred. death. at. London. Anno. 1418." Underneath: "In the latter Time fhall many be chofē, proued and purifyed

---

[b] The tranflator of Oforius's Pearl for a prince.
[c] To which is added in the head title before the preface, "out of the bokes and writtings of thofe Popyfh Prelates which were prefent both at his condempnatyon and iudgement."

by

by fyre yet fhall the vngodly lyue wickedly ftyll, and haue no vnderftanding. Daniel xij." On the back, " Imprinted——/ by Anthony Scoloker. And Wyllyā Seres, Dwelling wythout Alderſgate, Cum gratia & priuilegio—folum." I have two copies of this book, and have feen another, each of them in good condition, but wanting the title page.[4] G, in eights. " ¶ Thus endeth the brefe Chronicle concerning the examination and death of the bleffed martir of Chryſt, Sir Johan Oldecaſtel/ the Lord Cobham/ not canonifed of the pope/ but in the precioufe bloude of his lord Jefus Chriſt. Collected by Johan[e] Bale." On the laſt page: " ¶ Prophecies of Joachim, Abbas——.Ex compendario Guidonis Perpiniani/ de herefibus." W.H. Octavo.

The fame perſons printed another edition of this book, without date alſo, which was reprinted in the Harleian Miſcellany,[f] retaining the whole colophon of the original edition, printed in 1544. This probably induced the compiler of that miſcellany to ſuppoſe it to be the original; but that was printed without the printer's name, and reprinted in 1729, differing only in orthography from this edition by Scoloker and Seres. Octavo.

" The agreement of the holye Fathers—vpon the chiefeſt articles of the Chriſtian Religion," &c. as p. 580. Octavo.

" An aunfwere to the Proclamation of the rebells in 'the North." It is in verſe, and poſſibly may be by T. Norton. T. Baker's Maunfell, p. 76. Octavo.

" The Actes of the Apoſtles tranflated into Englyfhe metre," &c. as p. 690. " Imprinted ——in Powles churchyarde by Wyllyam Seres. Cum priuilegio——folum." Octavo.

" The ordenarye for all faythfull chriſtians to lead a vertuous and godly lyfe here in this vale of miferie. Tranflated out of Doutch into Inglyfh by Anthony Scoloker." Printed alfo by him, for " William Seres—in S. Botolphs parifh wythout Alderſgate. Cum gratia——per feptennium." With a number of pretty wood cuts.[e] Octavo.

" Certayne chapters of the prouerbs of Salomon drawen into metre by Thomas Sternholde[s] late grome of the kynges mageſties robes." Printed by John Cafe for him. Cum priuilegio——folum.[e] Octavo.

" The boke of Secretes of Albertus Magnus." M. S. account of ſome books omitted; ſent to Mr Ames. The ſize not mentioned.

[4] Theſe books and many others, have been thus defaced on account of the imaginary *portrait* on the title page.
[e] To which is added in the original edition in the ſecretary type, and without the printer's name, "and imprinted, Anno. Dom. 1544. and vi. die Auguſti." By which theſe editions may be diſtinguiſhed.
[f] Vol. 11. p. 233.
[s] See Warton's Hiſt. of Eng. Poetry, Vol. iii. p. 182.

HENRY

HENRY SMYTHE, or SMYTH

AT the fign of the Holy Trinity without Temble-bar, in faint Clement's parifh, was fon in law to Robert Redman, anno 1540. I find nothing printed either by or for him, but the following Lawbooks.

1545. " LYTYLTON TENVRES NEVVLY REVISED AND truly corrected Vuith a table (after the alphabete to fynde out brefely the cafes defyred in the fame) therto added uery neceffary to the readers. CVM PRIVILEGIO ad imprimendum folum per feptennium." Before the table is " A brefe deglaration of the table to the reders." At the end of it, " W.S." Then another title-page: " LYTYLTON TENVRES neulye impryntcd. An M.D. X L V." In a compartment with two boys kneeling on dolphins' heads, at the bottom. Contains, befides the table, 152 leaves. " ¶ Prynted at London by Henry Smyth dwellynge wythout Temple barre at the figne of the Trinitie. CVM priuilegio——folum." W.H. Octavo.

To this is annexed the " Olde tenures," without title-page or colophon, but the fame type.

1545. " NATVRA BREVIVM NEVVLY AND mooft treuly corrected, VVyth dyuers additions of ftatutes, boke cafes, plees in abatementes of the fayd wryttes and theyr declaratiōs, and barres to ý fame added & put in theyr places mooft conuenient. Imprynted in the yeare of our Lorde God M. CCCC. XLV. Cum priuilegio——folum." Contains Fol. 179, befides tables at the end. "¶ Imprynted——VVythout Temple barre in faint Clementes paryfhe by me Henry Smyth." W.H. Octavo.

1545. " Articuli ad narrationes nouas pertim formati." Begins on the back of the title-page; and contains D 2, in eights. " Hunc libellum excudebat Henricus Smythe, Anno M. CCCCC. XLV." W.H. Octavo.

1545. " INCIPIT PERVTILIS TRACTATVS magiftri Iohannis Parkins interioris Templi focii, fiue explanatio quorundam capitulorum in tabula huius libelli contentorum grandi cum diligentia et ftudio ac iuuenum informationem ualde neceffaria, nouiter edita magnaq; diligentia reuifa ac caftigata, in edibus Henrici Smyth. An. M. D. XLV." Contains y, in eights. " ¶ Imprinted——by me Henry Smyth dVVellyng VVythout Temple barre in faint Clementes paryfhe." W.H. Octavo.

1546. " The boke for a iuftyce of peace, neuer fo well and diligently fet forthe. Carta feodi. Returna Breuium. The manner of keeping a court lete.——Court Baron," &c. Octavo.

" ¶ Intra-

HENRY SMYTH.

"¶ Intrationum liber omnibus legum Anglie ftudiofis apprime necef- 1546.
farius, in fe complectens diuerfas formas placitorū, tam realium, perfo-
nalium, quam mixtorum, necnon multorum breuium tam executionum,
quam aliorum, valde vtilium ; nunc tandem in gratiam ftudioforum
maiori cura et diligentia quam ante hac reuifus ac emendatus, adiectis
etiam Jndice multo quam ante hac caftigatiore, cum nonnullus aliis
additamentis hactenus non excufis nec editis, cuius quidem Jndicis
ordinem feries alphabetica tibi demonftrabit. H. S." Beneath : " Excudebat Henricus Smythe, cōmorans extra Temple barre, in parochia
fancti Clementis, ad inter fignū fanctae Trinitatis, An. reftitutae falutis
M. D. XLVI." In an architecture compartment with W on the bafe of
one pedeftal, and R on the other. It had been Will. Raftal's; and
afterwards ufed by Tho. Gibfon, &c. On the back of the title-page. " A
briefe declaration of the table to the readers." At the end of the
table " W. S." Contains befides, " Fol. cc. xliiii." "¶ Explicit
opus excellentiffimum—— nouiter Jmpreffum Londoni in parochia fancti
Clemētis extra Temple barre impenfis honefti viri Henrici Smyth moram
fuam trahentis fub figno fancte Trinitatis Anno noftre redemptionis.
M. cccco. xLV. die vero primo menfis Nouembris." W.H.   Folio.

" Diversite de Courtes lour iurifdictions et alia neceffaria et ——
utilia." This has no printer's name or date ; but the title is in the fame
compartment as " Lytylton Tenures." W.H.   Octavo.

" Prenobilis militis cognomento Fortefcu," &c, as p. 549,   ——

Year books. 1——4, 7——11. Hen. vi. W.H.   Folio ——

****

ROGER CAR,

ABOUT the year 1548, printed " Herman, archbifhop of Colen, (1548)
Of the right inftitution of baptifm : alfo a treatife of matrimony,
and buriall of the dead. By Wolph. Mufculus. Tranflated by Richard
Rice."   Octavo.
Perhaps he printed; " Fiue Sermons of Bernadine Ochine of Sena :
Godlye, fruiteful, &c. Tranflated out of Italen into Englifhe Anno Do.
MDXCVIII. Jmprinted——by R.C. for William Beddell, at the fygne
of the George, in Pauls Church yarde."   Octavo.

NICHOLAS

NICHOLAS HYLL, or HILL

Dwelt in St. John's ftreet, near Clerkenwell.

1546. "Jnftitutions or principal grounds of the lawes and ftatutes of England." For Rob. Toy. Octavo.

1546. "John Clerke his declaration of certain Articles, with a recitall of the capitall errours againſt the ſame." Maunfell. p. 34. Octavo.

1546. "Three Godly and notable Sermons—by Wyllyā Peryn preeſt, bachelar of diuinite, & now ſet forth for the auauncemēt of goddes honor: the truthe of his worde, and edification of good chriſten people. ¶ Vos fratres preſciētes cuſtodite: ne inſipientiū errore traducti, &c. 2 Pe. vlti." In a compartment frequently uſed by him, with a cherubic head at top. The date " 1546." on the fell. Dedicated " Vnto the ryght reuerend father in god, and his ſpecial good lorde and mayſter Edmund (by the grace of god) byſſhope of London," &c. Next is an introductory addreſs " Vnto the Chriſtian reader." Then, " A Prayer vnto the Sacrament, in the Maſſe tyme." The text, " Math. 26. Marc. 14. Luc. 22. 1 Cor. 11. HOC EST CORPVS MEVM. Thys is my bodye." N, in eights, beſides the prefixes. Concludes with " Hec eſt fides catholica quam niſi quiſq; fideliter firmiterq; ſeruauerit abſq; dubio in eternum peribit. Jmprynted—in S. Johns ſtrete, by Nycolas Hyll, at the coſtes & charges of Robert Toye," &c. W.H. Octavo.

1548. " CATECHISMUS, That is to ſay a Shorte Inſtruction into Chriſtian Religion for the ſynguler commoditie and profyte of childrē and yong people. Set forth by the mooſte reuerende father in God Thomas Archbyſhop of Canterbury, primate of all England and Metropolitane. Gualterus Lynne excudebat." 1548." On the back is a neat wood-cut of K. Edward, ſitting on his throne, with a ſword in his right hand, delivering the bible with his left, to his clergy and nobility. It is dedicated " To the moſte excellent prince Edward vi.——kyng of Englande," &c. Concludes, " Youre Graces humble ſubiecte and Chaplayne Thomas Archbiſhoppe of Canterbury." Next is " The preface" with this head title " A ſhort Jnſtruction into Chriſtian religion——. Ouerſene & corrected[1] by the moſt reuerende father in God the Archbyſſhoppe of Canterburie." The work conſiſts of ſhort inſtructions, explaining the ten commandments, the creed, the Lord's prayer, baptiſm, the keys, and the Lord's ſupper; not by queſtion and anſwer, but

[1] In the edition without date, inſtead of " Ouerſene & corrected," is " Set forth."

in short sermons and homilies. Before each of these are neat cuts designed by Hans Holbein. Colophon: " ¶ Jmprynted——in S. Jhones strete by Nycolas Hyll. for Gwalter Lynne, dwellyng on Somers keye by Byllynges gate. Cum priuilegio——solum." Octavo.

The rates of the custom house, both inward and outward, newly corrected and imprinted for T. Petit. Octavo. 1550.

" A Declaration of the. x. holye commaundementes of almightye God, written. Exo. 20. Det. 5. Collectyd oute of the scripture Canonicall, by John Hopper. Cum and sc. John. 1. Anno M.D.L." In the compartment with a cherubic head at top, and N. H. on the sell. The address " Vnto the Christiane Reader" is dated " 5. Nouembris Anno. M. D. XLIX." The whole contains O 4, in eights. My copy wants the last leaf, on which, perhaps, was a colophon. W.H+. Octavo. 1550.

The English Bible was printed by him for Tomas Petit, and Robert Toy, and John Walley. See p. 556, and 587. Colophon: " Here endeth the whole bible after the translation of Thomas Mathew, with all hys prologues, that is to say, vpon the fiue bokes of Moses, the prophet Jonas, and to euery of the four euangelists, and before euery epistle of the New Testament, and after euery chapter of the boke are there added many playne annotations and expositions of such places, as vnto the symple vnlearned seame heard to vnderstand. With other diuers notable matters, as ye shall fınde noted next vnto the callender. Diligently perused and corrected. Imprinted at the coste and charges of certayne honest menne of the occupacyon, whose names be vpon their bokes." Folio 1551.

The Byble, &c. printed by Nich. Hyll. M. D. LII. Mr. T. Baker's Maunsell, p. 10. Quarto. 1552.

" Compendiosa totius anotomie delineatio aere exarata, per Thomam Geminum." Translated into English, by Nicolas Udall, in 1552, with 40 plates.[1] The preface is dated " at Windsor, 20th July." At the end: " Impryinted by Nycholas Hyll, dwellynge in Saynte Johns streate for Thomas Geminus." Folio. 1552.

" The Christen state of Matrymonye, wherein housebandes and wyues mayel erne to kepe house together wyth loue. ¶ The original of holy wedlock : whā, wher, how, and of whom it was instituted & ordeined : what it is : how it ought to proceade : what be the occasiōs, frute & commodities thereof. Contrarye wyse : how shameful & horrible a thīg whoredom and aduoutry is : How one ought also to chose him a mete 1552.

[1] This edition has the portrait of K. Edw. vi. on the title-page. See more under || Tho. Geminie.

NICHOLAS HILL

& conuenient fpoufe to kepe and increafe the mutual loue, trueth and dewty of wedloke: and how maryed folkes fhould bringe vp theyr children in the feare of God. Set forthe by Myles Couerdale. ¶ Wedlock is to be had in honoure amonge all men, and the bed vndefyled. As for whore kepers and aduouterers, God fhall iudge them. Hebre. xiii." In this edition Becon's preface is omitted; and the 7th, 8th, and 25th chapters have been added. Contains Fol. xci, and a table at the end. " Anno incarnationis Chrifti. M.D.LII. ¶ Jmprinted at London by Nycholas Hyll for Abraham Vele." W.H. See it before in p. 497, &c. Sixteens.
Some copies have " for Robert Toy;" others " for Rychard Kele."

1552. " A Chriften exhortation to cuftomable Swearers." Dr. Lort. See it without date. Sixteens.

1552. " The chriften rule, or ftate of the world from the hygheft to the loweft; and how every man fhould lyue to pleafe God in his callynge." Sixteens.

1553. " The Actes of the Apoftles, tranflated into Englyfhe Metre," &c. as p. 690.

1553. " The Pādectes of the Euangelicall Lawe." &c. as p. 691. Octavo.

1553· " Commō places of fcripture," &c. as p. 587. Printed alfo for John Walley. Octavo.

1553· " Catonis Diftica Moralia cum caftigat. D. Erafmi vna cum annatationibus & Scholiis Richardi Tauerneri Anglico Idiomate confcriptis in vfum Angliæ Juuentutis. Londini in ædibus Nicholai Montani. Anno falutis 1553." At the end, " Jmprinted——by Nicholas Hill for John Walley," &c. Octavo.

—— " The order of common prayer for mattins and evening fong throught the whole year." Printed for Abr. Vele. Quarto.

—— " Againft the tyrany of the bifhop of Rome." Quarto.

—— " The Statutes or ordinaunces concernynge Artificers Seruauntes, and Labourers, Journeymen and Prentyfes, drawen out of the common lawes of thys realme, fyth the tyme of Edwarde the fyrfte, vntyll the thyrde and fourth yeare of our dread foueraygne lorde Kynge Edwarde the vi. wyth the ftatute and order of the meafuryng of Landes. ¶ Jmprynted ——by Nycholas Hyll, in Saynt Johns ftreate." 30 leaves. W.H. Octavo.

—— " Of meafures and weights, and other commodities, very neceffary for all merchants." Octavo.

—— Cranmer's, catechifm. At the beginning, printed by G. Lynne; but at the end, for him. See it under G. Lynne. Octavo.

" A

A chriſtē exhortacion vnto cuſtomable ſwearers. ¶ What a righte and lawful othe is: whan, and before whom it ought to be. ¶ Jtem. The manner of ſaying grace, or geuyng thākes vnto God. Whoſoever heareth Goddes worde, beleue it, and do thereafter, ſhall be ſaued." In the ſame compartment as The declaration of the x commandments. At the end is "A ſhorte inſtructiō to the worlde," in ſeven-lined ſtanzas; addreſſed to kings, judges, counſellors, chamberlains, ſtewards, treaſurers, controllers, and prieſts. Contains E 4, in eights, the laſt blank. "¶ Jmprinted—by Nicholas Hyll, for Abraham Vele." W.H. Printed alſo for Richard Kele. Sixteens.

The Brittiſh proverbs, by Gruffyd Hiraethoe, a poet in North Wales, and lived about 1500. "Ou Synnwyr pen Kembero ygyd, Wedyrgynnull; ei gynnwys aegyfanſoddi mewn crynodab ddoſparthus a threfnodic awedrwy ddyual yſtryw. Gruffyd Hiraethoc prydydd o wynedd Is. Conwy," ends, "teruyn. Imprynted at London in St. Johns ſtreete, by Nycholas Hyll." This piece was in the poſſeſſion of W. Jones, eſq; Octavo.

RICHARD WYER.

I Cannot learn what relation he bore to Robert Wyer, who doubtleſs was the ſenior of this name; nor to John, who was his contemporary, and, if Mr. Ames was not miſinformed concerning the M. S. entry in his interleaved copy, they ſeem to have printed "The Ymage of both Churches" in partnerſhip. After theſe, there was Nicholas Wyer, but of whom i find no mention before 1565, nor after 1570. None of them appear to have been members of the Stationers' company.

"The rekenynge and declaracion of the fayth and belefe of Huldrike 1548. zwyngly, byſſhoppe of zuryk the chefe town of Heluitia, ſent to Charles. v. that nowe is Emperoure of Rome: Holdynge a parlement or Counſayll at Auſbrough with the chefe Lordes and lerned men of Germanye, The yere of our Lorde M. D. xxx. Jn the moneth of July. ¶ Come to me all that labour, &c. Mathe. xi. ¶ The veryte wyll haue the victory: Preſſe ye it downe neuer ſo ſtrongly. ¶ Tranſlated and Jmprynted at zuryk in Marche. Anno Do. M. D. XLVIII." E 2, in eigts. At the end, "Jmprynted by me Rycharde Wyer." W.H. Octavo. Though this book has the appearance of being printed at Zurick, yet as the type is the ſame that was uſed by Rob. Wyer, i apprehend it was printed at London, with the title-page cloſely copied from the Zurick edition, printed in 1543, except the date.

"The

1550. " The Debate Betwene the Heraldes of Englande and Fraunce, compyled by Jhōn Coke, clarke of the Kynges recognifaunce, or vulgerly called clarke of the Statutes of the ftaple of Weftmynfter, and fynyfhed the yere of our Lorde. M.D.L." On the back are three cuts, viz. " Lady Prudence" holding a lanthorn in her hand, over " The frenche Heralde," and " The Englyfhe Heralde." The fubject of this curious book, is a controverfy between the heralds of England and France, on a queftion fet forth by Lady Prudence, viz. " Which realme chriftened is moft worthy to be approched to honoure." The French herald claimeth preeminence for France on account of " Pleafure, Valyaunce & Ryches." This the Englifh herald confutes, and fhews that England excells therein; and at the end makes this requeft. " Right hygh and excellent princeffe Lady prudēce, pleafeth it you to haue in remēbraunce of the thynges which I haue refyted vnto you before. And how the forefayde hungarians callyng them felues Galles, and nowe frenchmen, be great bofters, blowers forth of fayned fables, & braggers, begynning alwayes warres & contencions," &c. M 6, in eights. Colophon, " Fynifhed by me John Coke, Le dernier Jour d'Octobre, Den yaer ons here duifent vij f. huadred negen en viertich. Finis Laudat opus. And Imprynted by me Rycharde Wyer, and be to be folde at his fhop in Poules churche yearde. Cum priuilegio——folum." On the back, a cut of St. John, like that ufed by Robert Wyer. Sixteens.

1550. " The Ymage of both Churches after the mofte wonderful and heauenly Reuelacion of Saincte John the Euangelyft, containing a very frutefull expofytion or Paraphrafe vpon the fame. Wherin it is cōferred with the other fcriptures & moft auctoryfed hiftories. Cōmpyled by by John Bale an exyle alfo in this lyfe for the faithfull teftimony of Jefu." This title, taken from Mr. Ames's interleaved copy, correfponding exactly with that printed by John Wyer the fame year, affords reafon to fuppofe the book was printed in like manner as the Great Bible, Chaucer's works, &c. See p. 515, and 557.

—— " A Dialogue wherein is contained the examination of the Meffe," &c. By W. Turner. Octavo.

JOHN WYER.

1550. " **T**HE YMAGE OF BOTH CHVRCHES AFTER The mofte wonderrful and heavenly Reuelacion of Saincte John the Euangelyft, &c. Go ye oute of Sodome," &c. as p. 674. This title is in the compartment

compartment of boys carrying one on the shoulders of four others. The first part contains 84, the second 88, and the third 83 leaves. " Jmprynted——Jn Fleteftrete By me John Wyer dwellyng alytle Aboue the Conduyte. 1550. Cum priuilegio——folum." Each part has a colophon to the fame purport. Mr. Ames, in his interleaved copy, has afcribed this to Richard Wyer: perhaps the fame edition, with their names changed after a certain number had been worked off. W.H. See p. 712. Quarto.

RICHARD JUGGE,

BRED to learning, and elected from Eaton to King's-College, in the year 1531, went from thence a fcholar. But being defirous to promote learning and virtue, about the time of the reformation acquired the art of printing, which he practifed in King Edward the fixth's time, and kept fhop, the fign of the Bible, at the north door of St. Paul's church, but dwelt in Newgate-market, next to Chrift's church. He had a licence from Government to print the New Teftament in Englifh, dated Jan. 1550. His editions of the old and new Teftaments are very neat, and adorned with many elegant initial letters and woodcuts. He was one of the original members of the Stationers' company, of which he was chofen warden in the years 1560, 1563, 1566; and mafter in 1568, 1569, 1573, and 1574. On the acceffion of Queen Elizabeth to the throne, he printed the proclamation, 17 Nov. 1558; and fome others afterwards: but the 7th Feb. following, John Cawood, who had been printer to the late queen, was conjoined with him in printing a proclamation for eating flefh; and they appear to have continued printing jointly the ftate papers from that time, though i do not find they had a patent for fo doing 'till 24th March, 1560; by which they were appointed printers to the queen's maiefty, with a falary of 6l. 13s. 4d, the fame as had been allowed to Cawood, by his patent from Q. Mary, for his life; and that feems to have been the reafon for his being joined with Jugge, in Q. Elifabeth's patent. He had licence from his Company to print the following books. viz. From July 1557, to July 1558. " The boke of Palmeftrye. The boke of Jofephus. The Kynge of Refte. The Kynge of Ryghtuoufnes. The fmall pfalter, in xvi, Englefhe. The fhorte dixtionary." In 1561. " The oration of Beze. Orders taken by my Lorde of Canterburye with the reft of the Cōmeffioners."

messioners." 1566—1567. " A defence of preeftis maryges." 1569—1570. *Directions* " for churchwardens and swornemen. Wether yt be mortal fynne to tranfgreffe Ciuil lawes. Dr. Storyes confeffion at his death." He furvived Cawood a few years, in which he enjoyed the privileges of the patent alone. The laft proclamation he printed is dated 16. Feb. 19 Eliz. 1576-7. He was in bufinefs about 30 years, and was fucceeded in it by one John Jugge, but what relation he was to him i cannot fay.

1547. " The booke of merchants, very profitable to all folkes to know of what wares they ought to beware of, for the beguiling of them. Newly perufed by the firft authore, well practifed in fuch Doynges. Read and Profite. Ad libelli repertorem. En fine mercatu merces, mercator, inemptas tolle tibi, merces has tibi fponte damno." Octavo.

1548. " A boke made by John Fryth pryfoner in the Tower of London, anfwerynge vnto M. Mores letter," &c. as p. 689. " Newly corrected and prynted after the fyrft copye, by Rychard Jugge, dwellynge in Powles church yarde, at the fygne of the Byble." 108 leaves. At the end " Per me Johan ffryth. ☞ Be wyfe as Serpentes, and innocent as Dooues. Jmprynted in the yare of oure Lorde. M.D.XLviii." W.H. Octavo.

1548. " A declaration of the twelue articles of the Chriften faythe with annotations of the holy fcripture, where they be grounded in, &c. By D. Vrbanum Regium." For Gwalter Lynne Octavo.

1550. " The Newe Teftament of oure Sauyour Jefus Chrift. Rom. xv. 14. Whatfoever things are wrytton afore tyme, are wrytten for oure learnynge. Tranflated by Wyllyam Tyndale, after the laft copye corrected, with fculptures," &c. A table of the epiftles and gofpels at the end. Mr. Bagford fays of this edition," The moft beautiful book i have feen of the black letter." Twenty-fours.

1550. " A declaratyon of the ten holy cōmaundementes of almyghtye God, wryten Exo. xx. Deu. v. Collected out of the fcripture canonycall, by John Houper, with certayne newe addifions made by the fame maifter Houper. Cum and fe: John. 1. Anno. M.D.L." In the compartment with the fun at top, and Whitchurche's mark at bottom. The epiftle " Unto the Chriftiane Reader," is dated " v. Nouembris. Anno. 1549." To which is added another, certifying that, " in the feuenth cōmaun dement thou fhalt fynde added more than was before for the confirmation of fuch diuorfement as many of late haue ben offended withal." Dated "From London. 28. Julii. 1550." Contains befides, Fol. cx, and a table at the end. " Jmprynted——in Paules church yarde at——ỹ Byble by Rycharde Jugge." WH. Octavo.

1550 The feconde books of Tertullian vnto his wyf tranflated into Englyfhe, wherī wherI is cõteined moſt goŭly cõuſel how thoſe that be vnmaryed, may choſe vnto them ſelfes godly cumpanyons, and ſo to liue quyetly in this worlde and bleſſedlye in the worlde to come. ¶ Let wedlocke be had in price in all poyntes, and let the Chaber be vndefiled, &c. Heb. 13." In N. Hill's compartment with the cherubic head at top, and N. H. in the ſell. It is introduced by an epiſtle from John Hoper, the tranſlator, "To the chirſtian Reader." D 10. in eights. "Jmprynted ——by Richarde Jugge, dwellynge in Paules churche yarde at——the Byble. M.D.L. Cum priuilegio——folum." W.H.   Octavo.

" The new teſtament of our ſauiour Jeſu Chriſt, faithfully tranſlated 1552. out of the greke, and peruſed by the commandement of the kings maieſtie, and his honorable councell and by them authoriſed." With the kings head printed in red ink; about it, " Edwardus Sextus aetatis *x." and beneath theſe words: " The pearle, which Chriſt commanded to be bought is here to be found, not els to be ſought." On the next leaf is the printers epiſtle " To the moſt puyſaunt and mightye Prince Edward the ſyxt,—Kyng of Englande," &c. An abſtract from which, &c. you may ſee in Lewis's Hiſt. of Eng. Tranſlations of the Bible, p. 195, &c. There are ſeveral wood-cuts interſperſed in the goſpels, and the revelation of St. John, ſaid to be done by Virgilio Sole, and others, but i find no mark to any of them. The text is printed in the Englifh, or black letter; the running titles, arguments, and notes in the Italic. At the end of the Acts is Jugge's device, a pelican in her neſt feeding her young ones, encompaſſed with this motto: " Pro lege, rege et grege." And about that: " Love kepyth the lawe, obeyeth the kynge, and is good for the commenwelthe." The whole ſupported by Prudence and Juſtice. At the bottom, his mark. Beneath, " Jmprynted——by ——Rychard Jugge, Vvith the king his moſt gratious priuilege." The epiſtles have a ſeparate title-page incloſed with 4 pieces, in the middle of the bottom one, R. I. encircled with this motto " Omnia deſuper." On the laſt page is Jugge's device, as at the end of the Acts. Underneath, " Jmprynted——in Paules churche yarde at the——byble. Vvith the kynge his mooſte gratious lycence, and priuilege, forbyddynge all other men to print or cauſe to be printed, this, or any other Teſtament in Englyſhe." ˙ W.H.+   Quarto.

" The newe Teſtament of oure Sauiour Jeſus Chriſte. Faythfully 1553. tranſlated oute of the Greke. ¶ With the Notes and expoſitions of the darke places therein." This title in red and black is over the king's portrait, as in the former edition, but printed all in black, with " Viuat" on one ſide, and " Rex" on the other. Beneath it, as in Lewis, p. 195. On the back, " The copye of the bill aſſigned by the kinges honorable counſell for the Auctoriſinge of this Teſtament. An

* Thus in Ames, but XV in Lewis, ‖ ſeen. My copy wants the Title-page. and in all the other editions which I have ‖

Abridgement

716   RICHARD JUGGE.

abridgement of which i have given below.¹ In the calender, the conversion of St. Paul is now inserted in black, and the Fish-days noted. The same translation, cuts, &c. as the foregoing article; but the arguments and notes are now in black letter. The colophon is at the end of the table to find the epistles, &c. to the same purport as the edition last year: his device on the back. WH.   Quarto.

1553.   "The whole Byble, that is the holye Scripture of the Olde and Newe Testament faithfullye translated into Englyshe by Myles Couerdale, and newly ouersene and correcte. M.D.LIII. ii. Tessa. iii. Praye for vs that the worde of God may haue free passage and be glorified. Prynted —at the North dore of Powles at the—Byble. Set forth with the Kinges mooft gratious Licence." This is the same edition as was printed at Zurich, 1550, for Andrew Hester." W.H.   Quarto.

1554.   An exhortation against rebellion.   Octavo.

1556.   "Fyue Homiles of late made by a ryght good and vertuous clerke, called master Leonarde Pollarde prebendary of the Cathedrall Churche of Woster, directed and dedicated to the ryght reuerende Father in God Rychard by the permissyon of God byshoppe of Woster his specyall good Lorde. Viewed, examined, and allowed by the right reuerende Father in God Edmonde Byshop of London, within whose diocese they are imprinted Cum priuilegio—solum." The privilege granted 1 July, 1555. With Cawood. Ames's M.SS.   Quarto.

1558.   A proclamation at the decease of Q. Mary; 17. Nov.—That the Epistle and gospel of the day should be read in English on 1. Jan. next; 13. Dec.—Against passing the seas; 21. Dec.——against disputes in Religion; f orbidding preaching, &c. 27. Dec. These he printed alone; the following ones jointly with John Cawood.

1558.   "Iohan. Indagines his pleasant Introductions vnto the art of Chiromancie, and Phisiognomie, with circumstances vppon the faces of the Signes. Also certaine Canons, vpon diseases and sicknes, where-

---

¹ Where as Richarde Jugge——hauinge the kynges——licence—to printe the newe Testamente in English, forbiddinge all other men——: was bounden by recognisance in a certen summe of money, that he shuld not sell—the same—when—prynted, but at such prices—as shuld be by vs appoynted. Forasmuch as the same bokes be nowe comen forth in printe, he hath made humble sute vnto vs, that we wolde cause the same to be perused—and that—we wold set vpon the same suche prices, as we shoulde thynke good—answerable with the charges & laboure that he hath susteyned——. Werfore, hauynge caused them to be ouerseen by persons mete—, who haue made relation vnto vs that the same bokes haue been printed with greate diligence, and care,—we haue estemed that the pryce of twentye and two pence for euerye boke in papers and vnbounde is a reasonable and conueniente price for the same——.The whiche price we haue agreed vpon with the sayde Richard, chargeinge him not to excede—the same. At Grenewhiche the. x. of June. M.D.Lii."

ᵐ See p. 568; also Lewis, p. 196, &c.

vnto

ynto is alfo annexed, as well the Artificiall, as natural Aftrologie, with the nature of the Planets, tranflated by Fabyan Wythers. Printed for R. Jugge. 1588." Maunfell, Part 11; p. 14. Octavo.

A proclamation concerning eating flefh; 7. Feb.——concerning the Sa- 1559. crament; 22 March.—of peace with France; 8 April. —againft players: the fame day. With John Cawood.

" Jniunctions Given By The Quenes Maieftie ¶ Imprinted——— 1559. in Powles Church yarde by Rychard Jugge and John Cawood, Printers to the Quenes Maieftie. Cum priuilegio Regiæ Maieftatis." Inclofed as the title to the Epiftles of the New Teftament, 1552. Contains, D 5, in fours. On the laft page, the colophon, as in the title-page. Alfo without date. W.H. Quarto.

Another edition entitled thus: " Jniunctions geuen by the Queenes 1556. Maieftie. Anno Domini. 1559. The firft yere of the raigne of our Soueraigne Lady Queene Elizabeth. Cum priuilegio" &c. W.H. Quarto.

Articles to be enquired in the vifitacion, in the firft yere of the 1559. raigne of our moft dread foueraigne Ladye, Elizabeth by the grace of God, of England, Fraunce, and Jreland Quene, defender of the fayth, &c. Anno 1559. "Inclofed as the Injunctions. Contains B 3, in fours. At the end, " God save the Qvene." On the back of the laft leaf " Jmprinted at London In Povvles Chvrch yarde by Rycharde Jugge and John Cawood Printers to the Quenes Maieftie. Cum priuilegio Regiæ Maieftatis." W.H. Quarto.

Another edition, with the title inclofed in like manner, but the 1559. colophon on the concluding page. W.H. Quarto.

Another, the title in a compartment with Cawood's mark on the fell; 1559. and the date M.D.LIX. inferted in the colophon. W.H. Quarto.

" The fourme and maner of makyng and confecratyng Bifhoppes, 1559. Prieftes, and Deacons. Anno domini 1559. Printed at London by Richard Jugge, printer to the Queenes Maieftie. Cum priuilegio." In the fame compartment as was ufed by Whitchurch to Erafmus's Paraphrafe on the New Teftament; but the arms of Q Katharine Parr and E. W. are taken out. Dr. Lort. Folio.

Mr. Ames mentions it printed with Cawood.

" The booke of Common Prayer,ª and adminiftracion of the Sa- 1559. cramentes, and other Rites and Ceremonies of the Churche of England.

ª The act for uniformity of Common Prayer paffed in the Houfe of Commons, 20th April; and by the Lords, on the 28th, 1559. To be in full force and effect from and after the feaft of St. John Baptift; but was performed in the queen's chapel on Sunday 12th May, four days after it was enacted. By this act K. Edward's Common Prayer-book was eftablifhed again, with certain alterations and additions made at the inftigation of M. Bucer. The alterations were in the form of the Litany: the additions were certain leffons to be ufed every Sunday in the year. See Strype's Annals, Vol. 1. p. 59, &c. Alfo Collier's Eccl.Hift. Vol. 11. Records, N. 77.

Londini

Londini in officina Richardi Jugge & Johannis Cawood. Cum Priuilegio Regiæ Maieftatis. 1559." The pfalms are printed feparately, with this title: " The Pfalter or Pfalmes of Dauid after the tranflation of the great Bible, poynted as it fhall be faid or fonge in churches. Londini. 1559. T. R." Folio.

Thefe no doubt were printed very frequently, and of all fizes.

1559. "Certayne Sermons appoynted by the Quenes Maieftie, to be declared and read, by all Perfones, Vycars, and Curates, euery Sonday and holy daye, in theyr Churches: And by her Grace's aduyfe perufed and ouerfene, for the better vnderftandyng of the fimple people. Newly Jmprynted in partes, accordyng as is mencioned in the booke of Commune prayers. Anno. M.D.Lix. Cum priuilegio Regiæ Maieftatis." In the fame compartment as the Injunctions and Articles. Contains A a, in fours. On the laft leaf, " Jmprinted—in Poules Church yarde by Richard Jugge and John Cawood prynters to the Quenes Maieftie. Cum priuilegio" &c. W.H. Quarto.

1559. "A Decree of the Priuye Counfell at Weftminfter. Anno. 1.5.5.9. xx. October. Articles agreed vppon by the Lordes and other of the Quenes Maiefties pryuy Counfayle for a Reformation of their feruauntes in certayne abufes of apparell, therby to gyue example to al other Lordes, noblemen and Gentlemen." With Cawood. Broad fheet.

1559. "ANNO PRIMO REGINÆ ELIZABETHE. At the Parliament begunne at Weftminfter, the. xxiij. of Januarie, in the fyrft yere of the raigne of our Soueraigne Lady Elizabeth,—And there proroged till the. xxv. of the fame moneth, and then and there holden, kept, and continued, vntyll the diffolution of the fame, being the eight day of May then next enfuyng, were enacted as foloweth. 1559." In a flourifhed compartment, with the queen's arms at top, Agriculture on one fide, and Pallas on the other. 52 leaves. " Imprinted——by R-Jugge and J-Cawood, Printers to the Queenes Maieftie. Cum priuilegio" &c. W.H. Folio.

1560. " The Pfalter or Pfalmes of Dauid. Corrected and pointed as they fhalbe fong in Churches, after the tranflation of the great Byble: With certayne addicions of Collectes and other the ordinary feruice, gathered out of the booke of Common prayer, confirmed by Acte of Parliament in the fyrft yere of the raigne of our foueraigne Lady Quene Elizabeth. Anno Chrifti. 1.5.60. Cum priuilegio." Inclofed like the Epiftles in the New Teftament, 1552; only W. S.° inftead of R. I. in the fell. On the back " An Almanacke for. xxii. yeres. M.D.lix.—M. D. lxxx." Contains the order for morning and evening prayer; the Litany; and the

* This W. S. feems to indicate its being printed by affignment from Will. Seres, || who had an exclufive privilege for printing the Pfalter.

collects,

collects, but without even references to the epistles and gospels. 39 leaves, and another perhaps with a colophon. To this is annexed
"The Pſalter, or Pſalmes of Dauid after the tranſlation of the great Bible, poincted as it ſhall be ſōg in Churches." In a compartment with the ſtory of Piramus and Thiſbe. At the end, "Imprinted at London by Richarde Jugge and John Cawood." W.H.   Quarto.

"A proclamation againſt breaking or defacing of Monuments of 1560. Antiquitie, being ſet vp in Churches, or other publike places, for memory and not for ſuperſtition.——Yeuen at Windſor the xix. of September, the ſecond yeare of her Maieſties raigne. God ſaue the Queene. Imprinted—by R-Jugge & J-Cawood, Printers to the Queenes Maieſtie. Cum priuilegio" &c. Weaver's Funeral Monuments, p. 52, ——54.

"Proteſtation FAICTE DE LA PART DV ROY TRESCHRESTIEN par ſon 1560. Ambaſſadeur, reſident pres la Royne d'Angleterra, a ſa Maieſté, & aux Seigneurs de ſon Conſeil. xx. April. Anno Domini. M. D. LX." Nine leaves.

"Reſponce a la proteſtation par L'ambaſſadeur du Roy Treſchreſtien 1560. de la part dudict Roy ſō Maiſtré, a la Royne d'angleterre, le vingtieſme Jour d'apuril. Anno. 1.5.60." Thirty one leaves. On the laſt page, "A Londres Par Rychard Jugge, & Jehan Cawood imprimeurs pour la Maieſtiè de la Royne. Lan M. D. LX." W.H.   Octavo.

"A Declaration of certain principal Articles of Religion, ſet out by 1561. the Order of both Archbiſhops, Metropolitans, and the reſt of the Biſhops, for the Unity of Doctrine to be taught and holden of all Parſons, &c. Imprinted——by R-Jugge, Printer to the Queenes Maieſtie. Cum privilegio" &c.   Hiſt. of the Reformation, Part II. Records, p. 336.

"A Compendious and moſte marueylous Hiſtory of the latter 1561. times of the Jewes communeweale, beginnyng where the Byble or Scriptures leaue, and continuyng to the vtter ſubuerſioñ and laſte deſtruction of that countrey and people. Wrytten in Hebrewe by Joſeph Ben Gorion, a nobleman of the ſame countrey, who ſaw the moſt thinges hym ſelfe, and was auctour and doer of a great part of the ſame. Tranſlated into Engliſhe by Peter Morvvyng of Magdalen Colledge in Oxford. And nowe newly corrected, and amended by the ſayde tranſlatour. Anno. 1561." At the end of "The Epiſtle to the Reader," on a ſpare leaf, is Jugge's device, reduced. Contains beſides, Fol. cclix. On the laſt page "Jmprinted——by Rycharde Jugge, Printer to the Quenes Maieſtie. 1561. Cum priuilegio" &c. W.H.   Octavo.

"The Arte of Nauigation, Conteynyng a compendious deſcription of 1561. the Sphere, with the makyng of certen Jnſtrumentes and Rules for Nauigations: and exemplified by manye Demonſtrations. Wrytten in Spanyſhe

Spanyſhe tongue by Martin Curtes, And directed to the Emperour Charles the fyfte. Tranſlated out of Spanyſhe into Englyſhe by Richard Eden. 1561." Incloſed like the Injunctions, &c. It is dedicated by the tranſlator " To the ryght worſhypfull ſyr VVyllyam Garrerd Knyght, and Maſter Thomas Lodge, Aldermen of the Citie of London, and Gouernours of the honorable felowſhyp or focietie,—,for the diſcouery of Landes—vnknowen," &c. Contains beſides, Fol. lxxxiiii, and a table at the end. " Jmprinted—in Powles Church yarde, by R-Jugge, Printer to the Quenes Maieſtie." W.H. Quarto.

(1561.) The New Teſtament. The ſame tranſlation and notes as in 1552. " An Almanacke for. xxiiii yeres. M.D. lxi.—M.D. lxxxiiii." W.H. Octavo.

(1561.) " An Oration made by Maſter Theodore de Beze, Miniſter of the word of God, accompanyed with. xi. other Miniſters and. xx. deputies of the reſourmed Churches of the Realme of Fraunce, in the preſence of the King, the Quene mother, the king of Nauarre, the Princes of Conde, and of La Roche ſur yon, Monſieur de Guiſe, the Conſtable, and other great Princes and Lordes of the kinges councel, being alſo preſente. vi. Cardinalles, xxxvi. Archbiſhoppes and Biſhoppes, beſydes, a great number of Abbots, Priours, Doctours of the Sorbone & other Scooles: Tueſday the. ix. day of September. 1561. in the Noonnery of Poyſſi. Truely gathered & ſet forth in ſuch ſorte as it was ſpoken by the ſaid de Beze. Wherunto is added a brief declaration exhibited by the ſayde Beze, to the Quene Mother, the next morowe after the makyng of the ſayd Oration, touchyng certain poyntes conteyned in the ſame." E 4, in eights. On the laſt page, " Jmprinted——in Powles Churchyarde by R-Jugge printer to the Quenes Maieſtie.. Cum priuilegio" &c. W.H. Octavo.

1562. A Proclamation for obſerving certain Statutes. With Cawood. W.H.. Quarto.

(1562.) " A Declaration of the Quenes Maieſtie, Elizabeth, by the Grace of God, Quene of England, &c. Conteyning the Cauſes which haue conſtrayned her to arme certeine of her Subiectes, for Defence both of her owne Eſtate, and of the moſt Chriſtian Kyng, Charles the Nynth, her good Brother, and his Subiectes. Septemb. 1562. Jmprinted—— in Powles Churchyarde, by R-Jugge and I-Cawood, Printers to the Quenes Maieſtie. Cum priuilegio" &c. Thirteen pages. Harleian Miſcel. Vol. III, p. 177. Quarto.

1563. " Certaine Sermons appoynted by the Quenes Maieſty, to be declared and read," &c. as in 1559. In a compartment with Cawood's mark' on the fell. Contains 12 diſcourſes; A 2, in fours. On the

ᵖ Theſe Homilies were frequently printed afterwards, and ſometimes with the titles in a compartment with Jugge's mark at top.

laſt

laſt leaf, " Imprinted——in Povvles Churcheyard, by R. Jugge, and J. Cawood, Printers to the Quenes Maieſtie. Cum priuilegio" &c.

" The Second Tome of Homilies, of ſuche matters as vvere promiſed 1563. and intituled in the former part of Homilies, ſet out by the aucthoritie of the Queenes Maieſtie. And to be read in euery Pariſhe Church agreablye. 1563." In the ſame compartment as the former part. Contains 20 diſcourſes; 276 leaves. A Colophon, as above, completes the laſt page. W.H. Quarto.

" A Fourme to be vſed in Common prayer twyſe aweke, and alſo an 1563. order of publique faſt, to be vſed every Wedneſday in the weeke, duryng this tyme of mortalitie, and other afflictions wherwith the Realme at this preſent is viſited. Set forth by the Quenes Maieſties ſpeciall cōmaundement, expreſſed in her letters hereafter folowyng in the next page. xxx. Julii. 1563." In a compartment with Jugge's mark at top. Contains C, in fours, the laſt leaf blank. To this is annexed, with the ſignatures continued,

" ¶ An Homyly, concerning the Juſtice of God, in punyſhing of impenitent ſynners, and of hys mercies towardes all ſuch as in theyr afflictions vnfaynedly turne vnto hym. Appoynted to be read in the tyme of ſickneſ." Ends on F 4. " Jmprinted——in Powels Churchyarde by R. Jugge and J. Cawood, Printers to the Quenes Maieſtie. Cum priuilegio" &c. W.H. Quarto.

" A ſhort fourme of thankſgeuyng to God for ceaſynge of the con- 1563. tagious Sickneſ of the Plague. Set forth by the Biſhop of London, to be vſed in the City of London and the reſt of his Dioceſſe: and in other places alſo at the diſcretion of the ordinary miniſters of the Churches." Printed Jan. 22. 1563, as noted in the margin. T. Baker's Maunſell, p. 84. With Cawood. (Quarto.)

" A brief Examination for the Tyme of a certain Deelaration, lately 1564. put in print in the name and Defence of certain Miniſters in London refuſing to wear the Apparel preſcribed by the Lawes of the Realme." Catal. Bibl. Harleianæ. Vol. III, No. 4206. Quarto.

" A fforme to be vſed in common prayer euery Sunday, Wedneſday 1565. and Friday through the whole realm: To excite and ſtirre vp all godly people to pray vnto God for the preſeruation of thoſe chriſtians, and their countreys, that are now inuaded by the Turke in Hungary, or elſewhere. Set foorth by the moſt reuerende father in God, Mathewe, archbyſhop of Canterbury, by the auſthoritie of the queenes Maieſtie." With Cawood. Again next year. Quarto.

" The birth of mankynde, otherwyſe named the womans booke. 1565 Newly ſet forth," &c. as p. 581. In the compartment with his mark.

at

at top. Neatly printed; with cuts. T 6, in eights, except A 4. At the end, " 1565."over his device. W.H. Quarto.

1566. " The Newe Teſtament of our Sauiour Jeſus Chriſte," &c. as p. 715. But the lips and feather in the king's portrait are printed in red; and the words " Vivat Rex" on the ſides are now omitted. On the back is " An Almanacke for. xxv. yeres.—M. D. lxvi——M. D. xc." Jugge's epiſtle dedicatory to king Edward is here retained; but the fiſh days are omitted in the Kalendar. The cuts to the Revelation of St. John are larger in this edition than in thoſe of 1552 and 1553. " Jm-printed——in Powles Churchyarde by R- Jugge, printer to the Queenes Maieſtie, Forbydding" &c. as in the former editions. " Cum priuilegio Regiæ Maieſtatis." W.H. Quarto.

1567. " A compendious——Hiſtorie of the latter tymes of the Jewes common weale," &c. as p. 719. Contains 257 leaves, but figured 357, beſides " The epiſtle to the Reader," and the colophon on a ſeparate leaf, as to the edition 1561. W.H. Octavo.

1568. " Propoſitions or Articles drawn out of holy Scripture, ſhewing the cauſe of continuall variance in the Duch Church in London: and thought meete to be publiſhed for ſtaying of other congregations which in theſe daies doe ſpring vp. Subſcribed vnto by Theod. Beza, and diuers other preachers beyond ſea. Printed in Engliſh and Latine by R- Jugge. 1568." Maunſell, p. 88. Octavo.

1568. " The Bible in Englyſhe, that is to ſaye, The content of all the holy Scripture both of the olde and newe Teſtament. According to the Tranſlation that is appointed to be read in Churches. Anno 1568." Lewis, p. 217. Quarto.

1568. " The holie Bible." This title is in a tablet over the portrait of queen Elizabeth, ſupported by Faith and Charity; the queen's arms over head, and a tablet underneath, with this motto, " Non me pudet Euangelij Chriſti, Virtus enim Dei eſt ad ſalutem Omni credenti. Rom 1." ſupported by the lion and dragon. The whole engraved on copper plate. The E. of Leiceſter's portrait is on the title page for the ſecond part, before the book of Joſhua, and that of Cecil lord Burleigh before the Pſalms; both alſo engraved on copper. This is the firſt edition of the Biſhops' Bible, ſet forth by archbiſhop Parker. The chapters are divided into verſes, which are numbered. See a very particular account of it in Lewis, p. 235, &c. Printed by Jugge alone. W.H. Folio.

1569. " The holy bible.—.The newe Teſtament of our Lorde and ſauiour Jeſus Chriſt. 1569. Cum priuilegio." In compartments with Jugge's mark or cypher at top. This edition has the numbers of the verſes placed in any part of the line, and ſo continued through the paragraphs.

The

The Bifhops' tranflation; with fome cuts. See more in Lewis's defcription of it, p. 253, &c. Again 1570, and 1573, &c. W.H. Quarto.

" Articles whereupon it was agreed by the Archbifhoppes and Bifhoppes 1571. of both prouinces and the whole cleargie in the Conuocation holden at London in the yere of our Lorde GOD, 1562. according to the computation of the Churche of Englande, for the auoiding of the diuerfities of opinions, and for the ftablifhyng of confent touching true religion. Put foorth by the queenes aucthoritie." In the compartment with his mark at top. The 20th article[q] begins, " It is not lawfull for the Church to ordayne any thing that is contrary to GOD's worde written," &c. D 2, in fours. " ¶ Imprinted——in Poules Churchyard, by R-Jugge and J-Cawood,—in Anno Domini. 1571. Cum priuilegio"&c. Jugge's device on the laft page. W.H. Alfo without date. Quarto. I have another edition of the fame tenour, date, &c. which differs only in compofing the lines, and the omiffion of the number to p. 3.

" Anno. xiij. Reginæ Elifabethe. At the parliament begunne and 1571. holden at Weftminfter the fecond of Apryll,——1571." In the fame compartment as Anno primo. With Cawood. W.H. Folio.

" Anno. xiiij.——begunne and holden at Weftminfter the eyght 1572. of May,——and there continued vntyll the laft of June follwyng. ——1572." In the fame compartment, but doubtlefs without Cawood, who died 1 April. My copy wants the laft leaf. W.H. Folio.

" A fourme of common prayer to be ufed, and fo commanded by 1572. aucthoritie of the queens maieftie, and neceffarie for the prefent tyme and ftate. 27 October." Quarto.

" The holie bible." The fecond folio edition of the Bifhops' Bible. 1572. The title page, and portraits of the E. of Leicefter and Lord Burleigh, the fame as to the firft edition, 1568. To this edition is added a map of the Holy Land, " engraven by Humphry Cole, goldfmith, an Englifhman, born in the North, and pertaining to the mint in the Tower, 1572." Alfo a double verfion of the pfalmes. And a genealogical table to prefix occafionally. For further particulars fee Lewis, p. 257, &c. Sir John Hawkins. Folio.

" The arte of Nauigation," as p. 719. " And now newly corrected and 1572. amended in dyuers places. 1572—Cum priuilegio" &c. W.H. Quarto.

A proclamation againft the defpifers or breakers of the orders pre- 1573. fcribed in the book of Common Prayer. Greenwich. 20. Octo. 1573. By Jugge alone.'

[q] Thofe who defire to be more particularly informed of the controverfy concerning this article, may confult the authors mentioned in note °, p. 651.

[r] Cawood died 1. April, 1572. Since that time the privileged books were printed with Jugge's name only.

The

1574. The Bishops' Bible reprinted, without the portraits that were in the former folio editions; but has a map of the Holy Land,* cut on wood, though without any mention by whom; and has the arms of Archb. Parker impaled with those of the deanery, instead of the archbishoprick, of Canterbury: these seem to have been let in where lord Burleigh's had been. W.H. Again next year, and frequently. Folio.

1575. " The calender of Scripture. VVhearin the Hebru, Challdian, Arabian, Phenician, Syrian, Persian, Greek and Latin names, of Nations, Cuntreys, Men, Weemen, Idols, Cities, Hils, Riuers, & of oother places in the holly byble mentioned, by order of Letters ar set, and turned into oour English toong. 1575. Psal. Lxxxviij. Nunquid narrabit aliquis in sepulchro misericordiam tuam? & veritatem tuam in perditione?" In a flourished compartment with the queen's arms at top, his device and rebus at bottom; viz. the pelican in the middle, and an angel holding the letter R, in one corner, a nightingale on a bush, with JVGGE on a label over it, in the other corner. Mr. Ames mentions William Patten as the author, but my copy is anonymous. It is introduced with an epistle from " The Printer vnto the gentle Reedar.——ix of Aprill 1575." C c c 2, in fours. At the end " Per Summum, Solum, Fortem, Potentem, Vnicum, Viuentem, Justum, Misericordem, Devm, præter quem non est Deus: Cui soli in Maiestate trina, honor, gloria, laus, & gratiarum actio." Annexed is a list of authors who have written on the sacred scriptures in whole or in part, on Ddd 8. On the last leaf a large table of " Fauts,—to be amended."* W.H. Quarto.

1575. " Articles whereupon it was agreede by the Archbishop of Cant. and the other Bishops of the same prouince in the Conuocation holden at Westminster. 1575. touching the admission of apt and fit persons to the ministery and establishing of good orders in the Church." Maunsell, p. 5. W.H. Quarto.

1575. " A compendious—Historie of the latter tymes of the Jewes common weale," &c. as p. 719. This edition has the pelican on a shield in the title-page, and the date at bottom 1575; but at the end, ¶ " Imprinted at London by John Wallie 1579. Cum Priuilegio." Over Jugge's device ornamented as p. 715. W.H. Octavo.

1575. " Anno. xviii. Reginæ Elisabethe. At this present Session of Parliament by prorogation holden at Westminster the viii. day of February ——and there continued vntyl xv. day of March folowyng.—1575." In the same compartment as Anno primo. K, in sixes. Imprinted—by R Jugge.——Cum priuilegio" &c. W.H. Folio.

1576. " Decade of Voyages. The navigation and voyages of Lewes Vertomanus, gentleman of the city of Rome, to the regions of Arabia, Egypt, Persia, Syria, Etheopia, and East India, both within and without the

---

* This map, on comparison, appears to be the same as was used to Couerdale's Bible, 1535.

Gangis, in the year of our lorde 1503. Tranflated out of Latin into Englifh, by Richard Eden." Included in the following article. Octavo.

THE Hiftory of Travayle in the Weft and Eaft Indies, and other coun- 1577. treys lying eyther way, towardes the fruitfull and ryche Moluccaes; as Mofcovia, Perfia, Arabia, Syria, Ægypte, Ethiopia, Guinea, China in Cathayo, and Giapan: vvith a difcourfe of the Northweft paffage. Gathered in parte, and done into Englyfhe by Richarde Eden. Newly fet in order, augmented and finifhed by Richarde Willes. ¶ Imprinted —by Richarde Iugge. 1577. Cum priuilegio. Dedicated " To the ryghte noble and excellent Lady, the Lady Brigit, Counteffe of Bedforde, &c.— At London the fourth day of July. 1577.—Richarde Willes." Then his preface " To the reader." Contains befides, 466 leaves; R. W's Aduifes for correcting errors, and a table of contents. W.H.   Quarto.

" The Holy Byble, conteyning the Olde Teftament, and the Newe. 1577. Set foorth by aucthoritie. Whereunto is ioyned the whole feruice vfed in the Churche of Englande. 1577. John. 5. Searche ye the Scriptures," &c. In a compartment with mafks at top and the fides; at bottom a nightingale in a bufh, underneath " Omne Bonū Supernæ." The old Teftament is divided as ufual into 4 parts, but with head titles only. Before Jofhua, " The feconde part of the Bible ;" before the Pfalms, " The thyrde part of the Bible; before the apocryphal books, " The fourth part of the Bible." The contents of thefe on 400 leaves. Then, " The Newe Teftament of our Saviour Jefus Chrifte. I am not afhamed of the Gofpel of Chrifte, &c. 1577. Cum Privilegio." In the fame compartment as the general title. Contains 103 leaves, including the title; and at the end, under his device, " Imprinted—by R. Jugge, Printer to the Queenes Maieftie. Cum priuilegio Regiæ Maieftatis." W.H.   Quarto.

" Anno Quinto Reg. Eliz. At the Parliament holden at Weftminfter the xii. of Januarie,—'1563." Proroged the 10th of April following. Fol. 71. To this is annexed, on a frefh fet of fignatures, " An act of a Subfidie" &c. C 7, in fixes, By Jugge alone. W.H.   Folio.

" Anno octauo——by prorogation holden at Weftminfter the laft day of September,——'1566." By Jugge alone. W.H.   • Folio.

" The holy byble, conteyning the old teftament and the newe, fet forth by aucthoritie." In an oval compartment, with the queen's arms at top; on one fide thereof the harp crowned, and on the other four lions quarterly, crowned alfo; thefe fuftained by Faith and Love. At bottom, " I am not afhamed of the Gofpel of Chrifte," &c. on a tablet fupported

---

† The epiftles and gofpels have only references to find them.

" Not the date of printing, but of prorogation. This accounts for Jugge's name alone in the colophon, and fhews it was printed after Cawood's deceafe. No doubt thefe acts had been printed jointly, foon after the rife of the parliament.

by the lion and dragon. Mr. Ames mentioned this without date; but his defcription anfwers in euery particular to the edition of 1574. Folio.

—— " A fhort fourme and order to be vfed in Common prayer thryfe a Weeke, for feafonable wether, and good fucceffe of the cōmon affayres of the Realme: meete to be vfed at this prefent and alfo hereafter when lyke occafion fhall aryfe. By the difcrefion of the ordinary within the prouince of Canterbury. Set forth by Mathew——Archbyfhop of Canterbury." Octavo.

—— A Prayer for Malta. Collier's Eccl. Hift. Vol. II. p. 505.

—— " A Declaracyon of the procedynge of a Conference begon at Weftminfter the lafte of Marche 1559, concerning certaine Articles of Religion; and the breaking vp of the faid Conference by default and contempt of certayne Bifhops, parties of the faid conference." With Cawood. Cum privilegio. Octavo.

—— " An Homilie againft Difobedience & wylfull Rebellion: With a thankefgeuing for the fuppreffion of the late Rebellion." With Cawood. Cum privilegio. Quarto.

—— " A collection of the fubftaunces of certayne neceffarye ftatutes to be by the Juftices of Peace executed within euery fheare of the Realme." Ten leaves, printed on one fide only. With Cawood. Mr. M. C. Tutet.

—— " A DEFENCE of prieftes mariages ftablyfhed by the imperiall lawes of the Realme of Englande, agaynft a Ciuilian, namyng hymfelfe Thomas Martin doctour of the Ciuile lawes," &c. Contains 359 leaves, exclufive of the contents, a preface to the reader, and a correction of faults, prefixed; and a table of the principal matters, at the end. "Cum priuilegio," &c. W.H. Quarto.

—— " The Piththy and moft notable fayinges of al Scripture, gathered by Tho. Paynel:" &c. as p. 359, &c.

—— " The Gratulation of the moofte famous Clerke M. Martin Bucer,——. And Hys anfwere vnto the two rayling epiftles of Steuē, Bifhoppe of Winchefter, concerning the vnmarried ftate of preeftes and cloyfterars, wherin is euidently declared, that it is againft the lawes of God, and of his churche, to require of all fuch as be, and muft be admitted to preefthood, to refrain from holpe matrimonie. Tranflated out of Latin in to Englifhe. Hebru. xiij. Wedlocke is to be had in price emonge al men," &c. Dedicated " To his right worfhypfull Brother Syr Philyppe Hobye knight, M. of ý Kinges maiefties ordinaunce, *by* Thomas Hobye," the tranflator. " At Argentyne. Kalendis Februarij." The running title " Bucer to the holy churche of Englande." Concludes, " Youre humble and daylie oratoure in the Lorde Martine Bucer. Jmprynted ad London by me R- Jugge dwelling at the nourth dore of Poules. Cum priuilegio—— folum." W.H. Octavo.

" Howe

"Howe we ought to take the death of the Godly. A sermon made in Cambridge at the Burial of the noble Clerck D. M. Bucer, by Matthewe Parkar D. of Diuinitie." His text, Wisdom iv. 7—19. "Imprynted— in Paules Churcheyarde at——the Bible. With the Kynges most gratious priuilege. And licenced accordyng vnto the meanyng of the late Proclamation." Octavo.

"Guilio of Millaine his 44 sermon touching the Lords supper, on 1. Cor. xi. 27." Maunsell, p. 99. Octavo.

".Whether it be mortall sinne to transgresse ciuil lawes, which be the commaundementes of ciuil Magistrates. The iudgement of Philip Melancton in his Epitome of morall philosophie. The resolution of. D. Hen. Bullinger, and. D. Rod. Gualter, of. D. Martin Bucer, and. D. Peter Martyr, concerning thapparrel of Ministers, and other indifferent thinges." In a border of metal flowers. On the back of p. 101, "Imprinted——in Powles Churchyarde by——Printer to the Queenes Maiestie. Cum priuilegio," &c. On p. 103 begins, with this head title, "¶ A briefe and lamentable consyderation, of the apparell now vsed by the Cleargie of England. Set out by a faithfull seruaunt of God, for the instruction of the weake. On p. 119 begins, "This letter sent by maister Bucer, to the letter of maister Hooper, before in the page. 59." which concludes on p. 134, "Youres most bounden in the Lorde. Mart. Bucer." W.H. Octavo.

"A newe worck concernyng both partes of the Sacrament to be receyued of the lay people as well vnder the kynd of wyne, as vnder the kynd of breade, with certen other articles, concernyng the masse and the auctorite of bysshops, the chapters wherof are conteyned in the next leafe: made by Philip Melanchton, and newly translated out of Latyn."* F, in eights; the last leaf blank. "Jmprinted by my R- Jugge." The same was printed again, page for page, and on the same type, but corrected. The running-title of this, for the first part, is, "The sacrament ought to be geuen to the layte under both kyndes." W.H. both. Octavo.

"A brief examination for the tyme of a certaine declaration," &c. as in 1564. "Jn the ende is reported the iudgement of two notable learned fathers, M. doctour Bucer, and M. doctour Martir, sometyme in eyther vniuersities here of England the kynges readers and professours of diuinitie, translated out of the originals, written by theyr owne handes, purposely debatyng this controuersye. Paul. Rom. 14. I besech you brethren marke them which cause diuision," &c. The examination contains ******* 2, in fours; the letters annexed, D, in fours. W.H. Quarto.

"A very necessarie and profitable Booke concerning Nauigation, compiled in Latin by Ioannes Taisnierus, a publike professor in Rome, Ferraria, & other Vniuersities in Italie of the Mathematicalles, named a treatise of continuall Motions. Translated into Englishe by Richarde Eden. The contentes of this booke you shall finde on the next page folowyng.
Imprinted—

RICHARD JUGGE.

Imprinted——by R- Jugge. Cum priuilegio." In the compartment formed of four pieces, as to the epiſtles of the New Teſtament 1552. "The Epiſtle Dedicatorie. To the ryght woorſhipfull Syr Wylliam Wynter, Knyght, Maiſter of the Ordinaunce of the Queenes Maieſties Shippes, and Surueyor of the ſayd Shippes. Richarde Eden wyſheth health," &c. Then, Taiſnier's preface to the Archbiſhop of Colen. Illuſtrated with cuts and ſchemes. 42 leaves. W.H.   Quarto.

——— "The Birth of mankynde, &c. as p. 721. In the flouriſhed compartment with his rebus at bottom. His device at the end. W.H. Quarto.

——— "The Image of bothe Churches," &c. as p. 674, &c. At the end, "Jmprynted——by R- Jugge,—in Paules churchyarde at the—Byble." Under which, perhaps, might be a date, but my copy is torn. W.H.+   Quarto.

JOHN JUGGE

Succeeded Richard in his buſineſs, and probably in his houſe; and though it does not poſitively appear what relation he was to him, it is highly probable that he was his ſon. We might perhaps have been ſatisfied in this particular, but that the regiſter-book, containing the Company's tranſactions from 1571 to 1576 is miſſing; in which period it is likely he took up his freedom, ſeeing he was brought on the livery in about 1574. May 20, 1577, he had licenſe to print "Fuller's farewell to Mr. Fourbouſier and other gentlemen adventurers who labour to diſcouer the right paſſage to Catay. The delectable and pleaſant hiſtorie of Gerillion of Englande." He ſeems not to have been living on the 6th of April, 1579, when one Miles Jennyngs claimed the copy-right of "a book entitled The hiſtorie of Gerillion of England. which he affirmed that he bought of Jhon Jugge." The buſineſs alſo from that time appears to have been carried on by Joan the widow of Richard Jugge.

1577.   "The aduiſe and anſwere of my lord the prince of Orange, countie Naſſau, &c. and of the ſtates of Holland and Zealand, &c. dedicated to the king of Spain. Tranſlated from Dutch. Imprinted at London by John Jugge and John Allde," but at the end, "Doon at Middleborough, the 19 Feb."*   Octavo.

——— P. Martir of the vſe and abuſe of dauncing, tranſlated by I. R. Octavo.

JOAN JUGGE,

The widow of Richard Jugge. By her not taking up the buſineſs till after the death of the abovementioned John, there is ſufficient reaſon to ſuppoſe he was their ſon.

1579.   Langham's Garden of Health.   Quarto.

"The

JOAN JUGGE.

" The Arte of Nauigation," &c. as p. 719. " And now newly cor- 1579.
rected, &c. 1579. Whereunto may be added, at the wyl of the byer,
another very fruitfull and neceffary booke of Nauigation, tranflated out
of Latine by the faide Eden. And Prynted by Richard Jugge." In the
compartment with his rebus at bottom. At the end, " Imprinted at
London by the Widowe of Richarde Jugge late printer to the Queenes
Maieftie. 1579. Cum priuilegio." Quarto.

The Book of Common Prayer. Dr. Lort. Folio. 1584.

" The Arte of Nauigation," &c. as in 1579. " and printed by Johan 1580.
Jugge wydowe. Cum priuilegio." Quarto.

There were certain Sermons appointed by the queen's majefty, printed
1587, with Jugge's rebus, probably by his widow, &c.

JOHN WALLEY, or WALEY,

WAS one of the original members of the Stationers' company before
they had their charter; and ferved renter, or collector of the quarterages, from 1554 to 1557; when he was chofen under-warden; he was
upper-warden in 1564, and again in 1569. He rented a chamber in the
Company's hall, for which he paid xiij. s. iiij. d. a year, in 1557; but in
1561, xx. s. His dwelling houfe was in Fofter-lane, at the fign of the Hart's-
horn. In 1558 he was fined ij. s. viij d. for keeping open fhop, and felling
books on a feftival day. Again, in 1564, for keeping open fhop on St.
Luke's day, with 18 others, xvj. s. viij. d. He had licenfe for printing:
viz. from July 1557 to July 1558, " Welth & Helthe. The Frere and the
boye. Stans puer ad menfam. Youghte, Charyte, and Humylyte. An
a b c for cheldren, in Engleshe, with fylables. An hundreth mery tales.—
the waye of God. The cronacle of yeres, in xvj." Alfo fundry ballads with
Mrs. Toy, as p. 588. 1559—1560, " Efopes fables in Englefhe. The
Shipman's Calender." 1562—1563, " An almanacke & prognoftication
of Mr. John Securys, for the year 1563—of Noftradamus, for this year,
Anno 1562." The Lateny in 'Welfhe." 1564—1565, " An. almanacke,
for xiij yeres from 1565." 1565—1566, " An almanacke & pronoftication.
of Mr. Buckmafter." 1566—1567, " The fecounde well a daye. The

---

* The entry of this feems to have been
omitted in its proper place.
ᵃ He was joined with William Salifbury,
of Llanroaft, Gent. in a patent for feven
years, to print the Bible in Welfh. Strype's
Annals, Vol. 1. p. 434.

Lamentynge

JOHN WALLEY.

Lamentynge of a yonge made, who by grace ys fully ftayde." 1567——1568, " Taveners poftell vpon the Gofpelles. An almanacke & pronoftication of Mychell Noftradamus for 1568." Aug. 3, 1579, " The fecond booke of Robyn Confcyence." Octo. 6, 1580, " iij balads: The Lord of lorne & the falfe fteward. Of going to market to buy the child fhoes. Of this fillie poore man." See other copies, declared to have been his, under his fon Robert, to whom they were accordingly allowed. In 1568, his fon *John* (as entered in the Company's regifter) was made free by patrimony; but as I find no farther mention of his name, fuppofe it a miftake for his fon Robert, of whom fee hereafter. Mr. John Waley, for fo he figned his name, died in the beginning of 1586, as appears by the following memorandum. " 27 Janij, 1586.——this day there was diftributed in the hall to the poore of the Companie, of the gift and legacie of John Walley Staconer deceafed the fome of Fyftie fhillinges by Agnes Walley executrix of his teftament by thandes of Robert Walley his fonne according to the faid Teftators teftament. This day of the diftribution thereof being the firft & next quarter day after the deceafe of the faid John Walley."

1546. " The boke for a Juftice of Peace. Kepynge a Court Baron. Carta feodi. Returna Brevium. Modus tenendi unum hundredum. Fees in the Kynges Efchequer." Octavo.

1547. " A dictionary in Englyfh and Welfhe, moche neceffary to all fuche Welfhemen, as wil fpedly learne the Englyfhe tongue, thought vnto the kynges maieftie very mete to be fette forthe, to the vfe of his graces fubiects in Wales: whereunto is prefixed, a little treatyfe of the Englifh pronounciation of the letters; by Wyllyam Salefbury." Dedicated to king Henry the VIII. Quarto.

1548. " A caueat for Chriftians agaynft the Archpapift." &c.

1550. " A new balade made by Nicholas Balthorp which fuffered in Calys the xv. daie of Marche M.D.L." Contains 12 fix-lined ftanzas.

1551. " The whole Byble after the tranflation of Tho. Mathew," &c. as p. 556. It was printed by Nich. Hill for him, and others. Folio.

1553. " Commō places of fcripture," &c. as p. 587. Nicholas Hill, for him and others. Octavo.

1553. " Catonis Diftica Moralia," &c. as p. 710. Printed alfo by Nich. Hill for him, " dwelling in Fofter-lane, at the fygne of the Hartfhorne."

1554. " A book of good maners for chyldren, by Erafmus Rot. With the interpretation of the fame, by Robert Wittinton, poet laureat. Cum Latino textu." See it printed by Nich. Bourman, p. 594. Octavo.

" A breuiate

JOHN WALLEY.

" A breuiate chronicle, containing all the kings from Brute to the 1554. year of Chrift, 1554." This he printed feveral times. Octavo.

" The hiftory of graunde amoure et la Bel Pucell." See it printed 1555. by Wayland, 1554. Quarto.

" Spirituall exercyfes and gooftly meditacions, and a neare waye to 1557. come to perfection and lyfe contemplatyue, very profytable for Religyous, and generally for all other that defyre to come to the perfecte loue of God, and to the contempte of the worlde. Collected and fet foorthe by the helpe of god, and diligente laboure of F. Wyllyam Peryn bacheler of diuinitie and Pryor of the Friers preachers of greate Sayncte Bartholomes in Smyth. fyelde. Jmprynted—in Fofter lane by Jhon Waley, Anno Domini: M. D. L. vij." Dedicated " Vnto the deuoute and very religious Sufter Katherin Palmer of the order of faincte Brigit in Dermount, and Sufter Dorothe Clement of the order of faincte Claire in Louaine.—The laft of December. Anno Do. M.D.L. iiij." Contains befides, V 6, in eights. W.H. Octavo.

" The confpiracie of Cataline, written by Conftancius, Felicius, Du- 1557. rantinus, and tranflated bi Thomas Paynell : with the Hiftorye of Jugurth, written by the famous Romaine Saluft, and tranflated into Englyfhe by Alexander Barcklaye." In a compartment with Atlas on each fide. De- dicated to K. Henry the eighth. Y 6, in fours; the laft leaf blank. " Thus endeth the confpiracy of Cataline Jmprinted—in Fofter-lane by John Waley."

" Here begynneth the famous Cronicle of warre, whyche the Ro- 1557. maynes hadde agaynft Jugurth vfurper of the kyngedome of Numidie: whiche Cronicle is compiled in Laten by the renowmed Romayne Salufte : and tranflated into englyfhe by fyr alexander Barklaye priefte. And nowe perufed and corrected by Thomas Paynell. Newely Jmprinted in the yere of oure Lorde God M. D. Lvij." On the back begins Paynel's dedication " To the ryghte honorable Lorde Antonye Vycounte Mountague, knyghte of the righte honorable order of the garter, and one of the Kynge & Queenes magefties pryuie counfayle." H h, in fours. " Thus endeth the famoufe Cronicle of the warre——againft Jugurth——tranflated——by fyr Alexander Barkeley priefte, at commaundemente of——Thomas duke of Northfolke, And imprinted—in Fofter-lane by Jhon Waley." W.H.
Octavo.

" The vvorkes of Sir Thomas More Knyght, fometvme Lord Chaun- 1557. cellour of England, wrytten in the Englyfhe tonge. Printed——at the coftes and charges of John Cawood, John Waly, and Richarde Tottell. Anno. 1557." Beneath is Cawood's mark. In a compartment with the arms of England crowned and gartered at top, and Peace on each fide. Dedicated " To the mofte hygh & vertuous Princeffe, MARY——quene of Englande, Spayne," &c. by " VVyllyam Raftell feriant at lawe." It has then prefixed, a table of the works, and another of the matters ; afterwards four pieces of poetry " wrote in his youth." Contains befides, 1458 pages, in double columns. " Jmprinted——in Fleteftrete at the
fygne

JOHN WALLEY.

fygne of the hande & ftarre, at the cofte and charge of John Cawood, John Walley, and Richard Tottle. Finifhed in Apryll—1557. Cum priuilegio—folum." W.H. Folio.

Mr. Ames fays he printed feveral books with Cawood and Tottle about this time.

1557. "The interlude of Youth." In verfe. See it without date. Quarto.

1558. "A TABLE collected of the yeres of our Lorde God, and of the yeres of the Kynges of England, from the fyrft yere of William Conquerour, fhewynge how the yeres of our Lord god and the yeares of the kynges of Englande concurre and agree togyther, by whyche Table it may be quickely accompted, howe many yeres, monthes, and dayes be pafte fyns the makynge of any euidences, inftrumentes or writynga that haue their dates of the yeres of the kynges reignes, and not the dates of the yeres of our lorde God. And alfo it fhalbe redyly fene, in what yere of our Lorde God thofe euidences, inftruments or writinges were made. Londini. 1558." Said to have been compiled by William Raftall. L, in eights.[b] "Londini. Ex officina Ioannis VValey typographi. Anno à virgineo partu M. D. LVIII. 22. die Menf. Decemb." W.H. Octavo. Again, 1562, 1563, 1565, 1567, and 1576.

1559. "Pfalmes or prayers taken out of holy fcripture," &c. as p. 357. Sixteens.

1560. Tho. Palfreyman's "Exhortation to the knowledge and loue of God," &c. as p. 570.

1561. He was fined xviii. d. for printing Jacob and his xii fons, without licenfe.

1569. "Articles of enquiry by John bifhop of Norwich, with Iniunctions to the Archdeacons, at the vifitation of the dioces of Norwich." Quarto.

1575. The Bible in Englifh. Lambeth Lift. Folio.

1575. "A compendious—Hiftorie of the latter tymes of the Jewes common weale," &c. as p. 724.

—— "An introductorie for to lerne to rede, to pronounce, and to fpeake Frenche trewly," &c. as p. 323. The firft book contains S ii, in fours. "Here foloweth the feconde booke of thys lytell worke, in the whych fhalbe treated of cōmunications, & other thynges neceffary to the lernynge of the fayde Frenche tonge. Newly corrected and amended." The fignatures proceed to E e 6; but x, y and z are omitted. "Printed at London by John Waley." W.H. Quarto.

—— "Thēterlude of Youth." This title over two whole-length figures of "Charitie" and "Youth:" the former with a long bow in his left hand, and an arrow in his right; the latter with an halberd in his right hand, a hat and feather on his head. C, in fours. "Imprinted—by John Waley—in Fofter lane." Quarto.

[b] This book appears evidently to have been printed by virtue of the abovementioned entry in the Hall Book, which though mentioned in xvj, on examination i find is printed on whole fheets.

"Hyckefcorner

JOHN WALLEY.

"Hyckefcorner." A morality, or interlude, reprinted in Hawkins's Origin of the Englifh Drama. "Imprinted by Johan Waley." See p. 215. Quarto.

"Syr Eglamore of Artoys." Over a cut of a knight on horfeback, with fword and buckler. E, in fours. "Jmprynted——in Fofter lane, at the Sygne of the Hartefhorne by John Walley." Quarto.

"The life of the glorious and bleffed virgin and martyr Saincte Katheryne." In the fame compartment as The Confpiracie of Cataline. D, in fours. "Jmprynted—in Fofter-lane by Jhon waley." W.H. Quarto.

"Here begynneth the feynge of Vrynes" &c. as p. 355. "Here endeth the boke of feynge of Waters. Jmprynted—in Fofter lane by me John Waley." Over a cut of S. John Baptift, W.H. Octavo.

"The Rutter of the See with the hauons, rodes, foundinges, kenninges, wyndes, flodes and ebbes, daungers & cooftes of dyuers regiōs, with ẏ lawes of the yele of Auleron, and the iudgementes of the fee. With a Rutter of the Northe added to the fame." On the back begins "The prologue of roberte Coplande the tranflatoure of this fayde Rutter." At the end of the laws, "Wytneffe the feale of the yle of Auleron eftablifhed by the contractes of the faide yle, ẏ Tuefday after the feafte of fainte Andrewe. The yeare of our Lorde. M. cc. lxvi." Then, "A newe Rutter of the fea, for the northe partyes: compyled by Rycharde Proude. M. D. XLI." F 3, in eights. "Jmprinted——by John waley——in Fofter lane." W.H. Octavo.

"This is the Myrrour or Glaffe of Helth," &c. as p. 375.

"A new mery balad of a maid that wold mary wyth a feruyng man." Containing 15 feven-lined ftanzas, each ending with "I will haue to my hufband a feruing man," a little varied. "FINIS quod Tho. Emley. Jmprinted——in Fofter lane by Jhon Waley." Broadfide.

"Interrogatories. For the Doctrine and Manners of Mynifters and for other Orders in the Churche.—Churchwardens. Scholemaifters. Clarkes. For the People." At the end "To thefe Interrogatories the Ordinarie requireth an Aunfwere accordingly by the lafte Daye of Auguft, or before if they maye. Imprynted—in Fofter-lane by Jhon Waley." See Strype's Annals, Vol. 1. p. 165, and Appendix, [N° XXI.] Quarto.

"The boke of Hawkynge, Huntyng and Fyfhing," &c. as p. 367. Mr. M. C. Tutet. Quarto.

"A brief treatife conteinyng many proper Tables and eafie rules, verie neceffary and needeful for the vfe and commoditie of all people, collected out of certaine learned mens woorkes.——.Newly fet fourth and allowed accordyng to the Queenes Maiefties Iniunctions. Imprinted——by Ihon VValey." Octavo.

"Here beginneth the kalender of fheepehards: newly augmented and corrected." See p. 208, &c. Folio.

"The copie of a letter fent into Scotlande" &c. as p. 562. Octavo.

ROBERT

ROBERT WALLEY, or WALEY,

WAS son of the forementioned John Waley, and seems to have been made free, by patrimony, in Aug. 1568, but entered in the Company's register, by mistake, under the name of John. However that be, he bound an apprentice in 1576; and was brought on the livery in 1585. He served renter in 1592. In 1594 he was taken into the court of assistants; so that probably he had fined for warden. Next year he was one of the three members who were annually appointed to dine at the lord mayor's feast, in Guildhall. His father seems to have quitted the trade to him in 1576. July 21, 1577, he had a reversionary licence from the company for printing a book entitled "An abstract of all the penall statutes," &c. after the death of "Raffe Newberye," who was not to enjoy it till after the death of "Richard Tottell." He had licence also for printing solely the following books. Feb. 20, 1577-8, "Cometographia quedam Lucis Aerij que circa decimū Nou. 1577, apparuit." Mar. 6, 1580-1, "Articles to be enquired, with D. Squiers visitation." May 4, "A true reporte of the strange connynge & breedinge of myse in the marshes of Dengie hundred, in Essex." Octo. 23, "The wonderfull aduentures of Don Symonydes a Spanish gent. Written by Barnaby Rich." April 23, 1582, "A lat practise enterprised by a papist with a yonge maide in Wales, taken amongest Catholikis for a prophetis, &c." Septem. 14, "A book of Engins for the destruction of vermyne, Crowes & Sparrowes, with the gouernement of Oxen, kyen, Calues, horse, shepe, hogis, mowles & doggis." Decem. 7, 1584, "The difference betweene the Auncient phisicke first taught by the godly fathers, consistinge in vnitie, peace & concorde; And the latter phisicke proceading from Idolatrie, &c." Septem. 4, 1586, jointly with John Charlewood, "A discourse of Englishe poetrye." March 22, 1586-7. "The pathway to Militarye practise, with a kalender for the ymbattelling of men, newly written by Barnabie Riche." March 1. 1590-1. "Allowed vnto him these copies, which were his fathers. viz. The Shepherdis Calender. Cato: Engl. & Latyn. The prouerbs of Salomon Jnglish. Salust, & bellū Jugurthium. Mr. Graftons computation. Esopes fables: Eng. Josephus, de bello Judico: Eng. Robyn Conscience." The 12th of October following he assigned all his copies to Tho. Adams.

It does not appear that he printed himself, seeing most of his copies that have been found were printed for him. They are as follows.

1579. "Jo. Hedlamb, his exposition on the 8. to the Rom. Printed for Rob. Walley." Maunsell, p. 100.   Octavo.

1581. "The straunge and wonderfull adventures of Simonides a gentilman Spaniarde. Conteyning uerie pleasaunte discourse. Gathered as well for the recreation of our noble yong gentilman as our honourable courtly ladies. By Barnaby Riche, gentleman. London, for Robert Walley. 1581." Warton's Hist. of Eng. Poetry, Vol. III. p. 482.   Quarto.

" The

ROBERT WALLEY.

" The difference between the auncient phificke, firft taught by the godly 1585.
forefathers, confifting in unitie. peace, and concord; and the latter phificke proceeding from idolaters, &c. by R. B. efq;" with the author's
obteftation to almighty God. See the licenfe for it above. Octavo.

" A difcourfe of Englifh Poetrie. Together with the authors iudgment 1586.
touching the reformation of our Englifh verfe. By William Webbe,
gradute." John Charlewood for him. See a circumftantial account of
this book in the Britifh Librarian, p. 86, &c. Mr. Warton mentions an
edition 1585. Quarto.

" A PATH-WAY to Military practife. Containinge Offices, Lawes, 1587.
Difciplines and orders to be obferved in an Army, with fundry Stratagems,
very beneficiall for young Gentlemen,——defirous to haue knowledge in
Martiall exercifes. Whereunto is annexed a Kalender of the Imbattelinge
of men: Newlie written by Barnabe Rich Souldiour, feruaunt to the
right honorable Sir Chriftopher Hatton Knight. Malui me diuitem effe
quam vocari. Perufed and allowed. At London. Printed by John Charlewood, for Robert Walley. 1587." In a border. On the back, the
queen's arms, crowned and gartered. An " Epiftle Dedicatorie, To the
moft High and mighty Princeffe Elizabeth,—Queene of England," &c.
Another, " To the moft noble Captaines & renowned Souldiours of
England." Another, " To the freendly Readers in generall. L, in
fours. W.H. Quarto.

WILLIAM POWELL, or POWEL,

DWELT in St. Dunftan's parifh in Fleet-ftreet, next to the church,
at the fign of the George, in the old fhop, that was laft William
Middleton's, formerly Rob. Redman's, and before that Rich. Pinfon's.
He was one of the original members of the Stationers' company. He
had licenfe to print the following books. Feb. 6, 1559-60, " The boke
of fortune, in folio." Nov. 30, 1561, " Raynolde the Foxe." Oct. 27,
1564, " A Crononicall table." 1565, " Ludlowes prayers."—1566,
" A petyous Lamentation of the miferable eftate of the churche of chrifte.
A warnyng for wydows that aged be/ how lufty yonge yough & age can
agree." He printed the following.

" The new book of Juftices of peace, made by Ant. Fitzherbarde," 1547.
&c. In a compartment with Will. Middleton's rebus. Octavo.

" Phill. Melangton, Of the Juftification of man by Faith onely, tranf- 1547.
lated by Nich. Leffe. Alfo an apologie of the Word of God, declaring how
neceffarie

neceffarie it is to be in all mens hands, the want whereof is the caufe of al vngodlines, written by the faid Nich. Leffe." Maunfell, p. 72. Again, 1549.   Octavo.

1547. The new teftament. My copy wants the title-page. The colophon, " Thus endeth the neweTeftament both in Englyfhe & in Laten/ of mayfter Erafmus tranflaciō, with the Pyftles takē out of ye Olde teftamēt. Set forthe with the Kynges mofte gracious lycēce, and Jmprynted by Wyllyam Powell dwellynge in Fleteftrete at ye fygne of the George, nexte vnto faynt Dunftons Churche. The yere of our Lorde. M.CCCCC.xlvii. and the fyrfte yere of the Kynges mofte gracious reygne. God faue the kynge." W.H.+   Quarto.

I have another copy of it with this title, " THE NEWE TESTAMENT in Englifhe and in Latin.'. NOVVM TESTAMENTVM ANGLICE ET LATINE. Anno Dnī. 1548." In the compartment of Mutius and Porfenna, formerly Pinfon's. On the back is a lift of " The bokes contayned in the newe Teftament."

1547. " Returna Breuium."   Octavo.

1547. " Articuli ad narrationes. In ÆDIBVS Wilhelmi Powell."   Octavo.

1548. " ENCHIRIDION MILITIS CHRISTIANI," &c. as p. 488. Colophon, " Here endethe——in the which boke is conteyned many godly leffons, very neceffary & profytable for the foule helth of al true chriften people. ¶ Jmprynted—M.D.xlviii. The xxv. daye of October. Cum priuilegio ——folum." W.H.   Octavo.

1548. " A boke of Prefidentes," &c. See p. 521.   Octavo.

1548. " Here begynneth the feynge of Vrynes" &c. as p. 355. At the end, " Imprinted—M.D.XLVIII. The xvi. day of September." W.H.   Octavo.

1549. The New Teftament in Englifh and Latin. The fame as in 1547; but the Latin printed with Roman type. " Cum priuilegio—folum." W.H.+   Quarto.

1549. " THE FAL OF THE LATE ARRIAN." Dedicated to the Lady Mary, by John Proctour. " Jmprinted——An. Dnī. M.CCCCC.XLIX. Mēf. Decemb. ix. Cum priuilegio—folum." W.H.   Octavo.

1549. " The Voice of the people to fuch Parfons, Vicars, Archdeacons, Deanes, Prebendaries, and others, as lie away from their Cures, and benefices. Alfo to fuch Lay men & vnlearned as couetoufly catch, and wrongfully withold Parfonages, Vicarages, Tithes, and other liuings, appointed to the miniftrie of Gods word and Sacraments." Maunfell, p. 119.   Octavo.

1549. " The office of Sheriffs. 9 March. John Baynes, Efq;   Octavo.

WILLIAM POWELL.

The book for a juſtice of Peace. "Londini in edibus Wilhelmi 1550. Povvell typis impreſſ. Cum priuilegio—folum. Anno. M. D. L." W.H.
Quarto.

"PROVERBES or adagies, gathered out of the chiliades of Eraſmus, 1550. by Rychard Taverner, with new additions as well of Latyne proverbes, as Engliſh. Anno 1550, xx day of Apryl." See p. 406. N. B. This Taverner (ſays T. Rawlinſon, Eſq; in his catalogue) was a lay-preacher under king Edward VI. got up with his gold chain in querpo Mr. Sherif, at St. Mary's, Oxon, to the Scholars. Sic fata ferebant. Octavo,

"The book of haukynge, huntynge, and fyſhing, with all the pro- 1550. perties, and medecynes, that are neceſſary to be kepte." Alſo without date. Quarto.

"A woorke of the holy biſhop S. Auguſtine concernyng adulterous 1550 mariages, written by him to Pollentius, diuided into two bookes, very neceſſary to be knowen of all men and women. Londini. Anno. 1550." L 4, in eights. W.H. Octavo.

A little herbal, by Ant. Aſkam. "Imprynted——M. D. L. the twelfe 1550. day of Marche." See it printed by John King. Octavo.

"Antho. Aſcham, his Treatiſe of Aſtronomie, declaring what herbs, 1550. and all kinde of medicines are appropriate, and vnder the influence of the Planets, ſignes & conſtellations: alſo howe ye ſhall bring the virtue of the heauens, and nature of the ſtarres to euery part of mans body being diſ- eaſed to the ſooner recouerie, &c." Maunſell, Part II. p. 2. Octavo.

"The Treaſure of pore men.——M. D. L. I. 25. July." Octavo. 1551.

"Inſtitutions, or principal groundes of the Lawes," &c. Octavo. 1551.

"A chronicle of yeres," &c. as p. 555. Octavo. 1551.

"Lytelton Tenures truely tranſlated" &c. 6. Septemb. Octavo. 1551.

The new great abridgement of the Statutes. Octavo. 1551.

"A prognoſtication made for the yere of our Lord God 1552, by 1552. Anth. Aſkam, Phiſition." Cum privilegio. Octavo.

"A treatiſe of Aſtronomy, declearing the leap year, and what is the 1552. cauſe thereof; and how to know St. Matthias day for euer, with the mar- uellous motion of the ſun, both in his proper circle, and by the mouing, that he hath of the 10th, 9th, and 8th ſphere. By Anthony Aſcham."
Octavo.

"The Order of The ſcheker." Running title. Octavo. 1552.

"The long quinto of Edward the IV. in a ſingle volume by itſelf." 1552.
Folio.

"The

1552. "The office of Shyryffes, Baylyffes," &c. See p. 574.   Octavo.

1552. "The Breuiary of Healthe, for all maner of sickneſſes & diſeaſes the which may be in man or woman, doth followe. Expreſſyng the obſcure termes of Greke, Araby, Latyn, and Barbary, in Engliſhe, concernyng Phiſicke & Chierurgerie, compyled by Andrewe Boorde of Phiſicke Doctour, an Engliſhe man. Anno. M.D.LII." 123 leaves. "Here endeth the firſt boke examined in Oxford in June——M. ccccc. xlvi.——And newly Imprinted & corrected—M.ccccc.LII."
"The ſeconde Boke of the Breuyary of Health, named the Extrauagantes foloweth. Compyled by Andrewe Boorde of Phiſicke doctor." Colophon: "Imprynted——. Cum priuilegio——ſolum." Again 1556, and 1557.   Quarto.

1552. "PETRI BELLOPOELII GALLI TARNACII DE PACE INTER INVICTISSIMOS HENRICVM GALLIARVM, ET EDVARDVM ANGLIAE, REGES ORATIO. LONDINI IN AEDIbus Guillelmi Povvell. CVM PRIVILEGIO. 1552." In the compartment with boys riding in paniers on an elephant, formerly uſed by Pinſon. Dedicated, "Potentiſſimo—Joanni duci Northumbriæ"—magno Regiæ Magiſtro, ac Illuſtriſſimi Garterij Ordinis aurato Militi." H 5, in fours. Roman letter. W.H.   Quarto.

1553. Natura brevium. 6th May. Cum privilegio.   Octavo.

1554. "ORATIO LEONHARDI GORETII EQVITIS Poloni de matrimonio ſereniſſimi ac potentiſſimi ſereniſſimæ, potentiſſimęq; Dei gratia Regis ac Reginæ Angliæ, Hiſpaniæ. &c. Ad populum principeſq; Angliæ. Londini. In ædibus Guilhelmi Powell. Anno 1554." In the ſame compartment as Petri Bellopoelii Oratio. K, in fours. Roman letter. W.H.   Quarto.

1555. "THE DECADES of the newe worlde," &c. as p. 587. He printed this book for Rob. Toy, Will. Seres, Edw. Sutton, and perhaps others, each having his name ſeparately inſerted in the colophon of his quota of copies; in like manner as to Chaucer's works, &c. See p. 515, and 557.   Quarto.

1558. "Anthony Aſkam's treatyſe, made 1547, of the ſtate and diſpoſition of the worlde, with the alterations and changings thereof thro' the hyeſt planets, called, Maxima, Maior, Media, & Minor, declaring the very tyme of the day, houre and minute, that God created the ſonne, moone and ſterres, and the places, where they were fyrſt ſet in the heavens, and the beginning of their mouings, and ſo contynued to this day, whereby the world hath receyved influence, as ſhall be declared by example from the creation vnto this preſent year of our Lord, an. M.D.LVIII to come, laſt daye Januarye." One ſheet.   Octavo.

1559. "The kalender of ſhepards, newly augmented and corrected." See p. 208, &c. Benet's College Library, Cambridge.   Folio.

"The

"The feeing of Vrines" &c. Maunfell, Part II. p. 25. Octavo. 1562.

"Certayne godly and comfortable letters of the conftant wytnefs of 1566. Chrift, by John Careles, written in the time of his imprifonment, and now fyrfte fet forth in printe."* Octavo.

"A compendyous regyment, or a dyetary of health, made in Mount- 1567. pyllyer, by Andrewe Boorde, of phyfycke doctour, newley corrected and imprynted, with diuers addycyons. Dedycated to the armypotent prince, and valyent lorde, Thomas, duke of Norfolke, the 1ft day of May 1547." See Warton's Hift. of Eng. Poetry, Vol. III. p. 77. Octavo.

"A Table Collected of the yeares of our Lord God" as p. 732. 1567.

The prymer, fet forth by the kinges highnefs in the 3d and 4th year of his reign. Quarto.

"The A, B, C, fet forth by the kynges maieftie, and his clergye, and commanded to be taught throughout all hys realme. All other utterly fet apart, as the teachers thereof tender his graces favour." Eight leaues. Octavo.

"A Pitious Lamentation of the miferable ftate of the Church of England, in the time of the late reuolt from the Gofpell : Wherein is contained a comparifon betweene the comfortable doctrine of the Gofpell, and the traditions of Popifh religion. With an inftruction how the true Chriftian ought to behaue himfelf in the time of triall. Whereunto are annexed certaine Letters of John Careleffe, written in the time of his imprifonment." By Nich. Ridley, late bifhop of London. Maunfell, p. 91. Octavo.

"The Ordinarie fafhion of good liuing." Maunfell, p. 77. Sixteens.

"Year book,—17. Edward, IV." Folio.

"An Exortation to Charite very needefull at this tyme for eche man and woman to inbrace, Compyled by Wyllyam Conway." B. 6, in eights. W.H. Octavo.

"Certayne godly Exercifes, Meditations and Prayers, &c. Set forthe by certayne godly lerned men : viz. T. Leuer, R. Coles, Ja. P." (James Pilkington.) Octavo.

"Here beginneth the Epyftles and Gofpels of euery Sonday, and holy day in the yeare." R, in fours. Quarto.

"Weftern Wyll vpon the debate betwyxte Churchyarde and Camell, with Dauid Dicars Dreame." In fix-lined ftanzas. Quarto.

Modus obferuandi Curiam Baronū, cum nouis adicionibus. Returna Breuium. &c. &c. Octavo.

HUGH SHYNGLETON, or SINGLETON,

APPEARS to have been free of the old Stationers' Company, as he is found binding an apprentice in 1562, though his name is not inferted among the freemen in their charter from Phil. and Mary, 4 May, 1556. He appears to have been very unfettled in his habitations. His name is moftly written Shyngleton in the hall book. Mr. Ames had heard that he printed in 1525, but he had met with no proof of it, neither have i. In 1562—1563, he was fined for fpeaking unfeemly words before Mrs. ——. In 1566—1567, he was authorifed with Tho. Purfoot to fearch for unlicenfed and diforderly books. About two years afterwards he received from the company X..s. Perhaps on the fame account as the year after he received ij. s. Viz. for taking up books at the waterfide. He appears to have been but an indifferent œconomift, and his principles rather loofe. "Sept. 17, 1577. Whereas Hu. Singleton is indebted to James Afkell lvj. s. It is ordered that he fhall pay the fame at v s. a week. And if default be made in anie payment, then Afkell hath libertie to feek his remedie by lawe. This money Mr. Daie to pay as long as Singleton workes with him, And after, Yt to be demaunded at Singletons houfe.—17. M'cij, 1577-8. Yt is ordered that Hu. Singleton fhall redeliuer vnto W^m. Dickens a pair of Shetis & a diaper towel at or before the 27th. day of this inftant M'che, vpon pain to be cōmitted to ward." Octo. 23. 1584, he borrowed of the Company 5 l. on bond; and for which John Charlewood was fecurity. In 1585—1586, he had xx s. given him by the Company, but no mention for what. In 1591, he and John Charlewood fupplicated the lord chancellor againft a privilege to John Wolf for printing books of lefs than fix fheets of paper. His device or rebus may be feen in the frontifpiece. Latterly he was printer to the city of London. He died between July 1592 and July 1593, in which year Rob. Robinfon difcharged his bond for 5 l. to the Company.

In 1561—1562, he had licenfe to print, "An inftruction full of heavenly confolation. The prefious perle. How a chriftian man oughte to behaue hymfelfe in the daunger of Deathe. Declarynge how God dothe calle vs to Repentaunce." In 1565—1566, "A complaynte betwene nede and pouerte." In 1566—1567. "Thre cōmandementes & leffon of olde Cato as he lay vpon his death bedd. A tretys which ys prouvyd that the fowle of man doth leve & wake after the departure of this world. The Couurt of Venus moralized by Thomas Bryce." In 1567—1568. "An hiftory of lyf & vertu, wherein ys towched the Couurce of mans lyf." In 1568—1569. "The Juftification of a chriftian Fayth. The Retorne of olde well fpoken no body." In 1579, "A neceffarie inftruccōn of the promifes of God. Tranflated from the Latin by Urb. Regius. An aunfwere to a Rebellious Libell. The Shepperds calender, conteyninge xij. eclogis, &c. In 1583, "xxvi fermons of Hen. Bullenger

HUGH SINGLETON.

lenger upon the Oration of the prophet Jeremye, &c. Which book he is appointed to print by his deputye." In 1586, " A thankſgyvinge for our deliv'y from the intended tyrānye of the Antichriſtian Pharao." In 1587, To Wyndet and him. " A brief inſtruccōn & manner howe to kepe bookis of accoumpts," &c. He printed the following.

" The vocacyon of Johā Bale to the biſhoprick of Oſſorie in Irelānde, 1553. his perſecuciōs in ẏ ſame, & finall delyueraunce." Colophon, " Imprinted in Rome before the caſtell of S. Angell/ at ẏ ſigne of S. Peter/ in Decembre Anno D. 1553." With Singleton's mark. * G, in eights. W.H. Octavo.

" The copie of a piſtel or letter ſent to Gilbard Potter in the tyme 1553. when he was in priſon, for ſpeakinge of our moſt true quenes part, the lady Mary, before he had his ears cut off, the 13 of Julye. Anno 1553. 1 Auguſt." Ends, " Quod Poore Pratte. Imprynted in Temſtrete, ouer agaynſte the Styliardes, at the ſigne of the Dobbelhood, by Hewghe Singelton." 8 leaves Octavo.

" De vera obedientia. An oration made in Latine/ by the right 1553. Reuerēde father in God Stephā biſhop of Wicheſtre/ now Lorde Chaūcelour of Englande. With the Preface of Edmonde Bonner than Archideacon of Leiceſtre/ and the kinges Maieſties Embaſſadour in Denmarke/ and now biſſhop of London: touching true obedience/ Printed at Hāburgh in Latine/ in officina Frāciſci Rhodi Menſe Januario/ 1536. And now tranſlated in to Engliſhe/ and printed eftſones/ in Rome/ before ẏ caſtle of. S. Angel/ at the ſigne of. S. Peter. Jn nouembre/ Anno do. M. D. Liij." Concludes with " A faire tale, a good tale, God quyte you for your tale. Very wele ſayed/ wele obeyed/ as ẏ is ſproken in ale." Then are added ſome texts of ſcripture againſt refiſting the goſpel; and a ſhort addreſs to the reader."——Pray that Englande for vnthankefulneſſe to God/ be not brought into ſuche a bondage and ſlauery as in Kynge Wyllyam Baſtardes tyme. Marke in the Chronyckles the prophecye of S. Edwarde/ concernynge the moſt vyle nacion. Fewe wurdes maye ſuffiſe a man that hath wytte/ a fole is neuer taught." &c. " At the end, his mark. See another edition, this year, in our General Hiſtory. W.H. Octavo.

"A godly letter ſent too the fayethfull in London, Newcaſtell, Barwyke, 1554. and to all other within the realme of England, that loue the cōminge of our Lorde Jeſus, by Jhon Knox.——Imprinted at Rome before the caſtle of S Angel, at the ſign of ſainct Peter, in the moneth of July, in the year of our Lord 1554. A confeſſion and declaration of praiers added thereto by Jhon Knox." His mark at the end. Octavo.

A ſpiritual and moſt precious pearle. See it without date. Sixteens. 1569.
M. Luther's " Expoſition on 130. Pſalme. Tranſlated by Tho. Potter." 1577.
Octavo.
Theod.

1578.   " Theod. Beza his little Catechifme."Maunfell, p. 29.   Octavo.

1578.   " A DYALL Of dayly Contemplacion, or deuine exercife of the mind : inftructing vs to liue vnto God, and to dye vnto the vvorld. Firft colected & publifhed in Latin, at the requeft of a godly Bifhop, and Reuerend Father Richard fometime Byfhop of Dirham, and Lorde Priuie Seale. Novv nevvly Tranflated into Englifhe, by Richard Robinfon. Citizen of London. Seene, and allowed.—Anno, 1578." See p. 135, &c. Dedicated by the tranflator to Dean Nowell ; then, the author's epiftle : contains befides, P. 7, in eights." Here endeth this woorke of Contemplacion, fyrft printed in Latine at Weftminfter,—1499, and nowe newly englifhed & printed at London by Hugh Singleton, dwelling in Creede Lane, at the figne of the gylden Tunne. Neare vnto Ludgate. Anno 1578." His mark, or rebus, larger than before. W.H.   Octavo.

1579.   " A Briefe Catechifme, which ftandeth on three parts."   Octavo.

1579.   Chrif. Watfon his Catechifme. Maunfell, p. 32.   Octavo.

1579.   " Neceffarie inftructions of faith and hope, for Chriftians to holde faft, and not to doubt of their Saluation." By Urbanus Regius. Tranflated by John Fox. Reprinted by Hugh Singleton. Maunfell, p.119.   Octavo.

1579.   " The Shepheardes Calender Conteyning twelue Æglogues proportionable to the twelue monethes. Entitled To the noble and vertuous Gentleman moft worthy of all titles both of learning and cheualrie M. Philip Sidney. At London. Printed—in Creede Lane neere vnto Ludgate, at the —gylden Tunne, and there to be folde. 1579." With neat wood-cuts of the 12 months. 56 leaves. George Mafon Efq.   Quarto.

1579.   " A Godly treatife, wherein is proued the true Juftification of a Chriftian man to come freely of the mercie of God. And alfo how good work ought to bee done, and what be true good works indeede. Whereunto is ioyned a conference betweene the lawe and the Gofpell : with a Dialogue of the faithfull and vnfaithfull. Tranflated out of high Almaine by Miles Couerdale." Maunfell, p. 63.   Sixteens.

1579.   " The Hope of the faithfull, declaring briefly & cleerely the refurrection of Jefus Chrifte paft, and of our true effentiall bodies to come. With an euident probatiō that there is an eternall life of the faithfull, & euerlafting damnation of the wicked. Tranflated out of high Almaine by Miles Couerdale." Maunfell, p. 59.   Sixteens.

1579.   " The Booke of Death, or, howe a Chriftian man ought to behaue himfelfe in the danger of Death : and how they are to be comforted whofe deare frendes are departed out of this world. Tranflated out of high Dutch by Miles Couerdale." Maunfell, p. 42.   Sixteens.

" THE

"The Discoverie of a Gaping Gvlf vvhereinto England is like 1579. to be fwallowed by another French mariage, if the Lord forbid not the banes, by letting her Maieftie fee the fin & punifhment thereof.—Menfe Augufti. Anno. 1579." John Stubbs, of Lincoln's Inn, the author, Will. Page the publifher, and Singleton the printer, were tried on the Statute, 1 and 2. of Phil. & Mary, againft the authors, difperfers, or printers of feditious words, or rumours; in confequence whereof Stubbs and Page had their right hands cut off with a butcher's knife and a mallet.— Singleton the printer was pardoned. See Camden's Elizabeth, 1581. F 4, in eights. Neat Roman Long Primer. W.H. Octavo.

"A Catechifme in 3 partes. 1. Of the mifery of men in themfelues. 1582. 2. Of the happinefs of thofe that belieue. 3. Of the duties we owe to God." Maunfell, p. 29. Octavo.

"A penfiue mans practife, very profitable for all perfons, wherein are 1585. conteyned very deuout and neceffary prayers, for fundry good purpofes. With requifite perfwafions, before euery prayer. Written by John Norden. Newly corrected and enlarged." Quarto.

"The order of my Lorde Maior, the Aldermen, & the Shiriffes, for 1586. their meetinges and wearing of their apparell throughout the yeere. Imprinted——by Hugh Singleton Printer to the right Honorable Citie of London. 1586." W.H.+ Octavo.

"The Abridgement of an Acte of Common counfel paffed at the 1586. Guildehal in London, the feconde day of July 1586.——And there at the fame time eftablifhed for a law, for the better feruice of hir Maieftie in hir cariage belonging to hir highneffe houfeholde, and for the better gouernement of Cartes, Carters, Carres & Carre men, and the gouernement thereof by the authority of the fame Act cōmitted to the Gouernours of Chriftes Hofpitall. WH.+ Octavo.

"A Briefe Inftruction and manner hovv to keepe bookes of Accompts 1588. after the order of Debitor and Creditor, & as well for proper Accompts partible, &c. By the three bookes named the Memoriall, Iournall, & Leager,—.Newely augmented & fet forth by John Mellis Scholemafter. 1588. Imprinted—by John Windet,—1588." To this is annexed "A Short and Plaine Treatife of Arithmeticke in whole numbers, comprifed into a briefer method than hetherto hath bin publifhed. Adioyned to this fmall Treatife for the ayde of Learners. By Iohn Mellis" &c. Colophon: "Printed by John Windet for Hugh Singleton, and are to be folde at his Shoppe at the North dore in Chriftes Hofpital, next vnto the Cloyfter, going into Smithfield. 1583." His device. W·H. Octavo.

"The fupplication of doctour Barnes vnto the mooft gracyous kynge Henrye the eyght, with the declaration of his articles condēned for herefy by the byfhops." It was printed by John Byddel, 1534. The colophon

HUGH SINGLETON.

colophon, "Imprinted at London in Poules church-yard, at the signe of St. Augustyne, by Hugh Syngelton. Cum priuilegio ad imprimendum solum." Octavo.

——— "An Jnstruccyon of Christen fayth howe to be bolde vpon the promyse of God and not to doubte of our saluacyon/ made by Vrbanus Regius. Traflated into englyshe. Prynted—at the singne of saynt Augustine in Paules churche yerde." Dedicated by J. Fox, the translator; "To his reuerende and singuler good father Ric. Melton." E 7, in Eights. W.H. Octavo.

——— "A spirituall purgation sent vnto al them that laboure of Luthers Errour, as touching the bodely presens of Christe our sauiour in the Sacrament, and to al them that haue espyed the libertie of the gospel, as touching theyr fleshe, yet seke not the lybertie to make free theyr spyrite from thys afore sayde errour. Jhon. xviii. Euery one that is of truth heareth my voyce. Cum. Priuilegio—Solum." Dedicated by the author "To the right worshipypfull & godlye knyght Syr Thomas Wyat." K, in eights. At the end, "God saue the Kynge.——T. C." On the last leaf, "Imprynted at the sygne of saynte Augustyne in paules churche yearde." W.H. Octavo.

——— M. Luther's Sermon on the angelses.. Octavo.

——— "A demonstratiue Oration of the resurrection of the deade compiled by Claudius Alberius Triuncurianus. Most necessary to be vsed of all Christians, which doo more regard Celestiall ioyes, then the infernall paynes. Printed at London by Hugh Singleton." Dedicated, "To the right worshipfull the Wardens and assistants of the Drapers societie, their humble scholler. W. M." the translator. 15 leaves. W.H. Octavo.

——— "Orders Appointed to be executed in the Cittie of London, for setting roges and idle persons to worke, and for releefe of the poore.—Printed—in Smithfielde, at the signe of the golden Tunne." 8 leaves. W.H. Quarto.

——— "A spirituall and most precious perle, teaching all men to loue and embrace the cross as a most sweet and necessary thing vnto the soul, &c. written for thy comfort, by a learned preacher, OTHO. WERMYLIERVS, and translated into English by MILES COVERDALE." The printer to the christian reader, thus: "Whereas (by the sinister dealing of some) this boke hath gone abroad, as imprinted by me, with mo and greater faultes then either hath or shall (I hope) be found by any boke imprinted by me, indeed I haue thought it good to set it forth once againe, according to the true copy of that translation, that I received at the hands of M. doctour Milo Couerdale, at whose hand I receiued also the copyes of three other workes of Otho Wermulerus, a German preacher in the city of Tigurie, who wrote them in the Germain tongue, as he did certain bookes moe, which are not as yet turned into the Englishe tongue. The

names

names of thofe bookes, which are tranflated, are thefe: the precious pearle, which the author calleth, of affliction, another of death, the third of iuftification, and the fourth of the hope of the faithful. Thefe I haue imprinted with as great diligence as I coulde, and I hope according to the copies, that I receiued from the tranflator of them, howfoever the fame haue been by others thruft out in my name, corruptly enough to my great difcredite, and fome hindrance alfo, but with the greateft difpleafure to the buyers. Mine hearty defire hath been (deare reader) alway to be fo occupied, that I might liue by profiting others, in fuch fort as that excellent prince, the late duke of Somerfet, and uncle to good king Edward the fixte, wifhed this pearle to be imprinted, after he in his trouble had felt the commodity of it; which was that other might reape the like commodity by the fame. Take therefore thys precious pearle, now brought to that brightnefs, that it had, when it firft came oute imprynted by me, and vfe it to thy comforte in all thy diftreffes. And for the comforte, that thou fhalt find therein, give God the whole prayfe. Farewell, quod Hugh Singleton." Then his mark and motto, GOD IS MY HELPER. At the end, his mark again, and is faid to dwell " at the Golden Tunn in Creed-lane, near Ludgate. Cum Privilegio." * Sixteens.

" THE SANCTVARIE OF SALVATION, HELMET OF HEALTH, AND MIRROVR OF MODESTIE AND GOOD MANERS. Wherein is conteined an exhortation vnto the inftitution of a Chriftian, vertuous, honeft, & laudable life, very behouefull, holfome, & fruitfull both to the higheft & loweft degrees of men, which defire either health of bodie, or faluation of foule. Written in Latin—by Leuinus Lemnius of Ziriziæa, Phifition, and englifhed by H. K."—His mark, and under it " Printed at London by Hugh Singleton." It is dedicated by the tranflator, Hen. Kinder, " To the right worfhipfull Maifter Stephen Thimelby Efq; Recorder of the Citie of Lincoln." 227 pages, befides the dedication, and preface prefixed, and a table at the end. White letter. W.H. Octavo.

" A brefe and a playne declaratyon of the dewty of maried folkes, gathered out of the holy fcriptures, and fet forth in the almayne tonge by Hermon Arcbyfhop of Colayne, whiche wylled all the houfholdes of his flocke to haue the fame in their bedchambers as a mirror or glaffe dayly to loke in, wherby they might know & do their dewties eche vnto others, and lede a godly, quiet & louing life togethers, and newly tranflated into y̆ Englifhe tonge by Hans Dekyn. Jacob. i. Se that ye be dores of the worde & not hearers, only difceyuing your felfes. Coloff. iij. Aboue al things put on loue which is the bond of perfectnes." C 3, in eights. Jmprynted—in Temeftrete by Hughe Syngleton, at the—dobbel hood, ouer agaynfte the Stylyarde." W.H. Octavo.

" Certayne Caufes gathered together, wherin is fhewed the decaye of England, onely by the great multitude of fhepe, to the vtter decay of houfeholde keeping, mayntenaunce of men, dearth of corne, and other
notable

HUGH SINGLETON.

—— notable dyfcommodityes, approued by fyxe olde prouerbes. Imprinted— in Pouls churchyeard, at the fygne of faynte Auften, by Heugh Syngleton." Dedicated to the king and parliament. 12 leaves. Harleian Pamphlets, N°. 174.　　　　　　　　　　　　　　　　Octavo.

—— John Bradford's " Fruitfull treatife, full of heauenly confolation, againft the feare of death : whereunto are annexed certaine fweet meditations of the knowledge of Chrift, of life euerlafting, and of the bleffed ftate & felicity of the fame. Maunfell, p. 23.　　　　　　Octavo.

—— " A farmon of Jhon Oecolampadius to yong men and maydens." Tranflated by John Fox, and printed by Hum. Powel. " And are to be f uld by Hugh Syngleton,—at the fygne of Saynt Auguftine, in Poules church yarde. W.H.　　　　　　　　　　　　　　　Octavo.

—— " Of Mifrules contending/ with Gods worde by name/
　　And then/ of ones Judgement/ that had heared the fame."
Contains 22 ftanzas of 4 lines each. " Finis. Quod Wyllym Kethe. Imprynted—in Temeftrete—ouer againft the Stiliardes."　Broad fide.

—— " A godlie and zealous prayer to bee vfed of euery Chriftian and dutifull Subiecte for the preferuation of our moft foueraigne Lady Elizabeth, —Queene : and in all caufes, as well ecclefiaftical as temporall, next vnder God, of the Church of England & Ireland fupreame Gouerneffe." I. P at the end of the prayer."—printer to the right honorable citie of London."　　　　　　　　　　　　　　　　　　　　Broad fide.

RICHARD KELE,

AT the long fhop in the Poultry, under St. Mildred's church, and in Lombard-ftreet at the fign of the Eagle, near unto Stocks Market.

1546.　" Modus tenendi vnum Hundredum fiue Curiam de recordo. Londini in edibus Richardi Kele." At the end, " Jmpreffum—per Richardum Kele. An. dnī. 1546. W.H.　　　　　　　　　　Octavo.

1546.　" Paruus libellus continens formam multarū rerum, prout patet in Kalendario in fine contento. Anno. 1546." 50 leaves. " Explicit Carta feodi. Impreffum—per Richardum Kele, An. dnī M.CCCCC.XLvi, Cum priuilegio—folum. R. K." W.H.　　　　　　　　Octavo.

" A faithful

RICHARD KELE. 747

"'A faithful and true prognoftication vpon the yeare M. CCCCC. XLVIII, 1548.
and perpetually after to the worldes end, gathered out of the prophefies
and fcriptures of God," &c. Tranflated out of high Almaine by Miles
Gouerdale." Octavo.

The fame. It was printed in like manner in 1536; and probably 1549.
printed frequently.

" The ymage of bothe paftoures, &c. as p. 690. 1550.

" Flores aliquot fententiarum" &c. as p. 354, &c. 1550.

" A godly and holfom preferuatyue againft difperacio," &c. as p. 355. 1551.

" The book of the properties of herbs," &c. as p. 355. Alfo without 1552.
date.

Richard Taverner's Adagies, gathered out of the chiliades of Erafmus. 1582.
Octavo.
If there be no miftake in the date, this book was probably printed by
a fucceffor of his name.

" The Garden of Wyfdome," &c. as p. 366. ———

" An expofition vpon the epiftle of Jude" &c. as p. 365. ———

" Admonycion of a certen trewe paftor & prophet fent vnto the Ger- ———
manes." Octavo.

" The workes of Geffray Chaucer" &c. as p. 557. ———

Several pieces compiled by mafter Skelton poet laureat. " Colyn ———
Cut. The boke of Phyllyp Sparowe." &c. Octavo.

" This is the Myrrour or Glaffe of Helth," &c. as p. 375. Colophon: ———
"Imprynted—by Richarde Kele, dwellinge at the longe fhope vnder faincte
Mildreds churche." W.H. Octavo.

" Here begynnethe a goode Booke of medecines called the treafure ———
of poore men." See p. 574. Octavo.

" A chrifte exhortation vnto cuftomable fwearers," &c. as p. 711. ———

" A proclamacyon of the hygh emperour Jefu Chrifte," &c. as p. 394. ———
B, in eights. Imprinted——by Rycharde Kele,——at the longe fhope
in the Poultry vnder faynte Mildredes church." W.H. Octavo.

" The boke of hufbandrye, verye profytable and neceffarye for all ———
manner of perfons, newlye corrected and amended by the auctor Fitzherbarde; with dyuers additions put thereunto." Contains H, fheets.
Sixteens.
Another edition containing Sheet I.

ANTHONY

ANTHONY SCHOLOKER.

PRINTED in London, Weftminfter, and at Ipfwich. He tranflated Viret's Collection of fcriptures serving for expofition of the Lord's Prayer, out of French; the Ordinary for all Faithful Chriftians, out of Dutch; and, A brief fum of the Bible, out of German. He ufed for his device, a touchftone with Verbum Dei engraved on it; over it is a hand extended from the clouds, rubbing a piece of money on the ftone, to try its genuinenefs; the wind blowing on it from another Quarter. Under this device, "Proue the fpyrites, whether they be of God. Jhon ỹ. iiij. i. Reg. viij. d. Mat. vij." Mr. Ames imagined, from a fimilitude of types, that he printed The complaint of Roderyck Mors; but when he printed, which feems to have been only in the time of Edw. the 6. there appears no reafon why he might not have put his name, or device to it. Some books he printed with Will. Seres, which are entered under his name. He printed alone the following.

1548. " A right goodly rule howe all faithfull chriftyans ought to occupie and exercyfe themfelues in their dayly prayers." With cuts. " Dwelling in St. Botolphs paryfhe." * Sixteens.

—— " The prayfe and commendacion of fuche as fought comenwelthes: and to the cōtrary, the ende and difcommendacion of fuch as fought priuate welthes. Gathered both out of the Scripture and Phylozophers. Prov. xiij .7. Imprinted at London,—Dwelling in the Savoyrēts Wythout Temple barre. Cum priuilegio.——folum."* Sixteens.

—— " The Right inftitucion of baptifme fett forth by the Reuerend Father in Chrift Herman Archebiffhop of Coleyne, Wherunto is alfo annexed a godly treatyfe of Matrimonie, compyled by the famous Clerke and faithfull Euangelift Wolfgangus Mufculus, no leffe frutefull thē neceffary for all godly minifters of chriftes church, tranflated by the vnprofitable feruaunt of chrift Richard Ryce.¶ Mark ¶ Suffer the infantes to come vnto me, and forbidde them not, for vnto fuch the kyngdome of God is due. Imprinted—by Anthony Scoloker. Dwelling in S. Botolphs Parifh wythout Alderfgate." c 6, in eights. W.H. Sixteens.

—— " The olde Fayth of greate Brittaygne and the newe learnynge of Inglande. Wherunto is added a fymple inftruction concernynge the Kinges Maiefties procedinges in the cōmunyon. Compyled by R. V. Nume. xvi. (verfe 28.) Imprinted——in the Sauoy rentes, Wythoute Templebarre. Cum priuilegio——folum." a 5, in eights, half fheets. W.H. Sixteens.

" The

"The Order of Matrimony. Hebre. xiij. (*verse* 4.) Let wedlocke be had in pryce in all poyntes, &c. Imprinted—in the Sauoy rentes. Cum priuilegio—folum." B 4, in eights, half sheets. W.H.   Sixteens.

"Pyers Plowmans exortation, vnto the lordes, knights, and Burgoysses of the parlyament house." In the reign of Edw. vi.   Octavo.

Mr. Warton mentions a collection of Skelton's poems printed for him 1582. Hist. of Eng. Poetry, Vol. II, p. 336, But query the date.

HUMPHREY POWELL, or POWEL,

DWELT in 1548 and 1549 above Holborn-conduit; but he appears soon afterwards to have gone over to Ireland, where some further notice will be taken of him. We find nothing of his printing after 1551: his name however stands in the list of freemen of the worshipful company of Stationers, inserted in their charter in 1556; but i find no further mention of him.

"The haruest is at hande wherein the Tares shal be bound and cast into the fire, &c. By John Champneis." Maunsell, p. 57.   Octavo.   1548.

"The fiue abominable blasphemies contayned in the Masse, by John Veron."   Octavo.   1548.

"AN HOLSOME Antidotus or counterpoyfen, agaynst the pestilent heresye and secte of Anabaptistes, newly translated out of latī into Englysh by John Veron Senonoys." Dedicated "To the most redoubted prynce Edwarde——Duke of Somerset, Lorde Protector," &c. Therein we are informed that this treatise was written originally in the German language, by Hen. Bullinger, and translated into Latin by Leo Jude, bishop of Tigure. The original consists of 4 parts; of which this is only the first, divided into ten dialogues. P 4, in eights. On the last leaf, "Thou shalt vnderstāde, good Cristen Reader, that as sone as may be thou shalte haue the seconde Booke," &c. Imprintyd——by Humfrey Powell, dwellyng Above Holburne Conduit. Cum priuilegio——solum. Anno Do. 1548." W.H.   Octavo.   1548.

"Certayne litel treatises set forth by John Veron Senonoys, for the erudition and learning of the symple and ignorant people."   Octavo.   1549.

U u u u   "The

— "The pfalter or pfalms of Dauid" &c. as p. 550.

— "A Sarmon of Jhon Oecolampadius to yong men and maydens." In a compartment with Will. Middleton's mark. Tranflated by John Fox. C 7, in eights. "Jmprinted—by me Humfrey Powell—aboue Holburne conduit. And——fould by Hugh Syngleton," &c. W.H. Octavo.

— "Here begynneth the Egloges of Alexander Barclay, prieft, wherof the firft thre conteineth the miferies of courters and courtes, of all Princes in generall. The mattier whereof was tranflated into Englyfhe by the faied Alexander, in forme of dialoges, out of a boke named in Latin, Miferie curialium," &c. P 2, in fours. Concludes, "Thus endeth the thyrde and laft Eglogue of the Mifery of Courte and Courters, Compofed by Alexander Barclay, preeft, in his youth. Jmprinted at London by Humfrey Powell." W.H.   Quarto.

ROBERT STOUGHTON,

Dwelling within Ludgate, at the fign of the bifhop's mitre.

1548. "Two Epiftles, one of Henry bullynger, wyth the confent of all the learned men of the Churche of Tygury: another of Johan Caluyne, chefe Preecher of the church of Geneue: whether it be lauful for a chryften man to communicate or be partaker of the maffe of the papyfts, without offendīg God and hys neyghbour, or not. i Corinth. x.——Imprinted—— Anno 1548." B 7, in eights. W.H.   Octavo.

1548. "THE OLDE LEARNYNG and the new compared together, wherby it may be eafely knowē which of them is better and more agreyng wyth the euerlafting word of God. Newly cerrected & augmented by Wyllyam Turner." Tranflated from Urban Regius. H 6, in eights. "Imprinted——M. D. XLVIII." W.H.   Octavo.

1548. "A Godly Newe fhort treatyfe inftructyng euery parfon howe they fhulde trade theyr lyues in ẏ Jmytacyon of Vertu/ and the fhewyng of vyce, and declaryng alfo what benefyte man hath receaued by chrift, through the effufyon of his moft precyous bloude. Jmprīted" &c. 17 leaves. By Hen. Harte. "Anno. 1548. the 23. October." W.H.   Sixteens.

1548. "A bryef & fhort declaracyon made wherbye euerye chryften man may knowe what is a facrament. Of what partes a facramente confyfteth
&

ROBERT STOUGHTON.

& is made, for what intent facramentes were inftituted, and what is the pryncypal effect of facramentes, and finally of the abufe of the facrament of chryftes body and bloud. Jmprinted" &c. 15 leaves. " Compyled by Rychard Tracye.——Anno. 1548. the. 10. of Nouembre." W.H.
 Sixteens.

" A compendious, or fhort treatife gathered out of the chief authors 1551. of phyfic, containing certen precepts neceffary to the preferuation of health, and long continuance of the fame. by Hen. Wingfeld." Dedicated to *Sir* William Cecil. Alfo without date. Octavo.

" William Samuel, minifter, his Abridgement of Gods ftatutes in 1551. meeter." Printed for him. Maunfell, p. 95. Octavo.

" A Briefe declaration of the facraments, expreffing the fyrft oryginall how they came by, ãd were inftitute with the—meaning and vnderftand-yng of the fame, very neceffarye for all men, that wyl not erre in the true vfe and receauing therof, Compyled by the godly learned man Wyllyam Tyndall. Jmprinted" &c. 40 leaves. W.H. Sixteens.

" The trew Judgemẽt and declaration of a faithful Chryftyan, vppon the facrament of the body and bloud of Chryft, agreeng wyth the fcrypture and mooft Catholycke and trew dyffinicyon of the doctours. Jmprinted" &c. 11 leaves. W.H. Sixteens.

" The Cenfure and iudgement of the famous clark Erafmus of Rot-erodam : Whyther dyuorfemente betwene man and wyfe ftondeth with the lawe of god." &c. as p. 580.

" A difcourfe, or traictife of Petur Martyr Vermilla, Florẽtine, the publyque reader of diuinitee in the Vniuerfitee of Oxford, wherin he openly declared his whole and determinate iudgement of the Sacrament of the Lordes fupper, in the fayde Vniuerfitee." In the compartment with Whitchurch's cypher on the fell. It is dedicated " To the right honorable Sir Wylliam Parre Knight, Lord Parre, Erle of Effex, Marqueffe of Northampton, Lorde great Chamberlaine of Englande, and Knight of the mofte noble ordre of ỹ Garter," by Nic. Udall the tranflator. Contains befides, Fol. cx. " Jmprinted—for Nycolas Vdall. Cum priuilegio—folum." W.H. Quarto.

" A notable treatife fhewing, that by the word of God we may eat at al times, fuch meats as God has created for the fubfiftaunce of man," &c. Octavo.

"•A bryfe fumme of the whole Byble. A Chryftian Jnftruction for all parfons, younge and old. To the which is annexed, The ordinary for all grees. Tranflated out of the Doutche—by Anth. Scholoker." Octavo.

" Here beginneth a fong of the Lordes Supper." &c. as p. 362.

Uuuu 2 GUALTER

[752]

GUALTER or WALTER LYNNE,

A Scholar and author, as well as printer of several books, dwelt on Sommers Key, near Billingsgate; and he seems to have kept shop at the Eagle next S. Paul's School. His device was a ram, marked with W, and a goat marked with L, as in the frontispiece.

1547. " The begininge and endinge of all popery," &c. as p. 579.

1548. " Richard Bonner priest, his treatise of the right worshipping of Christ in the sacrament of the bread and wine, when it is ministred with thankesgiuing in the holy supper. printed for Gualther Lin." Maunsell, p. 22.
    Octavo.

1548. " Catechismus.—set forth by—Thomas Archbishop of Canterbury," &c. as p. 708.
The same without date. Printed for him. W.H.   Octavo.

1548. " A declaration of the twelue Articles of the christen faythe," &c. as p. 714.

1548. " The diuision of the law and the gospele: gathered out of the holy scriptures, by Petrum Artopoeum."   Octavo.

1548. " A frutefull and godly Exposition and declaration of the kyngdom of Christ, and of the christen lybertye, made vpō the wordes of the Prophete Jeremye in the xxiij. Chapter with an exposycyon of the viii. Psalme, intreatyng of the same matter, by the famous clerke Doctor Martyn Luther, whereunto is annexed a Godly sermon of Doctor Vrbanus Regius, vpon the ix. Chapyter of Mathewe, of the woman that had an issew of blood & of the rulers daughter, newly translated oute of hyghe Almayne. Imprinted for Gwalter Lynne. Anno. M. D. Xlviii." Dedicated by G. L. " To the most gratiouse & vertuouse Lady Elizabeth syster to the Kynges moost Royal Maiestie." K, in eights. Under his device, " Jmprinted for——dwellyng vpon Somers kaye, by Byllynges gate.—Cum gratia & priuilegio—solum." W.H.   Octavo.

1548. " The chiefe and pryncypall Articles of the Christen faythe, to holde againste the Pope, and al Papistes, and the gates of hell, with other thre very profitable & necessary bokes, the names or tyttels, whereof are conteyneth in the leafe next followynge. Made by Doctor Marten Luther. ¶ To the Reader. Jn thys boke shal you fynde Christian Reader the ryght probation of the righte Olde Catholyke Churche, and of the newe false Churche, whereby eyther of them is to be knowen. Reade and iudge." Q 4, in eights. His device and colophon as to the last article. W.H.   Octavo.

" A lytle

GUALTER LYNNE

" A lytle Treatife after the maner of an Epiftle wryten by the fa- 1548.
moufe Clerk Doctor Vrbanus Regius to his friend, about the caufes of
the great controuerfy, that hath been, and is yet in the chriftian religion." Neatly printed, with a cut by H. Holbein. Octavo.

" A tragoedie or Dialoge of the vniufte vfurped primacie of the Bifhop 1549.
of Rome, and of all the iuft abolifhyng of the fame, made by mafter
Bernardine Oehine, an Italian, and tranflated out of Latine into Englifhe
by Mafter John Ponet Doctor of Diuinitie, neuer printed before in any
language. Anno Do. 1549." At the end, an epiftle of G. Lynne's
to K. Edw. 6. W. H. Octavo.

" A briefe collection of all fuch textes of the fcripture as do declare ỹ 1549.
moft bleffed and happie eftate of the that be vyfeted wyth fycknes & other
vifitations of God, and of the that be departinge out of this lyfe,
wyth moft godly prayers & general confeffions, verie expedient &
mete to be read to all ficke perfones, to make the wyllynge to dye. Wherunto are added two fruitfull & comfortable fermõs made by the famoufe
clarke doctor Martin Luther, verye mete alfo to be reade at the buri-
alles. Ecclefiaftes, vij. Jt is better to go into an houfe of mournyng,"
&c. M, in eights. " Jmprynted on Somers kaye M. D. XLJX.
Cum priuilegio folum." W.H. Octavo.

" A treatife or fermon of Henry Bullynger, much fruitfull and necef- 1549.
fary for this tyme, concernynge magiftrates and obedience of fubjects.
Alfo concerning the affayres of warre, and what fcryptures make mention
thereof," &c. With the epiftle of the printer W. Lynne. Octavo.

" A notable Sermon concerninge the ryght vfe of the lordes fupper, 1550.
and other thynges very profitable for all men to knowe, preached before
the Kinges moft excellent Mayeftye, and hys moft honorable counfel, in
hys courte at Weftmynfter, the 14. daye of Marche [on Mat. xiv. 4.]
by Mayfter John Ponet, Doctor of dyuinity." 1550. At the end, A
prayer againft the Pope and Turks. " Jmprynted—for Gwalter Lynne,
—on Somers Kaye—. ¶ And they be to be folde in Paules church yarde,
next ethe great Schole, at the figne of the fprede Egle. Cum priuilegio
——folum." W.H. Octavo.

" The thre bokes of Cronicles, whyche John Carion (a man fyngularly 1550.
well fene in the Mathematycall fciences) gathered wyth great diligence
of the befte Authors, that haue writtten in Hebrue, Greke, or Latine.
Wherunto is added an Appendix, conteynyng all fuch notable thynges
as be mentyoned in Cronicles to haue chaunced in fundry partes of the
worlde, from the yeare of Chrift. 1532. To thys prefent yeare of. 1550.
Gathered by John Funcke of Nurenborough. Whyche was neuer afore
prynted in Englyfh. G. Lynne dedicates it to king Edward the vith, and
fays, " I haue thought it my duetye (beynge one that fpendeth all hys
tyme in the fettynge forth of bokes in the Englyfhe tounge) emongeft
all

all other to set forth thys shorte Cronicle;" and intimates that he had translated it from the Latin, &c. Then a preface concerning " The vse of readynge hystoryes." The Chronicle contains Fol. cclxxix, besides an index ; at the end of which is his rebus, and a colophon, as the last article. W.H. Quarto.

1550. " The true beliefe in Christ and his sacramentes, set forth in a Dialogue betwene a Christen father and his sonne, very necessary to be learned of all men, of what estate soeuer they be. ¶ My sonnes, heare the enstruccions of youre father, &c. Prouerbiorum iiij. Jmprinted——for Gwalter Lynne,——On Somers kaye,——.Anno Domini. M. D. L. Cum Priuilegio——solum." Dedicated "To the moste gracioufe Lady, Lady Ann doucheffe of Somerset." In which he says,"——.The author of the boke J know not. Only this J finde that it was fyrste written in the duche tong, and then translated into latine.——Jt declareth effectuously——the. xij. Articles of the Christian faith &c.—J woulde wyshe therfore that al men, women, and chyldren, would read it. Not as they haue bene here tofore accustomed to reade the fained storyes of Robinhode, Clem of the cloughe, wyth such lyke to passe the tyme wythal, &c. Geuen at London——.M. D. &. L. The xx. daye of Januarye. Your Graces dayly Orator Gwalter Lynne." G 4, in eights. W.H. Octavo.

1550. " The spyrytual and precious pearle. To which is added, A humble peticyon to the Lord, practysed in the common prayer of the whole famylye at Shene, durying the trouble of their lord and master the duke of Somerset his grace. Gathered and set forth by Thomas Becon, minister there. Which trouble began the vi of October, the year of our Lord M. D. XLIX, and ended the vi of Febuarye then next ensuyng." Sixteens.

1550. " An Epistle written by D. Peter Martir to the Duke of Somerset. Translated by Thom. Norton." Maunsell, p. 66. Octavo.

1550. " A briefe and compendiouse table, in a maner of a concordaunce, openyng the waye to the principall histories of the whole Bible, and the moste comon Articles grounded and comprehended in the newe Testament and olde, in maner as amply as doeth the great concordauce of the Bible. Gathered and set furth by Henry Bullynger, Leo Jude, Conrade Pellicane, and by the other Ministers of the churche of Tigurie. And nowe first imprinted in Englyshe. M. D. L. The third boke of the Machabees a booke of the Bible also prynted vnto this boke which was neuer before Translated or prynted in any Englyshe Bible." Mr. Ames's copy wanting this general title, he printed from the head titles. The dedication " To the right noble and vertuouse Lady, Lady Anne, doucheffe of Somerset," begins : " Beinge credibly enformed (mooste gracioufe Lady) by such, as be nygh about youre grace, that youre graces chiefe and daylye study is the holy Byble : J thought J coulde not presente your grace wyth any Juwell more acceptable and pleasaunt to your grace than suche a bryefe concordaunce. Whereby youre grace myghte wyth more

redynes

redynes fynde all fuche thynges, as you fhoulde be defyroufe to fee. Then as for gold or filuer, J am not able to prefente vnto your grace, but my ruyde laboures, and that J do wyth all my herte. When thys lyttle boke therfore came to my hand (beynge wrytten in the hygh Almayne tongue) and J perceiued it to be mooft diligently gathered out of the holy Byble, and fet in order, that wyth in it lyghtelly as much maye be founde by it, as by a great concordaunce, J ceaffed not tyll J had it tranflated and fett in lyke order in the Englyfh tounge, that your grace myghte not be deftytute of fo neceffarye an inftrumente in your godlye ftudie. And that all other whych be ftudyoufe of godly knowledge myghte be the more defyroufe to haue thys lyttle boke and embrace it wyth better wyll (knowynge your graces feruent zeale for the furtherynge of Goddis trueth) J haue byn (as my dutie byndeth me) fo boulde as to dedycate vnto your grace for to forther your grace in your godly ftudye, truftinge that your grace, according to your acuftomable gentilnes, will accept my good wyll therein, and then fhall J thynke my labours well employed, & my Paynes well recompenfed.——Moreouer, it behoueth, that J lett youre grace knowe the caufe whye J haue annexed the thyrde boke of the Machabees vnto this table. Whych is for that it is verie often fpoken of in thys little table, and is not to be found in any Byble in Englyfhe, faueynge only in one, whych John Daye the prynter hath now in pryntynge. Lefte your grace therfore (or any other that fhall chaunce to haue thys lyttle boke) fhould thinke that there were no fuch boke of the Byble: I haue caufed thys thyrde boke of the Machabees to be tranflated, and haue imprynted it with this table, &c. Gwalter Lynne." * W.H. Octavo. The Lambeth lift of Bibles, &c. mentions it alfo in Quarto.

" A catechifme, that is to faie, a familiar introduction and training of the fimple."

" S. Auguftine his two bookes: the one of Predeftination of Saints, the otherof Perfeuerance vnto the ende. With the determination of 2 general Councelles concerning that matter. Tranflated by Nich. Leffe." Maunfell, p. 5. Octavo.
" The Diuifyon of the places of the lawe & of the gofpell gathered out of the hooly Scriptures by Petrum Artopœum. Whereunto is added two orations of praying to God made by S. John Chryfoftome, and no leffe neceffary then lerned. Tranflated into Englifh. Math 26. Watche & pray, &c. Imprinted by my Gwalter Lynne." Octavo.

WILLIAM HYLL or HILL,

AT the figne of the green hill, at the Weft door of S. Paul's church. One of both thefe names is inferted in the lift of freemen belonging to the Stationers' company, in their charter of Phil. & Mary, 1556. If this

this was the same person, he appears to have then left off printing, and followed Book-binding only; for i find no copies licenced to him, but about 1560 he was fined one shilling for binding Primers in parchment, contrary to the company's orders.

1548. "Here begynneth a godly newe story of. xii. men that moyses (by the Cōmaundement of god) sent to spye out the land of canaan: of whiche xii. onely Josua and Caleb, wer found faythful messengers. Psalme. cv. O that me woulde praise y goodnes of god, &c. By Phyllyp Nycolls." 51 leaves. "Jmprented——the Tenthe daye of Maye: Anno dominice incarnationis M·C XLViii.* By william hill remaynyng at the Signe of the hill in Paules Church Yarde: and be there by him to be solde." W.H.
Octavo.

1548. "A new Dialoge called the endightment againſt mother Meſſe." &c. as p. 690.

1548. "The summe of the holy Scripture, and ordinarie of the true Chryſtian, teaching of the Chryſtian Faith, by which wee are all iustified: and of the vertue of Baptisme, with an information howe all Chriſtian people should liue according to the Gospell." See p. 616.
Octavo.

1548. "Certein places gathered out of Austens boke intituled, De essentia diuinitatis, by Hermon Bodius, and now tranflated into Engliſh. 13th of December."
Octavo.

1548. "Solace of the soule againſt the bitter stormes of Sicknes and death," &c. Maunsell, p. 107.
Octavo.

1548. "A table of the principall matters conteined in the Bible" &c. as p. 583.

1549. "The Byble," &c. as p. 583.

1549. "The Physicke of the soule, wherin thou ſhalt finde many Godly emplastures & comfortable salues agaynſt al spiritual diseases, very necessary to be red of the true chriſtians in these last and perilous dayes. Set forth by Tho. Becon. Math. xi. Come vnto me all ye that are sycke, &c. Jmprinted——the tenth day of Julii.——M. D. xlix. And are to be solde by——remainīg at the sygne of the Hyll in Paules Churche Yarde. B 3, in eights.
Octavo.

——— "A sermon declaringe how vue ar iustified by faith.
Octavo.

——— "An Expoſicion vpon the. v. vi. vii. chapters of Mathew, which thre chapters are y keye and the dore of the scriptures, &c. Compyled by Wyllyam Tindal. Newly set forth and corrected according to his firſt copye." Fol. c. and a table at the end. "Jmprynted——at the signe of the Hyll, at the weſt dore of Paules. By Wyllyam Hyll. And there to be sold." W.H.
Octavo.

The

Thus in the original.

" **The** Obedyence of A Chriftian man: and how chriften Rulers ought to gouerne: wherin alfo (yf thou marke dylygentlye) thou fhalte finde eyes to perceyue the craftye conueighaunce of all Jugglers. Reade (whenfoeuer thou Readefte good Chriften Reader) with a pure affection, and vprighte Judgemente to Godes mofte holy Booke." In a compartment, having T on a tablet at the bottom of one jamb, and P on another tablet at the bottom of the other jamb. 140 leaves. No printer's name appears to this, but it feems to have been printed together with

" The parable Of the wycked mammon taken out of the. xvi. Ca. of Luke, with an expoficyon thervpon lately corrected & prynted. Luce. xvi. Facite vobis amicos de mammon iniquitatis " In the fame compartment as the former article. K, in fours. " Jmprynted—The. xv. day of September." W.H. Quarto.

The obedience of a chriftian man. See p. 585. Octavo.

" The Pyftles and Gofpeles." The late Dr. Freind, dean of Canterbury. Quarto.
A ballade made againft the pope & popery, by Wm. Punt.

ROBERT CROLE, CROLEUS, or CROWLEY,

WAS born in Glocefterfhire, became a ftudent in the Univerfity of Oxford about 1534, and was foon after made demy of Magdalen-college. In 1542, being bachelor of arts, he was made probationer-fellow of the faid houfe, by the name of Robert Crole. In the reign of Edward vi. he exercifed the profeffion of printing, in Ely-Rents, Holborn, London; fold books, and preached in the city and elfewhere; but upon the fucceffion of queen Mary, he, among feveral Englifh Proteftants, went to Frankfort, in Germany. After her deceafe he returned, and had feveral benefices beftowed upon him, among which was St. Giles by Cripplegate, London, of which church he wrote himfelf vicar in 1566. He lived to a fair age, was buried at St. Giles, Cripplegate, and over his grave a ftone was laid, with this infcription engraven on a brafs plate: "Here lieth the body of Robert Crowley, clerk, late vicar of this parifh, who departed this life the 18th of June, anno Domini 1588." He was admitted a freeman of the worfhipful company of Stationers by redemption (*gratis*) 27. Sept. 1578. Soon after, he is found, jointly with the wardens, licencing copies; as alfo are feveral other perfons; but by what authority does not appear. Though he printed two or three books after the date of the company's charter, there are none entered as licenced to him. Being very zealous for a reformation from popifh errors, he exerted his abilities againft them, both by writing and printing. His writings may be feen in bifhop Tanner's Bibliotheca, p. 210. But be pleafed to obferve that though "The fupper of the Lorde," and "The Vifion of Piers Plowman," are inferted among the reft of his writings, he wrote only the prefixes to them; And that The fchool of Vertue and good nurture is afcribed to F. S. in p. 692.

Mr. Ames fays he printed "The confutation of Nich. Shaxton; Explicatio petitoria ad Parliamentum; and The confutation of Miles Haggard's ballad. Crowley indeed was the author of them, but they were printed by Day and Seres. See p. 616, 619, and 620. Probably he might correct the prefs there, and learn the art of printing, which he afterwards practifed himfelf.

1549. "The Pfalter of Dauid newly tranflated in Englyfhe metre in fuch fort that it may more decently and wyth more delight of the mynde be read and fonge of al men. Whereunto is added a note of four parts wyth other thinges as fhall appeare in the Epiftle to the reader. Tranflated & imprinted by Robert Crowley in the year of our Lord 1549. the 20. day of September." Mr. Tutet. Quarto.

1549. "The Voyce of the lafte trumpet blowen bi the feueth Angel (as is metioned in the eleuenth of the Apocalips) callynge al the eftates of menne to the right path of their vocation, wherin are contayned xii. leffons to twelue feueral eftates of menne, whych if they learne and followe, al fhal be well, and nothynge amife. ¶ The Voyce of one criyng in the deferte. Luke iii. ¶ Make ready the Lordes way. &c. Efai xi." On the back of the title, "The contentes of thys boke. The Beggers leffone. The feruauntes—The Yeomans—The Lewde prieftes—The Scolers—The Learned mans—The Phificians—The Lawiers—The Marchauntes—The Gentilmans—The Maieftrates—The Womans." In metre. D 5, in eights. "Jmprinted—in Elie rentes—M. D. XLIX. the xxix daie of Nouembre. ☞ Authore eodem Roberto Croleo. ¶ Cum priuilegio—folum." Again 1550. Octavo.

1549. "A new yeres gyfte, wherein is taught the knowledge of ourfelf, and the fear of God: worthy to be giuen and thankfullye recyued of all chriften men. M DXLIX, the laft daye of December. Authore eodem Roberto Croleo." This efcaped the Notice of bifhop Tanner. Octavo.

1550. "THE VISION of Pierce Plowman, now fyrfte imprynted by Roberte Crowley, dwellyng in Ely rentes in Holburne. Anno Domini 1505. (*a miftake for* 1550.) Cum priuilegio—folum." In a compartment with the fun at top, and Edw. Whitchurch's cypher at bottom. The printer addreffes the reader in thefe words: "Beynge defyerous to knowe the name of the Autoure of this moft worthy worke (gentle reader) and the tyme of the writynge of the fame: J did not onely gather togyther fuche aunciente

ciente copies as J could come by, but alſo conſult ſuch mē, as J knew to be more exerciſed in the ſtudie of antiquities, then J my ſelfe haue ben. And by ſome of them J haue learned that the Autour was named Roberte langelande, a Shropſhere man, borne in Cleybirie, aboute viii. myles from Maluerne hilles. For the time when it was written : it chaunced me to ſe an auncient copye, in the later ende wherof was noted, that the ſame copye was written in the yere of oure Lorde. M. iiii. C. and nyne, which was before thys preſente yere, an hundred and xli. yeres. And in the ſeconde ſide of the. lxviii. leafe of thys printed copye, J finde mētion of a dere yere, that was in the yere of oure Lorde, M. iii. hundred and. L. John Chicheſter than beynge mayre of London. So that this J may be bold to reporte, that it was fyrſte made and wrytten after the yeare of our lord. M. iii. C. L. and before the yere M, iiii C, and. ix. which meane ſpaſe was. lix. yeres. We may iuſtly cōiect therefore, ẏ it was firſte written about two hundred yeres paſte, in the tyme of Kynge Edwarde the thyrde. Jn whoſe tyme it pleaſed God to open the eyes of many to ſe hys truth, geuing them boldenes of herte, to open their mouthes, and crye oute agaynſte the worckes of darckenes, as did John wickleſe, who alſo in thoſe dayes tranſlated the holye Bible into the Engliſhe tonge, and this writer, who in reportynge certaine viſions and dreames, that he ſayned him ſelfe to haue dreamed : doeth moſte chriſtainlye enſtruct the weake, and ſharply rebuke the obſtinate blynde. There is no maner of vice, that reigneth in anye eſtate of men, whiche this wryter hath not godly, learnedlye, and wittilye rebuked. He wrote altogyther in miter : but not after ẏ maner of our rimers that write nowe adayes (for his verſes ende not alike) but the nature of hys miter is, to haue thre wordes at the leaſte in euery verſe, whiche beginne with ſome one letter. As for enſample, the firſte two verſes of the boke renne vpon. ſ. as thus.

 Jn a ſomer ſeaſon whan ſette was the Sunne,
 J ſhope me into ſhrobbes, as J a ſhepe were.
  The next runneth vpon. H. as thus.
 Jn habite as an Hermite vnholy of werckes, &c.

This thinge noted, the miter ſhal be very pleaſaunt to read. The Englyſhe is according to the time it was written in, and the ſence ſomewhat darcke, but not ſo harde, but that it may be vnderſtande of ſuche, as will not ſticke to breake the ſhell of the nutte for the kernelles ſake." &c. Colophon : " Jmprinted at London——.M D. L." Contains 117 leaves, beſides the prefixes. W.H.   Quarto.

"¶ The viſion of Pierce Plowman, nowe the ſeconde time imprinted 1550. byRoberte Crowley dwellynge in Elye rentes in Holburne, Whereunto are added certayne notes and cotations in the mergyne, geuynge light to the Reader. And in the begynning is ſet a briefe ſumme of all the principall matters ſpoken of in the boke. And as the boke is deuided into twenty partes called Paſſus: ſo is the ſummary diuided, for euery parte hys ſummarie, rehearſynge the matters ſpoken of in euerye parte, euen in ſuche
            order

order as they ftande there. ¶ Jmprinted——,The yere of our Lord. M. D. L. ¶ Cum priuilegio——folum." W.H. Quarto.

Dr. Percy (Ancient Poems, Vol. II; p. 272.) mentions three editions printed by Crowley in 1550; but on examining my two copies, this laft mentioned is the very fame as the firft, with only a new title-page; the prologue reprinted; and the fummary now added. However, there doubtlefs was another edition, feeing the doctor found evident variations in every page, of which it might with more propriety be faid "nowe the feconde tyme imprinted" &c. And as my enlarged edition has its laft leaf fupplied with M. S. copied from a different edition than either the firft, or Owen Rogers's, it was probably taken from the real fecond edition, and if fo has "Cum priuilegio——folum" under the colophon, by which it may be known.

1550. "A metrical fermon on Pleafure and pain, Heauen and Hell, remember thefe four, and all fhall be well." Written in metre. See Warton's Hift. of Eng. Poetry. Vol. III; p. 188. Octavo.

1550. "The way to wealth wherein is plainly taught a moft prefent remedy for Sedicion. Wrytten and imprinted by Robert Crowley the. vij. Of Februaryo. in the yere of our LORDE A thoufand fiue hūdred & fiftie. Jn Elie rentes Holburne. Cum priuilegio——folum.

"Who fo thou be that doeft defyre, to liue and good dayes fe.
Loke ẏ in thy tonge and lyps, none yl or difceite be.
Fle from yl and do that good is, whereof commeth no blame.
Seke thou for peace diligently, and then enfue the fame."

The head-title, "By what meanes Sedicion maye be put a waye, and what diftruccion will follow it if it be not put away fpedely. Confultatio Roberti Crolei." B, in eights. W.H. Octavo.

1550. "One and thirty epigrams, wherein are briefly touched fo many abufes that may, and ought to be put away." In Ely-rentes. This bifhop Tanner afcribes to him as the author. Mr. Strype has given 15 of them. See Eccl. Memor. Vol. II; p. (266.) and Repof. O. O. p. 132. Again 1551.

1550. The lady Elizabeth Fane's 21 pfalms, and 102 proverbs. Octavo.

1550. "The Path way to perfect knowledge. The true copye of a prologue, wrytten about two C. yeres pafte, by John Wycklyffe. (as maye iuftlye be gatherid bi that, that John Bale hath wrytten of him in his Boke entitled the fūmarie of famoufe writers of the Ile of great Britan) the original whereof is founde written in an olde Englifh Bible bitwixt the olde Teftament, and the newe. Which Bible remainith now in the Kyng hys Maiefties Chāber. Imprinted—in Elie rents—Anno Do. M D L. Cum priuilegio—folum." It begins, "25 books of the old teftament ben books of faith, and fully books of holy writ." This was printed by John Gough in 1536, and 1540; then

entitled

entitled " The Dore of holy fcripture." In thofe editions the author is not named, only in the printer's epiftle we are told that it " is the Prologue of the fyrft tranflatoure of the byble out of latyn into Englyfhe," &c. See p. 493, &c. Hereupon Mr. T. Baker remarked, This Pathway was not Wickliff's, but the author's of the other old tranflation of the Bible. See Ant. Harmer's fpecimen, p. 16." Alfo Lewis's Hift. of Eng. Tranfl. of the Bible, p. 35, &c. W. Tindall wrote a treatife entitled " A Pathway into the holy Scripture." See it in his works, p. 377, &c.
Octavo.

" Ban wedy 1 dynny air yngair allan o hen gyfreith Howel da vap Cadell brenhin Kymbry ynghylch chwechant mlyned aeth heibio wrth yr hwn van y gellir deall bot yr offeiriait y pryd hynny yn priodi gwraged yn dichwith ac yn kyttal ac wynt in gyfreith lawn. A certain cafe extracte out of the aunciert law of Hoel da, king of Wales, in the yere of our Lorde, nyne hundred and fourtene paffed; whereby it maye be gathered, that prieftes had lawfully maried wyues at that tyme. R. Crowley." 4 leaves.
Quarto.

" The baterie of the Popes Botereulx, commmonlye called the high Altare. Compiled by W. S. *(William Salefbury)* in the yere of oure Lorde 1550. ii. Cor. x. The wapons of our war are not flefhly," &c. Dedicated "To—fyr Richarde Ryche, Lorde Ryche, & Lord Chauncelloure of Englande." G 6, in eights. " Jmprinted——in Elye rentes——.M. D. L. Cum priuilegio——folum." W.H. 1550. Octavo.

" A briefe and playne introduction, teachyng how to pronounce the letters in the Britifh tong (now commonly called Walfh) wherby an Englifhman fhal not only with eafe read the faid tong rightly, but marking the fame wel, it fhall be a meane for him with one labour and diligence to attain to the true and natural pronounciation of other expediente and moft excellent languages. Set forth by W. Salefburye. R. Crowley. Cum priuilegio——folum." 18 leaves. 1550. Quarto.

" KIYNNIVER Uith aban or yfcry thur lan ac a ỏarlleir yr eccleis pryd Commun, y Sulieu a'r Gwilieu trwy'r vlwy ỏyn : o Camberciciat, W. S." With the king's privilege "to William Salefbury and Jhon Waley, or affigns, for feven years, to print, or caufe to be printed, the book intitled, a dictionarie both in Englifh and Welfhe, whereby our beloued fubiectes in Wales may the fooner attayne vnto the Englifhe tongue. Given at Weftmifter, 13 Dec. 37 Hen. VIII. 1546." Colophon, " Imprinted by Robert Crowley for William Salefbury." See p. 730. 1551. Quarto.

" The fable of Philargyrie, the great gigant of Great Britain, what houfes were builded, and lands appointed for his prouifions, and how al the fame is wafted to contente his greedy gut wythall, and yet he rageth for honger." Mr. Warton mentions it in 4 and afcribes it to Crowley as the author. Hift. of Eng. Poetry, Vol. II; p. 361, note z. 1551. Octavo.

A Myroure

1551. "A Myroure or glaſſe for all ſpiritual miniſters to beholde themſelues in, wherein they may learne theyr office and duitie towardis the flocke cōmitted to their charg. Gathered out of holy Scripture and catholyke doctours, by Peter Pykeryng, feruant to the ryght worſhipful ſyr Anthonie Neuell, Knight, and one of the Kingis maieſties councel, eſtabliſhed in the northe, and ſent to ſyr Jhon Todkyll, vycar of ſouth Teuerton, and other his cōplifis in Notyngham ſhyre, for a newe yeres gyfte." B 7, in eights. "Jmprinted——in Elye rentes——.M. D. L. J." W.H. Octavo.

1551. "Poore Shakerley his knowledge of good and euil, called otherwiſe Eccleſiaſtes: by him turned into meeter." For John Caſe. Octavo.

—— "The ſupper of the Lorde After the meanyng of the Sixte of John and the. xi. of the fyrſt Epiſtle to the Corintiās, whervnto is added an Epyſtle to the reader, And incidently in the expoſition of the ſupper: is cōfuted the letter of maſter More againſt John Fryth. i. Corinth. xi. Whoſoeuer ſhall eate of this bread & drinke of this cuppe of the Lorde vnworthely ſhal be gyltye of thebody & bloud of the Lorde. Anno. M.CCCCC. xxxiii. v. daye of Apryll." Though this treatiſe is anonymous, Will. Tindall is allowed to have been the author; Crowley wrote only the preface. It was originally printed at Nornburg, and dated as above. Bearing no printer's name, nor date of printing, i have placed it to Crowley, being a printer, as having the juſteſt claim to it. W.H. Octavo.

—— "An information and petition againſt the oppreſſours of the pore commons of this realme. Compiled and imprinted for them, that haue to do in the parliamente," &c. By Robert Crowley." Dedicated " to the honourable lordes of the parliament." This is a book written with a bold freedom. Octavo.

Mr. Ames has given two other Books as printed by him, one in 1566, the other in 1567; but that is not very probable, being then vicar of St. Giles's Cripplegate. I have the former of them, printed by Hen. Binneman. He was however the author of both treatiſes.

—— Perhaps he printed, "Five ſermons of Bern. Ochine" &c. for Will. Beddell.

✣✣✣✣✣✣✣✣✣✣✣✣✣✣✣✣✣✣✣✣✣✣✣✣✣✣

JOHN KYNGE

WAS a freeman of the old company of Stationers, dwelt in Creed-lane, and kept ſhop in Paul's church-yard at the ſign of the Swan. He had licenſe to print, 1557——1558, "The defence of women. Adam Bell, &c. The brevyat cronacle of the kyngs, in viij°. A Jeſte

JOHN KYNGE.

A Jefte of fyr Gawayne/ The boke of Carvynge, & fewynge/ Syr Lamwell. The boke of Cokerye. The boke of nurture for mens farvaunts." 1558 —1559. He was fined " for that he ded prynt The nut browne mayde without licenfe. ij.s. vj.d." Feb. 6. 1559, 60. He had licenfe to print " Salomons proverbis, in viij." June, 10. " Lucas vrialis/ Nyce wanton/ Impaciens poverte/ The proude wyves pater nofter. The fquyre of Low degre/ Syr Deggre." Aug. 14. " A play called Juventus." Aug. 30. " A boke called Albertus magnus." Septemb. 20. " Lupfetts workes." Octob. 30. " The lyttle herball. The greate Herball. The medyfine for horfes." He probably died about the latter end of the yeare 1561 ; for then Tho. Marfhe had licenfe to print " The Cronacle in viij, which he bought of John Kynges wyfe." The following books only are found of his printing.

" A litle Herball of the properties of Herbes, newly amended & cor- 1550. rected, wyth certayn Additions at the ende of the boke, declaring what Herbes hath influence of certain Sterres and conftellations, wherby maye be chofen the beft and moft lucky tymes and days of their miniftracion, according to the Moone beyng in the fygnes of heauē, the which is daily appoīted in the Almanacke, made and gathered in the yeare of our Lorde God. M. D. L. the xii. daye of February, by Anthony Afkhā Phyfycyon." K 7, in eights. " Jmprynted———, in Paules churcheyarde, at the figne of the Swanne by Jhon kynge." It was printed alfo this yeare by Will. Powell. W.H. Octavo.

" A breuiat chronicle contayning all the kynges, from Brute to this 1555. day, and many notable actes, gathered out of diuers chronicles, from William the conqueroure, vnto the year of Chrift, M. V. C. LV. with the mayors, and fhryffes of the citie of London, newly corrected and amended." P. ii, in eights. In this book it is faid, " printing began at Mens by John Fauft. 1457." Again 1557,* and 1559, W.H. Octavo.

Robert Crowley's Epigrames. Maunfell, p. 41. Octavo. 1559.

" Tho. Lupfets workes. Anno domini. M. D. LX." D 4, in eights. 1560. " Jmprynted———in Paules churcheyearde, at the fign of the fwanne, by Jhon kynge." W.H. Octavo.

" A Lytle and bryefe treatyfe, called the defence of women, and efpe- 1560. cially of Englyfhe women, made agaynft the Schole howfe of women.

" Yf the turtle doue, Be true in loue, Voyde of reafon, than, What fhame is it, of man hath wyt, And hateth a woman ? Anno Domini. M. D. LX." In the fame compartment as The Obedience of A Chriftian man, printed by Will. Hill ; but the T. P. are left out in the tablets at the bottom of the jambs, and TR inferted in the tablet in the fell. On the back begins the dedication " To hys fynguler and efpecyall frende mayfter Wyllyam Page, Secretary to Syr Phillyip hobdy, Edward

JOHN KYNGE.

ward More fendeth gretyng.——frō Hambleden the xx. day of Julye M.D.L.V. JJ. Finis. qud. E. M." In Alexandrine verſe. C, in fours. " Jmprinted—in Paules Churche yearde," &c. W.H. Quarto.

1560. " The proude wyues Pater noſter that wolde go gaye, and vndyd her huſbonde and went her waye. Anno Domini. M.D.LX." In the octave ſtanza. C, in fours. " Jmprinted——in Paules Churche yearde," In the ſame compartment as the preceding article, but with an odd piece in place of the ſell. W.H. Quarto.

1561. " The great Herball." Without the cuts. Folio.

—— " The caſtle of loue, tranſlated oute of Spanyſſhe into Engliſſh by John Bowrchier knyght, lorde Bernes, at the inſtaunce of the lady Elizabeth Carewe, late wyfe to ſyr Nicholas Carewe, knyght. The which booke treateth of the loue betwene Leriano and Laureola, doughter to the kynge of Macedonia." R 4, in eights. " Imprinted at London by John Kynge." Octavo.

—— " A ſermon no leſs frutefull, then famous, Made in the yere of oure Lorde God, M.CCC.lxxxviii. In theſe our later dayes mooſt neceſſary to be knowen. Neyther addynge to nor dymynyſhyng from: ſaue the old and rude Englyſhe thereof mended here and there. Impirnted——by John Kynge, in Paules Church yearde, at the ſygne of the Swanne." See it reprinted in Morgan's Phoenix Britannicus, p. 1.

—— " Here after foloweth certaine bokes cōpyled by mayſter Skeltō, Poet Laureat, whoſe names here after ſhall appere. Speake Parot. The death of the noble Prynce Kynge Edwarde the fourth. A treatyſe of the Scottes. Ware the Hawke. The Tunnynge of Elynoure Rummyng." In a border of four pieces. D, in eights. " Jmprynted——in Crede Lane, by John kynge and Tho. Marche." This has the portrait of " Skelton Poete" ſitting at a deſk for the frontiſpiece. W.H. Octavo.

WILLIAM TILLY.

1549. The New Teſtament, after the tranſlation of M. Coverdale: printed by " Wylliam Tilly, dwellynge in St. Anne and Agnes pariſhe at Aldriſhgate." Dr. Gifford. Quarto.

THOMAS

THOMAS GALTIER, or GUALTIER,

HAD a licenfe to print in French all fuch church books as fhould be fet forth, &c. See Strype's Eccl. Memor. Vol. II; p. 518.

" Here beginneth the booke of RAYNARDE THE FOXE, conteining diuers goodlye hiftoryes and Parables, with other dyuers pointes necefsarye for al men to be marked, by the whiche pointes men maye lerne to come vnto the fubtyll knowledge of fuch thinges as daily ben vfed and had in the counfeyles of Lordes and Prelates, both ghoftely and worldely, and alfo among marchauntes and other comen people. ¶ Imprinted in London, in Saint Martens by Thomas Gualtier, 1550." * See p. 27, and 282. 1550.
Octavo.

" The new Teftament in Englifhe after the greeke trâflation annexed wyth the tranflation of Erafmus in Latin. Wherunto is added, a Kalendar, and an exhortation to the readyng of the holy fcriptures made by the fame Erafmus, wyth the Epiftles taken out of the olde teftamēt, both in Latin and Englyfh, a table neceffary to finde the Epiftles and Gofpels, for euery fonday and holy day throughout the yere, after the vfe of the churche of England nowe.—in officina Thomæ Gualtier. pro I. C. *John Cheek*. Pridie kalendas Decembris anno Domini. M. D. L." This title is in the compartment with the fun at top, and Edw. Whitchurch's cypher at bottom. On the back is " An almanacke for. xxii. yeares." Then, before the kalendar, " J. C. vnto the Chriften reders." The Latin printed on Brevier Roman, N°. 2. The whole contains J i 2, in eights, befides the prefixes.* The rev. Mr. Worfley, and W H. 1550.
Octavo.
This Englifh tranflation is afcribed to Sir John Cheke, in Catal. Bibl. Bodleianæ Vol. II; p. 576. It was commonly underftood for his in 1583, when Mr. Barker yielded certain of his copies to the Stationers' company for the relief of their poor. See our General Hiftory in that year. See alfo Lewis's Hift. of Eng. Tranfl. of the Bible, p. 184, &c.

" The Piththy and mooft notable fayinges of al Scripture, gathered by Thomas Paynell : after the manner of common places, very neceffary for al thofe that delite in the confolacions of the Scriptures. 1550." In Whitchurch's fmall compartment with his cypher at bottom. The preface begins on the back, infcribed " To the right excellent and mofte gracious Lady, my Ladye Maryes good grace." The old teftament contains Fol. lxxxvii, and an alphabetical table; the new teftament, Fol. lxxiiii, and a table alfo. " Jmprinted——at Flietebridge, ——at the coftes & charges of Robert Toye,——in Paules churcheyard at the figne of the Bel." W.H. 1550.
Octavo.

" The newe greate abredgement, brefly conteynyng, all thactes and ftatutes of this realme of England, vntyl the xxxv. yere of the reigne of our late noble kynge of mofte worthye and famous memorye Henry VIII. (whofe foule God pardone) newly reuyfed, trulye corrected and amended, to the greate pleafure and commoditie of all the readers thereof." The fame compartment as the laft article. 1551.
Octavo.

" Novum

THOMAS GUALTIER.

1551. "Novum teſtamentum gallicè, ex Joannis Calvini recognitione: cui præfixit iconem Moſis legis tabulas à Deo accipientis, cum his verbis, Legem pone michi domine, pſalm cxix. Is anno 1551, typis Italicis minutis ſed nitidiſſimis, excudit. See Mattaire's index.

1551. "LE LIVRE des Priers Communes, De l'adminiſtration des Sacremens & autres Ceremonies en l'Egliſe d'Angleterre. Traduit en Francoys par Francoys Philippe, ſeruiteur de Monſieur le grand Chancelier d'Angleterre. De l'imprimerie de Thomas Gaultier, Imprimeur du Roy en la langue Francoiſe, pour les Iſles de ſa Mageſté. Auec le priuilege general dudit ſeigneur. 1553." In the compartment with a male and female head on a medalion at top. The dedication inſcribed, "A tres reverend pere en Dieu Thomas Goodrik, Eueſque d'Ely & Chancelier d'Angleterre, Francoys Philippe, threſhumble Salut." Z z 4, in eights, beſides the prefixes. "Almanach pour. xix. ans," begins "M. D. lii." W.H. Octavo.

※※※※※※※※※※※※※※※※※※※※※※※※※※※※※

JOHN TYSDALL, or TYSDALE,

WAS an original member of the Stationers' company, dwelt in Knight-Riders-ſtreet, and had a ſhop in Lombard-ſtreet, in All-Hallows church-yard, near unto Grace-church.

1550. "An ouerſighte and deliberacion vppon the holy prophet Jonas; made, and vttered before the kings maieſtie, and his moſt honourable councell, by Jhon Hoper in the lent laſt paſt. Comprehended in ſeuen ſermons, anno M.D.L." Licenſed, to print again, in 1560. Octavo.

1551. A treatiſe of predeſtination. See it without date. Octavo.

1560. "Precepts of Cato, with annotations of D. Eraſmus of Roterdame, very profitable for all menne, newlye imprinted. With, Sage and Prudente ſayengs of the ſeuen wyſe men." Licenſed. Sixteens.

1560. "The firſt two partes of the Actes or vnchaſte examples of the Engliſhe Votaryes, gathered out of theyr owne legendes and Chronycles by Jhon Bale, and dedicated to our moſte redoubted foueraigne kyng Edwarde the ſyxte. ¶ Beware of the leuen of the phariſees &c. Luke. xii." Each part has a ſeparate title-page. The firſt part comprehends examples from the world's beginning to the year of our Lord 1000. Collected An. 1546. N 4, in eights, beſides the dedication. "Jmprinted ―― by John Tyſdale, ―― in Knight Rider ſtrete nere to the Quenes Waredrop. Anno. 1560." The ſecond part comprehends examples for for CC years, being continued to the reign of K. John. Collected Anno. 1550. V, in eights. "Jmprinted ―― by Jhon Tiſdale, ―― in Knight riders ſtreate." Licenſed. W.H. Octavo.

"The

JOHN TISDALE.

" The Firſte thre Bokes of the moſt Chriſtiā Poet Marcellus Palin- 1560.
genius called the Zodyake of Lyfe. Newly tranſlated out of Latin into
Engliſh by Barnabe Googe. Imprinted——by John Tiſdale for Rafe
Newberye. An. Do. 1560." On the back, Googe's coat of arms. Then an
epiſtle dedicatory " To the righte Woorſhipfull and his Eſpeciall good
graundmother my Lady Hales B. G. wiſheth long Lyfe and helth to the
pleaſure of God." Therein he ſtiles this piece " the firſt frutes of
his ſtudy." Next follows a Latin dedication " Clariſſimis ſimul ac
ſtudioſiſſimis Guli. Cromero, Th. Honiuodo, Ra. Heimundo Armigeris
B. Gogœus Aluinghamus, S. D.——Valete, ex Muſæo noſtro, Decimo
Martij, Anno Chriſti, 1560. ætatis noſtræ. xx." Then follows an
acroſtic of Latin verſes " (BARNABAS GOGEVS) by Gi. Duke." G 4, in
eights.   Octavo.
"The firſt ſix books: by the ſame."   Octavo. 1561.
In 1565 the whole was printed by Hen. Denham, for R. Newbery.
See Warton's Hiſt. of Eng. Poetry, Vol. III; p. 449.

" The enſamples of Virtue and vice, gathered out of holye ſcripture. 1561.
By Nicolas Hanape patriarch of Jeruſalem. Very neceſſarye for all
chriſten men and women to loke vpon. And Englyſhed by Thomas Paynell. Anno 1561." In the compartment with a cherub's head on each jamb;
formerly T. Berthelet's. Dedicated " To the moſte noble, moſt excellent,
and mooſte vertuous Lady Elizabeth. Quene of England," &c. It is
divided into 134 chapters, with a table at the end, " Jmprinted——in
Knyght Riders ſtreat. Cum priuilegio——ſolum." Licenſed. W.H.
                                                               Octavo.

" THE HVNtynge of Purgatorye to death, made Dialoge wyſe, by 1561
Jhon Veron Senonoys. Newly ſet foorth and alowed accordinge to the
order appoynted in the Quenes Maieſties Jniunctions. The bloudde of
the Lorde doeth clenſe vs from all ſynne. 1. Joan. 1." Dedicated " To
the righte honorable lord, my lord Ruſſell, erle of Bedforde." Contains
beſides, 397 leaves, and a table at the end. " Jmprynted——by Jhon
Tyſdale, and are to be ſolde at his ſhoppe in the vpper ende of Lombard
ſtrete, in Allhallowes churchyard nere vnto grace churche. 1561." On
the back is his device, an angel driving Adam and Eve out of Paradiſe.
Licenſed. W.H.   Octavo.

" A declaration of Edmonde Bonners articles, concerning the clear- 1561.
gye of Lōdon dyoceſe, whereby that execrable Antychriſte is in his righte
colours reueled in the yeare of our Lord a. 1554. By John Bale. Newlye
ſet fourth & allowed, according to the order appointed in the Quenes
Maieſties Jniunctions. ¶ Woo to them whiche builde in bloude & iniquity. Mich. iii. All thinges whan they are rebuked of the lyght are
manyfeſt. Ephe. v." Contains Fol. 70, beſides the preface. " Jmprynted——by Jhon Tyſdale, for Frauncys Coldocke, dwellinge in
Lombard ſtrete, ouer agaynſte the Cardinalles hatte, and are there to be
ſold at his ſhoppe. 1561." On the back, his device. W.H.   Octavo.

JOHN TISDALE.

1561. " The Ouerthrow of Iuftification of workes, and of the vain doctrine of merits of men : with the true affertion of the Iuftification of faith, and of the good workes that come of the fame, and in what refpect our good workes are crowned. By John Veron." Maunfell, p. 118. Licenfed. Octavo.

1562. " A Frvtefull booke of the comon places of all S. Pauls Epiftles right neceffarye for all fortes of people, but efpecially for thofe of the minifterye dyligentlye fette foorthe by Thomas Paniell. Anno. 1562." In the fame compartment as The enfamples of Virtue & vice. Dedicated "To the right worfhipfull mafter Thomas Argall." Z, in eights. " Imprynted ——by John Tifdale, and are to be folde at his fhoppe in the vpper end of Lombarde ftrete——1562." Licenfed. W.H. Octavo.

1562. " An expofition vpon the 23d pfalme of Dauid, full of frutefull and comfortable doctrin written to the cityc of London, by John Hooper bifhop of Gloceter and Worceter, and holy martyr of God for the teftimonye of his truth. Whereunto is annexed an apology of his, againft fuch as reported, that he curfed queen Mary," &c.* Octavo.

1562. " An apologye made by the reuerende father and conftante Martyr of Chrifte, John Hooper late Bifhop of Gloceter & Worceter, againfte the vntrue and fclaunderous report that he fhould be a maintainer & encorager of fuche as curfed the Quenes highnes that then was, Quene Marye. Wherein thou fhalte fee this Godly mannes innocency & modeft behauiour, and the falfhode & fubtilty of the aduerfaries of gods truth. Newelye fet foorth & allowed——1562." It is introduced with a preface "To the Godly reader" by Hen. Bull, in which he fays : "——beholde the prouidence of God, who hath brought this worke to lighte, which otherwyfe by the negligence of fome was like to perifhe. And here J haue iuft occafion to difcommend thofe men, whiche do defraude the congregation of fuch worthy monumentes," &c. At the end of the Apology are "Certayne letters. ¶ The copy of the letter wherby M. Hooper was certifyed of the takyng of a Godly companye in bowe churcheyearde at prayer.—3 Jan. 1554. M. Chamber, M. Monger, M. Sh. and the reft in the counter do praye for you," &c.——Mafter Hooper's aunfwer——4. Jan. 1554.——M. Hoopers Letter delyuered in the counter at breade ftreete. ——4. Jan. 1554." The fignatures continued to L, in eights. " Imprynted——by John Tifdale, and Thomas Hacket, and are to be folde at their fhoppes in Lombarde ftrete. Anno. 1562." * W.H. Octavo.

1562. " The ftatutes or ordinaunces concernynge artificers, feruantes, and labourers, iournemen and prentifes, drawn out of the common lawes of this realme, fith the time of Edward the fyrft, vntyll the thyrd and fourth yeare of oure dreade foueragne lord kyng Edward the vi. Wyth the meafuryng of lands." Octavo.

" A briefe

JOHN TISDALE.

"A briefe and compendyoufe table, in a maner of a concordaunce," &c. as p. 754. And nowe imprinted in Englifh. M.D.LX.JJJ." T2, in eights. "Jmprinted—by John Tyfdale, and—folde at hys fhoppe, in the vpper ende of Lombard ftrete,——at the fygne of the Eagles foote." W.H.   Octavo. 1563.

"An abridgement of the Notable worke of Polidore Virgile," &c. as p. 520. In a border of odd pieces. Contains Fol. c. lii, befides the dedication, and table at the end. "Jmprynted——in Knight riders ftreate, neare the Quenes Wardrop." On a feparate leaf, a cut of Abraham offering Ifaac, which feems to have been his device whilft he lived there. Licenfed. W.H.   Octavo.

"Beza's fecond Oration——at Poyffy in the prefence of the Quene Mother, &c." T. Baker's interleaved, Maunfell, p. 9.   Octavo.

"An Epitaph vpon the Deth of King Edward." With John Charlewood.   A fheet.

A memento mori. In 16 ftanzas of four lines. With John Charlewood.   A broadfide.

A FRVTEFVL treatife of predeftination and of the diuine prouidence of god, with an apology of the fame, againft the fwynyfhe gruntinge of the Epicures & Atheyftes of oure time. Whereunto are added, as depending of it a very neceffary boke againfte the free wyll men, and another of the true iuftification of faith, and of the good workes proceadynge of the fame, made Dialoge wyfe by John Veron. Jmprinted—by John Tifdale, and are to be folde at his fhop in the vpper end of Lombard ftrete," &c. Dedicated "To the moft Godlye vertuous & mightye Princes Elizabeth—— Queene of England," &c. Contains befides, Fo. 115, and a table at the end. To this is added "¶ An Apology or defence of the doctryne of Predeftination fette foorth by the Quenes hyghneffe hyr moft humble & obedyente fubiect John Veron," &c. Fol. 43, and a table. Licenfed. W.H.   Octavo.

Sermons. Without title, or numbers to the leaves.   Octavo.

"A mofte neceffary treatife of free wil, not only againft the Papiftes, but alfo againft the Anabaptiftes, which in thefe our daies go about to renue the deteftable herefies of Pelagius, & of the Luciferians, whiche fay and affirm that we be able by our own natural ftrength to fulfil the law & commaundementes of God. Made dialoge wyfe by John Veron, in a manner word by woorde, as he did fet it forth in his lectures at Paules." Dedicated "To—Lorde Robert Dudly, mafter of the horfes, and Knight of the——Garter." K, in eights. "Jmprinted——and fold at hys fhop in Lumbard ftreate. Cum priuilegio.—folum." W.H.   Octavo.

Befides

"Besides the books before mentioned to be licensed, in 1558 he had license to print "An abco in Laten for R. Jugge, John Judson and Anth. Smythe"; and yet he was fined " for pryntinge without lycense, The A. b. C. and another suche lyke, iiij. s. viij. d." But it was not paid. Also, in 1560, "Kynge Joseas." In 1561. "Bevys of Hampton." In 1562. " A pronostication of the scripture." In 1563. "The defence of women made agaynste the schole howse of women." Besides about 10 ballads.

STEPHEN MYERDMAN, or MIERDMAN.

1550. "THE market or fayre of vsurers. A new pasquillus or dialogue against vsurye, &c. translated from the high Almaigne, by William Harrys. Cum priuilegio——ad quinquennium." Octavo.

1551. Liturgia sacra, seu ritus ministerii in ecclesia peregrinorum profugorum propter euangelium Christi argentinae. Adjecta est ad finem breuis apologia pro hac liturgia, per Valerandum Pollanum Flandrum. Cum &c. Ad finem. s. d. s. m.* Octavo.

1551. " Absoluta De Christi Domini Et Catholicæ eius Ecclesiæ Sacramentis, tractatio, Authore Henrico Bullengero. Cui adiecta est eiusdem argumenti Epistola, per Ioannem à Lasco, Baronem Poloniæ ante quinquennium scripta. Accessit rerum ac verborum copiosus index, 1 Cor. 10. Non potestis mensæ Domini participes esse, & mensæ Dæmoniorum. Londini excudebat Stephanus Myerdmannus An. 1551. Men. Apri. Cum priuilegio——solum." Dedicated " Serenissimae Principi & Dominæ, Dominæ Elizabethæ——Regiæ Maiestatis Angliæ, &c. Sorori dignissimæ.——Tuæ Maiest. Serenissimæ addictissimus cliens Ioannes à Lasco.' 'The Epistle by A Lasco is dated Æmdæ Frisiorum. Mense Aprili. 1545. Contains, besides the prefixes, 123 leaves. W.H. Sixteens.

1551. " Compendium doctrinae de vera vnicaque Dei et Christi ecclesia, eiusque fide et confessione pura: in qua peregrinorum ecclesia Londini, autoritate atque assensu sacrae maiestatis regiae. Quem Deus opt. max. ad singulare ecclesiae suae decus, ornamentum, ac defensionem (per gratiam suam) seruet, gubernet, et fortunet, amen. Cum priv. ad imprimendum solum." Octavo.

1551. " A new herball, wherin are conteyned the names of herbes in Greeke, Latin, Englysh, Duch, Frenche, and in the potecaries, and
herbaries

herbaries Latin, with the properties, degrees, and naturell places of the same, gathered and made by William Turner, phificion vnto the duke of Somerfet's, grace." To whom it is dedicated. " And are to be fold by John Gybken." Folio.

" Breuis et dilucida de facramentis ecclefiae Chrifti tractatio, in qua 1552. et fons ipfe, et ratio, totius facramentariae noftri temporis controuerfiae paucis exponitur: naturaq; ac ius facramentorum compendio et perfpicuè explicatur: per Joannem a Lafco, Baronem Poloniae, fuperintendentem ecclefiae peregrinorum Londini." Octavo.

JOHN CASE

DWELT in Peter-college rents. He, or one of his name, was an original member of the Stationers' company.

" Poore Shakerley, his knowledge of good and evil," &c. as p. 762. 1551. Octavo.

" French Hoode, and newe apparel for ladies and gentlewomen wher- 1551. unto is added, a froffe pafte to lie in a nights." Maunfell afcribes this to the fame author; as alfo,

" A Dredge for defenders of womens apparel." Octavo.

" Certayne chapters of the prouerbs of Salomon," &c. as p. 705.

RICHARD CHARLTON

HAD printed for him " A treatife of the argumentes of the old and 1550. new Teftament, by John Brentius; tranflated by John Calcafkie." Maunfell, p. 23. Octavo

JOHN TURKE, or TURK,

WAS an original member of the Stationers' company, and ferved under-warden in 1558; dwelt at the Rofe in Pater-nofter-row, when Richard Banks feems to have printed for him. See p. 411. He dwelt afterwards at the Cock, in Paul's church yard.

A fheet

See p. 978, note g.

JOHN TURK.

1550. A sheet almanack and prognostication for the year of our lord 1551, Simonis Heuringii, Saelicedenfis, doctor in Phyfyck and aftronomy at Hagenaw.

1553. A comment on the LXXXII. pfalm.

—— "The ryght and trew vnderftandynge of the Supper of the Lord, and the vfe thereof, faythfully gathered out of the holy fcriptures, worthyly to be embrafed of all chriftian people. Perufed and allowed by dyuerfe godly men to the comfort of all the trew congregation of Chrift. By Thomas Lancafter." Dedicated to king Edw. VI. At the end, "Imprynted at London by John Turke." Dr. Lort. Octavo.

Between 1558 and 1559, he had licenfe to print "The Kynges & quenes Pfalmes, fett furth by Kynge Henry the viijth, & quene Katheryn."

ABRAHAM VELE, or VEALE,

WAS a member of the worfhipful company of Drapers, and feems to have been alfo admitted a brother of the Stationers' company. The firft mention i find of him in their books, is on the 20th of May 1558, when he was fined xx. s. without intimating for what. Probably it might be for his admiffion as a brother, the preceding article being for making one free: the fum however is much larger than was ufually paid on the like occafions; it may therefore be prefumed that he had alfo offended againft their ordinances, by printing without licence from the mafter and wardens. I do not find any licence for his printing The golden book of Marcus Aurelius; Xenophon's Treatife of houfehold; or, Regimen fanitatis Salerni; which he printed about that time. In 1565, he was fined, with 16 others, "for ftechen bookes, which is contrary to the orders of this houfe." In 1567, he was fined, with 5 others, "for bokes taken by Thomas Purfoote & Hewgh Shyngleton in ferten places, ij. l." He appears to have been in bufinefs 35 years, dwelling all the time at the fign of the Lamb, in St. Paul's Church-yard.

1551. "The firft two partes of the Actes, or vnchaft examples of the Englyfh votaryes, gathered out of their owne legeandes and chronycles by Johan Bale, and dedycated to our moft redoubted foueraigne kynge Edwarde the fyxte." Over a cut of the author prefenting his book to the king. Underneath, "Beware of the leuen of the pharifees, &c. Lu. xii." The firft part contains 79 leaves befides the dedication. "Jmprinted——by Abraham Vele, dwellyng in Paules churcheyarde at the fygne of the Lambe. Anno. 1551. Cum priuilegio——folum."

"The

ABRAHAM VEALE.

"The second part or contynuacyon of the Englifh votaries, compre- 1551.
hendynge theyr vnchaft examples for. CC. yeares fpace, from the yeare a
thoufande from Chriftes incarnacyon to the reigne of kyng Johan, collected
——by Johan Bale. ¶ Beholde (thou idolatrous churche) J wyll gather
together all thy louers, &c. Ezech. xvi. Jmprinted at London for
Johan Bale, in the yeare of our Lorde a M. D. & LI. and are to be folde
within Paules chayne, at the fygne of S. John Baptift. Cum priuilegio
——folum." On the fame type as the firft part. 124 leaves. W.H.
Octavo.

"A fruteful and pleafaunt worke of the befte ftate of a publyque weale, 1551.
and of the newe yle called Utopia: written in Latine by Syr Thomas
More knyght, and tranflated into Englyfhe by Raphe Robynfon, Citizein
and Goldfmythe of London, at the procurement and earneft requeft of
George Tadlowe, Citezien & Haberdaffher of the fame Citie. Jmprinted
—— in Pauls churcheyarde at the——Lambe. Anno. 1551." Dedicated
"To the right honourable,——maifter William Cecylle efquiere, one of
the twoo principal fecretaries to the kyng his mofte excellent maieftie."
144 leaves. W.H. Sixteens.

"PROVERBES or adagies, gathered out of the chiliades of Erafmus by 1552.
Richard Tauerner, with new additions, as well of Latin prouerbs as of
Englifh." See p. 406. Octavo.

"Here beginneth the feinge of Vrynes" &c. as p. 355, &c. 1552.

"The Chriften ftate of Matrymonye," &c. as p. 709. 1552.

"The Pādectes of the Euangelycall Lawe." &c. as p. 691. 1553.

"A frutefull pleafaunt, & wittie worke,——called Vtopia:——by 1556.
the right worthie & famous Syr Thomas More knyght, and tranflated
——by Raphe Robynfon, fometime fellowe of Corpus Chrifti College in
Oxford, and nowe by him at this feconde edition newlie perufed & corrected, and alfo with diuers notes in the margent augmented. Jmprinted" &c. S, in eights. "Jmprinted——in Paules Churche yarde, at the
——Lambe, by Abr. Veale. M. D. LVI." W.H. Again next year.
Octavo.

"The golden Boke of Marcus Aurelius." See p. 425. Octavo. 1557.

"Xenophons Treatife of Houfholde. Anno. M. D. LVII." In the 1557.
compartment with a cherubic head on each jamb. 64 leaves. "Jmprinted——by Abr. Vcle." W.H. Again 1577. Sixteens.

Regimen fanitatis Salerni. Englifhed by Tho. Paynel. Sixteens and 1557.
Octavo.
The fame reprinted. (See it in 1575.) Octavo and Quarto. 1558.

Sleidanes Commentaries. See it p. 633. Folio. 1560.

1569. "The Gouernance and preseruation of them that feare the Plage. Set forth by John Vandernote Phisicion & Surgion, admitted by the Kynge his highnesse. Now newly set forth at the request of William Barnard, of London Draper. 1569. Imprinted—by Wyllyam How, for Abr. Veale," &c. C 4, in eights. W.H. Sixteens.

1573. "A christian instruction, conteying the Law and the Gospel: also a summarie of the Christian faith, and of the abuse & errors contrarie to the same. By M. Peter Viret, sometime Minister of the word of God at Nymes in Prouence, translated by I. S. (John Shute)" Printed for him. See the first part, p. 640. Octavo.

1573. "The whole. xii. Bookes of the Æneidos of Virgill. Whereof the first. ix. and part of the tenth, were conuerted into English Meeter by Thomas. Phaer Esquier, and the residue supplied, and the whole worke together newly set forth, by Thomas Twyne Gentleman. There is added moreouer to this edition, Virgils life out of Donatus, and the Argument before euery booke. Imprinted——by Wyllyam How, for Abr. Veale,—at the—Lambe, 1573." In a border. Dedicated "To—Sir Nicholas Bacon Knight,—Lorde keeper of the great Seale of England." O o 2, in fours. W.H. Quarto.

1574. "Direction for the health of Magistrates & students, namelie, such as be in their consistent age, or neere there vnto: written in Latine by Guil. Gratarolus: translated by Tho. Newton. Printed for Abrah. Veale." Maunsell, Part 11; p. 8. Octavo.

1575. "THE TRAVELIER of Jerome Turler, deuided into two Bookes. The first conteining a notable discourse of the manner and order of traueiling ouersea, or into straunge & forein Countreys. The second comprehending an excellent description of the most delicious Realme of Naples in Italy. A woorke very pleasaunt for all persons to reade, and right profitable & necessarie vnto all such as are minded to Traueyll. Imprinted by William How, for Abr. Veale. 1575." In a border. It has the following paradoxical inscription prefixed. "At Bononie, the firste Stone from the Citie, in the villedge of Marke Antonie de L'auolta, a senatour, standeth this Monument to bee seene. Alia Lœlia Crispis neither Man, nor Woman, nor Mungrell, neither Maide, nor younge Woman, nor olde Wife, neither Chast, nor Harlot, nor Honest, But all. Dyinge neither by Famine, nor Sworde, nor Poyson, But by all. Lying neither in the open Ayre, nor in the Water, nor in the Earth, But euery where. Lucius Agatho Priscius vnto her neyther Husband, nor Louer, nor Kinseman, neither Sorowful, nor Glad, nor Weeping: this Monument beeyng neyrher Heape, nor Piller, nor Sepulcher, But all. He knoweth & yet hee cannot tell, for whom hee hath erected." Contains 192 pages, besides the prefixes. W.H. Sixteens.

"REGIMEN

ABRAHAM VEALE.

"REGIMEN Sanitatis Salerni. This booke teachyngall people to 1575. gouerne thē in health is tranflated out of the Latine tongue into Englifhe by Thomas Paynell: whiche booke is amended, & diligently imprinted 1575.——by Wyllyam How, for Abr. Veale." In a border. Dedicated " To the Right excellent & hon. Wyllyam Paulet, of the order of the Gartir knyght, lorde S. John, Erle of wilfhere, Marques of Winchefter, and lorde Treafurour of England." C. lxvii leaves, befides the prefixes. W.H.   Octavo.

"A NOTABLE Hiftorie of the Saracens. Briefly and faithfully defcrybing 1575. the originall beginning, continuaunce & fucceffe afwell of the Saracens, as alfo of Turkes, Souldans, Mamalukes, Affaffines, Tartarians & Sophians. With a difcourfe of their Affaires & Actes from the byrthe of Mahomet their firft peeuifh Prophet & founder, for 700 yeeres fpace. Whereunto is annexed a Compendious Chronycle of all their yeerely exploytes, from the fayde Mahomets tyme tyll this prefent yeere of grace 1575. Drawen out of Auguftine Curio & fundry other good Authours by Thomas Newton. Imprinted——by Will. How, for Abr. Veale. 1575." In a border. On the back is the arms of "Lorde Charles Howarde, Baron of Effynghain, and Knight of——the Garter," to whom this book is dedicated. Qq 2, in fours. W.H.   Quarto.

"Tragedies of Tyrants. Exercifed vpon the church of God from 1575. the birth of Chrift vnto this prefent yeere. 1572. Containing the caufes of them, and the iuft vengeance of God vpon the authors. &c. Written in Latine by Henry Bullinger, and tranflated by Thomas Twine." Printed by William How for him. W.H.   Octavo.

"Comfort againft all kinde of Calamitie. Written in Spanifh by 1576. John Peres. Tranflated by John Daniel." Dedicated "To—Edmonde —Archbifhop of Canterburie." 167 leaves. Printed for him. Octavo.

"ENCHIRIdion militis Chriftiani," &c. by Erafmus. W. How, for 1576. him. W.H.   Octavo.

"The Ouerthrow of theGowte, written in Latin verfe by Chr. 1577. Balifta. Tranflated by B. G. Printed for him."   Octavo.

"A Booke of Chriftian Queftions & anfwers. Wherin are fet forth 1577. the cheef points of the Chriftian religion in manner of an abridgement. A worke right neceffary & profitable for al fuch as fhal haue to deale with the captious quarelinges of the wrangling aduerfaries of Gods truth. Written in Latin by the lerned clarke Theodore Beza Vezelius, and newly tranflated into Englifhe by Arthur Goldinge. Jmprinted——by W. How, for Abr. Veale,——at the——Lambe. Ano. 1577." In a border. Dedicated "To——Lorde Henry Earle of Huntingdon, Baron

ABRAHAM VEALE.

Baron Haftinges, Knight of——the Garter.——London, the 12. of June 1572." Contains, befides, 90 leaves. W.H.　　　　　Octavo.
The fame next year, differing only in the date.

1578. "The Expofition, and Readynges of John Keltridge: Mayfter of the Artes: Student of late in Trinitie Colledge in Cambridge, Minifter, Preacher, & Paftor of the Church of Dedham, that is in Effex: Upon the wordes of our Sauiour Chrifte, that bee written in the. xi. of Luke. ——Imprinted——by Will. How, for Abr. Veale. 1578." In a border. Dedicated "To the right hon and reuerend——John Elmer, Bifhop of London.——In Dedham, this xxi. of June. 1578." Next is an epiftle "To him that readeth, & vnderftandeth.——I. K." Then a commendatory epiftle "Viro docto, mihique amicifsimo, Johanni Keltridgo: Thomas Numannus artium magifter, Magiftro Artium. Salut.——Vale. Tuus, T. N. Cantabrigia. Junij 2. 1578." Alfo, in 15 Latin diftichs. "J. R. Catabrigienfis, Artium Magifter, Authori." The expofition contains 218 pages. Then follows, (not mentioned in the title-page, but the pages continued to p. 251.) "A Sermon[a] made before the reuerend —John Bifhop of London (by J. Keltridge, Preacher) at his Mannor at

---

[a] This is a fevere Satyr, holding a whip with two thongs; lafhing, on one hand, the rotten part of the papiftical clergy, who for lucre's fake retained their benefices in the Church, living in luxury and debauchery; and, on the other, the puritanical fchifmaticks, who troubled the Commonwealth. The author appears to have been no velvet-mouthed, but a fincere, impartial preacher; for, before his diocefan, and others of the clergy, treating of the hofpitality recommended in his text, and cenfuring the covetoufnefs of the clergy, efpecially as an occafion thereof in the laity, frankly announces to them "J muft fpeak the trueth before the Lorde, & before you all—Jn a cleere teftimonie of my confcience, J vtter the trueth vnto you, for J have feene it in fome, Jknowe it in many, it is common in all.——J once was at Pauls Croffe, when as a learned & graue father preaching there, among many other thinges fpake much of this vice: He called to memorie at that time, for the fpace of many yeares before, that there had been no one Alder man in al London, whofe fonne was remembered to haue ufed that honeftly, which the father in many yeares had fcraped up couetoufly: Jt was a fore & heauie faying for thofe wife & gray headed men that were then prefent: And the faying of that wife Prophet moued the heartes of many that heard it. There is at this daye, nowe in London, in mofte florifhinge ftate, two fonnes of two Aldermen, gouerninge by the wealth their fathers left them, the happie and good ftate of the Citie. But if that olde and gray headed father were now aliue, and fpeaking of the minifters of the word of God, fhould turne his eyes to fee the iolitie, wherin their children be: that the fame difcreete father, for the two famous & moft worthy Shrifes (*Will. Kampfon & Geo. Barn*) could not (fince that England was inhabited & knowen of men) picke out fo much as two fonnes of two Bifhoppes, to haue fet in the feate of their fathers:——The Lord wil haue their fonnes to fpend it wantonly, for that their fathers haue rackte it vp fo vnkindly." Again, "Jt is a pitifull thinge, and J cannot but vtter it, take it as you will: we difagree among ourfelues, & be at difcord within our owne houfe: and there is harte burning euen among thofe that minifter in the Ephod, and ferue at the table of the Lord.—Followe not the cuftome of our Prefice men,——fo nice as may be, if they fee a furpleffe or a Coope: yet they will thruft them of their accorde into other mens charge, & take vpon them gouernment in the iurifdiction of others,—yet cannot find in their heart to beftowe them in any other place, where the worde of God hath not been heard, to do good there: And to conclude, let there be one agreement." &c. &c.

Fulham,

ABRAHAM VEALE.

Fulham, before them of the Clergie, at the making of Minifters: in the yeare—1577. and nowe fet out in Printe." His text, 1 Tim, III; 1, 2, 3. He had licenfe for this. W.H. Quarto.

In 1578, there was a difpute between Tho. Vautrollier and him about the French Littleton, which they agreed fhould be decided by the Court of Affiftants; who ordered, that T.V. fhould deliver to A.V. 100 books of every impreffion of 1250, during the life of the faid A.V, he allowing only paper for the printing thereof.

"A COMMENtarie or Expofition vpon the twoo Epiftles generall of Sainct 1581. Peter, and that of Sainct Jude. Firft faithfullie gathered out of the Lectures & Preachinges of that worthie Jnftrumente in Goddes Churche, Doctour Martine Luther. And now out of Latine, for the finguler benefite & comfort of the Godlie, familiarlie tranflated into Englifhe by Thomas Newton. Jmprinted——for Abr. Veale——in Paules Churchyard, at the—Lābe. 1581." In a compartment with David and Mōfes on the fides, ufed by Tho. Marfh. Dedicated "To the right hon. Sir Thomas Bromeley Knight,—Lorde Chauncelor of Englande.——From Butley in Chefshire, this firft of October, 1581. Your L. moft humble, T.N." Contains befides, 172 leaves. He had licenfe for this the 25th of June. W.H. Quarto.

" The. xiii. Bookes of ÆNEIDOS. The firft twelue beeinge the woorke 1584. of the diuine Poet Virgil Maro, and the thirteenth the fupplement of Maphæus Vegius. Tranflated into Englifh verfe to the fyrft third part of the tenth Booke, by Thomas Phaër Efquire: and the refidue finifhed, and now the fecond time new fet forth for the delite of fuch as are ftudious in Poetrie: By Thomas Twyne, Doctor in Phyficke. Imprinted—by Will.How, for Abr. Veale,—at the—Lambe 1584." Dedicated "To the right worfhipfull Maifter Robert Sackeuill Efq; moft worthie Sonne & heire apparant to the Right hon. Syr Thomas Sackeuill Knight, Lorde Buckehurft.—At my houfe in Lewis, this 1 Januarie. 1584." Contains X 3, in eights, yet the fame fize as in 1573. W.H. Quarto.

" The good admonifhion of the fage Ifocrates, to young Demonicus; 1585. tranflated from Greek by Richard Nuttall." Octavo.

" A breef declaration of the lordes fupper written by the fingular 1586. learned man, and moft conftant martyr of Chrifte, Nicholas Ridley, bifhop of London, prifoner n Oxford," &c. Octavo.

" Of Chriftian frendfhip with all the braunches, members, partes and 1586. circumftances thereof: together with an Inuective againft Dice play and other prophane games. Writ firft in Latin by Lambertus Danæus, and tranflated into Englifh by Tho. Newton. "Dedicated to Mr. Will. Bromley, and Mr. Reginald Skreuen, Secretaries to the Lord Chancellor. For him. Octavo.

" A manifeft

ABRAHAM VEALE.

—— "A manifeſt detection of the moſte vyle & deteſtable vſe of Diceplay, and other practiſes lyke the ſame, a Myrrour very neceſſary for all yonge Gentilmen & others ſodenly enabled by wordly abūdance to loke in. Newly ſet forth for their behoufe." With the names of the falſe dice, in a dialogue between R, and M. By Gilb. Walker. D, in eights. Catal. T. Beauclerk, S. R. S. N°. 4137. Octavo.

—— "Enchiridion Militis Chriſtiani." See it p. 588. Octavo.

—— "Introduction to wiſedome, Banket of Sapience, and precepts of Agapetus. For him. Sixteens.

—— "The fountaine, or well of life, out of the which doth ſpring moſt ſweete conſolations right neceſſary for troubled conſciences. See p. 325. Sixteens.

—— A chriſtian exhortacion vnto cuſtomable ſwearers." &c. See p. 711.

—— A mery dialogue, declaring the properties of ſhrowde ſhrewes and honeſt wifes." Octavo.

—— The Common Prayer of K. Edw. vi. For him. Willliam Bayntan Eq;

—— Tho. Phaer his Regiment of lyfe: Treatiſe of the peſtilence: With the booke of Children. He was fined for printing this before he had licenſe; and his licenſe was on condition, that if it had been granted to any other, it ſhould then be void. For him. Octavo.

—— "The Myrrour or Glaſſe of Health," &c. See p. 375, and 404. 16°.

—— "The parlement of Byrdes." Harl. Miſcel. Vol. v: p. 479. Quarto.

—— "Here after foloweth a litle boke of Phillip Sparow. compyled by mayſter Skelton Poete Laureate." Octavo.

—— "An Enterlude called LUSTY JUVENTUS: liuely deſcribing the Frailtie of youth: of Nature prone to Vyce: by Grace and Good Councell traynable to vertue." Colophon, "Finis quoth R. Weuer. Imprinted—in Paules churche yeard," &c. Lincoln Cathedral Library. Reprinted in Hawkins's Origin of the Engliſh Drama. See it characterized, Warton's Hiſt. of Eng. Poetry, Vol. III; p. 200, &c. Quarto.
Another edition of this Morality is among Garrick's Plays, in the Brit. Muſeum, but being imperfect at the end, it is uncertain by whom it was printed. It has been ſuppoſed to have been printed by R. Pinſon; but, according to Ames, he died before 1529. Quarto.

—— "Here beginneth the Piſtles & Goſpels of euery Sonday & holy day in the yeare." Printed in double columns. Contains Fol. lxvii, beſides a table at the end. " Jmprinted in Paules Churche yarde," &c. Richard Gough Eſq. and W.H. Quarto.

—— "The Booke of Caruyng. Imprinted—in Powles church yarde," &c. John Fenn Eſq; Octavo.

"Colyn

"Colyn Clout. Compiled by Mafter Skelton Poet Laureat." Ath. Oxon. Vol. I; c. 23. Octavo.

"Deathes generall proclamation, or, a generall Proclamation, Set forth by the inuincible, famous, renowned, & moſt mightie conqueror Death, his high maieſtie, Emperour of the wide world terreſtiall, & Supreme Lord ouer each creature bearing life: Directed to all people, Nations, kinreds & tongues, &c. By Valent. Leigh." Maunſell, p. 42. This he printed with licenſe. Octavo.

"Prognoſtication of the Huſbandman for euer." Octavo.

"The knyght of the ſwanne." &c. as p. 363.

Beſides the abovementioned, he had licenſe to print "An almanak & pronoſtication of M. Buliens" for 1564. "An almanake & pronoſtication of M. Wylliam Cōnyngham, for 1565." Alſo, "An almanake & pronoſtic'on of M. Johnſon, for 1568." June 25. 1581. "A myrrour for Courtyers written in the Caſtalian tonge by Ant. de Guevara, bp. of Moadovet."

JHON WYGHTE, or JOHN WIGHT,

WAS a member of the Drapers' company; and though not even a brother with the Stationers', appears, on account of his trade, to have been under their juriſdiction. He was once fined by them for keeping his ſhop open on St. Luke's day; another time for that certain books (which ſeem to have been Primers) illicitly printed, were found in his cuſtody, for which he was fined iiij. l.

"The

1551. "The Byble," &c. as p. 592. "Imprinted——by John Wyghte, dwellynge in Paules Churche Yarde, at the fygne of the Rofe. Cum gratia & priuilegio—folum. vi. day of May M. D. LI." See Lewis, p. 192, &c. W.H.+ Folio.

1558. "The Armour of patience, with the feuen tribulations ioyning thereunto: tranflated out of French by Edw. Tefs." For him. Octavo.

1559. "A VVOORKE OF JOANNES FERRARIVS Montanus, touchynge the good orderynge of a common weale: wherein afwell magiftrates, as priuate perfones, bee put in remembraunce of their dueties, not as the Philofophers in their vaine tradicions haue deuifed, but according to the godlie inftitutions and founde doctrine of chriftianitie. Englifhed by William Bauande. 1559. Jmprinted—by Jhon Kyngfton, for Jhon Wight—in Poules Churcheyearde." In the compartment with two heads in a medallion at top. Dedicated "To the moft high & vertuous Princeffe Elizabeth,—— Quene of Englande, &c.—At the middle Temple the. 20. daie of December. 1559.—w. Bauande." Contains befides, Fol. 212. And at the end "A brief Collection of the chiefeft matters," and a table of "Fauts efcaped in the Printing." He had licenfe to print this, for which he paid iiij. d. the cuftomary price. "And at the fynyfshynge of the fayde boke he fhall paye for euery iij leves a pannye." This condition i do not find impofed on any other. W.H. Quarto.

1561. Chaucer's works, with the fiege of Thebes. John Kingfton for him. W.H. Folio.

1566. "The thyrde and laft parte of the Secretes of the reuerende Maifter Alexis of Piemont, by him collected out of diuers excellent Authors, with a neceffary table in the ende, contayning all the matters treated of in this prefent worke. Englifhed by Wylliam Warde." This title over a neat wood-cut of the Genius of Englande, crowned with laurel, ftanding on a pedeftal, and holding the reins of two wild horfes. The motto "Armipotenti Angliæ. Under it, "Imprinted—by Henry Denham, for John Wyght." 75 leaves and the table. "Anno. 1566." W.H. Quarto.

1574. "Two notable Sermons. Made by that worthy Martyr of Chrift, Maifter John Bradford, the one of Repentance, and the other of the Lordes fupper, neuer before imprinted. Perufed & allowed according to the Queenes Maiefties Iniunction. 1574. Jmprinted at London by John Awdely, and John Wyght." They are introduced with a preface by Tho. Samfon, and M. Bradford's Epiftle, dated "The. xij. of July, 1553." The fame that was publifhed with the fermon on repentance in his life-time. W.H.+ Octavo.

1578. "The art of graffing and planting trees," &c. For him. See it in 1580, and 1582. Quarto.

"The third—part of the Secretes of—M. Alexis" &c. as in 1566. 1578. The title over a cut of Mr. Wight at full length, dreffed in his livery gown, with a bonnet on his head, holding in his hand a clafped book, with the word SCIENTIA on the cover; and about the whole, WELCOM THE WIGHT THAT BRINGETH SUCH LIGHT. Under it "Imprinted at London by Tho. Dawfon, for John Wyght. 1578." The whole in a border of metal flowers. Contains M iiii. W.H. Quarto.

" A verye excellent and profitable Booke conteyning fixe hundred 1578. foure fcore and odde experienced Medicines, appertayning vnto Phyficke & Surgerie, long time practyfed of the expert and reuerende Mayfter Alexis, which he termeth the fourth & finall Booke of his fecretes, and which in his latter dayes he did publifhe vnto a vniuerfall benefite, hauing vnto that time referued it onely vnto himfelfe, as a moft priuate & precious treafure. Tranflated out of Italian into Englifhe by Richard Androfe.———. Imprinted at London by John Wyght. 1578." On the back is the coat armour of " The right hon Fraunces Lorde Ruffell, Earle of Bedforde," &c. to whom it is dedicated by the tranflator. Then an epiftle of M. Alexis to the reader; and a table of contents. It is divided into three books: the 1. contains 52 pages; the 2. p. 57; and the 3. p. 50. At the end a cut of M. Wight, as to the third part. "Imprinted for John Wight———in Paules Churchyard, at the North doore of Paules. Anno Domini. 1578." W.H. Quarto.

The firft part of " The Secretes of the reuerend Maifter Alexis of Pie- 1580. mont: containyng excellente remedies againft diuerfe difeafes, woundes, & other accidentes, with the maner to make diftillations, Parfumes, Confitures, Diynges, Colours, Fufions, & meltynges. A woorke well approued, verie profitable and neceffarie for euery man. Newlie corrected & amended, and alfo fomewhat enlarged in certaine places, whiche wanted in the firfte edition. Tranflated out of Frenche into Englifhe, by Willyam Warde." The cut of England's Genius, as to the third part in 1566. " Jmprinted———by Jhon Kyngfton, for John Wight. ———1580." Dedicated " To———the lorde Ruffel, Erle of Bedforde," by the tranflator. Wherein he fays, "I haue taken in hand to tranflate this noble & excellente woorke,———firft written in the Italian tongue, and after tourned into Frenche, and of late into Dutche, and now lafte of all into Englifhe,—beeyng a worke—of fo famous a man as Alexis is, and dedicated firfte to fuche a noble Prince as the Duke of Sauoye is, to whom trifles or fables are not to be prefented, nor ———vnder whofe name lies or vaine inuentions ought to be fet forthe." Then, an epiftle of " Don Alexis" to the reader. In which he acquaints us that he was above 82 years of age when he firft publifhed this work, &c. Contains befides, 117 leaves, and a table, at the end of which is Mr. Wight's effigy, as to the third part, 1578. See the 2d part without date. W.H. Quarto.

Zzzz " The

JOHN WIGHT.

1580. " The Art of, Graffing and Planting all sorts of trees, how to set stones and sowe Pepins, to make wild trees to Graffe on, as also remedies & Medicines, with other new practises, &c. By Leonard Mascall." For him. Maunsell, Part II. p. 10.   Quarto.

1581. " Two NOTABLE Sermons. Made by—M. John Bradford," &c. as in 1574. " Imprinted at London by John Charlewood, and John Wight. 1581." J 7, in eights. W.H.   Octavo.

1582. " Theatrum Mundi. Theatre or Rule of the worlde: Wherein may bee seene the running race and course of euerie mannes lyfe, as touching miserie & felicitie: whereunto is added a learned worke of the excellencie of man. Written in French by Peter Boastuau. Translated by John Alday." For him.   Octavo.
  This copy was assigned to him by Tho. Hacket, at a Court of Assistants. 8 Jan. 1576-7.

1582. " A book of the arte and manner how to plant and graffe all sortes of trees, &c. by one of the abbey of Saint Vincent in France, &c. with the addition of certaine Dutch practises, set forth and Englished by Leonard Mascall." Dedicated to sir John Paulet, knight, lord St. John. This book is said to be printed by him, and has his picture cut in wood, as above. See it printed by T. East, 1592.   Quarto.

1584. " THE SCEPTER OF IVDAH: Or, what maner of Government it was, that unto the Common-wealth or Church of Israel was by the Law of God appointed. By Edm. Bunny. Deut. 4. verse. 8. What nation is so great, &c. Imprinted at London by N. Newton, and A. Hatfield, for John Wight. 1584." In a neat border. On the back is a griffin inclosed in a circle, with this motto, " INTEGRA. LEX. ÆQUI. CUSTOS; RECTIQVE. MAGISTRA; NON. HABET. AFFECTVS; SED. CAVSAS. IVRE. GVBERNAT. The whole in an octagon; over which is " Griphus Graienfis," and under it this Sapphic stanza.
   " Quo truci es vultu, Gryphe, parce rectis:
    In malos totus rapiare sævus:
    Sic tuis semper genuina fies
     Juris imago."
Dedicated to——" the Gentlemen and Students of Graies In.——From Bolton-Percie in the ancientie of York, the fift of September, 1584." Then, a short address to the reader; and an analytical table. The treatise contains 160 pages besides, and two tables at the end, one of the Scriptures alledged, the other of contents. The whole is printed on neat Roman and Italic type, of different sizes: the i and the j, the u and the v, are properly used in the small letter, though not in the capitals; the semicolon and the hyphen are here introduced, the first i have noticed, though perhaps we may hereafter meet with them of an earlier date. This printed by licence to him. W.H.   Octavo.
                   " A Booke

JOHN WIGHT.

" A Booke of chriftian exercife appertaining to refolution, that is 1584. fhewing how that we fhould refolue ourfelues to become chriftians indeed by R. P. *(Rob. Parfons)* Perufed and accompanied now with a treatife tending to pacification, by Edm. Bunny." N. Newton and A. Hatfield, for him. For this he had licenfe. - Octavo.

" A Regiment for the Sea, containing very neceffary matters for all 1584. forts of Sea-men and Trauailers, as Mafters of fhips, Pilots, Marriners, and Marchants, Newly corrected and amended by the Author. Wherevnto is added a Hidrographicall difcourfe to go vnto Cattay, fiue feuerall wayes. Written by William Bourne." The cut of a fhip of war, with St. George's flag at the Fore-top-maft head; the Royal ftandard at the Main-top-maft head; the lord high admiral's arms at the Mizen-top; the enfign the fame, in the garter, with a coronet. " Imprinted——by T. Efte for John Wight." Dedicated " To the right hon. Edward Earle of Lincolne, Baron of Clinton & Say, Knight of the Noble Order of the Garter, Lord high Admirall of England, Ireland & Wales, and of the Dominions & Iles thereof, of the Towne of Calice & Marches of the fame, Normandie, Gafcoyne, & Guyone, and Captaine generall of the Queenes Maiefties Seas & Nauie Royall." Next is a generall preface to the reader, as to the firft edition; then another concerning this. Contains V, in fours. On the laft page is Mr. Wight's effigy, fet in a neat compartment, with this colophon, " Imprinted at London by T. Eaft for John Wight," in a tablet, on which Grammar and Rhetoric ftand as fupporters to Mr. Wight; over thefe is a tablet with the date, " 1584." This copy was affigned to him by Tho. Hacket at a Court of Affiftants, P. Jan. 1576-7
Quarto.

" A Booke of Chriftian Exercife," &c. as in 1584. " Heb. 12. 8. 1586. Jefus Chrift yefterday, and to day, and the fame for ever. Imprinted— by I. Iackfon and Ed. Bollifant, for John Wight, and are to be folde at the great North doore of Paules. 1586." In a border. Dedicated " To —Edwin,—Archbifhop of Yorke." Next, a preface, dated " At Bolton-Percie, &c. 9. of Julie, 1584." Then, a table of contents. E e 10, in twelves; White letter, the i and j, the u and v, properly ufed in the Roman, but not in the Italic. W.H. Twelves.

" Foure Bookes of Hufbandrie, collected by Conradus Herefbachius, 1586. Councellor to the high & mightie Prince, the Duke of Cleue: containing the whole art & trade of Hufbandrie, Gardening, Graffing, and planting, with the antiquitie & commendation thereof. Newly Englifhed, and increafed by Barnabe Googe Efquire. Genefis 3. 19. In the fweate of thy face fhalt thou eate thy bread, &c. At London, Printed for Iohn Wight. 1586." On the back is the author's coat of arms. Dedicated " To the right worfhipfull his very good freend Syr William Fitz Williams, Knight. ——Kingftone, the firft of Februarie. 1577." Then, a preface; a lift of authors cited; and a table of contents. Contains befides, 194 leaves;

On the laft, "Olde Englifh rules for purchafing Lande." In verfe, as inferted in Ames, p. 279. But he was miftaken in intimating that they were in Mafcall's Art of Grafting and Planting. On the back is Mr. Wight's effigy, and under it "Imprinted——for John Wight,——in Paules.Church-yarde, &c. 1586." He had licenfe 30. March 1579, for "The firft bocke of Hufbandrye. Sold and fet ouer vnto him by Mr. Watkins, 29. July, 1577." W.H.   Quarto.

1588.   "The Arte of Warre, Written in Italian by Nicholas Machiauel, and fet foorth in Englifh by Peter Withorne, ftudient at Graies Inne: with other like Martiall feates and experiments, as in a Table in the ende of of the booke may appeare. New imprinted with other additions. 1588." In an elegant compartment, fuitable to the fubject. Dedicated "To the moft high & excellent Princes Elizabeth—Queene of England," &c. by the tranflator. Then, "The proheme of N. Machiauel, Citezen & Secretary of Florence." This treatife is divided into feven books, with fchemes, &c. It has neither printer's nor publifher's name; but by comparing the type with the two following tracts, with which my copy is bound, it appears to have been printed by, and perhaps for the fame perfon. G g, in fours. W.H.   Quarto.

1581.   "CERTAINE VVAIES for the ordering of Souldiours in battelray, and fetting of battailes, after diuers fafhions, with their manner of marching: And alfo Figures of certain new plattes for fortification of townes: And moreouer, how to make Saltpeter, Gunpowder & diuers fortes of Fireworkes or wilde Fire, with other things appertayning to the warres. Gathered & fet foorth by Peter Whitehorne. Jmprinted—by Tho. Eaft: for Ihon Wight. 1588." A line of Flowers at top and bottom. N 3, in fours, with cuts. W.H.   Quarto.

1588.   "Moft BRIEFE TABLES to know redily how manieranckes of footemen armed with Corfletts, as vnarmed, go to the making of a iuft battaile, from a hundred vnto twentie thoufand. Next a very eafie, and approued waye to arme a battaile with Harkabuzers, and winges of Horfemen, according to the vfe at thefe daies. Newlie increafed, and largelie amplified both in the tables, as in the declarations of the fame, by the Aucthour himfelfe, Girolamo Cataneo Novarefe. Tourned out of Italian into Englifh by H. G. ¶ Jmprinted——by Tho. Eaft: for Iohn Wight. 1588." Dedicated by the author "To the Earle Aloigi Auogardo.——From Brefcia the 5. day of Julie 1563." Then, a fhort addrefs to the reader. I 1, in fours; A 6, E 2. With fchemes. W.H.   Quarto.

1589.   "A Booke of Chriftian exercife," &c. as in 1584, and 1586. John Charlewood for him.   Octavo.

1589.   The art of Navigation &c. See p. 719.   Quarto.

A com-

JOHN WIGHT.

A compendious Treatife of the ten commandments with godlie Prayers. For him. Small

" The feconde parte of the Secrets of maifter Alexis of Piemont, by him collected out of diuers excellent authors, and nevvly tranflated out of French into Englifh. With a generall Table of all the matters &c. By Will. Warde." Device, the Genius of England, as p. 780. " Imprinted by Henry Bynneman, for John Wyght." 80 leaves. W.H. Quarto.

Another edition of this 2d. part, page for page, " Jmprinted by Jhon Kyngfton, for Jhon Wight." On the laft page Mr. Wight's effigy as p. 781. W.H. Quarto.

" A Lamentation that Ladye Jane made, faying for my fathers proclamation, Now muft I lofe my head." For him. Bagford's Papers. A ballad for which Tifdale had licenfe in 1561.

" A Godlye treatyfe of Prayer, tranflated into Englifhe, By John Bradforde." From the Latin of P. Melanchton. J 6, in eights. " Imprinted ———— in Paules Churche yearde, at the Rofe, By John Wight." W.H. Octavo.

In 1563—1564 he had licenfe alfo to print " The Dyall of prynnces." 1568—1569. " Yraffyn and Playntinge." 6. April 1579. " Plutarks lives," with Tho. Vautrollier.

JOHN CAWOOD, efq;

WAS of an ancient family in Yorkfhire, as appears from a book at the Herald's-Office. W. Grafton vi. A, B, c. Lond. Wherein are thefe words: " CAWOOD, Typographus regius reginae Mariae; his arms are, fable and argent parte per cheveron, embatteled between 3 harts heads cabofed, counter-changed within a border per feffe, counter-changed as before, with verdoy de trefyles fleped, numbered 10. Thefe Cawoods were once lords of the manor of Cawood near the citie of York, although the caftle had aunciently been the archbifhops fee. And it appears among the inquifitiones of the brethren in the time of King John throughout England, (that is to fay, in the 12th and 13th year of his reign, in the county of York, concerning knights fervice, and others held by him in chief, or capite, in the treafury rolls for the aforefaid liberty, by the hands of the fhireef of that time:) that John Cawood held by grand fergentie (fcilt. per fore ftantem inter Darwent & Owfe) one

JOHN CAWOOD.

one plowed land in Cawood. Which John, father of Peter, and Robert, clerk of the pipe, who had John, who had Margaret," &c. Thus it seems he was of that family in Yorkshire. He was an original member of the Stationers' company, and a bountiful benefactor to them; was appointed upper warden in their charter granted by Phil. and Mary, 4. May, 1556; and was chosen master in 1561, again in 1562, and 1566. He appears to have learned the art of John Raynes. See p. 414; however, he exercised the art three or four years before a patent was granted him by queen Mary, when Richard Grafton was set aside, and had a narrow escape for his life. The chief import of the patent, which you may see at length in Rymer, vol. xv. p. 125, is abstracted below.[f] In the next reign he dwelt in St. Paul's church-yard, at the sign of the Holy Ghost, and became partner with Richard Jugge, in queen Elizabeth's patent; and for their joint concern rented a room in Stationers' hall, at xx. s. per annum. His name is found but once on the black list, and that in 1565 "for stechen of bookes, which ys contrary to the orders of this howse, &c." when he with 16 others were fined viij. s. x. d. The mark he used, may be seen in the plate. He was buried in St. Faith's,

[f] The queen, to all whom it may concern, sends greeting. Know ye, that of our special favour, &c. for the good, true, and acceptable service of our beloved John Cawood, already performed, by these presents for us, our heirs, and successors, we do give and grant to the said John Cawood, the office of our printer of all and singular our statute books, acts, proclamations, injunctions, and other volumes, and things, under what name or title soever, either already, or hereafter to be published in the English language. Which office is now vacant, and in our disposal, for as much as R. Grafton, who lately had and exercised that office, had forfeited it by printing a a certain proclamation setting forth, that one Jane, wife of Gilford Dudley, was queen of England, which Jane is indeed a false traitor, and not queen of England; and by these presents, we constitute the said John Cawood our printer in the premises, to have and exercise, by himself, or sufficient deputies, the said office, with all the profits and advantages any way appertaining thereunto, during his natural life, in as ample manner as R. Grafton, or any others have, or ought to have enjoyed it heretofore.

Wherefore, we prohibit all our subjects whatsoeuer, and wheresoever, and all other persons whatsoever, to print, or cause to be printed, either by themselves, or others, in our dominions, or out of them, any books or volumes, the printing of which is granted to the aforesaid John Cawood; and that none cause to be reprinted, import, or cause to be imported, or sell within our kingdom, any books printed in our dominions by the said John Cawood, or hereafter to be printed by him in foreign parts, under the penalty of forfeiting all such books, &c.

And we do grant power unto John Cawood, and his assigns, to seize and consiscate to our use, all such books, &c. as he or they shall find so prohibited, without let or hindrance; and to enjoy the sum of 6l. 13s. 4d. per annum, during life, to be received out of our treasury. And whereas our dear brother Edward VI. &c. did grant unto Reginald Wolf, the office of printer and bookseller, in Latin, Greek, and Hebrew; we out of our abundant grace, &c. for ourselves, heirs, and successors, do give and grant to the said John Cawood the said office, with the fee of 16s. 8d. per annum, and all other profits and advantages thereto belonging, to be entered upon immediately after the death of the aforesaid Reginald, and to be by him enjoyed after his natural life, in as full and ample a manner, as the said Reginald now has, and exercises that office, &c. Given at Westminster, 29 Dec. 1553.

under

JOHN CAWOOD.

under St. Paul's, London, and his epitaph, preferved by Mr. Dugdale,[g] is thus: "John Cawood, citizen and ftationer of London, printer to the moft renowned queen's majefty Elizabeth; married three wives, and had iffue by Joane, the firft wife only, as followeth, three fons, four daughters. John his eldeft fon, being bachelor of law, and fellow in New Colledge in Oxenford, died 1570; Mary married to George Bifchoppe, ftationer; Ifabell married to Thomas Woodcock, ftationer; Gabrael, his fecond fon, beftowed this dutiful remembrance of his deare parents, 1591. then church-warden; Sufanna married to Robert Bullok; Barbara married to Mark Norton; Edmund third fon died 1570. He died 1 of April, 1572. being of age then, 58."

A Bible, and a New Teftament, are inferted in the Lambeth lift of 1549. Bibles, &c. as printed by John Cawood, in quarto; but no mention on what authority, or where to be found.

"The new Teftament" &c.. Printed for I. C. Who this I. C. was, 1550. does not clearly appear. See p. 765. W.H. Octavo.

"A fpiritual and moft precious perle," &c. as p. 744. This, Mr. 1550. Ames fays, after Palmer, was dedicated *to* Edward duke of Somerfet; but they fhould rather have faid prefaced *by* him. Dated "Somerfet Place, 6. May, 1550." They alfo added the words "A moft fruitful treatife of behaviour in the danger of death," as explanatory of the precious perle, whereas this latter is a feparate tract, indeed by the fame author, and ufually printed together. Mr. Ames intimates Miles Coverdale to be the author, but Singleton's edition informs us that it was written by Otho Wermylierus, and tranflated by M. Coverdale.[h] If this be fact, an author whom i much admire for his great abilities and impartiality, was mifled in afcribing to the duke of Somerfet the whole book, who wrote only the preface[i] to it, addreffed to the Chriftian Reader; and that not during his firft imprifonment, but after it; and to whom the author appears to have been unknown. This is evident from the extract in the notes, and comparing the date of the preface, as above, with the time when the duke was releafed[k] from the tower.

A Declaration of Q. Mary of her profeffion of the true religion and 1553. forbidding the names of diftinction (Papifts and Hereticks) among her

---

[g] Hiftory of St. Paul's, p. 125.
[h] Bp. Tanner rightly includes this treatife among the tranflations of M. Coverdale, but afcribes it to Calvin as the author; wherein he was miftaken, for that is a different tract. See p. 547.
[i] In which he fays exprefly, "This man whofoeuer he be that was the firft author of this boke, gocth the right way to worke: he bryngeth hys ground from Gods word.

—. Jn our greate trouble, which of late dyd happen vnto vs (as al the worlde doothe know) when it pleafed God for a tyme to attempt vs with his fcourge, & to proue if we loved him: Jn readyng this boke we did finde great comforte, &c ——. And hereupon we have required hym, of whom we had the copy of thys boke, to fette it forth in pryrt."
[k] The 16 Feb. 1549-50.

JOHN CAWOOD.

subjects. "Gyuen at Rychmond. xviii. Aug. Reg. 1. Excusum in Ædibus Johannis Cawodi." T. Baker's Maunsell, p. 88.

1553 "A dialoge describing the originall ground of these Lutheran faccions, and many of their abuses, Compyled by syr William Barlowe chanon, late byshop of Bathe. Anno. 1553." In a border. It begins with a preface intimating that whereas this book had been "long agoo set forth & printed, it hath ben thought good——(the prynte beynge nowe worne out) yet eftesones at this present to be reuiued," &c. Next is "A preface of the auctoute to the readers." This is supposed to be the same as was prefixed to the first ¹ edition. Contains.L, in eights, half sheets. See Strype's Eccl. Memor. Vol. III ; p. 153. Prynted—in Paules Churcheyard, —Cum priuilegio—solum." W.H. Sixteens.

1553. Concio quædam admodum elegans, docta salubris & pia magistri Iohannis Harpsfeldi sacræ Theologiæ baccalaurei, habita coram patribus & clero in Ecclesia Paulina Londini. 26. Octobris. 1553. Cui accedunt & sequentia, uidelicet Wilhelmi Pij Decani Cicestrensis, & Johannis VVymslei Archidiaconi Londini, Orationes laudatoriæ. Item, Magistri Hugonis VVestoni, Decani VVestmonasterij, uiri longe doctissimi & eloquentissimi, ac cleri referendarij. Oratio coram patribus & clero habita, miré elegans ac ualde docta, cum responso et exhortatione reuerendi patris, domini Edmundi Londinensis Episcopi. Romæ. 15. Quæcunq; scripta sunt," &c. D 4, in eights, half sheets. " Excusum Londini in ædibus Johannis Cawodi typographi Regiæ Maiestatis. Anno. M. D. LIII. Mense Decembri." The whole in neat Italic types. W.H. Sixteens.

1553. "The saying of John, late duke of Northumberland," vpon the scaffolde, at the tyme of his execution, the 22. August." Again, without date.
Octavo.

1553. "An exhortation to all menne to take hede, and beware of rebellion," &c. See it in 1554. Sixteens.

1553. "A Proclamation against raisers of sedition, printers, players of interludes. Given at our manour of Rychemonde, the 18. August, in the first yeare of our moost prosperous reygne. Londini in aedibus Johannis Cawodi typographi reginae, excusum anno M. D. LIII. Cum priuilegio. ——solum."

¹ When or where that edition was printed i cannot find ; but according to the author of Yet a courfe at the Romyshe foxe, fol. 55, these dialogues were commanded by bp. Stokesley to be preached of the curates throughout all his diocesse, at such time as he encouraged H. Pepwell to import a new edition of The Enchyridion of Eckius. See p. 316. As this edition mentions the casting out of images, &c. it seems to have been modelled to the times.

ᵐ Many pamphlets, &c. were published by the late D. of Northumberland to colour his proceedings. See The answer to the execution of Justice. Pref. p. 5th.

"—for

JOHN CAWOOD.

"——for the newe feuerall monies, and coines of fyne fterlynge fyluer 1553. and golde, and the valuation of euery of the fame, new fet furth by her highnes. Rychemonde." 20 Auguft, &c.

"——againft the malicioufe forfe of the moft errande traytour, fyr 1553. John Dudley, late duke of Northumberlande, and his complyces." 1 Sept.

——for fewell, for the cities of London and Weftminfter. 20 Nov. 1553.
——feditious talking. 1553.
——againft rebellion. 17 Febr. 1554.
"——touching coygnes." 4 May. 1554.
——or Summons to hold a parliament.
——of concluding a marriage with the prince of Spain. 1554.
"—— of centeyne moneyes and coynes of fyne gold and fine filuer, with 1554. the valuation of the fame, newlye fet forth by theyre highneffe. 26 Dec. 1 and 2 years of our raygnes."

——to put the lawes in execution againft tranfgreffors. God faue king 1554. and queen.

"ANNO MARIÆ PRIMO. Actes made in the Parliament begun and 1554. holden at Weftminfter, the. v. day of October, in the fyrft yeere of the reigne of our mofte gratious foueraigne Lady Mary,——Queene of Englande, Fraunce, & Ireland, defender of the Faith, and of the church of England, & of Ireland in earth the Supreme head. And there continued to the. xxi. day of the fame moneth, that is to fay, in the fyrft feffion of the fame Parliament, as foloweth. Cum priuilegio Regiæ Maieftatis." In a compartment with the queen's arms at top, and on the fides 2 female figures, each holding an olive branch in their left hand. One leaf befides the title. To this is annexed,

——"Actes made in the feconde & laft Seffion of this prefent Parlia- 1554. ment, holden vpon prorogation at Weftminfter, the xxiiii. day of October, in the firft yere &c. and there continued & kept to the diffolution of the fame,——vi. day of December then next enfuing." Contains together E 3, in fixes. " Excufum——in Ædibus——Typographi Regiæ Maieftatis.——M. D. LIIII. Cum priuilegio——folum." W.H. Folio.

" Philippe and Marye by the grace of God Kynge and Queene of Eng- 1554. land, Fraunce, Naples, Jerufalem, and Irelande, defendours of the fayth, Princes of Spayne and Cycilie, Archdukes of Auftria, Dukes of Myllayne, Burgondie & Brabande, counties of Hafpurge, Flaunders, and Tyroll." The fame printed beneath in Latin. " Anno. M. D. LIIII." A fheet.

" An exhortation to all menne to take hede and beware of rebellion: 1554. wherein are fet forth the caufes that commonlye moue men to rebellion, and that no caufe is there that ought to moue any man thereunto, with a difcourfe of the miferable effects, that enfue thereof, and of the wretched ende that all rebelles come to, mofte neceffary to be redde in this fediti-
& & & &   oufe

ouse & troublesome tyme; made by John Christoferson. At the ende whereof are ioyned two godlye Prayers, one for the Quenes highnes, verye conuenient to be sayd dayly of all her louing & faithfull Subiectes, and another for the good & quiete estate of the whole realme. Reade the whole, and then Judge." Dedicated "To the mooste excellent and vertuouse Queene Marye—Quene of England, &c.——Your hyghnefs Chapleyne and daylye oratoure John Christoferson." Afterwards dean of Norwich. G 4, in eights, half sheets. " Jmprynted——in Paules churcheyarde, at the signe of the holy Ghost, by—, Prynter to the Quenes highnes. Anno Domini. 1. 5. 5. 4. 24. Julij. Cum priuilegio Reginæ Mariæ." Sixteens.

1554. "Articles to be enquired of in the generall visitation of Edmonde Bisshoppe of London, exercised by him the yeare of oure Lorde, 1. 5. 5 4. in the citie and Diocese of London, and set forthe by the same for his owne discharge towardes God & the worlde, to the honour of God, & his catholyke churche, and to the commoditie & profite of all those that eyther are good (which he woulde were all) or deliteth in goodnesse (which he wissheth to be many) without anye particular grudge or displeasure to anye one good or badde within this Realme: whiche articles he desyreth all men of their charitie, especiallye those that are of his Diocese, to take wyth as good intente & minde, as he the sayde bysshoppe wyssheth & desyreth, whyche is to the beste. And the sayde Bysshoppe wythall desyreth all people to vnderstande, that whatsoeuer opinion, good or badde, *they* have conceyued of him, or whatsoeuer vsage or custome hath bene heretofore, hys onlye intente & purpose is, to doe his dutie charitablye, and wyth that loue, fauoure & respecte, both towardes God, and euery Christen person, whyche anye Bysshoppe should shewe to hys flocke in any wise." In a border, with Cawood's mark at the bottom. The first 37 of these articles concern the clergy, "because they shuld of duetie giue good example," &c. The next 11 are "concernyng archedeacons their officialles & ministers.—the thynges of the Churche & ornamentes of the same, 16.——the Laytye, 41——scholemaysters & teachers of chyldren, men or women, 8.——middewiues & and such as come to the trauayle of women beyng wyth chylde, 6.——the original patrones of benefices, and other that haue aduousons of the sayd benefices, 5. The tenour, forme & effect of the othe giuen by the sayd bisshop of London, to ȳ inquisitors & searchers for knowledge of thinges amisse," &c. E 3, in fours. "Excusum Londini in ædibus——Typographi Regiæ Maiestatis. Anno. M. D. LIIII. Mense Septembri. Cum priuilegio Regiæ Maiestatis.", See Bale's Reflexions on these articles, p. 767. W.H.
Quarto.

1554. "A treatise of S. John Chrisostome, concerning the restitution of a sinner, which is chiefly made against desperacyon, newly translated out of Greek into English." 10th May. Sixteens.

"An

JOHN CAWOOD.

"An exhortation to Margaret, wife of John Burges, clothier of Kingswood in the county of Wilts. By Paul Bush, bishop of Bristol." Eccl. Memor. Vol. III; p. 172. Also without date.   Octavo.   1554.

"EPITAPHIA et inscriptiones lugubres, a Gulielmo Berchero, cum in Italia animi causa peregrinaretur, collecta." See it in 1566. Quarto.   1554.

"Two Notable Sermons made the 3d, and 5th frydays in Lent last past before the Queens higgness concerning the real presence, &c. by Tho. Watson, D.D."   Quarto.
The same; 10. May. Contains & 7, in eights, half sheets. W.H. Sixteens.   1554.

"The epistle of Erasmus Roterodamus sente unto Conradius Pellicanus, concerning his opinion of the blessed Sacrament of Christes body and bloude."   Sixteens.   1554.

"Pro instauratione republicae Angl. proque reditu reverendissimi atque, illustrissimi domini, Reginaldi Pole, sanctae Romanae ecclesiae, titulo sanctae Mariae Cosmedim, diaconi cardinalis, sedis apostolicae legati a latere. Oratio ad prudentissimum senatum Angl. authore Jodoco Harchio^m Montensi." See Eccl. Memor. Vol. III; p. 157.   1554.

"Bulla Plenariae Indulgentiae per S. D. N. Julium Diuina Prouidentia papam III. concessae omnibus Christi fidelibus, qui Deo optimo pro vnione Regni Angliæ Sanctæ matri Ecclesiæ iam facta gratias egerint, ac pro cæteris qui adhuc in errore remanent, nec non pro pace inter Principes Christianos obtinenda humiliter supplicauerint. Datum Romae, 9 Kal. Jan.   Broadside.   1554.

"The Declaration of the bishop of London, to be published to the lay people of his diocesse, concerning the reconciliation." Dated Lond. 19 Feb.   Broadside.   1554.

"Actes made——and holden at Westminster the 2d day of Aprill, in the first yere of the raigne of our—Ladye Marie—Quene of Englande,—defender of the fayth, and there continued & kept to the dissolution of the same,—5th day of Maye, next ensuing,——Cum priuilegio," &c. In the same compartment as to Erasmus's Paraphrase, p 545; but the arms, &c. at the bottom obliterated.^a A and B in sixes, C and D in fours.
"Excusum—in edibus—M. D. LIIII, Mense Maio." W.H. Folio.   1554.

Bishop Bonner's Cathechism.   Quarto.   1555.

"The Song of the Chyld Byshop, as it was songe before the queenes maiestie in her priuie chamber at her manour of saynt James in the ffeildes   1555.

---

^m He was a learned physician of Mons in Germany, and in 1553, and 1576, published two tracts in Latin, on the subject of the Eucharist.   ^a See p. 717.

on faynt Nicholas day and Innocents day this yeare nowe prefent, by the chylde byfhope of Poules churche with his company. Londini, in ædibus——typographi reginæ, 1555." See Warton's Hift. of Eng. Poetry, Vol. III; p. 321. Quarto.

1555.   " S. Ambrofe his Deuout praier, expedient for thofe that prepare themfelues to fay Maffe, &c. tranflated by Tho. Paynell." Octavo.

1555.   Twelve fermons by S. Auguftin, tranflated by Tho. Paynell, and by him dedicated " To the mooft vertuous Lady & mooft gratious Quene Marye, daughter vnto the mooft victorious & moofte noble prynce, kinge Henry the viii." &c. L, in eights. My copy wants the title. "Jmprinted—at the fygne of the holye gooft by——Prynter to the Quenes Maieftie. :Cum priuilegio——folum." W.H.+ Octavo.

1555.   " A profitable and neceffary doctrine, with certayne homelies adioyned thervnto fet forth by the reuerende father in God Edmonde Byfhop of London, for the inftruction of the people beynge within his Dioceffe of London, and of his cure and charge. Declina a malo, & fac bonum. Prefis, vt profis." In a compartment with I. C. at bottom. C c c, in fours. " Excufum——in ædibus——Anno 1555. Menfis uero Septembris. 17." Other editions without date in the place of J. C. have his mark. W.H. Quarto.

1555.   " Homelies fette forth by the righte reuerende father in God, Edmunde Byfhop of London, not onely promifed before in his boke, intituled, A neceffary doctrine, but alfo now of late adioyned, and added therevnto, to be read within his dioceffe of London, of all perfons, vycars, & curates vnto theyr parifhioners, vpon fondayes, & holydayes. Anno. M. D. LV." In the compartment with I. C. The bifhop's epiftle to the clergy is dated, 1 July, 1555. T 2, in fours. At the end " Domine falue fac regem & reginam, & omnes qui eis bene uolunt. ¶ Jmprinted——in Poules churcheyarde,—by—Prynter to the Kynge and Queenes Maiefties. Cum priuilegio" &c. W.H. Quarto.
Another edition, at the end of which is a " Tetraftichon, in immodicam præfentis temporis pluuiam." Alfo, " A Dialogue betwene man and the Ayre of lyke effecte." Then, " Domine falue fac" &c. Quarto.

1555.   " Jniunctions geuen in the vifitatiō of the Reuerend father in God Edmunde, bifhop of London, begunne & continued in his Cathedral churche and dioces of London, from the thyrd day of September, in the of our Lorde god, a thoufand fiue hundreth and fifty foure, vntil the viij. daye of October, the yeare of our Lorde a thoufand fiue hundreth fifty and fiue then nexte° enfuyng." Nine leaves. "Jmprinted——in

₴ Thus in my copy.

JOHN CAWOOD.

Paules churchyeard, at the figne of the holy goft, by—Printer to the Kyng & Queenes highneffes. Anno Domini. M. D. LV. 4. Octobris. Cum priuilegio Regiæ maieftatis." W.H.   Quarto.
Mr. Ames mentions it in   Octavo.

" A fupplicacyō to the quenes Maieftie. Impryntid at London/ by (1555.) John Cawoode' Prynter to the quenes Mayeftie wyth here moft gracyus lycence." Twenty feven leaves. Ends with " Praye, Praye, Praye. ——Anno.¹ M. D. L." W.H.   Octavo,

" Anno primo & fecundo Philippi & Marie. Acts made at a Parlia- 1555, ment begun and holden at Weftminfter the eleuenth day of Nouember, in the firft & fecond yere of the raigne of our Soueraigne Lorde & Ladye Philip and Mary,——Kyng and Queene of Englande, Fraunce, Naples, Hierufalem, & Irelande, defenders of the fayth, Princes of Spaine, &c. there continued and kept vntill the diffolution of the fame,—xvi. day of Januarie then next enfuing, &c. Cum priuilegio" &c. In the fame compartment as the 2d. parliament of this queen. F', in fixes. " Excufum—— in Ædibus——Typographi Regiæ Maieftatis.—1555." W.H. Folio.

"——Actes made at a Parliament begunne——at Weftminfter, the 1555. xxi. day of October, in the feconde & thirde yere of——Philip & Marie, ——: And there continued——vntyll the diffolution of the fame, ——the ix. day of December then next enfuing,——Cum priuilegio" &c. In the fame compartment as Anno primo Reg. Eliz. p. 718. K 4, in fixes. " God fave the Kyng and the Quene. Excufum——in ædibus &c. 1555." W.H.   Folio.

A proclamation for fuppreffing heretical books, &c. 13 June.	1555.
——againft confpirators. Grenewich, 1 April.	1556.
——concerning the Coin. Grenewych, 3 April.	1556.
————, the fame. St. James. ·27 April.	1556.
————. the fame, for Ireland. 19 Sept.	1556.

" BOETIVS DE CONSOLATIONAE' philofophiæ. ¶ The Boke of Boecius, 1556, called the comforte of philofophye, &c.——in maner of a dialoge——

> ᵖ Though i have given this book a place among Cawood's performances, according to the print, it cannot rationally be fuppofed to be his, being fuch a ftinging Satyr on the clergy, and obliquely on the queen herfelf. " Your fine fetches in putting the names of Jhon Cawode the quenes Printer, and others (who with harte detefteth your doynges) to your beggerly libelles as to bee imprinter thereof, euery man nowe fpieth." Difplaying of the Proteftants, (by Miles Huggard) folio 118, b.
> ᵠ This date muft be an error of the prefs, omitting a V at the end, as it is evident by many circumftances mentioned in the treatife. See Oldys's Catal. of pamphlets in the Harleian Library, Nᵒ· 220- r Thus in my copy.

betwene

betwene———Boecius & Philofophy, whofe difputations———do playnly declare the diuerfitie of the lyfe actiue———and the contemplatyue,———. Tranflated out of latin into the Englyfhe tounge by George Coluile, alias Coldewel, to thintent that fuch as be ignoraunt in the Latin tounge, & can rede Englyfhe, maye vnderftande the fame. And to the mergentes is added the Latin, to the ende that fuch as delyghte in the Latin tonge may rede the Latin, accordynge to the boke of the tranflatour, which was a very olde prynt. Anno. M. D. LVJ." On the back begins the dedication " To the hygh and myghty pryncefſe, our Souereigne Ladye & Quene, Marye———Quene of Englande, Spayne," &c. F f 2, in fours. The Latin in Italics, on the inner margin. " Jmprynted———Prynter to the Kynge & Quenes Maiefties. Cum priuilegio———folum." W.H. Again in 1561; and without date. Quarto.

1556. " ALL THE SVBMYffyons, and recantations of Thomas Cranmer, late Archebyfhop of Canterburye, truely fet forth both in Latyn and Englyfh, agreable to the Originalles, wrytten and fubfcribed with his owne hande. Vifum & examinatum per reuerendum patrem & Dominum, Dominum Edmundum Epifcopum London. Anno, M. D. LVI." Six leaves. Excufum———in ædibus—Typographi Regiæ Maieftatis. Anno, M. D. LVI. Cum priuilegio." W.H. Quarto.

1556. " The folovving of Chrift," &c. as p. 323. The xviij. of September, Anno M. D. LVI. Cum priuilegio———folum." Lambeth Library. W.H.+ Octavo.

1556. " Fiue Homiles" &c. as p. 716.

1556. " John Churchfon his treatife of the Church, declaring what & where the Church is, that it is knowne, and whereby it is tried & knowne. Cum priuilegio———folum." Octavo.

1557. " CIRCES Of John Baptifta Gello, Florentine. Tranflated out of Jtalion into Englifhe by Henry Jden. Anno Domini. M. D. LVII Cum priuilegio———folum." In a border. Dedicated "To the ryght honorable ỹ lord Herbert of Cardiffe, Maifter Edwarde Herbert, & Mafter Henry Compton, his bretheren.—From London, the xv. day of March, ———H. Jden." Alfo, " To the mofte myghtie & excellent Prince Cofimo de Medici, Duke of Florence.———From Florence the firft of Marche. 1548. Giouanbaptifta Gello." This treatife is divided into ten dialogues, between Circes, Ulyffes, and his companions transformed by Circes into brutes. T 4, in eights, half fheets. " Imprinted———printer to the Quenes Maieftie. Cum priuilegio———folum." W.H. Sixteens.

" St.

JOHN CAWOOD.

" St. Bafil the great his Exhortation to his kinfmen to the ftudie of the 1557.
Scriptures: tranflated by Wiliam Berker." Octavo.

" An Homelye of Bafilius magnus, Howe younge men oughte to reade 1557.
Poetes & Oratours. Tranflated out of Greke. Anno M. D. LVII. Im-
printed——Printer to the King & Quenes Maiefties. Cum priuilegio"
Octavo.

" The VVorkes of Sir Thomas More Knyght," &c. as p. 731. W.H. 1557.
Folio.

" The acts made in the 4th and 5th of Philip and' Marie." At the end, 1557.
" God fave the kynge and queene.——Cum priuilegio." Folio.

" Anno quarto & quinto Philippi & Mariæ. Acts made at a Parlia- 1558.
ment begun——at Weftminfter, the xx. day of Januarie, in the fourth
& fifth yere of——Philip & Mary,——.And there continued vntyl the
vii. day of Marche, then next folowyng,——.Cum priuilegio" &c. In
the fame compartment as the 2d. parliament, the 1. of Mary. G 3, in
fixes. " Excufum——in Ædibus——Typographi Regiæ Maieftatis.
Anno Dom. 1555.' Cum priuilegio" &c. W.H. Folio.

An act' for hauing of horfe, armour, and weapons. On 7 fides. 1558.

" Certain godly and deuout prayers made in Latin by the Reuerend 1558.
Father in God Cuthbert Tüftall Bifhop of Durham, and tranflated into
Englifh by Tho. Paynell Clerke." Beautifully printed in red and black,
double columns, Latin and Englifh. " Imprinted——at the Sygne of
the holye Ghofte, by——Printer to the Kynge and Quens Maieftyes. Anno
1558. Cum priuilegio——folum." William Bayntun Efq.

Among the MSS. proclamations in the colection of the late Martin
Folkes" Efq; were " Ordinances devifed by the king and queenes majefties, for thordre of the poftes and hacqueny men betweene London
and Douer." Mr. Ames, after he had publifhed his Typographical
Antiquities, met with a compleat fet of thefe, printed in the time of Q.
Elizabeth, by Chr. Barker, in the poffeffion of Dr. R. Rawlinfon. As the
royal proclamations had been ufually printed, fo no doubt thefe were alfo at
their firft publication. This is the firft notice of pofts & ftages we have
met with.

* This feems to be the fame as the following article, but printed before the 25th
of March.

* Thus in my copy, but it muft be a
mifprint for 1558, feeing this parliament
did not end 'till the 7 March, 1557-8. The

late Mr. Ratcliff's copy was dated 1558.
t This was the 2d. act in the laft parliament.
u Now in the Library of the Society of
Antiquaries.

A proclamation

JOHN CAWOOD.

1558. A proclamation of the king and queen againſt divers books. "Geuen at our Manor of Saint Jameſes, the ſixt day of June." Fox's Martyr, p. 2235. Edit. 1570.

He printed ſeveral books with R- Jugge as Queen's printer, which i muſt intreat the reader to ſee under his name to avoid repetitions, this work having ſwelled largely under my hand; but as ſome articles were omitted there, they ſhall be inſerted here, with his name.

1559. "The complaint of peace. Written in Latin by the famous Clerke Eraſmus Roterodamus. And nuely tranſlated into Englyſhe by Thos Paynell.——1559." At the end, "Imprinted——in Paules Church-yard by——one of the Printers to the Quenes Maieſtye. Cum priuilegio Regiæ Maieſtatis." Octavo.

1559. A proclamation for preparing a navy of Ships. This no doubt with R- Jugge.

1560. "A Godly Treatiſe declaring the benefits fruites & great commodities of Praier, alſo the true vſe thereof, written in latin 40 yeares paſt by an engliſhman of great vertue & learning, tranſlated." Maunſell, p. 83. Octavo.

1561. The Bible, (Cranmer's) with notes and figures. Lambeth liſt, from Smith's catalogue, 1682. Folio.

1561. The ſame without notes or cuts. Job ends on Fol. cciii. The third parte ——contaynyge——The Pſalter" &c. The compartment with his mark at bottom. Malachy ends on Fol. cxxxiij. The numbers are continued through the Hagiographa to Fol. ccxiiij. "The Nevve Teſtament in Engliſh, tranſlated after the Greke, containynge theſe Bokes. Mathewe," &c. This title encloſed with pieces: a ſmall cut of our Lord's ſupper, at top; another of Judas betraying him, at bottom. At the end, A table to the epiſtles and goſpells. Fol. cij. "Imprinted——by Jhon Cawoode. Prynter to the Quenes Maieſtie. Anno. M. D. LXI. Cum priuilegio Regiæ Maieſtatis." W.H. Quarto.

1561. "The boke of Boecius, called the comforte of philoſophye." See it in 1556. Quarto.

1563. "A ryght fruitfull Monition concernyng the ordre of a good Chriſtian mans Lyfe, &c. Made by the famouſe Dr. Colete, ſometime Deane of Powles. Anno 1.5.6.3. I. C." At the end, "Imprinted in Powles Churche yard by Jhon Cawood, Printer to the Queens Maieſtie," T. Baker's Maunſell, p. 35. Octavo.

1564. The book of Common Prayer, &c. as p. 717, &c. With R- Jugge. W.H. Folio.

" Epitaphia

JOHN CAWOOD.

"Epitaphia et inscriptiones Lvovenses. A Gulielmo Berchero, cum in Italia, animi causa, peregrinaretur, collecta. Excusum——in ædibus Johannis Cavvodi Regiæ Maiestatis Tipographi. Anno Domini M. D. LX. VI. Cum priuilegio——folum." E, in fours. W.H. Quarto. 1566.

"Delectable demandes, and pleasaunt questions, with their seuerall aunswers in matters of loue, naturall causes, with morall and politique deuises. Newly translated out of Frenche into Englishe, this present year of our Lord God." Quarto. 1566.

The holy bible, &c. as p. 723. With Cawood's mark.* Quarto. 1569.

"Stultifera Nauis, qua omnium mortalium narratur stultitia, admodum vtilis & necessaria ab omnibus ad suam salutem perlegenda, è Latino sermone in nostrum vulgarem versa, et jam diligenter impressa. An. Do. 1570." A cut of several vessels laden with fools. "The Ship of Fooles, wherin is shewed the folly of all States, with diuers other workes adioyned vnto the same, very profitable and fruitfull for all men. Translated out of Latin into Englishe by Alexander Barclay Priest." This edition contains the Latin and English. The English alone was printed by Pinson in 1509. It was translated also into English prose, and printed by W. de Worde in 1517. See p. 156, and p. 253. The prefixes are "Venerandissimo in Christo Patri ac Domino, domino Thomæ Cornish, Tenemensis pontifici, ac diocesis Badonensis Suffraganio vigilantissimo, suæ paternitatis, capellanus humilimus Alexander Barclay, suiipsius recommendationem cum omni summissione & reuerentia. Narragonia latine facta à Iacobo Locher Philomuso Sueuo, eiusdem Epigramma ad Lectorem. Epistola Iacobi Locher Philomusi ad eruditissimum virum Seb. Brant Iurisconsultum & Poetam argutissimum, præceptorem suum dilectissimum.—.Datum Friburgi. Kal. Februariis.—1497. Carmen eiusdem ad Seb. Brant. Saphicos eiusdem——, excusantis ingenij sui paruitatem. Ad Iohannem Bergmannum de Olpe Iacobi Locher Decatasticon. Ad Iacobum Philomusum——. Exhortatio Seb. Brant. Prologus Iac. Locher Philomusi in Narragoniam. The Prologe of Iames Locher;" translated by Alex. Barclay, who adds at the conclusion "I haue ouersene the fyrst inuention in Doche, & after that the two translations in Latin & Frenche, whiche——agreeth in sentence, threefolde in language: wherefore willing to redresse the errours & vices of this our Realme of Englande,——I have taken vpon me,—to drawe into our Englishe tongue the saide booke named The Ship of Fooles, so nere to the saide three Languages as the parcitie of my witte will suffer me, &c. Hecatasticon in proludium auctoris & Libelli Narragonici. The Proeme:" A translation of the Hecatasticon, in 22 stanzas. " Argumentum in Narragoniam. The Argument. Epigramma—Iac. Locher—ad Lectores. In Narragonicam profectionem, Celeusma Seb. Brant. The clamour to the fooles." In 6 stanzas. Then 1570.

5 A begins

JOHN CAWOOD.

begins The Ship of Fools; to every Satyr of which is a suitable cut; the same seemingly as to the German edition, the date 1494, in antique figures, being on one of them. On the front page of leaf 259, "Thus endeth the Ship of Fooles, Tranflated——by Alex. Barclay Prieft, at that time Chaplen in the Colledge of S. Mary Otery in the Countie of Deuon.—— 1508." On the back, "Excufatio Iac. Locher," in Sapphic verfe: "Plaudite Mufæ" being the burden of every one, except the laft, "Lector amice." Then five ftanzas by "Alex. Barclay excufing the rudenes of his Tranflation." Laftly an Index in Latin, and then in Englifh. This book was licenfed to him 1567——1568; and is the only book he had licenfe for. Hereunto is annexed,

"The Mirrour of good Maners. Conteining the foure Cardinal Vertues, compiled in Latin by Dominike Mancin, and tranflated into Englifh by Alex. Barclay prieft, & Monke of Ely." The Latin on the fide of the Englifh. G, in fixes. See p. 293. Alfo,

"Certayne Egloges of Alex. Barclay", &c. as p. 750. D, in fixes. "Thus endeth the fifth & laft Egloge of Alex. Barclay of the Citizen and the man the countrey. Imprinted—in Paules Churchyarde by—Printer to the Queenes Maieftie. Cum priuilegio——folum." See Warton's Hift. of Eng. Poetry, Vol. II; p. 240, &c. W.H. Folio.

1571. Anno. xiij.——At the parliament begunne & holden at Weftminfter the fecond of Apryll in the. xiij. yere of the raigne of our mofte gratious foueraigne Lady Elizabeth,——and there continued vntyll the diffolution of the fame, &c. 1571." M 4, in fixes. "Imprinted—in Poules Church-yarde by R- Jugge & John Cawood, Printers to the Queenes Maieftie. Cum priuilegio" &c. W.H. Folio.

—— "A godly and deuout prayer for the Quenes hyghnes delyueraunce, and for the quietnes & wealth of this realme." In which the expected child is called an Ympe. "Excudit." Broadfide.

—— The Prices and Rates that euery particular perfon oweth to pay for his fayre or paffage vnto Watermen or Whyrrymen from London to Grauefende, and likewife from Grauefende to London, and to euery common landyng place betwene the fayd two places: and the Bote, or tyde Bote and to and from any of the faid places hereafter breyfelye appeareth." Annexed is

—— "The Rates and Prices——from London Brydge to Windefore, & to euery landing place betwene." Quarto.

—— Bifhop Bonner's works. Quarto.

—— The hiftory of Guy earle of Warwick. In verfe. Quarto.

—— "The office and duetie of a hufband, made by the excellent Philofopher Lodouicus Viues, and tranflated into Englifhe by Thomas Paynell. Jmprinted——

JOHN CAWOOD.

Jmprinted——in Pouls Churcheyarde by——prynter vnto the Quenes hyghnes. Cum priuilegio—folum." Dedicated " To—Syr Anthony Browne Knyght." D d 4, in eights. W.H. Sixteens.

" A brief declaration of the notable victory, giuen of God to oure foueraygne lady, quene Marye, made in the church of Luton, by John Gwinneth vicar there, the 23 July, in the first yere of her gracious reign." Thirty-two leaves. See a catalogue of Pamphlets in the Harleian Library. N°. 41: Sixteens:

" Articles" whereuvpon it was agreed by the Archbishops" &c. as p. 723. The 20th article is without the controverted clause; and the 29th article is wholly omitted, according to a Latin copy printed by Reg. Wolfe in 1563. B, in eights. " Jmprinted—in Powles Church yarde by R- Jugge & J- Cawood, Printers to the Queenes Maiestie. Cum priuilegio Regiæ Maiestatis." Dr. Lort. Octavo.

WILLIAM RIDDELL, or BEDDELL,

APPEARS to have been one of the original members of the worshipful company of Stationers, and lived at the George in Paul's churchyard. As he used a compartment of John Day's, it was supposed he might have been his servant. However he appears afterwards to have been in good circumstances, joining in most of the subscriptions set on foot in the company. He had licenfe for 7 ballads which R. Lant was to print for him. See p. 591.

" Abecedarium Anglico Latinum, pro tyrunculis, Richardo Huloeto excriptore." At the end is, " a peroration to the English reader," shewing he had been ten years about the work; had dedicated it to the bishop of Ely, lord chancellor of England; and that it should be better, if it came to another impression. " Londini ex officina Gulielmi Riddell. Cum priuilegio ad imprimendum folum. Mense Sept." Reprinted by T. March 1572, and entitled Huloet's Dictionary, &c. 1552.

ⁿ This book was placed by Mr. Ames to R. Jugge's account under the year 1562, but without a date in the margin; as indeed it has no date of printing. I therefore passed it over there, intending to have inserted it among his other books without date: however as Cawood was jointly concerned in the printing it, i have inserted it here.

" Two

WILLIAM BEDDELL.

1553. " Two epistles, wherin is declared the brainsick headines of the Lutherans, &c. Translated by Henry lord Strafford." Sixteens.

—— Fiue sermons of Bernardine Ochine of Sena. Godlye, frutefull, &c. Translated out of Italian into Englishe, Anno Do. MDXLVIII Imprinted at London by R. C. for William Beddell, at the sygne of the George in Pauls churchyarde. Mr. Alchorne. Octavo.

—— " E. Schambler, Vicar of Rie, and one of Peter house in Cambridge his Medicine prooued for a desperate Conscience. Print. by William Riddle." Maunsell, p. 95. Octavo.

—— " A Balade specifienge partly the maner, partly the matter, in the most excellent meetyng, and lyke mariage betwene our Soueraigne Lord & our Soueraigne Lady, the Kynges and Queenes highnes. Pende by John Heywood." Large black letter. A sheet.

ROWLAND HALL,

AT the death of king Edward vi, with seueral refugees, went and resided at Geneva, where he printed two books, as below. Whether he learnt the art of printing there, or before he went thither, with me is matter of doubt; especially that he printed any book here with his own name: for though Mr. Ames mentions The laws of Geneva, as printed by him in 1552, which he seems to have taken from Maunsell, p. 66; yet that date appears evidently a misprint for 1562, not only as the book is dedicated to Lord Robert Duddeley, Master of the Horse, and Knight of the Garter, who had not those titles before Q. Elizabeth's reign; but as the translator therein expressly says, "They charge us with libertie & licenciousnesse moste vniustly, reportinge that we departed out of this realme in the late tyme of banishement of Godds churche, onelye to this ende, to enjoye more vnchastised freedome of sensuall lyfe."

1559. " The Boke of Psalmes & Godly prayers. Printed at Geneva." Octavo.

1560. " The Bible and Holy Scriptures conteyned in the Olde and Newe Testament. Translated according to the Ebrue & Greeke, and confetred with the best translations in diuers langages. With most profitable Annotations vpon all the hard places, and other things of great importance

ROWLAND HALL.

importance as may appeare in the Epiſtle to the Reader." This title over a neat wood-cut of the Iſraelites going to paſs the Red Sea, the cloudy pillar on the oppoſite ſhore, and the Egyptians purſuing them, having theſe Scriptures about it. At top, "Feare ye not, ſtand ſtill and beholde the ſaluacion of the Lord, which he will ſhew to you this day. Exod. 14, 13." On the ſides, "Great are the troubles of the righteous: but the Lord deliuereth them out of all. Pſal. 34, 19." At bottom, "The Lord ſhal fight for you: therefore holde you your peace. Exod. 14, 14. At Geneva. Printed by Rouland Hall. M.D.L.X." On the back, are "the names and order of the books of the olde and newe Teſtamēt with the nombre of their chapters, and the leaf where they begin." Dedicated "To the moſt vertuous & noble quene Eliſabet, &c. Your humble ſubjects of the Engliſh Churche at Geneva wiſh grace & peace from God through Chriſt Ieſus our Lord.——From Geneva. 10. April. 1560." Then an epiſtle "To our beloued in the Lord, the brethren of England, Scotland, Ireland, &c." Here are no diviſions into parts, but the leaves are numbered throughout the Old Teſtament, and the apocryphal books, progreſſively to 474.

"The Newe Teſtament of our Lord Ieſus Chriſt, Conferred diligently with the Greke, and beſt approued tranſlacions in diuers languages." The ſame cut as before. "At Geneva. Printed by Rouland Hall. M.D.LX." On neat Roman types; 122 leaves. This is the firſt edition of the whole Bible of the Genevan tranſlation; has a few cuts of the tabernacle, the temple, the utenſils, maps, &c. W.H. Quarto.

This Bible was printed again 1561, at Geneva, in Folio; but not by R. Hall.

After Hall returned from Geneva, he dwelt firſt in Golden-lane near Cripplegate, at the ſign of the three arrows; then remoued to Gutter lane, at the ſign of the Geneva arms, the half-eagle and key, which he uſed alſo for his device, with this motto round it "POST TENEBRAS LVX." Sometimes he uſed the device of a boy in a looſe garment, lifting up his right leg; his right arm winged, ſtretched out towards heauen; and taking his left hand from off a ball upon the ground. In the clouds is a repreſentation of the Deity, as a royal perſonage, and as ſaying, Set your affections on things above, &c. Encloſed in an oval, broadways.

"The Confeſſion of fayhe and doctrine, beloued and profeſſed by the Proteſtants of the Realme of Scotlande, exhibited to the eſtates of the ſame in parliament, and by their publicke voices authoriſed as a doctrine, grounded upon the infallible worde of God. Set forth, &c. according to the quenes maieſties injunctions.——dwellyng in Goldyng lane, at the ſygne of the thre arrowes." At the end, "From Edenburghe, the 17 of Auguſt 1560, theſe actes and articles were red in the face of the parliament, and ratified by the thre eſtates." Licenſed. Octavo. 1561.

"FOVRE Godlye ſermons agaynſt the pollvtion of idolatries, comforting men in perſecutions, and teachyng them what commodities thei ſhal 1561.

find in Chriftes church, which were preached in French, by the moft famous clarke Jhon Caluyne, and tranflated fyrft into Latine, and afterward into Englifh by diuers godly learned men. Pfal. 16. I wyl not take the names of the Idols in my mouth. Printed—by—dwelling in Golding lane at the——thre arrowes. 1561." With an epiftle of R. Hall's to the reader.* Q 4, in eights. Licenfed. W.H. Sixteens.

(1561.) A very profitable treatife, made by M. Jhon Caluyne, declarynge what great profit might come to all Chriftendome, yf there were a regefter made of all Sainctes bodies and other reliques, which are afwell in Italy, as in Fraunce, Dutchland, Spaine, and other kingdomes and countreys. Tranflated out of Frenche into Englifh, by Steuen Wythers. 1561. Set furth and authorifed according to the queenes maiefties Jniunctions—At the—thre arrowes. The running title is a treatife of Reliques. H, in eights. Licenfed. W.H. Sixteens.

1561. "An Epiftle or letter of Exhortation vvritten in Latyne by Marcus Tullius Cicero, to his brother Quintus the Proconfull or Deputy of Afia, wherin the office of a Magiftrate is connyngly & wifely defcribed. Tranflated into englyfhe by G. G. Set furth——according to the—— Jniunctions. Prynted——at the——three arrowes. 1561." An epiftle is prefixed, "Goddred Gylby to the reader.——.At London the vii. of July—1561." C, in eights. Licenfed. W.H. Sixteens.

1562. The Lavves and Statutes of Geneua, as well concerning ecclefiaftical Difcipline, as ciuill regiment, with certeine Proclamations duly executed, whereby Gods religion is moft purelie mainteined, and their common wealth quietli gouerned: Tranflated out of Frenche into Englifh by Robert Fills. Except the Lorde kepe the Citie, &c. Pfal. 127." His device, as aboue, " Printed—in Gutter Lane, at the fygne of the halfe Egle and the Keye. 1562." Dedicated, " To—Lorde Robert Duddeley Maifter of the—horfe, and knight of the—Garter." A table of contents. Contains befides, 86 leaves: on the laft, "Printed—by R-Hall and Tho. Hacket, the 16. of Aprill,—1562." Licenfed. W.H. Sixteens.

1562. " The Treatye of thaffocation made by the Prince of Condee, together wyth the Princes, Knyghtes of the order, Lordes, Capitaines, Gentlemen, & others of al eftates which be entred or hereafter fhall entre into the Affociation for to mainteine the honour of God, the quiet of the Realme of France, and the ftate & Libertie of the king under the gouernment of the quene his mother, who is authorifed thereunto, and eftablifhed by the Eftates." For Edw. Sutton. Octavo.

1562. "A Declaration made by the Prince of Conde for to fhew & declare the caufes that haue conftrained him to take upon him the defence of the Kinges authoritie, &c." T. Baker's Maunfell, p. 38. Octavo.

" The

"The Caftle of Memorie: wherein is conteyned the reftoring, aug- 1562.
menting, & conferu ng of the Memorye & Remembraunce, with the
fafeft remedies & beft precepts thereunto in any wife appertayning: made
by Gulielmus Gratarolus Bergomalis Doctor of Arts & Phyficke, Englifhed by William Fulwood." Some pretty verfes on the fubject, at the
beginning and end. Licenfed. Again 1563. Sixteens.

"Thre Notable fermones made by the godly and famous Clerke 1562.
Maifter John Caluyn, on thre feuerall Sondayes in Maye, the yere 1561.
vpon the Pfalm 46. Teaching vs conftantly to cleaue vnto God's truth in time
of aduerfitie & trouble; and neuer to fhrinke for any rage of the wicked
but to fuffer all thynges in fayth & hope in Jefus Chrift. Englifhed by
Will. Warde." His device. "Printed——in Gutter Lane, &c. 1562."
On the back a fhort epiftle by "The Printer to the reader." F, in
eights. W.H. Sixteens.

"A fhort and pithie defence of the doctrine of the holy election & 1562.
predeftination of God, gathered out of the firft Chapter of St. Paules
epiftle to the Ephefians. By J. Bradford." Alfo annexed to

"Godlie meditations vpon the Lordes Prayer, the beleefe, &' ten 1562.
commaundementes, with other comfortable meditations, praiers & exercifes. Whereunto is annexed a defence of the doctrine of gods eternall
election & predeftination, gathered by the conftant martyr of God, John
Bradford in the tyme of his imprifonment. The contents wherof appeare
in the page nexte folowyng." His device, and under it, "Nowe fyrft
prynted by——in gutter lane at the figne of the halfe Egle and key, the
12. of October 1562." Q 6, in eights. Licenfed. W.H. Sixteens.

"Tvvo very notable Commentaries, the one of the originall of the 1562.
Turcks and Empire of the houfe of Ottomanno, written by Andrewe Cam-
bine, & thother, of the warres of the Turcke againft George Scanderbeg,
prince of Epiro, and of the great victories obteyned by the fayd George,
afwell againft the Emperour of Turkie, as other princes, and of his other
rare force and vertues, worthye of memorye, tranflated oute of Italian into
Englifh by John Shute. Prov. xxi. The horfe is prepared againft the daye
of battayle but the lord giueth the victorie. Printed——for Humfrey Toye
dwelling in paules Churche yearde at the figne of the Helmette. 1562."
Dedicated "To—fyr Edw. Fynes lorde Clynton & Say, Knight of the
order and highe Admirall of England & Ireland." Shewing the efficacy
of fkilful commanders, and difciplined fouldiers. The firft treatife contains
befides, 100 leaves; the fecond, 45. W.H. Quarto.
Mr. Ames mentions an edition in Twelves.

"The Pleaufaunt and vvittie Playe of the Cheaftes renewed, with 1562.
Inftructions both to learne it eafely, and to play it well. Lately tran-
flated out of Italian into French: And now fet furth in Englifhe by
James

James Rowbothum. Printed—for James Rowbothum, and are to be sold at hys shoppe vnder Bowe churche in Cheapsyde. 1562." On the back 12 distichs of Latin verses, "Liber ad Lectorem—W. Ward." Probably the translator. It is dedicated by J. Rowbothum " To—Lorde Robert Duddeley, Maister of the—horse, and Knight of the——Garter." Contains besides, G 4, in eights. With several schemes of the Play. W.H. Sixteens.

1562. " The Nyne fyrst Bookes of the Eneidos of Virgil converted into Englishe vearse by Tho. Phaer Doctour of Phisike, with so muche of the tenthe Booke, as since his death coulde be founde in vnperfit papers at his house in Kilgarran forest in Penbroke shire." A cut of the Genius of England, as p. 780. " Printed——for Nicholas England, 1562." Dedicated To—Syr Nicholas Bacon knyght, Lorde Keeper of the great Seale of England.—From London the vi. of July, 1562.—Will. Wightman." An intimate of Dr. Phaer's with whom he deposited the 8th and 9th books before his death. The first 7 books were printed by John Kingston in 1558. Then a general account of all the 12 books of Virgil. Contains besides, G g, in fours. W.H. Quarto.

1562. " The secretes of the reuerend Maister Alexis," &c. as p. 781. " Prynted——for Nycolas England, dwellyng in Pater noster rowe. 1562. Contains 122 leaves, besides the prefixes, and a table at the end.

1563. " The second part" &c. as p. 785. " Printed—for Nicholas Englande. 1563." Contains 88 leaves. Licensed. W.H. Quarto.

1563. " The Catechisme or maner to teache Children the Christian Religion. Made by the excellent Doctour and Pastour in Christes Churche, John Calvin. Wherein the minister demandeth the question, and the Childe maketh answere." Licensed. Sixteens.

1563. " The historie of Leonard Aretine, concerning the warres betwene the Imperialls & the Gothes for the possession of Italy. a worke very pleasant & profitable. Translated out of Latin into English by Arthur Goldyng. Printed—, for George Bucke. 1563." Dedicated " To—Sir William Sicill Knighte principall Secretarie to the Queenes Maiestie, and Maister of her hyghnesse Court of wardes & liueries.—Finished at your house in y Strond the second of Aprill. 1563.——Arthur Goldyng." An epistle To the reader. Then, The Preface of Leon. Aretine. Contains besides, 180 leaves. This was Licensed to Geo. Buck. W.H. Sixteens.

1563. " A Summe or a briefe collection of holy Signes, Sacrifices & Sacramentes instituted to God euen since the beginning of the world, And of the true originall of the Sacrifice of the Masse." This is the head title, my copy wanting the head title-page, and prefixes. Translated out of French by N. Ling. Maunsell, p. 23. Licensed. W.H. Octavo.

" An

"An excellent Treatife of wounds made with Gunfhot, in which is 1563.
confuted bothe the grofe errour of Jerome of Brunfwicke, John Vigo,
Alfonfe, Ferrius, and others; in that they make the wound venemous,
whiche cometh through the common pouder and fhotte: And alfo there
is fet out a perfect and trew methode of curyng thefe woundes. Newly
compiled and publifhed by Thomas Gale, Maifter in Chirurgerie." His
device of the boy with a winged arm, &c. as p. 801. "Printed——by
—for Tho. Gale. 1563." On the back is a cut of a wounded man, with
fundry inftruments. Then, verfes in praife of M. Gale by John Field
Chirurgian; and his portrait "Ætatis fue 56." At the end are cuts of
inftruments, neatly done. Octavo.

"An Enchiridion of Chirurgerie, conteyning the exacte & perfect cure 1563.
of woundes, fractures, and diflocations, newly compiled and publifhed by
Thomas Gale Maifter in Chirurgerie." His device of the boy, &c.
"Printed———for Tho. Gale. 1563." On the back is a cut of a
wounded fouldier, fupported by another, while a Surgeon is drawing out
an arrow from his breaft. Prefixed is an Epiftle from "Richarde Ferris
Seargeant Chirurgian vnto the Queenes—Maieftie, vnto his louing frende
Maifter Gale.——Paddinton, 2. July, 1563." Some commendatory
verfes by "Jhon Hall Chirurgian." Then "Tho. Gale Chirurgian
vnto the yonge men of his company," &c. Contains befides, 58 leaves.
W.H. Octavo.

He printed alfo The inftitution of a Chirurgian; and The antidotary: 1563.
both by T. Gale; and perhaps his other chirurgical pieces.

T. Beza's brief and pithy fum of the Chriftian Faith, &c. See it by 1563.
Rob. Walde-graue in 1585. Sixteens.

"Enarratio in Deuteronomium." &c. Octavo.

"An anfwere to the examination, that is fayde to haue bene made of 1563.
one named John de Poltrot, calling himfelf the Lord of mercy, vpon
the death of the late duke of Guyfe; by the lord of Chaftillion, admyrall
of Fraunce, and others named in the faide examination—At Caen. 1562."
Printed by him for Edward Sutton.* Octavo.

"The moft ancient and learned play, called the philofophers game, 1563.
invented for the honeft recreation of ftudients, and other fober perfons,
in paffing the tedioufnefs of tyme, to the releafe of their labours, and
the exercife of their wittes. Set forth with fuch playne precepts, rules,
and tables, that all men with care may underftand it, and moft men with
pleafure practife it. By W.F. It is dedicated by J. Robothum to lord
Dudley, whofe head is at the back of the title, and at the end of the
epiftles are thefe lines:

5 B "All

" All things belonging to this game
for reafon you may bye,
At the bonke fhop vnder Bochurch,
in Chepefyde redilye."
With wooden cuts. " Printed the 21 May." See it in 1562. Octavo.

" AN ADmonition againſt Aſtrology Iudiciall, and other curioſities, that raigne now in the world." Written in French, by J. Calvine, Engliſhed by G. G. (Goddred Gylby). At the three arrowes. Licenſed. Sixteens.

He had licenſe beſides, in 1561, for " The ſtrange newes." In 1562, " A letter of Nycholas Nemo. The ſome of the pryncypall poynts of the Chreſtian faythe by Peter verett. An exhortation made to a Certen Cytye to leaue Papiſtrye. Creſtenmas Carroles, &c. The iiijor mouages of John ſlydon. Morritio Detromon, auctoryſed by my lorde of london. A poofye in forme of a vyſion agaynſte wytche Crafte & Sofyrye in myter by John Hall." In 1563, " A ſpretuall preſeruiture for the plage and alſo for the fowle: made by doctour Rychardſon."

RICHARD TOTTEL

HAD his name ſpelt as different as poſſible, was a very conſiderable printer of law, and an original member of the Stationers' company. He dwelt in Fleet-ſtreet, within Temple Bar, at the ſign of the Hand and Star.

In Dugd. Orig. Jurid. p. 59. and 60 are found the following licenſes. A ſpecial licenſe to Richard Tathille (which I ſuppoſe Tottel) citizen, ſtationer, and printer of London, for him and his aſſigns, to imprint, for the ſpace of ſeven years next enſuing the date hereof, all manner of books of the temporal law, called the common law; ſo as the copies be allowed, and adjudged mete to be printed, by one of the Juſtices of the law, or two ſerjeants, or three apprentices of the law; whereof the one to be a reader in court. And that none other ſhall imprint any book, which the ſaid Richard Totell ſhall firſt take and imprint, during the ſaid term, upon pain of forfeiture of all ſuch books. T. R. apud Weſtm. 12 April, 7 Edward VI. p. 3. A licenſe[a] to Richard Tottle, ſtationer of London,

---

[a] Rychard Tottle brought in a patente for prynting of bokes of Lawe, to be confyrmed & allowed by this howſe: and the ſayd patente ys for vii yeares. Dated anno ij & iij. Phil. & Marie." Stat. Rigiſter. 1558. ——1559.

RICHARD TOTTEL.

to imprint, or cause to be imprinted, for the space of seven years next ensuing, all manner of books, which touch or concern the common law, whether already imprinted, or not. T. R. apud Westm. 1 Maii. Pat. 2, and 3 Phil. and Mary, p. 1. Again, a licenfe to Richard Tottell, citizen, printer and stationer of London, to print all manner of books, touching the common laws of England, for his life. T. R. 12 Jan. Pat. 1 Eliz. p. 4. Also, there was a patent ready drawn for queen Elizabeth's signing for seven years, priviledging Richard Tothill, stationer, to imprint all manner of books, or tables, whatsoever, which touched, or concerned cosmography, or any part thereof; as geography, or topography, writ in the English tongue, or translated out of any other language into English, of whatsoever countries they treated, and whosoever was the author. But whether this was ever actually signed or not, is uncertain. At the decease of archbishop Parker, Strype says, there was due to him for books, 1l. 11s 6d.

He was renter, or collector of the quarterages in 1559, and 1560; under warden in 1561, upper warden in 1567, 1568, and 1574; master in 1578, and 1584.

January 8. 1583, he yielded up to the Stationers' company, seven copies of books, for the relief of their poor. What they were you may see in our General History, under that year.

Latterly he retired into the country for the benefit of his health, whereby the business of the company was sometimes at a stand, as appears by an ordinance[y] made by the Court of Assistants, 30. Sept. 1589. His own business however appears to have been carried on for him untill 1593, after which time i find no more concerning him.

"Io Vaffe his iudgment of Urines, translated by Humfr. Lloyd." 1551. Octavo.

"A Manifest detection of the most vile and detestable vse of Dice 1552. play" &c. as p. 778. 32 leaves. "Imprinted——in Flete strete betwyne the two Temple gates, by R- Tottyl. An. 1552." Octavo.

---

[y] "Forasmuche as the affayres of the Company are often hindred by reason of the continuall absence of Mr. Richard Tottell, who dwellyth in the furtheft partes of the Realme: and of Garret Dewce & Richard Greene. Yt is now therefore for the furtherance & more spedy execution of the busyneffe of the Company ordered at a full Court, holden this day (beinge the quarter daye) That the said Mr. Tottell, Mr. Dewce & Mr. Greene from henceforthe shall stand discharged & removed from their Affistantship in this Company, and thereof they are discharged & removed by this Court. Nevertheles as touching the said Mr. Tottell (havinge bene alwayes a lovinge & orderly brother in the Company, and now absent, not for any cause savinge his infyrmytie, & farre dwellinge from the cyty) Yt is agreed that alwaies whensoever he shall resort hither, he shall, in regard of the offices he hath borne in the Company, syt with the Affistants of the Court at their metings. And his name shall be still entred in the booke amongft the said Affistants. And Mr. Dewce & Mr. Greene to be wrytten in the book amongft the rest of the lyuery.

"LA

1553. " LA NOVVELLE Natura breuium de iuge trefreuerend Monfieur An-
thione Fitzherbert, dernierement reneue et corigee par laucteur, auecques
vne table parfaicte de chofes notables contenues en ycelle nouuellement
compofee par Guilliaulme Raftell, et iammais par cy devant imprimee.
Cum priv." In Whitchurch's compartment. Contains 271 leaves,
befides 33 leaves in the table. Printed again the fame year; alfo 1567,.
and frequently.   Octavo.

1553. The hiftory of Quintus Curtius. See it in 1561.   Quarto.

1553. " A dialogue of comfort againft tribulation, made by fir Thomas
More, knight, &c. Cum priuilegio——folum." In his works p. 1139.
See it printed at Antwerp in cur General Hiftory, 1573.   Folio.
Mr. William White has the fame in   Quarto.

1553. " Marcus Tullius Ciceroes thre bookes of duties, to Marcus his
fonne, turned out of Latine into Englifh, by Nicholas Grimald.[a] Whereunto the Latin is adioined. Dedicated to Thomas, bifhop of Elie."
Printed frequently.   Octavo.

1553. " Regiftrum omnium breuium tam originalium quam iudicialium,
nouiter Jmpreffum, et quam exacte correctum et emendatum. Excufum
Londini, apud Richardum Tottelum. Anno Domini. 1553. Cum priuilegio——folum." In the compartment of boys riding on an elephant.
Both the Regifters however are the very fame as printed by W. Raftell,
in 1531; and the tables only are reprinted by R. Tottel. See
p. 475. W.H.   Folio.

1554. " A TREATISE excellent and compedious, fhewing and declaring, in
maner of Tragedye the falles of fondry moft notable Princes & Princeffes
with other Nobles, through ÿ mutabilitie and change of vnftedfaft Fortune,
together with their moft deteftable and wicked vices. Firft compyled
in Latin by the excellent Clerke Boetius an Jtalian borne. And fence
that tyme tranflated into our Englifh & vulgare tong, by Dan John
Lidgate Monke of Burye. And nowe newly imprynted, corrected &
augmented out of diuerfe & fundry olde writen copies in parchment. In
ædibus Richardi Tottelli. Cum priuilegio." In the fame compartment
as was ufed by Whitchurch to The Paraphrafe of Erafmus. See p. 544.
It has, prefixed, a table of contents, and " The Prologe of John Lydgate,"
at the end of which is a cut of the author, in a pofture of adoration before
the wheel of Fortune, turned about by Providence, reprefented by a royal
perfonage with expanded wings. The poem is divided into 9 books,
with a cut before each of them, and ends on Fol. ccxix. To this edition
is annexed,

[a] Of whom fee Warton's Hift. of Eng. Poetry, Vol. III. p. 60.

" The

"The daunce of Machabree, wherin is liuely expreffed & fhewed the ftate of manne, and how he is called at vncertayne tymes by death, and when he thinketh leaft theron made by thaforefayde Dan John Lydgate, Monke of Burye." A cut of Death's dance at the head, with this motto, " Cunctis mortalibus mors debetur;" and another at the conclufion of a royal corps laying on a tomb, eaten by worms, and 3 fpectators contemplating it, with this motto,
  "Nil ita fupreme eft, fupraq; pericula rendit:
  Non fit vt inferius, fuppofitumq; Deo."
The number of the leaves continued to Fol. CCxxiiii. Jmprinted—— in Flete ftrete within Temple barre at the fygne of the hande & ftarre, by Ric. Tottel, the. x. day of September——1554. Cum priuilegio—— folum." W.H. Folio.

"The newe boke of Juftices of the Peace," &c. Newly corrected. 1554. Octavo.

"The dyaloges in Englifhe, betwene a Doctour of diuinitie and a 1554. ftudēt in the lawes of Englād, newly corrected & imprinted, with new addicions." In a compartment with Grafton's mark at bottom. Contains 182 leaves, and tables to both books. " Londini in aedibus—— An. 1554. Cum priuilegio—folum." W.H. Printed frequently. Octavo.

"Anni regis Henrici feptimi. Quibus accefferunt annus primus et fe- 1555. cundus de noua, et valde bona collatione. Ac etiam, annus decimus, vndecimus, decimus tertius, decimus fextus, et vigefimus, nunquam antehac editi. Octavo.

"The Comentaries of Don Lewes de Auela, and Suniga, great Mafter 1555. of Acanter, which treateth of the great vvars in Germany by Charles the fifth Maximo Emperoure of Rome, King of Spain, againft John Frederike Duke of Saxon, and Philip the Lantgraue of Heffon, with other gret princes & Cities of the Lutherans, wherin you may fee how god hath preferued this vvorthie & victorious Emperor in al his affayres againft his enemyes. tranflated out of Spanifh into Englifh An. Do. 1555. Londini in Aedibus Richardi Totteli." On the back is the dedication "To ——Edwarde Earle of Darby, Lord Stanley & Strainge, Lord of Man & the Jles, knyght of the—Garter," by John Wilkinfon, probably the tranflator. The author's preface is dated 1546. Contains U 6, in eights. W.H. Octavo.

"The Hiftory of graund Amoure and la bel Pucell," &c. as p. 561. 1555. In the compartment ufed by Will. Copland, in 1553, for Douglas's tranflation of Virgil's Æneids. With " Anno domini. 1555," in the tablet at the bottom. On the back begins the table of contents, at the end of which is a note, that this work was compiled in the 21ft year of the reign of king Henry vii. Contains D d, in fours, and has wooden cuts
like

like that by W. de Worde in 1517.' " Jmprinted——in Fleteſtreate, at the——hande and ſtarre," &c. W.H. Quarto.

1555. "THE ABRIDGEMENT of the booke of Aſſiſes, lately peruſed ouer, and corrected, & nowe newelye imprinted by Ric. Tottill the laſt day of September. Anno do. 1555. Cum priuilegio." In the compartment with cherubic heads, formerly T. Berthelet's. Fol. 165, and a table. W.H. Octavo.

1555. "A Treatiſe of the Figures of Grammer and Rhetorike, profitable for al that be ſtudious of Eloquence, and in eſpeciall for ſuche as in grammer ſcholes doe reade moſte eloquente Poetes, and Oratours: Whereunto is ioygned the oration, which Cicero made to Ceſar, geuing thankes vnto him for pardonyng, and reſtoring again of that noble mā Marcus Marcellus: ſette foorth by Richarde Sherrye, Londonar. Londini in ædibus Richardi Totteli. Cum priuilegio——folum." Dedicated " Honoratiſſimo domino Guilielmo Pagetto, nobiliſſimi ordinis Garterii equiti aurato, domino Beudeſert, & illuſtriſſi. noſtræ reginæ Mariæ cōſiliario." Contains beſides, 75 leaves. "Imprinted——in Flete ſtrete——by R- Tottill, the iiii. daye of Maye——MDLV." W.H. Octavo.

1555. Year Books. 1.Edw. v. Alſo, 1—16. 20. 21. Hen.VII. Thoſe of Hen. VII. have this title, "Anni Regis Henrici Septimi. Quibus acceſſerunt Annus primus & ſecundus de noua & valde bona collatione, Ac etiam, Annus decimus, vndecimus, decimus tertius, decimus ſextus. & vigeſimus, nunquam ante hac æditi. An. Do. 1555. Londini in ædibus R. Totteli. Cum priuilegio——folum." In the compartment uſed by R. Grafton to the ſtatutes, 1 Edw. VI. Then, "Præfatio in laudem legum Anglie, printed on neat Roman letter, as are ſeveral of the caſes. The numbers of the leaves begin afreſh every year; but the ſignatures are continued throughout. "Jmprinted——the xij. daye of September, &c. W.H. Folio.

1556. "MAGNA CHARTA, CVM STATVTIS quæ Antiqua vocantur, iam recens excuſa, & ſumma fide emendata, iuxta vetuſta exemplaria ad parliamenti rotulos examinata: quibus acceſſerunt nonnulla nunc primum typis edita: apud R- Totelum. 12 June 1556. Conferre and then preferre. Cum priuilegio——folum." On the back begins the printer's preface, addreſſed "To gentlemen ſtudious of the lawes of Englande." Some extracts from which may be ſeen below.' Then a table of the ſtatutes and another of

"thinges

---

* "To ſtuffe a preface with praiſe of this boke, or with exhorting you to reding of it, were in a matter not doubtfull to take fond peine not nedefull,——.Onely touching myſelfe, and my labours in this & other, J praye ye ſuffer that J maye ſomewhat vſe your pacience, as ye ſhall alway vſe my diligence.——How vnperfit the bokes of the lawes of England were before, what price the ſcarcenes had raiſed, the moſt marueilouſly mangled, & no ſmall part no where to be gotten, ther be enow, though J rehorſe it not, that do freſhlye remembre, & can truely witnes. Likewiſe how,

ſubens

"thinges moſt notable." This firſt part ends on fo. 170, where begin and follow the tables for the ſecond part, notwithſtanding that is printed with ſeparate folios and ſignatures.

"Secvnda Pars Veterum Statutorum. Anno. m.d.lv." Contains fo. 72. " Londini in ædibus R- Tottelli. Anno do. 1556. Cum priuilegio——ſolum." W.H. Octavo.

"Marcvs Tvllius Ciceroes thre bokes of duties, to Marcus his ſonne, turned out of latine into engliſh, by Nicholas Grimalde. Cum priuilegio. ——ſolum. Anno domini 1556." In a compartment late T. Berthelet's, with 1534 on the ſell. Dedicated "To the right reuerend——Thomas Biſſhop of Elie one of the King & Quenes Maieſties moſte honorable priuie Counſell." Then, "N. G. to the reader." Contains beſides theſe, Fol. 158, and a table at the end. "Jmprinted &c. by R- Tottel." W.H. Octavo. 1556.

A year book. At the end, "Explicit annus iiii Henrici ſexti." With his ſign, "Cum priuilegio" about it, his cypher, and his name at length, beneath. And year books. 40——50. Edw. iii. W.H. Folio. 1556.

Littleton's Tenures in Engliſh. Cum priuilegio. Again 1576. Sixteens. 1556.

"Littletons Tenures. (In French) Conferred with diuers true wrytten copies and purged of ſondry caſes, hauing in ſome places more then ỹ autour wrote, and leſſe in other ſome. Apud Richardum Tottel. Cum priuilegio. 1557." On the back " A figure of the diviſion of poſſeſſions." 173 leaves and a table at the end. "Jmprinted——within Temple barre ——, the xxviij. daie of October." &c. W.H. Sixteens. 1557.

ſithens J toke in hand to ſerue your vſes, the imperfections haue ben ſuplied, the price ſo eaſed as the ſcarcenes no more hindreth but that ye haue them as chepe (notwithſtanding the common dearth of theſe times) as when they were moſt plentifol, the print much pleaſanter to the eye in the bokes of yeres than any that ye haue ben yet ſerued with, paper & margine as good & as fayre as the beſt, but much better and fairer then the moſt, no ſmall nomber by me ſet forth newly in print that before were ſcant to be found in writing, J nede not my ſelf to report it.——.But now to ſay alſo ſomewhat of this preſent work: albeit it might ſeme ſuperfluous and nedleſſe to haue emprinted it again ſo ſodeinly, being ſo lately done in ſo fair paper & letter by an other: (The Marſh) yet when ye ſhal wey how in ſondry places, much here is added out of bokes of good credit, as examined by the roules of parliament, how eche were the truth euen of the beſt printes is ouer matched by their faultes not fewe not a litle reformed, the light of pointing adioined, the chapiters of ſtatutes truly diuided, and noted with their due nombers, the alphabetical table iuſtly ordred & quoted, the leaues not one falſly marked, with many other helps to correct it & further you, when (J ſaie) ye ſhal haue weyed bothe all theſe by me performed, & the want of theſe in al other heretofore, J hope your wiſedoms wil ſone eſpie that nether J haue newe printed it for you cauſeleſſe, nor ye ſhal bye it of me fruteleſſe. This thought J fit in mine own behalf firſt to haue ſaid vnto you: and ſo now J ceſſe further to trouble you from your more earneſt ſtudies: &c. R. T.

I have

RICHARD TOTTEL.

1557. ,I have another edition with the fame number of leaves, and the fame date in the colophon, but it wants the title-page. W.H.+ Sixteens.

1557. " Les plees del coron deuife in plufieurs titles et common lieux. Per queux home plus redement et plenairement, trouera quelque chofe, que il quira touchant lez ditz pleez compofees Ian du Grace." Printed frequently. Quarto.

1557. " The vvorkes of Sir Thomas More" &c. as p. 731. W.H. Folio.

1557. " Songes and Sonnettes of Henry earle of Surrey," & others. " Imrinted —— the fifte day of June. An. 1557. Cum priuilegio —— folum." Frequently printed, and yet very fcarce. See Warton's Hift. of Eng. Poetry, Vol. III; p 11, 12, 60, 69. Sixteens.

1557. Raftell's Collection of all the Statutes from Magna Charta. Again frequently. Quarto.

1557. "Natura breuium in Englifhe, newelye corrected: with diuers addicions of ftatutes, booke cafes, plees in abatementes of the faide writtes: and theire declaraciōs: and batres to the fame added & put in their places mofte conueniēt. Cum priuilegio——folum." 180 leaves, and a table at the end. " Jmprinted——in Flete ftret——ỹ. xxvii. daie of Februarie.——1557." W.H Sixteens.

1557. The fame in French. " In Ædibus R- Tottell, 1557. Cum priuilegio——folum." Sixteens.

1558. " Marcus Tullius Ciceroes thre bookes of duties," &c. as in 1553. The Latin on the Italic type. Fol. 168. W.H. Sixteens.

1558. The paffage of Queen Elizabeth through the city of London. Licenfed Quarto.

He had licenfe alfo for printing " The frute of foes/ and a treatife of Seneca."

1559. " A Boke of Prefidentes exactly written, in maner of a Regifter. Newly correcfed with addicions of dyuers neceffarye Prefidentes, mete for all fuch has defyre to learne the fourme and maner howe to make all maner of evidēces & inftrumentes as in the table of this boke more playnely appeareth. With alfo the begynning & ending of the termes. Anno do. 1559. Cum priuilegio " On the back is " An Almanake for. xvii. yeres." beginning 1548. Then follow, A Calendar. " An abridgement of the actes made for the abrogation of certaine Holydaies." A table, " When the terme beginneth & endeth." Thefe all printed in red and black. Then, " The Table of this booke." Contains befides, 159 leaves. Again frequently. W.H. Sixteens.

Troas,

RICHARD TOTTEL.

Troas, a tragedy by Seneca; tranflated by Jafper Heywood. See Warton's Hift. of Eng. Poetry, Vol. III. p. 388, note *. Sixteens. 1560.

Abftracts of all the penal Satutes——which do threaten to the offenders thereof, the lofs of life, member, &c. In French, with remarks thereon in Englifh. Collected by Ferdinando Pulton. Again 1577. See it without date. Quarto. 1560.

Fitzherbert's New Book of Juftices of the peace. See it in 1566. Sixteens. 1560.

"Herein is conteined the booke called Nouæ Narrationes, the booke called Articuli ad Nouas Narrationes, and the booke of diuerfitees of courtes. Newly imprinted. 1561. In aedibus R- Tottell. Cum priuilegio." Fo. 120. "Jmprinted——by R- Tottill, the firfte daye of Aprill." &c. W.H. S.xteens. 1561.

"THE HISTORIE OF Quintus Curtius, conteyning the Actes of the greate Alexander, trãflated out of Latin into Englifhe by John Brende. In ædibus R- Tottell. Anno Domini. 1561. Cum priuilegio." In a compartment with 2 ftags couchant at the bottom. Dedicated "To the right hyghe &myghtye Prince, John Duke of Northumberlande, Earle marfhall of Englande." Contains befides, fo. 229. "Jmprinted—— the v. day of Apryl," &c.* W.H. Quarto. 1561.

Year books. 17, 18, 21, 29, 30, 38, 39 Edw. III. W.H. Folio. 1561.

"The Accedens of Armory." This title is in a tablet at the bottom of a device, or viniet, defcribed at large on 4½ pages preceding the treatife. On the back is an octave ftanza to caution againft cenfuring the the book. The author Gerard Legh, or Leigh, addreffes his preface "To the honorable affembly of gentlemen in the Innes of Court and Chauncery. Next follows, a recommendatory epiftle by "Richarde Argall of thinner Temple." Then, "The Defcription of the Viniet with the circumftaunce therof, contayned in the fyrft Page of the booke." Contains befides, Fol. 232, on Italic types, with cuts, and a table at the end. "Jmprinted——the laft day of December, An. Do. 1562." On the laft leaf is a cut of Æfop, &c. and on the back thereof "The waye to vnderftande Trycking." Licenfed. W.H. Again 1591. Octavo. 1562.

In this Booke is contayned the Office of Shiriffes, Bayliffes of Libertyes, Efcheatours, and Coroners. Octavo. 1562.

"An abridgement of the Chronicles of England, gathered by Richard Grafton citizen of London. Anno. Do. 1563. Perufed & allowed, according to an order taken. In ædibus R- Tottyll. Cum priuilegio." On the back, "The contentes of this Booke." This title-page with the calendar 1562.

5 C

calendar, &c. feem to have been added after the firft publication of the book, which by the colophon appears to have been printed " the 21. daye of February——1562." Dedicated " To the right honorable—— Lord Robert Dudley, knyght of the Gartier, mafter of the horfe, and one of the Queenes maiefties preuy counfayll." Contains, befides the prefixes and table at the end, Fo. 172. Concerning the controverfy between Grafton and Stow, fee p. 502. W.H. Octavo.

1562. " THE TRAGICALL HYSTORY OF ROMEUS AND JULIET: Contayning in it a rare example of true Conftancie, with the fubtill Counfells and practifes of an old fryer and ther ill event. Imprinted——the xix. day of November. Ann. Do. 1562." A metrical paraphrafe from the Italian of Bandello by Arthur Brooke. See Warton's Hift. of Eng. Poetry. Vol. III; p. 471. By licenfe: with Sonnets.

1563. " THE EXPOSICIONS of the termes of the lawes of England, with diuers proper rules & principles of the lawe, afwell out of the bookes of maifter Littleton, as of other. Gathered both in French & Englifh, for yongmen very neceffary. Whereunto are added the olde tenures. 1563. In ædibusR-Tottell. Cum priuilegio." It is introduced with " Prologus Iohannis Raftell," in Englifh, which you may fee with little variation, in p. 331, &c. Contains befides, Fo. 140, and a table at the end. The French on Italic types. " Jmprinted——the. xxix. day of Aprill." &c. W.H. Again 1579, and 1592. Octavo.

1564. A treatyce of Moral philofophy containing the fayinges of the wife. Wherin you maye fee the worthye and pithye fayinges of y Philofophers, Emperors, kinges, and oratours, of their liues, their aunfwers, of what lignage they came of, and of what coûtrey they were, whofe worthy & notable prefepts, counfailes, parables & femblables doth hereafter folow. : Firft gathered & englifhed by Williã Baldwin, after that, twife augmented by Thomas Paulfreyman, one of the gentlemen of the Queenes maiefties chaple, and now once againe enlarged by the firft aucthor. Cum priuilegio. 1564." This is greatly enlarged from Baldwin's editions; and before the fecond book has a dedication " To the vertuous & right honorable Lorde Henrye Haftinges," and a prologue to the reader. Contains Fol. 224. " Jmprinted—The firft day of December," &c. W.H. Alfo without date. Octavo.

1565. " The tragical hiftory of two Englifh louers, 1563, written by Bar. Gar." 95 leaves in verfe. This Mr. Ames has given alfo in his General Hiftory, under the fame year, as " containing 59 leaves." Licenfed. Octavo.

About the fame time he had licenfe for " A Cronenacle lately called Mr. Graftons Cronenacle." Stat. Regifter A. Fol. 118. This was a.

fevere-

f:vere flur upon Mr. Grafton, and feems to point at his Abridgement of
the Chronicles above mentioned, in 1562; for his large Chronicle was
not publifhed till 1569. His Manuell of the Chronicles indeed was printed this year by John Kingfton, which has Grafton's name; as has his
Abridgement printed again in 1572, by Tottel. So that this article ftands
in need of further explication than i am able to give it.

" La table cōteynant en fommarie les chofes notables en la graunde 1565.
Abridgement, compofeè par le Iudge trefreuerend monfieur Anthony Fytz-
herbert, dernierment renue et corige: Au quell eft nouelment adiouftee
les nombres des cafes, auecques afcuns diuifions iammes deuant imprimee.
In ædibus R- Tottell decimo Nouembris. (1565) Cum priuilegio." In
the compartment with Grafton's Rebus on the fell. Contains Fol. 210.
Firft printed by John Raftall in 1517.   W.H   Small folio.

" La Graunde Abridgement Collect par le Judge trefreuerende monfieur 1565.
Anthony Fitzherbert, dernierment Conferre auefq; la Copy Efcript, et
per ceo Correct: Aueques le nombre del fueil, per quel facilement poies
trouer les Cafes cy Abrydges en les Lyuers dans, nouelment annote: iammais deuaunt imprimee. Auxi vous troues les refiduums de lauter liuer
places icy in ceo liuer en le fyne de lour apte titles. In Ædibus R- Tottell
duodecimo Nouembris. (1565) Cum priuilegio." In a grand architective,
compartment with a tablet on the fell containing " Ne moy Reproues
fauns caufe, car mon entent eft de bon amour." Under it, R.T. This
book is divided into three parts, bound in two volumes. The firft part contains Fol. 379; the fecond part, Fol. 128; the laft part, fol. 207. It
was firft printed by Pinfon in 1514, then by W. de Worde in 1516. W.H.
Again 1577.   Large folio.

" A colleccion of entrees, Of declaracions, barres, replicacions, reioin- 1566.
ders, iffues, verdits, iudgements, executions, proces, contynuances, ef-
foynes, and diuers others matters. And fyrft an epiftle, with certayn ein-
ftructions neceffarye to bee redde for the redy fyndīge of the matters in
thys booke. In ædibus R- Tottell. Cum Priuilegio." In the fame
compartment as La Graunde Abridgement. In the tablet, "Anno falutis M. D. LXVI. decimo quarto die Maij." Under it, R.T. It is introduced with a preface by W. Raftell, giving not only directions for
readily finding the matters, but a more comprehenfive view of its defign
and ufe; and how much of it he claims as[b] his own. Contains fol. 627.
W.H. Again 1574.   Large folio.

" THE NEVVE BOKE OF IVSTICES OF peace made by Anthonie Fitz 1566.
Herbert iudge, lately tranflated out of Frēch into Englifhe and newlye
                                                        corrected.

[b] " ——vnderftand this good reader, | wordes in the margent, be mine & of mye
that al the notes & refermentes, and al that | ftudye. But none of the declaracions, &c.
is in frenche in this booke, and al the | that be in latin—be of mye makinge or
                                    5 C, 2   compilinge:

corrected. The yere of our Lorde. 1554. Cum priuilegio——folum."
Contains Fo. 173, and a table at the end. "Imprinted——the xiij.
day of July,——1566." W.H. Octavo.

1567. " The expoficions of the termes of the lawes of England," &c. as
in 1563.——17. Feb.* Octavo.

1567. "La Novvelle Natura breuiū" &c. as in 1553. "In ædibus." On the
back, "La preface fuis ceft liure, compofe par le reuerende Juftice Antonie Fitzherbard." Then, an alphabetical table. W.H. Octavo.

1567. " A learned commendation of the politique lawes of England, &c.
written in Latin above a 100 yeares paft by the learned, and right honourable maifter Fortefcue, knight, lord chancellor of England, and newly
tranflated by Robert Mulcafter, Latin and Englifh." * In was printed in
Latin only by Edw. Whitchurch, without date. W.H. Again 1573.
Sixteens.

1567. Surveying, by Mafter Fitzherbert; and his book of Hufbandry.
Octavo.

1567. " A profitable Booke of Mafter John Perkins, felow of the Inner
Temple, treating of the lawes of England. Apud R- Tottell." At the
end, Tottil. In French; and fo it was printed by Redman in 1532, but
then had a Latin title; and alfo by Hen. Smyth, in 1545.

1567. " Les plees del Coron : diuifees in plufiours titles & common lieux.
Per queux home plus redement & plenairemēt trouera, quelq; chofe
que il quira, touchant les dits plees. Compofees per le trefreuerend
Iudge Monfieur Guilliaulme Staunforde Chiuauler dernierment corrigee
auecques vn table parfaicte des chofes notables contenues en ycelle, et
iammais per cy deuant imprimee. Anno domini. 1567. In ædibus
——.Cum priuilegio." In a light airy compartment. There are prefixed, a fhort addrefs to the reader in Latin, a table as to the former

compilinge: For al them (except a fewe which I gathered together while I was a prentice of the law, Seriant & Juftice) haue I taken & gathered out of fower feueral bookes of good prefidents. Firft the olde prented booke of entrees, the feconde, a booke of prefidents of matters of the comon place, dylygentlye gathered toge-ther, & written by mafter Edw. Stubbis, ——one of the prothonotaryes of the comon place, the thyrd a booke of good prefidentes of matters of the kinges benche, written & gathered bye John Lucas fe-cundary to mafter william Roper protho-notary of the kings bench. The fourth a booke of good prefidents which was my grandfathers fyr Jhon More, fometime one of the iuftices of the kinges benche (but not of his colleccion) And al the prefidents that be in al thefe fower bookes haue I collected into this booke——which (with fuch copies as I had, being out of England, & lacking conferens with learned men) to the furtherance of the practife of the law, I haue finifhed the eight & twenty day of Marche, in the yeare of our lorde god a thowfand fyue hundred threefcore & fower."

edition,

edition, and " The newe table to this booke." Contains befides, 198, leaves. With this is ufually bound,

" An expoficion of the kinges prerogatiue collected out of the great 1568. abridgement of Juftice Fitzherbert and other olde writers of the lawes of Englande, by the right woorfhipfull fir William Staunford Knight, lately one of the Juftices of the Queenes maiefties court of comon pleas: Whereunto is annexed the Proces to the faine Prerogatiue appertaining. 1568." This had been dedicated by the author to fir Nicholas Bacon, the kings attorney of his court of Wards and Liveries, dated, " from Greys Jnne the vi. of Nouember. Anno Do. 1548;" which dedication is here inferted both in Latin and Englifh, with an apology by the printer for fo doing, in the fame ftile, &c. as before, though now advanced by the queen to be Lord Keeper of the great feal of England, in which character he now addreffes him. Contains Fol. 85. " Imprynted—An. 1568. Cum priuilegio." W.H. Both thefe books were frequently printed.
Quarto.

" M-T- Ciceroes three bookes of duefies," &c. The Latin on Roman 1568. type. Fol. 168. Again 1574. W.H. Octavo.

The Diall of princes, &c, as p. 564. Corrected, &c. with the ad- 1568. edition of a fourth Book, omitted in the French edition, from which the former Englifh edition, printed by John Wayland, 1557, was tranflated. My copy wants the title-page; but it was printed with Tho. Marfhe, and licenfed to them. John Wight had licenfe for printing this book in 1563-4, but i do not find he ever printed it. See it in 1582. W.H.+ Folio. 1568.

" The Accedens of Armory," &c. as in 1562. Quarto. 1568.

" A Chronicle at large and and meere Hiftory of the affayres of Eng- 1569. lande and Kinges of the fame, deduced from the Creation of the vvorlde, vnto the firft habitation of thys Jflande: and fo by contynuance vnto the firft yere of the reigne of our moft deere & fouereigne Lady Queene Elizabeth: collected out of fundry Aucthors, whofe names are expreffed in the next page of this leafe. Anno Domini. 1569. Cum priuilegio." In a compartment with Mofes and Brute at top; queen Elizabeth and attendants at bottom; Saul, David, Solomon, and William the conqueror, on one fide; Locrine, Albenact, Camber, and Henry the 8th, on the other. It is dedicated by Rich. Grafton "To the Right Hon. Sir Wylliam Cecill Knight, principall Secretarij to the Queenes Maiefty, and of hir priuie Counfayle, Mayfter of the Courtes of VVardes & Lyueries, and Chauncelour of the Vniuerfitie of Cambridge." Then, his epiftle to the reader, and another by Thomas N. The firft volume begins with Brute, and proceeds to William the conqueror; containing 192 pages befides the prefixes, and a table at the end.

" This

RICHARD TOTTEL.

"This feconde Volume, beginning at William the Conquerour, endeth wyth our mofte dread & foueraigne Lady Queene Elizabeth. Seene & alowed according to the order apointed. Cum priuilegio Regiæ M. ieftatis. Anno 1568." In a compartment of boys blowing horns, &c. Cawood's mark at the bottom. Contains Pages 1368, befides a table of the bailiffs, fheriffs, and mayors; and another of the principal matters. On the laft page is Grafton's rebus; and under it "Jmprinted——by Henry Denham,——for Richarde Tottle and Humffrey Toye. Anno 1569 the laft of March.——Cum priuilegio——folum." Licenfed. W.H. Folio.

1569. "THE CONTENTES of this booke. Fyrfte, the booke for a Iuftice of peace. The booke that teacheth to keepe a court Baron, or a Lete. The booke teaching to kepe a court hundred. The booke called returna Breuium. The booke called Charta feodi, conteininge the fourme of deedes, releafes, Judentures, obligacions, acquitaunces, letters of atturney, letters of permutacion, teftaments, & other thingz. And the booke of the ordinance to be obferued by the officers of the kinges Efcheker for fees taking." 195 leaves, & a table at the end. "Imprinted——1569. Cum pruilegio——folum." W.H. Sixteens.

1569. "Henrici de Bracton de Legibus & confuetudinibus Angliæ Libri quinq; in varios tractatus diftincti, ad diuerforum et vetuftifsimorum codicum collationem, ingenti cura, nunc primū typis vulgati: quorum quid cuiq; infit, proxima pagina demonftrabit.——An. do. 1569. Cum priuilegio." It has prefixed, "T. N. Candido lectori. S." Next, "Varietates lectionis huius operis ex fide duodecem libroium antiquorum defcriptæ." Then, "Index eorum omnium quæ in hoc opere continentur." Contains befides, Fol. 444. W.H. Folio.

(1570.) "A Table to al the Statutes made from the beginning of the raigne of Kyng Edwarde the. vi. vnto this prefent. xii. yeare of the reigne of our mofte gratious & foueraigne Ladye Queene Elizabeth. In ædibus——. Cum priuilegio." In the compartment with Grafton's rebus on the fell. 6 leaves. W.H. Folio.

1570. "Inftitutions or principall groundes of the Lawes and ftatutes of England, newly & very truely corrected & amended, with many newe & goodlye addicions, verye profitable for all fortes of people to knowe, lately augmented & imprinted.——by Rycharde Tottyl. 1570." Fol. 66, and a table; at the end of which is W. S. W.H. Octavo.

1570. "Graftons abridgement of the chronicles of Englande. Newly and diligrently corrected, and finifhed the laft of October 1570. The contents whereof appeareth in the next page of this lefe. Seen and allowed, according to an order taken." This book has a remarkable prologue againft
John

John Stow, which met with an anfwer from him in his fummary, anno 1573.* Again 1572. Octavo.

"Les comentaries, ou les reportes de Edmundi Plowden, vn apprentice de le comen ley, de dyuers cafes efteantes matters en ley, et de les argumentes fur yceax, en les temps des raygnes les roye Edwarde le fize, le roigne Mary, le roy et roigne Phillipp et Mary, et le roigne Elizabeth." Oct. 24. Folio. 1571.

" A litle treatife, conteyning many proper Tables & rules, very neceffary for the vfe of al men, the contentes wherof appere in the next page folowing. Collected & fet forthe by Richard Grafton. 1571.—— In ædibus———.Cum priuilegio———folum." G 6, in eights. W.H. Octavo. 1571.

A Treatife concerning Impropriations of Benefices. Quarto. 1571.

A fermon by Dr, Smythe, with which he entertained his congregation in Q. Mary's reign, was publifhed by R. T who affirmed he was both eye and ear witnefs. See Morgan's Phœnix Brit. p. 18. 1572.

" Les Tenures du monfier Littleton, ouefq certein cafes addes per auters de puifne temps, queux cafes trouers fignes ouefque ceft figne—al cōmencement & al fin de chefcun deux, au fine que ne poies eux mifprender pour les cafes de monfieur Littleton: pur quel inconuenience, ils fuerent dernierment tolles de ceft liuer, & cy vn foytz plus admotes al requefte des gentil homes ftudients en le ley dengleterre. Cum priuilegio. 1572." Contains y 4, in eights. At the end " Imprinted ———by R- Tottyl. 1574." I apprehend the laft 4 leaves only are of the edition 1574, as all the other fignatures are capitals. W.H. Sixteens. 1572.

" The Expofitions of the Termes of the Lawes of Englande," &c. as in 1563. Contains Fo. 138, befides the prologue and table. The French on Roman types. W.H. Again 1579. Sixteens. 1572.

A book of prefidents, &c. as in 1559. Sixteens. 1572.

" A difcourfe vpon vfurye by way of Dialogue & Oracions, by Tho. Wilfon Dr. of the Civil Lawes, and one of the Mafters of her Maiefties courte of Requeftes. Seene & allowed according to—Iniunctions. Dedicated " To the right hon. hyghe & myghty Earle,———,the lord Robert Duddeley, Erle of Leycefter, Barō of Denbigh, mafter of the horfe———, knight of———the garter, Chauncelloure of the vniuerfitie of Oxforde, & one of her hyghnes moft hon. preuy counfell.———From the Queenes maiefties hofpital at faincte Katherynes, thys twenty of Julye 1569." Next " A Chriftian Prologue to the Chriftian reader." Then, " A letter found 1572.

found in the ftudie of——the late bifhop of Salifburie,——fent to the author——by Ihon Garbrande M. A. in Oxforde, & prebendary of Salifbury, who had by legacy——al his papers, &c.——From Salifbury this 20. of Auguft. 1569." Alfo fome Latin commendatory verfes, by Will. Wickham, John Garbrand, and John Cook. Contains befides, 204 leaves. "Londini in ædibus——.1572." W.H.   Octavo.

1572.   "La vieux natura breuium, dernierment corrigee et amend. et cy nouelment imprimee." Printed frequently.   Sixteens.

1572.   "Graftons Abridgement of the Chronicles of Englande, newely corrected, &c. 1572. And in thende of thys Abridgement is added a propre & neceffary Treatife, conteynyng many good Rules, and fpecially one excellent manner of Computacion of yeres, wherby you maye readely finde the date & yeres of any euidēce. The particuler contentes of this Booke appereth in the next page folowing." This edition is dedicated "To——the Lorde Robert Dudley, Earle of Leycefter, &c. Knyght of——the Garter, one of the Queenes Maiefties priuie Counfaill, and Mafter of her hyghnes Horffe;" and has Grafton's addrefs to the reader, which you may fee at large in p. 504, &c. The chronicle contains 216 leaves. "In ædibus——Cum priuilegio." Licenfed. W.H.   Octavo.

1572.   "Workes of armorie, diuided into three bookes, entituled, the conꞏꞏrds of armorie, the armorie of honor, and of coats, and creftes, collected and gathered, by John Bofwell, gentleman.——In ædibus."   Quarto.

1573.   "Fiue hundreth points of good hufbandry vnited to as many of good hufwiferie, firft deuifed, & nowe lately augmented with diuerfe approued leffons concerning hopps & gardening, and other needeful matters, together with an abftract before euery moneth contelling the whole effect of the fayd moneth, with a table & a preface in the beginning both neceffary to be reade, for the better vnderftanding of the booke. Set forth by Thomas Tuffer gentleman, feruant to the honorable Lord Paget of Beudefert. Imprinted——anno. 1573." In a compartment with Midas on one fide, and Venus on the other. For this he had licenfe in 1561. George Mafon Efq;   Quarto.

1573.   Ricardi Willeii poematum liber. Dedicated to Secretary Cecil.   Octavo.

1573.   "La graunde abridgement, collecte et efcrie per le Judge tres reuerend fyr Robert Brooke, chiualier, nadgairs chiefe Juftice del comon banke.— 5 Jan." Again 1576, 1586.   Folio.

1575.   "A Summarie of the Chronicles of England, from the firft comming of Brute into this land, vnto this prefent yeare of Chrift 1575. Diligently collected

collected, corrected & enlarged by Iohn Stowe Citizen of London. Imprinted——by R. Tottle and Henry Binneman. Cum priuilegio." In a compartment with a naked boy bearing a basket of fruit on his head, on each side. On the back is the earl of Leicester's crest gartered, to whom it is dedicated, as were the former editions printed by Tho. Marsh. The preface is much the same as to the edition 1570, of which you may find some account in p. 504. This edition is adorned with cuts, designed for portraits, at half length, of the kings of England from William the Conqueror. Some of them had been used to Reyn. Wolfe's edition of Mat. Paris, but placed differently. The Summarie contains 570 pages. At the end is annexed a short account of the universities of England; the distance of the principal towns from London; the time for the fairs; the counties, the cities, &c. sending members to the parliament, 439 in number; lastly, a table of contents. W.H. Again 1579.—* Octavo.

Leigh's Accedents of Armory. See it in 1562.   Quarto. 1576.

" Regis pie memorie Edwardi tertij a quadragesimo ad quinquagesimum. 1576. Anni omnes a mendis quibus miserrime scatebant repurgati & suo nitori restituti. Anno Domini. 1576.——in ædibus——.Cum priuilegio—— folum. ¶ Ne moy reproues sauns cause, car mon entent est de bone amour." In the compartment with Grafton's rebus on the sell. The leaves of every year are numbered separately. At the end, " Jmprinted ——the xiii. daye of Januarye, in the yere of oure Lorde. 1556." This seems to be the true date of the book, and the title-page to have been prefixed since. W.H.   Folio.

" Magna charta, cum statutis, tum antiquis, tum recentibus, maxi- 1576. mopere animo tenendis, nunc demum ad vnum, typis edita." The reader is informed that this edition " cōteyneth the most necessarie——olde statutes, and diuers later & newe statutes most conuenient to bee had perfect & ready." There are two tables prefixed; one, of the statutes in the order they are printed, the other, alphabeticall, agreeing with Rastall's collection. Contains besides, fo. 247. " Jmprinted——the 8. day of March. 1576." W.H.   Sixteens.

" Ascvn Nouell cases de les ans et temps le Roy, H. 8. Edw. 6. et la 1578. Roygne Mary, Escrie ex la graund Abridgement, compose per Sir Robert Brooke Chieualer &c. la disperse en les Titles. Mes icy collect sub ans. Anno Do. 1578. In ædibus——Cum Priuilegio." In the same compartment as Stowe's Summarie, 1575. Contains 116 leaves, and a table at the end. " Imprinted——the xv. of October, 1578." W.H. Again 1587.   Sixteens.

" Of the knowledge and conducte of warres, two bookes, latelye 1578. wrytten & sett foorth, profitable for suche as delight in Hystoryes, or martyal martyall affayres, and neceffarye for this prefent tyme.——In ædibus ——vij. die Iunij. Anno Domini. 1578. Cum priuilegio——folum." On the next leaf is the author's coat of arms, under which are the initials T.P. and this motto "Virtus fuperat ardua." On the back fome verfes by the author to his book, figned T. P. Then, the preface, and contents of the 2 books; the firft treating of the captain & foldiers, the fecond of the dicipline, obfervations and admonitions of war. Contains 48 leaves, befides the prefixes. Licenfed. W.H. Quarto.

1578. "Les Commentaries, ou Reportes de Edmunde Plowden vn apprentice de le comen ley, de diuers cafes efteants matters en ley, & de les Arguments fur yceux, en les temps des Raygnes le Roye Edwarde de le fize, le Roigne Mary, le Roy & Roine Phillip & Mary, & le Roigne Elizabeth. Ouefque vn Table perfect——per VVilliam Fletevvoode Recorder de Loundres, & iammes cy deuaunt imprime. Auxy vous aues in ceft impreffion plufures bone notes en le mergent, &c. In Ædibus ——Octobris 20. Cum Priuilegio." In the fame compartment as Fitzherbert's Grand Abridgement. On the back is a table of "The principall cafes," &c. It has prefixed, Edm. Plowden's prologue or preface, and the fame " yelded in Englifh by E. M." Then, "The Table——by W- Fletewoode." Contains befides, Fol. 401.

1579, "Cy enfuont certeyne Cafes Reportes per Edmunde Plowden——, puis le primer imprimier de fes Commentaries, & ore a le feconde imprimter de les dits Commentaries a ceo addes. Ouefque vn Table en fine de ceft Lieur des toutez les principall cafes, cibien en le dift primier Lieur des Commentaries, come de les cafes icy de nouel addes, iammes deuaunt imprimie. Cum priuilegio. Anno. 1579." In a compartment of the fame defign as the firft part, but reduced. The folios continued to Fol. 565. Then the table abovementioned, and another by W. Fletewood. To thefe is added.

1579. "Vn Report fait per vn vncerteine author del part de vn argument del Edm. Plowden de Milieu Temple" &c. Contains Fol. 15. "In Ædibus ——1579. Cum Priuilegio." W.H. Short folio.

1579. "Le digeft des Briefes originals, et des chofes concernants eux, compofe per Simon Theloall.——in Ædibus——Octobris decimo quarto. 1579. Cum priuilegio——folum." Dedicated "To the right Hon. Sir Thomas Bromley Knight Lorde Chauncellour of Englande.——From my poore houfe neere Ruthin in Wales, the firft of October. 1579." Then, a table. Contains 424 leaves befides. W.H. Octavo.

1579. "ANNALIVM tam Regum Edwardi quinti, Richardi tertij, & Henrici feptimi, quàm Henrici octaui Titulorum Ordine Alphabetico digeftorum, Elenchus. Studio & Labore Guilhelmi Fletewoodi Recordatoris Londinenfis. In Ædibus—1579. Cum Priuilegio." Dedicated "Clariffimo,

RICHARD TOTTEL.

ac Illuſtriſsimo viro, Domino Thomæ Brumleio Summo Angliæ Cancellario, Sereniſſimæ Reginæ Elizabethæ à conſiliis.——Londini, ex domo Baconica, Decimo Calendas Septembris, Anno 1579." BB 4, in eights, half ſheets. " Imprinted——the x. day of September." W.H. Sixteens.

Kitchyn on Courts. See it in 1585. Sixteens. 1580.

" Le liuer des aſſiſes et plees del corone, &c. oueſque deux nouels 1580. tables, l'un de touts les principal caſes contenus in ceſt liuer, laut monſtrant ſouth queux titles, ſir Robert Brooke, eux ad abridge." Folio.

"EIRENARCHA: or of The Office of Juſtices of Peace, in two Bookes: 1581. Gathered. 1579. and now reuiſed, and firſte publiſhed, in the. 24. yeare of the peaceable reigne of our gratious Queene Elizabeth: By William Lambard of Lincolnes Inne Gent. Hæ tibi artes erunt, paciq; imponere morem. At London: Imprinted[b] by Ra: Newbery, and H. Bynneman, by the aſſ. of Ri. Tot. & Chr. Bar." In a compartment with Barker's arms on one ſide, and his creſt on the other. On the back, the arms of " Syr Thomas Bromley, Knight, Lord Chauncelour of England," to whom the book is dedicated. This dedication is omitted in the ſubſequent editions, but i know not how to omit the author's pious concluſion: " The Lord of Lordes bleſſe you (for his Chriſtes ſake) with the ſpirit of godly & couragious wiſedom, and make you a long & happie Counſailour of this Eſtate, to the furtherance of his holy goſpell, the weale of the Engliſh nation, the true ſeruice of the Queenes Maieſtie, and your owne Honorable & long laſting memorie. Amen. From Lincoln's Inne, this 27. day of Januarie. 1581." Then, a table of contents. Beſides them, 511 pages; and at the end, " A table of the imprinted Statutes,—— wherewith Iuſtices of the Peace haue in any ſorte to deale." Binneman's device on a ſeparate leaf. W.H. Octavo.

" The Dial of Princes (&c. as p. 564,) Engliſhed out of the French by 1582. Tho. North, ſonne of Sir Edward North knight, L. North of Kirtheling. And nowe newly reuiſed & corrected by hym, reſourned of faultes eſcaped

---

[b] "xvto die Januarij. At a courte holden this daie it is ordered & agreed by the maiſter, wardens & aſſiſtants of this companie (Stationers) with thaſſente of the parties, concerning the printinge of a booke of Mr. Lamberdes compiling, called Eirenarche, That the ſaid Raffe & Henrye ſhall printe this firſte ympreſſion of 1500 books thereof, which they are nowe in hande withall, to their owne vſes giuing, vnto the ſaid Mr. Tottell & Mr. Barker freelie halfe a hundred of the ſame bookes equallie betwene them. And———that alwaies after the finiſhinge of the ſaid firſt ympreſſion.... as often as it ſhalbe reprinted, augmented or altered, ſhall be printed to the equall benefittes of the ſaid Mr. Barker, Mr. Tottell, Raffe & Henry, and at the equall charges of them foure. Provided———— that the ſaid firſt ympreſſion & all other ympreſſions thereof hereafter ſhallbe onlie publiſhed as printed by the ſaid Raffe & Henry, at the aſſignement of the ſaid Chr. Barker & R. Tottell." Regiſter B. fol. 435.

in the firſt edition: with an amplification alſo of a fourth booke annexed to the ſame, Entituled The fauoured Courtier, neuer heretofore imprinted in our vulgare tongue. Right neceſſarie & pleaſaunt to all noble & vertuous perſons. Nowe newly imprinted——1582. Cum priuilegio." In the compartment with Midas and Venus. Contains 476 leaves beſides the prefixes. W.H. Quarto.

April the 6th 1582, he had licenſe to print "A book of Judgmentes and ſuche like matters, which he perpoſes to annexe to the booke of Preſidentes." On the 28th of January following, it was licenſed again under the title of "The booke of Preſidentes with addicōns newlie put thereto." February the 18th he enters, and has licenſe for theſe 36 books, viz. "L'Office & Authoritie de Juſtices de peas, vicountes, &c. in parte collecte per le Juge——Anth. ffytzherbert,——et enlarge——per R. Crompton,——et publie L'an du grace 1582. The bookes of yeres & termes, viz. from Henr. 3, to Henr. 8, incluſive. Bothe the volumes of Mr. Plowdens Cōmentaries: ffitzherberts Abridgement. Brookes Abridgement. Bracton de legibus Angliæ. The booke of Entries. The booke of Aſſiſes. The Regiſter of Writtes. Plees of the crowne, with prerogative Regis. Mr. Brookes Newe caſes. The court Baron, with Returna b'viū. Littleton's Tenures. ffitzherbertes Natura b'viū. The olde Natura b'viū. Magna Charta. The Digeſtes of wryttes. Mr. Perkins booke. Doctor & Student. The termes of the lawe. The olde Juſtice of peace, with others. fforteſcues booke of the cōmendation of the lawes of England. A table to Henr. the 7th. &c. yeres. Nove narrationes. Inſtitutions & groundes of the law. The office of Shriues. cūſtables, &c. The Accidence of Armorie. The concordes of Armorie. The art of lymninge. Morall philoſophie. Songes & Sonettes. Tullies offices, Lat. & Eng. Quintus Curtius in Eng. Engliſhe lovers. Romeo & Juletta. And, Dr. Wilſon's booke of vſurie." May the 9th, 1583, he had licenſe for "Les Trois mondes, par Le Seigneur de la popellince. To be tranſlated into Engliſh. vpon condicōn he obtain the Ordinaries hand thereto."

1583. ' " De Legibus Angliæ municipalibus Liber, ordine locorum communium diſpoſitus. Laborem queras, vt laborem vites." Folio 371. "Imprinted——the ix. day of October, Anno 1583. Cum priuilegio—ſolum." W.H. Octavo.

1583. Year books. xx. and xxi. Henry VII. W.H. Folio.

1584. " Loffice Et aucthoritie de Juſtices de peace, in part collect per le trefreuerende Monſier Antho. Fitzherbert, iades, vn de les Juſtices del common Banke, & inlarge per R. Crompton, vn Apprentice de le common ley, & ore luy reuyſe, corrygie, & augment, 1584. A que eſt
annex

annex Loffice de Vicountes, Baylifes, Efcheators, Conftables, Coroners, &c. collect per le dit Mounfier Fitzherb.——At London By R- Tottell." In a border. Dedicated "Al Trefhonorable Sir Thomas Bromley Chiualer, Seignior Chauncelour Dengleterre——.Del Melieu Temple le 30. iour de Nouemb. 1584." Alfo, an epiftle, "As toutes mes companions del myddle Temple que funt Iuftices de peace. Then, " Le Table tanque al Office de Vicounts &c. que commence fol. 156." And, " An Exhortation to the Jurye." At the end, " Placita, Proceffus, &c., Indictamenta, & Appella." Contains 227 leaves. "Imprinted—An. 1584." On a feparate leaf, the Errata. W.H. Again 1593. Quarto.

Year books. 1 Edw. v. Alfo, 1—21. Henry vii. W.H. Folio. 1585.

" Le Court leete, et court Baron, collect, per John Kitchyn de Grayes 1585. Inne, vn apprentice en le Ley. Et les cafes & matters neceffaries pur Senefchalles de eaux Courtes afcier, pur les Studentes de les meafons de Chauncerie. Ore nouelment imprimee, & per laucthour mefme Corrigee, ouefque diuers nouel Additions, come Court de Marfhalfey, Auncient demefne, Court de Pipowder, Effoines, Imparlance, View, Actions, Contracts, Pleadings, Maintenance, & diuers auters matters. In ædibus ——Maij fexto, Anno Do. 1585. Cum priuilegio—folum." Dedicated " Al Trefhonorable Sir Thomas Sackuile Chiualer, Seignior Buckhurft." The Preface, " All Studentes de les meafons del Chauncery——De Graies Inne 10. Nouembris Anno falutis. 1581." The table; and Errata. Contains 364 leaves befides. To this is commonly annexed,

" Retourna Brevivm Nouelment corrigee, ouefque dyuers auters 1585. bones retournes, & multes cafes del common ley a ceo addes,—collectes per Iohannem Kitchin—Iammais per ce deuant emprimes. In ædibus —1585. Cum priuilegio." Contains 58 leaves, and a table at the end. W.H. Again 1587. Sixteens.

" Ce enfuon afcuns nouels Cafes collectes per le iades trefreuerend 1585. Iudge, Mounfieur Iafques Dyer, chiefe Iuftice del common banke, ore primierment publies & imprimies.——in Ædibus—Ianuarij primo. 1585. Cum priuilegio Regiæ Maieftatis." In the fame compartment as the fecond part of Plowden's Commentaries. On the back begins an epiftle from R. F. and I. D. the author's nephews "To the Students of the common lawes of this Realme, and efpecially to our Maifters the Benchers & felow ftudents of the Middle Temple." Then, fome latin verfes to the reader, and others in commendation of the author. 377 leaves. " Imprinted——by R- Tottyll." On another leaf, " Nomina Capitalium, Iuftic' de Banco Regis, ab anno xxij. Edwardi tertij, vfque Annum 16. Reg. Elizab." Alfo, " Nomina Capitalium Iuftic' de Communi Banco ab anno domini 1399.——vfque Annum 24. Reg. Elizab. Et quibus Terminis & Annis inierunt officium, et quandiu in eodem permanferunt:

Et

"Et quando ſtilus Militis-incepit in Rotulis placitorum." N.B: Theſe liſts differ greatly from the Chronica Juridicialia. Licenſed. W.H. Folio.

1587. Year books. 1, and 2. Richard III. W.H. Alſo without date. Folio.

1588. " La Table al lieur Reports del treſreuerend Judge Sir James Dyer Chiualer, iades chiefe Iuſtice del common bank: per quel facilment cy troueron toutes choſes conteinus in icel ore tarde compoſe per T. A. Anno Domini 1588.——In Ædibus R-Tottelli." It has prefixed " A note to know the yere wherein any caſe is contained"; with a table for that purpoſe: Alſo a table of the general titles. Contains beſides, 167 leaves. W.H. Sixteens.

1590. " Symbolaeographia. Which may be termed the art, deſcription, or image of inſtrumentes, couenants, contracts, &c. or the notarie, or ſcriuener. Collected and diſpoſed by William Weſt, of the Inner Temple, gentleman, atturney at the common law, in fower ſeuerall bookes." Again 1592. Quarto.

1591. Year books. The whole reign of Henry VIII. W.H. Folio.

1592. " An Expoſition of certain difficult & obſcure words & termes of the Lawes of this Realme, newly ſet forth & augmented both in Frenche & Engliſh, &c. In ædibus——Cum priuilegio. 1592." A table prefixed, and contains beſides, 196 leaves. W.H. Octavo.

1594. " Giacomo Di Graſsi his true Arte of Defence, plainlie teaching by infallable Demonſtrations, apt Figures, and perfect Rules, the manner & forme how a man without other Teacher or Maſter may ſafelie handle all ſortes of Weapons, aſwell offenſiue as defenſiue: VVith a Treatiſe Of Diſceit or Falſinge: And with a waie or meane by priuate Induſtrie to obtaine Strength, Iudgement & Actiuitie. Firſt written in Italian by the foreſaid Author, And Engliſhed by I. G. gentleman. Printed at London for I. I. and are to be ſold within Temple Barre at the Sign of the Hand & Starre. 1594." It is dedicated " To the——L. Borrow, Lord Gouernor of the Breil, and Knight of——the Garter," by T. C. (Tho. Churchyard) who appears to have been the publiſher. Next is " The Authors Epiſtle vnto diuers Noble men and Gentle-men." Another, " To the Reader." Then, " An Aduertiſement," by the publiſher; and a liſt of " The Sortes of Weapons handled in this Treatiſe." Contains beſides theſe, E e 3, in fours; but the firſt alphabet proceeds only to R 2. Printed on Roman type, with cuts. W.H. Quarto.

" An abſtract of all the penall ſtatutes, which be in general in force & vſe, wherein is conteined the effect of all thoſe ſtatutes, which do threaten to the offenders thereof, the loſſe of life, member, lands, goods, or other puniſhment, or forfaiture whatſoever. Whereunto is alſo added, in their apt titles, the effect of ſuch other ſtatutes, wherein there is anie thing

thing materiall, and neceffarie for each fubiect to knowe. Moreouer, the authoritie & duetie of all iuftices, fheriffes, coroners, efchetors, maiors, baylifes, cuftomers, comptrollers of cuftome, ftewardes of leetes & liberties, aulnegers, & purueyors, and what things by the letter of feuerall ftatutes in force, they may, ought, or are compellable to doe. Collected by Fardinando Pulton, of Lincolnes Inne, gentleman. Deut. xvii. If there rife a matter to hard for thee in judgment, in matter of controverfe," &c. Octavo.

This book had a reverfionary licenfe, 21. July, 1577. Firft to Ri- Tottel for his life; then to Ra- Newbery for his life; and laftly to Ro- Walley for his life.

"TRACTATUS de legibus, et confuetudinibus regni Anglie, tempore regis Henrici fecundi compofitus, Iufticie gubernacula tenente illuftri viro Ranulpho de Glanuilla iuris regni, & antiquarum confuetudinu eo tempore peritiffimo. Et illas folu leges continet et confuetudines, fecundum quas placitatur in Curia regis ad fcaccarium, et coram Iufticiis vbicunque fuerint. Huic adjectæ funt a quodam legum ftudiofo adnotationes aliquot marginales non inutiles. W.H. Sixteens.

" A Bouclier of the catholike fayth of Chriftes church conteynyng diuers matters now of late called into controuerfy by the newe gofpellers. Made by Richard Smith doctour of diuinitee, & the Quenes hyghnes Reader of the fame ī her graces vniuerfitie of oxford." On the back, " The contentes." Dedicated " To the moft gratious & vertuous Ladye, oure fouereigne Mayftres, Quene Mary, Quene of England, &c. and defenders[c] of the faith." Next is a preface to the reader; at the end of which, "The fecond[d] boke fhalbe prynted fhortely by Gods grace." Then, " A Cōfutaciō of an errour fet forth by the autours of the Catechifme made in king Edward the fixte tyme, & publifhed in 'hys name, which is, that no man can obferue gods cōmaundementes." Contains fol. 80. Lambeth Library. Octavo.

[c] Perhaps defigned for Defendrefs. [d] Printed by Rob. Caly, 1555.

ROGER MADELEY,

OF him I have only found a copy of verfes intitled, " An inuectiue agaynft treafon." In two columns on a half fheet, fignifying the joy of the people, &c. on the 19th of July 1553. At the end, " Finis qd̃ T. W. Imprynted at London by Roger Madeley; and are to be fold in Paules Churche yearde at the fign of the Starre." A M.S. copy, W.H.
Folio.

1553.

ROBERT CALY, or CALEY,

DWELT in the precinct of Christ's-hospital, and may be supposed to have succeeded Richard Grafton, esq; in his house, though his office of king's printer was given to Cawood, when queen Mary had put him aside for printing the proclamation mentioned before, in p. 537. Soon after the parliament had renewed the statutes for the punishment of heretics, meaning all such as opposed the pope's supremacy, " fyr Roger Cholmeley knyght, mayster Willyam Roper, & Doctour Story" were in commission for the examination of them; and to bring them in,——— " Bearde, John Vales, and Robert Caley, two taylors & a prynter." Dec. 17. 1557, he was fined by the Stationers' company " iiii. s. for pryntinge a boke contrary to the ordinances, that ys not hauyng lycense from the master & wardyns for the same." About the beginning of the year 1559, he paid " at his makinge fre, and for his good will to the Hall, viij. s. iiij. d." Sometime before this, he was fined with 9 others " for that thay came not to the hall vpon the quarter daye." He with Henry Caley had a license, or patent, for printing for 7 years, granted them the 30 April, 1557.

1553.   " Oratio coram patribus et clero, anno primo Mariae. Hugo Weston, decanus Westmonast." See p. 788.   Octavo.

1553.   The restitution of a sinner. Translated out of Greek.   Octavo.

1553.   " DIACOSIO MARTYRION. id est Ducentorum virorum testimonium de veritate corporis et sanguinis Christi, in eucharistia, ante triennium, aduersus Petrum Martyrem, ex professo conscriptum. Sed nunc primum in lucem æditum. Ioanne Whito Anglo Collœgij Wicamensis apud inclytam Wintoniam præside Authore. Math. xviii. In ore duorum aut trium testium stet omne uerbum. Excusum Londini, in aedibus Roberti Cali, Typographus, Mense Decembris, Anno 1553. Cum priuilegio—solum." On the back, " Tipographus Lectori." Wherein we learn that the copy above 3 years before had been sent to Louain to be printed, but on the knowledge thereof, the author was committed to prison. The dedication in verse designed for it then is retained, being addressed " Ad Sereniffim. Illustriffim. que principem Mariam, Edouardi sexti Angliæ, &c. sororem." Next is " Epistola. Petro Martyri sanam mentem.——— Ex collegio nostro wintoniæ Ipsis Kalendis Ianuarijs. Anno domini. 1553." Then a summary of the witnesses in the following manner.
 " VI. Quattur euangelistæ, Apostolio duo,
  XXIII. Vnus propheta, martyres bis vndecem.
                                                                                                                                   " VII

VII. Septem rabinis ex vetuſtioribus,
V. Papæ, quot vna promicant fratres roſa,"
&c. to the number of 211. Laſtly, queen Mary's approbation and injunction for it to be taught in the ſchools, &c. "Datum Vveſtmonaſterii, Idibus Martiis, regni noſtri primo." The teſtimony of each witneſs is given in a ſhort poem; but they are of different metres; on 100 leaves. A recapitulation on 4 leaves, and a table on 2 leaves. The type chiefly Roman. W.H. See Eccl. Memor. Vol. II; p. (270.) Quarto.

"James Brokis D. D. and maſter of Baliol colledge, Oxon. his ſermon at Pauls croſs. 12. Nou. 1553." His text, "Math. ix. the latter part of the 18th verſe. Lord my daughter is even now deceaſed," &c. Theſe words he applied to the kingdom and church of England upon its late defection from the pope; but the proteſtants cenſured it, ſaying that he made himſelf to be Jairus; England, his daughter; and the queen, Chriſt. Eccl. Memor. Vol. III; p. 74. 1554.

The ſame, peruſed by the author, and newly imprinted. Octavo. 1554.

"A Sermon very notable fruictefull and Godlie made at Paules Croſſe in London, Anno Domini 1521. within the Octaues of the Aſcenſion by that famous & great Clerk John Fiſher, Biſhop of Rocheſter, concerning the hereſies of Martin Luther. Excuſum Londini in ædibus—— Menſe Nouembris, Anno 1554. Cum priuilegio." Dr. Lort. Again 1556. See p. 219. Octavo. 1554.

"The aſſault of the ſacrament of the altar, containyng as well ſix ſeverall aſſaults, made from tyme to tyme againſt the ſayd bleſſed ſacrament; as alſo the names and opinions of all the heretical captains of the ſame aſſaults: written in the year of our Lord 1549, by Miles Huggarde, and dedicated to the queɴes moſt excellent maieſtie, being then lady Marie: in whiche tyme (hereſie then reigning) it could take no place." 20 Sept. In verſe. Octavo. 1554.

"A diſcourſe wherin is debated whether it be expedient that the ſcripture ſhould be in Engliſh for al men to reade that wyll, Fyrſt reade this booke with an indifferent eye, and then approue or condemne as God ſhall moue your heart. Excuſum——in ædibus——Menſe Decembris, Anno. 1554. Cum priuilegio." The head title, "A queſtion to be moued to the highe courte of Parliament." John Standiſh was the author, as appears by a prayer at the end in an acroſtic. L 4, in eights. W.H. Octavo. 1554.

"The hiſtorie of Wyates rebellion, with the order and manner of reſiſting the ſame. Wherunto in the ende is added, an earneſt conſerence with the degenerate and ſedicious rebelles, for the ſerche of the cauſe of their daily diſorder. Made and compyled by John Proctor. Menſe Decembris 1554." At the end, "Jmprinted——by Rob. Caley, within the precincte of the late Houſe of the grey Friers, nowe conuerted into
an 1554.

ROBERT CALEY.

an Hofpital, called Chrift Hofpitall. The xxii daye of December. 1554.' Cum priuilegio.——folum," Again, " Menfe Januarij. Anno 1555.' *W.H. Sixteens.

1554. " A TRAICTISE declaryng and plainly prouyng, that the pretenfed marriage of Prieftes, and profeffed perfones, is no marriage, but altogether vnlawful, and all ages, and in al counteries in Chriftendome, bothe forbidden, and alfo punyfhed. Herewith is comprifed in the later chapiters, a full confutation of Doctour Poynetes boke intitled, a defenfe for the marriage of Prieftes. By Thomas Martin, Doctour of the Ciuile Lavves. Excufum——in ædibus——Menfe Maij. Anno 1554. Cum priuilegio." On the back, " The contentes of this woorke whiche is deuided into xiiii. chapiters." Dedicated " To the moft high and moft vertuous Ladie,—Marie—Quene of Englande, &c. defendour of the faieth, &c." Mm, in fours. W.H. Quarto.

1554. The way home to Chrift, by——Proctor. 22 Octo. Again 20. Jan. 1556. Octavo.

1555. " Reverendi in Chrifto Patris Domini Epifcopi Wyntonienfis doctoris Gardineri Angliæ Cancellarii Epitaphium, Joanne Morrenno Collegii Corporis Chrifti focio authore. Londini. Ex ædibus——Menfe Novembris. Anno falutis. 1555." See it in Hearne's Coll. of Difcourfes, &c. Quarto.

1555. Catonis difticha moralia, cum notis et fcholiis Taverneri. 6. Nov. Quarto.

1555. " The feconde parte of the booke called a Bucklar of the Catholyke fayeth, conteyninge feuen chapiters: Made by Rychard Smyth doctoure of diuinitie of Oxforde, & reader of the fame there. Fxcufum——Menfe Ianuarii, Anno. 1555. Cum priuilegio——folum." L 4, in eights. " Imprinted——by Rob. Caly within the precinct of——the graye Freers,——called Chriftes hofpitall. M. D. LV." Lambeth Library. See the firft part, p. 827. Octavo.

1555. " Hugh Glafier his fermon at paules Croffe the 25 of Auguft. 1555. on Luke 18. verf 10 &c. being the Gofpell on the 11 Sonday after Trinitie." Maunfell,.p. 99. Octavo.

1555. " A notable fermon made wichin S. Pauls church in Londō : in the prefence of certen of the Kinges and Quenes honorable priuie Counfell, at the celebration of Exequies of the Right Excellent & famous princeffe, Lady Jone Quene of Spayne, Sicilie & Nauarre, &c. the 18. of June, Anno 1555. by Maifter John Feckenham, Deane of the faid Churche of Paules. Set forth at the requeft of fome in authoritie whofe requeft could not be denayed. Excufum—Menfe Augufti, an. 1555. Cum priuilegio. Octavo.

1556. " The difplaying of the proteftantes and fundry their practifes, with a difcription of diuers of their abufes. Perufed & fet furth with thaffent of authoritie, &c. Excufum——in ædibus——An. 1556." No author's

nam

name in this copy; but at the end, in the hand of the late Mr. T. Baker. "This book was wrote by Miles Huggard." Mr. Warton remarks, "This piece soon acquired importance by being anſwered by Lawrence Humphries, & other eminent reformers." Hiſt. of Eng. Poetry, Vol. II; p. 459, note k. Folios 130, and a table. " Jmprynted——within the precinct of the late diſſolued houſe of the graye Freers, &c. Menſe Iulii. Anno. 1556. W.H.+ Sixteens.

" A ſhort treatiſe in meter vpon the cxxix pſalme of Dauid, called 1556. De profundis. ¶ Compiled and ſet forth by Miles Huggarde, ſeruante to the queues maieſtie." With Grafton's compartment. 4 Jan. Quarto.

" The Complaint of Grace, compiled by John Redman, D. D. late 1556. Preſident of Trinitie Colledge in Cambridge, containing in it much godly learning & veritie of matter, publiſhed by Tho. Smith ſeruant to Queene Marie." Maunſell, p. 36. Octavo.

" The primer in Engliſh and Latin, after Saliſbury Uſe, ſet out at 1556. length with many praiers, and godly pictures. Newly imprinted this preſent yeare. 1 Oct." Twelves.
Mr. Gough has a copy with a written title, dated, 1 Octob. 1557. 4to. Query, whether there be a miſtake in the print, or that they are different editions See Britiſh Topography, Vol. II; p. 361.

" A notable and learned Sermon or Homelie vpon St. Andrewes day 1556. laſt paſt 1556, in the Cathedral Church of S. Paul in London, by mayſter Jhon Harpesfield D. D. and Canon Reſidentiary of the ſaid Church. Set furth by the Biſhop of London, vltimo Decembris. 1556. Cum priuilegio——ſolum." Nineteen pages. " Imprinted——within the precinct," &c. Dr. Lort. Sixteens.
" Henry Pendleton' D. D. his Declaration in his ſickneſs of his faith & 1557. belief in all points as the Catholick Church teacheth againſt ſclanderous reports againſt him." See Eccl. Memor. Vol. III; p. 384. Quarto.

" Sermons very fruitfull and learned, preached and ſette foorth by 1557. maiſter Roger Edgworth, D. D. canon of the cathedral Churches of Sariſburie, Welles, and Briſtow, Reſidentiary of Wells & chancellor of the ſame : wherein is contained, 1. Of the 7 gifts of the Holy Ghoſt. 2. Of the Articles of the Chriſtian Faith. 3. Of ceremonies, and of Mans Laws. 4. An expoſition on the 1ſt Epiſtle of Peter. Menſe Sept."
Quarto and Octavo.

" Holſome and Catholyke doctryne concerning the ſeuen Sacramentes 1558. of Chryſtes Church, expedient to be knowen of all men, ſet forth in maner of ſhorte Sermons to bee made to the people, by the reuerend father in God Thomas byſhop of Lincolne, Anno 1558 Menſe Februarij. Excuſum

f A gun was fired at him whilſt he was preaching at Paul's Croſs, 10 June, 1554.

ROBERT CALEY.

———in ædibus———. Cum priuilegio———folum." In the fame compartment as was ufed by Grafton to the firft book of Homilies in 1548. Contains X, in eights. This feems to be the reafon of its being denominated Octavo, in Ames, having been probably fent him by fome friend; but it is the fame fize as thofe in general termed Quarto. When books of this fize have only 4 leaves to each fignature, they are printed on half fheets. I have 2 or 3 other editions of this fize, & never faw or heard of its being printed fmaller. One of them is only a copy of the faid edition, with a title of the fame purport, without any compartment, the table of contents beginning on the back. Thefe have no colophon. Two other editions have their leaves numbered, but very incorrectly. One of them has the fame title-page and table as the firft mentioned, which after Fol. lxxvi has 4 leaves unnumbered, and proceeds with Fol. lxxvii to Fol. cxc. "Imprynted———within the precinct of Chriftes Hofpitall. The. x. day of February, M. D. LVIII." The leaves of the other edition are all numbered to Fol. cxciiii. "Imprinted.———The. vii. day of June." He with Hen. Caley, had a patent for printing this book for 7 years, dated 30 April, the 4th and 5th of Philip and Mary. W.H.   Quarto.

——— "Orationes tres dicendae in miffis, pro agendis Deo gratiis de reconciliatione regni cum ecclefia catholica."   Broadfide.

——— "Two homilies vpon the firft, fecond & third articles of the Crede, made by maifter John Fecknam Deane of Paules." 8 leaves. "Excufum ———in ædibus—Cum priuilegio." Mr. William White.   Quarto.

JOHN KINGSTON

PUT a Y for an I, and an E at the end of his name, or fometimes wrote Jhon Kyngftone, according to the ufage of thofe times, when they were negligent in fpelling. He appears to have been connected in trade with Hen. Sutton all Q. Mary's reign, efpecially in the Church books. He had a fhop at the Weft door of St. Paul's. The latter part of his time he printed chiefly for others.

1553. "The regiment of life, whereunto is added a treatife of the peftilence, with the Booke of children, newly corrected and enlarged by Thomas Phaire." B b 7, in eights. "Imprinted———by Ihon kyngfton and Henry Sutton, dwelling in Paules Churchyarde. Anno Domini. 1553." W.H.   Small.

"Manuale

"Manuale ad vſum per celebris eccleſie Sariſburiēſis. Londini recēter 1554.
impreſſum, nec non multis mēdis terſum atq; emendatum. Londini. Anno
Domini 1554." In the ſame compartment as Manuale, &c. p. 561, to which
this is very ſimilar, being printed page for page alike. They both contain 168 leaves, but mentioned there only 148, by miſtake. "Explicit
Manuale &c. Impreſſum Londini ex officinæ Iohannis Kingſton. & Henricus* Sutton impreſſ. Anno. 1554." W.H. Quarto.

"Proceſſionale", &c. as p. 415. In the ſame compartment. Dr. Lort. 1555.
Quarto.

"Miſſale ad vſum eccleſie Sariſburienſis. 1555." Over a cut of the root of 1555.
Jeſſe; under which are three diſtichs, "Ad ſacerdotem." On the back, "Ex-
tracta e compoto." The kalendar, in which "Thome cant. archiepi. et
mar." is again reſtored, after having been expunged above 15 years. Colophon, "Miſſale ad vſum Sariſburienſis explicit, optimis formulis (vt
res ipſa indicat) diligentiſſime reuiſum ad correctum, cum multis annotatiunculis, ac litteris alphabeticis Euangeliorum atq; Epiſtolarum origi-
nem indicantibus. Lōdini impreſſum. Per Johannem Kyngſtō et Henricum Sutton typographos." W.H. Quarto.

"Hymnorum cum notis opuſculum, vſui inſignis eccleſie Sarum ſub- 1555.
ſeruiens." With Henry Sutton. Quarto.

"Portiforiū ſeu Breuiarium ad vſum eccleſie Sariſburienſis caſtigatum, ſup- 1555.
pletum, marginalibus quotationibus adornatum, ac nunc primum ad veriſſi-
mum ordinalis exemplar in ſuum ordinem a peritiſſimis viris redactum.
Pars Hyemalis. Londini. 1555." In the ſame compartment as Har-
dyng's Chronicle printed by Grafton.—"Pars Eſtiualis." Theſe have
the "Pſalterium Dauidicum," and the "Proprium de Sanctis," annexed
to them both, as appendages. Colophon, "Breuiarium ſeu Por-
tiforium—Londini impreſſum, Per Johannem Kyngſton et Henricum Sutton typographos.—Die vero ſeptima menſis Martij." W.H. Again 1556.
Quarto.

"See an account of theſe Roman church-books in Gough's Britiſh
Topography, Vol. II. p. 319, &c.

"The Fardle of facions conteining the aunciente maners, cuſtomes, & 1555.
Lawes of the peoples, enhabiting the two partes of the earth, called Affrike
& Aſie. Printed at London, by Jhon Kingſtone, & Henry Sutton." In a
compartment with a cherubic head at top, and the date, 1555, on the
fell. It is dedicated "To——the Erle of Arundel, Knight of the ordre,
& Lorde Stewarde of the—houſeholde." by William Waterman. Contains Z 3, in eights." Imprinted——by John Kyngſton and Henrie Sutton. The. xxii. daye of December. Anno Domini. M.D.LV." W.H.
Octavo.

"The Primer in Engliſhe and Latine, ſet out all along after the vſe of 1557.
Sarum: with many godlie and deuoute praiers: as it appeareth in the
table. Jmprinted—by Jhon Kyngſton and Henry Sutton. 1557. Cum
priuilegio

* Thus in the original.

1557. "The whetſtone of witte, whiche is the ſeconde parte of Arithmetike: containing thextraction of Rootes: The Coſſike practiſe, with the rule of Equation: and the woorkes of Surde Nombers. Theſe Bookes are to bee ſolde at the Weſte doore of Poules by Jhon Kyngſtone." Dedicated "To the right worſhipfull, the gouerners, Conſulles, and the reſte of the companie of venturers into Moſcouia, *by* Robert Recorde Phiſitian.—At London the. xii. daie of Nouember. 1557." Then, the preface; after which are two octave ſtanzas, one "Of the rule of Coſe," the other "To the curiouſe ſcanner." Contains beſides, R, in fours. "Jmprinted,—by Jhon Kynſton. Anno domini. 1557. W.H." Quarto.

1548. "The ſeuen firſt bookes of the eneidos of Virgill, conuerted into Engliſhe meter by Thomas Phaer eſquier, ſollicitour to the king and quenes maieſties, attending to their honorable conſaile in the marchies of Wales. 28 Mai." Dedicated to queen Mary. See Warton's Hiſt. of Eng. Poetry, Vol. III. p. 396, note *. Quarto.

1558. The ſecrets of Alexis, &c. See them printed for John Wight. Quarto.

1559. "The Chronicle of Fabian, whiche he nameth the concordaunce of of hiſtories, newly peruſed. And continued from the beginnyng of Kyng Henry the feuenth to thende of Quene Mary. 1559. Menſe Aprilis. Jmprinted—by Jhon Kyngſton." In the compartment with king Edw. vi. in council at top, and Grafton's rebus at bottom. On the back, "The Printer to the ᵇReader. Then, "The Table for the Firſte parte," which contains 369 pages, beſides.

ᵇ "Becauſe the laſt printe of Fabians Chronicle was in many places altered from the firſt copie, J haue cauſed it to be conferred with the firſt print of all, and ſet it foorthe in all pointes accordyng to the aucthours meanyng. Alſo al through the ſtorie of the Britons, wherin he followeth Geffrey of Monmouth, J haue cauſed his ſtorie to be conferred with Gefferies, & noted the chapiters in the margine, where out the matter is taken, and ſuch thinges as he (J wote not for what cauſe) omitted, J haue cauſed to be tranſlated, & duely placed under the peruſers *(Rob. Record)* name. The like haue J doen, for the moſte parte, all the boke through, notyng the places of ſoche aucthours as he allegeth. And bicauſe the controuerſie & varieties is greate emong writers aboute the nomber of yeres from Adams creation to Chriſtes incarnacion, therfore as maſter Fabian followeth the Septuaginta & ſainct Bede. So haue J in the margine added the accompt of Jhon Functius & other all through the Storie, till the beginnyng of our Sauiour, to the ende you may know the diuerſities of them, and lacke nothing neceſſarie for the truthe: J haue alſo continued the ſtorie from Fabians tyme till the ende of our late ſoueraigne quene Maries. Briefly touchyng the ſpeciall matters that haue happened therein. And if J haue in any place miſtaken ought, or one letter be ſkaped for another, J beſeche you of your gentlenes to amende it. Thus J praie God farther you in all good ſtudies."

The

JOHN KINGSTON.

"The feconde volume of Fabians Chronicle, conteinyng the Chronicles of Englande & of Fraunce from the beginning of the reigne of king Richard the firſte, vntill the eande of the reigne of Charles the nineth." In the fame compartment as the firſt part. On the back is the author's "Lenuoie," in three ſeven-lined ſtanzas. This part is introduced with an account of "The wardes of London," wherein the pariſh churches are enumerated 112; the cathedral, chapels, and other churches within the city, 19; without the city, 24; in all, 155. This part contains 571 pages. W.H. See the former editions, p. 265, 415, 480, 592.   Folio.

"A VVoorke of Joannes Ferrarivs Montanus, touchynge the good orderynge of a common weale:" &c. as p. 780.   1559.

"The Arte of Rhetorique by Thomas Wilſon, 1553. And now newlie ſet forth again, with a prologue to the Reader. Anno Domini M.D.LX." See p. 537. Again 1567, and 1584.   Quarto.   1560.

"The Arte of warre", &c. as p. 784. "Anno M.D.LX." At the end, the cut of the Genius of England, as p. 780. To this is annexed, "Certain waies for orderyng Souldiers in battelray," &c. as p. 784. Colophon, "Jmprinted—by Ihon Kingſton: for Nicolas Englande. Anno ſalutis, M.D.LXII." On the back, the Genius of England, which ſeems to have been N. England's device. W.H.   Quarto.   1560.

"The woorkes of Geffrey Chaucer, newlie printed with diuers addicions, whiche were neuer in print before: With the ſiege and deſtruction of the worthy Citee of Thebes, compiled by Jhon Lidgate, Monk of Berie. As in the table more plainly doeth appere.[1] 1561." Chaucer's works end on Fol. ccclxxviij. "Jmprinted—by Jhon Kyngſton for Jhon Wight, dwellyng in Poules Churchyarde. Anno. 1561." W.H. Folio.   1561.

"Bulleins Bulwarke of defẽce againſte all Sicknes, Sornes, and woundes that dooe daily aſſaulte mankinde, whiche Bulwarke is kepte with Hillarius the Gardiner, Health the Phiſician, with their Chyrurgian, to helpe the wounded ſoldiors. Gathered and practiſed frõ the moſte worthie learned, bothe old and newe: to the greate comforte of mankinde: Doen by Williyam Bulleyn, and ended this Marche. Anno ſalutis, 1562. Jmprinted— by Jhon Kyngſton." In the ſame compartment as Eraſmus's Paraphraſe, p. 544; but the arms of queen Catharine Parr, and the E. W. are taken out. Dedicated "To the right hon. Lorde Henry Cary, Baron of Hunſdon, Knight of the moſte noble order of the Garter;" whoſe arms are on the back of the title. Then, an epiſtle to the reader; and a liſt of "The Aucthours, Capitaines, and Souldiours of this Bulwarke." The firſt part,   1562.

---

[1] This title differing in orthography from that inſerted by Mr. Ames, ſeems to corroberate Mr. Bagford's aſſertion, that Chaucer's works were printed by John Kingſton and George Biſhop, in 1561; and by Adam Iſlip the year following: but Biſhop was not made free till Apr. 1562; nor Iſlip till June, 1585.

or

or " The boke of Simples" contains Fol. xc, befides an index, with cuts of plants and ftills. " ¶ Here after infueth a little Dialogue, between twoo men, the one called Sorenes, and the other Chyrurgj : Concernyng Apoftumacions, & woundes, their caufes, and alfo their cures," &c. This part contains Fol. xlviij, and an index, with " The Anatomie" of the bones. Next is " The booke of Compoundes." This contains 48 leaves, befides a table, and " the Apoticaries rules," at the end. " Hereafter foloweth the booke of the vfe of fickmen, and wholfome medicen." A cut of an infirm man, with a walking ftick, which fome fuppofe to be Bullein's portrait. The leaves of this part are numbered from xlviij. to Fol. lxxxij ; then, the index. It does not appear that he had licenfe for this before 2 Mar. 1578-9. It was reprinted by Tho. Marfh, in 1579. W.H. Folio.

1562. Regiment againft the Plurifie ; by Will. Bullein. Printed by Joh. Kingftone. Maunfell, Part II. p. 5. Octavo.

1563. The rule of reafon, containing the art of logick. Again 1567. Quarto.

1563. " A booke called the Foundacion of Rhetorike, becaufe all other partes of Rhetorike are grounded thereupon, euery parte fette forthe in an Oracion vpon queftions, verie profitable to bee knowen & redde : Made by Richard Rainolde Maifter of Arte, of the Vniuerfitie of Cambridge. 1563. Menfe Marcij. vj. Jmprinted—by Jhon Kingfton." In the compartment with a medallion at top. Dedicated " To—Lorde Robert Dudley Maifter of the Queenes Maiefties horfe, one of her highnes priuie Counfaile, and knight of—the Garter." An epiftle " To the Reader." In which, "—becaufe as yet the verie grounde of Rhetorike is not heretofore intreated of, as concernyng thefe exercifes, though in fewe yeres paft a learned woorke of Rhetorike is compiled & made in the Englifhe tongue, of one, who floweth in all excellencie of arte, who in iudgement is profounde, in wifedome & eloquence mofte famous." &c. Then, " The contentes of this Booke." Contains befides, Fol lxij. Licenfed. W.H. Quarto.

1564. " APOPHTHEGMES, that is to faie, prompte, quicke, wittie & fentecious faiynges, of certaine Emperours, Kynges, Capitaines, Philofophers & Oratours, afwell Grekes, as Romaines, bothe verie pleafaunt & profitable to reade, partely for all maner of perfones, and efpecially Gentlemenne. Firft gathered & compiled in Latin by the right famous clerke Maifter Erafmus of Roterodame. And now tranflated into Englifhe by Nicolas Vdall. Jmprinted—Menf. Februarij. 1564." Firft is an epiftle by " Nicolas Vdall vnto the gentle & honefte harted readers,—Written in the yeare of our Lorde God. M.D.XLII." Then, " The Preface of Def. Erafmus Roter. vnto a Duke's foonne of his Countrey.—Yeuen at Friburge, the. 26. daie of Februarie,——M.D.XXXJ." Contains befides, Fol. 245, and a table. " Thefe bookes are to bee folde at his Shoppe, at the Wefte doore of Paules." Licenfed. W.H. Octavo.

" The

JOHN KINGSTON.

" The Hope of Health wherin is conteined a goodlie regimente of life: 1564.
as medicine, good diet, and the goodlie vertues of fondrie Herbes, doen
by Philip Moore. Jmprinted—Maii, Anno falutis M D lxiiii." Dedicated
to fir Owen Hopton, knight. On the back of the leaf which concludes
the dedication, is a letter, in Latin, from Will. Bullein to the author.
William Bayntun Efq; Again, November, 1565. * Octavo.

" A Manuell of the Chronicles of Englande. From the creacion of the 1565.
worlde, to the yere of our Lorde 1565. Abridged & collected by Richard
Grafton. Jmprinted at London by J- Kingfton." On the back is " The contentes of this booke." It has a kalendar prefixed, in which the evil and
unfortunate days, and fuch as are not altogether fo evil are noted; alfo " An
Almanacke for xviii. yeres." Then Grafton's dedication " To his loving
frendes the Mafter & wardens of the companie of the mofte excellent
Arte & fcience of Jmprintyng." Of which, and the epiftle to the reader,
fee p. 503. The Manuell on 100 leaves. At the end, a lift of the principal fairs, and an index. W.H. Small.

" A brief requeft or declaracion, prefented vnto madame the ducheffe 1566.
of Parme, &c. regente of the Lowe-Countrie of Flaunders, by the lords
and nobilitie of the fame: with the anfwere and reply. Englifhed
by W. F." *

" The Rule of Reafon, conteinyng the Arte of Logike. Sette foorthe 1567.
in Englifhe, and newlie corrected by Thomas wilfon. Anno Domini.
M. D. LXVIJ. Menf. Feb. Imprinted", &c. In a compartment with Mofes
and David in the fides. Dedicated to king Edw. VI. Then five diftichs
of Latin verfes, by Dr. Gual. Haddon, and two diftichs by the author.
Contains befides, 90 leaves, including a table at the end. W.H.
Quarto.

" The Arte of Rhetorike, for the vfe of all fuche as are ftudious of 1567.
Eloquence, fette forthe in Englifhe by Thomas wilfon. 1553. And now
newlie fette foorthe againe, with a Prologue to the Reader.—1567. Imprinted", &c. In the fame compartment as the laft article. Dedicated,
&c. as p. 537. Afterwards, " A Prologue to the Reader.—This feuenth of
December Anno Dnī. 1560." In which the author gives an account of
his being imprifoned at Rome, about 2 years before, charged with being an
heretic for writing thefe two books; and fays, " If others neuer gette more
by bookes then J haue doen: it wer better be a Carter then a Scholer,
for worldlie profite." Next, " The Preface," with this title, " Eloquence
firfte given by God, and after lofte by manne, and lafte repaired by God
again." Then, a page of Latin verfes by Dr. Haddon, and the author.
The treatife, with a table, on 116 leaves. W.H. Both frequently printed.
Quarto.

" A Briefe and eafye inftruction to learne the tableture, to conducte 1568.
and difpofe thy hande vnto the Lute, englifhed by J. Alford Londenor."
A cut of the Lute. " Imprinted—for Iames Roubothum, and are to be
folde at hys fhop in pater nofter rowe. Lycenfed accordynge to the order,
&c.

5 F

JOHN KINGSTON.

&c. 1568." A short preface by the author, and another by the translator, dated 24 Sept. 1568. W.H. Again, with additions, in 1574. Broad quarto.

1569. Bullein's Regiment against the plague. See it in 1578.. Octavo.

1571. C. Crispi Salustii—historia. Octavo.

1571. "The Foreste or collection of Histories,—by Tho. Fortescue". For W. Jones. See p. 659. Quarto.

1571. "Spiritus est Vicarius Christi in terra. A breefe & pithie summe of the christian faith made in fourme of a confession, with a confutation of the papistes obiections & argumentes in sundry pointes of religion, repugnaunt to the christian faith: made by John Northbrooke, Minister & Preacher of the worde of God.—Seene & allowed, according to the order appointed in the Queenes Iniunctions. Printed—for W. Williamson,—.. Anno. 1571." In a border. Dedicated to "Gylbert—Bishop of Bathe & Welles.—From Redcliffe in Bristoll." Then, a copious preface, and a table of the contents. Contains besides, 143 leaves, and an index. At the end, a cut of the sun, with M at top, I and S on the sides, and W at bottom. W.H. Quarto.

1571. "A confutation of a Popishe, & sclanderous libelle, in forme of an apologie: geuen oute into the courte, and spread abrode in diuerse other places of the Realme. Written by William Fulke, Bachelor in Diuinitie, and felowe of S. Johns Colledge in Cambridge. Imprinted—, for Will. Jones,—1571." Dedicated "To the right hon. & vertuous Ladie, the Ladie Margaret Strange." Contains besides, 116 leaves. W H. Octavo.

1572. "Three faithfull sermons made by Thomas Lever, anno Domini 1550, and now newlie perused by the author: first made in the shroudes in London: second, before the king and councell: third, at Pauls Crofs." See p. 623, &c. Octavo.

1573. M. T. Ciceronis de oratore. Octavo.

1574. "A briefe and plaine Instruction to set all Musicke of eight diuers tunes in Tableture for the Lute. With a briefe Instruction how to play on the Lute by Tablature, to conduct and dispose thy hand vnto the Lute, with certaine easie lessons for that purpose. And also a third Booke containing diuers new excellent tunes. All first vvritten in French by Adrian Le Roy, and now translated into English by F. Ke. Gentelman. Imprinted at London by Iames Rowbothome, and are to be sold in Pater noster row at the signe of the Lute. Anno. 1574." In a border. Dedicated "To the Right Hon.—Lord Edward Seamour, Viscount Beauchamp, Erle of Hertford," by J. R. Also, "To—the Countesse of Retz, by Adr. le Roy." Then, "The Preface of Jacques Gohory," in praise of the author, his work, the lute, and music of Orlando de Lassis. The whole 78 leaves. On the last page " Imprinted—by Ihon Kyngston for James Robothome. Anno. 1574." W.H. Broad quarto.

"Principia

" Principia Latine loquendi scribendique, siue selecta quaedam ex 1575.
Ciceronis epistolis, ad pueros in Latina lingua exercendos, adiecta interpretatione Anglica, et (vbi opus esse visum est) Latina declaratione. Ad
rationem, quam nuper suis sedulitate summa Gallicè conscripsit Maturinus
Corderius." Translated by T. W. anno domini 1575.—for Oliver Wilkes.
Octavo.
" Petri Martyris Vermilii, Florentini, praestantissimi nostra aeta:e the- 1576.
ologi, loci communes. Ex variis ipsius auctthoris libris in vnum volumen
collecti, & in quatuor classes distributi." A burning bush within a neat
compartment adorns the title-page.   Folio.

Virgilii Opera.   1577.

" M. T. Ciceronis quaestiones Tusculanae per Erasinum.   Octavo. 1577.

" A Dialogue bothe pleasante & pittifull, wherin is shewed a goodlie 1578.
regimente against the feuer of pestilence, with a consolation & comfort
against death. Newlie corrected by W. Bullein, the author thereof." Licensed in 1563.   Octavo.
" An INVECTIVE Againste vices, taken for Vertue. Gathered out of 1579.
the scriptures, by the vnprofitable seruant of Jesus Christe, Richard Rice.
Also certaine necessary instructions, meet to be taught the younger sort,
before thei come to be partakers of the holy Communion. Doen by
D. w. Arch.——Seen & allowed.—Reade & iudge. But condeme not
before ye reade. Imprinted—for Henry Kirkham. 1579." In a border. It
is introduced with an address " To the Christian Reader, by Rob. Crowley"; and another by the author. E 5, in eights. W.H.   Sixteens.

This year a dispute, which began in 1573, was renewed between him 1579.
and Tho. Marshe, concerning the copy-right of " Esopes fables, Holie
Dialoges, and thepitome of Colloquies," which was submitted to the
Master, Wardens, and Court of Assistants. Their case, and the decision
may be seen below.[k]

" A blazyng Starre or burnyng Beacon, seene the 10. of October laste 1580.
(and yet continewyng) set on fire by Gods prouidence, to call all sinners

[k] By a Court held 22 May 1573, it appears that Kingston was then appointed to print the said three books, until such time as a certain sum of money was paid him, " and now pretendyd by the said Marshe to be released to him by meanes of a generall quitance, which he affyrmeth to haue obteyned at the handes of Kingston. For the appeasinge of which debates & controuersies, and for an vnitie & frendship to be had & contynued betwixt the said parties, Yt, is this last day of August 1579 ordered & decreed by the Master, Wardens, and Assistants of this cumpanye, by & with the assent & submission of the said parties in this behalf, that the said former order shall from henceforth stand voyd, And that the bondes wherein the said parties are bound for the performance thereof shall be cancelled. Jtem, that the said Tho. Marshe shall from henceforth enioye the printing of the said three bookes without leit of Kingston, Jtem, that the said Tho. Marshe shall pay vnto the Wardens of this Cumpany to the vse of the said Kingston L.s. ymedyately, and other L. s. before Hellentyde next." Both which sums were paid accordingly.

5 F 2   to

to earneſt & ſpeedie repentance. Written by Francis Shakelton, Miniſter & preacher of the worde of God." The cut of a blazing ſtar, or comet. "I will ſhewe wonders in heauen, &c. Joel 2. 30. Imprinted—for Henry Kirkham, and are to bee ſolde at the ſigne of the blacke boye, adioynyng to the little Northdore of ſainct Paule. 1580." Dedicated "To—ſir Thomas Bromley knight, lord Chauncellour of Englande," &c. Then, "In hanc Franciſci Shakeltonj Cometologiam Ogdoaſtichon.—Tho. Newton, Ceſtreſhyrius." Contains beſides, D 7, in eights. W.H. Sixteens.

1580. The 1ſt and 2d parts of the ſecrets of M. Alexis of Piemont. Quarto.

1580. The catechiſm to teach children, &c. By John Calvin. He bought the copy of W. Copland for 5 s. Licenſed in 1565. See it in 1582. Octavo.

1580. "The rule of reaſon,—the art of logike," &c. Pages 221, beſides the table. See it in 1567. Quarto.

1580. "A pollitique platt for the honour of the prince, the great profitt of the publique ſtate, relief of the poore, preſeruation of the rich, reformation of roges and idle perſons, and the wealth of thouſands, that know not how to liue. Written for a new yeres gift to England, and the inhabitants thereof, by Robert Hitchcok, late of Cauerfeld in the countie of Buckyngham, gentleman. Addreſſed to the parliament, 1ſt of Januarie." With Hitchcok's arms on the back of the title, and large wooden cuts. Licenſed. Quarto.

1581. "A briefe and pleaſaunt treatiſe, entituled Naturall & Artificiall concluſions: Written firſt by ſundrie ſcholers of the Vniuerſitie of Padua in Italie, at the inſtant requeſt of one Barthelmewe a Tuſcane: And now Engliſhed by Tho. Hill Londoner, as well for the commoditie of ſundrie Artificers, as for the matters of pleaſure to recreate wittes at vacant tymes." The cut of a candle burning in the mouth of a bottle. The ſame is uſed again on leaf C 7, with this title, "How to make a candle to be maruailed at, a proper ſecrete." Under the cut in the title-page, "Imprinted—for Abraham Kitſon, 1581." On the back is "The Preface to the Reader," ſeemingly by the bookſeller: wherein he ſays that this book had been promiſed in a former publication of "The interpretation of dreames, bothe of Joſephes & Salomons together," and on taking this in good part, promiſes ſhortly to publiſh his "treatiſe of Problems, with anſwers annexed." Alſo, "a breif Herbal verie rare for the meruailous matters, &c. And a treatiſe of the rare & ſtraunge wonders ſeen in the aire for theſe many yeres paſte in ſundrie Realmes," &c. Contains D, in eights. W.H. Sixteens.

1582. THE Catechiſme, or maner to teache Children the Chriſtian Religion. Made by the excellent doctor and paſtor in Chriſtes Churche Jhon Caluin. wherein the miniſter demaundeth the Queſtion, and the Childe maketh Aunſwere. Ephes. 2. The doctrine of the Apoſtles & Prophetes is the foundation of Chriſtes Churche. Imprinted at London by Ihon Kyngſton."

ston." In the compartment with a cherubic head at top, and the date 1582 in the fell. This catechism is divided into "55 Sondais." After which is "The maner to examine children before they be admitted to the Supper of the Lorde." Then, sundry forms of prayer, confessions, &c. K, in eights. W.H. Octavo.

"A COMPENDium of the rationall Secretes of the worthie Knight and 1582. moste excellent Doctour of Phisicke & Chirurgerie, Leonardo Phiorauante Bolognese, deuided into three Bookes. Jn the first is shewed many secretes apperteining vnto Phisicke. Jn the seconde is shewed many secretes apperteining vnto Chirurgerie, with their vses. Jn the third is shewed diuers compositions, apperteinyng bothe to Phisicke & Chirurgerie, with the hidden vertues of sondrie vegitables, animalles, & mineralls, and proued well by this author, heretofore neuer set out before. Jmprinted—for George Pen, and I. H. 1582." Dedicated "To the worshipfull & vertuous louer of learnyng Maister Richard Garth Esquire, one of the Clarkes of the Pettie Bagges in the Chauncerie —Your old & poore I. Hester." Next, his addresse "To the Courteous Reader," concluding thus: "For the receiptes in this Booke specified, as also for many other rare thynges mentioned els where: if any be disposed to vse them, let them repaire to my house at Poules wharfe, where they shall either finde them readie made, or me at reasonable warnyng readie to make them—without sophistication.—I. H." Then, tables for the three books. Contains besides, 142 pages, the last closing with a colophon, as in the title-page; then, on another leaf, "The vertue of this Balme"; meaning probably the last recipe in the book. At the end, "This Balme is to bee solde at Poules Wharfe, at the Signe of the Furnace, by Ihon Hester." W.H. Sixteens.

Injunctions¹—by the Queen's majesty. 1583.

"A Tragicall Historie of the troubles & Ciuile warres of the lowe (1583.) Countries, otherwise called Flanders.—And there withall, the Estate and cause of Religion, especially from the yere 1559. vnto the yere 1581. Besides many Letters, Commissions, Contracts of Peace, &c. published & Proclamed in the saied Prouinces. Translated out of French into Englishe, by T. S. gēt. Jmprinted—for Tobie Smith, dwelling in Paules Churchyarde, at the signe of the Crane." Dedicated "To—the Lorde Robert Dudley, Erle of Leicester, Baron of Denbigh, Knight of the moste noble Order of either Garter, and of sainct Michaels, Maister of her Maiesties Horses, & one of her Highnesse moste honorable Priuie Counsell.—London the xv. of Marche 1583.—Thomas Stocker." Then, the dedi-

¹ "7. October. Whereas John Kingston hath printed of her maiesties Jniunctions to the number of XXX. c. as he affyrmeth vpon his credyt, and not aboue. Yt is ordered that he shall bringe in all the first leaues thereof to Mr. Barker, who will print the same again for Kingston with the vinyate & marke of Mr. Barker to yt. And this impression finished & vttered, the said K. neuer to reprint the said booke. By me Ihon Kyngston." Stat. Reg. B. fol. 436.

cation.

cation of the original " To the high, noble, hon. & wife Lordes—of the Eftates, the Deputies, Prefidentes, and Counfelles, Burroughmaifters, Scoutes or Marfhalles, Maiors, Bailiefes, and to al other Officers, &c. vnited to the lowe Countries.—Theophile D.L." The hiftory is divided into four books; the third book ends on folio 139; the fourth book contains 67 leaves. At the end, a table of the whole. W.H. Quarto.

1584. " The Arte of Rhetorike, &c. By Tho. Wilfon." 225 pages. See it in 1567. Quarto.

1584. " Monardo: The Tritameron of Loue: wherein certain pleafant Conceits vttered by diuers worthy Perfonages are perfectly dyfcourfed, and three doubtfull Queftions of Loue pleafantly difcuffed: fhewing to the Wife, how to vfe Loue, and to the Fond, how to efchew Luft, and yielding to all both Pleafure & Profit. By Rob. Greene, Mafter of Arts, in Cambridge." Dedicated to Phillip, Earl of Arundell. In 23 leaves. Catal. of Pamphlets in the Harleian Library, No. 10. Quarto.

—— " An inuectiue againfte vices taken for Vertue," as in 1579. L 2, in fours, half fheets. On the laft leaf " Thefe bookes are to bee folde at the figne of the blacke Boye at the little Northe doore of Poules." On the back, H K, the initials of Henry Kirkham, and the fun with radiant beams over them; alfo fprigs of a rofe and a fun-flower. W.H. Octavo.

—— " Iniunctions giuen by the Queenes Maieftie—the firft year of our foueraigne Lady Queene Elizabeth." Alfo,

—— " Articles to be enquired in the vifitation, in the firft year of—Elizabeth, &c. Jmprinted at London by Jhon Kyngfton, by the affent of Chr. Barker." John Fenn, Efq; Quarto.

—— " The Commentaries of M. Jhon Caluin vpon the firft Epiftle of Sainct Jhon, and vpon the Epiftle of Jude: wherein accordyng to the truthe of the woordes of the holie Ghoft, he moft excellently openeth & cleareth the poinct of our iuftification with God, and fanctification by the Spirit of Chrift, by the effects that he bryngeth forthe in the regineration. Tranflated into Englifhe by W.H. Pfal. 129. 5. Prou. 19. 14. Pfal. 32. 11. Jmprinted—for Jhon Harrifon the yonger." In a border. Dedicated " To the Worfhipfull—Maifter VVilliam Swan of Wye in Kent, and to the right Vertuous & Chriftian Gentlewoman Miftreffe Amy Swan his Wife, with all thofe that in the truth of a fingle hart loue the Lord Jefus in that Congregation." The comment on S. John in 108 leaves; that on S. Jude in 25. W.H. Octavo.

—— " The feconde part of the Secretes of M. Alexis," &c. as p. 785. Jmprinted—by Jhon Kyngfton for Ihon Wight. W.H. Quarto.

—— " A defence of prieftes mariages," &c. as p. 726. See Strype's Life of Archb. Parker, p. 504, &c.

" The

JOHN KINGSTON.

"Tho. Brice his regifter of the Martirs that fuffered in queene Maries time.—for Richard Adams." Maunfell, p. 23. Octavo.

Befides the books abovementioned, he had licenfe to print the following. In 1562, "An almanacke of M. Doctor harycok, fometyme pryor of f. auguftyens fryer in norwyche." In 1564, "The ftory of Jobe, the faythfull feruaunte of God." In 1565, "An Epytaph of Mrs. Afthelay, made by Henry Towers." In 1566, he was fined for printing "a boke intitled The voyce of God." He had licenfe for printing "An expofition of faynt John baptefte." In 1567, "An almanacke & pronofticcation of Tho. Jenkynfon, for 1568." In 1569, "A godly meditation in myter for the preferuation of the quenes maieftie, *and* for peace." Sept. 3, 1579, "Mofelius & Sufinbrotus figures." Nov. 10, "Churchyardes chaunce." Apr. 11, 1580, "A thinge in verfe of thearthquake." July 13, "An epitaph vpon the deathe of Sir Alex. Auenon."

HENRY SUTTON

APPEARS to have been an original member of the Stationers' company, as he prefented an apprentice 1 Oct, 1556, though his name is not in the lift annexed to their charter, a copy of which Mr. Ames procured from the Rolls. He had a fhop in S. Paul's churchyard, and dwelt in Paternofter row at the fign of the Black Boy, and other places. During Q. Mary's reign he printed chiefly in connection with John Kingfton, efpecially the Romifh church books, under whofe name they are placed. In 1561, he was fined "iiij.s. iiij.d. for ferten bokes taken hawking aboute the ftretes—contrary to the orders of the Citye of London." In 1571, he bound an apprentice, after which i find no more of him.

"BREVIS ET PERSPICVA RATIO IVDICANDI GENITVRAS EX PHYSICIS 1558. CAVSIS, et varia experientia extructa: & ea Methodo tradita, vt quiuis facilè, in genere omnium Thematum iuditia inde colligere poffit: Cypriano Leouitio à Leonicia excellente Mathematico Authore. Praefixa eft Admonitio de vero & licito Aftrologiæ vfu: per Hieronymum VVolfium, virum in omni humaniore literatura, linguarum artiumq; Mathematicarum cognitione præftantem, in Dialogo confcripta. Adiectus eft præterea libellus de præftantioribus quibufdam Naturæ virtutibus: Ioanne Dee Londinenfe Authore. Londini. Anno. M.D.LVIII. Menfe Julio." The characters of all the planets, and their afpects, are ufed in this treatife, which contains R 2, in fours. The additional tract feems to have been defigned for a feparate publication, having its peculiar title-page, and

and series of signatures. This title-page is engraved on copper, the only one I have met with besides The Compendium Anatomiæ, p. 577.

"ΠΡΟΠΑΙΔΕΎΜΑΤΑ ΑΦΟΡΙΣΤΙΚ'Α ΙΟΑΝΝΙS DEE LONDINENSIS, de Præstantioribus quibusdam NATURÆ virtutibus, ad GERARDVM MERCATOREM RVPELMVNdanum, Mathamaticū, & Philosophum insignem." This title is within an architective compartment, over a shield with the character of Mercury, and I. D. on it. A waved ribband on the left side, with "Est in hac Monade quicquid quærunt sapientes," another on the right with "ΣΤΙΑΒΩΝ, acumine præditus, est instar omnium Planetarum." On the lintel, "Qui non intelligit, aut taceat aut discat." On the jambs, (pilasters of the Corinthian order,) are the figure of the sun on the one, and that of the moon on the other: over the former pilaster is the word, *Calidum*; and over the other, *Humidum*: on the base of the former, *Terra*; and on the other base, *Aer*. On the sell, " Erunt signa in Sole, & Luna, & Stellis. Luca. 21." It is dedicated, as the title intimates, " Clarissimo viro D. Gerardo Mercatori, Rupelmundano, Philosopho. & Mathematico illustri: ac amico suo longè charissimo." Herein he enumerates eleven books written by him, ten of which are inserted in the list of Dee's unpublished pieces, in Biographia Brit. p. 1642, viz. articles 6, 10, 15, 16, 17, 18, 19, 20, 22, 23; but the tenth here enumerated, " De Religione Christiana lib. 6. demōst," seems to have been unknown, or overlooked. This Epistle dedicatory is dated "Londini: à nostro nato Redemptore 1558. Julii. 20." The treatise consists of 120 aphorisms. " Londini. Excudebat Henricus Suttonus, Impensis Nicholai England, Anno à virgineo partu, 1558. Mense Julio." Alexander Dalrymple Esq; Quarto. A copy of this last tract without " Impensis N- England. W.H.

Mr. Ames says, several more of Dee's pieces were printed by him.

1559. " Almanacke for the yeare 1559. composed by Mayster Mych. Nostrodamus Dr. of Phisicke.—by him, for Luc. Haryson, the xx. of Feb. MDLIX." Mr. T. Baker mentions this as the first he had seen. In 1561, he was fined iiij. s. for printing it without licence. His Prognostication for the same year, with predictions and presages for every month, was printed at Antwerp. Octavo.

1559. " THE SECRETES OF THE REVERENDE MAYSTER ALEXIS OF PIEMOUNT," &c. as p. 781. "A worke wel approued, verye profitable and necessary for euery man. Translated out of Frenche—by Wyllyam Warde. Londini. Anno—M.D.LIX. XII. die Mens. Nouemb." This is the first edition in English, containing only the first part. At the end, " Imprinted—By H. Sutton, dvvellyng in Pater noster rovve, at the signe of the blacke Moryan. Anno—1559." W.H. Quarto.

1559. " The examination of the constante martir of Christ, John Philpot, archdeacon of Winchester, at sundry seasons, in the tyme of hys sore imprisonement, conuented and baited, as in these particular tragedies folowynge it maye (not only to the christen instruction, but also to the mery recreation of the indifferent reader) moste manifestly appeare." Quarto.

An

An edition of this without printer's name, or date; at the end of which is "An Apologie of Johan Philpot, written for fpitting vpon an Arrian, with an inuectiue againft the Arrians," &c. Colophon, " Imprynted at London by Henry Sutton." For printing this without licenfe he was fined vs. Octavo.

" A Treatife containing A Declaration of the Popes vfurped Primacie: 1560. written in Greeke aboue feuen hundred yeres fince by Nilus Archbifhop of Theffalonica. Tranflated by Thomas Greffop ftudent in Oxford. Printed —for Raph Newbery." Octavo.

" Antiprognofticon contra inutiles aftrologorum praedictiones, Noftro- 1560. dami, Cuninghami, Loui, Hilli, Vaghami, et reliquorum omnium. Authore Gulielmo Fulcane. Authoritate Londinenfis epifcopi, juxta formam in edictis reginae prefcriptam. Sexto die Septembris." At the end, " impenfis Humfredi Toii." Octavo.

" Antiprognofticon, that is to faye, an inuectiue agaynft the vayne and 1561. vnprofitable predictions of the aftrologians, as Noftrodame, &c .Tranflated out of Latine into Englifhe. Whereunto is added by the author, a fhort treatife in Englifhe, as well for the better fubuerfion of that fained arte, as alfo for the better vnderftanding of the common people, vnto whom the fyrft labour feemeth not fufficient." Octavo.

" A ftrong battery againft the Jdolatrous inuocation of the dead Saintes, 1562. and againft the hauyng or fetting vp of Jmages in the houfe of prayer, or in any other place where there is any peril of Jdolatrye, made dialoguewife by John Veron, Pfal. l. Calle vppon me in the daye of trouble, & I wil hear thee. Iohn. xi. I am the way,—no man comes to the father but by me." In the compartment with W. Seres's mark on the fides. Contains P 6, in eights, befides a preface, infcribed " To—fir Nicholas Bacon knight Lord keper of the great Seale of Englande," and a table, prefixed. Colophon, " Imprinted—for Tho. Hacket. The x. daye of Marche. —1562." W.H. Octavo.

" The Deftruction and facke cruelly committed by the duke of Guyfe 1562. and his company, in the towne of Vaffy, the fyrfte of Marche, M.D.LXII. Imprinted for Edw. Sutton.——The firft day of May." Octavo.

" Toxaris, or the frendfhyp of Lucian, tranflated out of Greke into 1565; Englifh. With a dedication to his frende A. S. from A. O." For Edward Sutton. Octavo.

" A Difcourfe vpon the Libertie, or Captiuitie of the King." (of France.) For Edw. Sutton. Octavo.

" Camelles Reioindre to Churchyarde, *or* Camelles Conclufion.
    " Camelles Conclufion, and laft farewell then,
    To Churchyarde and thofe that defende his when.
  " A man that hath mo thynges then two, to put him vnto paines,
  Hath euen fo many cares the mo to worke hym wery braynes.

So I that late haue laboured harde, and plucked at my pw,
Am come to towne, where nowe I fynde mo matters the yno we.
Mo then I looked for by muche, mo matters to then needes,
Mo makynges & mo medlynges far, then I haue herbes or wedes.
And all agaynſt me one alone, a ſory ſymple man,
That toyles and trauailes for my foode, to erne it as I can."

And ſo proceeds in this ſort of verſe on four ſides folio; and ends, " T. Camel.

" The hartburne I owe you is, yf you come to Lynne,
I pray you to take my poore howſe for your ynne.
Jmprinted by Hary Sutton dwelling in Poules Churchyarde at the ſign of the blacke boye."

His anſwer " To Goodman Chappels ſupplication," is contained in 20 lines of a very odd kind of Engliſh Poetry, ſpelt ſtrangely, and ſigned, " Thomas Camell." M.S. Joſ. Ames. Broadſide.

Beſides the books abovementioned, he had licenſe to print the following, viz. In 1557, " An entertude vpon the hiſtorye of Jacobe & Eſawe. The Courte of venus." In 1558, "Queſtions for Children, *deſigned to partake of the Lords ſupper.*" See p. 840. In 1559, " An almanack & pronoſtication of Lowe." In 1560, " An almanack & pronoſtication of Fulkes. iiijor ſtoryes of the ſcripture in myter." In 1562, " Taverners Poſtell. The brefe dyxcyonary. The boke of medycyne."

※※※*※※※※※※※※*※※*※※※*※※※※※※※※*※※※

THOMAS MARSHE, or MARSH,

AN original member of the Stationers' company, was taken into the Livery in 1562, ſerved Junior Collector in 1567, and Renter in 1568, Under warden in 1575, Upper warden in 1581. His behaviour in general ſeems to have been diſorderly, diſregarding the Company's ordinances, for which he was frequently fined. His character likewiſe does not appear to advantage in the controverſy between him and Kingſton. See p. 839, note ª. Nor in that with Vautrollier's wife, while her huſband was abroad on ſpecial affairs, which was compromiſed as' below. He dwelt in Fleet-ſtreet, at the ſign of the

" 17 July, 1581. Yt is agreed that Tho. Vautrollier his wife ſhall finiſhe this preſent ympreſſion which ſhee is in hand withall in her huſbands abſenſe, of Tullies epiſtles with Lambines annotations, and deliver to thoſe that haue partes therein with the ſaid Thomas. Yf his title be found vnſufficient to the ſaid booke, then the ſaid Thomas & his partners to yield ſuche Recompence to Mr. Tho. Marſhe for this impreſſion as the Table ſhall think good: for that the ſaid Tho. Marſhe nowe pretendeth title thereto." Stat. Regiſter, B. fol. 434, b.

Prince,

THOMAS MARSHE.

Prince's arms, according to some of his colophons, but in Aſkham's almanac it is ſtiled the "Kyngs arms." Whether this was near St. Dunſtan's church does not appear quite clear; for in 1556, &c. when he gives his direction, "nere to S. Dunſtones church," he does not mention any ſign. Strype, in his edition of Stow's Survey, B 5, p. 222, mentions his having a great licenſe to print Latin books uſed in the grammar ſchools of England, (ſuppoſed to be the patent granted the 29 Sept. 14 Eliz, 1572.) Againſt this, and indeed againſt all of like purport, the poor and unprivileged Stationers complained to the Lord Treaſurer: the reſult of which was by compromiſe, for the opulent and privileged Stationers to yield up their right in ſeveral copies to the Company, for the benefit of their poor, as will be further particularized in our General Hiſtory. In June 1591, Thomas Orwin had "granted unto him, by the conſent of Edw. Marſhe, theſe" copies, which did belong to Tho. Marſhe deceaſed."

"A ryghte

* "In Engliſh, 8°.

The mariage of wyt & wiſdome.
Arcandum.
Dehortation from papiſtrie.
Caſtle of Health.
Tuchſtone of Complexions.
A proued medecynes.
Keeping of a Goſhawke.
Myrrour of madnes.
Biles of Epemerides.
Brief ſome of the Bible.
Mariage of prieſtes.
Detection of Seaues.
Prouerbes of Salomon.
Tullies Old age.
Jnſtitution of a gentleman.
Ayde to Schollers.
Collection of munſer.
Jndentures and obligations.
Baniſters Surgerie.
Flowers of Terence.
Jdle Jnuentions.
Heywooddes woorkes.
Dyall of Deſteny.
Watchword for wilfull women.
Booke of Cheſſt plaie.
Skeltons woorkes.
Hills Dreames.
Lyne of liberalitie.
Nobilitie of Dr. Humfrey.
Tuſculan queſtions. In Engliſh.
French grammer.
Confeſſion of faithe.
Warres betwene the Romaine, & Catholiques.
Seeinge of Vrynes.
Tom tell trouthe.
Scottis cathemiſme.
Letters out of Holland.
Sipions Dreames.
Offices of bailifes, & ſheriffes.

In folio.

Bullins Bulwark.
Holettes dictionary.
Flores Hiſtories.
Arcetecture, in folio.
Perpetual Kalender.
Deſtruction of Troy, in meeter.
Marlorat in Mathew.
———on John. Folio.
———on Mark and Luke. Quarto.
Brocard in engliſh.
Palace of pleaſure. j part.
Palace of pleaſure. 2 part.
Tragicall diſcourſes.
Herodotus. In Engliſh.
Ovid de triſtibuz. In engliſh.
Seneca his tragedies.
Engliſh pollicie.
Ciuill pollicy.
Oſorius of nobilitie.
Diggs Tectonicon.
Digges pronoſtication.
Leaden goddes.
Mirror of magiſtrates. j & laſt part.
Schoole of ſhootinge.
Churchyardes Chippes.
Cardanus comfort.
Juſtine in engliſh.
Spider & the Flie.
Geneſis, in verſe.
Horace epiſtles.
A diſcourſe of complexions.
Horace Saturs.
Pageant of popes.
Lyſes of the Emperors.
Funerall of K. Edw. 6.
Jeſuits challenge.
Cornelius Agrippa.
Hiſtory of Julie.
The art of gardeninge.

Halls

THOMAS MARSHE.

1554. "A ryghte, excellente treatife of Aftronomie, called in Latine Facies Cœli, the face of the Heauens, for 1554 & 1555, made in the Tufcane or Italian tongue, by maifter Ant. de Montulmo: tranflated by Frederick van Brunfwicke."　　　　　　　　　　　　　　　　　　　　　Octavo.

(1554.) "An almanacke or prognoftication, made for the yeare of our Lord God MVCLV. made by maifter Anthony Afkham, phyfician and prefte," in columns as now, with the figns of the zodiac at the top of each month, "a declaration of almanacke," and the human figure with the 12 figns. "Imprinted—by Thomas Marche, dwellyng in Fleteftrete at the fign of the Kyngs arms." Brit. Topography, Vol. II. p. 355, &c, Richard Gough, Efq;　　　　　　　　　　　　　　　　　　　　　　　　Broadfide.

He had licenfe for this, and alfo for Henry Lowe's, in 1557.

1555. "The Jnftitution of a gentleman. Anno Domini. M.D.L.V. Jmprinted—at the—Princes armes, by Thomas Marfhe." Dedicated "To——the Lorde Fitzwater, Sonne & heire to the Erle of "Suffex." Then, "The prologe of the Booke." Contains befides, M 3, in eights. W.H. Again 1568.　　　　　　　　　　　　　　　　　　　　　　　　Octavo.

1555. Pfalterium Davidicum ad vfum ecclefiæ Sarifburienfis.　Octavo.

1555. "THIS TREATYfe concernynge the fruytfull fayinges of Dauyd, &c. as p. 148. An. M.D.L.V." Contains & 6, in eights. "Jmprinted—at the—Prynces armes," &c. W.H.　　　　　　　　　　　　　Octavo.

1555. "A Boke of precidents," &c. as p. 521. William Bayntun, Efq;　　　　　　　　　　　　　　　　　　　　　　　　　　　Octavo.

1555. "THE AVNCIENT HISTORIE AND onely trew and fyncere Chronicle of the warres betwixte the Grecians and the Troyans, and fubfequently of the fyrft euercyon of the auncient and famoufe Cytye of Troye, vnder Lamedon the king, and of the lafte and fynall dyftruction of the fame vnder Pryam: wrytten by Daretus a Troyan, and Dictus a Grecian, both fouldiours, and prefent in all the fayde warres: and digefted in Latyn by the learned Guydo de Columpnis, and fythes tranflated into Englyfhe verfe, by John Lydgate, Monke of Burye. And newly imprinted. An. M.D.L.V." In the fame compartment as Hall's chronicle, p. 530. See it printed by Pinfon in 1513. This edition has prefixed, an "epiftle "To the

Halls Surgerie.
The 4 books of the diall of princes.
　　　Jn latine, 8°.
Carters logick.
Buckeys arethmetick.
Terrana moftope
Conftitutiones prouinciales.
Chriftiane fidei.
The lyne of liberalitie.
Watfons Amyntas.
　　　Jn Jnglifh & latine, 16°.
Erafmus teftament.
Feare of deathe
Defenfe of the Soule.

Fame of the faithfull.
Ordinarie for all degrees.
Right godlie rules, in 16 & 32."
ⁿ Thus in my copy; but Mr. Ames, who had a copy alfo, has "Effex."
º Wherein, after mendoning the partiality of the Greek and Roman poets, and that therefore truth was not to be expected from their relations, efpecially from the tranflators of their works, he proceeds, "Of whome although fome J do confeffe haue learnedlye—perfourmed theyr enterpryfe therin. Yet hath there ben other fome fo beaftly bolde to vndertake, without
　　　　　　　　　　　　　　　eyther

the reader," by Robert Braham. See Warton's Hift. of Eng. Poetry,
Vol. I. p. 127. Vol. II. p. 81. W.H. Folio.
" Jnftitutions or princypal groundes of the lawes and ftatutes of En- 1555.
glande, newly and very truely corrected," &c. as p. 818. In the fame
compartment as The treafure of poor men, p. 563. It is introduced with
a fhort but pithy prologue by the author. Contains Fol. lxvi, and a table
at the end. " Jmprynted——in Fleteftrete at the fygne of the Pryncles
armes——Anno. M.D.L.V." W.H. Octavo.
Again 1556; but in the compartment with W. Middleton's rebus. 1556.

" Magna Charta, cum veteribus ftatutis.———Firft proue, and then re- 1556.
proue." At the end, " ———near St. Dunftans church." Tottel printed
it this year alfo, in oppofition to this. See p. 810. Octavo.

" The triall of the fupremacy wherein is fet fourth y̆ vnitie of chriftes 1556.
ehurch militāt geuē to St. Peter and his fucceffoures by Chrifte: And that
there ought to be one head Bifhop in earth, Chriftes Vicar generall ouer
all his churche militant: wyth anfweres to the blafphemous obiections
made agaynfte the fame in the late miferable yeres now pafte. Aug. li. 3.
de trinitate. ca. 3. Vtile eft plures libros a pluribus fieri, diuerfo ftilo, nō
diuerfa fide," &c. Dedicated " To the mofte holye & godly Prince,
Reginalde Pole, Cardinall & Legate," &c. Contains V 4, in eights.
" Jmprinted——nere to S. Dunftones church—Menfe Julii, 1556. W.H.
Again 1576. Octavo.

eyther wyt or any learning, to tranflate— the Eanedes of Virgyle into englyfhe,—. As by example, if a man ftudyoufe of that hiftorye, fhoulde feke to fynde the fame in the doynges of Wyllyam Caxton, in his leawde recueil of Troye: what fhoulde he then fynde thyncke you? affuredlye none other thynge, but a longe tedious & braynles bablyng, tendyng to no ende, nor hauing any certayne begynnynge: but proceadynge therin as an ydyot in his follye, that can not make an ende tyll he be bydden. Much like the foolyfhe & vnfauerye doynges of Oreftes, whom Juuinall remembreth,—. whych Caxtons recueil, who fo lyft wyth iudgement perufe, fhall rather thyncke his doynges worthye to be numbred amongeft the trifelinge tales & barrayne luerdries of Robyn Hode, & Beuys of Hampton, then remaine as a monument of fo worthy an hiftory." This harfh cenfure upon Caxton would have appeared with rather a better face againft his Eneydos; (p. 64.) though Caxton pretended to nothing more in either, than to give a tranflation of his French authors; therefore the reflection was unjuft. He afterwards commends the diligence of Lydgate, who " may worthyly be numbred amongeft thofe that haue chefelye deferued & imitator of the great Chaucer, the onelye glorye & beauty of the fame. Neuertheles, lykewyfe as it hapned the fame Chaucer to leafe the prayfe of that tyme wherin he wrote, beyng then when in dede al good letters were almoft aflepe,—. That if it had not bene—by the dylygence of one willyam Thime, (Thynne) a gentilman, who laudably ftudyoufe to the polyfhing of fo great a Jewell, with ryghte good iudgement, trauail & great paynes caufing the fame to be perfected & ftamped as it is nowe read, the fayde Chaucers workes had utterly peryfhed, or at the left bin fo depraued by corrupcion of copies, that at the laft, there fhoulde no parte of hys meaning haue ben founde in any of them. Euen the fame iniurye almoft hathe happened to this wryter in this his Pamphlite of the euercion of Troye :—J haue therefore taken vpon me as one ftudyous of the language of my countreyth,— to bring again this hiftorian into lyght, fomewhat J trufte more perfecte & polifhed then before," &c.

" CONSTI-

1557. " CONSTITVTIONES ANGLIÆ PRouinciales ex diuerfis Cantuarienfium Archiepifcoporum Synodalibus decretis, per Guilielmum Lyndewode Anglum iam olim colleɛtæ. Conftitutiones item Legatinæ, quas alii legitimas vocant, Reuerendifs. in Chrifto patrū, Othonis & Othoboni, quondam fedis Apoftolicæ in Anglia Legatorum, nunc demum accuratius quam antehac alias, in ftudioforum gratiam impreffa. Acceffit Cantuarienfium Archiepifcoporum quorum in hoc libro Conftitutiones continentur Catalogus, nunc primum confcriptus, atq; in lucem editus. Londini Excudebat Thomas Marfhe. '2557. The dedication is thus infcribed : " Reuerendiffimo in Chrifto Patri ac Domino, D. Reginaldo Polo Cardinali S. D. N. Papæ, & fedis Apoftolicæ Legato, Cantuarienfi Archiepifcopo, & totius Angliæ primati. T. Marfhus humilis Typographus perpetuam felicitatem optat." Next, " De Guilielmo Lynvvod, ex Anglorum fcriptorum Catalogo ." Then, " Cantuarienfium Archiepifcoporum—catalogus. - - D. Guilielmi Lyndewod—prefatio. - - - Titulorum index alphabeticus." The Conftitutiones provinciales on 282 pages. The Conftitutiones Othonis have only the leaves numbered, and end on folio 21. The Conftitutiones Othoboni proceed, and end on folio 76. Then a table to each of thefe Legatine Conftitutions. The fignatures are continued throughout to G g 4, in eights. See p. 392. W.H. Oɛtavo.

(1557.) " A Prayer fayd by the Lorde Sturton, being on his knees, before he went up the ladder, and alfo his confeffion before his death, the vi. day of March, 1557."

1557. The golden book of Marcus Aurelius. See p. 425. Oɛtavo.

1558. " Of the life Aɛtive & contemplatiue, entitled, The pearle of perfeɛtion. By James Cancellar." Maunfell, p. 28. Licenfed. Oɛtavo.

1559. " A MYRROVR FOR MAGISTRATES, Wherein may be feen by example of others, with howe greuous plages vices are punifhed, and howe frayl and vnftable worldly profperitie is founde, euen of thofe whom Fortvne feemeth moft highly to favour. Felix quem faciunt aliena pericula cautum. Anno 1559.——in ædibus," &c. Contains, befides Baldwin's dedication to the nobility, folio c. lx. and at the end, a table of the contents, for both parts, and another of " Faultes efcaped in the Printing." In this ftate it feem's to have been reprinted in 1563, 1571, and 1574. See Warton's Hift. of Eng. Poetry, Vol. III. p. 216, and 259. Licenfed. W.H.+ Quarto.

1559. " An Epitome of Cronicles," &c. as p. 692. In 1452, it is faid, " one named Johannes Fauftius, firft found the craft of printinge, in the citee of Mens in Germanie." * W.H. Quarto.

1560. " A pleafaunt Almanacke feruinge for thre yeres, as 1560, 1561, 1562. teachinge not only worthie leffons in the letting of bloud, and taking of

* Thus printed in my copy.

purgacions

THOMAS MARSHE. 851

purgacions, but extraordinarie rules for the weather, matter profittable for the meaner people & hufbandmen to know. Imprinted——neare to S. Dunftones churche, 1560." By Tho. Hill. Licenfed. Quarto.

" The Funeralles of King Edward the Sixt. Anno 1553. Wherin 1560. are declared the caufers & caufes of his death." With king Edward's head, finely cut in wood; and William Baldwin's dedication to the reader. Anno 1560. The remainder of the book is in verfe. Twelve leaves. Licenfed. Quarto.

" A compound Manuell, or Compoft of the hand, &c. Tranflated by 1560. W. Wuddus." See it in 1566. Octavo.

" A briefe Cronicle, contaynyng the accoumpte of the raygnes of all (1561.) *the* Kynges in this realme, from the entring of Brutus to this prefente yeare, with all the moft notable actes done by eche of theym, gathered oute of the moft trufty writers, whereunto is added, a perpetuall Kalender, for the readier findinge of dayes & tymes herein mentioned. Jmprinted— nere to faynct Dunftons church." The date is taken from the concluding 'article. Contains N, in eights, befides the kalendar, &c. prefixed. He bought the copy of John King's wife. Licenfed.' W.H. Octavo.

" Thofe fyue Queftions, which Marke Tullye Cicero difputed in his 1561. Manor of Tufculanum: Written afterwards by him, in as many bookes, to his frende and familiar Brutus, in the Latine tonnge. And nowe out of the fame tranflated & englifhed by John Dolman, Studente and fellowe of the Inner Temple. 1561. Jmprinted——in Fleteftrete nere S. Dunftones church." Licenfed. Octavo.

" THE HISTORYE OF ITALYE." &c. as p. 452. " Jmprinted——nere 1561. to S. Dunftons Church—1561." In a very rude compartment. Licenfed. Quarto.
Again 1562, with cuts. M. C. Tutet. Quarto. 1562.

" Anno xxv. Henrici Octaui. Actes made in the fiffion of this prefent 1562. Parliament, &c. Excudebat Londini Anno M.D.LXII." In a compartment with an alphabet of capitals in a tablet at bottom. W.H. Folio.

¶ " On Wednefday the fourth day of June, in this yere. 1561.—betwene one & two of the clock at after noone was fene a maruaylous great fyrye lightninge, and immediately enfued a mofte terrible cracke of thunder at which inftaunt the corner of a turret of S. Martins church within ludgate was torne, and diuers gret ftones caft downe: and diuers beinge in the feildes adioyning to the citie affirmed that they fawe a long efpeare poynted flame of fyer runne th roughe the toppe of the Broche of Paules fteple from the eaft, weftward. Betwene 4 and 5 of the clock a fmoke was efpied to breake oute vnder the bowle of the fayd fhafte, & fodenly after,—the flame brake fortho in a circle like a garlande rounde about the broche, aboute 2 yardes to theftimation of fyghte—, and increafed in fuche wife, that in one houres fpace the broche of the fteeple was brent down to the battilmentks, & the moft part of higheft roofe of the church was likewife confumed." See p. 696.
' He reprinted this, or another chronicle, in fixteens, without licenfe; for which he was fined v. s.

" A briefe

THOMAS MARSHE.

1562. A briefe treatife conecerning the burnynge of Bucer and Phagius at Cambrydge, in the tyme of Quene Mary, with theyr reftitution in the time of our mofte gracious fouerayne Lady that nowe is. Wherein is expreffed the fantafticall & tirannous dealynges of the Romifhe Church, togither with the godly & modeft regimet of the true Chriftian Church, moft flaunderouflye diffamed in thofe dayes of herefye. Tranflated into Englyfhe by Arthur Goldyng. Anno. 1562. ☞ Read and iudge indifferently, accordinge to the rule of Gods worde. Jmprinted—nere to faynct Dunftons Churche." Contains M 4, in eights, befides a fhort addrefs " To the reader" prefixed. W.H. Sixteens.

1563. "The Nobles, or of Nobilitye. The original nature, duties, ryght, and, chriftian Jnftitucion thereof, in three Bookes. Fyrfte eloquentlye writtē in Latiae, by Lawrence Humfrey, D. of Diuinity, and Prefidente of Magdaleine Colledge in Oxforde, late englifhed. Whereto, for the readers commoditítye, and matters affinítye, is coupled the fmall treatyfe of Philo, a Jewe. By the fame Author out of the Greeke Latined, nowe alfo Englifhed. 1563. Jmprinted," &c. Dedicated " To the mofte Chriftian Princeffe Elizabeth Queene of Englande,——&c. The nobleft protectour & defendour of the true fayth." Alfo, " To the ryghte honorable & worfhipfull of the Inner Temple." Then, fome verfes on the fubject of the book. The treatife on nobility contains Z 4 in eights, befides the prefixes; the tranflation of Philo begins on A a. i. and ends on A a 6. On a feparate leaf, " Scapes in Prynting." Colophon on the back. Licenfed.* W.H. Sixteens.

1563. " The Firft and chief grounds of Architecture vfed in all the auncient & famous monyments: with a father and more ample difcourfe vppon the fame, than hitherto hath been fet out by any other. Publifhed by Jhon Shute, Paynter & Archytecte." With cuts. Licenfed. Again 1584. Folio.

1563. " Petri Carteri Cantabrigienfis in Johannis Setoni dialecticam annotationes, vt clariffimae, ita breuiffimae. Ad illuftriffimum et clariffimum dominum Edouardum, Durouentanae, quam Darbiam vocant, comitem obiliffimum." Again 1568. Octavo.

1564. " A briefe chronicle, wherein are defcribed fhortlye the originall, &c. of the Romain wele publique, &c. collected and gathered by Eutropius, and Englifhed by Nicholas Hawarde. Dedicated to mayfter Henry Compton efq; (praifed for his learning.) From Thauies Jnne, 6 Octob. Licenfed. W.H.+ Octavo.

1563. " A pleafaunt Treatife of the interpretacion of fundrie dreames gathered parte out of the woorcke of the Learned Phylofopher Ponzettus, and part out of Artemidorus." By Tho. Hill. Licenfed.

1563. " A briefe Treatyfe of gardeninge teaching the apt dreffing, fowing & fetting of a Garden, with the remedies againft fuch beaftes, wormes, flyes, &c.

&c. that commonly annoye Gardens, encreafed by me the feconde tyme. Anno. 1563." By Tho. Hill. Licenfed. See it in 1568.

Juftin tranflated by Arthur Goldinge. Licenfed. See it in 1570. 1564.
Quarto.

" The Boke named the Gouernour, deuifed by fir Thomas Elyot, knyght. 1565. —In ædibus." Licenfed. W.H. See p. 455. Sixteens.

" A Summarie of Englyfh Chronicles,' Conteyning the true accompt 1565. of yeres, wherein euery kyng of this Realme of England began theyr reigne, howe long they reigned, and what notable thynges hath bene doone durynge theyr Reygnes. Wyth alfo the names and yeares of all the Bayfyffes, Cuftos, maiors, and fheriffes of the Citie of London, fens the Conquefte, diligently Collected by John Stow, citifen of London, in the yere of our Lorde God 1565. Whervnto is added a Table in the end, conteynyng all the principall matters of this Booke. Perufed and allowed accordyng to the queenes maiefties Jniunctions. In ædibus Thomæ Marfhi." The firft letter only of this title is printed in red. This book has prefixed, a kalender, and an almanac for 30 years, from 1564

* In this book he fays, folio 152, " The art of printing was found out by John Cuthenbergus, at Mentz 1458, and 16 years after their ink; and that William Caxton firft brought it to England about the year 1471, and practifed the fame at Weftminfter."

Among the catalogue of fuch unlawful books, as were found in the ftudy of John Stow, of London, February 24, 1568, N°. xvii, in Strype's life of archbp. Grindal, you will find this book in M:S. 1563. So that i conclude this the firft edition of it, which was afterward printed almoft annually as almanacs. Thofe Mr. Ames obferved, were

Another edition this fame year by Thomas Marfhe, Twelves, 1565.
Thomas Marfhe two, 16°. 12°. 1566.
By the fame perfon, 16°. 1567.
Richard Bradocke, 12°. 1570.

My copy of this edition wants the titlepage, and latter end; but it appears to me to be printed by T. Marfhe, having the fame blooming letters, &c. as the edition 1565; and has the characters of authors, but not of R- Grafton. Contains 413 leaves, exclufive of prefixes and affixes. Befides, Ric. Bradocke was not made free till 14 Octo. 1577.

Thomas March, a very thick 12°. 1573. This is the edition, that has the characters of authors. Amongft them Richard Grafton.
Richard Tothill and Binneman, 16°. 1575.

Richard Tothill and Binneman 16°. 1579.
Ralph Newbery and H. Denham 16°. 1584.
By the fame perfons 16°. 1587.
Ralph Newbery 12°. 1590.
Richard Bradocke 16°. 1598.
John Harrifon 12°. 1604.

The fizes of the books in this note i have given exactly from Mr. Ames; but be pleafed to obferve, that but very few of thefe early printed books can properly be filled 12°. Thofe that are really fuch i have noticed where they occurred; they are chiefly fome of the firft editions of Magna Charta. Moft of the books denominated 12°. by Mr. Ames, are rather 16°. or 8°. and fo they are termed by Maunfell. I followed Mr. Ames in the former part of this work, in order to avoid confufion, that fuch books might not have the appearance of different editions; but when I found Maunfell named the fizes more confiftently, i followed his method. Indeed when thefe fmall books are bound up, it is difficult to determine whether the fignatures are whole or half fheets. This fummarie in particular Mr. Ames calls twelves, but the form of it is larger than that of his foregoing article, called Octavo. Much is to be allowed for the condition the binder has left the book in, as fome have been cropt more than others. In fhort, no dependence can be relied on, as to the denomination of the fixes.

to 1593, inclufive, printed in red and black; a dedication " To—Lorde Robert Dudley, Earle of Leicefter," &c. an epiftle " To the Reader;" and " The names of the Authours—alledged." The Summarie contains befides, Fol. 248; the ages of the world; the diftance of places, in miles; and a table of contents. " Jmprinted—nygh vnto S. Dunftones churche,— Anno Salutis. 1565." See p. 502. Licenfed, being " auctorilfhed by my lorde of Canterbury." *W.H. Octavo.

1565. " A moft excellent and learned vvorke of Chirurgerie, called Chirurgia parua Lanfranci, Lanfranke of Mylayne his briefe: reduced from dyuers tranflations to our vulgar or vfuall frafe, and now firft publifhed in the Englyfhe prynte by Iohn Halle Chirurgien. Who hath thervnto neceffarily annexed. A Table, as well of the names of difeafes and fimples with their vertues, as alfo of all other termes of the arte opened. Very profitable for the better vnderftanding of the fame, or other like workes. And in the ende a compendious worke of Anatomie, more vtile & profitable, than any here tofore in the Englyfhe tongue publyfhed. An Hiftoricall expoftulation alfo againft the beaftly abufers, both of Chyrurgerie & Phificke in our tyme: With a goodly doctrine & inftruction, neceffary to be marked & folowed of all true Chirurges. All thefe faithfully gathered, and diligently fet forth, by the fayde Iohn Halle. Imprinted—, nyghe unto faint Dunftones churche,—An. 1565." On the back is Halle's portrait, " anno ætatis fuæ, 35—1564." There are prefixed, " The bookes verdict." An epiftle " Vnto the worfhipful the maifters, VVardens, and confequently to all the whole company & brotherhod of Chirurgiens of London, Iohn Halle, one of the lefte of them, fendeth hartie & Iouynge falutation." Another by " VV. Cuningham, Doctor in Phifique, vnto the profeffors of Chirurgerie—Fare hartely well, at my howfe in Colmaftrete, this, xviii. daye of Aprill. Anno, M.D.lxv." Another by " Tho. Gale maifter in Chirurgerie unto his welbeloved friende Iohn Halle.——London, the 14. daye of May. Anno. 1565." Four octave ftanzas " To the louynge Readers." Then, Halle's preface. The expofitive table, The treatife of Anatomy, and The Expoftulation, are printed fo as to fell feparately. Licenfed. W.H. Quarto.

1565. " The Courte of Vertue, contayning many holy or fprotuall fonges, fonnettes, pfalmes, balletts, and fhorte fentences, as well of holy fcripture, as others." With mufic notes. By John Hall, as appears by two acroftics under " Nomen authoris" at Fol. 99 and 172. That he was a phyfician appears by " A Ditie made to the prayfe of God by the Author for a pacient to vfe after helth attayned, who contrary to all mens expectation, was in hys handes by the goodnes of God cured." By the 'pro-

" A booke alfo of fonges they haue, | And in fuche fonges is all their game.
And Venus court they doe it name. | Wherof ryght diuers books be made,
No filthy mynde a fonge can craue, | To narythe that mofte filthy trade."
But therein he may finde the fame

logue

logue, which is in seven-lined stanzas, this book was written in contrast to one named The Court of "Venus. See Warton's Hist. of Eng. Poetry, Vol. III. p. 181, and 424. Licensed. W.H.+   Sixteens.

"A compound manuell, or compost of the hand, in Englishe, faith-   1566. fully translated, whereby you may easily, and with small trauaile finde out (by the art of the hand) all thinges pertaining to the use of the common almanacks: As the golden numbre, the circle of the sonne, the change of the moone, festiuall dayes, bothe fixe and mouable, dominical letter, and such other needful thynges, mete for all such, as may not conueniently cary any bokes with them. Whereunto is annexed, a practise of the hand diall, whereby to know the houre of the daye at all tymes (the sonne shynyng) without any difficultie, or any great studye; mete for the commoditie of al men, by W. W. [*W. Wuddus.*] Licensed.   Octavo.

"A Medicinable Morall, that is, the two Bookes of Horace his Satyres,   1566. Englyshed accordyng to the prescription of saint Hierome. Episto. ad Ruffin.
Quod malum est, muta,
Quod bonum est, prode.
The Wailyngs of the Prophet Hieremiah, done into Englyshe verse. Also Epigrammes. T. Drant. Antidoti salutaris amator. Perused and allowed &c. Imprinted—in Fleteftrete—M.D.LXVI." On the back, "To the Right Honorable my Lady Bacon, and my lady Cicell, sisters, fauourers of learnyng and vertue." only. It is introduced with an addres "To the Reader," concluding with a distich of Greek verses. Then, a poetical definition of a satire. Contains sheets "M 2. Licensed. " auctoryshed by my Lorde of London. W.H.+   Quarto.

"A new Postil Conteinyng most Godly and learned sermons vpon all   1566. the Sonday Gospelles, that be redde in the Church thorowout the yeare: Lately set foorth vnto the great profite not onely of al Curates, and spiritual Ministers, but also of all other godly and Faythfull Readers. Perused & allowed &c. Imprinted—nere to S. Dunstons church—M.D.LXVI." In a compartment with T. R. on a tablet at bottom. On the next leaf are "Certayne Sentences of holy scripture." Next is the preface of Tho. Becon[z] the author "To his faithfull felow labourers in the

---

[u] See Mr. Tyrwhitt's Appendix to the preface to The Canterbury Tales, p. xv.

[v] It may be necessary to add here, to what has been mentioned in note (s) above, concerning the sizes of books, that though this book, and many others of this size and shape have eight leaves to a signature, yet it would confound too much to call them octavos, being of the same squarish form with those which have but 4 leaves to a signature, and much larger than those denominated octavos in this work; i therefore call them quartos. When the leaves have been numbered, or paged, i have given the contents accordingly, and not by the signatures; but when the leaves are not numbered, as is the case with this book, i put the word sheets before the signature, supposing such as have 8 leaves to be whole sheets, and those with only 4 leaves to a signature to be half sheets.

[z] See p. 636, &c. This pious author appears to have laboured more abundantly, in forwarding and establishing the Reformation,

the Lordes harueſt, the Miniſters & Preachers of Gods moſt holy word.——From my houſe at Cantorbury, the. xvi. of July.——M.D.lxvi." Then, "A Prayer to be ſayde before the Sermon. A ſhorter prayer—. A thankeſgeuynge after the Sermon." This book is divided into two parts; the firſt endeth with Trinity Sunday, on Fol. 312, the ſecond is concluded on Fol. 195. See Strype's Life of Abp. Parker, p. 228. Licenſed. W.H. Again 1567. Quarto.

1566. Painter's Palace of Pleaſure. Mr. William White. Quarto.

1566. "John Securis his Detection of ſundrie Errors and Abuſes of vnlearned Phiſicions, Apothecaries & Surgions." Licenſed. Octavo.

1566. Bartlet's "Pedegrewe of Popiſh Heretiques." See it by H. Denham. Quarto.

1567. "HORACE His arte of Poetrie, piſtles, and Satyrs, Engliſhed, and to the earle of Ormounte by Tho. Drant addreſſed. Imprinted nere to S. Dunſtones Churche,——1567." In a compartment with the Stationers' arms at top, and T. M. joined, at bottom. On the back, "De ſeipſo," in 17 hexameters. Dedicated "To the right hon. and verye noble Lord Thomas Earle of Ormounte, and Oſſorye, Lord Butler, Vicounte Thurles, Lord of the libertie of Typparye, and highe Treaſurer of Ireland," with his arms prefixed. Then, an addreſs "To the Reader," Contains beſides, ſheets R. W.H. Quarto.

1567. "Certaine Tragicall Diſcourſes written oute of Frenche and Latin, by Geffraie Fenton, no leſſe profitable then pleaſaunt, and of like neceſſitye to al degrees that take pleaſure in antiquityes or foreine reapportes. Mon heur viendra. Jmprinted——1567." In the ſame compartment as the foregoing article. Dedicated "To the right hon. and vertuous Ladie, the Ladye Marye Sidney.——: at my chamber at Paris, xxii. Iunij, 1567." Then, ſundry verſes in praiſe of the tranſlator, by Sir John Conway, M. H. George Turberville, and Peter Beverley. Contains beſides, Fol. 306, and a table of contents. Licenſed: authoriſed by my lorde of Canterbury. W.H. Again 1579. Quarto.

1568. "Flowers for latin ſpeaking, ſelected and gathered out of Terence by Nich. Vdall." See p. 447. Theſe are ſelected from the firſt three comedies only. Licenſed. See the whole in 1581. Octavo.

1568. "The dial of princes," &c. as p. 817. This edition ſeems to have been printed at two preſſes, the folios and ſignatures beginning afreſh at the third book. The firſt and ſecond books contain Fol. 165, beſides the

ation, by his writings, than any of his cotemporaries, it is hoped therefore he will one day find a place in the Biogr. Britan. Much matter may be collected from Strype's Life of Abp. Cranmer, whoſe chaplain he is ſuppoſed to have been, p. 423; bp. Tanner's Bibliotheca, p. 85, &c. and from the ſeveral prefaces and dedications prefixed to his own works.

prefixes

prefixes which are the same as to the first edition, p. 564. The third and fourth books contain Fol. 173, besides the letters at the end. With R. Tottel. Licensed. W.H.+ Folio.

" Pithy, Pleasaunt, and Profitable workes of Maister Skelton, poete 1568. Laureate. Nowe collected, and newly published. Anno 1568." Containing the following pieces, viz. " The crowne Lawrell. The Bouge of Courte. How the douty duke of Albany, &c. Speake Parrot. On the death of Edw. IV. Against the Scottes. Warre the hawke. Elynour Rūmyng. Why come ye not to Court. Colin Clout. Philip Sparrow. Against Coystrowne, &c. The boke of 3 fooles. On the death of the E. of Northumberland." A new collection of his works was printed, 1736. See Warton's Hist. of Eng. Poetry, Vol. II. p. 130, and 336, &c. Octavo.

" The proffitable Arte of Gardening, now the third tyme set fourth: 1568. to whiche is added much necessarie matter, and a number of secretes with the Phisick helpes belonging to eche herbe, and that easily prepared. To this is annexed 2 propre treatises, the one intituled The marueilous gouernment, propertie, & benefite of the Bees, with the rare secrets of the honny & the waxe. And the other, the yerely coniectures, meet for husbandmen to know: englished by Tho. Hill, Londoner. 1568." In a border. Dedicated " To the Righte Worshipful Sirre Henry Seamer knight." Next is " The Preface to the Reader," in which is promised his Art of planting and graffing, &c. as appears in a table at the end. Then, some verses by " A friende to the Reader;" a table of contents; and a list of authors from whose works this was gathered. Contains besides, 195 leaves. Hereunto is annexed a table of " The bookes and treatises all readie printed; The bookes of mine, now in readynesse to be imprinted, &c. The bookes which remaine with mee fullye ended, halfe done," &c. The particulars of the two former heads are noticed in their proper places; those of the latter, which remained by him, are abridged 'below. The tracts on Bees, and Husbandry, are printed with a separate title-page, in the same border as the Gardening, viz.

" A pleasaunt

? " 1. A treatise of the Sphere, for Mariners & seafaringe men, gathered part out of the large Commentarie of Stoeflerus vpon the Sphere of Proclus, and parte of others, and this in a readynesse to the printing. 2. A Treatyse of Physicke, conteyninge manye woorthye Lessons and Secretes, in the drawyne Oyles of the 7 metalls, and other simples, gathered out of the beste practisioners, and this half writen, in sundrye papers. 3. A briefe Herbal wryten first by a singuler Phylosopher & skylful practisioner of Bolognia in Italye, and part begunne of the same. 4. A Treatyse of the Judgement of vrines, much helping young practitioners of Phisicke, and that lacke the Lattine tungè, vnto whyche is annexed sundrie pleasaunte Cautells helpinge greatelye the furtheraunce in iudgement, and this in a maner readie for printing. 5. The greate woorcke of the arte of Paulmestrie, dyuided into two bookes, the fyrste prouynge it by solempne argumentes, alledged oute of auncientc Phylosophers, &c. that the same be an arte founde oute, of longe experience, & this so reasoned of in dyologue forme. And the other proueth it by apt demonstrations accordyng to the instructions of the art, and by sundrie examples, whyche not a lytle further the confirminge of the former, and besydes added to such number of handes as maye gyue an Euydente lyghte,

to

THOMAS MARSHE.

1568. "A pleasaunt Instruction of the parfit ordering of Bees, with the marueilous nature, propertie & gouernement of theim: and the myraculous vses, bothe of their honny & waxe (seruing diuersly) as well in inwarde as outwarde causes: gathered out of the beste writers. To whiche is annexed a profitable treatise, intituled Certaine husbandly coniectures of dearth & plentie for euer, and other matters also meete for husbandmen to knowe, &c. Which is now Englished by Thomas Hyll Londyner. 1568." His portrait, "Ætatis suæ 28," on the back. Dedicated "To the worshipfull mayster M. Gentilman." Then a preface, and a table of contents. These together contain 83 leaues. "Imprinted—neare to S. Dunstones Churche——1568." W.H.   Sixteens.

1568. "Greg. Nazianzen his Epigrams, and spirituall sentences. Translated by Tho. Drant." Maunsell, p. 46, and 75.   Octavo.

1569. "The worthy booke of old age, otherwise entituled, the elder Cato, containing a learned defence and praise of age, and aged men, written by M. T. Cicero, and now Englished: whereunto is added, a recital of diuers men, that liued long, with a declaration of sundry sortes of yeares, and the diuersitie betwene the yeres in the old time, and our years now a dayes." Dedicated to the lord William Paulet, marquis of Winchester, and lord treasurer, then 96 years of age. 64 leaues. Licensed.   Small.

1569. "The palys of Pleasure." In two tomes.   Quarto.

1569. "The Pleasaunt and vvittie Plaie of the Cheastes renewed with Instructions both to learne it easely & to plaie it wel. Lately translated out of Jtalie into Frenche: And nowe set furth into English By J. R. (*James Rowbothum*) 1569. Jmprinted" &c. In a border. On the back "Liber ad Lectorem." in 12 distichs, signed "w. ward." Dedicated "To the right honorable; the Lorde R. D. Maistre of the Quenes horse, & knight of the noble order of garter. *by* I. R. citisin of London." Then, The preface to the reader. It has seueral schemes of the board and men; and contains together G 3, in eights. See p. 803. W.H.   Octavo.

1569. "The Line of Liberalitie dulie directinge the wel bestowing of benefites, and reprehending the comonly vsed vice of Jngratitude. Anno. 1569.

to the arte. All whyche by greate dyligence gathered out of most authors auncient or of late yeares, the lyke not extante in the common shoppes, and this in a maner fynished by me. 6. A Treatyse intytuled the Ecclesiasticall counte, in the whyche is conteyned muche necessarie matter, for all soortes of people to reade, and thys in a readynesse to the pryntynge. 7. The last parte of Alexis Piemont not yet extant in the Englishe tongue as the same shall appeare at the comming furth of the Booke,

vnto which J haue annexed (here & there) sundrie newe inuentions about the drawing of costly Oyles & waters, not only for preseruation of healthe, but for the liuely garnishing of the face, colouring of the heare, yellow or flaxine, &c. 8. A Treatyse of the daungerous tymes of the sicke, wryten by Hypocrates, & this in a readynesse to the printinge."

ˢ This is the only semicolon i have met with since that mentioned in p. 512, note ⁹. Both seem to have been used accidentally.

Imprinted

THOMAS MARSHE.

"Imprinted" &c. In a border. Dedicated " To the ryght woorſhipfull Sir Chiſtopher Heydon knight, his moſt courteouſe Creditour of many bounties & benefites.—Your worſhips depe dettour Nicolas Haward." Contains beſides, Fol. 131. Licenſed. W.H. Sixteens.

" FLORES HISTORIARVM PER MATTHÆum Weſtmonaſterienſem collecti, Præcipuè de rebus Britannicis ab exordio mundi vſque ad Annum Domini 1307. Londini, Ex officina Thomæ Marſhij,—1570." In a grand compartment, having in an oval at top the following verſes from 1570.
      " Marcel. Paling.
  " Hiſtorię placeant celebres, mihi credite, vitā
  Inſtituūt, quæ ſint fugiēda ſequēdaq; mōſtrāt.
  Fabula nō omnis ſpernēda eſt recta legatur,
  Iſta iuuāt, eadē pariter ſine crimine profunt."
And in an oblong tablet at bottom, " Cicero de Oratore. Hiſtoria teſtis temporum, lux veritatis, vita memoriæ, Magiſtra vitæ, et vetuſtatis nuntia." It is introduced with a preface, moſt probably by archbiſhop Parker. Then, at the head of the author's prologue, " Flores Hiſtoriarum Matthæi Weſtmonaſterienſis monachi, in vtraque litteratura eruditi. Prologus." The firſt letter T is a blooming one, with a boy riding on a ſwan, not an uncommon one. The firſt book concludes with Haraldus ſecundus on Pag. 410. Then, on Page. 1. " Jncipit Liber ſecundus, de coronatione regis Gulihelmi Primi, conquæſtoris." Pag. 1001, is printed for 101, and ſo in like manner to Pag. 1008 incluſive; but afterwards properly to Pag. 466, which concludes with this colophon, " Londini, Excudebat Thomas Marſhius ſecundo die Junij, anno gratiæ 1570." Then, an index, and the errata. W.H. Folio.

" This book was publiſhed again with the tilte new compoſed, and 1573, put in the place of 1570; alſo the firſt ſignature after the preface was new compoſed, and inſtead of the former head title was this, " In librum, qui Flores hiſtoriarum intitulatur, Prologus;" and inſtead of the firſt blooming letter of 5 lines, a magnificent one of 14 lines, with archbp. Parker's arms impaled with thoſe of the deanery of Canterbury, the ſame that had been uſed to the preface of the New Teſtament, in the firſt edition of The Biſhops' Bible, 1568. The Harleian catalogue mentions an edition in 1567.

" A CATHOLIKE AND ECCLESIASTICAll expoſition of the holy Goſpell 1570. after S. Mathewe, gathered out of all the ſinguler approued Deuines (whiche the Lorde hath geuen to his Churche) by Auguſtine Marlorate. And tranſlated out of Latine into Engliſhe by Thomas Tymme Myniſter. Sene & allowed &c. Jmprinted—1570." In the ſame compartment as the foregoing article. In the tablet at top, " Auguſt. lib. 11. Super Gene. Maior eſt huius ſcripturæ authoritas quam omnis humani ingenij perſpicacitas." In the tablet at bottom, " Quanto, magis quiſq; in ſacris eloquijs aſſiduus fuerit, tanto ex eis vberiorem intelligentiam capit. Iſidor. De ſummo bo. lib. 3." Dedicated " To the right hon. Sir William
                  Brooke,

Brooke, Lorde Cobham, and Lorde Warden of the cinque portes.—
Your humble orator Thomas Tymme." Then, the preface of Aug.
Marlorate addreſſed "To all the brethren diſperſed here and there, that
fauoure and loue the Goſpell.—At Viuiacum, in the Calendes of January,
1559." Contains beſides, "Fol." (*pages*) 759, and an alphabetical table
at the end. Licenſed: "auctoryſſed by my lorde of London." W.H.
Folio.

1570. "Thabridgement of the Hiſtories of Trogus Pompeius, gathered
& written in the Laten tung, by the famous Hiſtoriographer Iuſtine, and
tranſlated into Engliſhe by Arthur Goldinge: a worke conteyning brefly
great plentye of moſte delectable Hiſtoryes, and notable examples, worthy not only to be Read, but alſo to bee embraced & followed of al men.
Newlie conferred with the Latin Copye, and corrected by the Tranſlator.
Anno Domini. 1570. Jmprinted at London by Tho. Marſhe." In the
ſame compartment as Horace's Art of Poetry, 1567. Dedicated "To
the Right Hon.—Edward de Veer, Erle of Oxinforde L. great Chamberlayne of England, Vicount Bulbeck, &c." Next is an epiſtle "To the
Reader"; and a table of "The Succeſſion of the Kings of the 3 Monarchies mentioned in this Booke." Then, "A preface of Simon Grineus—
concerning the profite of reading Hiſtories." Another by Juſtine. Contains beſides, Fol. 200. "Imprinted nere vnto Sainct Dunſtons churche,"
&c. W.H. Again 1578. Quarto.

1570. "A boke name Tectonicon,* briefly ſhewinge the exacte meaſuring,
and ſpedye reckonynge all manner of lande, ſquares, tymber, ſtone, ſtea-
ples, pyllers, globes, &c. Further declarynge the perfecte makinge and
large vſe of the carpenters ruler, contayninge a quadrante geometricall,
comprehendinge alſo the rare vſe of the ſquare. And in thende, a lytle
treatiſe adioyned, openinge the compoſition and appliancie of an inſtrument, called the profitable ſtaffe, with other thinges pleaſaunt and neceſſarye, moſt conducible for ſurveyers, landmeaters, jointers, carpenters,
and maſons. Publiſhed by Leonarde Digges, gentleman, in 1556."
W.H.4. Again 1585.

1571. Toxophylus. The ſchoole or partitions of ſhooting, contayned in ii.
bookes, written by Roger Aſcham, 1544, and now newly peruſed, anno
1571. See it p. 541. Quarto.

1571. "A Chronicle of all the noble Emperours of the Romaines from Iulius
Cæſar, orderly to this moſte victorious Emperour Maximilian, that now
gouerneth, with the great warres of Iulius Cæſar & Pompeius Magnus:
Setting forth the great power, and deuine prouidence of almighty God,
in preſeruing the godly Princes and common wealthes. Set forth by
Richard Reynoldes, Doctor in Phiſicke. Anno. 1571. Imprinted", &c.

* This copy, with that of the Prognoſtication, he bought of Luke Harriſon; and
they were licenſed to him in 1564.

In

In the compartment with his mark at bottom: the same as to Justin, 1570. Dedicated "Honorando domino, D. Gulihelmo Cecillo, domino Baroni de Burgley, Senatori Regiæ Maieſtatis prudentiſsimo." Next is "An epiſtle to the Reader." Then, "A table of all the Emperours"; and an alphabetical table of the principal matters. Contains beſides, Folio 230; with medallions of the emperors. "Jmprinted——1571." Licenſed. W.H. Quarto.

"Tho. Hill his pleaſant Art of interpretatiō of dreames: wherevnto 1571. is annexed ſundrie problemes, with apt aunſweres neare agreeing to the matter." Maunſell, Part II. p. 13.

The 29th of Sept. 14th of Eliz. a licenſe was granted him to print Catonis dyſticha de moribus, Marci Tull. epiſt. familiares, Æſopi fabulae, and other claſſick authors, for 12 years; and none to print any of his copies; with privilege to enter any houſe, or warehouſe, to ſearch for, and ſeize any books printed and brought into the realm, contrary to the tenour of theſe our letters patent, and the ſame to ſeize to the uſe of us, and our heirs and ſucceſſors.

"Hvloet's Dictionarie newelye corrected, amended, ſet in order, and 1572. enlarged, vvith many names of Men, Tovvnes, Beaſtes, Foules, Fiſhes, Trees, Shrubbes, Herbes, Fruites, Places, Jnſtrumentes, &c. And in eche place fit Phraſes, gathered out of the beſt Latin Authors. Alſo the Frenche therevnto annexed, by vvhich you may finde the Latin or Frenche, of anye Engliſhe woorde you will. By Iohn Higgins late ſtudent in Oxeford. Liber ad Lectorem.

Simplice iampridem qui ſum tantum ore loquutus,
Bino contentus reddere verba ſono,
Nunc ego triglottos trinis en prodeo linguis
Illuſtris, mira commoditate nouus.
Higginus claro vos iſtoc munere donat
Heus iuuenes, grato me accipetote ſinu.

Londini, In ædibus——1572." In the ſame compartment as the Flores hiſtoriarum. In the oval at top, and continued in the tablet at bottom.

"To write & many pleaſe is much, to pleaſe not write is paine:
Than rather write & pleaſe the good, then ſpende thy time in vaine.
The wyſe will vewe & deeme the beſt, vnwiſe the worſt will marke:
The wyſe do vſe reprouinge leaſt, the fooliſh euer carke.
Then care the leſſe be thou at reſt, though Momus baule & barke."

On the back is the coat armour of Sir George Peckham knight and baronet, to whom the book is dedicated. Then, an addreſs "To the Reader"; and certain commendatory verſes, "Domini Buggans. Henrici Torquati. A. W. E. B. and Tho. Churchyarde. Contains beſides, A a, in ſixes.* Bought of Mr. Reddell. See p. 799. Licenſed. W.H. Folio.

"Dialecta Joannis Setoni Cantabrigienſis, annotationibus Petri Car- 1572. teri, ut clariſſimis, ita breviſſimis, explicata. Huic acceſſit, ob artium ingenuarum

ingenuarum inter fe cognationem, Gulielmi Buclaei arithmetica." Again·
1574, 1577. Octavo.

1572. " A Fort againſt the feare of death, and loſſe of frends, and all other
commodities of this world." Maunſell, p. 51. Sixteens.

1572. " The Fovre bookes of Flauius Vegetius Ranatus, briefelye contayninge a plaine forme, and perfect knowledge of Martiall policye, feates
of Chiualrie, and whateuer pertayneth to warre. Tranſlated out of lattine,
into Engliſhe, by Iohn Sadler. Anno 1572. Seene & allowed &c. Imprinted" &c. In the compartment with his mark at bottom. Dedicated
" To—Lorde Ruſſell, Earle of Bedforde, one of the—priuye coun
ſell, and Knighte of the——Garter.——From Oundell, 1. Octob. 1571.
——Iohn Sadler." Prefixed alſo are " A preface to the Reader"; and commendatory verſes, " Chr. Carlili Saphphica ; Tho. Drante; Guliel. Jacobi; Gul. Charci; Will. Bulleyne; and John Higgins; alſo,. Sadleri
carmen ad præcedentia, reflecting the praiſes, aſcribed to him, on Sir
Edm. Brudenell, at whoſe requeſt he undertook the tranſlation. Contains beſides, Fol. 66 ; cuts of engines, four leaves ;. and a table of contents. W.H. Quarto.

1573. " Ultimo Marcii 1573," he changed the copy-right of Stowe's chronicle " with H. Bynneman for Terence ; and T. Hill's Gardening, for
Cato, per licentiam Mag'ri, et gardianorum."

1573. " Alæ sev Scalæ Mathematicæ, quibus viſibilium remotiſsima Cœlorum Theatra conſcendi, & Planetarum omnium itinera nouis & inauditis Methodis explorari : tùm huius portentoſi Syderis in Mundi Boreali
plaga inſolito fulgore coruſcantis, Diſtantia, & Magnitudo immenſa,
Situſq'; protinùs tremendus indagari, Deiq'; ſtupendum oſtentum, Terricolis expoſitum,. cognoſci liquidiſsimè poſsit. Thoma Diggeſeo, Cantienſi, Stemmatis Generoſi, Authore. Londini. Anno Domini. 1573." On
the back is a repreſentation of Caſſiopeia, and its 13 ſtars, according to
Ptolemy, ſhewing the place of the new ſtar, which appeared near that
conſtellation in 1572 ; of which ſee Chambers's Dictionary, &c. Facing
the figure is a table of the Lat. Long. and Magnitude of each of thoſe
ſtars On the back of this table is the coat armour of Sir William Cecil,
Lord Burghley, to whom the book is dedicated. Contains beſides, L 3,
in fours. The whole well printed, on Roman and Italic types of different ſizes, and the ſchemes neatly cut. " Londini. Apud Thomam
Marſh." W.H. Again 1581. Quarto.

1573. Novum Teſtamentum Deſ. Eraſmi Roterodami. Octavo.

1573. Dialogorum ſacrorum libri quatuor. Authore Sebaſtiano Caſtalione.
Again 1580. Octavo.

1574. " Marcelli Palengenii, ſtellati poetae doctiſſimi, Zodiacus vitae.
Hoc eſt, de hominis vita, ſtudio, ac moribus optimè inſtituendis, libri
xii." With a cut of Mars, or a marſhall, alluding to his name. Octavo.

" The

"THE PREACHER, or Methode of preaching, vvrytten in Latine by 1574. Nicholas Hemminge, and tranflated into Englifhe by I. H. *(John Horsfall)* Verye neceffarie for al thofe that by the true preaching of the word of God, labour to pull downe the Sinagoge of Sathan, and to buylde vp the Temple of God. 1. Corinth. 1. 18. The preaching of the Croffe, &c. Scene and alowed &c. Imprinted—1574. Cum Priuilegio." Dedicated " To the right Hon. Douglas Lady Sheffeld, late wyfe of Lord Iohn Sheffeld difceafed." Next, in an epiftle " To his brethren & fellowe Minifters—, the interpretour wifheth peace," &c. Then, a table of the contents. Contains befides, 68 leaves. W.H. Again 1576. Octavo.

"THE PAGEANT OF POPES, contayninge the lyues of all the Bifhops 1574. of Rome, from the beginninge of them to the yeare of grace 1555. Deuided into iii. fortes, bifhops, Archbifhops, and Popes, vvhereof the two firft are contayned in two bookes, and the third forte in fiue. In the vvhich is manifeftly fhevved the beginning of Antichrifte, and increafing to his fulneffe, and alfo the vvayning of his povver againe, accordinge to the Prophecye of Iohn in the Apocalips. Shewing many ftraunge, notorious, outragious, and tragicall partes played by them, the like vvhereof hath not els bin hearde : both pleafant and profitable for this age. Written in Latin by Maifter Bale, and now Englifhed, with fondrye additions, by J. S. *(John Studley.)* Behold I come vpon thee, &c. Nahum. 3. (5, 6.) Come away from her my people, &c. Apoca. 18. (4. 6.) Anno 1574." Prefixed are the tranflator's dedication " To the right Hon. Thomas Earle of Suffex, Vicount Fitzwalter, Lorde of Egremont and of Burnel, one of the Queenes Maieftyes hon. priuye Counfaile, & Lord highe Chamberlaine of her houfe, Of the noble order of the Garter knighte, Juftice of Oyer, of the Forefts, Parkes, VVarraines and chafes from Trent Southward, and Captaine of the Gentlemen Pentioners.---The tranflator to the Reader." Bale's dedication " To the moft vvorthie and learned men maifter Simond Sulcer, Henry Bullenger, John Caluin, *and* Philip Melanthon.----- Iohn Bale to the Reader.---To the Reader, T. R. Gentleman." In verfe. The Pageant contains Fol. 198. At the end, " Diuers cafes wherin the Pope doth fell Difpenfation*s* contrarye to Gods Lawe and his owne Canons, and the price of the difpenfation according to the rate in his Courtes." Thefe i fhall not enumerate, but refer the inquifitive reader to a fimilar lift, annexed to Sir Richard Steel's Romifh Ecclef. Hiftory of late years, 1714. W.H. Quarto.

" A Briefe Collection and compendious extract of ftrange and memorable thinges, gathered out of the Cofmographye of Sebaftian Munfter. Wherein is made a plaine defcription of diuers and ftraunge Lawes, Rites, Maners and properties of fondrye Nations, and a fhort report of

---

ᵇ According to a M.S. note in my copy, he died 9 June, 1583. " A Remembraunce of the Life, Death, and Vertues of the moft noble Lord Thomas late Earl of Suffex," in verfe, was printed the fame year. Catal. Bibl. Harleianæ. Vol. III. No. 5962.

ftraunge

ſtraunge hiſtories of diuers men, and of the nature & properties of cer‐
taine Fowles, fiſhes, Beaſtes, Monſters, and ſondry Countryes & places.
Imprinted—1574." It is introduced with an addreſs " To the Reader."
Contains beſides, fol. 101.   Octavo.

1575.   " A Brief ſumme of the whole Bible," &c. as p. 751. Licenſed.
        Octavo.

1575.   " A CATHOLIKE and Eccleſiaſticall expoſition of the holy Goſpell
after S. Iohn. Gathered," &c. as to his expoſition of S. Mathewe. " Im‐
printed—1575." In the compartment with the alphabet on a tablet at
bottom. Dedicated " To——Lorde Thomas Earle of Suſſex, &c. And
to the right hon. Lady his wyfe." Then, an alphabetical table. Con‐
tains beſides, 619 pages. Licenſed. W.H.   Folio.

1575.   " THE FIRST parte of the Mirour for Magiſtrates, contayning the
falles of the firſt infortunate Princes of this lande: From the comming
of Brute to the incarnation of our ſauiour and redemer Jeſu Chriſte. Ad
Romanos. 13. 2. Quiſquis ſe opponit poteſtati, Dei ordinationi reſiſtit.
Imprinted——1575. Cum Priuilegio." In the compartment with his
mark at bottom. It begins with a table reciting the ſeveral hiſtories, 17
in number, ending with " The Tragœdy of Irenglas." Next is I. Hig‐
gin's epiſtle " To the Nobilitie and all other in office," ſuperſcribed
" Loue and liue." Then, another " To the Reader." Contains beſides,
Fol. 81." W.H.   Quarto.

In the author's introduction to this book is the firſt attempt i have ob‐
ſerved of the idea of diſtinguiſhing an apoſtrophe by a character. At
firſt the hyphen is uſed, as in the 6th line of the firſt ſtanza,
    " As when the time of yeare and wether-s fayre."
In the 10th ſtanza it is neglected in the word " Phaëtons"; but, as you
may obſerve, the diæreſis is uſed; the firſt i have obſerved. Afterwards
the comma is applied as in modern uſe, in the 14th, and 19th ſtanzas;
but in the 15th, and 16th, the comma is ſet at the bottom of the line,
thus, " th, ende."

1575.   " Iohn Baniſter, Maiſter in Chirurgery & Licentiate in Phiſick, his
neceſſary treatiſe of chirurgery, briefly comprehending the general & par‐
ticular curation of Vlcers: drawne eſpecially out of Antho. Calmetteus,
and Io. Tagaltius. Whereunto is added certaine experiments of his owne
inuention."   Octavo.

1575.   " THE FIRSTE parte of Churchyardes Chippes, contayninge twelue ſeueral
Labours, diuiſed and publiſhed onely by Thomas Churchyard, gentilman."
The contents are, " 1. The ſiege of Leeth   2. A farewell to the world.
3. A fayned fancie of a ſpider and the gowte. 4. A dolefull diſcourſe of
a lady and a knight. 5. The rode into Scotland, by ſir William Drury,
knight. 6. Sir Simond Burleis tragedie. 7. A tragecall diſcourſe of the
unhappie mans life. 8. A diſcourſe of vertue. 9. Churchyards dreame.
10. A tale of a frier and Shumakers wief. 11. The ſiege of Edenborough
caſtle. 12. The whole order of the receiuing of the queens maieſtle into
Briſtowe." Sheets O 1111. Again 1578. George Maſon, Eſq; Quarto.

" THE

"THE MIRROVR of Madnes, or a Paradoxe maintayning Madnes to 1576. be moſt excellent: done out of French into Engliſh, Ia. San. Gent. Imprinted——1576." Dedicated " To the right woſhipful Sir Arthur Champernon Knighte." Then, ſome verſes " To the Reader," ſigned, " Tutto per il Meglio." Contains together D 3, in eights, half ſheets. W.H.
Sixteens.

" The Touchſtone of Complexions. Generallye appliable, expedient 1576. and profitable for all ſuch, as be deſirous & carefull of theyr bodylye health. Contayning moſt eaſie rules & ready tokens whereby euery one may perfectly try, and thoroughly know, aſwell the exacte ſtate, habite, diſpoſition, & conſtitution of his owne Body outwardly: as alſo the inclinations, affections, motions & deſires of his mynd inwardly. Firſt written in Latine by Leuine Lemnie, and now engliſhed by Thomas Newton. Noſce teipſum. Imprinted—1576. Cum priuilegio." Dedicated " To—Sir William Brooke Knighte, Baron Cobham, and Lorde Warden of the Cinque Portes.——From Butley in Cheſſhire, 21 Sept. 1576." Contains beſides, 157 leaves, and a table at the end. W.H. Again 1581.
Octavo.

" CARDANVS Comforte, tranſlated into Engliſhe. And publiſhed by 1576. commaundement of the righte Honorable the Earle of Oxenforde. Newly peruſed, corrected, and augmented. Anno Domini 1576. Imprinted——. Cum Priuilegio." In the compartment with his mark. Dedicated " To the Righte Hon. & my good Lord the Earle of Oxenforde, Lorde great Chamberlayne of Englande.——1. Jan. 1571.——Tho. Bedingfeld." Then, " To my louinge frend Thomas Bedingfeld Eſquyer, one of her Maieſties gentlemen Pentioners.——From my newe contrye Muſes at Wiuenghole——By youre louinge & aſſured frende. E. Oxenford." Alſo, ſome commendatory verſes by " the Earle of Oxenforde, Thomas Churchyarde, and George Gaſcoigne." Contains beſides, 102 leaves. W.H.
Quarto.

" A MORAL METHODE of ciuile Policie, Contayninge learned and 1576. fruictful diſcourſe of the inſtitution, ſtate & gouernment of a common Weale. Abridged oute of the comentaries of the Reuerende and famous clerke Franciſcus Patricius Byſhop of Caieta in Italye. Done out of Latine into Engliſhe by Rycharde Robinſon, Citezen of London. Seene and allowed. &c. Anno Domini 1576. Imprinted", &c. as the foregoing article. Dedicated " To——his ſinguler good maiſter Sir William Allen, Knight. Aldermā of the City of London." Then, ſome Latin verſes by Tho. Newton, Cheſtreſhyrius. Contains beſides, 88 leaves, and a table; at the end thereof " Certaine notes ſelected oute of the Preface of Franciſcus Patricius Senenſis," &c. W.H.
Quarto.

" IOHN HEYWOODES WOORKES. A dialogue conteyning the number of 1576. the effectual Prouerbes in the Engliſh tongue, compact in a matter concerning two maner of Mariages. With one hūdreth Epigrammes: and three hundreth of Epigrammes vppon thre hundreth prouerbes: and a
fifth

fifth hundred of Epigrammes. Whereunto are newly added a fixte hundred of Epigrammes by the faide John Heywoode. 'Anno Domini 1576. Imprinted", &c. as the two foregoing articles. Contains sheets P 5. George Mason, Esq; Quarto.
Sir Thomas Elyot's Castle of Health. Octavo.

1576. "THE FIVE Bookes of the Famous, learned, and eloquent man Hieronimus Oforius, contayninge a difcourfe of Ciuill, and Chriftian Nobilitie. A worke no leffe pleafaunt then profitable for all, but efpeciallye the noble Gentlemen of England, to vievv their liues, their eftates, and conditions in. Tranflated out of Latine into Englifhe by VVilliam Blandie late of the Vniuerfitie of Oxeford, and novv fellovv of the middle Temple in London. Imprinted—1576. Cum Priuilegio." In the compartment with his mark. Dedicated "To—Lord Robert Dudley, Erle of Leycefter, Baron Denbigh, Mafter of the Horfe to the Queenes Maieftie, Knighte of the noble order of the Garter, highe Chancelour of the Vniuerfitye of Oxforde.——At Newberie, 6 Aprill, 1576.—W. Blandie." Next are commendatory verfes ; Henr. Ferrarius Badifleius Guilielmo Blandeio suo; Leonardius Louelaceus; Ioannes Butterwike ; Richardus VVarnefordus ; Ioannes VVakemanus; Thomas Newtonus ; *and* VVilliam Fofter." Then, the epiftle of Hier. Oforius to prince Lewis, fon to Emanuell king of PortugaL Contains befides, 110 leaves. W.H. Quarto.

1576. "Nicolai Carri Nouo caftrenfis Angli, de fcriptorum Britannicorum paucitate." Bibl. Publ. Cantab. Octavo.

1577. "The golden booke of the leaden goddes, wherein is defcribed the vayne imaginations of heathen Pagans, and counterfeit chriftians. Wyth a defcription of their feueral tables, what each of their pictures fignified. By Stephen Bateman," (or Batman). 36 leaves.* Licenfed. Quarto.

1577. "Dialectica Ioannis Setoni Cantab. annotationibus Petri Carteri, vt clariffimis ita breuiffimis explicata. Huic acceffit, ob artium ingenuarum inter fe cognationem, Guilielmi Buclæi Arithmetica." His device: Fortune ftanding on a ball, holding a palm in her right hand, and a fword in her left; the fun on one fide, and the moon on the other. "Londini, Apud Thomam Marfh. Anno Domini, 1577. Cum Priuilegio." On the back, are fome Greek verfes by John Cheek, and Richard Stephens. Then follow thefe prefixes: "Clariff. præfuli epifcopo Wintonienfi, fummo huius Academie Cancellario.—Decimo octavo calendas Februarij. Cantabrigiæ, ex collegio diuo Ioanni dicato.—Io. Setonus, dicti coll. focius.---Hon. & clariff. viro ac domino, D. Edouardo, Durouentanæ prouinciæ comiti, Monæ infulæ Hebridum celeberimæ regulo, nobiliff. Garteriorum ordinis equiti aurato, &c. literarum Patrono benigniffimo.—Cantabrigiæ. iiij. Calendas Ianuarij.—P. Carterus, focius Collegij S. Ioan. Euangeliftæ.---Ad lectorem, Petri Carteri Hexafticon.---In Annotationes, Carteri, Tho. Dranta.---Tho. Newtonus,
"Ceftrefhyrius."

THOMAS MARSHE.

Ceftrefhyrius." Contains befides, 17, in eights. Licenfed. W.H. Again 1584. Octavo.

" FOVVR SEueral Treatifes of M. Tullius Cicero: Conteyninge his 1577. moft learned and Eloquente Difcourfes of Frendfhippe: Oldage: Paradoxes: and Scipio his Dreame. All turned out of Latine into English, by Tho. Newton. Imprinted—Cum Priuilegio. 1577." Dedicated " To ——, his very good Lord, Fraūcis Earle of Bedford, Lord Rufsell, of the Noble order of the Garter Knight.—From Butley in Chefshyre, 4 Maye. 1577." Contains befides S, in eights, including a table at the end. Licenfed. W.H. Octavo.

"THE LAST part of the Mirour for Magiftrates", &c. The fame as was 1578. printed in 1559. " Newly corrected and enlarged", with the tale of Humphry duke of Glocefter. Contains Fol. 183, befides the prefixes. W.H. Alfo without date. Quarto.

" Iohn Gardener his confeffion of the Chriftian faith, tranflated out of 1578. French by Iohn Broke." Maunfell, p. 30. Again 1583. Octavo.

" A HYVE full of honye. contayning the Firft Booke of Mofes, called 1578. Genefis, turned into englifhe meter, by William Hunnis, with notes in the margin." Dedicated to R. Dudley, earl of Leycefter. " Cum Priuilegio Regiæ Maieftatis." Licenfed. Quarto & Octavo.

" A Prognoftication euerlaftinge of right good effecte, fruictfully aug- 1578. mented by the auctour, contayning plain, briefe, pleafaunte, chofen rules to iudge the Weather by the Sunne, Moone, Starres, Comets, Rainebow, Thunder, Cloudes, with other extraordinarye tokens, not omitting the Afpects of the Planets, with a briefe iudgement for euer, of Plenty, Lacke, Sickenes, Dearth, VVarres &c. opening alfo many naturall caufes worthy to be knowen. To thefe and other now at the laft are ioyned diuers General pleafaūt Tables, with many compendious Rules, eafye to be had in memory, manifold wayes profitable to all men of vnderftanding, Publifhed by Leonard Digges Gentleman. Lately corrected and augmented by Thomas Digges his Sonne." The cut of a naked man and the figns of the Zodiac: the engraver's initials, C. I. "Imprinted at London by Tho. Marfhe. Anno 1575." Dedicated "To the right Hon. Sir Edward Fines, Earle of Lincolne, Baron Clinton and Say, knight of the—Garter, Lord high Admiral of England, Ireland & Wales, and of the Dominions & Iles thereof. of the towne of Calice & Marches of the fame, Normandy, Gafcoigne & Guian. And Captain generall. of the Queenes Maiefties Seas and Nauie Royal.—Tho. Digges," Then, an epiftle " To the reader." Contains befides P 2, in fours. See the Tectonicon, p. 860. W.H. Quarto.

"THE MOST EXCELLENT PRofitable, and pleafaunt Booke of the fa- 1578. mous Doctor and expert Aftrologian Arcandam, or Aleandrin, to find the Fatall deftiny, conftellation, complexion, & naturall inclination of euery

man

man and childe by his birth. VVith an addition of Phifiognomy, very pleafat to read. Now newly tourned out of French into our vulgar tongue, by William Warde. Imprinted——1578." Contains Q 4. in eights. W.H. Octavo.

1578. " A perfite platforme of a hoppe garden, and neceffarie inftructions for the making and mayntenance thereof, with notes and rules for reformation of all abufes, commonly practifed therin, very neceffarie and expedient for all men to haue, which in any wife haue to do with hops. Now newly corrected and augmented, by Reynolde Scot." With cuts. 63 pages.* Quarto.

1578. " An Ordinary for all faithfull Chriftians", &c. as p. 705. To which is added " A briefe fumme of the Bible: alfo a Chriftian inftruction for all perfons, young and old." Sixteens.

1578. " The Defence of the Soul againft the ftrongeft affaults of Satan, by R. C." Sixteens.

1579. " Bvlleins Bulwarke of Defence againft all Sickneffe, &c. as p. 835. Imprinted—1579." In the fame compartment as Flores Hiftoriarum. In the oval at top, " Eccle. 38. Altiffimus creauit de terra medicinam, & vir prudens non abhorrebit illam." At bottom, " Cicero. 1. Offic. Eos quorū vita perfpecta eft in rebus honeftis, & bene de Rep. fentientes obferuare & colere folemus." This edition hath, added to the prefixes of the former edition, feven ftanzas by " Tho. Newton to the Freendly Reader." Within the firft letter D, is his own arms with T. N. over the fhield. The title for the Dialogue between forenefs and Chirurgi is in the compartment with the alphabet on a tablet at bottom. W.H. Folio.

1580. Æfopi Fabulæ. From which W. Bulloker tranflated his edition.

1580. " Sententiæ Ciceronis, Demosthenis, ac Terentij Dogmata Philofophica. Item, Apophthegmata quædam pia. Omnia Ex fere ducentis auctoribus, tam Græcis quàm Latinis, ad bene beatéq; viuendum diligentiffime collecta. Auctorum nomina fequentes pagellæ Indicabūt." His device, Mars at full length in a Roman habit, with fword and fhield. " Londini—1580. Cum Priuilegio Regiæ Maieftatis." On the back, " Hoc volumine continentur.---Ad Lectorem. Habes hunc librum, amice Lector, fi vnquam antea caftigatum, nūc fanè, &c. Lege igitur, vtere, fruere, ac vale." It has prefixed, " Nomina Auctorum." Alfo, " Index rerum & verborum." Contains befides, 482 pages. W.H. Octavo.

1580. " Approved Medicines and Cordiall Receiptes, with the natures, qualities, and operations of fundry Simples. Very Commodious & expedient for all that are ftudious of fuch knowledge. Imprinted—1580." In a border. Dedicated " To the worfhipful, the Maifter, wardens & generall Affiftantes of the fraternitye of Chirurgians in London.—From Butley,—19 Octob. 1580.—Tho. Newton," An addrefs to the reader, and a table. Contains befides, 95 leaves. W.H. Sixteens.

" The

THOMAS MARSHE.

"The Caſtle of Health, by ſyr Thomas Elyot knight." Octavo. 1580.

"FLOVVERS OR ELOQUENT Phraſes of the Latine ſpeach, gathered out 1581. of al the ſixe Comœdies of Terence. VVherof thoſe of the firſt thre were ſelected by Nicolas Vdall. And thoſe of the latter three, novv to them annexed, by I. Higgins, very profitable and neceſſary for the expedite knowledge of the Latine tounge. Imprinted—1581. Cum Priuilegio." In a border. On the back, "Nic. Vdalli carmen." It has prefixed alſo, "N. Vdallus ſuauiſſimo diſcipulorum ſuorum gregi ſalutem plurimam dicit.——In hos P. Terentij floſculos N. Vdalli & I. Higgini opera excerptos, Tho. Newtoni Δωδεκαϭιχον." See p. 447. Contains beſides, Dd 2, in eights. W.H. Octavo.

"A COMPENDIOUS or briefe examination of certayne ordinary com- 1581. plaints of diuers of our country men in theſe our dayes: which although they are in ſome part vniuſt and friuolous, yet are they all by way of dialogues throughly debated and diſcuſſed. By W. S. 'Gentleman. Imprinted—1581. Cum Priuilegio." In the compartment with his mark. On the back, the queen's arms, crowned and gartered. Dedicated " To the moſt vertuous and learned Lady, my moſt deare and Soueraigne Princeſſe Elizabeth, by the Grace of God Queene of England, &c. Your Maieſties moſt faithfull and louing Subiect W.S." Then, "A Table of thynges moſt notable," Contains beſides, Fol. 55. The running title, "A briefe Conceipte of Engliſh pollicy." W.H. Quarto.

Meredith Hanmer's " Confutation of the great bragge and challenge of 1581. M. Champion a Ieſuite, containing 9 articles by him directed to the Lords of the Councell." Maunſell, p. 57. Licenſed. Octavo.

"The maner and forme of examination before admiſſion to the table of 1581. the Lord,—uſed in Scotland." Licenſed. Octavo.

"SENECA HIS TENNE TRAGEDIES, TRANSLATED INTO Englyſh. Mer- 1581. curij nutrices, horæ. Imprinted—1581." In the compartment with his mark at bottom. Dedicated " To—Sir Thomas Heneage knight, Treaſurer of her Maieſties chamber.—From Butley in Cheſſhyre, 24. Aprill 1581.—Tho. Newton." Then, " The names of the Tragedies, and by whom each of them was tranſlated. Hercules Furens, Thyeſtes, and Troas, by Jaſper Heywood. Oedipus, by Alex. Neuile, 1560. Hippolitus, Medea, Agamemnon, Hercules Oetæus, by John Studley. Octauia, by T. Nuce. Thebais, by Tho. Newton." Contains beſides, 217 leaves. Licenſed. W.H. Quarto.

"The Diall of deſtinie, wherein may bee ſeene the continuall courſe, 1582. diſpoſition, qualities, effects, and influence of the 7 Planets. Compiled aſwell Aſtrologically, as poetically, and philoſophically. By John Maplet. Octavo.

⁎ See the prefixes to Shakeſpeare's Plays by Johnſon and Steevens, p. 205, &c. Edit. 1778.

1583. "A CATHolike and Ecclefiafticall Expofition of the Holy Gofpell after S. Marke and Luke. Gathered" &c. as to St. Mathew,-and S. John. "Im-printed—neare vnto S. Dunftanes church. 1583. Cum Priuilegio." In the compartment with his mark at bottom. Contains pages 341. Licenfed. W.H. Quarto.

1584. "THE Famous Hyftory of Herodotus. Conteyning the Difcourfe of dyuers Countreys, the fuccefsion of theyr Kynges: the actes and exploytes atchieued by them: the Lavves and cuftomes of euery Nation: with the true Defcription and Antiquitie of the fame." A medal "IMP. IVSTINUS IVN. AVGVST." Under it, "Deuided into nine Bookes, entituled with the names of the nine Mufes. At London. Printed by Tho. Marfhe, 1584." In the compartment with his mark at the bottom. Dedicated "To the right excellent and vertuous Gentleman Mayfter Robert Dormer, fonne to the noble Knight Sir Wyllyam Dormer.—B. R." Then, an epiftle, "To the Gentlemen Readers.—B. R." Perhaps Barnaby Rich. Contains befides, Fol. 119. Licenfed. W.H. Quarto.

1587. "Rules and Documentes touching the vfe and practife of the common Almanaches, which are called Ephemerides: alfo a brief Introduction vpon the iudiciall Aftrologie, to prognofticate things to come: with a treatife added, touching the Coniunction of the Planets in euery of the 12. Signes, tranflated by Humfrey Baker." Licenfed in 1557. Quarto.

1587. "THE Caftell of Health corrected, and in fome places Augmented by the firft Author—.And now newly imprinted,—1587. Imprinted—in Fleeteftreete by Tho. Marfh." In a border. Befides the prefixes, 96 leaves. W.H. Sixteens.

1587. "A Difcourfe of the fubtill Practifes of Deuilles by VVitches and Sorcerers. By which men are and haue bin greatly deluded: the antiquitie of them: their diuers forts & Names. With an Aunfwer vnto diuers friuolus Reafons which fome doe make to prooue that the Deuils did not make thofe Aperations in any bodily fhape. By G. Gyfford. Imprinted at London for Toby Cooke. 1587." In the compartment with David and Mofes on the fides. Dedicated "To—Maifter Richard Martin, Alder-man, and Warden of her Maiefties Mint." 34 leaves. W.H. Quarto.

1587. "The banifhment of Cupid. Tranflated out of Italian into Englifh by Thomas Hedly." Alfo without date. Octavo.

1587. "The Boke of Surueying and Improuementes, newly corrected and amended, very neceffary for all men. Imprynted—nere to f. Dunftanes Church." Prefixed are a table, and two ftanzas by the author, to his book. 54 leaves befides. Sixteens.

—— "The Booke of Hufbandrye, newly corrected and amended by the author, Fitzharbarde." Licenfed, 1559. Mr. Tutet. Octavo.

"The

"The Fan of the Faithfull to trie the truth in controuerſie: collected by
A. B. Dedicated by Iames Price." Maunſell, p. 49. Sixteens.

"IN THIS BOOKE IS CONTEINED the office of ſhiriffes," &c. as p. 574.
K 6, in eights. "Imprynted—nere to S. Dunſtones Church. W.H.
Octavo.
Another edition, with "Anno Domini" on the title-page, but no date.
Octavo.
"—— certaine bokes cōpyled by mayſter Skelton," &c. as p. 764.

"A ſtronge defence of the maryage of pryeſtes, agaynſte the pope,
Euſtachians, and Tatanites of our time, made dialogue wiſe, by John
Veron, betwixte Robin Papyſte, and the true Chriſtian." * Octavo.

"The offspring of the houſe of Ottomano, and officers belonging to
the great Turkes court. Whereunto is added, Bartholomeus Georgieuiz
epitome of the cuſtomes, rites, ceremonies, and religion of the Turkes,
&c. Engliſhed by Hugh Goughe." Dedicated "To the right worſhippeful
knight, Sir Thomas Greſsam. Hugh Goughe wiſheth all godly honour,
with Neſtors yeres, and Galens healthe." Wherein Sir Thomas is highly
commended for his building the Burſe, or Exchange of London. Licenſed.
W.H.+ Octavo.
Beſides the books abovementioned, he had licenſe for printing the
following, viz. In 1557, "An Almanacke & Pronoſtication of Hen.
Lowes. A Pronoſtication of Aſkams." In 1558, "An almanacke, &c.
of Lewes, Waughans, with Aſkams. Beuys of Hampton. The vij wyſe
maſters of Rome." In 1560, "The almanackes, &c. of Waughan,
Monflow, & of Kēnyngham. The Epythaſe of Bradfordes." In 1561,
"The pycture of a monſterous chylde, bourne in Suffolke." In 1562, "An
almanack, &c. of H. Lowe. The Myrrour of Mageſtrates. Alſo, The
ij parte. Ryght godly Rules how all deuoute and vertuous people ought to
occupye and exerciſe themſelves." In 1563, "An almanack, &c. of
H. Lows." In 1564, again. In 1565, "The almanacks, &c. of John
Securis, and of M. Dr. Lows. An epygrāme of the death of Cuthberte
Skotte sōme tyme beſſhoppe of Cheſter, by Roger Shacklocke, and replyed
agaynſte by Tho. Drant. The tru copye of the laſte aduertiſemente that
cam from Malta. An epytaphe of the deathe of the famous and renowned
knyghte Tho. Challenor. An almanacke with the names of the Kynges:
The hiſtory of Phil. Cōmines vpon the actes and deades of kyng Lewes xj.
auctoryſſhed by my lorde of London." In 1566, "The 4th book of
the diall of prēnces. An almanacke, &c. of M. Dr. Low; alſo, of Mr.
Securis." Both of them again in 1567. Alſo, "The book of Salomans
Prouerbs. Epygrames & Sentences ſprituall by Drant." In 1568, "The
almanacks &c. of Low, Securys, & of T. Stephens gent." In 1569,
"The maryage of wytt and Scyence. The almanack, &c. of H. Lowe.
The Jniunction of Edwarde byſſhope of Peterbroughte." In 1570,
"The

" The almanake, &c. of John Securis, *authorised* by my lorde of Canterbury." In 1581, " An anfwere to the calumnious letter and erronious propofitions of an Apoftata named John Hambleton, compofed by Mr. Will. Flower. A brief difcours of roiall monarchie as of the beft cōmon weale: ' written by Charles M'burie. Whereunto are annexed cerſaine Italian prouerbs. By his owne cōmandement, without anye handes. Nowe head warden. *Alfo*, .An Apologie *or* defence of—William prince of Orange, againſt the Proclamation of the K of Spaine, in Englifh. Neuer printed by him. *Alfo*, Tractatus de Eccleſia per Phil. Mornew, Plaſſen."

THOMAS GEMINIE, or GEMINI,

WAS an engraver on copper, and probably the firſt who exercifed that art in England; however his name appears on the plates of the firſt edition of Compendiofa totius anatomie delineatio, in 1545. See p. 577, &c. The 2d edition was printed by Nic. Hill, for him, in 1552. Of him i find the following entries in the Stationers' regifter book A. " 21 July, 1554. Rec. of Tho. Gemyne ſtranger for tranfgreſſynge the ordenaunce of this howfe, callynge a brother of the companye falſſe knave. xij. d." Alfo, in their collection " for the howfe of Brydewell by cōmaundement of the Lorde the maior & the Courte of Aldermen," he contributed xx. d. than which ſum few contributed more. He afterwards became a printer, when he dwelt within the Black-friars.

1556. " A prognoſtication euerlaſting of ryght good effecte, frutefully augmented by the author, conteyning playne, brief, pleaſant, chofe rules to iudge the weather. by the ſunne, moone, ſterres, &c. once again publiſhed by Leonard Digges, gentleman." Dedicated by him to ſir Edward Fines, knight of the garter. Containing 41 leaves. It had been printed, 1555. Again imprinted by Thomas Geminie, &c. Afterwards by T. Marſh.* See p. 867. Quarto.

1559. " Compendiofa totius anatomie delineatio, ære exarata per Thomam Geminum Londini. 1559." This is all the title in my copy, and differs from thofe of 1545, and 1552, only in the date, and having the portrait of the queen in the place of the king's arms in the edition of 1545, and the portrait of K. Edw. VI. in that of 1552. The fame copper-plate, thus occaſionally altered, ſerving for each edition. Mr. Ames i ſee took his title from Maunfell. Geminie dedicates this edition to the queen, as he did the former impreſſions firſt to her father, then to her brother, both which were wafted, and the fame then required at his hands; fays he

had

had *firſt* made uſe of the pains of Nicholas Udal, and now of Richard Eden, and hopes it may be of uſe to the publick. Contains K 2, in ſixes, beſides the copper-plate prints. " Imprinted at London, within the blacke-fryars by Tho. Gemini—1559. Menſe Septemb." W.H. Folio.

ANTHONY KYTSON

DWELT, or kept ſhop in Paul's church-yard, at the ſign of the Sun. Though he was no member of the Stationers' company he was fined by them, in 1558, for keeping his ſhop open on Sundaies: again, in 1566, for having certain books found in his cuſtody on a ſearch made by Hu. Singleton and Tho. Purfoot, appointed for that purpoſe; when he was fined iij. l. vj. s. viij. d. He put up a monument for his wife in the north aiſle of St. Faith's, with the inſcription⁴ below.

The book of Bertram the prieſt. See p. 583.   Octavo. 1549.

" Pſalterium Dauidicum, ad vſum eccleſie Sarisburienſis. Impreſſum (1555) per Antonium Kitſon." The almanack begins, 1555.

" This lytle Practice of Johannis de Vigo in medicine, is tranſlated out 1564. of Latin into Englyſhe, for the health of the Body of man. Now newly Imprinted the fyrſt day of May 1564."   Octavo.

Will. Williamſon printed an almanac for him; and Legat at Cam- 1573. bridge printed for him after.

" A little book whych hath to name, " Why came ye not to court, compyled by Mayſter Skelton, poet laureate." See p. 588.

" Hereafter followeth a litle boke called *Colyn* Clout compiled by maſter Skelton, poete laureate." See p. 779.   Sixteens.

" A little olde booke of Cookerie. For him". Maunſell, P. 2; p. 6. Sixteens.

" The huſband mans practiſe, or Pronoſtication for euer. As teacheth Albert, Alkind, Haly, & Ptolome." 2 ſheets.   Sixteens.

" ᵈHear lyeth the bodie, taken from lyfe, | She ys gone before, he is yet behind,
Of Margaret, Anthony Kytſon's wyf; | And hoopes in Heaven his wyfe to fynde:
Whoſe vertues every where were ſuch, | Whoſe leeke on earthe, for his degree,
As his great want bewayleth much. | He never lookes alive to ſee.
T'en fair bab:s ſhe brought to blys, | Obiit xxi. November 1567."
And of th' eleventh now departed ſhe ys: | Dugdale's St. Paul's, p. 119.

" A

ANTHONY KYTSON.

—— " A Godly dyfputacyon very expedyent for euery man to read. Compofed by the holly Dyonifius, and lately tranflated into Englyſhe, for the fortherance of them in vertue, that are not lerned in latyne."
Octavo.

—— " A booke of the properties of Herbes, called an Herball. Whereunto is added the tyme that Herbes, Floures & Seedes ſhould bee gathered to bee kept the whole yeare, wyth the vertue of the Herbes when they are ſtylled. Alſo a generall rule of all maner of Herbes, drawen out of an auncient booke of Phyficke by VV. C." *Walter Carey*. On the back begins " The tyme of gatherynge Seedes," &c. Contains befides, X 4, in eights. For him. W.H.
Octavo.

—— " Spare your good." Alſo, " The Parlament of byrdes." Quarto.

THOMAS POWELL, or POWEL,

WAS made free of the Stationers' company 21 July, 1556, and his name is inſerted in their charter. I do not find him once fined, and yet he has licenſe for only one book. He appears to have done the printing buſineſs for T. Berthelet, and to have dwelt in his houſe in Fleet-ſtreet. See p. 465.

1547. The Attorney's Academy. Middle Temple Library.   Octavo.

1556. " THE SPIDER and the Flie. A parable of the Spider and the Flie, made by Iohn Heywood. Imprinted at London in Fleteſtrete by Tho. Povvell. Anno 1556." On the back is the author's portrait at full length. Prefixed are " The preface," in verſe; and " The Table" of contents. This poem is compoſed in the ſeven-lined ſtanza, and is divided into 98 chapters, with a cut to each of them. Then, " The concluſion,* with an expoſiſſion of the Auctor touching one peece of the latter part of this parable." At the end " Imprinted—Cum priuilegio—folum." George Maſon Eſq; and W.H.   Quarto.

1557. " A PLAYNE DEMONSTRATION OF JOHN Frithes lacke of witte a d

* In which we are informed that by the ſpiders we are to underſtand the proteſtants; the flies, the catholicks; the maid, queen Mary; her broom, the civil ſword; her maſter, Chriſt; her miſtreſs, mother church. How juſtly the characters are ſupported i do not pretend to determine. The author was profeſſedly a papiſt, and as ſuch attempts vindicating their cauſe in the adminiſtration of that reign.

learnyge,

learnyge, in his vnderftandynge of holie Scripture, and of the olde holy doctours, in the bleſſed Sacrament of the Aulter, Newly ſet foorthe by John Gwynneth, Clerke. Londini. 1557." C c 3, in fours. " Cum priuilegio—folum."* W.H.   Quarto.

The Italian Grammar and Dictionary, by W. Thomas. See p. 453, &c. 1561.
Quarto.
Statutes. 27, 28, 31, 34, 35 Hen. VIII. W.H.   Folio. 1562.

" The Booke of freendeſhip of Marcus Tullie Cicero. Anno dnĩ. 1562. 1562." Dedicated " To the right vertuous, and my ſinguler good Lady Katharine duches of Suffolke.—John Harryngton." Then, an epiſtle " To the Reader." Contains beſides, 65 leaves, and a table at the end. See p. 466. W.H.   Small.
" An Apologie of priuate Maſſe, ſpred abroade in writing, without 1562. name of the authour: as it ſeemeth againſt the offer and proteſtacion made in certayne Sermons, by the reuerent father, Byſſhop of Salſburie : with an anſwer to the ſame Apologie, ſet foorth for the maintenance and defence of the trueth. Peruſed and allowed, by the reuerent father in God, Edmonde, Biſſhop of London, according to the Queenes maieſties Jniunctions. Londini Mens. Nouemb. 1562." The orthography of this differing ſo conſiderably from Mr. Ames's copy has the appearance of two editions in the ſame month. The apology on 31 leaves. Then, on a ſeparate title page, " An Anſwere in defence of the truth. Againſte the Apologie of priuate Maſſe. Londini Menſ. Nouĕb. 1562." In a compartment with Lucretia in a medallion at bottom. On the back " The cheeſe poinctees touched in this defence of the trueth." Then, " The preface to the Reader." Contains beſides, 121 leaves. The whole on one ſeries of ſignatures. Licenſed.* W.H.   Octavo.

" John Heywoodes VVorkes." &c. as p. 865. Again, by Felix 1562. Kingſton, 1598.   Quarto.
" A breeſe balet touching the trayterous takyng of Scarborow Caſtle. ——
—Cum priuilegio—folum."   A broadſide.

" The Caſtell of Helth corrected, and in ſome places augmented, by the —— firſt author—1541." In Berthelet's compartment with 1534 on the fell. The Proheme begins on the back ; then, the table. Contains beſides, 97 leaves. On a ſeparate leaf, " Jmprinted—in Fleteſtrete,—Cum priuilegio—folum." W.H.   Sixteens.

JAMES BURREL

PRINTED " A Godly and wholſome preſeruatiue againſt deſpera- 1559. tion," &c. See p. 355. At the end, " Imprinted—by James Burrel, dwelling without the north gate of Paules, in the corner houſe of Paternoſter-row, opening into Cheapſide. Cum priuilegio—folum."* Octavo.

OWEN

OWEN ROGERS, or AP-ROGERS,

WAS made free of the Stationers' company, 8 Octob. 1555; but a disorderly member, printing other men's copies; and without Licenſe; for which he was ſeveral times fined. He dwelt at the Spread Eagle between both St. Bartholomews, in Smithfield.

1559. "The ſerpent of diuiſion. Whych hath euer bene yet the chefeſt vndoer of any Region or Citie, ſet forth after the Auctours old copy by J. S. Anno M. D. L. J X. the iiii. of May." In a neat architectiue compartment with O. R. on the ſell. D 7, in eights. "Thus endeth this little treatiſe entituled: the Serpent of diuiſion, made by John Lydgate." Then, "The declaracion of thys tragical Hiſtory" &c. in three octave ſtanzas. On the back, "Jmprinted—in Smithfielde, by the Hoſpital in litle S. Bartelmewes." W.H.   Sixteens.

1561. "A dialogue conteynyng the number in effecte of all the prouerbs in the Engliſh tunge compact, in a matter concerning twoo manner of maryages, made and ſet foorth by Jhon Heywood. Newly overſene, and ſomewhat augmented, by the ſayde John Heywood." In verſe. Octavo.

1561. "The viſion of Pierce Plowman, newlye imprynted after the authours olde copy, with a brefe ſummary of the principall matters, ſet before euery part, called Paſſus. Wherevnto is alſo annexed, the Crede of Pierce Plowman, neuer imprinted with the booke before. Jmprinted—neare vnto great ſaint Bartelmewes gate, at the ſygne of the ſpred Egle. The yere of our Lorde God, a thouſand, fyue hundred, threſcore and one. The xxi daye of the Moneth of Februarye. Cum privilegio—ſolum." See p. 758, &c. This book is often found without the creed. W.H. At the end of the creed, which contains 15 leaves, are the explanations of ſeveral old words * Licenſed.   Quarto.

—— "Pierce Plowman in proſe. (I did not ſee the beginning of this booke, but it endeth thus)

"God ſaue the King, and ſpeede the plough,
And ſend the prelats care Inough,
Inough, Inough, Inough."

Maunſell, p. 80, 81. Octavo.

—— "The troubled mans medicine, very profitable to be read of all men, wherein they may learn patiently to ſuffer all kindes of aduerſitie, made by William Hughe to a friend of his." Licenſed, 1558.   Small.

—— The ſeditious and blaſphemous Oration of Cardinal Pole, both againſt god & his Coüntry, which he directed to themperour in his booke intytuled, the defence of the ecleſiaſtical vnity, mouing the emperour therein
to

to feke the deftruction of England and all thofe whiche had profeffed the gofpele. Tranflated into englyfh by Fabyane Wythers. Reede and than Judge." On the back begins " Fabyan wythers to the gentle reader." On fignature C.i, begins " The glofe of Athanafius vpō the oration" &c. which ends on E. iiii. This fignature put under the colophon, " Jmprinted—betwene both S. Bartelmews, at the Spread Eagle." W.H. Sixteens.

" A New balet entituled howe to Wyue well.—Finis. quod Lewys Euans. Imprinted—at the fpread Egle betwyxte both the Saynct Bartholomews."

He had licenfes alfo for the following books, viz. In 1558, " The complante of verite. Inftruction for chyldren." In 1559, " Efocrats to Demonicus." In 1560, "The oration of cardēnall Poole. The epyftells of Bradfordes & Fylpottes, vpon predeftination. An almanack and pronoftication of Henry Rotherforthes." In 1561, " A new yeres gyfte made by Leues Euans." In 1562, " An almanacke for the monythes." In 1565, " A tru dyfcription of twoo cheldren borne at Herne in Kente, 26 Aug. 1565. An epytaph vpon the death of fyr John Mafon, knyghte." He printed without licenfe, " Epiftills & gofpills. The Regefter of all them that were burned. The booke of hufboundry, beynge Mr. Tottles." Alfo, feveral ballads with, and without, licenfe.

※※※※※※※※※※※※※※※※※※※※※※※※※※※※※※

WILLIAM NORTON

WAS an original member of the Stationers' company, and one of the firft fix, who came on the livery when it was revived again in 1561, after their charter from Phil. and Mary. He ferved collector in 1563 and 1564; was chofen under warden in 1569 and 1570; upper warden in 1573 and 1577; mafter in 1580, 1586, and 1593, in which year he died,[f] and George Bifhop was chofen in his ftead. Dec. 24, 1593, his will, which had been depofited with the company, was opened and read; then delivered to Mr. Tho. Woodcock, warden, and Mr. Ric. Watkins, to deliver to Mr. Bonham Norton, the executor. It contained divers fheets of paper, each fubfcribed with the teftator's name, as alfo thofe of Raffe Jackfon and Will. Yonge as witneffes.[g] He gave 6l. 13s. 4d. yearly to his company, to be lent to young men, free of the fame fociety; and the

[f] Aged 66 years; and had iffue only one fon, Bonham Norton. Dugdale's St. Paules, p. 123. [g] Regifter B. fol. 458.

like fum yearly for euer to Chrift's hofpital.[b] He dwelt at the King's arms in Paul's church yard, and used his rebus for a device as in the frontispiece. He was fined, not only for keeping his fhop open upon S. Luke's day, but on Sundays, and in fermon time, &c. Though he had a licenfe for printing near ten years fooner, i find nothing by or for him before.

1570. " The Tranquillitie of the minde: an excellent Oration directing euery man and woman to the true tranquillity and quietnefs of the minde, written in Latin by John Bernard, ftudent in Cambridge: tranflated by Anth. Marten, Gentleman Sewer of her Maiefties chamber." For him, Octavo.

1571. " Actes of Conference in religion, or Difputations holden at Paris betweene two papiftes of Sorbon, and two godly minifters of the Church: tranflated by Geffrey Fenton." For him and Humf. Toy. Dedicated to Lady Hobbie, 4 July, 1571. Quarto.

1572. " A Difcourfe, Wherein is plainly proued by the order of time and place, that Peter was neuer at Rome. Furthermore, that neither Peter nor the Pope is the head of Chriftes Church. Alfo an interpretation vpon the fecond Epiftle of S. Paul to the Theffalonians, the fecond Chapter. Seene and alowed, &c. Imprinted—by Tho. Eaft and H. Myddleton: for VVilliam Norton.—1572." Dedicated " To—Sir Henry Sidney, Knight of the—Garter, Lord Prefident of the Marches of VVales, and Lord deputie of Ireland.—R. T." Then, " A defcription of the Pope." In verfe. The third difcourfe ends on O, iii. On the back, begins a dialogue, in Latin diftichs, between the emperor Frederick, and Pope Innocent. " Sic quidam de Papa." Alfo, " De Roma meretrice Babilonica." W.H. Quarto.

1574. " Q. HORATII FLACCI VENVSINI, Poetae Lyrici, Poëmata omnia, doctiffimis fcholijs illuftrata." His rebus of a tun, with the letters NOR upon it, and a fweet-william growing out of it, alluding to his name. " Excufum apud Guil. Norton.—1574." Octavo.

1574. Juuenalis et Perfius. Octavo.

1574. " Diuine meditations of the milde Chriftiã," &c. For him. Sixteens.

1575. The Bible. Lambeth lift. Folio.

1575. " A Looking Glafs for the Court: compofed in the Caftilian Tongue, by the Lord Anthony of Gueuarra," &c. as p. 526. " And now newly printed, corrected & fet forth with fundry apt notes in the Margent, by T. Tymme, Minifter." By whom it is dedicated to John Lord Ruffel, fon and heir of Francis, earl of Bedford. Then, a poem of the editor's

---

[b] Stow's Survey of London, B. 1. p. 266. Edit. 1733.

in praife of the author, and his Englifh tranflator, Sir Francis Briant.[1] Contains befides, 76 leaves. Octavo.

"Joyfvll Nevves ovt of the newe founde worlde, wherein is declared the rare and finguler vertues of diuerfe & fundrie Hearbes, Trees, Oyles, Plantes & Stones, with their aplications afwell for Phificke as Chirurgerie, the faied beyng well applied bryngeth fuche prefent remedie for all defeafes, as maie feme altogether incredible: notwithftandvng by practize founde out to bee true: Alfo the portrature of the faied Hearbes, very aptly difcribed: Englifhed by Jhon Frampton Marchaunt. Jmprinted at London in Poules Churche-yarde by Willyam Norton.—1577." Dedicated "To the right worfhipfull Maifter Edvvarde Dier Efquire." Wherein we learn that it was tranflated from the Spanifh of Doctor Monardes, a learned phyfician of Seville. It is divided into three parts, each with a feparate title-page. The fecond part begins at Fol. 33. The third at 87, which concludes on Fol. 109. Then, a table of contents for the three books. W.H. Again 1580. Quarto. 1577.

"Quinti Horatii" &c. as 1574, " et nouis aliquot annotatiunculis, illuftrata. Cum Joann. Harifon." Octavo. 1578.

"Heauenly Philofophie, containing not onely the moft pithie fentences of Gods moft holy fcriptures, but alfo the fayings of ancient fathers, and other writers." By Tho. Palfreyman. For him. Quarto. 1578.

"Patrickes places, containing fruitfull gatheringes of fcripture, declaring faith and workes. For him." Maunfell, p. 79. See p. 365. Octavo. 1578.

"John Gibfon his Catechifme. For him." Maunfell, p. 30. Octavo. 1579.

"The Historie of Guiccardin, containing the warres of Italie, and other partes, continued for many yeares vnder fundry Kings and Princes, together with the variation *and accidents* of the fame, deuided into twenty bookes: *&c.* Reduced into Englifh by Geffray Fenton. Mon heur viendra." His rebus. Printed for him; probably by Tho. Vautroullier, as a joint concern: however, he printed it this year with the fame title, literatim, only ufing his own device, name, &c. in the place of Norton's. Licenfed to them jointly. Folio. 1579.

"Edm. Bunnie his abridgemente of Caluins Inftitutions: tranflated by Edw. May. For him." Maunfell, p. 25. Octavo. 1580.

"The Defperation of Francis Spira, who for feare of men denyed Chrift, and the knowen veritie: written by Math. Gribald D. of the 1582.

[1] See Oldys's Catal. of the Harleian pamphlets, No. 177. Alfo, Warton's Hift. of Englifh Poetry. Vol. III; p. 42.

5 L 2   Law.

WILLIAM NORTON.

Law. With a preface of M. Io. Caluin: whereunto is added a preseruatorie againſt Deſperation. For him." Maunſell, p. 43. Alſo without date. Licenſed 1569. Octavo.

1585. "Qvincti Horatii—poemata omnia. Quibus adiunximus I. Iuuenalis & A. Perſij opera: doctiſſima etiam in vnumquemq; Scholia & annotationes quàm maximè idoneas coniecimus. Londini, Apud Nin. Newtonum, impenſis Gulielm. Norton, & Joan. Hariſon. 1585." Horace ends on p. 316. A ſeparate title-page to Juvenal & Perſius: "Apud Nin. Newtonum & Arnoldum Hatfildum." The paging proceeds to p. 476. W.H. Sixteens.

1586. "Dionis Gray, *of London Goldſmith*, his Storehouſe of Breuitie in woorkes of Arithmetick, containing as well the ſundrie parts of ſcience, in whole and broken numbers, with the rules of proportion, as alſo ſundrie rules of Breuitie: a worke of rare pleaſant & commodious effecte; corrected & printed 1586. For him and John Hariſon." Maunſell, P. II; p. 2. Alſo without date. Licenſed 18 Mar. 1576-7. Octavo.

1587. "PARABOLÆ Siue Similia Des. Eras. Rot. ex diligenti auctorum collatione nouiſſimum recognita, cum vocabulorum aliquot non ita vulgarium explicatione. Acceſſerunt annotationes longè vtiliſſimæ, vna cum indice, quæ adoleſcentiæ vſum manifeſtè commonſtrabunt, auctore Ioanne Artopæo ſpirenſe. Similitudines aliæ etiam collectaneæ ex Cicerone, aliiſque ſcriptoribus additæ." His rebus. "Londini, Impenſis Guilielmi Nortoni, 1587." The dedication thus inſcribed, "Eraſmus Rot. cum primis erudito Petro Ægidio celebratiſſimæ ciuitatis Antuerpienſis à libellis, S.D.—Baſileæ, Anno a Chriſto nato M. D. xiiii. Idibus Octobris." 216 pages, and a table at the end. Licenſed. W.H. Octavo.

1587. "THE Inſtitution of Chriſtian Religion, written in Latine by M. John Caluine, and tranſlated into Engliſh according to the Authors laſt edition, with ſundry Tables—by Tho. Norton. Whereunto there are newly added in the margent—notes—of the matter handled in each ſection." His rebus. "—Printed by H. Midleton, for W. Norton.—1587." There are prefixed, the preface by T. N.; Calvin's dedication to the French King, Baſil, 1 Aug. 1536; Calvin's epiſtle to the reader, Geneva, 1 Aug. 1559; and a table of contents. The Inſtitution on 507 leaves. At the end are four other copious tables. W.H. Quarto.

1589. "THE HAVEN OF HEALTH: Chiefly made for the Comfort of Students, and conſequently for all thoſe that haue a care of their health, amplified vppon fiue wordes of Hippocrates, written Epid. 6. Labour, Meate, Drinke, Sleepe, Venus: By Tho. Cogan, Maiſter of Artes & Bacheler of Phiſicke: And now of late corrected and augmented. Hereunto is added a Preſeruation from the Peſtilence: With a ſhort Cenſure of the late ſickneſſe at Oxford. Ecclus. 37; 30. By ſurfet haue many periſhed: but he that dieteth himſelfe prolongeth his life." His rebus.
"Imprinted—

WILLIAM NORTON.

" Imprinted—by Tho. Orwin, for W. Norton. 1589." Dedicated " To—Syr Edward Seymor Knight, Baron Bewchamp, and Earle of Hertford." Whofe arms are on the back of the title-page. Then, " Authoris carmen Sapphicum—ex Ecclefiaftico, Cap. 30. ver. 14—17." And an epiftle to the reader. The Haven on 276 pages. An alphabetical table at the end. Maunfell mentions an edition in 1586; but as our dedication is dated 1588, imagine the laft figure of his date was turned the wrong way. W.H. Quarto.

"TESTAMENTI VETERIS BIBLIA SACRA, five Libri Canonici Prifcæ Iudæorum Ecclefiæ à Deo traditi, Latini ex Hebræo facti brevibufq; Scholiis illuftrati ab Immanuele Tremellio & Francifco Junio. Accefferunt Libri qui vulgo dicuntur Apocryphi, Latine redditi & notis quibufdam aucti à Fr. Junio. Multo omnes quàm antè emendatiùs editi & aucti locis innumeris: quibus etiam adjunximus Noui Teftamenti libros ex fermone Syro ab eodem Tremellio, & ex Græco à Theodoro Beza in Latinum verfos notifque itidem illuftratos. Secunda cura Fr. Junii. Londini, Impenfis Guliel. N.—1593." In a magnificent compartment cut on wood, having, in a tablet at top, a lamb laying on a ftool, with its legs bound together, and a knife at its throat. On a fcroll above, Poffidete animas veftras. The fides are twifted columns with vine branches about them. At the bottom, between their bafes, a vafe, out of which the vine branches iffue, with the date, 1574. The engraver's mark NH. To the Old Teftament are prefixed a dedication, " Illuftriffimo Principi—Frederico IIII. Comiti Palatino ad Rhenum, &c.—Heidelbergæ pridie Kal Aprileis, 1587.—Fr. Junius." A preface, " Illuft. Potentiffimoq; Principi—Frederico III. &c.—Immanuel Tremellius, Fr. Junius." Then, " Lectori falutem in Chrifto Iefu." The prophetic and the apocryphal books have the device of Arnold Hatfield, and the date M.D.XCII 1592. 1593.

" D. N. Jefu Chrifti Novum Teftamentum, &c. Excudebant Reg. Typograph. Anno falutis humanæ, 1592." Licenfed in 1578, to John Harrifon, fen. Geo. Bifhop, W. Norton, & Chr. Barker. W.H. Folio.

" Tabulæ Analyticæ, quibus exemplar illud fanorum fermonum de Fide, Charitate et Patientia, quod olim Prophetæ, Euangeliftæ, Apoftoli literis memoriæque mandauerunt, fideliter declaratur. Authore Steph. Szegedino Pannonio. De operis huius, nunquam anteà editi ratione & vfu longè maximo, ad ampliff. Senatum Schaphufianum Epiftola Dedicatoria. Authores tam veteris quam noui Teftamenti, his Tabulis illuftratos, feptima pagina continet." His rebus. " Londini, Excufum in ædibus Richardi Fieldi, impenfis G. Nortoni. 1593." Contains 377 pages befides the prefixes. W.H. Quarto. 1593.

" Adam Hill his fermon on Gen. 18; 21, 22. For him." Maunfell, p. 100. Licenfed. Octavo. 1593.

" Tho. Palfreyman his Paraphrafe on the Romans: alfo, certain little tracts of Mart. Cellarius. For him." Maunfell, p. 78. Quarto.

" Of

WILLIAM NORTON.

—— "Of the arrivall of the 3 graces in Anglia, lamenting the abuses of the present age." By Steph. Batman. In five sheets. Quarto.

—— "A Briefe and Compendious exposition vpon the Psalme called De profundis, which haue bene. And presentlye is, horrible and detestable, Abused in the Churche of God. And now translated to the trew sens: to Gods glorie & to the Edification and comfort of his Church. By M. Rob. Richardson, B. D. And Minister in London." B, in eights. "Jmprinted—by Tho. Purfoote for W. Norton." Licensed, 1569. W.H.
Octavo.

Besides these he had Licenses for the following, viz. In 1561, "Stans puer ad mensam." In 1567, "A Godly medicene which hath done moche good, by Rob. Rychardson." In 1569, "Of ij louers, Euryalus & Lucreffia, pleasaunt & delectable." In 1576, "The well of wisdome. Recordes Castle of knowledge." In 1577, "Andoniari Talei rethorica. Rethorica Rami." The two last with John Harrison, senior. Again in 1586. In 1578, "Crispin's Lexicon, 4°. Apophthegmatū ex optimis utriusq; linguæ scriptoribus per Conradū Lycosthenem Rubeaquensem collectorem Loci cōmunes denuo aucti & recogniti : Cum Parabolis— in locos cōmunes digestis." These with John Harrison senior, and Geo. Bishop. "An easie entrance into the principall pointes of Christian Religion: verie shorte & plaine, for the simpler sorte." In 1581, "A copie of the great pardon, generally granted, without terme of tyme, for full remission of all synnes whateuer, vnto all vnfained repentant Synners." In 1588, "A prayer against the Enemyes of Gods Truthe." In 1589, "The marchauntes Auizo."

RICHARD ADAMS.

WAS presented by the executor of Richard Kele to be made free of the Stationers' company in the year 1558. He printed

1559. "A register in metre, containing the names, and patient sufferings, of the members of Jesus Christ, afflicted, tormented, and cruelly burned, here in England, in the time of queen Mary, gathered by Tho. Brice." This was printed before Fox's account: but printing it without licenfe he was fined v. s. It was printed also by John Kingston for him, without date. Octavo.

1579. "A report of an assault against Mastricht, 26 Apr. 1579." By commandment from the Wardens, 26 June.

"Francisci Junii de peccato primo Adami, et genere caufae, qua ad peccandum adductus est, liber in questiones quatuor distributus."
Octavo.
RICHARD

RICHARD HARRISON

WAS an original member of the Stationers' company, and doubtlefs one of the old livery, as he was chofen under warden in 1562, without being called on the livery when new revived, or ferving collector: but he died before the expiration of his year. The company attended at his funeral fermon; and Mrs. Harrifon gave them in reward x.s. See p. 417, and 586.

The New Teftament.* Licenfed, Quarto. 1561.

"The Institution of Chriftian Religion, &c. as p. 606. Folio. 1561.

"The Bible in Englyfhe, that is to fay, the contentes of all the holy 1562. Scriptures,—according to the tranflation that is appointed to be read in Churches. Imprinted—in White Croffe ftreet by Richard Harryfon, An. Dom. 1562." See Lewis's Hift. of Eng. Tranflations of the Bible, p. 213. My copy wants the title-page; and differs from Mr. Lewis's account, in not having a preface to the Hagiographa. The title of the New Teftament in my copy runs thus: " The newe Teftament in Englifh trãflated after the Greke, wherin are conteined thefe bokes folowing." &c. Jmprinted—by Richarde Harrifon Anno Do. M.VC.LXII." This title in a fuitable compartment. On one fide, Adam and Eve at the tree of Knowledge; over them, Mofes receiving the two tables: on the other, the crucifixion of our Saviour; over it, the Angel appearing to the Virgin Mary. At the bottom, a finner fitting in diftrefs, converfing with a doctor of the law, behind whom is a clofed tomb with a dead carcafe on it; on the other hand is an evangelical minifter fhewing him Chrift crucified, and by whom is an open tomb, on which Chrift, rifen from the dead, is fitting, and triumphing over Death. This compartment was ufed to Matthews's Bible, printed in 1537; both to the old and new teftaments. The table of Epiftles and Gofpels begins on the back of this title-page. The New Teftament contains Fo. cxix. " Imprinted—in White Croffe ftrete—the yeare of oure Lorde, a thoufande fyue hundred thre fcore and two. Cum priuilegio—folum." Hence it appears that he printed 2 editions, at leaft of the New Teftament. But, doing this without licenfe, he was fined viij. s. W.H.+ Folio.

"The Institution of Chriftian Religion—By maifter John Caluine, 1562. and tranflated—accordyng to the Authors laft edition, by T. N. &c. Seen and allowed" &c. His rebus, as in the frontifpiece: John Harrifon ufed the fame; and it is one of thofe noted by Camden in his Remains. " Imprinted—in White Croffe ftrete—1562. Cum priuilegio—folum." The Inftitution on 502 leaves. The prefixes and affixes, fee page 880. Licenfed. W.H. Folio.

" De neutralibus et mediis. Grofly Englyfhed, Jack of both fydes. 1562.

A

A godly and neceſſary admonition, touching thoſe that be neutres, holding vpon no certayne religion, nor doctryne, and ſuch as holde with both partes, or rather of no parte, very neceſſary to ſtaye and ſtablyſh Gods elect in the true catholike faith, againſt thys preſent wicked world." Some perſon had written on the title of Mr. Ames's copy, " Wigand authore." But Maunſell has, " Written in latine by Rodolph Gualter. Tranſlated."
Octavo.

He had licenſe, a little before his death, to print " The Dyxcionary of Mr. Tho. Elyott, and Mr. Cowper."

DAVID MOPTID, and JOHN MATHER,

SEEM to have been partners together, and dwelt in Red-croſs-ſtreet, next adjoining to St. Giles church, without Cripplegate. Moptid was bound apprentice to Tho. Heaſte, for ſeven years from the feaſt of St. Michael, An. 1566; and Mather to Hen. Binneman for the ſame time.

—— " A briefe declaration of the chiefe poyntes of Chriſtian Religion, ſet foorth in a Table. Made by M. Theodore Beza. Gal. 3. d. The Scripture hath ſhut up all vnder ſinne, to the intent, that the promiſe by faith in Ieſus Chriſt, ſhould be giuen to them that beleeue. Seene and allowed &c. With a ſcheem, and a briefe declaration of the table of predeſtination. Engliſhed by W. Whittingham, together with a treatiſe of election and reprobation, with certain anſwers to the obiections of the aduerſaries of this doctrine: written by Anthonie Gylbie." This is a general title for the two following books reprinted from the Geneva edition, 1556; as may be ſeen in our General Hiſtory.

—— " A BRIEFE declaration, &c. Seene & allowed, &c. Imprinted by Dauid Moptid and John Mather." D 4, in eights. W.H. Sixteens.

—— " A BRIEFE Treatice of Election, &c. Written by Anthonie Gylbie. Rom. 9. I will ſhew mercy" &c. E 4, in eights. " Imprinted—by Dauid Moptid and John Mather, Dwelling in Redcroſſe ſtreete, next adioyning to S. Gylfes Church, without Criplegate." W.H. Sixteens.

JOHN SAMPSON, alias AWDLEY,

APPEARS to have been an original member of the Stationers' company, though his name is not inſerted in the charter liſt, as he bound an apprentice with them in 1559, and others afterwards. He is
mentioned

JOHN AWDELEY.

mentioned in the company's books sometimes by one name, sometimes by the other, and sometimes by both, as above. He dwelt in Little Britain street, by Great St. Bartholomew's, without Alderſgate, and always printed his name Awdeley, or Awdely. He printed a Catechiſm in 1559, for which he was fined ij.s. Alſo, in 1563, he was " fined xx.d. for printing other mens copies ;" probably ballads.

"The ſume of diuinitie. Tranſlated from the Latin, by Robert Hutten, 1560, 24 Oct." Licenſed. Again in 1561, and 1567.   Sixteens.

" The Deſcription of Swedland, Gotland & Finland, the auncient Eſtate 1561. of theyr Kynges, the moſte horrible & incredible tiranny of the ſecond Chriſtiern kyng of Denmarke agaynſt the Swecians, the politicke attaynyng to the Crowne of Goſtaue &c. Collected &c. By Geo. North. Set forth according to the order &c. Imprinted—in Litle Britaine ſtreete, by great St. Bartelmews. 1561. Octo. 28." On 28 leaves. At the end is the Lord's prayer, in the Swediſh language.*   Quarto.

" The vnfained retractation of Francis Cox, which he vttered at the 1561. Pillery in Chepeſyde, and els where, according to the Councels cōmandement. The 25. of June, 1561. Being accuſed for the vſe of certayne ſiniſtral & Diueliſh Artes. Imprinted—by great S. Bartelmews, The 7. of July, 1561." Concludes, " Thus as a true declaracion of my vnfayned repentaunce in theſe Diueliſh practices, I haue cauſed this my Retractacion to be put in print, to haue it openly publiſhed to all ſorts of People. Purpoſing (by Gods helpe) within theſe fewe dayes to ſet forth a ſmall peice of woorke to the utter defaſing of thoſe diueliſh ſciences, with the declaration of the horrible practiſes and deathes of ſuch as haue vſed thoſe Diabolical Artes, which (as I truſt) ſhall fear all other to practiſe the lyke." Copy in MS. by Mr. Ames. Licenſed thus: " A confeſſion made by a preſte which ſtode vpon the pyllorye with vij moo."
      A ſheet.

" A famous and godly hiſtory, contaynyng the lyues and actes of three 1561, renowned reformers of the chriſtian church, Martine Luther, John Ecolampadius, and Huldericke Zuinglius : the declaration of Martin Luthers faythe before the emperoure Charles the fyſt, and the illuſtre eſtates of the empyre of Germanye, wyth an oration of hys death : all ſet forth in Latin by Phillip Melancthon, Wolfangus Faber, Capito, Simon Grineus, and Oſwald Miconius.   Newly Engliſhed by Henry Bennet Calleſian."
      Octavo.

" The Booke of Huſbandry." Fitzherbert's. John Baynes, Eſq; 1562.
      Octavo.

" The fraternitie of vacabondes, as wel of ruſling vacabones, as of 1565, beggerly, as wel *of women as* of men, and as wel of gyrles as of boyes, with their proper names and qualityes. Alſo the xxv orders of knaues, otherwiſe called, A quartten of knaues. Confirmed this yere by Cocke Lorel, 13 December."   Quarto.
      " The

1566. "The examination of John Walſh before the biſhop of Exeter his cōmiſſary touching Witchcraft & Socery." 20 Aug. Octavo.

1566. "The golden Booke, of Marcus Aurelius &c. 1566. Jmprynted—in litle Britaine ſtreete, beyonde Alderſgate. Cum priuilegio—ſolum." See p. 455. Oo 4, in eights. W.H. Again 1573. Sixteens.

1567. "The ſumme of Diuinitie Drawen out of the holy Scripture, very neceſſarie for Curates & yong Studentes in Diuinitie, and alſo meete for al Chriſten men & women, what ſoeuer age they be of. Drawen out of Latin into Engliſh by Rob. Hutten. Jmprinted—The 15. of March, 1567." On the back, "William Turner to the Chriſtian Reader." Begins, "After that my Scholler ſometime & ſeruaunt Rob. Hutten had tranſlated this booke, miſtruſting hys owne iudgement to be ſufficient to iudge whether the Compiler of this booke had done althinges according to the vayne of holy Scripture, offered the booke vnto me that J ſhould examine it wyth the touchſtone of the Scripture." &c. T 4, in eights. W.H. Sixteens.

(1569.) "A ſermon preached before the Quenes Maieſtie, by Maiſter Edward Dering, the 25. day of February—1569." Sixteens.

1570. "Iohn Gough his anſwere to M. Fecknams obiections againſt his ſermon lately preached in the Tower, the 15. of January, 1570." Licenſed. Octavo.

1571. "A ſermon exhorting to pitie the poore. Preached the xv. of Nouember—1571, at Chriſtes Churche in London, By Henry Bedel Vicar there, which Treatiſe may well be called The Mouth of the poore." Again 1572, & 1573. Sixteens.

1571. "A Sermon preached at Hampton Court on Sunday the 12. day of Nouember—1570. VVherein is playnly proued Babylon to be Rome, both by Scriptures and Doctors. By William Fulke B. D. and fellow of S. Johns Colledge in Cambridge. Apoc. 14. She is fallen," &c. Dedicated "To—lord Ambroſe Dudley, Earle of VVarwick, Maiſter of the—Ordinance, and Knight of the—Garter." H j, in eights. Licenſed. Again frequently. W.H. Sixteens.

1572. "A briefe & neceſſary Inſtruction, Verye needefull to bee knowen of all Houſholders, Whereby they may the better teach & inſtruct their Families in ſuch points of Chriſtian Religion as is moſt meete. Not onely of them throughly to be vnderſtood, but alſo requiſite to be learned by hart of all ſuche as ſhall be admitted vnto the Lordes Supper. Jmprinted—1572." It has prefixed a copious epiſtle "To the Chriſtian Reader,

JOHN AWDELEY.

Reader, by E. D. *(Edw. Dering)* decrying the vain and lewd[x] books of that age." It was printed again in 1575, but the catechifm greatly varied. W.H.+ Octavo.

" A comfortable Sermon of Faith in temptations & afflictions. Preached 1574: at S. Botulphes wythout Alderfgate in London, the xv. of Februarye, 1573. By Maifter VVilliam Fulke, D. D. 1 Iohn, 5, 4. All that is born of God ouercometh the world: &c. Imprinted—1574." G, in fours, half fheets. W.H. Octavo.

" A Lecture or expofition vpon a part of the v. chapter of the Epiftle 1574. to the Hebrues. Set forth as it was read in Paules Church in London, the vj. of December, 1573. By Edw. Deryng. Geuen for a Newyeares gift to the godly in London, and els where. Perufed & alowed by authoritie. Imprinted—1574." G. 2, in fours. W.H. Octavo.

" An introduction of algorifme, to learn to recken wyth the pen, or 1574. wyth the counters, in whole numbers, or in broken. Newly ouerfene, and corrected: Whereto is annexed, certain notable and pleafant rules of falfe poffitions, not before fene in our Englifh toung, by which all manner of difficil queftions may eafely be diffolued and affoyled." *˙ See p. 579. Octavo.

" Gregory Scots Treatife againft certain Errors of the Romifh 1574. Church." In verfe. Octavo.

" Two notable fermons. Made by—John Bradford," &c. as p. 780. 1574.

" Certayne godly learned, and comfortable conferences. Betwene the 1374. two reuerende Fathers, & holy Martyrs of Chrift D. Nicolas Rydley, late Bifhop of London, and M. Hugh Latimer, fometyme Bifhop of Worcefter, during the tyme of their imprifonments. Whereunto is added a treatife of the Lordes Supper, made by the fayd reuerende Father D. Nic. Rydley, a little before he fuffered death, 1555. Now newly again imprinted. 1574." K6, in eights. " The 14. day of October.". W.H. Again without date. Octavo.

" The Chriftian ftate of Matrimony," &c. as p. 709. Licenfed. 1575.
" A Chriftian exhortation vnto cuftomable fwearers," as p. 711. 1575.

---

*̽ " Beuis of Hampton, Guy of Warwicke, Arthur of the round table, Huon of Bourdeaux, Oliver of the Caftle, the foure fonnes of Amond, the witles deuices of Gargantua, Howleglas, Efop, Robyn Hoode, Adam Bel, Frier Rufhe, the Fooles of Gotham, and a thoufand fuch other. And yet of all the refidue the moft dronken imaginations, with which they fo defiled their Feftival and high holy daies, their Legendawry, theyr Saintes lyues, their tales of Robyn Goodfellow, &c. and were warranted vnto fale vnder the Popes priuiledge to kindle in mens hartes the fparkes of fuperftition, that at laft it might flame out into the fire of Purgatorie.—] trow we haue multiplied to ourfelues fo many new delightes that we might iuftify the idolatrous fuperftition of the elder world. To this purpofe we haue printed vs many bawdy fonges (I am lothe to vfe fuch a lothfome woord, faue that it is not fyt inough for fo vile endeuours) to this purpofe we haue gotten our Songes & Sonets, our Pallaces of pleafure, our vnchaft Fables and Tragedies, and fuch like Sorceries,—and haue not bene afhamed to entitle their bookes The Court of Venus, The Caftle of Loue. O that there were among vs fome zealous Ephefians, that bookes of fo great vanity might be burned up."

JOHN AWDELEY.

575. "A Sermon made in the Chappel at the Gylde Halle in London, the xxix. day of September, 1574. before the Lord Maior and the whole state of the Citie, then assembled for the chusing of their Maior that shuld then succede in the gouernmēt of the same Citie. Concionatore Roberto Croleo. Perused & licensed, &c. Jmprinted &c. 1575." The text, Pf. 139, 21, &c. An Appendix is added, "Notyng the duties both of them to whom it appertayneth to chuse, and also of them that shall be chosen to anye gouernment, in either the Ecclesiastical or Ciuile state." G 3, in fours. W.H.   Octavo.

1575. "A Sermon no lesse fruitful than famous. Made in the yeare of our Lord God M.ccc lxxxviij, and founde out hyd in a wall. Which Sermon is here set forth by the old copy without adding or diminishing, saue the old & rude English here and there amended. Jmprinted—1575." F 4, in eights. W.H.   Octavo.

1575. The same in a very small edition.*

1576. "A sermon preached at Pauls crosse on Trinity sunday, 1571. By E. B." *(Edw. Bush.)* H, in fours. W.H.   Octavo.

576. "Godlye priuate Prayers for Housholders to meditate vpon, and to say in their Families. Mark, XI, 24. VVhat soeuer ye desyre when ye pray, &c. Imprinted—in litle Brittaine streete—1576." C, in eights. W.H.   Octavo.

—— "Tho. Knell his answere to the most heretical trayterous papistical Bill, cast in the streets of Northampton, and brought before the Judges," &c. In verse. Licensed, 1570.

He had licenses also to print the following, viz. In 1559, "A mornyng & Euening prayer." In 1560, "The proude vyues pater noster. A pany worth of wytt. The plowmans pater noster. The gouernaunce of Vertu. Haue an eye to your Conscyence. The Epestilles & Gospelles. The boke of Repentauns." In 1561, "Coxes *treatise* agaynste sofferers & Cōngerers, with an Almanacke:" with John Alde. In 1562, "The Epitaphe of Mr. Veron." In 1564, "A copye of a letter dyscrybinge the wonderfull workes of God." In 1566, "The cruell inquysetion & plaquet which the papestes wolde brynge into Antiwarpe, &c. The newes from Veenna, 5 Aug. of the strong towne & castell of Juba, in Hungarye. An other newes from Veenna, the laste of August. The Somonynge of dame popery." In 1567, "A dyscription of vij pryncipall vices, with the devyce." In 1568, "A monsterous chylde which was borne at Maydestone. The endes of ij presoners lately pressed to death in Newgate. The daunce & songe of Death." In 1569, "A shorte expostulation of madame Popery to hyr chaplayne Mr. Bankes." In 1570, "The bewalynge of tru subiectes. An Epytaphe of Dr. Haddon. An Epitaphe of Mr. Frauncis Benyson." Besides several ballads,

JOHN

JOHN ALDAYE, ALLDE, or ALDE,

AT the long shop adjoining to St. Mildred's church, in the Poultry, had a servant named Hugh Mores, who died in the Stocks of the plague, (a room so called, as appears from the register book of that parish) and was buried the 5th of May, 1570. He is the first man who appears on the register to have taken up his freedom in the Stationers' company. This was about Christmas 1554.

" A short treatise declaring the detestable wickedness of magical 1561. sciences: as, necromancie, coniurations, curious astrologie, and such like, made by Fr. Cox." Licensed. Also, without date. Octavo.

" The lAmentatiō of a Sinner made by the most vertuous Lady Queen 1563. Katherin, bewailing the ignorance of her blinde life: first set foorth and put in Print at the instaunt desire of the right gratious Lady Katherin Duches of Suffolke, and the ernest request of the right honorable Lord William Parre, Marquesse of Northamton. And now again newly Jmprinted, 1563." The preface thus superscribed " Wylliam Cicill hauing taken much profite by the readyng of this treatise folowyng, wisheth vnto euery Christian by the readinge therof like profite with increase from God." H 4, in eights. " Jmprinted at Lonon at the long Shopp adioining vnto S. Mildreds Churche in the Pultrie by John Alde. 1563." See Walpole's Royal & Noble authors, vol. I; p. 19. W.H. Octavo.

" A good and godly prayer" &c. as p. 570. Again 1565. 1563.

" The troubled mans medicine, very profitable to bee red of all men 1567. where in they may learn paciently to suffer all kindes of aduersity: made and written by William Hughe vnto a freed of his." H 7, in eights. " Oxford, the xv. day of Marche."

A sweet consolation, & the second booke of the troubled mans medicine, made & pronounced by William Hughe, to his freend lying on his death bed. Watche for yee know no day nor houre.—1567." Dedicated " To—maistres Lady Deny." The signatures proceed to P, in eights. " Jmprinted—at the long shop &c. by John Allde.—1567." W.H. 1567. Octavo.

" An Epitaph, or rather a short discourse made vpon the life & death of 1569. D. Boner, sometimes vnworthy Bisshop of London, whiche dyed the v, of September in the Marshalsie. Jmprinted &c. An. do. 1569. Sep. 14." On eight leaves. Ends " quoth T. Knell, Iu." W.H. Harl. Miscell. Vol. I. p. 595. Octavo.

To this Mr. Ames has annexed, " A Commemoration or Dirge of Bastarde Edmonde Boner, alias Sauage, vsurped Bisshoppe of London. Compiled by Lemeke Auale. Episcopatum eius accipiet alter,—1599.

Imprinted

JOHN ALLDE.

Imprinted by P. O." And then added John Allde, as if he had been the real printer. Though i apprehend P. O. to be fictitious, yet the types are not like any of Allde's that i have seen. However, as i know not to whom else to ascribe it, i have followed Mr. Ames in placing it here. It is a most severe, and indeed profane burlesque, in the Skeltonic manner. W.H. Octavo.

1569. "The play called the foure P. a very merry enterlude of a palmer, a pardoner, a potecary, a pedlar." In verse: ends, "quoth John Heywood." 14th September. Again 1596, quarto. Garrick's Plays, in the Brit. Museum. See p. 576. Octavo.

1569. "A wonder and true history of pope Alexander VI." Being a large picture of the pope: turn up the head, and a devil appears: round it is a description of the wickedness of the pope. Broadside.

1571. "A Declaration of such tempestous & ouragious Fluddes, as hath been in diuers places of England. 1570." Enumerating the loss and damage in the several counties, particularly in the bishopric of Ely. Printed by Will. How for him, and Will. Pickering. Octavo.

1573. "Xenophons treatise of Housholde—1573." See p. 424. W.H. Octavo

1573. "—the offices of Shiriffes, Bayliffes of liberties," &c. as p. 574. W.H. Octavo.

1574. "The Composition or making of the moste excellent pretious Oil, called Oleum Magistrale. First published by the comandement of the King of Spain, vvith the maner how to apply it particulerly. The which Oyl cureth these disseases folowig. That is to say, Woūds, Contusiōs, Hargubush shot, Cankeris, pain of the Raines, Apostumes, Hermerhoids, olde Vlcers, pain of the Joints & Gout, and indifferently all maner of disseases. Also the third book of Galen of curing of pricks & wounds of Sinowes. A method for curing of wounds in the ioynts, and the maner how to place them. A breefe gathering together of certain errours which the comon Chirurgians dayly vse. Very profitable & necessary for all Chirurgians, &c. Faithfully gathered & translated into English by Geo. Baker Chirurgian. 1574." Dedicated "To the right hon. Edwarde de Vere Earle of Oxford," &c. whose arms are on the back of the title-page. It has prefixed some commendatory verses, Latin & English; an epistle to the reader; and a preface. Contains 52 leaves besides. W.H. Octavo.

Also, his treatise of the nature & properties of Quicksilver.

1575. "Exhortation to the ministers of Gods worde in the Church of Christ, that they set aside all mutuall discord, and in these later daies & dangerous times, purely and with one consent, preach vnto the world the onely true faith in Christ, and amendment of life. By Hen. Bullinger: translated by John Cox." Octavo.

"Joyful

JOHN AILDE.

" Joyfull newes out of Heluetia, from Theophr. Paracelfum, declaring 1575. the ruinate fall of the papall dignitie: alfo a treatife againft Vfury. By Steph. Batman." For him.   Octavo.

" Fran. Coxe his Treatife of the making & vfe of diuers Oyles, vn- 1575. guents, Emplaifters, & ftilled waters."   Octavo.

" THE DEFENCE of Death. Containing a mofte excellent difcourfe of 1576. life and death, vvritten in Frenche by Philip de Mornaye Gent. And doone into Englifh by E. A." A wyvern rifing out of a ducal coronet. The creft of Lord Clinton of Chudleigh. " Imprinted—for Edw. Aggas —1576." Dedicated " To—Margaret Counteffe of Darby." H6, in eights. W.H. Again 1577.   Octavo.

" Celius fecundus Curio to his deer freend Fuluius the ftraunger 1576. Morato. S.P.D." This is the head title, my copy wanting the prefixes. It is an epiftle for the godly bringing up of children. Concludes. " From Luce 1542. the iiij. of the Jdes of June. quoth VV. Liuing.—Imprinted &c. 1576. the 28. of Iune." W.H.   Octavo.

" The advife and anfwere of—the prince of Orange" &c. as p. 728. 1577.

" The Proclamation of Jefus Chrift concerning the caftle of Faith, 1577. which is like to be woone into Chriftian mens hands."   Octavo.

" Newes from the North. Otherwife called the Conference between 1579. Simon Certain, and Pierce Plowman, faithfully collected & gathered by T. F. Student. Aut bibe aut Abi. Imprinted &c. 1579." Dedicated " To—Sir Henry Sidney Knight of the—Garter, Lord Prefident of Wales, & Marches of the fame—26. Nov. 1579." An epiftle from the author; another from the printer; then, fome commendatory verfes. Concludes with " The Apologie of the Author," in 6 fix-lined ftanzas. L, in fours. Licenfed. W.H.   Quarto.

" The horrible acts of Eliz. Style, alias Rockingham, Mother Dutton, 1579. Mother Douell, and Mother Margaret, 4 witches executed at Abington, 26. Feb. vpon Richard Galis." For Edw. White. Licenfed.

" A warning to the wife, a feare to the fond, a bridle to the lewde, and 1580. a glaffe to the good. Written of the late earthquake chanced in London, and other places, the 6th of April 1580, for the glory of God, and benefite of men, that warely can walke, and wifely can iudge. Set forth in verfe and profe, by Thomas Churchyard, gentleman. Seen and allowed." 15 leaves. With Nich. Lyng. 8th April. See Oldys's Catalogue of Pamphlets in the Harleian Libr. N°. 222.   Octavo.

" A golden chaine, or the defcription of theologie, containing the 1591. order of the caufes of faluation and damnation, according to Gods woord. Written

Written in Latin by William Perkins, and tranflated by an other. Hereun o is adioyned the order, which M. Beza vfed in comforting troubled confciences." Octavo.

1596. " A difcourfe of the variation of the compaffe, or magnetical needle, wherein is fhewed, the manner of the obfervation, effects, and application thereof, made by W. B.[1] With a preface to the trauailers, feamen, and mariners of England." Quarto.

1596. " Edward the fourth and the tanner of Tamworth." Quarto.

—— " A merry Jeft of Dane New, munk of Leiceftre: and how he was foure times flain, and once hanged." Quarto.

—— " The Batayll of Egynge courte & the great fege of Rone." Quarto.

—— Leonard Staveley's " Breef difcourfes on the miferies of life. Cicero 1. Tufcul. Moriendum eft omnibus: eftque finis miferiae in Morte." Octavo.

—— " A Spirituall and moft precious parle, teaching all men to loue & embrace the croffe as a moft fweet & neceffary thing vnto the foule, what comfort is to be taken thereof, where and how bothe confolation & ayd in al maner of afflictions is to be fought: and again how all men fhould behaue themfelves therein, according to the word of God.—By me John Alde, for Hugh Singleton." Small.

—— " A Lamentable Tragedy, mixed ful of pleafant Mirth, conteyning the life of Cambifes king of Percia, from the beginning of his kingdome vnto his death, his one good deed of execution, after that many wicked deeds & tiranous Murders committed by and through him: and laft of all, his odious death, by Gods Juftice appointed, doon in fuch order as foloweth, by Thomas Prefton." It was reprinted by Edw. Allde; alfo, in Hawkins's Origin of the Englifh Drama, Vol. I. Licenfed.

—— " Hen. Bullinger his 2 fermons on the ende of the world, tranflated by Tho. Potter, expoundinge Math. 24; 29, &c. and Dan. c. 7. concerning the moft troublefome kingdome of the pope, &c." Octavo.

—— " Meditations of true & perfect confolation, declared in two tables: in the firft, 7 confiderations of the euills which happen vnto ys; in the fecond, 7 confiderations of the good we receiue. Tranflated out of French by Rob. Fills." Licenfed, 1563. Octavo.

—— " A DIALOGVE between Cuftome & Veritie, concerning the vfe and

[1] Mr. Ames fuppofed this B to mean Burrough. I have not feen the book; but unlefs that name be found in the dedication, &c. i am rather inclined to believe thefe initials W. B. ftand for William Barlow, author of the Navigator's Supply, 1597, quarto. See Ath. Oxon. Vol. I; c. 495.

abufe

abuſe of Daucing & Minſtrelſie. Rom. 13; 12—14." Dedicated " To the godly & faithfull miniſters &c. Maiſter Rob. Crowley & Maiſter Tho. Braſbridge.—Tho. Louell." In verſe. D, in eights. Licenſed, 1581. Octavo.
" The book of Wiſdom tranſlated into Engliſh metre by Peter Tie." Licenſed, 1562. Octavo.
" An exhortacion to be giuen to the Priſoners lying in Newgate & other places, for Roberies & Murthers." Octavo.

" A treatiſe of the Sin againſt the Holy Ghoſt, by Aug. Marlorat: tranſlated out of French. For Luc. Harriſon." Octavo.

" Yues Rouſpeau, his Dialogue of preparation to the Supper of our Lord: tranſlated from French into Engliſh by R. B." Octavo.

Beſides theſe he had licenſes to print the following, viz. In 1561, " The Rates of the Cuſtome howſe. An almanacke & pronoſtication of Keningham. An admonition to Elderton, &c. The picture of a monſterus pygge." In 1562, " An almanake & pronoſt. of Fran. Coxe. The brefe Poſtyll." In 1564, " An almanake, &c. of Fr. Coxe." In 1565, " An hundreth poyntes of euell buſwyfrye. The Scyence of Lutynge. An almanake &c. of Fr. Coxe. A copye of the treates of the Confederation betwene the prence & the Lordes of the Lowe Country." In 1566, " A tru diſcription of a chylde borne with Ruffes in the paryſhe of Myttcham, —Surry. Almanake, &c. of Fr. Coxe." In 1567, " The complaynt of John a Neale. Hiſtory of pope Alexander. A dyſcryption of a monſterus gyante." In 1568, " The prayſe of Good Women. Pro. ch. xiij. A profitable & pleaſaunt Fayrynge. A play: lyke wyll to lyke, q. the Deuell to the Collyer. An epytaphe of M. Couerdayle. Freer Ruſht. The Freer & the boye. The defeate of women. Beues of Hampton. *A vocabulary:* Frynſhe, Engleſſhe & Duche." In 1569, " The repentance of the Nenyuetes. An enterlude: a lamentable tragedy full of pleaſaunt myrth. A neceſſary wepon, or ſharpe ſpirituall Sworde. ij ſhorte ſpeaches agaynſte Rebellion." In 1570, " A very lamentable & wofull dyſcource of Floudes. A dyſcourſe of Floudes & waters: with W. Poekering. A New yeres gyfte. An Epytaph of Mr. Bryce preacher. An admonition of Dr. Storye." In 1576, " A new fayringe for Bartilmew fayre. A newe yeres gifte for—1577." In 1577, " Stanbridges vocabularie. A diſcription of thoſe ſtrange kinde of people whiche Mr. Fourboſier brought into England. Godly meditations vppon the Lordes prayer, &c. whereunto is annexed a defence of Gods eternall election, &c. by Jhon Bradford, in the tyme of his impriſonment." In 1578, " Tarlton's deuiſe vpon this vnlooked for great ſnowe. The vj. newe yeres gift, and iiijth proclamation of outlawrye. An oration in Latin, made by Mr. Malem ſcholemaſter of Paules ſcole, to Duke Jhon Caſſimer, &c." In 1579, " A Bartilmewe fayringe for this yere, 1579. The myrror of mutability. A godly

JOHN ALLDE.

godly hymne, or caroll for Chriftmas. A newe yeres gifte for 1580, againſt the ſilence of Ric. Briſtowe, prieſte. A true report of this earthquake in London. Alarm for Londoners, ſettinge forthe the thunderinge peales of Gods Mercye & Juſtice vnto vs, by the comfortable preaching of the Woord, and laſte by extraordynarye ſignes & tokens." In 1580, "Newes out of Eſſex, conteyninge a preſent cure for the newe ſickneſs. Godlie carolles, hymnes & ſpirituall ſonges. Newes out of Jrelande. A prophecie wonderfull & ſtrange, which ſhall happen in the yere 1581. A newe yeres gifte. Certaine true markes whereby to know a papiſte." In 1581, "A pleaſaūt recreacion for an indifferent mynde." In 1582, "A Rate for expences.. On condition that no other man be intereſſed in yt." Alſo a great many ballads.

THOMAS HACKET

SEEMS to have been an original member of the Stationers' company, though his name is not found in the Charter liſt, as i do not find any entry of his being made free, and yet he was called on the livery in 1569, and ſerved collector in 1574, and 1575. In February 1559-60, he was fined ij.s. "for that he ſett his name to a booke, which ys cōtrary to the orders of the howſe." What thoſe orders were does not appear, this being the only inſtance of the kind i have obſerved. Jan 8. 1576-7, he ſeems to have been abroad, or under ſome confinement, as his wife acted by a Letter of Attorney from him, at a Court of Aſſiſtants, in aſſigning over to John Wight the copies of The regiment for the ſea; and the Theatre, or Rule of the world: And thus " the ſaid John and Thomas diſcharge either to other all maner of Debtes, Duties & Matters, from the begining of the world to this day." Though the former book does not appear to have been licenſed to Hacket himſelf, yet it might have been between July 1571, and July 1576, as the clerk's book for the entry of copies during that interval is ſuppoſed to be loſt. He was never choſen warden, nor came into the Court of Aſſiſtants. Indeed it looks as if he was of a quarrelſome diſpoſition, ſeeing he was twice fined on that account. He dwelt in Lombard ſtreet, at the ſign of the Pope's-head; and kept ſhop in the Royal Exchange, at the ſign of the Green Dragon; alſo, at the ſign of the Key, in St. Paul's Church-yard.

1560. "The fable of Ouid treting of Narciſſus, tranſlated out of Latin into Engliſh mytre, with a moral thereunto, very pleaſaunte to rede."
Quarto.

1562. John Hooper's Apologie. With John Tiſdale. See it in p. 768.
"The

"The moſt wonderfull, and pleaſaunt Hiſtory of Titus, and Giſippus, 1562. whereby is fully declared the figure of perfect frendſhyp, drawen into Engliſh metre. By Edwarde Lewicke. Anno 1562." C 3, in eights. "Imprinted at London by Tho. Hacket, and are to be ſolde at hys ſhop in Lumbarde ſtreete." Licenſed.

"The Lawes and Statutes of Geneua," &c. as p. 802. 1561.

"A ſtrong battery againſt the Jdolatrous inuocation of the dead 1562. Saintes," &c. as p. 845.

"A true & perfect deſcription of the laſt voyage or Nauigation, at-(1566.) tempted by Cap. John Ribaut, deputie & generall for the French men, 1565. Truely ſet forth by thoſe that returned from thence, wherein are contayned things as lamentable to heare as they haue bene cruelly executed. Imprinted—by H. Denham for Tho. Hacket, and are to be ſold at his ſhop in Lumbart ſtreate." It is introduced by an epiſtle from "The Author to his friend.—Deepe the xxv. day of May, 1566. Your louing brother & friend N. le Shalleux." Running title, "The laſt voyage to Terra Florida." D, in eights. Licenſed. Octavo.

"The Hyſtories of the moſt famous & worthy Cronographer Poly- 1568. bius: Diſcourſing of the warres betwixt the Romanes & Carthagineſes, a riche and goodly Worke, conteining holſome counſels & wonderfull de- uiſes againſt the incombrances of fickle Fortune. Engliſhed by C. W. Whereunto is annexed an Abſtract, compendiouſly coarcted out of the life & worthy acts perpetrate by oure puiſſaunt Prince King Henry the fift. Imprinted—by Hen. Bynneman for Tho. Hacket. And are to be ſold at his ſhoppe in Paules Churchyard at the ſigne of the Key." On the back, "The Phiſnomie of Polybius." A medallion, with eight alexandrines under it by B. G. Next is "The life of Polybius collected out of his Hiſtorie." The dedication thus inſcribed, "To the right worſhipful Thomas Gaudy Eſquier, Chr. Watſon wyſheth Argantos age, Policrates proſperitie, Au- guſtus amitie, and after the conſummation of this terreſtriall Tragedie, a ſeate amongeſt the celeſtial Hierarchie, &c." Concludes, "Thus then, al vain words ſet apart, J deſire your worſhip benignely to accept this as a token of the intier affection J bear towards you, which taken as J meane it, ſhal deſerue to furniſhe ſome voyde corner in the loweſt parte of your Librarie. From my chamber in your houſe at Gaudy Hall," in Norfolk. His epiſtle "To the Reader.---Franciſcus Aſulanus Lectori ſalutem.--- R. W. in laudem Hiſtor. Polybij, Anglico Lectori:" in five ſhort ſtanzas. The hiſtories of Polybius end on folio 100, on the back of which begins the author's epiſtle "To the Queſtioners:" thoſe deſirous to know why he joyned the abridgement of K. Henry the fifth's life to this foreign hiſtory. Then, "The victorious actes of king Henry the fift," from Hall's chronicle, which ends on folio 130. On another leaf, "Imprinted —Anno. 1568." Licenſed. W.H. Octavo.

1568. " THE NEw found vvorlde or Antarctike, wherin is contained wõderful & ſtrange things, as well of humaine creatures, as Beaſtes, Fiſhes, Foules, & Serpents, Trees, Plants, Mines of Golde & Siluer: garniſhed with many learned aucthorities, trauailed and written in the French tong by that excellent learned man, maſter Andrewe Theuet. And now newly tranſlated into Engliſhe, wherein is reformed the errours of the auncient Coſmographers." H. Bynneman for him: " And are to be ſold at his ſhop in Poules Church-yard, at the ſigne of the Key." Dedicated by T. Hacket " To—Sir Henrie Sidney, Knight of the—Garter, Lorde Preſident of Wales, &c. Lord Deputie Generall of the Queenes Maieſties Realme of Ireland." Then, " An Admonition to the Reader."---Some verſes " In prayſe of the Author."---Six diſtichs " In Theuetum Noui Orbis peragratorem & deſcriptorem, Io. Auratus literarum Græcarum Regius profeſſor."---Andr. Thevet's dedication " To my Lord the Right reuerend Cardinallof Sens, keper of the great ſeales of France.---A Preface to the Reader." Beſides theſe, 138 leaves, and a table. " Imprinted—1568." Licenſed. W. H. Quarto.

(1570.) " Tho. Buckminſter, miniſter; his right Chriſtian Calendar, or ſpirituall Prognoſtications, made for the year 1570." For him. Licenſed. Octavo.

1574. " A TOUCHESTONE for this time preſent, expreſsly declaring ſuch ruines, enormities & abuſes as trouble the church of God, and our Chriſtian commonwealth at this daye, &c. Newly ſett foorth by E. H. (*Edw. Hake.*) Imprinted—by Tho. Hacket, and are to be ſolde at his ſhop, at the Greene Dragon in the Royall Exchange, 1574." Annexed is, " A Compendious fourme of education to be diligently obſerued of all parentes & ſcolemaſters in the training vp of their children & ſchollers in learning. Gathered into Engliſhe meeter by Edw. Hake. It is an epitome of a Latin tract, " De pueris ſtatim ac liberaliter inſtituendis." See Warton's Hiſt. of Eng. Poetry, Vol. III; p. 275. Octavo.

1574. Theatrum Mundi. The Theatre or rule of the vvorlde, vvherein may be ſeene the running race and courſe of euery mans life, as touching miſerie and felicitie, wherein be contained wonderfull examples and learned deuiſes, to the ouerthrow of vice and exalting of vertue. VVherevnto is added a learned and pitthie worke of the excellencie of mankynd. Written in the French and Latin tongues by Peter Boayſtuau, Engliſhed by John Alday." H. Bynneman for him, and ſold in the Royal Exchange, at the Green Dragon, 1574. On the back are verſes " In prayſe of the Booke." Dedicated " To—Sir William Cheſter Knight, Alderman of the Citie of London, and Merchant of the Staple." Then, " The Printer to the Reader.---Peter Boayſtuau to the Reader.---The Table of principall matters," &c. Contains beſides, 287 pages. W. H. Again without date. Licenſed, 1566. Sixteens.

1584. " A VVatch-vvoord to Englande To beware of traytors and tretcherous practiſes, which haue beene the ouerthrowe of many famous Kingdomes

THOMAS HACKET.

dômes and common weales. Written by a faithfull affected freend to his Country: who defireth God long to bleffe it from Traytours, and their fecret confpiracyes, Seene and allowed, &c. Iofua 1. verfe 5. Take a good hart, and be ftrong, &c. London, Printed for Tho. Hacket, and are to be folde at his fhop in Lumberd ftreete, vnder the figne of the Popes head, 1584." It has prefixed, the queen's arms crowned, with E and R on the fides, and over it two diftichs of Latin verfes. On the back fix others, " In laudem Elyzabethæ Reginæ" by R. W. A dedication by A. M. *(Anth. Munday)* the author, " To the high, mightie, and right excellent Princeffe Elyzabeth, by the grace of God, of England, Fraunce and Ireland Queene, defendreffe of the true auncient, Catholique and Apoftolique Faith &c." Another, " To the right Hon. Mafter Thomas Pullifon Lord Mayor elected of the famous Citie of London, and to the worfhipfull Maifter Stephen Slanie, and Maifter Henrie Billingfley, Sheriffes," &c. In this tract are recited moft of the treafons practifed from the reign of Richard I. to the year 1584. Contains 47 leaves. At the end, " God long preferue and bleffe our Queene Elizabeth, and confound all her enimies. Honos alit Artes. A. M." W. H. Quarto.

An extract from Pliny. Probably the Polyhiftor of Solinus. Tranflated from the French. 1585. Quarto.

He had a book licenfed to him in 1565, intitled " The Somary of Plinie."

The Worke of Pomponius Mela the Cofmographer, &c. See it in 1590. 1585. Quarto.

" Of the Ladder to repentance, the third ftepp, which is of faith, by L. Simpfon. For him." Maunfell, p. 65. 1585. Octavo.

" The Harborough of Chriftianitie, containing many godly praiers and meditations, collected for the benefit and profit of the penitent finner, by Iohn Seuell Gent." For him. Licenfed. 1585. Octavo.

" Ant. Monday, his godly exercife for Chriftian famelies, containing an order of Praiers for Morning and Euening, with a little Catechifme betweene the man & his wife. For him." Maunfell, p. 86. 1586. Octavo.

" The excellent and pleafant worke of Iulius Solinus Polyhiftor. Contayning the noble actions of humaine creatures, the fecretes & prouidence of nature, the defcription of Countries, the maners of the people: with many maruailous things and ftrange antiquities, feruing for the benefitt and recreation of all forts of perfons. Tranflated out of Latin into Englifh, by Arth. Golding, Gent." A medallion of a royal perfonage, at half length, crowned, with a fceptre in his right hand, and a fword in his left. " At London, Printed by I. Charlewood for Tho. Hacket. 1587." It has prefixed, " The life of Solinus,—by Iohn Camertes.---C. Iulius Solinus fendeth hartie commendations to his freende Autius.---The epiftle dedicatorie of the Author to the fame Autius." The whole Gg 2, in fours. 1587.

This

THOMAS HACKET.

This Solinus lived after Pliny, and does little more than tranfcribe him. W.H. Quarto.

1587. "Secretes and wonders of the worlde, containing many wonderfull properties giuen to Man, Beafts, Fowles, Fifhes, and Serpents, Trees and Plants, abftracted out of Plinie." For him. Quarto.

1587. "A MIRROVR of Monfters: Wherein is plainely defcribed the manifold vices, & fpotted enormities, that are caufed by the infectious fight of Playes, with the defcription of the fubtile flights of Sathan, making them his inftruments. Compiled by Wil. Rankins. Magna fpes eft inferni. Seene and allowed." The fame medallion as to Polybius, 1568. "At London, Printed by I. C. (*John Charlewood*) for T. H. in Anno Do. 1587." Without any prefix. 24 leaves. Licenfed. W.H. Quarto.

1588. "The houfholders philofophie, wherein is perfectly, and profitably difcribed, the true oeconomia, and forme of houfekeeping. Firft written in Italian, by that excellent orator and poet, fignior Torquato Taffo, and now tranflated by T. K. Whereunto is annexed, a dairie booke for all good hufwiues. Dedicated to them by Bartholomew Dowe." Printed by I. C. for him.* Quarto.

1588. "A maruailous combat of Contrarieties, malignantlie werking in the members of humaine creatures, wherin the extreame vices of this prefent tyme are difplaied againfte Traytors & Treafons, &c. by W, A." (*William Auerell.*) Dedicated "To Maifter George Bond, Lord Mayor of London." Licenfed. William Bayntun, Efq; Quarto.

1590. "Four notable hiftories, applyed to foure worthy examples: as, 1. A diall for daintie darlings. 2. A fpectacle for negligent parents. 3. A glafs for difobedient fonnes. 4. And a myrrour for virtuous maydes. Wherunto is added a dialogue, expreffing the corruptions of this age. Written by W. A." Printed for him. Quarto.

1590. "The Rare and Singuler worke of Pomponius Mela, That excellent and worthy Cofmographer, of the fituation of the world, moft orderly prepared, and deuided euery parte by it felfe: with the Longitude and Latitude of euerie Kingdome, Regent, Prouince, Riuers, Mountaines, Citties and Countries. Wherevnto is added, that learned worke of Iulius Solinus Polyhiftor, with a neceffarie Table for thys Booke: Right pleafant and profitable for Gentlemen, Marchaunts, Mariners, and Trauellers. Tranflated into Englyfhe, By Arthur Golding Gentleman." For him, and fold "in Lumbertftreete, vnder the figne of the Popes head. 1590." Dedicated "To—Syr William Cicill, of the Noble order of the Garter, Knight, Barron of Burgley, Lord high Treafurer of England, &c.—written thys fixt of February, 1584." Then "A Table of the principall matters—in Pomp. Mela." Which tract, on 124 pages, feems to be the fame edition as printed in 1585, with a general title-page, annexing the Work of Solinus; which evidently had been printed and publifhed feparate, in 1587. W.H. Quarto.

"ENGLANDS

THOMAS HACKET.

"ENGLANDS Parnaffus: or The choyfeft Flowers of our Moderne 1600. Poets, with their Poeticall comparifons, Defcriptions of Bewties, Perfonages, Caftles, Pallaces, Mountains, Groues, Seas, Springes, Riuers, &c. Wherevnto are annexed other various difcourfes both pleafaunt and profitable." A Ling entangled in the branches of a honeyfuckle,—N and L in the lower corners; the device of Nic. Ling. "Imprinted at London for N. L. C. B. and T. H. 1600." Dedicated "To—Syr Thomas Mounfon, Knight." In verfe, by R. A. (*Rob. Allot*) Alfo, "To the Reader by the fame." See Warton's Hift. of Eng. Poetry, Vol. III; p. 280. Then "A Table of the fpeciall matters." Contains befides, 510 pages. George Mafon Efq; and W. H.+   Octavo.

"THE TREAfurie of health" &c. as p. 360. Contains *H h* 4, in eights.

"A REGIMENT for the Sea: Conteyning moft profitable Rules, Mathematical experiences, and perfect knovvledge of Navigation for all Coaftes and Countreys: moft needefull & neceffarie for all Seafaring men, and Trauellers, as Pilotes, Mariners, Marchants, &c. Exactly deuifed and made by VVilliam Bourne." A Sea aftorolob or ring. "Imprinted—by Tho. Hacket, and are to be folde at his fhop in the Royall Exchange, at the —Greene Dragon." On the back is a fhip, as defcribed in p. 783. This i take to be the firft edition, and to have been printed about 1574; the table of Primes, &c. beginning that year. In this edition after the preface are verfes by "I. H. In commendation of the Booke.---T. H. In prayfe of the Author.---A. R. To the Author." The whole, Q, in fours. W. H.   Quarto.

"An A, B, C. with the Pater nofter, Aue Maria, the crede, and ten commandements. With certain inftructions, that fchoolmafters ought to bring vp children in."

"Pyramus and Thifbe." For him. See Warton's Hift. of Englifh Poetry, Vol. III; p. 417.   Quarto.

He had licenfes alfo for printing, in 1560, "An almanack of Noftradamus. A playe of wytts. Of Cranmer, Ridley, & Latimer. Titus & Jofephus." In 1561, "The vnyuerfall Coffemographe. Phelantropofe. The entrynge of Chrifte into Englande. An alm. & pronoft. of Notradamus. A dyalogue of wyuynge & thryuynge, of Tuffhers, with ij leffons for olde & yonge. A cronacle of all the Emperours. The armour of pafyence. A new interlude of the ij fynnes of kynge Dauid." In 1565, "How to attayne to an honefte lyf, by Audryan Heffie. The wondors in Italy, aboute Naples & Rome. The hiftory of the moofte noble kynge Plafadas. Of the tow moofte noble prynces of the worlde, Aftromax & Polipena, of Troy. An inuectiue agaynfte the horryable ende & miferye of Mr. Grene." In 1566, "A play intitled Rauf Ruyfter Dufter. A play, Farre fetched & deare bought ys good for lades. An alm. & pronoft. of Barnabe Gaynsforth. The banquett of danties for all fuche geftes that

loue

loue moderatt dyate." In 1567, " An alm. & pronoſt. of Barnardyn. A letter ſente by the maydes of London to the vertuous matrons and miſtreſſes of the ſame Cetie. The traſurye a mydyce, cōtanyng Eloquente oracions, made by Tho. Pannell." In 1570, " Fawltes facoultes by—Boreman." In 1583, " The Turkiſſhe eſtate of this tyme preſent, with the confeſſion of Germadius, patriarke of Jeruſalem, and of the battle fought with the Perſians. A hiſtorye of an vſurer that hanged himſelf in Hell ſtreete in Fraūce, betwene the cittie Neuers & a place called St. Peter le monſtier, on xpmas Eue, 1782. A diall for daintie Darlinges. The banquet of dayntie conceiptes." In 1584, " Fedele & Fortim: The deceiptes in Loue, diſcourſed in a comedie of ij Italian gent. and tranſlated into Engleſhe." In 1585, " A panople of deuiſes: authoriſed under the hand of Mr. Fletewood, recorder." In 1586, " Thexploites & enterpriſes of Sir Francis Drake, at St. Domingo, &c." In 1587, " The philoſophical diſcourſe of the Houſeholder. A poſtyll of Math. Judex vppon the Epiſtles of all ſundaies and daies feſtiuall throughout the whole yere.—Abp. of Cant." In 1588, " A derie booke for huſwyfes: vpon condition. The anatomie of Abſurdities." In 1589, " Twoo epitaphs, or cōmemorations vpon the deathes of Sir Walter Myldmay, & Sir John Calthorp." Beſides ſome ballads, one in particular

" In the prayſe of worthy ladyes herein by name,
An eſpecyally of quene Elyſabeth ſo worthy of Fame."

RALPH, or RAFE NEWBERY,

WAS made free of the worſhipful company of Stationers, 21 Jan. 1559-60, but their regiſter is ſilent by whom. He had books printed for him the ſame year, but they give no information where he then dwelt. In 1563 he ſucceeded Tho. Powell in the houſe late Tho. Berthelet's, and i don't find him dwelling any where elſe. He was admitted into the livery 29 June, 1570; ſerved collector or renter in 1579, and 1580; under warden in 1583, and 1584; upper warden in 1589, and 1590; maſter in 1598, and 1601. In 1583, when the lower claſs of printers petitioned the privy-council againſt the privileged copies (ſee p. 688; &c.) he being joint aſſignee with Hen. Denham to execute the privilege which belonged to Hen. Bynneman, then deceaſed, yielded up to the company the copy-right of certain books for the benefit of their poor. They gave others alſo of their own; as did Mr. Day, Mr. Tottel, and Mr. Barker, as may be ſeen more particularly in our general hiſtory. Stow ſays he gave a ſtock of books, and privileges of printing to be ſold for the benefit of Chriſt's hoſpital and Bridewell.

" A

"A treatife, containing a Declaration of the Pope's vfurped Primacie:" 1560.
&c. as p. 845.
"The Firfte thre Bookes of—the Zodyake of Lyfe." &c. as p. 767. 1560.
Licenfed.
"Egloges, epitaphes, and fonnets, newly written by Barnabe Googe· 1563.
15. Marche.—Tho. Colwell for Raue Newbury, dwelling in Flete ftrete
a little aboue the Conduit, in the late fhop of Tho. Barthelet." Licenfed.
Octavo.
"The ZODIAKE OF LIFE, written by the godly and learned poet Marcellus 1565.
Pallingenius Stellatus, wherein are conteyned twelue bookes difclofing
the haynous crymes and wicked vices of our corrupt nature: And plainlye
declaring the pleafaunt & perfit pathway vnto eternal life, befides a number of digreffions both pleafaunt and profitable. Newly tranflated into
Englifhe verfe by Barnabee Googe. Probitas laudatur et alget. Imprinted—by Henry Denham for Rafe Newberye—in Fleete ftreate. Anno
1565. Aprilis 18." On the front of the next leaf is Googe's coat of arms;
and on the back thereof begin commendatory verfes in Latin and Greek:
viz. An acroftic (Barnabas Gogeus) by Gilb. Duke, and three Sapphics
by the fame; fome hexameters and pentameters by Chr. Carlile; James
Itzwert; G. Chaterton; and David Bell; alfo an Octaftich in Greek,
"Εισ τον ζωγιον, Ριχαρδοσ ο ςιΦανοσ," with a Latin tranflation. After them,
an epiftle dedicatory "To—Sir William Cecill Knighte, principall Secretary to the Queenes highneffe, & Maifter of her Maiefties Courte of
Wardes & Liueryes." Then, a preface "To the vertuous and frendely
Reader." See Warton's Hift. of Eng. Poetry, Vol. III; p. 449, &c.
Yy 4, in eights. W.H. Sixteens.

"An Expofition of the Hymne commonly called Benedictus: with 1573.
an ample & comfortable application of the fame to our age and
people. By A. Anderfon, Preacher. Stay prophane & vain bablings, &c.
2 Tim. 2 16. But if you haue bitter enuying &c. Iames 3. 14, 15, 16.
Imprinted—by H. Middelton, for Raufe Newbery." On the back, "The
booke to the Reader." In verfe. A. A. Dedicated to Thomas bp. of
Lincoln: From Medborne, 15 Jan. 1573. Contains Fol. 77. W.H.
Octavo.
Gualter's Homelies on Abdias and Ionas. Tranflated by Rob. Norton. 1573.
Octavo.
"A briefe expofition of fuch Chapters of the olde teftament as vfually 1573.
are red in the Church at common praier on the Sondayes, fet forth for the
better helpe and inftruction of the vnlearned. By Thomas Cooper
Bifhoppe of Lincolne. Whatfoeuer things are writté before time, &c.
Rom. 15. Imprinted—by H. D. for Rafe Newbery—in Fleeteftreete."
In a rich architective compartment. On the back begins "An Epiftle
to the Reader." 392 leaves befides. "Imprinted—by Henrie Denham
for him a little aboue the Conduite. Anno 1573." W.H. Quarto.
An edition, prior to this, was printed the fame year. Mr. Jennings.

RALPH NEWBERY.

1574. "The Familiar Epiftles of Sir Antony of Gueuara, Preacher, Chronicler, and Councellour, to the Emperour Charles the fifth. Tranflated out of the Spanifh toung by Edward Hellows, Groome of the Leafh. Wherin are contained very notable letters, excellent difcourfes, curious fayings, and moft natural reafons. Wherein are contained expofitions of certaine figures, authorities of holy Scripture; very good to be preached, and better to be followed.—declarations of ancient ftampes, of writings vpon ftones. Epitaphes of Sepulchres, Lawes & cuftomes of Gentils.—Doctrines, Examples, and Counfels, for Princes, for noble men, for Lawyers, and Churchmen: very profitable to be followed, and pleafant to be read. Printed—for Raufe Newbery, in Fleetftreet, a little aboue the Conduit." Dedicated "To—Sir Henry Lee Knight, Maifter of the Leafhe; and has an epiftle "To the Reader." Contains befides, 513 pages, and a table. "Jmprinted—by Henrie Middleton for Rafe Newbery;—a little aboue the Conduit." Over, a device of Chrift carrying a lamb on his fhoulder, with this motto about it, "Periit et inventa eft." An elephant under it. W.H. Again 1577, for him, and 1584, by him. Alfo without date, H. Bynneman for him. Quarto.

1574. "A forme of Chriftian pollicie, gathered out of French by Geffray Fenton. A worke very neceffary to al forts of people generally, as wherein is contayned doctrine, both vniuerfall & fpecial touching the inftitution of al Chriftian profeffion: and alfo conuenient particularly for all Magiftrates & gouernours of common weales, for their more happy Regiment according to God. Mon heur viendra. Jmprinted—by H. Middelton for *him*, Anno 1574." Dedicated "To—Sir William Cecill Knight—Lord high Treafurer of England.—At my chamber in the Blacke Friers, this xvj of May, 1574." Then, a table of contents. Contains befides, 352 pages. W.H. Quarto.

1575. "The true and perfect copie of a godly Sermon, preached in the Minfter at Lincolne, by the reuerend Father in God, Thomas (*Cooper*) L. Bifhop of Lincolne, the 28 of Auguft. Anno 1575. Jmprinted—by Henrie Middleton for Rafe Newberie," &c. On the back, an addrefs To the Reader, which concludes, "The excellency of the thing made manie men to requeft it; and the defire of many made me print it, hoping that among manie fome wil profite by it." The head title, "A fermon no leffe godly then neceffarie, preached, &c. Treating on the xvi. Chapter of Matthewe, verfe 26, 27." E 5, in eights. W.H. Alfo without date.
Octavo.

1575. Golden Epiftles, &c. Printed by H. Middleton for him. See it in 1577. Quarto.

1576. "A Perambulation of Kent: Conteining the defcription, Hyftorie, and Cuftomes of that Shyre. Collected and written (for the moft part) in the yeare 1570. By William Lambard of Lincolnes Inne Gent. and nowe increafed by the addition of fome things which the Authour himfelfe hath obferued fince that time. Iuuat immemorata ferentem Ingeniis,

geniis, oculifq; legi, manibufq; teneri. Imprinted—for Ralphe Nevvberie,—a litle aboue the Conduit, Anno 1576." This book is introduced with an addrefs " To his Countriemen, the Gentlemen of the Countie of Kent.—The xvj of Aprill, 1576. Your Countrey man and very louing friende T. W. ---Gulielmus Fletewodus, vrbis Londinenfis Recordator, ad candidum Lectorem.---The Saxon Characters, and their values.--- Sundry faultes *corrected*.---Angliæ Heptarchia." The expofition of the faid map begins on page 1. Contains 429 pages, and an alphabetical table. W.H. Quarto.
It was printed this year alfo for H. Middleton. Q. if not jointly?

" Galateo of maifter John Della Cafe, archbifhop of Beneuenta. Or 1576. rather, a treatife of the manners and behauiors it behoueth a man to vfe, and efchew, in his familiar conuerfation. A work very neceffary and profitable for all gentlemen, and others. Firft written in the Italion tongue, and now done into Englifh, by Robert Peterfon of Lincolnes Inne, gentleman." Dedicated " To—Lord Robert Dudley, Earle of Leycefter," &c. Prefixed alfo are fome commendatory verfes by Francefco Pucci, and Alefandro Citolini, in Italian; Edw. Cradock, D. D. in Latin Sapphics; Tho. Drant, Archedeacon, I. Stoughton, and Tho. Brown, in Englifh, Then, a lift of errors corrected. Contains befides, 122 pages. W.H.; Quarto.

" A Panoplie of Epiftles, Or, a looking Glaffe for the vnlearned. 1576. Conteyning a perfecte plattforme of inditing letters of all forts, to perfons of al eftates and degrees, as well our fuperiours, as alfo our equalls and inferiours: vfed of the beft and the eloquenteft Rhetoricians, that haue liued in all ages, and haue beene famous in that facultie. Gathered and tranflated out of Latine into Englifh, by Abraham Flemming. Armat fpina rofas, mella tegunt apes. Imprinted—for Ralph Newberie,—Anno à Virgineo partu, 1576." On the back begins, " A Catalogue of fuch Authours as haue written in this Panoplie." Next, " In Abr. Flemingi Londin. epiftolarum Panoplian, Th. Sp. Ebor. Octaftichon." Dedicated " To —fyr William Cordell knight, Maifter of the—Rolles, &c.---To the learned and vnlearned Reader." Then, " An Epitome of precepts, whereby the ignoraunt may learne to indite, acoording to fkill & order, reduced into a Dialogue betweene the Maifter & Scholer." The firft collection is " of certaine felected Epiftles out of M. T. Cicero, the moft famous Rhetorician & eloquent Orator among the auncient Romanes." Thefe end on p. 153. Then begin others out of Ifocrates, &c. They end on p. 240.---out of C. Plinius, &c.; which end on p. 313.---out of Paulus Manutius, and other late writers, moft meet for imitation; which end on p. 412.---" out of M. Gualter Haddon, and M. Roger Afchame, Gentlemen of late memorie, for their rare learning & knowledge verie famous." Thefe conclude with the book on p. 448. Licenfed.* W.H. Quarto.

" The

RALPH NEWBERY.

1576. " The Zodiake of life written by the excellent and Chriſtian Poet, Marcellus Palingenius Stellatus. Wherein are conteined twelue feuerall labours, painting out moſte liuely the whole compaſſe of the world, the reformation of manners, the miſeries of mankinde, the pathway to vertue and vice, the eternitie of the Soule, the courſe of the Heauens, the myſteries of nature, and diuers other circuſtances of great learning and no leſſe iudgement. Tranſlated out of Latine into Engliſhe, By Barnabie Googe, and by him newly recogniſhed. Probitas laudatur & alget. Herevnto is annexed (for the Readers aduantage) a large Table, as well of wordes as of matters mentioned in this whole work. Imprinted—for *him*—1576." On the back is the coat-armour of Sir William Cecill, Lord high Treaſurer of England, to whom this edition is alſo dedicated, but differently from that in 1565. Next, inſtead of the former preface to the reader, is a tranſlation of that of M. Palingenius Stellatus to Hercules the ſecond, Duke of Ferrar. Then, the commendatory verſes as before. Contains beſides, 242 pages. At the end, other verſes by Abr. Fleming; and an alphabetical table. W.H.+ Quarto.

1577. " Fiftie Godlie and Learned Sermons, diuided into fiue Decades, conteyning the chiefe and principall pointes of Chriſtian Religion, written in three ſeuerall Tomes or Sections, by Henrie Bullinger, miniſter of the Church of Tigure, in Swicerlande. Wherevnto is adioyned a triple or threefolde Table, verie fruitfull and neceſſarie. Tranſlated out of Latine—by H. I. ſtudent in Diuinitie." The loſt ſheep, as p. 902, without the elephant. " Matthewe 17. This is my beloued Sonne in whom I am well pleaſed: Heare him. Imprinted—by Ralphe Newberrie,—Anno Gratiæ 1577." The prefixes to the firſt two decades are " A Preface to the Miniſtrie of the Church of England, and other wel diſpoſed Readers of Gods woorde. —Of the foure generall Synodes or Counſels, &c.—The firſt table, conteyning the arguments—of euery Sermon," in their order.—" The ſecond table—ſuch places of Scripture as are vſed of the Authour," &c.—" The thirde and fourth Decade of Sermons,—. The ſecond Tome." Dedicated " To the moſt renowmed Prince Edward the ſixt, King of England and Fraunce, Lord of Jreland, Prince of Wales, and Cornewall, defender of the Chriſtian faith.—Tigure. March, 1550." The laſt eight ſermons of the fourth Decade were not publiſhed till the Auguſt following, when they were alſo dedicated as the former twelve.—" The fift and laſt Decade of Sermons,—. The thirde Tome." In a ſmall tract, entitled " The Judgement of—Hen. Bullinger—1566," mention is made of " hys preface of the fyft Decade to my Lord Gray"; but neither my copy of this, nor the edition 1584, have any preface; though the chaſm in the paging of both editions ſeem to indicate ſome deficiency; the ſignatures however are regular; ſo that probably it was never printed, if tranſlated. Printed in double columns, with lines; 1142 pages. W.H. Again 1584, and 1587. Quarto.

1577. " A Chronicle, conteyning the liues of tenne Emperours of Rome. Wherin are diſcouered their beginnings, proceedings, and endings, worthie

to

to be read, marked, and remembred. Wherein are alfo conteyned Lawes of fpeciall profite and policie. Sentences of fingular fhortneffe and fweetneffe. Orations of great grauitie and Wifedome. Letters of rare learning and eloquence. Examples of vices carefully to be auoyded, and notable paternes of vertue fruitfull to be followed. Compiled by the moft famous Syr Anthonie Gueuara, Bifhop of Mondonnedo, Preacher, Chronicler, & counfellour to the Emperour Charles the fift: and tranflated out of Spanifh into Englifh by Edw. Hellowes, Groome of her Maiefties Leafhe.—. Imprinted—for *him*—Anno Gratiæ 1577." Dedicated " To the moft excellent & vertuous Princeffe—Elizabeth, by the grace of God, of Englande—Queene: defender of the Chriftian Faith. &c." Then, " The Prologue of the famous Syr Anth. Gueuara." He begins with Trajan and ends with Alexander Severus. 484 pages, including the preface. A general table of the titles of every chapter is annexed. Licenfed. W.H: Quarto.

" Golden Epiftles, conteyning varietie of difcourfe, both Morall, 1577. Philofophicall, and Diuine: gathered, as well out of the remaynder of Gueuaraes woorkes, as other Authours, Latine, French, and Jtalian. By Geffrey Fenton. Newly corrected & amended. Mon heur viendra. Imprinted—for Raph Newberie,—1577." Dedicated " To—Ladie Anne Counteffe of Oxenford.—At my chamber in the Blacke Friers, London, 4 Feb. 1575." Contains befides, 185 leaves, and a table at the end. W.H. Again 1582. Quarto.

" A BOOKE OF THE Inuention of the Art of Nauigation, and of the 1578. greate trauelles whiche they paffe that faile in Gallies : Compiled by the famous Sir Anthonie of Gueuara, bifhop &c. Dedicated by the faid Authour vnto the famous Sir Frances de la Cobos, great Comptroller of Leon, and Counfeller vnto the faid Emperour Charles the fift. Wherin are touched moft excellent antiquities, and notable aduertifements for fuch as faile in Gallies. Imprinted—for *him*;—1578." Dedicated " To —Lord Charles Haward Baron of Effingham, and Knight of the—Garter. ---To the Reader.---A Letter Miffiue, or Dedicatorie of the Authour, vnto the renowmed fir Frances de la Cobos." The whole 30 leaves. W.H. Quarto.

" AN AVNCIENT Hiftorie and exquifite Chronicle of the Romanes 1578. warres, both Ciuile and Foren. Written in Greeke by the noble Orator and Hiftoriographer Appian of Alexandria, one of the learned Counfell to the moft mightie Emperoures, Traiane and Adriane. In which is declared: Their greedy defire to conquere others. Their mortall malice to deftroy themfelues. Their feeking of matters to make warre abroad. Their picking of quarels to fall out at home. All the degrees of Sedition, and all the effects of Ambition. A firme determination of Fate, thorough all the changes of Fortune. And finally, an euident demonftration, That peoples rule muft giue place, and Princes power preuayle. With a continuation, bicaufe that part of Appian is not extant, from the death of Sextus Pompeius, fecond fonne to Pompey the Great, till the

ouerthrow

ouerthrow of Antonie & Cleopatra, after the which time, Octauianus Cæſar had the Lordſhip of all alone. Βασιλίδι Κάτιση διεκότιδι τῇ ἐπιικισίατη. Imprinted—by Raufe Newbery and Henrie Bynneman. Anno 1578." This title taken from a copy of the late T. Beauclerk's Eſq; my own wanting the title-page, and all before " The Preface of the authour." The fifth book ends on p. 354, as printed, but ſhould be 370; and has Bynneman's device of the Brazen Serpent. Then, a leaf of the Faults corrected; but they are all on the front page. The continuation has a ſeparate title-page, and dedication " To—his ſingular good Mayſter Sir Chr. Hatton, Knight, Capitaine of the Queenes Maieſties Gards, Vizchamberlaine to hir Highneſſe, and one of hir—priuie Counſayle.—H. B." The ſignatures and paging continued to p. 394. Then, an alphabetical table for the whole, in double columns; and another leaf of faults corrected, full on both ſides, including the former, which proceeded no further than for p. 161. Licenſed. W.H.+ Quarto.

Mr. Beauclerk's copy contains 445 pages, with the errata on the back of the laſt; and then a table, in double columns, on four leaves: mine is on five. So, there were two editions this year.

The ſecond part was printed by H. Bynneman, this year; but as my copy wants the latter end, i cannot ſay whether it was not printed for Newbery. The licenſe for printing " Appius Alexandrinus of the Romaine Ciuill warres" was granted to both jointly; and though the entry makes no mention of their foreign wars, the former title includes both.

1578. " A Lamentable, and pitifull Deſcription, of the wofull warres in Flaunders, ſince the foure laſt yeares of the Emperor Charles the fifth his raigne. With a briefe rehearſall of many things done ſince that ſeaſon, vntill this preſent yeare, and death of Don Iohn. Written by Thomas Churchyarde Gentleman. Imprinted—by him. Anno 1578." Dedicated " To—Sir Frauncis Walſingham Knight, Principall Secretarie to the Queenes Maieſtie." Then, " The lamentation of Flaunders," in verſe; which with the narrative occupy 72 pages. At the end, an epilogue " To the VVorlde," on two leaves more. Licenſed.* W.H. Quarto.

1578. " A rich Store howſe, or treaſury for the Sicke, full of Chriſtian Counſailes, & godly meditations: whereunto is added a comfort for the poor priſoners condemned to die, with an exhortation to repentance. Tranſlated out of Dutch by Tho. Godfrey Eſq;" Licenſed. Octavo.

1578. Beza's " Diſplay of popiſh practiſes, or patched pelagianiſme, wherein is cleered the truth of Gods eternall Predeſtination, tranſlated by Will. Hopkinſon." For him. Quarto.

1579. " The Arte & Science of preſeruing Bodie & Soule in Healthe, Wiſedome & Catholike Religion: Phiſically, Philoſophically, and Diuinely deuiſed: By Iohn Iones Phiſition. Right profitable for all perſones: but chiefly for Princes, Rulers, Nobles, Byſhoppes, Preachers, Parents, and
them

them of the Parliament houſe. Pro. 4. (20—22) Hearken vnto my wordes:" &c. Printed alſo, if not the ſame, by H. Bynneman. Quarto.

" The moſt noble and moſt famous trauels of Marcus Paulus of the   1579.
Nobilitie of the State of Venice into the Eaſt parts of the Worlde, as Armenia, Perſia, Arabia, Tartaria, with many other Kingdoms and prouinces, no leſs pleaſant than profitable as appeareth by the table or Contents of this booke, moſt neceſſary for all ſorts of perſons, and eſpecially for trauellers, tranſlated (from the Spaniſh of Roderigo, by Iohn Frampton) into Engliſhe." For him. 167 pages. Quarto.
He had licence about this time to print " A deſcription of the Eaſt Indies, tranſlated out of Italion." Q. if this be not the book intended?

" A Treatiſe Touching the Libertie of a Chriſtian man. Written in   1579.
Latin by Doctor Martine Luther. And Tranſlated by Iames Bell.—Imprinted by Ra. Newbery and H. Bynneman.—1579." Dedicated " To —Lady Anne, Counteſſe of Warwicke." Then, an epiſtle from M. Luther to pope Leo X. Contains beſides, 92 pages. Licenſed. W.H. Octavo.

" Certaine Sermons vvherein is contained the Defenſe of the Goſpell   1580.
nowe preached, againſt Cauils and falſe accuſations as are obiected both againſt the Doctrine it ſelfe, and the Preachers and Profeſſors thereof, by the friendes and fauourers of the Church of Rome. Preached of late by Thomas, by Gods ſufferance Byſhop of Lincolne. Imprinted—by Ralphe Newbery—1580." In the ſame compartment as A briefe expoſition, &c. 1573. It has prefixed a ſhort addreſs " To the Godly diſpoſed Reader. —T. N."---A liſt of the 12 ſermons, their texts and folios; another, of the faults corrected: all on one leaf. The ſermons on 241 pages. At the end, a table of Common places. Mr. Ames, who had a copy, ſays they were gathered by Abr. Fleming: my copy has not the leaſt intimation thereof; but that certain zealous & godly miniſters craved that they might be publiſhed. Licenſed.* W.H. Quarto.

" A comfortable Treatiſe for a troubled Conſcience. Alſo a   1580.
briefe Catechiſme, wyth a forme of prayer for houſeholders." By Tho. Sparke, D. D. Octavo.

" The Chronicles of England, from Brute vnto this preſent yeare of   1580.
Chriſt, 1580. Collected by Iohn Stow Citizen of London. Printed—by Ralphe Newberie, at the aſſignment of Henrie Bynneman. Cum Priuilegio Regiæ Maieſtatis." In a compartment exhibiting the genealogy of Q. Elizabeth from K. Edw. III. On the back begins an alphabetical liſt of " Authours out of whom theſe Chronicles are collected. Dedicated " To—Lord Robert Dudley, Earle of Leiceſter,—Knight of the moſt noble Orders of the Garter, & of St. Michaell, one of hir Maieſties—priuie Counſell, & Maiſter of hir Horſſe:" whoſe arms are on the oppoſite page. Next is
" The

RALPH NEWBERY.

"The Preface to the Reader," varying* only occaſionally from what he had publiſhed in his Summaries of 1570, and 1575. Then, "A Table of the principall matters."---A liſt of "Faultes eſcaped in the Printing," on one page. The Chronicle on 1215 pages. At the end, an account of our univerſities, which is concluded on p. 1223. Again 1592, 1600, &c. W.H. Quarto.

1581. "AN EXPOSITION Vpon the Epiſtle of S. Paule the Apoſtle to the Epheſians. By S. Iohn Chryſoſtome, Archbiſhop of Conſtantinople. Peruſed, & auctorized, &c.—Printed by Hen. Binneman and Ra. Newberie. 1581. Cum Priuilegio Regiæ Maieſtatis." In a compartment with H. B. at bottom, and a tablet with "Decembris 24." It is introduced with an epiſtle "To the Reader," by the anonymous tranſlator; and dedicated "To—Anne Counteſſe of Oxenforde," by the ſame. It conſiſts of 24 ſermons or homelies, and 21 moral diſcourſes, which occupy 341 pages; on the back of the laſt is Bynneman's device of the brazen ſerpent, over a colophon of the ſame purport as the title-page. Then, a table of Common-places, alphabetically. W.H. Quarto.

1581. "THE DOOME warning all men to the Iudgement : Wherein are contayned for the moſt parte all the ſtraunge Prodigies hapned in the Worlde, with divers ſecrete figures of Reuelations tending to mannes ſtayed conuerſion towardes God : In maner of a generall Chronicle, gathered out of ſundrie approued authors by St. Batman profeſſor in Diuinitie. Imprinted by R- Nubery, aſſigned by H- Bynneman. Cum priuilegio Regali." On the ſell "Anno Domini 1581." Dedicated "To—Sir Thomas Bromley Knight,—Lord Chauncellor of Englande.---To the gentle Reader." Then, commendatory verſes in Greek, Latin, Italian, French and Engliſh, by James Sanford; an hexaſtich by Chr. Carlile; and ſome Engliſh verſes by H. H.---A catalogue of authors.---" The antiquitie of Englande firſt called Brutaine."---The author's coat of arms, viz. three ſtars, the lower one iſſuant out of a creſcent. See Guillim's Heraldry p. 96, and 430.---Te Deum, &c. Contains beſides, 437 pages, with many cuts of prodigies, monſters, &c. W.H. Quarto.

1581. "A Sermon preached at Paules Croſſe, the 23. of Aprill, being the Lords day, called Sonday. 1581. By Anth. Anderſon." Dedicated "To —Edm. Anderſon Eſq; Sergeaunt at Law to the Queene : and Tho. Fanſhawe Eſq; hir Maieſties Remembrauncer in hir hon. Court of the exchequer." Text, Luke 13. 6—9. H 6, in eights. On the laſt leaf, a phœnix in flames, and the ſun. Lambeth Library. Licenſed. Octavo.

* The chief difference conſiſts in this, that whereas, in the Summary 1570, he had ſaid, It is now five years, and in that of 1575, It is now ten years, ſince I (ſeeing the confuſed order of our late Engliſh Chronicles, &c.) leaving my own peculiar gains, conſecrated myſelf to the ſearch of our famous Antiquities, agreeing with his firſt publication, 1565; in this edition he ſays, "It is now ſeuentene yeres ſince," &c. (1563) whereby he muſt mean the time of leaving off his buſineſs in order to apply himſelf to his ſtudies. The Catal. Bibl. Harleianæ N°. 7675. mentions an edition 1560, but it is evidently a miſprint for 1580.

Ten

RALPH NEWBERY.

"Ten books of Homer's Iliad translated from a metrical French verſion 1581. into Engliſh by A. H. (*Arthur Hill, Eſq.*) of Grantham, a member of parliament. Nov. 25, 1580, it was licenſed to H. Binneman. Warton's Hiſt. of Eng. Poetry, Vol. III; p. 440. Quarto.

"EIRENARCHA: or of The Office of the Iuſtices of Peace: in two 1581. Bookes: Gathered, 1579, and now reuiſed, and firſte publiſhed, in the 24 yeare of the peaceable reigne of our gratious Queene Elizabeth. By William Lambard of Lincolnes Inne Gent. Hæ tibi artes erunt, paciq; imponere morem. At London: Imprinted by Ra. Newbery, and H. Bynneman, by the aſſ. of Ri. Tot. & Chr. Bar.—1581." In a compartment frequently uſed by Chr. Barker. Dedicated "To—Syr Thomas Bromley Knight, Lord Chauncelour of England," whoſe coat-armour is on the back of the title-page. Dated, "From Lincolnes Inne, 17 Jan. 1581." Then, a table of contents. The Eirenarcha on 511 pages. At the end, "A Table, conteining (verie neare) all the imprinted Statutes."--- The Faults corrected, and the brazen ſerpent on a ſeparate leaf. Of this edition 1500 books were printed. Licenſed to him and Binneman. Again 1582, 1588, 1592, 1594. Theſe have an Appendix containing ſundry Precedents. W.H. Octavo.

"A VIEWE OF A SEDITIOVS Bul ſente into Englande, from Pius 1582. Quintus Biſhop of Rome, Anno 1569. Taken by the reuerende Father in God, Iohn Iewel, late Biſhop of Saliſburie. Wherevnto is added A ſhort Treatiſe of the holie Scriptures. Both which hee deliuered in diuers Sermons in his Cathedral Church of Saliſburie, Anno 1570." His device of the loſt ſheep; but no elephant under it."—Printed by R. Newberie, & H. Bynneman.—1582. Cum gratia & Priuilegio." Prefixed is a preface by Jo. Garbrand, dated North Crowley, 27 Jan. 1582. in which we learn that biſhop Jewel " rendred vp his ſoule to God, the 23 of Sept. 1571." The view of the bull on 88 pages; the treatiſe proceeds to p. 175. W.H. Octavo.

"ANGLORVM PRÆLIA ab anno Domini 1327, anno nimirùm primo 1582. inclytiſsimi Principis Eduardi eius nominis tertij, vſque ad annũ Domini 1558. Carmine ſummatim perſtricta. Item, De pacatiſſimo Angliæ ſtatu, imperante Elizabetha, compendioſa Narratio. Authore Chriſtophoro Oclando, primo Scholæ Southwarkienſis prope Londinum, dein Cheltenhamenſis, quæ ſunt à ſereniſſima ſua Maieſtate fundatæ, Moderatore. Hæc duo Poëmata, tàm ob argumenti grauitatem quàm Carminis facilitatem, Nobiliſſimi Regiæ Maieſtatis Conſiliarij in omnibus huius regni Scholis prælegenda pueris præſcripſerunt. Hijs Alexandri Neuelli Kettum: tùm propter argumenti ſimilitudinem, tùm propter orationis elegantiam adiunximus." The device, a hind ſtanding on a wreath, with " Cerva chariſſima et gratiſſimus Hinnulus. Pro. 5." around it, in a compartment, " Londini Apud Radulphum Nubery, ex aſſignatione H-Bynneman Typographi,—1582. Cum priuilegio Regiæ Maieſtatis." On the back, "Regum Angliæ nomina, quorum aufpicijs hæc bella gefta funt." Then, "The Tenour of the Letters* directed by the Lords of hir highneffe priuie Counfell to hir Maiefties high cōmiffioners in caufes Ecclefiafticall, for the publike receyuing & teaching of Chr. Ocklandes Booke in ail Grammer & freefchooles within this Realme."---" A Copie of the Letters, directed by hir Maiefties High Commiffioners in caufes Ecclefiafticall, to all the Bifhops throughout hir Highneffe dominions of Englande and Wales, by efpeciall order from the Lordes of the Priuie Counsel, for the publike reading and teaching of this booke in all grammer and Free fchooles within their Diocefes." On the back of this is the queen's arms on a fhield, gartered and crowned. On the oppofite page " Ad illuftriffimam potentiffimamq; principem Elizabetham Angliae Franciæ, & Hiberniæ Reginam propugnatricem.—Chr. Ocklandus." Then, " In hæc Chr. Oclandi Anglorum Prælia, πανηγυρικὸν.—

ᵃ " After our right hartie commendations. Whereas there hath beene of late a booke written in Latine verfe by one Chr. Ocklande, entituled Anglorum prælia, aboute halfe a yeare fithence imprinted & publifhed, and nowe againe lately reprinted with the addition of a fhort treatife or appendix concerning the peaceable gouernement of the Queenes Maieftie: Forafmuche as his trauaile therein, with the qualitie of the verfe, hath receiued good commendation, and that the fubiect or matter of the faid Booke is fuch, as is worthie to be read of all men, & efpecially in common fchooles, where diuers heathen Poets are ordinarily read & taught, from the which the youth of the realme doth rather receiue infection in manners than aduancement in vertue: in place of fome of whiche Poets we thinke this Booke fit to be read & taught in the Grammer fchooles: We haue therefore thought good, for the encouraging of the faide Ockland & others, that are learned to beftowe their trauell & ftudies to fo good purpofes, as alfo for the benefit of the youth, and the remouing of fuch lafciuious Poets as are commonly read & taught in the faide Grammer fchooles (the matter of thys Book being heroicall & of good inftruction) to praye & require you vpon the fight hereof, as by our fpeciall order, to write your letters vnto al the Bifhops throughout this realme, requiring them to giue commandement, that in all the Grammer & free fchooles within their feuerall Dioceffes, the faide Bookes, de Anglorum prælijs, & peaceable gouernement of hir maieftie, may be in place of fome of the heathen Poets receyued & publiquely read & taught by the Scholemaifters vnto their Scholers in fome one of the fourmes in their fchoole, fitteft for that matter. Whereof praying you there may be no defaulte, fo as this our direction mays take place accordingly, wee bidde you hartily farewell: from the Court at Greenewich, the 21 of Aprill, 1582. Your louing friends

Edw. Lincolne.   Rob. Leicefter.
Amb. Warwicke.   Fran. Knollys.
Ja. Croft.   Chr. Hatton.
Fran. Walfingham."

ᵇ " After our heartie commendations, &c. Wheras we of hir Maiefties high Commiffion Ecclefiaftical, haue receiued Letters, from the Lords of hir Highneffe mofte honourable priuie Counfell, THAT we fhould direct order to all the Byfhops of the Realme, to caufe to be receiued, and publiquely read and taught, in all Grammer and Free fcholes, within their feuerall diocefes, a Booke in Latine verfe of late imprinted, entituled Anglorum Prælia, fet foorth by one Chriftopher Ocklande, as by the true Copie of their Honours Letters, which we fend you here inclofed, it may appere vnto you. THESE are therefore to require you, according to their Honours pleafures fignifyed to vs in that behalfe, forthwith vppon receipt hereof to take prefent order within your Dioces, for the due accomplifhment of their faid Letters accordingly. AND fo we bidde you heartely farewel. From London the feauenth of May, 1582. Your louing Friendes.

John London.   Owin Hopton.
Da. Lewes.   W. Fleetewoode.
Bar. Clerke.   Pet. Ofborne.
W. Lewyn.   Tho. Fanfhaw."

Tho.

Tho. Newton Ceftrefhyrius." The prælia end on I. iiij. Each of thefe pieces have particular title-pages fo as to fell feparate occafionally.

"EIPHNAPXIA Siue ELIZABETHA. De pacatifsimo Angliæ ftatu, imperante Elifabetha, compendiofa narratio. Huc accedit illuftrifsimorum virorum, qui aut iam mortui, fuerunt, aut hodie funt Elifabethæ Reginæ à confilijs perbreuis Catalogus. Authore Chriftophero Oclando" Device, as to the general title-page. "Londini: Apud Ra: Nubeirie, ex afsignatione H- Bynneman, 1582. Cum fereniff. Regiæ Ma. priuilegio." It is dedicated in hexameters "Ad prænobilem, & in primis eruditam fœminam vtriufq; literaturæ & Græcæ & Latinæ peritiffimam, Dominam Mildredam, Dynaftæ Burglæi magni Angliæ Thefaurarij coniugem laudatifsimam.---Ad Lectorem Candidum," in feven diftichs.--- "In Chr. Oclandi Elifabetham.—Ric. Mulcafter.---Ad Oclandum, de Eulogijs fereniffimæ noftræ Elifabethæ poft Anglorum prælia cantatis, Decaftichon.—Tho. Watfonus." The fignatures proceed to M. iij. See this laft tract under Bynneman. W.H. Octavo. 1582.

"The duties of conftables, borsholders, tithingmen, and fuch other lowe minifters of the peace, by W. Lambarde of Lincolns Inn, gent." The device of the loft fheep. Licenfed, with Henry Middleton. Again 1594. Octavo. 1583.

"An EXPOSITION vpon the two Epiftles of the Apoftle Sainct Paule to the Theffalonians, By the reuerende Father Iohn Iewel late Byfhop of Sarifburie." The loft fheep. "At London: Printed by R. Newberie, and H. Bynneman, Anno Salutis 1583." Dedicated "To—Sir Francis Walfingham, Knight, principal Secretarie &c.—Iohn Garbrand." Contains befides, 424 pages. Licenfed. W.H. Octavo.
The fame with an index, by Newbery alone. Cum privilegio. 1583.

"A BRIEFE Difcourfe, declaring and approuing the neceffarie and inuiolable maintenance of the laudable Cuftomes of London: Namely, of that one, whereby a reafonable partition of the goods of hufbands among their wiues & children is prouided: With an anfwer to fuch obiections and pretenfed reafons, as are by perfons vnaduifed, or euil perfuaded, vfed againft the fame.—Printed by Henrie Midleton for him. 1584." On the back, "Iuris ciuilis de confuetudine Axiomata, fiue maximæ." 48 pages. W.H. Octavo. 1584.

"The hiftorie of Cambria, now called Wales: A part of the moft famous Yland of Brytaine, written in the Brytifh language aboue two hundreth yeares paft: tranflated into Englifh by H. Lhoyd Gentleman: Corrected, augmented, and continued out of Records, and beft approued Authors, by Dauid Powel, Doctor in diuinitie. Cum Priuilegio." In a compartment with H. D. at the bottom. Dedicated "To—Sir Philip Sydney knight —From my lodging in London, 25 March, 1584. Dauid Powel.---To the Reader." Then, "A defcription of Cambria—: Drawne firft by Sir Iohn Prife knight, and augmented & made perfect by Humf. Lhoyd Gent." On 22 pages. The hiftory, 401 pages, exclufive of the table; 1584.

table; at the end of which is Denham's device of the ſtar, &c. "Imprinted by Raſe Newberie & H. Denham." W.H. Quarto.

1584. "A Dictionarie in Latine and Engliſh, heretofore ſet forth by Maſter Iohn Veron, and now newlie corrected and enlarged, For the vtilitie and profit of all yoong' ſtudents in the Latine toong', as by further ſearch therein they ſhall find: By R. W. Imprinted—by R. Newberie & H. Denham. Cum Priuilegio Regiæ Maieſtatis. 1584." It is introduced with an Epiſtle "To the Reader." At the end of which, "Rodolphus Waddingtonus Ludimagiſter Hoſpitij Chriſti ſcholæ, in renouatam Veroni Bibliothecam: ad ſtudioſam Anglorum Iuuentutem." A decaſtichon. Contains ſheets Vv. "Finis propoſiti, laus Chriſto neſcia Finis." W.H.
Quarto.

1585. "The Nomenclator, or Remembrancer of Adrianus Iunius Phyſician, diuided in two Tomes, conteining proper names and apt termes for all thinges vnder their conuenient Titles which within a few leaues doe follow: VVriten by the ſaid Ad. Iu. in Latine, Greeke, French, and other forrein tongues: and now in Engliſh by Iohn Higins: VVith a full ſupplie of all ſuch vvords as the laſt inlarged edition affoorded; and a dictional Index, conteining aboue fourteene hundred principall words, with their numbers directly leading to their interpretations: Of ſpecial vſe for all ſcholars & learners of the ſame languages. &c. Imprinted— for R. Newberie, & H. Denham. 1585." Dedicated "Dom. Doctori Valentino Dale, Regiæ Maieſtati a libellis ſupplicibus, Mecoenati ſuo benigniſſimo,—Winſamiæ, Idibus Nouembris, 1584.—Ioannes Higins. ---Then, Auctorum—catalogus.---The Contents." Contains beſides, 539 pages. "Finis propoſiti," &c. Licenſed, with H. Denham. W.H.
Octavo.

1585. "Pontici Virvnnii Viri Doctissimi Britannicæ Historiæ Libri Sex, Magna Et Fide Et Diligentia Conscripti: Ad Britannici codicis fidem correcti, & ab infinitis mendis liberati: quibus præfixus eſt catalogus Regum Britanniæ: per Dauidem Pouelum, S. Theolog. profeſſorem." The device, Abraham going to offer up Iſaac, with this motto, Deus providebit. "Londini Apud Edmundum Bolliſantum, impenſis H. Denhami & R. Nuberij, 1585." Dedicated "Ampliſſimo viro D. Henrico Sydneo,—Ruabonæ, pridie Calend. Iunij, 1585.—Dauid Pouelus." Then, the catalogue of kings from Brute to Cadvalader. The hiſtory on 44 pages. W.H. Octavo.

1585. "Itinerarium Cambriæ: Sev Laboriosæ Baldvini Cantuar. Archiepiſcopi per Walliam legationis, accurata deſcriptio, Auctore Sil. Giraldo Cambrenſe. Cum Annotationibus Dauidis Poueli ſacræ Theologiæ profeſſoris." Device &c. as above. Dedicated "Illuſtri viro omni tàm virtutis quàm doctrinæ laude clariſſimo Philippo Sidnaeo,

ᵖ Thᵉ oo in this title are joined in a ſingle. It had been in uſe ſome time before in the type, in Roman Letter; as they are aſſⁿ in black letter. the dedication of The Hiſt. of Cambria, &c.

auratæ

suratæ militiæ equiti, D. Pouelus S. D.—Ruabonæ 3 Calend. Julij, 1585." The life and prefaces of Girald are alfo prefixed. The paging continued from the hiftory to p. 284. Then, an index. W.H. Octavo.

"Politique Discovrses vpon Trveth and Lying. An inftruction 1586, to Princes to keepe their Faith and Promife: Containing the fumme of Chriftian and Morall Philofophie, and the duetie of a good man, in fundrie politique difcourfes vpon trueth and Lying. Firft compofed by Sir Martyn Cognet, Knight, one of the *French* Kings priuie Councell, mafter of requeftes of his houfhold, and lately Embaffadour to the Cantons of Zwitzers & Grizons. Newly tranflated out of French into Englifh, by Sir Edward Hoby, Knight. Ἀγαθον ἐλπίζειν ἐπὶ κυριον ἢ ἐλπίζειν ἐπ᾽ ἄρχυσι. It is better to truft in the Lord, then to put confidence in Princes. Pfalm 118. 9." The fame device as to Britannia. "At London, Printed by R. Newberie, Cum gratia & Priuilegio Regiæ Maieftatis, 1586." On the back, the coat-armour of Sir Edw. Hoby. Dedicated "To—Sir VVilliam Cecill, of the moft noble order of the Garter, knight, &c.—From the Ifle of Shepey, 10. Dec. 1585. Your honors moft bound Neuewe, Edw. Hobye." Then, "A commendation of this worke. By Tho. Digges." And a table of contents. Contains befides. 246 pages. Licenfed. W.H. Quarto.

"The Brvtish Thvnderbolt; or rather Feeble Fier-Flafh of Pope 1586, Sixtus the fift, againft Henrie the moft excellent King of Nauarre, and the moft noble Henrie Borbon, Prince of Condie. Togither with a declaration of the manifold infufficiencie of the fame. Tranflated out of Latin jnto Englifh by Chr. Fetherftone. Minifter of Gods word. Nahum, 3. I wil reueale thy filthines vpon thy face, &c. Imprinted—by Arnold Harfield, for G. B. and R. Newbery, 1586." Dedicated "To—Lord Robert Dudley, Earle of Leicefter,—and Lord Lieutenant, and Captaine general of hir Maiefties forces in the Low countries.—London, 3 Octo. 1586.---To the Reader." Then, "The bleffings of—Pope Sixtus the fift againft the King of Nauarre, & the Prince of Condie.—Pfalm..109. O Lord, they fhall curfe, and thou wilt bleffe:" &c. Contains 321 pages befides, and "The Declaration" &c. at the end. Licenfed. W.H.
Octavo.

"Britannia" in a feparate compartment, "Sive Florentissimorvm 1586. Regnorvm Angliæ, Scotiæ, Hiberniae, E.t Insvlarvm Adiacentium ex intima antiquitate Chorographica defcriptio, Authore Guilielmo Camdeno." The device; an anchor, on the ftock of which is a lion crowned paffant; about it DE-SIR-N'A-RE-POS. 1586. "Londini Per R. Newbery. Cum gratia & priuilegio Regiæ Maieftatis,⁴ 1586." "Dedicated

---

⁴ Bifhop Nicholfon, Hift. Libr. p. 4. note h, Edit. 1714. mentions two other editions, viz. 1582 and 1585, and has omitted this of 1586; which has all the appearance of being the firft edition, the dedication being dated 2. May, 1586, in this and the fubfequent editions.

"Clariffimo,

"Clariffimo, et honoratiffimo viro D. Guilielmo Cecilio, Baroni Burghlæo, Georgiani ordinis Equiti aurato, Summo Angliæ Thefaurario,—bonarum literarum, & Collegij Weftmonaft. Patrono fingulari. S. P.—Weftm. 2. Maij. Amplitudini tuæ deuotiffimus, G. Camdenus.---Beneuolo Lectori. S." Then, feveral commendàtory verfes. England on 476 pages; Scotland continued to p. 485. Ireland has a feparate title-page; and dedicated " Clarif.—Domino Edwardo Hobeio, equiti aurato," &c. Continued to p. 556; and then, tables of the antique and modern names. Licenfed. W.H. Octavo.

1587. The fame, " Nunc denuo recognita & plurimis locis adaucta." The Saxon alphabet illuftrated; and an additional index. England on 550 pages; Scotland continued to p. 559. Ireland to p. 648. W.H. Octavo.

1587. Holinfhed's Chronicles. The large edition. See it under H. Denham. This was licenfed to John Harrifon, fenior, Geo. Bifhop, Ra. Newbery, Hen. Denham, and Tho. Woodcock. Harrifon feems to have been the fole proprietor of the edition 1577. However, 30 December following, it is with other' books entered in the Stationers' regifter as belonging to Newbery and Denham, " by vertue of the queenes pryuiledge & grante." W.H. Folio.

1589. " Antimartinvs, Sive Monitio cuiufdam Londinenfis ad Adolefcentes vtriúfq Academiæ, contra perfonatum quendam rabulam, qui fe Anglicè Mantin Marprelat, Hoc eft Martinum Μαςλιγάρχον, ἢ μισάρχον, vocat. Rom. 16. verf. 17. Παρακαλῶ δὴ ὑμᾶς, &c. Londini, Excudebant G. Bifhop & R. Newbery,—1589." Contains 60 pages. Totus vefter, A. L. Licenfed. W.H. Quarto.

1589. "The Principall Navigations, Voiages and Discoveries of the Englifh nation, made by Sea or ouer Land, to the moft remote and fartheft diftant Quarters of the Earth, at any time within the compaffe of thefe 1500 yeeres: Deuided into three feuerall partes according to the pofitions of the Regions whereunto they were directed.—.Whereunto is added the laft moft renowned Englifh Nauigation round about the whole Globe of the Earth. By Richard Hakluyt, M. A. and ftudent fometime of Chrift-church in Oxford. Jmprinted—by G. Bifhop, and R. Newberie, Deputies to Chr. Barker, Printer to the Queenes moft excellent Maieftie, 1589." Dedicated " To—Sir Francis Walfingham, Knight, Principall Secretarie,—Chancellor of the Duchie of Lancafter, &c.—

ᵗ " Jmprimis Hollinfheddes Chronicle in folio, & all other Chronicles in the faid volume of folio. Item, Stowes Chronicle in iiijto. viij, and xvi. and all other Chronicles in iiijto. viijvo. and xvjvo. Item, Mr. Dr. Cowpers Dictionarie in folio, & all other dictionaries in folio. Item, Morelius Dictionarie in Latyn & Greek, and all other dictionaries, Latin & Greeke. Item, Verons Dictionarie in iiijto. and all other dictionaries in iiijto. Item, The Dictionarie in Frenche & Englifhe in iiijto. and all other dictionaries Fr. & Eng. in quarto. Item, The Dictionarie in Italian & Englifhe in iiijto. and all other dictionaries Italian & Eng. in quarto."

London,

London, 17 Nouember.---To the Reader." Some commendatory verses; and a table of the order of the voyages. 825 pages besides, and an index. Licensed. W.H. Folio.

He printed in Greek types, "D." in a compartment by itself, "IOANNIS 1589. CHRYSOSTOMI ARCHIEPISCOPI CONSTANTINOPOLITANI HOMILIAE ad populum Antiochenum, cum presbyter esset Antiochiæ, habitæ, duæ & viginti. Omnes excepta prima, nunc primùm in lucem editæ, ex manuscriptis Noui Collegij Oxoniensis codicibus. Opera & studio Ioannis Harmari Collegij prope Winton Magistri Informatoris. Cum Latina versione eiusdem, Homiliæ decimæ nonæ, quæ in Latinis etiam exemplaribus hactenus desiderata est.—Excudebant G. Bishop & R. Newberie. Anno salutis humanæ cɔ ɔ xc." Dedicated "Clariss. & Hono. Heroi, D. Christoph. Hatton, illustriss. ordinis Periscelidis Equiti aurato, summo Angliæ Cancellario, &c.—E Collegio propè Winton, 10 Kal. Decembris, Anno vltimi temporis cɔ ɔ xc." 381 pages; then, "Beneuolo Lectori. —Tuus Io. Harmar.---Errata, &c. Licensed in 1581. W.H. Octavo.

"A REMONSTRANCE:" in a compartment, "Or plaine Detection of 1590. some of the Faults, and hideous Sores of such sillie Syllogismes and impertinent Allegations out of sundrie factious Pamphlets & Rhapsodies, as are cobled vp together in a Booke, Entituled, A Demonstration of Discipline: Wherein also, The true state of the Controuersie of most of the points in variance, is (by the way) declared. 2 Tim. 3. ver. 6 & 7. They haue a shew of godlinesse, &c.—Imprinted—by G. Bishop & R. Newberie,—1590." The preface addressed "To the factious and turbulent T. C. W. T. I. P. and to the rest of that anarchicall disordered Alphabet, which trouble the quiet & peace of the Church of England." Contains besides, 210 pages, and the Errata. Licensed. W.H. Octavo.

Under the queen's arms "Academiarum quæ aliquando fuere et hodie 1590. sunt in Europa, Catalogus & enumeratio breuis." The vniversity arms, with "Academia Oxoniensis" around it —Excudebant G. Bishop & R. Newberie.—1590." On the back, the arms of Sir Chr. Hatton, to whom the book is dedicated, viz. "Clariss. viro, Sacræ Maiestati a melioribus consiliis D. Christophoro Hattono, summo Angliæ Cancellario, Georgiani ordinis equiti aurato, & florentissimæ Oxoniensis Academiæ Archicancellario Honoratissimo S. P.—Oxon, 15 Maij, 1590.—M. W." The running title, "Europaei Orbis Academiae." The academies are described alphabetically. 60 pages. Licensed. W.H. Quarto.

Stow's Summarie. Containing 759 pages. See p. 853. Octavo. 1590.

"Laur. Deios, his 2 sermons against Antichrist, on Apocal. 19; 19." 1590. Licensed. With G. Bishop. Octavo.

"THE," in a compartment, "CONSENT OF TIME, DISCIPHERING the 1590. errors of the Grecians in their Olympiads, the vncertain computation of
the

the Romanes in their Penteterydes and building of Rome, of the Persians in their accompt of Cyrus, and of the vanities of the Gentiles in fables of antiquities, disagreeing with the Hebrewes, and with the Sacred Histories in consent of time. Wherein also is set downe the beginning, continuance, succession, and ouerthrowes of kings, kingdomes, States and gouernments. By Lodovvik Lloid Esquire. Prov. 24. Vir sapiens est fortis, & vir doctus robustus. Imprinted—by G. Bishop, and R. Nevvberie, 1590." Dedicated " To the most reuerend Iohn Archbishop of Canterburie. Primite, &c.---To the Reader."---A table of the arguments. Contains besides, 722 pages. W.H. Quarto.

1590. " A Treatise of Ecc'esiasticall Discipline: Wherein that confused forme of gouernment, which certeine vnder false pretence, and title of reformation, and true discipline, do striue to bring into the Church of England, is examined and confuted: By Matth. Sutcliffe.—Printed by G. Bishop and R. Newberie, 1590." Dedicated " To—the Earle of Bathe. – London, 1 Jan. 1590.---The Epistle to the Reader." Contains besides, 230 pages. " Imprinted &c. 1591." W.H. Quarto.

1590. " A sermon—at Pawles, 9 Nouemb. 1589, on 1 Cor. 12; 25—27. by William James, D. D. and Deane of Christes Church, Oxford." Licensed, with G. Bishop. Quarto.

1591. Virgilii Georgica. A prose Latin paraphrase, by Nich. Grimoald. With G. Bishop. See Hist. of Eng. Poetry, Vol. III; p. 60. Quarto.

1593. Testamenti veteris Biblia Sacra, &c. Excudebant G. B. R. N. & R. B. See it under G. Bishop. Again 1597. Folio.

1594. " THE SECOND PART OF THE FRENCH ACADEMIE. VVherein, as it were by naturall historie of the bodie and soule of man, the creation, matter, composition, forme, nature, profite and vse of all the partes of the frame of man are handled, with the naturall causes of all affections, vertues and vices, and chiefly the nature, powers, workes and immortalitie of the Soule. By Peter de la Primaudaye Esquier, Lord of the same place, and of Barre. And translated out of the second Edition, which was reuiewed and augmented by the Author. At London, Printed by G. B. R. N. R. B. 1594." Dedicated " To—Sir Iohn Puckering knight, Lorde Keeper of the great Seale of England," By T. B. (Tho. Bowes) the translator. Then, a copious epistle to the reader; and a table of contents. Besides, 600 pages. Licensed, with G. Bishop, in 1588. The first part was printed in 1589, " Impensis G. Bishop. Licensed to both in 1586. W.H. Quarto.

1597. " The Navigators Svpply, Conteyning many things of principall importance belonging to Nauigation, with the description and vse of diuerse Instruments framed chiefly for that purpose; but seruing also for sundry other of Cosmography in generall: the particular Instruments are specified
or

on the next page." A copper-plate cut" of the travellers' jewell : "They that goe downe to the Sea in Ships, &c. Pfal. 107. Imprinted—by G. Bishop, R. Newbery, & R. Barker, 1597." On the back, " Of the Compasse in generall. Of the Compasse of Variation. Of the Trauailors Iewell. Of the Pantometer. Of the Hemisphere. Of the Trauerse-boorde. A friendly Aduertisement to the Nauigators of England." Dedicated "To—Lord Robert, Earle of Essex,—Master of her Maiesties Ordinance & Horse, &c.—William" Barlowe." Then some commendatory verses. Contains besides, L 2, in fours. Alex. Dalrymple Esq; Quarto.

" The principal Navigations," &c. as in 1589. " within the compasse 1599. of these 1600 yeres: Diuided into three Volumes, &c. By Richard Hakluyt Preacher, and sometime Student of Christ-Church in Oxford. Imprinted—by G. Bishop, R. Newberie, and Rob. Barker, 1599." Dedicated "To—Lord Charles Howard, Erle of Notingham,—Lord high Admirall of England, Ireland, Wales, &c." The first volume, 606 pages besides prefixes. The 2d and 3d volumes have each separate title-pages, &c. The 2d vol. is dedicated " To—Sir Robert Cecil knight, principall Secretarie to her Maiestie, master of the Court of Wardes & Liueries," &c. 204 pages.

The third volume, printed 1600; and dedicated also to Sir Robert 1600. Cecil. 868 pages. W.H. Folio.

" A Summary of the Sacraments, as well of the old as of the new law, gathered out of the most learned writers of our dayes, by Anth. Foord, preacher: with an exhortation to the communicantes, truely to examine themselues." For him. Octavo.

A sermon against Discord, by Will. Overton, D. D. on Rom. 16. vers. 17. Octavo.
He had licenses besides to print the following, viz. In 1562, "An accidence in Laten & Frensshe." In 1564, " An almanake & pronostic. of Mr. Buckemaister." In 1565, " An alm. and pronost. of Mr. Gayle; towching Surgery." Again, in 1566. In 1567, " An alm. and pronost. of Mr. Startoppe." In 1569, the like of Joachem Hubryghte. " An Exhortation vnto batchelers. Manipulus Curatorum." In 1570, " Medytations vpon the Psalmes." In 1576, he had a reversionary licence for " An abstract of all the penall Statutes in force." See p. 734. In 1577,

* On it is engraved very small, " If any man desire more ample instructions concerning the vse of these instruments hee may repayre vnto Ihon Goodwin dwellinge in Bucklarsburye teacher of the grownds of these artes. The instruments are made by Charles Whitwell, ouer against Essex howse, maker of all sortes of mathematicall instruments, and the grauer of these portraitures."

w " This booke was written by a Bishops sonne,
And by affinitie to many Bishops kinne:
Himselfe a godly Pastour, prayse hath wonne,
In being diligent to conquer sinne. J. R."

"A difcourfe of the life of a Seruinge man, By Will. Dorrell." In 1578, "Guido his Surgery: with H. Bynneman." In 1584, "Orders for the dyoceffe of Wynchefter. Will. Parries voluntarye cōfeffion. A true report of Willm Parries arainement. The office of Juftices of peace, by John Goldwell of Graies inne Efq;" In 1585, "Tables of Surgery, by Horati a Florentine. Britania antiqua, in Englifhe." In 1586, "Bullengers Decades, in Latin, with Hu. Jaxfon. Articles to be enquired of in the dioces of Winchefter." In 1588, "Mr. Dr. Fulkes Anfwere againft the Remifhe Teftament, with G. Bifhop. An admonition to the People of England." In 1589, "A Declaration of the caufes wherewith the nauie of the noble quene of England being moued, did in their voiage to Portingal take certen fhips furnifhed with Corne, &c. Both in Lat. & Eng." In 1590, "De diuerfis miniftrorū euangelij gradibus &c. Cui duo alij additi: alter de honore qui debetur eccl'iarum paftoribus: Alter de facrilegijs & facrilegiorum pœnis. Authore Hadriano Sarauia, Belga. Dr. Syklif De Prefbiteriū. Efopes fables, in Greeke." Thefe, with G. Bifhop. In 1594, he "affigned to Ja. Newberie Stowes Chronicles in 8vo and 4to; Lamberdes Eirenarche, & Duty of Conftables; Jewell vpon the Theffalonians; The Panoplie of Epiftles; The familiar & Golden Epiftles." In 1595, "An anfwere vnto a certen columnious letter publifhed by Mr. Job Throkmorton—by Math. Sutcliffe. With Bifhop and Barker."

FRANCIS COLDOCK

WAS prefented by William Bonham, and made free of the Stationers' company, 2 December 1557; came on the livery 29 June 1570; ferved renter in 1575 and 1576; under warden in 1580 and 1582; upper warden in 1587 and 1588; mafter in 1591, and 1595. He lived in Lombard ftreet, over againft the Cardinal's Hat; and afterwards at the Green Dragon in St. Paul's church-yard. According to his monumental infcription in St. Faith's, he married Alice the widow of Richard Waterfon, and had iffue by her, two daughters; Joane, married to William Ponfonby, ftationer, and Anne who died young: he departed this life the 13th day of January, 1602. See Dugdale's St. Paul's, p. 126. He was fined for keeping open fhop on St. Luke's day; and another time, the fermon time.

1561.   "A declaration of Edmonde Bonners articles," &c. as p. 767. Licenfed.
1572.   "Andr. Hiperius his regiment of pouertie, tranflated by Hen. Trippe. For him." Maunfell, p. 58.   Octavo.

" A

FRANCIS COLDOCK. 919

"A fermon preached before the Queenes Maieftie, at Greenwich, 14   1573.
of March, by Richard Coorteffe bifhop of Chichefter, on Eccles. 12.
1—7." H. Bynneman for him. John Fenn Efq; Again 1586. Octavo.

"The fyrft parte of Commentaries, Concerning the ftate of Religion,   1573.
and the Common vvealthe of Fraunce, vnder the reignes of Henry the
fecond, Frauncis the fecond, and Charles the ninth. Tranflated out of
Latine into Englifhe by Thomas Tymme Minifter. Seene & allowed.
Deut. 32. (7.) Remember the dayes of olde, &c. Jmprinted—by Henrie
Bynneman for *him*, 1573." Dedicated "To—Sir Richard Baker, knight,
one of the Queenes Maiefties Iuftices of peace in the countie of Kent,"
on whofe name is an acroftic on the back of the title-page, figned T. T.
---"The Authours Preface to the Reader.---Edw. Grant Schoolemafter
of VVeftminfter to the Booke:" in verfe. "In hiftoriam de Gallæ Ecclefiæ ftatu recens editam, Rob. Rolli carmen:" in 7 diftichs. The commentaries on 271 pages. Then, "The tranflator to the reader;" and an
index. This was publifhed again next year, with the remainder.

"THE THREE PARTES of Commentaries, Containing the whole and   1574.
perfect difcourfe of the Ciuill warres of Fraunce,—. With an Addition of
the cruell Murther of the Admirall Chastilion, and diuers other Nobles,
committed the 24 daye of Auguft. Anno 1572. Tranflated, &c. Imprinted—by Frances Coldocke, 1574." Each part has a feparate title-
page.

"The feconde parte of Commentaries, &c. Imprinted—by Frances   1574.
Coldock. And are to be fold at his fhop in Pawles churchyard at the
figne of the greene Dragon, 1574." Dedicated "To—Sir Richard
Baker, Knight."---15 Latin diftichs by E. G.---An epiftle "To the
Chriftian Reader:" by the author.---The table. The Commentaries on
267 pages.

"The thirde parte" &c. The table begins on the back of the title-   1574.
page. 494 pages. "Imprinted—by Hen. Middleton; for Frauncis
Coldocke,—1574." Each of thefe parts contains three books. Pet.
Ramus the author of them was murthered in the bloody maffacre at Paris,
of which a particular account is annexed in

"The tenth Book treating of the furious outrages of Fraunce, vvith
the flaughter of the Admiral, and diuers other noble and excellent men,
—24 Auguft, 1572." On 38 leaves. W.H. Quarto.

"ΤΗΣ ΕΛΛΗΝΙΚΗΣ ΓΛΩΣΣΗΣ ςαχυυλογία. Græcæ Linguæ Spicilegium,   1575.
Ex præftantiffimis Grammaticis, in quatuor Horrea collectum, breuifsimis
quæftiunculis & intellectu facilimis, ad puerorum intelligentiam difpofitum, & in Scholæ Weftmonafterien: Progymnafmata diuulgatum Collectore E. G. Scholæ eiufdem moderatore. Quintilianus. Nifi Gram-
matices fundamenta fideliter ieceris, quicquid fuperftruxeris corruet. Ex
officina H. Binnemani pro Francifco Coldock." Dedicated "Illuftriff.
viro virtute, doctrina confilio præftantifsimo, D. Guiliel. Cæcilio, aureæ
perifcelidis Equiti aurato,—fummo Angliæ Thefaurario,—Academiæ
Cantab.

FRANCIS COLDOCK.

Cantab. Cancellario, & Weſtmon. Collegij ſcholæq; benigniſsimo pa‑ trono." 199 leaves, beſides prefixes and affixes. At the end, under the device of the brazen ſerpent, "Excuſum in ædibus H. Binemani Typographi, impenſis F. Coldock, Anno à virgineo partu, 1575." W.H. Quarto.

1577. "Two ſermons at Paules in the time of the plague: the 1, on Sophonie, 3. 1, 2, 3. The 2, on Ieremie, 23. 5, 6. By Dr. Tho. White." For him. Octavo.

1577. "A Supplication exhibited to the moſte Mightie Prince Philip king of Spain &c. Wherein is contained the ſumme of our Chriſtian Religion, for the profeſſion whereof the Proteſtants in the lowe countries of Flanders, &c. doe ſuffer perſecution : with the meanes to acquit and appeaſe the troubles in thoſe partes. Annexed is The confeſſion of Auſpurge, &c. Written in Latin & French by Ant. Corranus." With H. Bynneman. Licenſed. Octavo.

1577. "The praiſe of ſolitarineſs ſet down in the form of a dialogue, wherin is contained, a diſcourſe philoſophical of the life active and contemplatiue." With H. Bynneman. Quarto.

1577. "Apologia Doctissimi Viri Rogeri Aſchami, Angli, pro cæna Dominica, contra Miſſam & eius prǣſtigias: in Academia olim Cantabrigienſi exercitationis gratia inchoata. Cui acceſſerunt themata quædam Theologica, debita diſputandi ratione in Collegio D. Ioan. pronunciata. Expoſitionis itèm antiquę in epiſtola Diui Pauli ad Titum & Philemonem, ex diuerſis ſanctorum Patrum Gręcè ſcriptis commentarijs ab Oecumenio collectę, & à R. A. Latinè verſę. Excuſum Londini pro F. Coldocko, 1577." Dedicated "Illuſtriſſ. ac nobiliſſ D. Roberto Dudleio, Leceſtriæ comiti, &c. Has doctiſſ. viri Rogeri Aſchami lucubrationis Theologicas, nunc primum collectas, & æditas, gratitudinis ergò debitiq; officij ratione E. G. dedicat conſecratq;." Then, ſome hexameters " In Symbolium Gentilitium Honoratiſſ. Domini, Comitis Leceſtrenſis," with the earl's creſt gartered; and an epiſtle "Benevolo Lectori. Contains beſides 296 pages. "Londini, ex Typographia H. Middletoni, 1577." W.H. Octavo.

1578. "Dissertissimi Viri Rogeri Aschami Angli, Regiæ olim Maieſtati à Latinis Epiſtolis familiarium Epiſtolarum libri tres, magna orationis elegantia conſcripti, nunc denuò emendati & aucti. Quibus adiunctus eſt Commendatitiarum, Petitoriarum, & aliarum huius generis ſimilium Epiſtolarum, ad alios Principes & Magnates conſcriptarum, Liber vnus. Huc acceſſerunt pauca quædam eiuſdem R. A. Poemata. Item, Oratio E. G. De vita & obitu R. A. & eius dictionis elegantia. Londini Pro F. Coldocko, 1578." On the back, 20 diſtichs "Ad illuſtriſſimam Reginam Elizabetham. Liber de ſe." Dedicated alſo "Sereniſſ. Potentiſſimæq; Divæ Elizabethæ, Angliæ,—Reginæ, &c. Collegij Scholæq; Weſtmon. liberaliſſimæ fundatrici.—E Schola tuæ Maieſtatis Weſtmon. Februarij 7, An. Do. 1578.—E. Grant." Then, "Liber

FRANCIS COLDOCK.

"Liber de fuo Domino vita defuncto," 82 diftichs.---" Candido Lectori, præfatio.---Amplifs. viri Thomæ Wilfoni—Carmen Encomiafticum.---In doctiff. viri R. A. laudem Sylua.—Gul. Camden." Afcham's poems end on folio 258. His life is on 49 pages; then, an index. "Londini, in officina H. Middletoni Typographi, 1578." Licenfed. W.H. Again 1581, H. Bynneman, for him; and 1590, Arn. Hatfield for him. Octavo.

"A Godlye and fruitefull Sermon againft Idolatrie, &c. preached the xv. daye of Ianuarie, 1581. in the parrifhe church of Eaton Sooken within the countie of Bedforde, by P. W. *(Peter White)* Minifter, &c. At London, Imprinted by F. Coldocke, 1581." Licenfed. Octavo. 1581.

"The Figure of Antichrift, with the tokens of the end of the world: being an expofition on the 2d epiftle to the Theffalonians. By Tho. Timme." For him. Licenfed. Octavo. 1586.

The Duties of Conftables, Borfholers, &c. With H. Middleton. Octavo. 1587.

"An Æthiopian Hiftorie, written in Greeke by Heliodorus, no leffe wittie then pleafaunt: Englifhed by Tho. Vnderdowne, & newly corrected, and augmented with diuers & fundry newe additions by the faid Authour. Whereunto is alfo annexed the argument of euery booke in the beginning of the fame, for the better vnderftanding of the ftorie. Imprinted—for Frauncis Coldocke, &—fold—in Paules church yeard, at —the greene Dragon, 1587." Dedicated "To—Edwarde Deuiere Lord Boulbecke, Earle of Oxenford, Lord great Chamberlain of England." Then, an addrefs "To the gentle reader" acknowledging the incorrectnefs of a former edition, &c.---An anecdote "of the author out of the Latine tranflation." Befides, 152 leaves. Licenfed in 1568. W.H. Quarto. 1587.

"A BRIEFE DISCOVERIE OF DOCTOR ALLENS feditious drifts, contriued in a Pamphlet written by him Concerning the yeelding vp of the towne of Deuenter, (in Ouerriffel) vnto the king of Spain, by Sir William Stanley. The contentes whereof are particularly fet downe in the page following. Reuel. 17. 3. And I fawe a woman fit upon a fkarlet-coloured beaft, &c. Matth. 15. 6. Thus haue ye made the commandement of God of no authoritie by your traditions. and Chap. 23. ver. 13. Woe therefore be vnto you Scribes, &c. Matth 7. 15. Beware of falfe prophets, &c. London, Imprinted by I. W. *(John Wolfe)* for F. Coldock, 1588." On the back, is a Summarie of the contents. Then, a prefatory epiftle "To the Reader," by G. D. Contains befides, 128 pages. Licenfed: the Counfelles hands. W.H. Quarto. 1588.

"IOANNIS STVRMII, HIERONYMI OSORII, ALIORVMQVE EPISTOLÆ, Ad Rogerum Afchamum aliofque nobiles Anglos miffæ, ab E. G. collectæ nunc primum editæ. Londini Impenfis F. Coldock, 1589." This is inferted between Afcham's letters and poems, printed by Arn. Hatfield, 1590, and paged accordingly. W.H. Octavo. 1589.

"A

1591. " A Siluer Watch-bell to waken all estates from the drousy sleep of Sin &c. By Tho. Timme." For him. Octavo.

—— " All the famous Battels that haue been fought in our age throughout the worlde, as well by sea as lande, set foorth at large liuely descrībed, beautified and enriched with sundry eloquent Orations and the declaratiōs of the causes, with the fruites of them. Collected out of sundry good Authors," &c. With H. Bynneman. In a rich military compartment. Dedicated " To the right hon. Christopher Hatton, Captaine of the Queenes Maiesties Garde attending vpon hir most royall person, Vicechamberlaine to hir Highnesse," &c. By Henry Bynneman, who had caused it to be translated. Besides, 337 pages, and a table. W.H. Quarto.

He had licenses also for printing, in 1566, " A new Revinge for an olde grudge, by Sutton." In 1567, " The brefe dyscource of Rob. Baker, in Gynney, India, Portyngale & Fraunce, &c." In 1577, " A courtly Controuersie of Cupids cautels:" with H. Bynneman. In 1582, " Sermons preached before the quene, and at Paules crosse by John Jewell, late Bp. of Salisburie: With a treatise on the Sacramentes, gathered out of his sermons, in his Cathedrall." In 1588, " The psalmes of Dauid in meter: in all volumes & notes & tunes: in the Scottysshe, Frenche, Dutche, & Thitalian; or any of thése iointly & feuerallye. Granted to Mr. Harrison, master; Mr. Coldock, & Mr. Denham, wardens, to the vse of this corporation."

WILLIAM GRIFFITH, or GREFFYN,

WAS an original member of the Stationers' company, but it does not appear he ever came on the livery. He dwelt in Fleet-street, at the sign of the Falcon, and kept shop in St. Dunstan's Church-yard, in the West of London. He used for his rebus a griffin sitting, holding an escutcheon with his mark or cypher, and a sweet-william in its mouth. In 1558, he was fined " for that he printed, A medisine made by Dr. Owyn, without lycense, xij.d." And Will. Peckeryng and he, " for cōtensious words betwene them had for cōvaynge awaye of a copye or Dr. Owyns medysine ys fined vi.d. le pece."

1561. " All the examinacions of the Constante martir of God, M. Iohn Bradforde, before the Lorde Chauncellor, B. of Winchester, the B. of London, & other cōmissioners: whervnto ar annexed his priuate talk & conflictes in prison after his condemnacion, with the Archb. of York, the B. of Chichester, Alphonsus, and King Philips confessour, two Spanishe freers, and sundry others. With his modest learned and godly answeres. Anno

WILLIAM GRIFFITH.

Anno Domini, 1561. Cum Priuilegio—folum." It is introduced with "The Oryginall of his Lyfe." 114 leaves. "Imprinted—in Fleetſtrete, at the Signe of the Faucon by William Griffith, and are to be ſold at the litle ſhop in ſaincte Dunſtones churchyard.—The xiii daie of Maye." On the laſt page, a demi-angel, with a Roman G beneath. Licenſed. W.H.
Octavo.

A goodly gallery &c. See it in 1571. 1563.

"The tragedie of Gorboduc, whereof three actes were wrytten by Thomas Nortone, and the two laſte by Thomas Sackvyle. ¶ Sett forthe as the ſame was ſhewed before the queenes moſt excellent maieſtie, in her highnes court of Whitehall, the 18 Jan. 1561. By the gentlemen of ſhynner Temple in London. Sept. 22." See Royal and Noble Authors, Vol. I. p. 163. Dodſley's Old Plays, Vol. II. and Hawkins's Orig. of the Eng. Drama, Vol. II. Licenſed. 1565.

"The heavy horrable hiſtory of the dreadfull death of the Righte Reuerente Roode of Weſt Cheſter." Licenſed. Octavo. 1565.

"A goodly Gallery vvith a moſt pleaſaunt Proſpect, into the garden of naturall contemplation, to beholde the naturall cauſes of all kind of Meteors. As well fyery and ayery, as watry annd earthly, of which ſorte be blaſing Starres, ſhootinge Starres, flames in the ayre &c. thonder, Lightninge, Earthquakes &c. Rayne, Dew, ſnowe, Cloudes, Springes, &c. Stones, Metalles, Earthes. To the glory of God, and the profitte of his creatures. Pſalme. 148; (7, 8.) Prayſe the Lord vpon Earth, &c. Jmprinted—in Fleteſtrete by Wylliam Gryffith, 1571." Dedicated "To —Lord Robart Dudley, Maſter of the Queenes maieſties horſe," &c. by Will. Fulce, who therein claims his being the author of the Philoſophers game, which James Robothum had dedicated to this lord in 1562, and again in 1563. See p. 802, and 804. Contains Fol. 71. On the back, a demi-angel; about it "Thinke and thanke, lyve in fere, knowe thy ſelfe, far and nere." Under it, "Jmprynted—in Fleteſtreate, at the ſigne of the Faucone,—And they are to be ſold at his ſhop in S. Dunſtones churchyarde in the Weſte. 1563." I take this to be the date of printing the book, and that in 1571 the dedication was added, with a freſh title-page. Licenſed. W.H.   Octavo. 1571.

"A Detection of the malice and miſcheefe that is in Hereſie, declaring the cauſes why heretickes bee brent. Made by Theoph. Miſoheriſon."
Octavo.

"A playn and ſmall confutation: Of Cammells corlyke oblatracion." In verſe, by Tho. Churchyard. "Imprinted in Fletſtrit by Wyllyam Gryffyth, a lyttle aboue the condit, at the ſyne of the Gryffin." 200 lines, on one ſheet.   Folio.

He had licenſes for printing alſo, in 1557, "An almanake & pronoſtication of Geo. Williams doynge." In 1559, "A Catecheſme in Laten,
Frynſhe

WILLIAM GRIFFITH.

Frynſhe & Engleſſhe." In 1562, " The cōmendation of Muſycke by Churchyarde. Perymus & Theſbye." In 1563, "A comfortable drynke for the Plage, to be taken at all tymes." In 1564, "The ſtory of K. Henry IV, and the tanner of Tamouthe." See p. 892. " An almanake &c. of Geo. Wylliams. A quere entitled, A godly new dyalogue betwene Chryſte & a ſynner: meyte for all ages. A pycture of a chylde borne in the Ile of Wyghte with a cluſter of grapes aboute yt nauell. A monſterus chylde which was in Anwarpe." In 1565, " A new dyalogue or dyſputation betwene Day & Nighte, &c. A pleaſaunte Recytall of iiij worthy Squyers of Darius kynge of Perſia. A catecheſme in myter. A fayre well called Churchyardes Ronde from the Couurte to the Coūtry Groūde." In 1566, " An almanake &c. of Geo. Wylliams. A mooſte delectable cōference betwene the wedde lyfe & the ſingle by Hen. Hake. Eraſmus Rot. cōtaynynge a moſte pleaſaunt Dialogue towchynge the entertaynment & vſage in cōmen Jnnes, &c. A caueat for cōmen Torſetors, vulgarly called Vagabons, by Tho. Harman. Tempora labuntur, otherwyſe the Jmage of tyme. An Ephethappe of Capt. Randall." In 1567, " An almanake &c. of Simonde Pembroke. A newe yeres yefte. The x cōmandementes in Welſhe, peruſed by M. Dr. Yeale." In 1568, " The tragecall hiſt. of Floredicus. A tragecall dyſcourſe of ij Engleſſhe Louers." In 1569, " A godly gardyn out of which the moſt cōfortable hirbis may be gathered. The fortriſſe of Fayth, by Edw. Crane. A dyſcourſe of rebelles drawen fourth to warre, by Churchyarde. A letter with ſpede ſent to the pope, declarynge the rebelles. A ſpedy remyde for the peſtelence by a byſſhope of Denmarke. An Epytaphe of the Lorde of Pembroke by Mr. Edwardes. A geliflower or ſwete marygolde, wherein the frutes of teranny you may beholde/ by Tho. Preſton." In 1570, " Very Godly pſalmes and prayers. An Epytaph on Mr. Oneflowe." Alſo for many ballads.

LUCAS, or LUKE HARRISON,

WAS made free of the worſhipful company of Stationers 21 May 1556; came on the Livery in 1568; and ſerved the office of Renter in 1570, and 1571. He dwelt at the ſign of the Crane in S. Paul's church-yard. His name is always printed Lucas after 1561.

1559. An almanacke & pronoſtication. by M. Mych. Noſtrōdamus. Hen. Sutton for him. See p. 844. Licenſed. Octavo;

1561. " A briefe reherſal of the doings at Poyſſye in Fraūce, betwixt the lordes of ſplritualty, and the miniſters of the goſple. Sett forth by Nicholas Galaſius, one of the diſputers there." Another edition without date, dedicated

dedicated to Francis Ruffel, earl of Bedford.* My copy wants the title-page, but conclude it to be the edition without date, as it has the dedication. After which is " The proclamation of fummons and faue conduit to the generall confultation of the clergy in Fraunce:—Geuen at Sainct Germaines in Laye ỹ xxv. day of July in the yeare of grace M.D.LX.J." The rehearfal contains H 3, in eights. " Jmprinted—in Paules Churchyarde, by Luke Harryfon." W.H.+ Sixteens.

" THE Agreemente of Sondry places of Scripture, feeming in fhew 1563. to Iarre, Seruing in ftead of Commentaryes, not only for thefe, but others lyke. Tranflated out of French, and nowe fyrft publyfhed by Arthure Broke. Seene & allowed, &c. Imprynted—in Paules Churchyard, at the figne of the Crane, by Lucas Harrifon, 1563." On the back begins an apology from " The Prynter to the Reader," on account of the author's, or rather tranflator's abfence. Then, the tranflator's Epiftle to the Reader, which concludes with " Thys is but only a profe offered in haft, at whiche if you fhal beginne to take profite, let vs vnderftand it, & you fhal geue him ỹ hath thus begon ỹ occafion to go further forward," &c. Fol. 907. On the back, 6 four-lined ftanzas by " Tho. Broke the yonger," wherein he mentions the fhipwreck of the tranflator. Licenfed. W.H. Octavo.

Beza's Life and Death of M. John Calvin. H. Denham for him. 1564. Licenfed. Octavo.

" AN AVNSWERE TO THE TREATISE OF THE CROSSE: wherin ye fhal 1565. fee by the plaine & vndoubted word of God, the vanities of men difproued: by the true and Godly Fathers of the Church, the dreames & dotages of other controlled: and by lavvful Counfels, confpiracies ouerthrowen. Reade & Regarde," H. Denham for him, 1565. Prefixed, is an epiftle " To Iohn Martiall: ftudent in Diuinitie: James Calfhill Bacheler of the fame." Contains befides, 187 leaves, of which the preface occupies 20. At the end, a table, and fome Latin verfes, entitled " Mortis & Crucis collatio." Alfo, " The fame in Englifhe." Colophon, " Imprinted &c. Nouembris 3." Licenfed. W.H. Quarto.

" The Pedegrewe of Heretiques. Wherein is truely & plainely fet out 1566. the firft roote of Heretiques begon in the Church, fince the time & paffage of the Gofpell, together with an example of the ofspring of the fame. Efay. 47. O Babylon fapientia &c. O Babilon thy wifdome & conning hath deceiued thee. Perufed & alowed" &c. The earl of Leicefter's creft, to whom the book is dedicated by J. Barthlet, minifter. Then, " L. G. Cantabrigienfis ad Lectorem. S. D.----Ad eundem Dicolon Tetraftrophon." The running title, " The Pedegrewe of Popifh Heretiques." 89 leaves. Imprinted by H. Denham for him, 1566. W.H. Quarto.

" A SHORT Difcourfe of the meanes that the Cardinal of Loraine vfeth, 1568. to hinder the ftablifhing of peace, & to moue new troubles in Fraunce. Imprinted—by H. Denham for him, 1568." F, in fours. Licenfed. W.H. Sixteens.

5 R " To

LUKE HARRISON.

1569. "To the Quenes Maiesties poore deceiued Subjectes of the North Countrey, drawen into rebellion by the Earles of Northumberland and Westmerland. Written by Tho. Norton. And newly perused and encreased. Seen & allowed," &c. D 4, in eights. Imprinted—by H. Bynneman, for him, 1569. Licensed. W.H.   Octavo.

1569. "A true Declaratiō of the troublesome voyage *(the second)* of Mr. John Hawkins to the parties of Guynea, & the West Indies, 1567, and 1568," Printed by Tho. Purfoot for him. Licensed.   Octavo.

1570. A Dictionary French & English. H. Bynneman for him. Licensed.   Quarto.

1570. "A Postil, or orderly disposing of certeine Epistles vsually red in the Church of God, vppon the Sundayes & Holydayes throughout the whole yeere. Written in Latin by Dauid Chytræus, and translated into English by Arthur Golding. Seen & allowed" *&c.* Dedicated "To—Sir Walter Myldmay Knight, Chancelour of the—Exchequer, *&c.*—Finished at Powles Belchamp, the last day of March, 1570." The postil on 489 pages; then, a table and exposition of words. At the end, Bynneman's Sign. "Imprinted—by H. Bynneman,—for Lucas Haryson & G. Bishop, 1570. Cum priuilegio." W.H. Again 1577.   Quarto.

1571. A politike discourse &c. Tho. Purfoot for him.   Octavo.

1571. "The Psalmes of Dauid and others. With M. John Caluins Commentaries. Anno Do. MDLXXI." In a magnificent compartment, neatly cut on wood: at the bottom, a dragon sprawling on its back; a smaller beast seizing it by the throat, with this label, "Non vi sed virtute." The engraver's initials C. T. Dedicated "To—Lord Edw. De Vere Erle of Oxinford," by Arth. Golding, the translator, 20 Octo. 1571. Then, "John Caluin to the godly Readers—At Geneva, 23 Julij, 1557." In 2 partes; the first on 287 pages, the second, 259. Printed by Tho. East and Hen. Middleton, for him and Geo. Bishop. Licensed. W.H. Again 1576.   Quarto.

1571. "A Confession of Faith, made by common consent of diuers reformed Churches beyond the seas: With an Exhortation to the Reformation of the church. Perused & allowed" &c. There are prefixed an epistle "Vnto the Christian Reader," by I O.---An exhortation to, or for the reformation of the church, addressed "To the moste renowned Prince, Lewes of Burbon, Prince of Conde, &c. and other famous & noble Dukes, *&c.* that embrace the true Gospell of Chryst in the Realme of Fraunce." By Theod. Beza.—"At Geneua, the tenth of the Calendes of Marche. The yere from Chrysts incarnation for vs. 1565.---An epistle "To all the faithfull Christians which are in Germany & other forayne Nations, by the Ministers of the Churches throughout Heluetia, whose names are subscribed.—1 March 1566." To the general confession is annexed "A Confession of faythe, made by cōmon agremēt of such Frēchmen,

"Frenchmen, as desire to liue accordyng to the puritie of the Gospel of oure Lorde Iesus Christ." With an expostulation to the king. At the end, "This confession of faithe was publiquely presented ageyne to the Kings Maiestie, Charles the ix. of that name, at Poissy, ix. of Septemb. 1561." S, in eights. "Imprinted by H. Bynneman, for *him*, M.D.LXXI." Licensed. W.H.   Octavo.

"The Fall of Hugh Sureau from the Gospell, with his penitent sub-   1573. mission, with confession of his fault." For him and G. Bishop. Octavo.

"A Postill, or Expositions of the Gospels read in the Churches of God   1574. on Sundayes & feast days of Saincts. Written by Nich. Heminge, and translated into English by Arth. Goldinge." 'My copy wants the title-page. Dedicated "To—Sir Walter Myldmay Knight, *&c*—London, the xij. of October, 1569." Then, A warning, " Too all the seruants of God, and Ministers of Iesu Chryst,—within the famous Realmes of Denmarke & Norwey *by* Nich. Heminge, Minister of the Gospell in the Vaiuersitie of Hasnie.—the xxx. of Marche, The yere since Chryst was borne, 1561." The postill on 345 leaues; then, a table, and " An exposition of certaine words." Imprinted by H. Bynneman for him and G. Byshop, 1574. Cum Priuilegio. Licensed. W.H+ Again 1577.   Quarto.

" A Catholike exposition vpon the Reuelation of Sainct Iohn. Col-   1574. lected by M. Augustine Marlorate, out of diuers notable Writers, whose names ye shal find in the page following. Printed by H. Binneman for *him*, and G. Bishop." In his grand compartment. Dedicated by Arth. Golding " To—Sir Walter Mildmay knyghte," &c.—Finished at my lodging in London the last day of August, 1574." Then, an alphabetical table. The exposition on 318 leaues. W.H.   Quarto.

" Sermons by M. John Caluin vpon the booke of Job." Printed by   1574. H. Binneman for him, and G. Bishop.   Folio.

" SERMONS of M. Iohn Caluine vpon the Epistle of Saincte Paule to   1574. the Galathians. Jmprinted—by *him*, and G. Bishop, 1574. Dedicated " To—Sir William Cecill knight, *&c.*—VVritten at my lodging in the forestreete vvithout Cripplegate the 14. of Nouember, 1574.—Arthur Golding." Then, the argument, and the table. The sermons on 329 leaues. At the end, Mr. Calvin's prayers before and after his sermons. W.H. Again without date by H. Middleton for them.   Quarto.

" THE Philosopher of the Court, written by Philbert of Vienne in   1575. Champaigne, And Englished by Geo. North, Gentlemā." Dedicated " To—Master Chr. Hatton Esquier, Captaine of the Queenes Maiesties Garde, and Gentleman of hir highnesse priuie Chamber." Some verses by Io. Daniell, and Will. Hitchcockes, Gent. The Argument of the Booke, in verse also. Then, " The Prologue, or Authors Epistle to his Ladie, a louer of vertue." The treatise, 114 pages. At the end, the tree of Charity,

LUKE HARRISON.

Charity, as ufed by Reg. Wolfe. Imprinted—by H. Binneman for him, and G. Bifhop, 1575. W.H. Octavo.

1576. "A GODLIE Sermon, Preached on Newe yeeres day laft befor Sir Will. Fitzwilliam Knight, late Lord deputie of Irelande, Sir Iames Harrington Knight, their Ladyes & Children, vvith many others, at Burghley. By the minifter of God, Anth. Anderfon. Hereto is added a very profitable forme of prayer, good for all fuch as paffe the feas: by the fame author framed, and vfed in his aduentured iourney. Pfal. 37. Marke the vpright man," &c. Dedicated "To the right worfhipfull Sir William Fitzwilliam &c.—Ian. 3, 1575." The form of prayer is dedicated alfo "To—Syr Will. Fitzwilliam, &c.—At Holme Patricke, in Irelande, 17. Octob. 1575. E 4, in eights. Tho. Purfoote for him. W.H. Octavo.

1576. "The vvhole Summe of Chriftian Religion, giuen forth by two feuerall Methodes or Formes: the one higher, for the better learned; the other applyed to the capacitie of the common multitude, and meete for all: yet both of them fuch as in fome refpect do knit themfelues together in one. By Edmund Bunny. B. D.—Imprinted—by Tho. Purfoote for Lucas Harifon and Geo. Bifhop, dwelling in Paules Churchyarde. 1576." Dedicated "To—Edmunde—Archbifhop of Canterbury." Then the preface, and an advertifement of fome additions and corrections. Each part has a feparate title-page: the former, "A fhort fumme of Chriftian Religion vnder the confideration of the three perfons in the Trinitie," &c. Prefixed is the cut of a tree-root, with letters of reference, which are explained in letter-prefs on the back. At the end of this part are two other cuts, in like manner, of the tree of Death, and the tree of Life. The 2d title-page is at Fol. 30. "A fhort fumme of Chriftian Religion vnder the confideration of the Ten-Commandementes." This part proceeds to Fol. 71. On a feparate leaf, by way of Epilogue, "Deut. 4, a. 5—9. Behold I haue taught you ordinances," &c. W.H. Octavo.

1576. "XXVII LECTVRES, or readings, vpon part of the Epiftle written to the Hebrues. Made by Maifter Edw. Deering, B. D. Imprinted by him, 1576." In his grand compartment. It has prefixed, an epiftle 'To the Reader, dated the xxiiij. of Nouember, 1576; and 5 fix-lined ftanzas by T. N. "Aucthorized by the lord Archbyffhops grace of Canterbury." The readings on fheets Hh, in eights. See p. 855, note w. W.H. Quarto.

1576. "A VIEWE OF mans eftate, wherein the great mercie of God in mans free juftification by Chrift is very comfortably declared. By Andrewe Kingefmill. Diuided into Chapters—. Where vnto is annexed a godly aduife—touching marriage. Seene & allowed, &c. Imprinted—by H. Bynneman for him & G. Bifhop, 1576." Prefixed is an epiftle "To the Reader," giving fome account of the author. M 4, in eights. On the laft leaf Bynneman's fign. Licenfed, 1570. W.H. Octavo.

" Will.

"Will. Fulke his sermon at Alphages, on Gallath. 4: 21—31." For 1577. him and G. Bishop. Octavo.

"The Sermons of M. Iohn Caluin vpon the Epiftle of S. Paule too the 1577. Ephefians. Tranflated out of French into Englifh by Arth. Golding. Imprinted—for him & G. Byfhop, 1577." In the grand compartment. Dedicated "To—Edmund—Archbifhop of Canterbury. &c.—At Clare in Suffolke, the vii. of January, 1576." Then, an addrefs "To all Chriftians baptized in the name of the Father, and of the Sonne, and of the holy Ghoft, dwelling or abyding in Fraunce.—Your brethren in our Lord, the caufers of thefe fermons too bee brought to lyght."---The Argument, &c. The fermons on 347 leaves. W.H. Quarto.

"The Lectvres of John Knewftub, vpon the twentith Chapter of 1577. Exodus, and certeine other places of Scripture. Seene & allowed &c. Imprinted by him, 1577." In his grand compartment. Dedicated "To —the Ladie Anne, Counteffe of Warwick." The lectures on 355 pages. "Authorifed by the Bifhop of London." Again 1578. W.H. Quarto.

"The benefit that Chriftians receyue by Iefus Chrift crucified. Tranflated out of French, into Englifh, by A. G. Imprinted—nigh vnto the three Cranes in the Vintree for him & G. Byfhop." It has two epiftles prefixed: one, To the Englifh Reader; in which we are told that this treatife was firft written in Italian, and printed at Venice, after that tranflated into French, and printed at Lions: the other, to all Chriftians vnder Heaven. H 3, in eights. W.H. Octavo.

"A treatife of the Sin againft the Holy Ghoft. &c. as p: 893.

"A shorte Difcourfe of the Ciuill warres and laft troubles in Fraunce, vnder Charles the ninth." Head-title; my copy wanting the title-page, and all prefixed, except the laft leaf of the table. In three books. The licenfe fays by Jeffray Fenton. 218 pages. Imprinted—by H. Bynneman—For him & G. Byfhop. Licenfed, 1569. W.H. Octavo.

"Yues Roupeau his Dialogue" &c. as p. 893.

He had licenfes alfo for the following, viz. In 1559, "A generall pardon for euer." In 1562, "A Declaration of the prynce of Conde & his affociats to the quene, vpon the Jugement of Rebellion fett out againft them by theyre enymees. Dygges pronoffic. and his Tyctonycon." In the 1564, "A dyaloge betwene the hed & the cappe. An expofition vpon fyrfte chap. of ỹ prouerbis of Salomon by Mygchell Coupe." In 1565, "An hiftory of Sampfon. The vertues & properties of tryaucles. The ofsprynge of Symonde Magus, or iij of the popeffhe heryfees." In 1568, "The game of iij whetftones. A dyfcourfe of thynges happened in Fraunce." In 1569, "An aunfwere to fertayne obiections of Mr. Feckham, fometyme abbott of Weftm. agaynfte a fermon which Mr. Gough made in the tower, &c." In 1570, "An Edict for the peace of Fraunce." In 1576,
" to

" to prynt in Englyſhe, one book which in Latyn is intitled Quæſtionū & reſponſionū Chriſtianarū pars altera, quæ eſt de Sacramentis." In 1577, "A Declaration of the iuſt cauſes whiche haue conſtrained the Kinge of Nauarre & thoſe of the religion to betake them to their weapons. Neuer printed."

15 July, 1578, Mrs. Harriſon ſold to Tho. Woodcock all the copies and ſhares, which belonged to the ſaid Mr. Luke Harriſon.

THOMAS COLWELL

WAS preſented to the Stationers' company, with Richard Watkyns, and another, as apprentices with Will. Powell, 13 Octob. 1556; and was made free 30 Aug. 1560; but never came upon the livery. He ſucceeded Rob. Wyer, at the ſign of St. John Evangeliſt in St. Martin's pariſh, beſide Charing croſs; and made uſe of ſeveral of Wyer's wood-cuts. He dwelt alſo in Fleet ſtreet, beneath the Conduit, at the ſame ſign; alſo, in St. Bride's Church-yard. In 1561, he "prented the Diatory of helthe; the aſſyce of breade and ale; with Arra pater, without licenſe:" for which he was fined xij.d. Soon after he printed two ballads without licenſe alſo, and was fined v.s. Again in 1563, he was fined xx.d. for printing other men's copies. Alſo in 1565, and in 1569, for printing ij ballads againſt Bonner.

1562. " A NEVVE Comedy or Enterlude; concerning thre lawes, of Nature, Moiſes, and Chriſte, corrupted by the Sodomytes, Pharyſies, and Papiſtes: Compoſed by John Bale :* and nowe newly Imprynted. The yere of our Lord M.D.LXJJ. ¶ The Players names. Deus pater. Moſeh lex. Jnfidelitas. Jdolatria. Ambitio. Pſeudodoctrina. Vindicta Dei. Nature lex. Chriſti lex, vel Euangelium. Sodomiſmus. Auaricia. Hypocriſis. Fides Chriſtiana. Baleus Prolocutor." On the back, " Jnto ſyue Perſonages maye the partes of this Cōmedy be deuyded. 1. The Prolocutour. Chryſten fayth. Inſydelyte. 2. The lawe of Nature. Coueytouſneſſe. Falſe doctryne. 3. The lawe of Moſes. Jdolatrye. Hypocryſye. 4. The lawe of Chriſt.. Ambycion. Sodomye. 5. Deus pater. Vindicta Dei. ¶ The apparellynge of the ſyxe vyces, or fruytes of Jnſydelytie. Let Jdolatrye be decked lyke an olde Wytche. Sodomye like a Monke of all ſectes. Ambycion lyke a Byſſhop. Coueteouſnes lyke a Pharyſie, or ſpyrytual lawer. Falſe doctryne lyke a popyſſhe Doctour. And Hypocryſye lyke a graye fryer. The reſt of the partes are eaſye ynough to coniecture." My copy imperfect at the end. Licenſed. Quarto.

1562. "Here Foloweth a Compendyous Regimente or Dietary of health, &c. as p. 379. Anno Domini M.D.LXJJ. XJJ Die Menſis Januarij." H,

* Anno 1538; and was firſt printed in the time of K. Edw. VI. See it in our General Hiſtory, between the years 1549, and 1550.

in

THOMAS COLWELL.

in eights: "Jmprinted by me Thomas Colwel, Dwellynge in the houfe of Roberte Wyer, at the Signe of S. John befyde Charynge Croffe." W.H. Sixteens.

"John de Vigo his little practice in chyrurgery. Tranflated out of Latin." Sixteens. 1562.

"Egloges, epitaphes, and fonnets," &c. as p. 901. 1563.

The famous hiftory of the vertuous and godly woman Judith. Octavo. 1565.

"Here begynneth a goode booke of medicines called, the treafure of poore men." Sixteens. 1565.

"The boke of wifdome, otherwife called, the flower of vertue, folowing the auctorities of auncient doctours and philofophers, deuiding and fpeaking of vices and vertues, wyth many goodly examples, wherby a man may be praifed or difpraifed, wyth the maner to fpeake well and wyfelie to al folkes, of what eftate fo euer they bee. Tranflated fyrft out of Italion into French, and out of French into Englifh, by John Larke." Alfo without date. W.H. Sixteens. 1565.

"The Prognoftication For euery of Erra Pater:" &c. as p. 380. Over his portrait, "Erra Pater." On the fides thereof, "This Prognofticatiō ferueth for al the world ouer." B.7; in eights. "Imprinted—in Fleteftrete, beneth the Conduyt; at the Sygne of S. John Euangelyft by Tho. Colwell, 1565." W.H. Alfo without date. Octavo. 1565.

"The fyght Tragedie of Seneca, entituled AGAMEMNON, tranflated out of Latin into Englifh by John Studley ftudent in Trinitie college Cambridge. Imprinted—A. D. M.D.LXVI." See Warton's Hift. of Eng. Poetry, Vol. III. p. 383. Licenfed. Octavo. 1566.

"THE CLOSET of Counfells, conteining The aduice of diuers wyfe Philofophers, touchinge fundry morall matters, in Poefies, Preceptes, Prouerbes, & Parrables, tranflated, and collected out of diuers aucthors into Englifh Verfe; by Edmond Eluiden, Gent. Wherunto is anexed a pithy & pleafant difcription of the abufes and vanities of the vvorlde. 1569. Imprynted—in Fleetftreat," &c. Dedicated "To his Neuew william Bufher." Then, an epiftle to the reader. Contains befides, Fo. 99, and on another leaf his colophon, and a cut of St. John. Licenfed. W.H. Octavo. 1569.

"The end and confeffion of John Felton, the rank traytor, who fet vp the traytorous bull on the bifhop of Londons gate." With Richard Johnes. Quarto. 1570.

"A declaration of the lyfe and Death of Iohn Story, Late a Romifh Canonicall Doctor by profeffyon, 1571." In a light compartment with Victory at top, Juftice and Prudence on the fides, and the queen's arms at bottom. D, in fours. At the end, "God faue the Queene, and confounde her enemies. Seene & allowed &c." St. John, &c. on the laft leaf. Licenfed. W.H. Octavo. 1571.

"The

THOMAS COLWELL.

1574. "The famooste & notable hyftory of two faithfull louers, named Abfagus and Archelaus, by Edward Jenynges." In myter." Licenfed, 1565. Quarto.

1575. "A ryght pithy, pleafaunt, and merie comedie, intytuled, Gammer Gurton's nedle. Played on the ftage not long agoe in Chrifts colledge, in Cambridge. Made by Mr. S. mafter of art." Quarto.

1575. "Here begynneth certayne Merye Tales of Skelton,* Poet Laureat." Licenfed, 1566. Alfo without date. Octavo.

—— "The compoft of Ptolomeus," &c. as p. 375. Licenfed. Mr. Will. White.

—— "The treafure of poor men, &c. as p. 355.

—— A ballad againft marriage by Will. Elderton, ballad maker. This is the myrrour," or glas of health.*

—— "The Boke of meafurynge of Lande afwell of woodland as plowland, and pafture in the felde: & to compt the true nobre of Acres of the fame. Newly inuented & compyled by Syr Richarde de Benefe." In a compartment ufed by John Day. It is introduced with "The Preface of Tho. Paynell, Chanon of Marton," as was the author. G, in eights. "Jmprinted in S. Brydes Churchyarde:—Sent & allowed; accordyng to the ordre appointed in our Hall." Licenfed, 1562. W.H. Sixteens.
I have another edition of this book, which feems to be of his printing, but is imperfect at the end, thus entitled: "This Boke, Newely Imprynted, fheweth the maner of meafutyng &c. Newely inuented & compyled by Syr Richard de Benefe, Chanon of Marton Abbay; befyde London."

—— New Sonets & pratie pamphlets, by Tho. Holwell, Gent. Licenfed. Quarto.

—— "Here begynneth the book named, the affife of bread, what it ought to wey after the pryfe of a quarter of wheet. And alfo the affyfe of ale,

---

γ " 1. How Skelton came late home to Oxford from Abington. 2. How S. dreft the Kendall man, in the fweat time. 3. Howe S. tolde the man that Chryft was very bufye in the woodes with them that made fagots. 4. Howe the Welfhman dyd defyre S. to ayde hym in hys fute to the kynge, for a patent to fell drynke. 5. Of Swanborne the knaue, that was buried vnder St. Peters wall in Oxford. 6. Howe S. was complayned on to the Bifhop of Norwich. 7. Howe S when hee came from the Bifhop made a Sermon. 8. How the fryer afked S. to preach at Dys, which S. wold not grant. 9. How S. handled the fryer that woulde needes lye with him in his Jnne. 10. Howe the Cardynall defyred S. to make an Epitaphe vpon his graue. 11. How the Hoftler dyd byte Skeltons Mare vnder the tale, for biting him on the arme. 12. Howe the Cobler tolde maifter S. it is good fleeping in a whole fkinne. 13. How Mafter Skeltons Miller deceyued hym manye times, by playinge the theefe, & howe he was pardoned by Mafter S. after the ftealinge awaye of a Preeft oute of his bed at midnight. 14. Howe S. was in prifon at commaundement of the Cardinall, 15. How the vinteners wife put water into Skeltons wine."

* Mr. Ames's printer has added this in the fame paragraph, as if it belonged to the ballad; but it feems rather to be another edition of the book with a like title, in p. 375.

with

with all manner of wood and cole, lath, bowrde, and tymber. And the weyght of butter and cheefe." With wooden cuts of the fhapes of the " quarter of wheat, farthynge waftell, farthynge fymnell, farthynge whyte lofe, a halfpenny whyte lofe, a halfpenny wheten lofe, a penny wheten lofe, and a halfpenny houfeholde lofe." Quarto.

He had licenfes alfo for the following, viz. In 1562, " A play entitled Dyccon of Bedlam, &c. A lamentable hiftory of the prynce Oedipus." In 1563, " The boke of knowledge of thynges vnknowen, appertayning to Aftronymey. An epytaphe on the deathe of the worthye pryncess Margarete late duches of Norfolk." In 1564, " A pycture of a chylde. The fyrfte twoo Satars or poyfes of Orace, englifhed by Lewes Euans, fcholemaifter." In 1565, " A mery play both pytthy & pleafaunt, of Albyon Knyghte. A play of kyng Daryous. A Song or Pfalme for the delyuerance of his people from the handes of the Turke, & all heathen infideles. To the tune of the xix pfalme. The geftes of Skoggan, gathered together in this volume. The pleafaunte fable of Ouide, intituled Hermaphroditus & Salmacis. The mofte notable hiftory of the lorde Mandozze. An hiftory of meke & pacyent Grefell. The confeffion of parfon Darfy vpon his death. A picture of a monfterus pygge at Salufbury. A tragedy of Seneca, Media, by John Studley, of Trenety Coll. in Cambryge." In 1566, " The dyfprayces of ingratitude &c. and prayces of Fryndefhippe, &c." In 1567, " An epytaphe on the worthy lady Elyfabeth, countes of Shrewfbury." In 1568, " The maruelus dedes & the lyf of Lazaro de Tormes. A mounflerus fyffhe which was taken at Ipwyche. The playe of Sufanna. The hift. of payciente Grefell." In 1569, " An enterlude for boyes to handle & to paffe time at Chriftinmas. An epytaphe on my lorde of Pembroke, made by Dauid Rowlande." In 1570, " A tru report of the newes in Heryfordfhyre. The iij parte of Hercules Oote. The new newes of Doctor Story." Alfo for many ballads, from firft to laft.

HUMPHREY TOY,

THE fon of Rob. Toy, or Toye, was made free of the Stationers' company, by his father's copy, 11 March 1557-8; and came on the livery at the firft reviving thereof, in 1560. He ferved renter in 1561, and 1562; under warden in 1571, and 1752. In 1566, he was one of the delinquents, who were fined " for bokes taken by Tho. Purfoote & Hu. Shingleton in ferten places;" and was fined ij.l. Only x.s. brought to account. In 1568, he paid to the company iiij.l. the bequeft of Mrs. Eliz. Toy, his mother, late deceafed; which fhould have been mentioned at the end of her memoirs, p. 589. He dwelt at the fign of the Helmet, in St. Paul's church-yard.

1550. John Fab. Montanus oration againſt the councel of Trent. Again 1562. Licenſed. Quarto.

1560. "Antiprognoſticon" &c. Henry Sutton for him. See p. 845. Licenſed.

1562. "Tvvo very notable Commentaries," &c. as p. 803. Licenſed.

1564. "The Tranſlation of a letter—vpon the death of—Elenor of Roye, Princes of Conde," &c. See p. 638.

1567. "A playne and familiar introduction, teaching how to pronounce the letters in the Britiſhe tounge. By William Saleſbury, 1555, and now augmented." H. Denham for him, 17 May. Licenſed. Quarto.

1567. (N. T. in Welſh) "Teſtament newydd ein arglwydd Jeſu Chriſt. Gwedy ei dynnu, yd y gadei yr aney fiaith, 'air yn ei gylydd or Groec' a'r Llatin, gan newidio ffurf uythyren gairiae-dodi. Eb law hyng y mae pop gair dybiwyt y vot yn andeallus, ai o ran llediaith y' wlat, ai o ancyne-findery deunydd, wedy ei noli ai eglurhau ar' ledemyl y tu dalen gy-drchiol." Dedicated by William Saleſbury to Q. Elizabeth. Contains beſides, 39) leaves. Printed by H. Denham, at the coſts of Humphrey Toy; and licenſed to him. Bodl. Libr. Quarto.

1568. "Edw. Dering, B. D. his ſparing reſtraint of many lauiſh vntruthes, which M. Dr. Harding doth challenge in the firſt article of my Lord of Sariſburies reply: with an aunſweare to that long & vncurteous Epiſtle, entituled to Maſter Iewell, and ſet before Maiſter Hardings reioynder." For him. Licenſed. Alſo without date. Quarto.

1569. "A Chronicle at large" &c. as p. 817, 818. Grafton's.

1571. "Actes of conference in religion," &c. as p. 878.

1571. "An anſwere to a certen libell intituled An admonition to the parliament." For him. Again 1572, and 1573. "Newly augmented by the Authoure, as by conference ſhall appeare." On the back, "The titles of this Booke." The prefixes are, An epiſtle "To his louing Nurſe, the Chriſtian Church of England, I. VV. a member & miniſter of the ſame, wiſheth peace in Chriſt, &c.---A briefe examination of the reaſon vſed in the book called an Admonition to the Parliament.---To the Chriſtian Reader.---An Exhortation to ſuch as bee in authoritie, and haue the gouernement of the Churche committed vnto them, whether they be Ciuile or Eccleſiaſticall Magiſtrates.---Certayne notes & properties of Anabap-

---

* Apoc. or Rev. 5; 8. φιαλας, vials, the Engliſh; and that he did not diſtinguiſh Brit. Ffiolan N. B. Biſhop Morgan tranſ- between vials and viols, or violins. See lates it eryiban, crouds. Which ſhews that Warton's Hiſt. of Eng. Poetry, Vol. I. he had not the original before him, but only p. 147, note r.

tiſtes,

tiftes, & other perturbers of the Churche, collected out of Zuinglius, &c." Contains 344 pages, including the title, and colophon. "Imprinted—by H. Bynneman, for Humfrey Toy,—in Paules Church yard, at the figne of the Helmet, 1573." W.H. Quarto.

" HISTORIAE BRYTANNICAE DEFENSIO, IOANNE PRISEO EQVESTRIS 1573. ORDINIS BRYTANNO AVTHORE." The genius of England, as p. 780. " In ædibus H. Binneman typographi, impenfis Humfredi Toy, 1573." Dedicated "Honoratiſſimo viro, Gulielmo Cecilio Baroni de Burgley, &c.—Londini tertio Non. Martij, 1573." Then, fome commendatory verfes; an epiftle "Præpotenti et omnibus modis ornatiſſimo viro D. Gulielmo Penbrochiæ comiti," &c. A preface infcribed " Ad illuftriſſimum iuxta ac potentiſſimum principem Angliæ—Edouardum eius nominis fextum, in Hiftoriæ Brytannicæ defenfionem." Contains befides, 160 pages, and "Index capitum," and "Index rerum." Hereunto is annexed "De Mona Druidum Infula, Epiftola Humfredi Lhuyd.—Denbighiæ, Northvvaliæ fiue Guynedhiæ, quinto Aprilis, 1568." W.H.
Quarto.

" The Supremacie of Chriftian Princes, ouer all perfons throughout 1573. their dominions, in all caufes fo wel Ecclefiaftical as temporall, both againft the Counterblaft of Tho. Stapleton, replying on the Reuerend father in Chrifte, Robert Biſhop of VVincheſter: and alfo Againft Nicolas Sanders his Vifible Monarchie of the Romaine Church, touching his controuerfie of the Princes Supremacie. Anfvvered by Iohn Bridges. The Princes charge in his inftitution to ouerfee the direction of Gods lawe. Deut. 17. After he fhall be fettled in the throne of his kingdom, &c. Printed—by Hen. Bynneman for Humf. Toye, 1573." On the back is the queen's arms, to whom this book is dedicated. Then, a preface.--"Mafter Stapletons common places; and his beadroll of vntruthes." Contains befides, 1114 pages, and a lift of faults, at the end of which is Binneman's fign. W.H. Quarto.

" Bart. Traheron his Expofition on Rev. IV. in fundrie redings, in Ger- 1573. many," &c. H. Bynneman for him. Octavo.

" A DEFENSE of the Ecclefiafticall Regiment in Englande, defaced by 1574. T. C. in his Replie agaynft D. VVhitgifte. Seene & allowed" &c. H. Bynneman for him, 1574.

" The Defenfe of the Aunfwere to the Admonition, againft the Replie 1574. of T. C. By Iohn VVhitgift, D. D. In the beginning are added thefe 4 Tables. 1. Of dangerous doctrines in the Replie. 2. Of Falfifications & Vntruthes. 3. Of matters handled at large. 4. A Table generall. If any man be contentious, &c. 1 Cor. 11. 16. Printed—by H. Binneman for *him*, 1574." In a fcroll compartment, with the mermaid at bottom. In a partition at top, " 1 Cor. 8. 2. If any thinke that he knoweth any thing," &c. In another, at bottom, " Gal. 5. 26. Let vs not be defirous of vaine glorie," &c. The preface is addreſſed " To the godlie Reader."— The defenſe on 812 pages; then, " An examination of the place cited in the end of the Replie." W.H.   Folio.

1574.   " A ſermon preached before the queene at Greenwiche, 27. March, 1574. on John vi; 25—27. by Dr. John Whitgifte, dean of Lincoln."   Octavo.

1575.   " A Sermon preached before the Queenes Maieſtie at Richmond 6 March, by—the Biſhop of Chiceſter, on Judges 1. 1—13." For him.   Octavo.

1576.   " The fourth parte of the Cōmentaries of the Ciuill warres in Fraunce, and of the lovve countrie of Flaunders: Tranſlated out of Latine into Engliſh by Tho. Tymme, Miniſter. Seene & allowed." Dedicated " To—Lorde Ambroſe Dudley, Earle of Warwicke, &c. Generall of the Queenes Maieſties Ordinance, within hir highneſſe realmes & dominions." Then, a ſhort note to the reader, and the table. Contains beſides, 150 pages, printed 1410. With a plan of the ſiege of Rochelle. Then, after another ſhort note, " A Supplication to the Kings Maieſty of Spain, by the Prince of Orange, the States of Hollande & Zeland, &c. By which is declared the original of the troubles in the Lowe countrey." H. Binneman for him. W.H.   Quarto.

1576.   " Of the Conſcience. A Diſcourſe wherein is playnely declared, the vnſpeakeable ioye & comfort of a good Conſcience, and the intollerable griefe & diſcomfort of an euill Conſcience. Made by Iohn Woolton Miniſter of the Goſpell. 1576." H. Iackſon for *him.* Dedicated " To—Sir Iohn Jylbert, Knight.—From Exceter, 20 Marche." The whole on O 2, in fours. W.H.   Octavo.

———   " A Sermon preached at Paules Croſſe on the Monday in Whitſon weeke, 1571. Entreating on this Sentence. Sic Deus dilexit mundum, &c. John 3. (16) Preached & augmented by Iohn Bridges." Printed by H Binneman for him. Dedicated " To—Sir. William Cecill Knight, &c.—Iohn Brigges." The ſermon on 182 pages. W.H.   Quarto.

———   " A righte noble & pleaſant Hiſtory of the Succeſſors of Alexander ſurnamed the great, taken out of Diadorus Siculus: and ſome of their liues written by the wiſe Plutarch: tranſlated out of French into Engly̆ſh by Tho. Stocker." H. Bynneman for him. Licenſed, 1568. Mr. M. C. Tuter.   Quarto.

He had licenſes alſo for the following, viz. In 1566, " The ſervis boke in Welſhe:—my lorde of London." In 1569, " Jniunctions & articles gyuen by—Rycharde byſihoppe of Techeſter."

HENRY

HENRY WYKES, or WEKES,

WAS prefented as an apprentice with Mr. Berthelet, 15 Octo. (1556) but as Mr. Berthelet died before that time, 'tis probable he continued apprentice with the widow, &c. as the books printed in the houfe late Tho. Berthelet's after 1555 till 1561, give no intimation by whom, or on whofe account they were printed. See p. 464. He was made free of the Stationers' company, 15 Aug. 1565; but he printed books in his own name a confiderable time before. He was fined 7. Aug. 1564, for printing without licenfe, Confabulationes, and The banket of Sapience; iij.s. Again, for printing other mens copies, and for the forfiture of his obligation, xl. s. It does not appear what thofe copies were, or what the faid obligation was for; but the wardens paid about the fame time " to my Lorde mayres officer, & the fheriffes officer, waiting for Hen. Wekes, viij.s. viij.d." At his being made free, he gave befides the accuftomed fee, a benevolence of x.s. perhaps on account of the expence he had put the company to, as above. By the colophon of the title-page to Cooper's Thefaurus, 1565, he appears to have dwelt in the houfe that was Mr. Berthelet's; and yet R. Newbery appears to have lived there alfo. Wykes dwelt however in Fleet ftreet, at the fign of the black elephant; and frequently ufed Berthelet's compartments, and Powel's mafks, &c. One of Mr. Ames's papers mentions his being, in 1570, the fervant of Sir Francis Knowlis, whofe creft is a black elephant, and that therefore Wykes chofe the fame for his fign; but i have not found any book of his printing with fo late a date.

" A VERY FRVTEFVL AND PLEASANT BOOKE CALled the Jnftruction of 1557. a chriftian woman," &c. as p. 440. " Londini Anno M.D.LVII." In a heavy compartment ufed by Berthelet, with T. B. under it; afterward by Tho. Powel, with T. P. Some of Powel's mafks are ufed in the work. This edition begins with the preface of L. Vives, as to the former edition, 1541. Nn, in fours, " Jmprinted—in Fleteftrete, by Henry Wykes. Cum priuilegio—folum." W.H. Quarto.

" Principal Rules of the Jtalian Grammer," &c. as p. 453. " Newly 1560. corrected & imprinted,—In ædibus H. VVykes." Again 1567. W.H.
Quarto.

Pfalmes or Prayers taken out of the holy fcriptures. Octavo, 1562.

" In this volume are conteined the ftatutes" &c. as p. 443. " Jm- 1564. printed—in Fleetftreete—1564." W.H. Folio.

" THESAVRVS LINGVAE Romanæ & Britannicæ, tam accurate congeftus, 1565. vt nihil penè in eo defyderari pofsit, quod vel Latine complectatur ampliffimus Stephani Thefaurus, vel Anglicè, toties aucta Eliotæ Bibliotheca: opera & induftria Thomæ Cooperi Magdalenenfis. Quid fructus

ex

ex hoc Thefauro ftudiofi pofsint excerpere, & quam rationem fecutus author fit in vocabulorum interpretatione & difpofitione, poft epiftolam demonftratur. Acceffit Dictionarium hiftoricum & poëticum propria vocabula Virorum—& cæterorum locorum complectens, & in his iucundifsimas & omnium cognitione dignifsimas hiftorias." The earl of Leicefter's creft, gartered. Under it. " In Thefaurum—hexaftichon Ri. Stephani.---Excufum—in aedibus quondam Bertheleti, cum priuilegio Regiæ Maieftatis, per H. VVykes, 1565. 16 Martij." Prefixed are " Ad clariff. virum Robertum Dudleium Leiceftriæ comitem, &c. Tho. Cooperi de Thefauri dedicatione, epiftola.---Annotationes quibus ftudiofi lectores admonentur de ratione, & ordine iftius libri," &c.---Some directions for young fcholars, in Englifh.---Commendatory verfes, in Latin, and Greek. The Latin dictionary on SSSsss 5, in fixes, or 797 leaves; the hiftorical dictionary, R 5, in fixes. W.H. Folio.

It was printed again in 1573; but without the printer's name.

1565. " A Reprovfe, written by Alexander Nowell of a booke entituled, A proufe of certayne articles in Religion denied by M. Iuell, fet furth by Tho. Dorman, B. D. Prouerb. 19. a. Teftis falfus non erit impunitus: &c. Set foorth & allowed &c. Imprinted—in Fleeteftreete,—1565. 30 die Maij." Ji, in fours. W.H. Quarto.

1565. " A Replie vnto M. Hardinges Ansvveare: By perufinge whereof the difcrete & diligent Reader may eafily fee the weake & vnftable groundes of the Romaine Religion, which of late hath beene accompted Catholique. By Iohn Iewel Bifhoppe of Sarifburie. 3 Efdræ, 4. Magna eft veritas & præualet.—Ex Edicto Imperatorum Valentin. & Martiani in Concil. Chalcedon. Actione 3. Qui poft femel inuentam veritatem aliud quærit, Mendacium quærit, non veritatem.—Imprinted—in Fleeteftreate, at the figne of the Blacke Oliphante,—1565. VVith fpecial Priuilege." Prefixed are, An epiftle " Vnto the Chriftian Reader.—From London, 6 Aug. 1565. Iohn Ievvell Sarifburien.---An anfwere to M. Hardinges Preface.---The Table." Contains befides, 641 pages; and " An anfweare to M. Hardinges Conclufion." W.H. Again 1566. Folio.

1566. " The Reprovfe of M. Dorman his proufe of certaine Articles in Religion &c. continued by Alex. Nowell. With a defenfe of the chiefe authoritie & gouernment of Chriftian Princes as well in caufes Ecclefiafticall, as ciuill within their owne dominions, by M. Dorman malitiouflie oppugned. Imprinted &c. 1566." Prefixed are " An admonition to the Reader." And " The faultes—mended." 289 leaves. " Imprinted & allowed according to the order—in the Queenes Maiefties Iniunctions." W.H. Quarto.

1566. " The XI Bookes of the Golden Afse, conteining the Metamorphofie of Lucius Apuleius interlaced with fundrie pleafaunt & delectable Tales: with an excellent Narration of the marriage of Cupide & Pfyches, fet out in the 4, 5, and 6 Bookes. Tranflated out of Latine into Englifh by Will.

HENRY WYKES.

Will. Adlinton." Dedicated " To—Thomas Earl of Suffex, &c.—Vniuerfity Coll. Oxon. 18 Septemb. 1566." Licenfed. Dr. Lort. Quarto.

" An Ansvveare, Made by Rob. Bifhoppe of VVynchefter, to a 1566
Booke entituled, The Declaration of fuche Scruples & ftaies of Confcience, touchinge the Othe of Supremacy as M. Iohn Fakenham, by vvrytinge did deliuer vnto the L. Bifhop of VVinchefter, vvith his Refolutions made thereunto.—Imprinted &c. 1.556." The preface is dated 25 Feb. 1565. Contains befides, 130 leaves. Licenfed. W.H. Quarto.

" Iohn Heywoodes woorkes." &c. as p. 865. Quarto.

" A defence of the apologie of the churche of Englande, conteininge 1567.
an anfweare to a certaine booke, lately fet foorth by M. Hardinge, and entituled, A confutation of &c. By John Jewel bifhop of Salifburie." October 27. Again 16th of June 1570. Alfo in 1571. Folio.

" An almanacke and prognoftication, for the yeare of our Lorde God 1568.
1568, feruying for al Europe; wherein is fhewed the natures of the planets, very neceffary for all ftudents, marchaunts, mariners, and trauellers both by fea and lande, gathered by Joachim Hubrigh, doctor in phificke and aftronomie. Whereunto is annexed, a profitable rule for mariners to know their ebbes, and fluddes, courfes, landinges, foundinges, markes, and dangers, all alonge the coafte of Englande and Normandie. Alfo al the principall fayres and martes, where, and when they be holden; with new additions, meete for all thofe, that vfe the trade thereof." For William Pickringe.* Octavo.

" A fhorte dictionarie (*vocabulary*) for yonge beginners. Gathered of 1568.
good authours, fpecially of Columel, Grapald, and Plini. Anno M.D.LXVIII." The prologue addreffed " To—fyr Thomas Chaloner knight, Clerke of the Kynges maiefties priuie Counfayle." Y 3, in fours. " Compiled by J. Withals.—Cum priuilegio—folum." W.H. Quarto.

" Henrie Cornelius Agrippa, of the Vanitie & vncertaintie of Artes 1569.
and Sciences. Englifhed by Ja. San. Gent. Ecclef. 1. All is but mofte vaine Vanitie: and all is moft vaine, and but plaine Vanitie. Seene & allowed &c. Imprinted—in Fleete ftreat, at the figne of the blacke Elephant, 1569." Dedicated, by James Sanford, " To the Noble & Vertuous Prince Thomas Duke of Northfolke, Earle Marfhal of England," &c. Whofe arms are on the back of the title-page. An epiftle from the tranflator " To the Reader."---A table of common places.---" Cornelius Agrippa to the Reader." Contains befides, 187 leaves, and a lift of faults. Licenfed. W.H. Quarto.

" A True Shield and Buckler of faith, wherein is intreated 23 articles 1569.
of religion dialogue wife: written in French by Bart. Caufe. Tranflated by T. S." Licenfed. Quarto.

" The

"The fables of Esope in Englifhe, with all his life, and fortune; and how he was fubtyll, wyfe, and borne in Greece, not far from Troy the great, in a towne named Amoneo: how he was of all other men moft diffourmed and euil fhapen: for he had a great head, a large vifage, long iawes, fharp eyen, a fhort neck, crok-backed, great bely, great legs, large feet; and yet that which was worfe, he was dombe, and could not fpeake: but notwithftanding this, he had a fingular wit, and was greatly ingenious and fubtill in cauillations, and pleafaunt in woordes, after he came to his fpeache. Whereunto is added, the fables of Auian, and alfo the fables of Poge the Florentyne, very pleafaunte to reade." Contains 134 leaves, and a table. For Waley.* Octavo.

"Merie tales of the madmen of Golam, gathered together by A. B. (Andr. Borde) of phyfick doctour."

"Meditations and praiers, gathered out of the facred letters, and vertuous writers: difpofed in fourme of the alphabet of the queene her moft excellent maiefties name. Whereunto are added, comfortable confolations (drawen out of the Latin) to afflicted mindes."

He had licenfes alfo for the following books, viz. in 1565, "A tragicall & pleafaunte hiftory, Ariounder Jenenor the doughter vnto the kynge of (blank) by Peter Beuerlay." In 1566, "The Kynges pfalmes, and the quenes prayers. Orace epeftles in Englefhe." In 1567, "The myrror of the Laten tönge by Dauyd Rowlande." In 1567, "A trygecall hiftorye of Agathecles." In 1568, "Roffencis Pfalmes: in Laten."

GERRARD DEWES,

APPRENTICE with Andr. Hefter, was made free of the Stationers' company, 4 Octo. 1557; taken into the livery in 1568; ferved renter in 1572, and 1573; and under warden in 1581. In 1566, he was fined with others "for books taken by T. Purfoote and H. Shyngleton in ferten places, xx.s." In 1567, he was fined for printing "The boke of Rogges, contrary to the orders of this howfe, wh'ch was feffed in the tyme of Mr. Jugge and Mr. Daye, iij.l. vj.s. viij.d." He dwelt at the fign of the Swan in St. Paul's church-yard; and ufed for a device his rebus, which is one of thofe, mentioned among others by Camden in his remains, "And if you require more, I refer you to the witty inventions of fome Londoners; but that for Garret Dewes is moft remarkable: two in a garret cafting Dews at dice." This may be feen in the frontifpiece.

This Gerrard or Garret was the eldeft fon of Adrian D'Ewes, who came over from Guelderland when it was depopulated by inteftine wars. Gerrard

ard married Grace Hinde of Cambridgeshire, a Dutch woman, who died in 1583, and was buried in St. Faith's, under St. Paul's, London. Soon after her death, he left off trade, and probably then purchased the manor of Gaines, in Upminster, in the county of Essex. He died 12 April 1591, leaving Paul his son and heir, who was born in 1570, and was the father of Sir Simonds D'Ewes, noted for his collections of the Journals of Parliament during the reign of Q. Elizabeth. The inquisition taken after the death of Gerard is recorded, and his epitaph, with his figure in armour, in brass, on his grave stone, in the chapel at Gaines, delineated in Weever's Fun. Monuments, p. 653.

"Epitome troporum ac schematum, et gramaticorum et rhetorum, ad autores, tum prophanos tum sacros, intelligendos, non minus vtilis quam necessaria. Joanne Susenbroto Almangauus, Rauenspurgi ludimagistro, collectore." Octavo. 1552.

"TESTAMENTVM NOVVM ex des. Erasmi Roterodami versione, ac eiusdem recognitione postrema. Londini, Apud Gerardum Deuues, Anno 1568." E e e 7, in eights. Dr. Lort. Sixteens and Quarto. 1568.

"A Nievve Herball, or Historie of Plantes :—First set foorth in the Doutch or Almaigne toungue, by Rembert Dodoens : and now first translated out of French, by Henry Lyte, Esquyer." As this was printed at Antwerp, and only sold by G. Dewes here, a more particular account of it will be given in our General History. W.H. Folio. 1578.

"Baptistae Mantuani, carmelitae theologi, Adolescentia, seu Bucolica, breuibus Jodoci Badii commentariis illustrata. Excudebant Gerardus Dewes et Henricus Marsh, ex assignatione Thomae Marsh. Cum priuilegio." Octavo. 1584.

"A SHORTE Discourse of the most rare and excellent vertue of Nitre: Wherein is declared the sundrie & diuerse cures by the same effected, and how it may be aswell receiued in medicine inwardly as outwardly plaisterwise applied : seruing to the vse & commoditie aswell of the meaner people as of the delicater sorte. Ecclesiast. 38. 4. The Lord hath created medicines, &c. Imprinted—by Gerald Dewes,—in Paules churchyearde at the—Swanne, 1584." In the compartment with T. Marsh's mark at the bottom. It has prefixed, an address from "The Printer to the Reader," signed A. D. (Query if not a misprint for G. D.) Under it, T. Marsh's device of Fortune. This is the copy of a letter, written from the Isle of Lamby on the East coast of Ireland, by Tho. Chaloner, gent. to his cousin John Napper, apothecary, over against Soper-lane end in Cheap-side. 22 leaves. W.H. Quarto. 1584.

"A Catholicke and ecclesiastical Exposition on S. Jude, by Aug. Marlorat: translated by J. D. minister." With Hen. Marsh. Octavo. 1584.

5 T "Dialectica

GERARD DEWES.

1584. " Dialectica Joannis Setoni" &c. as p. 866. " Excudebant Gerardus Dewes & Henricus Marsh, ex assignatione Thomæ Marsh. 1584. Cum Priuilegio." R 2, in eights. W.H. Octavo.

1587. " Anwick his Meditations vpon Gods Monarchie, and the Deuill his Kingdome. And, Of the knowledge that Man in this life may obtaine of the almightie, eternal, and most glorious Godhed. With other thinges, not only worth the reading, but also the marking, and the retayning. —Imprinted—by Gerred Dewes,—in Powles Churchard, at the—Swan, 1587." Dedicated "To—Sir Fraunces Walsingham Knight," &c. Then, the preface and contents. Contains besides, 117 pages. W.H. Quarto.

He had licenses also for the following, viz. In 1562, " The pycture of a monsterus pygge at Hamsted." In 1566, " An almanake &c. of Mr. Buckmaster." The same again in 1567, and 1568. In 1576 " Thepitaphe of therle of Essex."

HENRY DENHAM

WAS presented 14 Oct. (1556,) as an apprentice with Mr. Tottel, and made free of the Stationers' Company, 30 Aug. 1560; was admitted to the livery in 1572; served renter in 1580 and 1584; under-warden in 1586 and 1588. He was an exceeding neat printer, and the first who used the semicolon with propriety. In 1564, he "prented premers without licenfe, contrary to the orders," for which he was fined xl. s. Also, in 1565, " for myfufing Mr. Warden, x. s." and, " for printing The vtter apparell for mynesters, x. s." Again, 10 Aug. 1579, " for arresting a freeman of the company without licenfe, xij. d." Also, 6 Apr. 1584, " for vfing vndecent speaches to the elder warden, xx. s." Remitted. He dwelt, at times, in Pater-noster-row, at the sign of the Star, which, with this motto, Os HOMINI SVBLIME DEDIT, about it, he used in his books: also, in White-cross street; and afterwards in Alderfgate street, at the same sign. When William Seres the father was grown infirm, and the son perhaps too gallant to mind business, he made our H. Denham, his assignee or agent, &c. as p. 688. In 1583, he assigned over certain of his copies to the company for the benefit of their poor; as may be seen in our General History.

He printed several books for or with Ri. Tottel, John Wight, Thomas Hacket, Ra. Newbery, L. Harrison, and Humf. Toy, which may be seen in the accounts of them.

HENRY DENHAM.

The Pſalter, with marginal notes. Very ſmall. 1559.

" This Booke is called the Treaſure of Gladneſſe, and ſemeth by the 1563. copy, being a very little manuell, and written in velam, to be made aboue Cc yeares paſt, at the leaſt. Whereby appeareth how God in olde time, and not of late onely, hath bené truely confeſſed & honored. The copy hereof, for the antiquitie of it, is preſerued and to be ſeen in the printers Hall. Set forth and allowed &c. And now fyrſt Imprinted, 1563." At the end, " Imprinted—by H. Denham for Iohn Charlewood, &c. the ſyxt day of Septembre—1564." On 70 leaues. Sixteens.

" The pitiful Eſtate of the Time preſent. A Chriſtian conſideration 1564. of the Miſeries of this time with an Exhortation to amendment of Life." In White-croſs ſtreet. Licenſed. Alſo without date. Sixteens.

" Tho. Cole, Archd. of Eſſex, his ſermon on 4 Kings, 10; 15. Is thine 1564. hart vpright." Licenſed. Octavo.

" The Zodiack of Life: tranſlated into Engliſh verſe, by Barnabie 1565. Googe." See p. 904. Quarto and Octavo.

" Ordinances decreed for the reformation of diuers diſorders in printing, 1566. and vttering of books." Strype's Life of Abp. Parker, p. 221. A ſheet.

" The great wonders, that are chaunced in the realme of Naples, with 1566. a great misfortune happened at Rome, and in other places, by an earthquake in the month of December laſt paſt. Tranſlated out of Frenche into Engliſhe by J. A." Sixteens.

" An Apologie or Defence of thoſe Engliſhe Writers and Preachers, 1566. which Cerberus, the three headed Dog of Hell, chargeth with falſe doctrine, vnder the name of Predeſtination. Written by Ro. Crowley, Clerke, & Vicare of St. Giles without Cripplegate,—14 October." See it by H. Binneman the ſame day. Quarto.

" The worthie Hiſtory of the moſt noble and valiant knight Plaſidacis." 1566. Alſo, " The notable hiſtorie of two famous princes Aſtianax and Polixona." Both in verſe by John Patridge. William Bayntun, Eſq; Octavo.

" A Greene Foreſt, or a naturall Hiſtorie, Wherein may bee ſeene 1567. firſt the moſt Sufferaigne Vertues in all the whole kinde of Stones and Mettals: next of Plants, as of Herbes, Trees & Shrubs: Laſtly of Brute Beaſts; Foules, Fiſhes, creeping wormes & Serpents, and that Alphabetically: ſo that a table ſhall not neede. Compiled by John Maplet, M. A. and Student in Cambridge." 112 leaues. Octavo.

" The Heroycall Epiſtles of the Learned Poet Publius Ouidius Naſo, 1567. Jn Engliſhe Verſe; ſet out and tranſlated by George Turberuile Gent.

with Aulus Sabinus Aunfweres to certaine of the fame. 1567." Dedicated "To—Lord Tho. Hovvarde,. Vicount Byndon, &c." Then, "The Tranflator to his Mufe," and his epiftle To the Reader. Contains befides 162 leaves. Arguments in verfe are prefixed to each of Ovid's epiftles. Licenfed. W.H. Again, 1569. Alfo without date; at the end of which are 11 fix-lined ftanzas from "The Tranflator to the captious fort of Sycophants;" and correction of "Faultes efcaped:" W.H. Octavo.

1567. "The imitation, or following of Chrift, and the contemning of worldly vanities: At the firft written by Thomas Kempife, a Dutchman, amended and polifhed by Sebaftianus Caftalio, an Italian; and Englifhed by E. H. [*Edward Hake.*] Seene and allowed." &c. Dedicated to Thomas, duke of Norfolk.* Octavo.

1568. Again 1568. "Whereunto as a fpring out of the fame roote, we haue adioined, A fhort pretie treatife touching the perpetuall reioyce of the godly euen in this life. Seene and allowed." * Again 1584; alfo without date. Octavo.

1568. "Phificke for the Soul, very neceffarie to be vfed in the agonie of death, and in thofe extreme and mofte perilous feafons, as well for thofe, which are in good health, as thofe which are endewed with bodily ficknesse. Tranflated out of Latine into Englifhe, by H. Thorne. Perufed and allowed," &c. Again 1570. W.H. Sixteens.

1568. "The Caftle of Chriftianitie, detecting the long erring eftate, as well of the Romaine Church, as of the Byfhop of Rome: together with the defence of the Catholique Faith. Set forth by Lewis Euans." Licenfed. Mr. Edward Jacob, of Feverfham. Octavo.

1568. "A modeft meane to Mariage, pleafauntly fet foorth by that famous Clarke, Frafmus Roterodamus, and tranflated into Englifhe by N. L. (*Nich. Leigh*) Anno 1568. Imprinted—in Pater nofter Rowe, at the Starre." Dij, in eights. W.H. Octavo.

1568. "THE ARBOR OF AMITIE, wherein is comprifed plefaunt poems, and pretie poefies, fet foorth by Thomas Howell, gentleman, anno 1568." Dedicated "To—ladie Anne Talbot." See Warton's Hift. of Eng. Poetry, Vol. III; p. 418, note, b. Licenfed. Again 1569. Octavo.

1569. "The trauayled pylgrime, bringing newes from all partes of the worlde, fuch like fcarfe harde of before." The dedication "To the right worfhipfull fir William Dounfell, knight, receiuer general of the queens maiefties wards and liueries." In verie; with many cuts. Licenfed. Quarto.

1569. "A fetting open of the fubtyle Sophiftrie of Thomas VVatfon, D.D. which he vfed in hys two Sermons made before Queene Mary, in the thirde and fift Frydayes in Lent, Anno 1553, to prooue the reall prefence of Chrifts body and bloud in the facrament, and the Maffe to be the facrafice of the newe

HENRY DENHAM.

newe Teſtament. Written by Robert Crowley Clearke."—Seene and allowed" &c. It has 3 epiſtles prefixed ; one in Latin, to the divines of our Univerſities, one to doctor Watſon, and one to the Chriſtian readers ; alſo a table for the notes of each ſermon. The firſt ſermon ends on p. 209, and the ſecond begins on the back thereof, paged 2, and ends on p. 188. " Imprinted—in Pater noſter Rovve, &c. 1569. Cum priuilegio." W.H. Quarto.

" The Miſeries of Schoole maiſters, vttered in a Latine Oration made by the famous Clearke, Philip Melanchthon." Licenſed. Octavo. 1569.

The fourth part of the ſecrets of Alexis. See p. 781. Licenſed. Quarto. 1569.

" A Chronicle at large and meere Hiſtory of the affayres of Englande and Kinges of the ſame, deducted from the Creation of the vvorlde, vnto the firſt habitation of thys Jſland : and ſo by contynuance vnto the firſt yere of the reigne of our moſt deere & ſouereigne Lady Queene Elizabeth : collected out of ſundry Authors,—1569." In 2 volumes. The ſecond volume appears to have been printed in 1568. At the end is Grafton's rebus : under it " Jmprinted—for Richarde Tottle, and Humffrey Toye, Anno 1569, the laſt of March. Cum priuilegio—ſolum." See p. 817, &c. W.H. Folio. 1569.

" A Sermon of gods fearfull threatnings for Idolatrye, &c. With a Treatiſe againſt vſurie. Preached in Paules Churche, the xv daye of Maye, 1570.—by Richarde Porder." Parſon of St. Peter's Cornhill. Text, " Sophonie 1. 1—6." Licenſed. Octavo. 1570.

" Newes from Niniue to Englande brought by the prophete Jonas.— By Brentius : tranſlated by Tho. Timme, Miniſter." Licenſed. Octavo. 1570.

" The Popiſh Kingdome, or reigne of Antichriſt, written in Latine verſe by Thomas Naogeorgus, and englyſhed by Barnabe Googe. 2 Tim. 3. (8,9.) Lyke as Jannes and Jambres withſtoode Moyſes, &c. Imprinted —for Richarde Watkins, Anno 1570." On the back begins Googe's dedication to Q. Elizabeth. Then, the author's epiſtle to Philip, Landgrave of Heſſe, &c.—Baſill, 20 Feb. 1553. Divided into four books. The tranſlation in Alexandrine verſe. 60 leaves, and a table at the end. To this is annexed, " The ſpirituall Huſbandrie," dedicated " To the— Lords, the gouernors & Senate of Bern.—From Campidun, 1 March, 1550." The folios continued from fol. 61 to 88. W.H. Quarto. 1570.

" A ritch Storehouſe or Treaſure for nobilltye & gentlemen, written in Latin by John Sturmius, aud tranſlated by T. B. gent." Octavo. 1570.

" Epitaphs, epigrams, ſongs, and ſonets, with a diſcourſe of the friendly affections of Tymets to Pyndara his ladie. Newly corrected, with additions, and ſett out by George Turberuile, gentleman." Licenſed. Octavo. 1570.

" The

1570. "The Morall Philosophie of 'Doni: drawne out of the auncient writers. A worke first compiled in the Indian tongue, and afterwardes reduced into diuers other languages: and now lastly englished out of Italian by Tho. North, Brother to the right Hon. Sir Roger North Knight, Lorde North of Kyrtheling." This title is over an emblematical cut, having this motto "The wisedome of this worlde is folly before God." Dedicated "To—Lorde Robert Dudley, Earle of Leycester, &c.---To the Reader."---Commendatory verses, in Italian, by G. B. and in English by T. N. and E. C. This curious book is divided into 4 parts, introduced by a prologue beginning on fol. 1, and adorned with several neat wood-cuts. Fol 111. "Here endeth the Treatise of the Morall Philosophie of Sendebar: In which is layd open many infinite examples for the health & life of reasonable men, shadowed under tales and similitudes of brute beastes without reason." His sign encircled with the motto. "Imprinted—in Pater noster Rowe,—1570. Cum Priuilegio." W.H.
Quarto.

1570. "The three Orations of Demosthenes chiefe Orator among the Grecians, in Fauour of the Olynthians, a people in Thracia, now called Romania: with those his fower Orations titled expreslely & by name against king Philip of Macedonie: most nedefull to be redde, in these daungerous dayen, of all them that loue their Countries libertie, and desire to take warning for their better auayle, by example of others. Englished 'out of the Greeke By Thomas Wylson Doctor of the ciuill lawes. After these Orations ended, Demosthenes lyfe is set foorth, and gathered out of Plutarch, Lucian, Suidas, &c. with a large table, declaring all the principall matters conteyned in euerye part of this booke. Seene & allowed," &c. To this book are prefixed the commendatory verses in Latin, Gual. Haddoni; Ægidij Laurentij; Tho. Bingi; Joannis Cooci; and J. M. Also, Interpres Lectori.---The dedication, "To—Sir William Cecill Knight, &c.——From the Queenes Maiesties Hospitall of St. Katherins nigh the Tower of London, 10 June, 1570."---The preface.----"Testimonies—of Demosthenes his worthynesse by diuers learned men.---The Othe that the yong men of Grecia did take, when they we are appointed Souldiers.---The description of Athens:" with a plan.---" The bounding of Greecelande according to Ptolomeus.---The hystorie of P. Sulpitius Consull, according to Titus Liuius.---The argument of the first Oration." Contains besides, 145 pages, and a table. "Imprinted—Cum priuilegio—folum. 1570." Licensed W.H. Quarto.

ᵇThis word Doni seems to be of like import with that of Magi, as appears by the head-title: "The first part of the Morall Philosophie of the auncient Sages, compiled by the great and learned Philosopher Sendebar, in the Indian tongue, who by sundrie and wonderfull examples bewrayeth the deceyts and daungers of this present worlde." The fables are linked together in like manner as those of Pilpay, which are also supposed to be of Eastern origin.

ᶜQ. Elizabeth, desirous of having this book translated, being then at war with Philip K. of Spain, was so pleased with it, that she rewarded the doctor, recommended by Lord Burliegh, with the great posts and places he afterwards enjoyed.

Demosthenis

HENRY DENHAM.

"Demosthenis Græcorum oratorum principis, Olynthiacæ orationes tres, 1571. et Philippicae quatuor, e Græco in Latinum conuersae, a Nicolao Carro, Anglo Nouocastrensi, doctore medico, et Græcarum literarum in Cantabrigiensi academia professore regio. Addita est etiam epistola de vita, et obitu eiusdem Nicolai Carri, et carmina, cum Græca, tum Latina, in eundem conscripta." * Quarto.

" An almanack published at large, in forme of a booke of memorie, ne- 1571. cessary for all such, as haue occasion daily to note sondrie affayres, eyther for recytes, payments, or such lyke. Newly set forth, by T. H. (Tho. Hill) Londoner." Quarto.

" A briefe and pleasant discourse of duties in Mariage, called the Flower 1571. of Friendshippe. Imprinted—in Pater noster Rowe,—1571. Cum priuilegio." The queen's armes on the back, to whom this book is dedicated by Edm. Tilney. E 7, eights. W.H. Again 1577. Octavo.

" An hundred, threescore & fiftene Homelyes or Sermons, vppon the 1572. Actes of the Apostles, written by Saint Luke: made by Radulpe Gualthere Tigurine, and translated out of Latine into our tongue for the commoditie of the Englishe reader. Iohn, 1. Beholde the Lambe of God, &c. Seene & allowed, &c. 1572." In the compartment with John Day's nebus. There are prefixed, a dedication " To—Fraunces Earle of Bedforde, Knight of the Garter," &c. (whose arms are on the back of the title page) by " Iohn Bridges Vicare of Herne," the translator, 21 April 1572.—To the Reader.—" Raufe Gualthere Tigurine, To the noble & hon. Consuls, & whole Senate of the famous Commonweale of Zurich.—1 Aug. 1567."—Some Latin verses by John Parkhurst, bishop of Norwich.—A table of matters expounded; and another of the places of Scripture. Contains besides, " Pag. 919." On the back " The Iudgement of S. Hierome, vppon the Actes of the Apostles.—Imprinted—in Pater noster rowe," &c. Licensed. W.H. Folio.

Mascall's Art of planting and graffing. See p. 782. Quarto. 1572.

" The olde Lawes and Statutes of the Stannarie of Deuon: as many 1574. as were in force, and heretofore imprinted. Whereunto are added certayne other newlie made, in the yeare of the reigne of our Soueraigne Ladie Queene Elizabeth, the xvj. 1574. Imprinted—in Pater noster Rowe. Cum priuilegio—solum." 56 leaves; on the last, the Bedford arms. W.H. Quarto.

" The lyues of holy Sainctes, Prophetes, Patriarches & others, con- 1574. tayned in holye Scripture,—with the interpretation of their names: Collected & gathered into an Alphabeticall order,—By Iohn Marbecke. Psalm 97. 10. O ye that loue the Lord" &c. Lord Burleigh's crest, incircled with his motto. " Seene & allowed, &c. Anno 1574." Dedicated " To—Lorde Burleigh, Lo..'s High Treasurer," &c. Then, an epistle " To the Reader," by R. M. 3.5 pages besides. " Imprinted—by H. Denham & R. Watkins, 1574." W.H. Quarto.
The same in Benet's College Library. Folio.

The

1576. "The newe Iewell of Health, wherein is contayned the moſt excellent Secretes of Phiſicke & Philoſophie, deuided into fower Bookes. In the which are the beſt approued remedies for the diſeaſes, as well inwarde as outwarde of all the partes of mans bodie: treating very amplye of all Dyſtillations of Waters, of Oyles, Balmes, Quinteſſences, with the extraction of artificiall Saltes, the vſe & preparation of Antimonie, and potable Gold. Gathered out of the beſt & moſt approued Authors, by that excellent Doctor Geſnerus. Alſo the pictures, and maner to make the Veſſels, Furnaces, and other Inſtrumentes thereunto belonging. Faithfully corrected & publiſhed in Engliſhe, by George Baker, Chirurgian." A cut of "Alchymya" with implements of chymiſtry. "Printed—1576." Dedicated "To—the Noble Counteſſe of Oxeforde, &c." whoſe arms are on the back of the title page.---"George Baker to the Reader.—From my houſe in Bartholomewe lane beſide the Royal exchange in London, this xxj day of February, 1576---The Table." Contains beſides, 258 leaves. W.H. Quarto.

1576. "An Alphabet of Prayers, wherein many prayers haue the firſt letter of them in Alphabetical order: and the initial letters of others form his patron's name, Robert Dudley. By James Cancellar." See Maunſell p. 84. Dr. Lort.† Sixteens.

1577. "The COVRTYER OF COVNT BALDESSAR CASTILIO Deuided into foure Bookes, Imprinted—in Pater noſter Row,—1577." In a heavy architective compartment. On the back are ſome verſes by "Tho. Sackeuyll in commendation of the worke." Contains ſheets Y, in eights. W.H. Quarto.

1577. "A Funerall Sermon preached the xxvi day of November—M.D.LXXVI. In the Pariſhe Church of Caermerthyn, By the Rev. Father in God, Richard by the permiſſion of God, Biſhoppe of St. Dauys, at the buriall of The Right Hon. VValter Earle of Eſſex and Ewe, Earle Marſhall of Irelande, Viſcounte Hereforde & Bourgcher, Lord Ferrers of Chartley, Bourgcher & Louein, of the moſt Noble order of the Garter, Knight. Jmprinted &c. 1577." On the back is the earl's coat armour, with coronet and garter Dedicated "To—L. Robert Earle of Eſſex, &c. hir Maieſties Ward. &c.—E.W." Next, is a genealogy of the family, with arms in the margin. Then, verſes on the death of the late earl, in Hebrew, Greek, Latin, Welſh, and French. The whole, F 2, in fours. Licenſed. John Fenn Eſq. and W.H. Quarto.

1577. Tuſſer's 500 pointes of good huſbandry. Again 1580. See it in 1586. Quarto.

1578. "A Perfite platforme of a Hoppe Garden, and neceſſarie Jnſtructions for the making & mayntenaunce thereof, with notes and rules for reformation of all abuſes, commonly practiſed therein, very neceſſarie & expedient for all men to haue, which in any wiſe haue to do with Hops. Now newly corrected & augmented By Reynold Scot. Prouerbs, 11. Whoſo laboureth after goodneſſe findeth his deſire. Sapien. 7. Wiſedom is
nymbler

nymbler than all nymble things. She goeth through & attayneth to all things. Imprinted—in Pater noster Rowe, 1578. Cum priuilegio—folum." Dedicated " To—Mayſter Willyam Louelace, Eſquire, Sergeaunt at Lawe.----To the Reader.----The Table." Contains befides, 60 pages. W.H.+ Quarto.

" Phiſicke for the Soule, neceffary to be vſed in the agonye of death, 1578. as alſo for thoſe which are in good health : tranſlated by Henry Thorne. Alſo, a Sermon of patience, and of the confummation of the worlde."
Sixteens.

" The Booke of Pſalmes,—with briefe and apt Annotations in the 1578. margent, &c. Imprinted—being the aſſigne of W. Seres. 1578." The ueen's arms on the back, to whom the book is dedicated.—" From Geneua this 10. of Febru. 1559."---An epiſtle " To the Reader.---The Argument vpon the Pſalmes of Dauid." 373 pages; and two tables at the end. W.H. Sixteens.

" A Golden Chaine taken out of the rich Treaſure houſe, the pſalmes 1579. of King Dauid. Alſo, The pretious Pearles of King Salomon. Publiſhed, &c. By Tho. Rogers. Printed—1579." The former tract has a cut of K. David, playing on his harp, and contains p. 218. The latter has a cut of K. Solomon, kneeling, and has pages 198. Licenſed.— Again 1587. Sixteens.

" Phil. Melangton his praiers, tranſlated by Richard Robinſon." 1579. Seres's aſſignee. Octavo.

" The Condyt of Comfort, by Abr. Fleming.". Licenſed. 1579.

" AN ALVEARIE or Quadruple Dictionarie, containing foure ſun- 1580. drie tongues; namelie, Engliſh, Latine, Greeke, & French. Newlie enriched with varietie of Wordes, Phraſes, Prouerbs, and diuers lightſome obſeruations of Grammar. By the Tables you may contrariwiſe find out the moſt neceſſarie wordes placed after the Alphabet, whatſoeuer are to be found in anie other Dictionarie. Which Tables alſo ſeruing for Lexicons, to lead the learner vnto the Engliſh of ſuch hard wordes as are often read in Authors, being faithfullie examined, are truelie numbered. Verie profitable for ſuch as be deſirous of anie of thoſe languages." In a neat compartment, with the queen's arms at top, and a bee-hive, with Lord Burleigh's creſt over it, at bottom. Underneath, " Cum Priuilegio Regiæ Maieſtatis." Dedicated " Ad illuſtriſsimum virum Gulielmum Cecilium præclariſsimi ordinis equitem auratum:" &c. whoſe arms, creſted, are encloſed in the blooming letter D.—" Io. Baretus Cantabrigienſis." Then, ſundry commendatory verſes.----An epiſtle to the reader.---A brief inſtruction, explaining the figures by numeral letters. " Anno 1580. Ianuarie 2." After the Alvary are tables for Latin & French words, but not for Greek. " Londini, Excudebat H. Denhamus Typographus, Gulielmi Sereſij vnicus aſſignatus. Anno ſalutis humanæ 1580." W.H. Folio.

5 U " Pſalmi

1580. "Pſalmi Dauidis ex Hebraeo in Latinum conuerſi, ſcholiiſque perneceſſariis illuſtrati ab imman. Tremello, et Franc. Junio. Excud. H. Denhamus, typographus Gul. Sereſii vnicus aſſignatus." Octavo.

1580. "The foure chiefeſt Offices belonging to Horſemanſhip, That is to ſaie. The office of the Breeder, of the Rider, of the Keeper, and of the Ferrer. In the firſt part whereof is declared the order of breeding of Horſes. In the ſecond, how to break them, and to make them Horſes of ſeruice. Conteining the whole Art of Riding latelie ſet forth, and now newlie corrected and amended of manie faults eſcaped in the firſt printing, as well touching the Bits, as otherwiſe. Thirdlie, how to diet them, as well when they reſt, as when they trauell by the way. Fourthlie, to what diſeaſes they be ſubiect, togither with the cauſes of ſuch diſeaſes, the ſignes how to knowe them, and finallie how to cure the ſame. Which Bookes are not onlie painfullie collected out of a number of Authors, but alſo orderlie diſpoſed and applied to the vſe of this our countrie. By Tho. Blundeuill of Newton Flotman in Norfolke. Imprinted—being the aſſigne of W- Seres, 1580. Cum priuilegio Regiæ Maieſtatis." The laſt three tracts have diſtinct title-pages, &c. ſo as to ſell ſeparate occaſionally. The firſt three are dedicated " To—Lord Robert Dudley, Earle of Leiceſter," &c.—The firſt treatiſe contains 22 leaves; the 2d, 80; the 3d, 22; the laſt, 86; beſides their ſeveral prefixes. W.H. Quarto.

1580. Three moral treatiſes, &c. as p. 693. Octavo.

1580. "A ſecond and third ᵈ blaſt of retrait from plaies & Theaters: the one whereof was founded by a reuerend Byſhop, dead long ſince; the other by a worſhipful & zealous Gentleman now aliue. One ſhewing the filthines of plaies in times paſt; ᵉ the other the abhomination of Theaters in the time preſent: both expreſly prouing that that Common-weale is nigh vnto the curſſe of God wherein either plaiers be made, or Theaters maintained. Set forth by Anglo-phile Eutheo. Epheſ. 5; 15, 16.—Allowed by auctoritie, 1580." On the back is the arms of the city of London. 128 pages beſides the preface. On a ſeparate leaf, his ſign, &c. with the city arms over, and the Stationers' under it. "Imprinted—in Pater noſter Row,—being the aſſigne of W- Seres. Cum priuilegio," &c. Licenſed. W.H. Octavo.

ᵈ The editor in his preface ſays, "The firſt blaſt in my compt is The Schoole of abuſe." See it printed by Tho. Woodcock, 1579. "The former of theſe two was written in Latine by that reuerend man Siluianus, biſhop of Maſſilia,—1100 yeeres ſithence.—Touching the Autor of the latter blaſt, thou maiſt coniecture who he was, but I maie not name him at this time, for my promiſe ſake: yet this do I ſaie of him, that he hath bine, to vſe his verie wordes, A great affecter of that vaine Art of plaie making, &c. Yea, which I ad, as excellent an author of thoſe vanities, as who was beſt. But the Lord of his goodnes hath called him home; ſo that he did not ſo much delight in plaies in times paſt, but he doth as much deteſt them now." &c. ᵉ This is the firſt ſemicolon i have met with properly uſed.

" Dauids

"Dauids Sling againſt great Goliah: A ſword againſt the feare of death. A battell between the Deuill & the Conſcience. The dead mans ſchoole. A lodge for Lazarus. A retraite from ſin, by E. H." Maunſell, p. 84. Licenſed.　　　　　　　　　　　　　Sixteens. 1580.

"The enemie of Securitie, or a daily exerciſe of godly meditations, drawn out of the holy ſcriptures by Iohn Auenar, tranſlated by Tho. Rogers." Licenſed.　　　　　　　　　　　　　Small. 1580.

"Feed, Nawſe, his generall Doctrine of Earthquakes, tranſlated by Abr. Fleming." Licenſed.　　　　　　　　　　　Octavo. 1580.

"Bullokars Booke at large for the Amendment of Orthographie[f] for Engliſh ſpeech: wherein a moſt perfect ſupplie is made, for the wantes & double ſounde of letters in the olde Orthographie, with Examples for the ſame. With the eaſie conference & vſe of both Orthographies, to ſaue expence in Bookes for a time, vntil this amendment grow to a generall vſe, for the eaſie, ſpeedie, and perfect reading & writing of Engliſh (the ſpeech not changed, as ſome vntruly & maliciouſly, or at the leaſt ignorantly blowe abroade) by the which amendment the ſame Author hath alſo framed a ruled Grammar, to be imprinted heereafter, for the ſame ſpeech, to *the* no ſmall commoditie of the Engliſh Nation, not only to come to eaſie, ſpeedie & perfect vſe of our owne language, but alſo to their eaſie, ſpeedie & readie entrance into the ſecretes of other Languages, and eaſie & ſpeedie pathway to all Strangers to vſe our Language, heeretofore very hard vnto them, to *the* no ſmall profite & credite to this our Nation, and ſtay therevnto in the weightieſt cauſes. There is alſo imprinted with this Orthographie a ſhort Pamphlet for all Learners, and a Primer agreeing to the ſame, and as learners ſhall go forward therein, other neceſſarie Bookes ſhall ſpedily be prouided with the ſame Orthographie. Herevnto are alſo ioyned written Copies with the ſame Orthographie. Giue God the praiſe, that teacheth alwaies. When truth trieth errour flieth. Seene & allowed &c. Imprinted—1580." To this curious book are prefixed, "Bullokar to his Countrie.---The Prologe," in Alexandrines. As a ſpecimen, in part, of the author's plan, be pleaſed to accept the following extract from it. 1580.

"Sith letters be chéefe ſtay of all, in ech time, in theſe points,
　let perfectneſſe in ſingles be, and concord in their ioints.

[f] He does not herein endeavour ſo much to reform the ſpelling of words, as to eſtabliſh an alphabet in ſuch "concord of the eye, voice, and ear, as to yield a pleaſant harmony to the mind;" giving to ſuch letters as have different ſounds a little mark of diſtinction only, not forming new characters for them, as he intimates had been done by Sir Thomas Smith, and Mr. Cheſter, which ſeems to have been the reaſon why their ſchemes did not take place. Indeed thereby old deeds and writings would become as unintelligible as a dead language. Mr. Bullokar's plan appears to have met with no better reception, ſince no auther beſides is found to have followed it; owing perhaps to his ſpinning the thread too fine. See more of this author in Warton's Hiſt. of Eng. Poetry, Vol. II; p. 171. Vol. III. p. 346, &c.

HENRY DENHAM.

"Of which default, complaine we may, in the old A. B. C:
wherein be letters twentie fower, whereof but sixe agree,
In perfect vse, of name, and sound, besides misplacing some,
other are written vnsounded, wherein concord is none.
But he that will in Inglish knowe, diuisions in voice,
shall find therein fortie and fower, without any more choice.
Whereof are Consonants, twentie sixe, of vowels eight there be,
and diphthongs seuen, and likewise, halfe vowels there be three:
Of seuerall sounds, and perfect vse: and letters for the same,
are now prouided in this worke, and none haue double name."
The amendment of orthography is on 54 pages. Then a table of the contents of the 13 chapters.---The names of the letters according to this amendment; with specimens of their use in Roman, Italian, chancery and secretary hands. These are on two leaves, printed on one side only: the whole pages seemingly cut on wood, like the first method of printing. Licensed. W.H. Quarto.

1581. "THE PSALMES OF Dauid truly opened & and explaned by Paraphrasis, &c. Translated from the Latin of Th. Beza, by Anth. Gilbie. And by him newlie purged from sundrie faultes escaped in the first print, &c. Printed—1581. Cum priuilegio Regiæ Maiestatis." Dedicated "To— Ladie Katherine, Countesse of Huntingdon.—Ashbie, 7 March, 1579.--- The Epistle to the Reader.---The Psalmes digested into a briefe table of principal heads, according to Beza, and Temellius." On the back is a cut of K. David on his knees. Contains besides, 357 pages, and a table of the purport of each psalm in its order. The whole by signatures, Q 11, in twelues. Licensed. W.H. Twenty-fours.

1581. Tho. Knell's treatise of the use and abuse of prayer. Sixteens.

1581. "A Manuell of Christian praiers made by diuers deuout and godly men, as Caluin, Luther, Melangton, &c. augmented and amended by Abr. Fleming." Maunsell, p. 86. Licensed. Sixteens.

1581. The foot path to felicitie, &c. by Abr. Fleming. See The Diamond of Deuotion, 1586. Twenty-fours.

1581. "A PRETIOVS BOOKE OF HEAVENLIE MEDITATIONS, called A priuate talke of the soule with God: which who so zealouslie vse and peruse, shall feele in his mind an vnspeakable sweetnes of the euerlasting happines: Written (as some thinke) by that reuerend, and religious father, S. Augustine, and not translated onlie, but purified also, and with most ample, and necessarie sentences of holie scripture adorned, by Thomas Rogers. Printed—in Pater noster Row,—1581. Cum priuilegio," &c. Dedicated "To the hon. Master Tho. Wilson, Doctor of the Ciuil lawes; and one of hir Maiesties principal Secretaries," &c. whose arms are on the back of the title-page. The Meditations on 210 pages. Another leaf with his device; under it, being the assigne of W. Seres." W.H. Twenty-fours.

"A

" A right Chriſtian Treatiſe, entituled S. Auguſtines Praiers: Publiſhed 1582. in more ample ſort than yet it hath bin in the Engliſh tong; purged from diuers ſuperſtitious points; and adorned with manifold places of the S. Scriptures by Tho. Rogers. Wherevnto is annexed S. Auguſtines Pſalter: tranſlated & quoted by the ſame. T. R. 1. Theſ. 5. 17. Praie continualie. Imprinted—in Pater noſter Row,—1581. Cum priuilegio," &c. Addreſſed " To the Chriſtian Reader.—10 June, 1581." L 6, in twelues. W.H. Twenty-fours.

" S. AVGVSTINES MANVEL. Conteining ſpecial and piked meditations, and godlie praiers:—Collected, tranſlated & adorned by Tho. Rogers.— Imprinted—. Cum priuilegio, &c. 1581." E 10, in [b] twelues. W.H. Twenty-fours.

" THE Hammer for the Stone: So named, for that it ſheweth the moſt 1581. excellent remedie that euer was knowne for the ſame. Latelie deuiſed by Walter Cary Maiſter of Art, and ſtudent in Phyſicke. Imprinted— 1581." On the back is a goat's head on a wreath: See Guillim's Heraldry, p. 152. It is introduced with an addreſs " To the Reader;" an abſtract of which you have in the notes.[b]---"The Author to thoſe that are vexed with the ſtone:" in verſe. B 4, in eights. Licenſed. W.H.
Sixteens.

[g] This and the three foregoing articles are the firſt books I have met with ſince the Magna Charta, &c. in p. 259, whoſe ſignatures have really 12 leaues; and theſe by their ſize ſeem to be only half-ſheets.

[h] " Conſidering how common a diſeaſe the ſtone is, and how little helpe the parties grieued haue by the vſuall meanes of Phyſicians in this our time, I did endeuor with all ſtudie & diligence to find our ſome ſpeciall thing, which might far excell the remedies now dailie vſed for the cure of that greuous diſeaſe. Whereupon taking mine inuention from Etius, who vſed verie much the powder of Goats bloud for cure of the ſame: alſo being further perſuaded therevnto by authoritie of diuerſe, writing of the nature of goats bloud, I did with my great charge drawe a pure and cleare liquor out of the bloud of the male Goat, which with the patience of the Phyſicians, I will be bold to call a Quinteſſence. And hauing made experience thereof now two yeares & better, I thought good to publiſh the ſame, to the reliefe of manie, in ſuch ſort, as it ſhall appeare I rather ſeeke herein to benefite my Countrie, than anie priuate gaine to my ſelfe. For whereas no man troubled with that diſeaſe can haue the helpe of the Phyſician, without his great charge: I haue deuiſed that meane that anie ſo diſeaſed may haue ſuch eaſe with verie ſmall charge, as I dare auouch, cannot be had by anie vſuall meanes. And that my Quinteſſence may take the better effect, this treatiſe is diuided into 4 chapters. 1. The cauſes of the ſtone. 2. The difference of ſtones ingendered in mans bodie. 3. The vſuall waie both to preuent & cure it. 4. The waie now late deuiſed for the ſame. In this laſt chapter the author ſays, " But the maner of making this Quinteſſence, the choice of the Goat, the time of the yeare, the diuerſitie of the bloud of the arterie & the veine, and the order of diſtilling & circulating the ſame, I will not here ſpeake of: but (vpon requeſt) will deliuer it in writing to the right worſhipfull & moſt learned companie of Phyſicians in London, wiſhing them to appoint certaine Apothecaries, which ſhall be ſworne for the iuſt & true making of this Quinteſſence. Which being ſo made, the Phyſicians may direct the ſame to be giuen as to them ſhall ſeeme moſt meete. But in the meane time you ſhal haue of this Quinteſſence as much as I can conuenientlie make, at maiſter Graies houſe the Apothecarie in Fanchurch ſtreet, whoſe honeſtie & approoued good dealing I dare boldlie commend vnto you: and alſo at my houſe in great Wickham, in the countie of Buckingham,—at fiue ſhillings the wine pint."

" A

1581. "A Ripping vp of the popes fardel. By John Marbecke." Licenſed. Octavo.

1582. "A Monomachie of Motives in the mind of man: Or a battell be-tweene Vertues & Vices of contrarie qualitie. Wherein the Imperfections & weakneſſes of Nature appeare ſo naked, that anie reaſonable ſoule may ſoone ſee by what ſpirit he is lead: Herevnto alſo, beſides ſundrie deuout praiers neceſſarilie interlaced, diuers golden ſentences of S. Barnard are annexed: and alſo a briefe concluſion of his vpon this Theame, that Victorie is obtained by reſiſting temptation. Newlie engliſhed by Abr. Fleming. Iames 4; 7, 8. Reſiſt the Diuell & he will flie from you. Drawe neere to God, and he will drawe neere to you. Imprinted—Cum priuilegio," &c. Dedicated "To—Sir George Carey, Knight, Knight Marſhall of hir Maieſties—houſhold, Sonne & heir apparent to—Lord Henrie Lord Hunſdon, &c." Whoſe arms are on the back of the title-page, in 20 coats. Then, "The names of ſuch Vices & Vertues as are ſpecified in this booke." Contains beſides, 339 pages, each encloſed in a ſimilar border. On the back of the laſt is his device; under it, "being the aſſigne of W. Seres, 1582." Licenſed. W.H. Twenty-fours.

1582. "The Monvment of Matrones: conteining ſeuen ſeuerall Lamps of Virginitie, of diſtinct treatiſes; whereof the firſt fiue concerne praier & meditation: the other two laſt, precepts & examples, as the woorthie works, partlie of men, partlie of women; compiled for the neceſſarie vſe of both ſexes, out of the ſacred Scriptures, and other approoued authors, by Tho. Bentley of Graies Jnne, Student. Luke 12; 35. Let your loines be girt about, and your lampes burn cleerelie. 2. Tim. 2; 19. Let euerie one that calleth vpon the name of the Lord depart from iniquitie. Printed by H. Denham." Each of theſe lamps has a diſtinct title-page, and a different compartment; the firſt five ſeemingly were done on purpoſe for this work: the two laſt lamps were printed by T. Dawſon. On the back of the firſt, or general, title-page is "A praier vpon the poeſie prefixed." Then follow in order, the dedication "To the moſt vertuous Ladie & Chriſtian Princeſſe, Queen Elizabeth.----Lampas Virginitatis:" in 12 diſtichs of Latin verſes.---A prefatory addreſs "To the Chriſtian Reader.---Facies militantis Eccleſiæ:" in hexameters, by L. S.----"Rob. Marbeck ad lectorem:" in 9 diſtichs.---"Argumentum libri:" in 16 diſtichs, by L. S.---"The names of ſundrie famous Queens, godlie Ladies,[1] & vertuous women, which were notablie learned, whereof ſome were the authors of a great part of this booke.---What ceremonie euerie woman ought by Gods word to vſe in the time of praier, publike or priuate. 1 Cor. xi. 4, &c.---The firſt Lampe of Vir-

[1] Among theſe, "Elizabeth Q. of England. Ladie Elizabeth Tirwit. Ladie Francis Aburgauennie. Ladie Iane Dudley. Katherine (Parre) Q. of England. Margaret Q. of Nauar."

ginitie,

ginitie, conteining the diuine Praiers, Hymnes, or Songs, made by sundrie holie women in the Scripture; something explained in the hardest places," &c. is on 49 pages.

" THE SECOND LAMPE OF VIRGINITIE: Conteining diuers godlie Meditations & Christian Praiers made by sundrie vertuous Queenes, and other deuout & godlie women in our time: and first, A Godlie Meditation of the inward loue of the soule towards Christ our Lord: composed first in French by the virtuous Ladie Margaret, Queene of Nauarre: aptlie, exactlie, and fruitfullie translated by our most gratious Souereigne Ladie Queene Elizabeth, in the tender & maidenlie yeeres of hir youth & virginitie, to the great benefit of Gods Church, and comfort of the godlie. Imprinted—in Pater noster Rowe,—1582. Cum priuilegio" &c. On 252 pages.

" THE THIRD LAMPE of VIRGINITIE; Conteining sundrie formes of diuine meditations & Christian praiers penned by the godlie learned, to be properly vsed of the Queenes most excellent Maiestie, as especiallie vpon the 17 daie of Nouember, being the daie of the gladnesse of hir hart, and memorable feast of hir coronation: so on all other daies & times at hir Graces pleasure. Wherevnto also is added a most heauenlie Heast, spoken as it were in the person of God vnto hir Maiestie, conteining his diuine will & commandement concerning gouernement: and a right godlie & Christian Vow, vttered againe by hir Grace vnto God, comprehending the heroicall office & dutie of a Prince: faithfullie compiled out of the holie Psalmes of that princelie Prophet King Dauid, as they are learnedlie explaned by T. Beza: verie profitable to be often read & meditated vpon of hir Maiestie, and all other Christian Rulers & Gouernours, to the glorie of God; the benefit of his Church, and their own euerlasting ioie & comfort in the holie Ghost. Psalme 45. Audi filia & vide, &c. 1582." This seems to have been printed with the second lamp, the paging being continued to p. 362. At the end is a neat cut of the last judgement comprised in an oval, and divided into two parts, separated by clouds, and a rainbow. Above, is Christ Jesus on a throne of glory, holding out in his right hand a crown, and in his left, a palm-branch. Before him, kneeling right under the crown, is the Virgin Mary, from whose mouth proceed " Hallelu-iah, Hallelu-iah." Behind her is K. David, playing on his harp, and a group of virgins in adoring postures, holding in their hands lighted lamps. On the other side, is K. Solomon holding an open book; also other virgins, as on the opposite side. About this upper part, " Be thow faithful vnto the death, and I wil giue thee a crowne of lyfe. Rev. 2." Between the two parts, on each side is an angel blowing a trumpet. Below, the dead are rising out of their graues; beyond these is " Q. Katherine" laid, in a robe, on a mattress. About this lower part, " We must all appeere before the iudgment seat of Christ.

[k] See the original edition of this translation in our General History, 1548.

Rom.

Rom. 14." In the 4 corners, without the oval, are Juſtice, Prudence, Authority, and Fortitude. On the back of this cut begins a table for theſe three lamps, which compoſe the firſt volume.

"The Fourth Lampe of Virginitie. Conteining the moſt pure ſacrifice of Euangelicall deuotion, or an exerciſe of holie praiers, and Chriſtian Meditations for ſundrie purpoſes, &c. Compiled out of ſeuerall works of the moſt approoued Authors in our age, to the glorie of God, and the profit of his Church. By Tho. Bentley, Gent. Epheſ. 6. 18.—Philip 4. 6.—Imprinted—in Pater noſter rowe." The paging of this lamp is continued from the third to p. 1000. Then, the ſame cut of the laſt judgement, and a table for this part, which alone makes the 2d volume.

"The Fift Lampe of Virginitie: Conteining ſundrie forms of chriſtian praiers & meditations, too bee vſed onlie of and for all ſorts & degrees of women, in their ſeucrall ages & callings; as namelie, of Virgins, Wiues, Women with child, Midwiues, Mothers, Daughters, Miſtreſſes, Maids, Widows, & old women. A Treatiſe verie needfull for this time, and profitable to the Church; now newlie compiled to the glorie of God, & comfort of al godlie women, by the ſaid T. B. Gent. Matth. 26. 41.—Imprinted &c. being the aſſigne of W. Seres. 1582." On 213 pages. A table at the end.

"The Sixt Lampe of Virginitie; Conteining a Mirrour for Maidens & Matrons: Or, The ſeuerall Duties & office of all ſorts of women in their vocation, out of Gods word, with their due praiſe & diſpraiſe by the ſame: together with the names liues & ſtories of all women mentioned in the holie Scriptures, either good or bad: verie neceſſarie, pleaſant & profitable for all women to read & vſe, both for inſtruction & imitation. Newlie collected & compiled to the glorie of God, by T. B. Gent. Eccluſ. 26. 19.—1582." In the ſame compartment as to Bp. Cooper's Expoſitions, &c. p. 901. On 115 pages.

"The Seuenth Lampe of virginitie, conteining the acts & hiſtories, liues & deaths of all maner of women,—as well by name, as without name, ſet forth in alphabeticall order, with the ſignification & interpretation of moſt of their names: and in ſome part paraphraſtically explained & enlarged, &c. Wherevnto are added—the liues & ſtoryes of ſundrie ſuche other women as are mentioned in the thirde booke of Macchabees, and Joſephus. A treatiſe very neceſſary &c. eſpecially to the true imitation of vertue & ſhunning of vice, by example of all women kinde. Newely collected, &c. by the ſaid T. B. G. Prouer. 31: 29, 30, 31. Many daughters (through the feare of the Lord) haue done vertuouſly, &c. 1582." In a border with T. D. in the lower corners. The paging of this is continued from the former lamp to p. 331. "Imprinted—at the the three Cranes in the Vintree, by Tho. Dawſon for the aſſignes of W. Seres. 1582." The 5th, 6th, and 7th lamps make the third volume. Licenſed. W. H. Alſo without date. Dr. Lort. Quarto.

"Mabs

" Mabs [1] remembrances, faithfullie printed out of his own hand writing; the true copie whereof was found carefullie wrapped vp with his laſt will and teſtament, and other writings of great weight; and by himſelfe thus intituled, A delaration of my faithe ; mine opinion of religion ; a thankſgiuing to God, for all his benefits; an exhortation to my children, wherein all ſuch are to learne a good leſſon, as the lord hath crowned with anie kind of bleſſing ; and ſpecially with bodilie iſſue." Licenſed. 1583.
Sixteens.

"A briefe Treatiſe, called Caries farewell to Phyſicke: wherein thou ſhalt find rare and ſpeciall helpe for manie common diſeaſes.[m] Herevnto alſo is to be referred a gentle remedie againſt the Collicke: named The Hammer for the Stone, by the ſame W. C. Eccluſ. 30. 15.—Imprinted— 1583." On the back, " The Authors verſe." Two ſix-lined ſtanzas. It is introduced with a ſhort addreſs " To the Reader." Contains beſides, 57 pages. Licenſed. W.H. Again 1587. 1583.
Sixteens.

" The Common Places of the moſt famous & renowmed Diuine Doctor Peter Martyr, diuided into foure principall parts: with a large addition of manie theologicall & neceſſarie diſcourſes, ſome neuer extant before. Tranſlated & partlie gathered by Anthonie Marten, one of the Sewers of hir Maieſties moſt Honourable Chamber. Meliora ſpero. In the end of the booke are annexed two tables of all the notable matters therein conteined. 1 Cor. 3. 11. Other foundation can no man laie, than Chriſt Ieſus, which is alreadie laid." In the magnificent comparrment deſcribed, p. 891. Dedicated " To the moſt excellent, mightie & religious Princeſſe, Elizabeth,—Queene of England, &c.—At your Maieſties Court in Greenewich, 6 Maie, 1583.—Anth. Marten."---An epiſtle To the Chriſtian Reader.---Another, To all godly miniſters, ſtudious in divinity.---Tables of the method and diſtribution of the Common Places.---Faults eſcaped corrected. This book is printed in double columns, and ornamented with neat head and tail pieces. The firſt two parts on 640 pages; the third part on 380 pages; the fourth part on 331 pages. The additions have a diſtinct title-page, in a broad border; and dedicated alſo " To hir Maieſtie." Theſe additions are in two diviſions; the firſt on 252 pages, the other on 165. To theſe is added The life and death of Peter Martyr, by Joſias Simlerus, prefaced to biſhop Jewel. Dated at Zurick, 4 Cal. of January 1562. On 14 leaves. At the end of the tables, " Imprinted—in Pater noſter Rowe, at the coſts & charges of H- Denham, Tho. Chard, W. Brome, & Andr. Maunſell, 1583. Allowed according to hir Maieſties Iniunctions." W.H.   Folio. 1583.

This John Mabs was chamberlain of London, and one of the goldſmiths company, 17th of January, 1572.

[m] For moſt of theſe the author preſcribes a general purging potion, to be had of " Maiſter Graie Apothecarie in Fanchurchſtreete, (to whom he had given the receipt for making it) for 6.s. the wine pint." It ſeems he did not chooſe to offer this noſtrum to the phyſicians, as he had his Hammer for the Stone. See p. 953.

5 W " The

HENRY DENHAM.

1583. "The Pſalter or Pſalmes of Dauid: Corrected & pointed as it ſhall be ſaid or ſoong in Churches, after the tranſlation of the great Bible, with certain additions of collectsⁿ & other ordinarie ſeruice, gathered out of the booke of Common praier: Confirmed by act of Parliament, in the firſt yeare of the reigne of our ſouereigne Ladie Queene Elizabeth. Printed by the aſſignes of W. Seres. Cum priuilegio Regiæ Maieſtatis." In Seres's compartment with his mark on the ſides. The Pſalter—1583, has its title in the ſame compartment. Annexed, are "Certaine godlie Praiers." At the end is his device. "Imprinted &c. being the aſſigne of W. Seres." W.H. Again 1586. Quarto.

1583. "The exerciſe of the faithfull ſoule, or praiers & meditations for one to comfort himſelfe in afflictions, and eſpecially to ſtrengthen himſelfe in faith. Set in order according to the articles of the Chriſtian faith. By Dan. Tauſſain; tranſlated by Ferd. Filding" Licenſed.. Octavo.

1584. "AN ANSWER. To the two firſt and principall Treatiſes of a certeine factious libell, put forth latelie, without name of Author or Printer, and without approbation by authoritie, vnder the the title of An Abſtract of certeine Acts of Parliament: of certeine hir Maieſties Iniunctions: of certeine Canons, &c. Galath. 5. (10) He that troubleth you, &c. Publiſhed by authoritie. Printed—for Thō. Chard, 1584." It is introduced with a preface "Concerning the Title, and the Epiſtle of the Booke." Contains beſides, 350 pages. The two treatiſes are prefixed. W.H.
Quarto.

1584. "A Catalog of the Biſhops of Exceſter, with the deſcription° of the antiquitie and firſt foundation of the Cathedrall Church of the ſame. Collected by Iohn Vowell alias Hoker, Gent.—Imprinted—1584." The Author's coat of arms is on the back. Dedicated to the biſhop, the dean & chapter of Exeter. The laſt biſhop mentioned is John Wolton, conſecrated in 1579. K. 1, in twos. W.H. Quarto.

1584. "A Pamphlet of the Offices, and duties of euerie particular ſworne Officer, of the Citie of Exceſter: Collected by Iohn Vowell alias Hoker, Gentleman & Chamberlaine of the ſame. Numb. 30. Who ſo euer ſweareth an oth to bind him ſelfe, he ſhall not breake his promiſe. Imprinted—1584." On the back, are the arms of the city, with creſt and ſupporters; over which, "Pſal. Cxxvij. 2. Except the Lord keepe the Citie," &c. Alſo, the author's ſingle arms under one of the ſupporters. Dedicated "To—the Maior, Bailiffes, Recorder, Aldermen, and all others, the ſworne officers of the Citie of Exceſter. Exon, the laſt of the old yeere, and the beginning of the New, 1583.—Iohn Hoker." On the back, the author's arms as to the laſt article, in ſix coats. I, in twos. Licenſed. W.H. Quarto.

ⁿ Theſe collects have not after them even the places where the Epiſtles and Goſpels are to be found.

° This deſcription was afterwards inſerted in Holinſhed's chronicle, with ſome amendments.

The

HENRY DENHAM.

The Summary of Chronicles by John Stow. With R. Newbery. 1584. Again 1587. Sixteens.

" The Confutation of Follie; Conteining certeine selected queſtions, 1584. pithie anſwers, and ſyllogiſticall obiections, touching places of holie ſcripture, ſome preſentlie confuted, and others taken out of Loſsius Latine queſtions. No leſſe profitable than neceſſarie to ſtop the cauilling mouthes of the aduerſaries & wilfull opiniatiue Papiſts. By Henrie Thorne, Miniſter. Prov. 26. (5) Reſponde ſtulto &c. 1584." Dedicated " To M. George Speke Eſq;" whoſe arms are oppoſite.---" To the vnlatined Reader.---Ad candidum lectorem libelli Proſopopœia." Contains beſides, 150 pages. Licenſed. W.H. Octavo.

" The Art of Riding, ſet foorth in a breefe treatiſe, with a due inter- 1584. pretation of certeine places alledged out of Xenophon, and Gryſon, verie expert & excellent Horſſemen: Wherein alſo the true vſe of the hand by the ſaid Gryſons rules & precepts is ſpeciallie touched: and how the Author of this preſent worke hath put the ſame in practiſe, alſo what profit men maie reape thereby: without the knowledge whereof, all the reſidue of the order of Riding is but vaine. Laſtlie is added a ſhort diſcourſe of the Chaine of Cauezzan, the Trench, and the Martingale: written by a Gentleman of great ſkill & long experience in the ſaid Art. Jmprinted— 1584." Prefixed are, " A letter miſſiue To the right worſhipfull Gentlemen Penſioners, M. Henrie Mackwilliam, and M. Wm. Fitzwilliams.— 30 Sept. 1583. G. B.°.---To our verie louing Companions & fellows in Armes, hir Maieſties Gentlemen Penſioners; and to the gentle Reader whoſoeuer.—From the Court at Whitehall, 31 March, 1584. H. Mackwilliam. W. Fitzwilliams." Contains beſides, 79 pages. Licenſed. W.H. Quarto.

" The Art of Riding, conteining diuerſe neceſſarie inſtructions, de- 1584. monſtrations, helps & corrections apperteining to horſſemanſhip, not heretofore expreſſed by anie other Author: Written at large in the Italian toong, by Maiſter Claudio Corte, a man moſt excellent in this Art. Here brieflie reduced into certeine Engliſh diſcourſes to the benefit of Gentlemen deſirous of ſuch knowledge.—1584." To this tract are prefixed an epiſtle " To—M. Hen. Mackwilliam,—From the Court at Greenewich, 18 Maie, 1584.—T. Bedingfield..--To—my verie louing companions— Gentlemen Penſioners.—From the Court at Richmund, 1 June, 1584.— H. Mackwilliam.---To the Reader.—T. B." Contains beſides, 12 pages. W.H. Quarto.

" The Imitation of Chriſt, three bookes, made 170 yeares ſince by 1584. Tho. de Kempis, and for the worthines thereof oft ſince tranſlated into

---

p By the enſuing epiſtle we learn, that || a gentleman of the queen's Privy Chamber, notwithſtanding theſe initials, G. B. Mr. || was the author of this treatiſe. John Aſtley, maſter of the Jewel houſe, & ||

ſundrie

fundrie languages, now newly tranflated, corrected, and with moft ample textes & fentences of holy fcripture illuftrated by Tho, Rogers." Again 1589. Sixteens.

1584. "A Handfull of holefome, though homely hearbes, gathered out of the godly garden of Gods moft holy word, dedicated to all religious Ladies, Gentlewomen & others, by Anne Wheathill, Gent." Licenfed. Sixteens.

1585. "Seuen Sobs of a Sorrowfull Soule for Sinne: Comprehending thofe feuen Pfalmes of the Princelie Prophet David commonlie called Pœnitentiall; framed into a forme of familiar praiers, and reduced into meeter by William Hunnis one of the Gentlemen of hir Maiefties Chapell, and maifter to the Children of the fame. Wherevnto are alfo annexed his Handfull of Honifuckles; the Poore Widowes Mite; a Dialog betweene Chrift and a finner; diuers godlie & pithie ditties, with a Chriftian confeffion of and to the Trinitie; newlie printed and augmented. 1535." With mufic notes. Dedicated, "To—Ladie Francis, Counteffe of Suffex, and one of the Ladies of hir Maiefties moft Hon. priuie chamber.---The Author to his Booke.---The booke to his readers." On the back of this laft is a cut of K. David on his knees. The feven fobs on 85 pages. The annexes have diftinct title-pages, though the fignatures are continued to H 10, in twelves. "Printed in the now dwelling houfe—in Alderfgate ftreete at the figne of the Starre. Cum priuilegio Regiæ Maieftatis ad imprimendum folum." Licenfed. W.H. Twenty-fours.

1586. "Horat Morus a florentine, his tables of Surgerie, briefly comprehending the whole Art and practife thereof, in a maruelous good Methode, tranflated by Rich. Caldwell, Doctor in Phificke." Folio.

1586. "The garland of godlie flowers, commonlie called Twines praiers, carefully collected, and beautifully adorned with the moft fragrant bloffoms, that flourifh in the comfortable garden of the right pure and facred fcriptures," &c. Every page is printed in borders. Again 1589. Seres's affignee. Twenty-fours.

1586. "Fiue hundreth pointes of good Hufbandrie, as well for the Champion or open countrie, as alfo for the Woodland or Seuerall, mixed in euerie month with Hufwiferie, ouer and befides the booke of Hufwiferie. Corrected, better ordered, and newlie augmented to a fourth part more, with diuers other leffons, as a diet for the farmer, of the properties of winds, planets, hops, herbs, bees, and approoued remedies for fheepe and cattell, with manie other matters both profitable and not vnpleafant for the Reader.—Newlie fet foorth by Tho. Tuffer Gent.—Printed—in Alderfgate ftreet." The Hufbandry is dedicated to Lord Thomas Paget of Beaudefert, as it had been before to his late father deceafed. The Hufwifery to Lady Paget. Contains 164 pages. W.H.+ Quarto.

"THE

HENRY DENHAM.

"THE DIAMOND of Deuotion: Cut and fquared into fixe feuerall 1586. pointes: Namelie, 1. The Footpath of Felicitie. 2 A Guide to Godlines. 3. The Schoole of Skill. 4. A fwarme of Bees. 5. A Plant of Pleafure. 6. A Groue of Graces. Full of manie fruitfull leffons auaileable vnto the leading of a godlie and reformed life. By Abr. Fleming.—Printed—in alderfgate ftreete—1586." Dedicated " To—Sir George Carey Knight," &c. as p. 954. Each of thefe tracts has its peculiar title-page, but the paging and fignatures are progreffive throughout. O 10, in twelves. Every page in a border. Licenfed, 1581. W.H. Twenty-fours.

" THE Firft and fecond volumes of Chronicles comprifing, 1 The de- 1587. fcription and hiftorie of England. 2 The defcription and hiftorie of Ireland. 3 The defcription and hiftorie of Scotland. Firft collected and publifhed by Raphaell Holinfhed, William Harrifon and others: Now newlie augmented and continued (with manifold matters of fingular note and worthie memorie) to the yeare 1586, by Iohn Hooker, alias Vowell, Gent. and others. With conuenient tables at the end of thefe volumes." In the magnificent compartment, defcribed p. 881. Each of thefe hiftories is printed fo as to fell feparate occafionally. That of England is dedicated " To—S. William Brooke Knight, Lord Warden of the cinque Ports, and Baron Cobham," by W. H. (*William Harrifon.*) After the names of the authors from whom this hiftory is collected, and a table of contents, are fome Latin verfes by Tho. Newton of Chefhire. The defcription of Britain, on 130 pages; that of England continued to p. 250. Then, " The Hiftorie of England, from the time that it was firft inhabited, vntill the time that it was conquered: &c. By Raphael Holinfhed. Now newlie—digefted &c. by Abr. Fleming." 202 pages.

" THE Second volume of Chronicles: conteining the defcription, conqueft, inhabitation, and troublefome eftate of Ireland; firft collected by Ra. Holinfhed; and now newlie recognifed, augmented & continued from the death of king Henrie the eight vntill this prefent time of fir Iohn Perot knight, lord deputie: &c. By Iohn Hooker alias Vowell gent. Wherevnto is annexed the defcription and hiftorie of Scotland, firft publifhed by the faid R. H. and now newlie reuifed, inlarged and continued to this prefent yeare, by F. T. (Francis Thin) 1586." The hiftory of Ireland is dedicated " To—fir Henrie Sidneie knight lord deputie generall of Ireland," &c. by R. Holinfhed. Another dedication to him by Rich. Stanihurft, who compiled the defcription of Ireland; containing 61 pages. Then, " The Irifh hiftorie—written by Giraldus Cambrenfis, and tranflated into Englifh (with fcholies to the fame;) together with the fupplie of the hiftorie from the death of K. Henrie the eight, vnto 1587. By Iohn Hooker." Dedicated " To—fir Walter Raleigh knight, fenefchall of the duchies of Cornewall and Excefter; and lord warden of the ftannaries in Deuon & Cornewall.—Exon. 12. Octob. 1586.—Iohn Hooker." Then, two prefaces by Giraldus Cambrenfis, and his dedication to king John.

John. This hiſtory is on 183 pages. Next is "The Deſcription of Scotland, Written at the firſt by Hector Boetius in Latine, and afterward tranſlated into the Scotiſh ſpeech by Iohn Bellendine, Archdeacon of Murrey, and now finallie into Engliſh by R. H." Dedicated "To—Tho. Secford Eſq; Maiſter of the Requeſts," by W. Hariſon. This part on 23 pages. "The Hiſtorie of Scotland, conteining the beginning, increaſe, proceedings, acts, and gouernment of the Scotiſh nation, from the originall thereof vnto the yeere 1571, gathered &c. by R. H. and continued to 1585 by others." Dedicated "To—Lord Robert Dudley, Earle of Leiceſter," &c. by R. H. The paging of this is continued from the deſcription to p. 464. Theſe together form what is now denominated the firſt volume.

"The third Volume of Chronicles, beginning at Duke William the Norman, commonly called the conqueror, and deſcending by degrees of yeares to all the Kings and Queens of England in their orderlie ſucceſſion." Dedicated "To—Sir William Cecill, Baron of Burghleygh, &c. —-The Preface." Holinſhed collected and compiled the moſt part of this hiſtory to the year 1576, which was printed for John Harriſon in 1577; but it was now enlarged by Abr. Fleming with interpolations from the collections of Fr. Thin, the abridgement of R. Grafton, and the ſummary of John Stow. On p. 1268 begins "The Continuation from 1576, to —1586," prefaced by Abr. Fleming. This part contained ſeveral curious particulars, ſome of which were ſuppreſſed at the firſt publication, whereby the paging from p. 1220 to p. 1275 is very irregular: the caſtrations however have been reprinted. The whole on 1592 pages, beſides the prefixes, and a table at the end. "Finiſhed in Ianuarie 1587,—at the expenſes of Iohn Hariſon, Geo. Biſhop, Rafe Newberie, Hen. Denham, and Tho. Woodcocke." Denham's device; under it, "At London, Printed in Alderſgate ſtreet at the ſigne of the Starre. Cùm priuilegio." See p. 914. W.H. Folio.

1588. "Recreations: containing Adams banniſhment, Chriſt his Cribbe, the loſt ſheepe, and the complaint of old age. By William Hunnis." Licenſed. Twenty-fours.

1590. "Daily exerciſe of a Chriſtian, gathered out of the Scripture, againſt the temptations of the Deuill, by Tho. Lant." Maunſell, p. 84. Licenſed. Sixteens.

1591. "An Enemie to Atheiſme, or chriſtian godly praiers for all degrees, &c. written in the German tongue by Iohn Auenar: tranſlated out of Latine by Tho. Rogers.—Printed for the aſſignes of W. Seres." Sixteens.

——— "The Gouernment of all Eſtates wherein is contayned the perfect way to an honeſt life, by Andrew Heſſe: tranſlated by N. Boorman." Sixteens.

——— "A contemplation of myſteries: contayning the rare effectes and ſignifications of certayne comets, and a briefe reherſall of ſundrie hyſtoricall examples,

HENRY DENHAM.

examples, as well diuine, as prophane, verie fruitfull to be read in this our age: with matter delectable both for the fayler, and husbandman, yea and all traueylers by the sea and lande, in knowing aforehande, how daungerous tempest will succeede, by the sight of the clowd coming ouer the head, and other matters fruitful to be read, as shall appere in the table next after the preface. Gathered and Englished by Thomas Hyll." Octavo.

"Meditations set forth after the alphabet of the queens name, dedicated to queen Elizabeth. Ad finem, Meditationes Margaretae reginae Navarrae, translat. per reginam Elizabetham." Licensed, 1564. Twenty-fours.

"The Pitifull estate of the time present. A Christian consideration of the miseries of this time, with an exhortation to amendment of life, compiled by I. S." Licensed, 1563. Octavo.

"Theatrum mundi." &c. as p. 896. For Tho. Hacket. Sixteens.

"A Iewell for Gentlewomen, containing (besides many godly exercises) a spiritual almanach, wherein euery Christian may see what hee ought dayly to doe or leaue vndone: also a perpetual prognost. &c. by Tho. Humfrey." Maunsell, p. 86. Licensed, 1585. Sixteens.

The New Testament, and Common Prayer. With Barker. Small.

He had licenses also for the following, viz. In 1564, "The Reproued: written by M. Nowell agaynst Dorman." See p. 938. In 1565, "Letters of Sommon sente backe agayne by the greate Turke/ Soltain Soliman, to the cetizens of the cetye of Malta: and also of the castell of S. Elme. The myrror of pollicy/, a work no lesse profitable than necessarye for all monarckes/ kynges, prences, lordes, magestrates, & other rulers of the comen welth: aucthoryffhed by my lorde of London. An alm. & pronost. of M. Browne.—of Mygehell Nostrodamus." In 1566, "The ix. and x tragede of Lucius Anneus (*Seneca*) oute of the Laten into Englesshe/, by T. N. (*T. Nuce*) Felowe of Pembrek hall in Cambryge. A presydent for a prynce. An alm. & pronost. of Rych. Raynolds. The pycture of the Crosse, which was auctoryff'hed by my lorde of London. Serten verses in Laten. A warnynge agaynste the day of Judgement. The iiijth parte (*tragedy*) of Senecas workes. Argemonie, or the pryncipall vertues of stones by John Maplett." In 1567, "A mery metynge of maydes in London. Newes out of Powles churcheyarde/ a trappe for syr monye. A godly medytation of ỹ christian sowle. Wyckcleffes wycked. The abrydgemēt of the arte of Reason. The flowre of fryndshippe. Serten medytations vpon the Cxlij psalme. Serten prayers to be vsed in scholles, &c. Pamphilus the Louer/ & Maria the woman beloued. Pleasaunt tayles of the lyf of Rych. Wolner. Pleasaunte letters. to be vsed in

fuetes,

fuetes of lawfull maryages." In 1568, " An exortation to Englonde to ioyne for defence of tru Religion. A table cōtaynynge many prety pleafaūt paftymes. The x cōmandemēts in Spanyſſhe." In 1569, " The ijde parte of Euonimus, or the fecret remydes of Conradus Gefnerus. The Jewell of helth." In 1570, " Naturall & artificiall cōclufions, which was Cowplandes. Rayre wonders & feyrefull fyghtes in earth as in Heauen."---In 1579, " A paradox, prouinge by reafon & example that baldnes is much better then bufhie heare." In 1580, " Praiers vpon thoccafion of thearthquake. The praiers of the Byble. xiiij holy pfalmes chofen forthe of the newe & old teftament by Th. Beza. A complaynte & fupplication of the Soule : with a treatife on praier neceffarie to be ioyned to prayer bookes." In 1581, " vij Steppes to heauen, alias, the vij pfalmes reduced into meter by Will. Hunnys: the Honny fuccles, & the Wydoes myte." In 1587, " The iiijth parte of thimitacion of Xpiſt. An abridgement of the booke of Martyrs, with the pyctures." In 1588, " Anthinolophi, or the Hills of Numbers." This is entered to Mr. Will. Denham; but i apprehend it to be a miftake for Henry, as that name does not occur elfewhere in the company's books. " Directions, declaring the difcharge of our duties towards God and man, with certen pfalmes ānexed." In 1589, " The Armes of all the cumpanys of the worfhipfull cytye of London."

❊❊❊❊❊❊❊❊❊❊❊❊❊❊❊❊❊❊❊❊❊❊❊❊❊❊❊❊

RICHARD SERLL, or SCERLLE,

IN Fleet-lane, at the fign of the Geneva-arms, or the Half-eagle and Key; the fame as was fet up by Rowland Hall.

1566. " A new almanack and prognoſtication feruing for the year of Chriſt our Lorde M.D.LX.VI, diligently calculated for the longitude of London, and pole articke of the fame, by William Cunyngham, doctour in Phiſicke." Printed for William Jhones. Licenfed, 1563. Octavo.

—— " A briefe and pithie fome of the chriftian faith, made in forme of a confeffion, vvith a confutacion of all fuch fuperſtitious errours, as are contrary therevnto. Made by Theodore de Beza. Tranflated out of Frenche by R. F." (*Robert Fyll.*) Dedicated " To the lorde Haftings, Earl of Huntington, &c." Then, a table made by Serll, as he has told us in a ſhort note immediately following; on the back of which are certain texts, exhorting chriftians to render a confeffion of their faith before men. Bb, in eights. This had been printed by Rowland Hall in 1563. W.H. Sixteens.

He had licenfes alfo for the following, viz. In 1563, " A diologe betwene

twene Experyence & a Courtiour. An expofition—by Agapetus menefter of the mooft holy & greate churche of God, and now tranflated mooft truly put of Greke into Engleffhe by Tho. Whyte." In 1564, " A godly medytation vpon the Lordes prayer. A catechefme in Laten of M. Caluyn." Alfo feveral ballads; fome in 1565.

HENRY BYNNEMAN, or BINNEMAN,

WAS bound apprentice to Richard Harrifon, for eight years, from the feaft of St. John Baptift, 1560; but as Mr. Harrifon died in 1562, he probably ferved the refidue of his time, till he was made free on the 15th of Aug. 1566, with Reg. Wolfe, whofe device of the Brazen Serpent he fometimes ufed. He was called on the Livery 30 June, 1578; but I do not find him ferving any office. Soon after he was made free he was fined iij.s. " for vndermynding & procurynge as moche as in hym ded lye a copye from Will. Greffeth, called the boke of Rogges." He dwelt, firft in Pater nofter row, and then in Knightrider ftreet, at the fign of the Mermaid; which he fometimes ufed for a device, with this motto, *Omnia tempus habent*. Laftly, in Thames ftreet, near Baynard's caftle. In 1580, he was obliged to appear at the bar of the Houfe of Commons, on account of a libellous book he had printed fometime before for Arth. Hall, Efq; member for Grantham in Lincolnfhire. This book was charged with being not only greatly reproachful againft Sir Robert Bell the late fpeaker, and other worthy members, but alfo flanderous and derogatory to the general authority &c. of the Houfe, and prejudicial to the validity of their determinations,—charging them with drunkennefs, and their proceedings to be *opera tenebrarum*. The book was dedicated to Sir Henry Knivett, without his knowledge or approbation. Only 80 or 100 of them were printed. The author, printer, and parties concerned, were examined, firft before the the Privy Council, and afterwards brought to the bar of the Houfe. Hall as a culprit, the reft as witneffes. In the event, Hall was committed to the Tower for fix months, and fo much longer as until he fhould make a Revocation or Retraction, under his hand in writing, of the errors and flanders contained in the faid book, to the fatisfaction of the Houfe; to pay a fine of 500 marks to the queen's ufe; and expelled the Houfe, during that parliament. It was alfo " refolved and ordered, that the faid book and libel was, and fhould be holden, deemed, taken, and adjudged to be, for fo much as doth concern the errors aforefaid, condemned." See D'Ewes's Journals of the Parliaments of Q. Elizabeth, p. 291—309. Bynneman appointed H. Denham and Ra. Newbery his affignees, or as they are elfewhere called, the executors of his privilege, who in 1583 yielded up to the Stationers' company certain copies of Bynneman's privilege,ᵖ then deceased, for the relief of the poor of the said company. What Mr. Ames has said of the great encouragement he met with from Archb. Parker, is rather applicable to John Day.ᑫ He printed also several books for and with R. Newbery, L. Harrison, and H. Toy, which may be seen under their memoirs.

1566. "An Apologie, or Defence," &c. as p. 943. "Imprinted—in Paternoster Rowe at the signe of the blacke boy, by H. Binneman, Anno 1566- Octobris 14." Hen. Sutton lived here; but as nothing of his printing appears after 1565, our printer seems to have succeeded him here, for a while. As Denham printed this book also the same day, query whether it be the same edition with this, each having his own name to his proper quota? Crowley's epistle to the reader, is dated 1 March, 1565. This apology was written against an anonymous piece, without the printer's name, entitled "The Copie of an Aunswere, made vnto a certaine letter wherein the Aunswerer purgeth himselfe and other, from Pelagius errours, and fro the errour of free will, or iustification of workes: wherewithall he semeth to be charged, by the sayde letter: And further he sheweth wherin he differeth in iudgment from certaine Englishe writers & preachers, whome he chargeth with teaching of false doctrine, vnder the name of Predestination." This tract is printed verbatim, and answered paragraph by paragraph, on 104 leaves. At the end, "Seene & allowed according to the order appointed." Licensed. W.H.   Quarto.

1567. "The Manuell of Epictetus, Translated out of Greeke into French, and now into English, conferred with two Latine Translations. Hereunto are annexed Annotations, and also the Apothegs of the same Author. By Ia. Sanford." His device: a mermaid in an oval, with this motto, Omnia tempus habent; the Stationers' arms at top, and his mark at bottom. "Imprinted at London by H. Bynneman for Leonard Maylard,—1567." On the back, "Tetrasticon ad Reginam." Prefixed are, an epistle dedicatory to Q. Elizabeth; another to the reader; some Latin verses by Edm. Lewkener; others in English, against curious carpers; and the life of Epictetus. The Manuell on 37 leaves, the first not numbered. Then, some verses by the translator on Epictetus, " hys badge or Cognisaunce, 'Ανέχυ καὶ ἀπέχυ," on one leaf. "The Apophthegs" on two leaves. The colophon on another; "Imprinted—in Pater Noster Rowe, —And are to be sold in Paules Churchyarde at the signe of the Cock, 1567." W.H.   Octavo.

ᵖ Aug. 3, 1584, two of his apprentices were turned over to John Windet and Tho. Judson, who were then in partnership; and for which his widow was to have vjl. vjs. viijd. viz. iijl. vis. viijd. by the 24th instant, and the other iijll. by the sixth of Dec. next.

ᑫ See Strype's Life of Arohbp. Parker, p. 540.

" Iames

"Iames Sandfordes Amorous Tales and Sentences of the Greeke Philo- 1567. sophers." Dedicated To—Sir Hugh Pawlet, and has his arms. 40 leaves. At the end, " Printed in Pater noster Roe, at the signe of the Marmayde—for Leonard Maylarde—1567. Thefe bokes are to be folde in Paules Churchyarde at the figne of the Cocke." This was in the poffeffion of the late Thomas Martin of Palgrave, in Suffolk; but his copy wanted the title-page. Octavo.

" The fecond Tome of the Palace of Pleafure, conteyning ftore of goodly 1567. Hiftories, Tragicall matters, and other morall argument, very requifite for delighte & profit. Chofen & felected out of diuers good and commendable Authors: By William Painter, Clerke of the Ordinance & Armarie—1567. Imprinted—in Pater Nofter Rowe,—for Nich. England." Dedicated " To—Sir George Howarde Knighte, Mafter of the Queues Maiefties Armarie.—From my poore houfe befides the Toure of London, 4 Nov. 1567.---A Summarie of the Nouels enfuing.---The Preface to the Reader.—Authorities from whence thefe Nouels be collected: and in the fame auouched." From the dedication we learn that the firft tome had been publifhed almoft two years paft; and dedicated to the Earl of Warwick. See it in p. 856. The preface concludes thus: " Wherefore as in this I haue continued what erft I partly promifed in the firft: So vpon intelligence of the fecond fign of thy good will, a third (by Gods affiftance) fhall come forth." But none fuch has yet appeared. This tome contains 34 novels, on 427 leaves. " Imprinted—M.D.LXVII. Nouembris 8." On another leaf, the Faults with their corrections; at the end, N. England's device, as p. 780, and p. 855. W.H. 4 Quarto.

" A Confutation as wel of M. Dormans laft Boke entituled A Dif- 1567. proufe, &c. as alfo of D. Sander his caufes of Tranfubftantiation, by Alexander Nowel. Whereby our Cuntreymen (fpecially the fimple & vnlearned) may vnderftand, how fhamefully they are abufed by thofe & like Bokes, pretended to be written for their inftruction. Imprinted—1567. Cum priuilegio." It is introduced with a copious preface; and contains befides, 456 leaves, including Dorman's treatife, which is introduced, and anfwered paragraph by paragraph; as alfo part of Dr. Sander's book of The fupper of our Lord. Licenfed. W.H. Quarto.

" A very rich Lotterie general, without any Blankes, contayning a great number of good Prices, as well of redy money as of Plate, and certaine fortes of Merchaundizes, hauing ben valued & priced, by the comaundment of the Queenes moft excellent Maieftie, by men expert & fkilfull: and the fame Lotterie is erected by Her Maiefties order, to the entent that fuch comodities as may chaunce to arife thereof, after the charges borne, may be conuerted towards the reparation of the Hauens, and ftrength of the Realm, and towards fuch other good works. The num of Lotts fhall be foure hundreth thoufand, and no more: and every Lott fhall shall be the summe of Tenne shillings sterling onely & no more. To be ready the feast of St. Bartholmew, 1567. The shew of the Prises, &c. to be seen in Cheapside, at the signe of the Queenes armes, the house of Mr. Dericke, Goldsmith, seruant to the Queen." Another order 3 Jan. 1567; another, 9 Jan. 1568; and another, 13 July 1568, to finish the affair of the Lottery.

1568. "De antiquitate Cantabrigiensis Academiæ Libri duo. In quorum secundo de Oxoniensis quoq; Gymnasij antiquitate disseritur, & Cantabrigiense longe eo antiquius esse definitur. Londinensi Authore. Adiunximus assertionem antiquitatis Oxoniensis Academiæ, ab Oxoniensi quodam annis iam elapsis duobus ad Reginam conscriptam," &c. as p. 657. "Excusum Londini.—1568. Mense Augusto. Per Henricum Bynneman." The antiquity of Cambridge on 360 pages, exclusive of a catalogue of Authors, an index, and table of errors. The antiquity of Oxford contains E, in fours. On the last leaf is the printer's device only. Licensed; W.H. Octavo.

1568. "The Enimie of Idlenesse: Teaching the maner & stile how to endite, compose, and write all sorts of Epistles & Letters: as well by answer as otherwise. Deuided into foure Bokes, no lesse pleasaunt than profitable. Set forth in English by Will. Fulwood Marchaunt, &c.—Imprinted at London—for Leonard Maylard,—1568." Dedicated, in verse, "To—the Maister, Wardens, & Company of the Marchant Taylors of London.---The bokes verdict:" in verse also.---" To the reader." Contains besides, 145 leaues, and a table of contents. "Imprinted—dwelling in Knightrider strete, at the signe of the Mermaide, for L· Maylerd—1568." W.H. Octavo.

1568. "The secretes of—Maister Alexis of Piemont:" &c. See p. 781. Quarto.

1568. "An auncient order of knighthoode, called, the order of the band, instituted by Don Alponsus, king of Spain, in the year 1368, from Caesar Augustus, to wear a red ribbon of three fingers breadth, and subject to xxxv rules, the knights whereof were called by the same name; first translated out of Spanish into French, by Don Anthonie de Guauares and now Englished by Henry D." Dedicated "To—Sir Henry Sidney." Licensed. Octavo.

1568. "A newe mery and wittie Comedie or Enterlude, newely imprinted, treating vpon the Historie of Iacob & Esau, taken out of the xxvij Chap. of the first booke of Moses entituled Genesis. The partes and names of the Players, who are to be consydered to be Hebrews, and so should be apparailed with attire. 1 The Prologue, a Poete. 2 Isaac, an olde man, father to Iacob and Esau. 3 Rebecca, an olde woman, wife to Isaac. 4 Esau, a yong man and a hunter. 5 Iacob, a yong mā of godly conuersation. 6 Zethar, a neighbour. 7 Hanan, a neighbour to Isaac also. 8 Ragau,

8 Ragau, feruant vnto Efau. 9 Mido, a little Boy, leading Ifaac. 10 Debora, the nurfe of Ifaacs tente. 11 Abra, a little wench, feruaūt to Rebecca. Imprinted—in Knightrider ftreate,—1568." The prologue, of three feven-lined ftanzas, is on the back. The interlude confifts of five acts, which are divided into fcenes. It concludes with a fong. " Then entreth the Poete, and the reft ftand ftill, til he haue done" fpeaking the epilogue, though it has no name, any more than the prologue. When it is finifhed, " All the reft of the actours anfwer Amen. Then followeth the prayer, fpoken in parts, viz.

Ifaac.   Now vnto God let vs pray for all the whole clergy,
         To giue them grace to auaunce Gods honor & glory.
Reb.   Then for the Quenes maiefty let vs pray
         Vnto God, to kepe her in helth & welth night & day:
         And that of his mere mercy and great benignitie
         He will defend and maintaine hir eftate and dignitie,
         That fhe, being greeued with any outward hoftilitie,
         May againft her enemies alwaye haue victorie.
Jacob.   God faue the Quenes counfailours moft noble and true,
         And with all godlineffe their noble heartes endue.
Efau.   Lord faue the nobilitie, and preferue them all:
         And profper the Quenes fubiects vniuerfall. Amen.
Thus endeth this Comedie or Enterlude of Jacob & Efau."   Quarto.

" AN EPISTLE or godlie admonition of a learned Minifter of the Gofpel of our Sauiour Chrift, Sent to the Paftours of the Flemifh Church in Antwerp (who name themfelves of the confeffion of Aufpurge) exhorting them to concord with the other Minifters of the Gofpell. Tranflated out of French by Geffray Fenton. Here may the chriftian reader lerne to know what is the true participatiō of the body of Chrift, & what is the lauful vfe of the holy Supper." By Anth. de Corro, alias Belle Rive. " Imprinted—1569. Cum priuilegio." Dedicated " To Iohn Byron Efq;—10 Dec. 1569." Then, " A prayer of the Author for the concorde of doctrine & vnitie in wils in diuine things, &c.---A prayer to Jefus Chrift for peace & vnitie in the Church by Geff. Fenton." Contains befides, 56 leaves. On the laft, the printer's device, &c. The running title " A moft chriftian and godly Sermon." W.H. Again 1570. Octavo.   1569.

" Certaine fecretes and wonders of nature, containing a defcription of fundry ftraunge things, feeming monftrous in our eyes and iudgement, becaufe we are not priuie to the reafons of them. Gathered out of diuers learned authors, as well Greeke as Latine, facred as prophane, by Edward Fenton." With cuts.* Licenfed. W.H.+   Quarto.   1569.

" A difcourfe of fuch things, as are happened in the armie of my lordes, the princes of Nauarre and Condey, fince the moneth of September laft, 1568."*   Sixteens.   1569.

" A

1569. "A Theatre wherein be represented as well the Miseries & Calamities that follow the voluptuous worldlings, &c. Deuised by S. Iohn Vander Noodt. Imprinted—1569." Dedicated To Q. Elizabeth. Prefixed also are seven epigrams, and fifteen sonnets, with a cut to each. The running-title is "A Theatre for Worldlings." Licensed. Octavo.

1570. "Margarita theologica" &c. See it in 1573. Octavo.

1570. "Virgilii opera omnia, culex, dirae," &c. Licensed. Octavo.

1570. "An edict set forth by the French king for appeasing of troubles in his kingdome. Proclaymed in the court of parliament at Rowen, 16 August."° Sixteens.

1570. "DE ARTE Concionandi formulæ, vt breues, ita doctæ & piæ. Joanne Reuclino Phorcensi, Anonymo quodam rhapsodo, Philippo Melanchone, D. Ioanne Hepino, Autoribus. Eiusdem Melanchonis discendæ Theologiæ rationem, ad calcem adiecimus. Londini,—Pro Io. Waley.—1570." On the back are some Latin verses, Ad lectorem, &c. It is dedicated by I. Reuchlin "Ad Reuerendū patrem D. Petrum præpositum in Denckendorff, &c.—Kal. Januarias 1503." K. 2, in eights. W.H. Octavo.

1570. "The Arraignment & execution of Iohn Felton, hanged and quartered for Treason in Paules Churchyard, Aug. 8." In verse. Licensed. Octavo.

1571. "A Geometrical Practise, named Pantometria, diuided into three Bookes, Longimetra, Planimetra and Stereometria, containing Rules manifolde for mensuration of all lines, Superficies & Solides: with sundry straunge conclusions both by instrument and without, and also by Perspectiue glasses to set forth the true description or exact plat of an whole Region: framed by Leonard Digges Gentleman, lately finished by Thomas Digges his sonne. Who hath also thereunto adioyned a Mathematicall treatise of the fiue regulare Platonicall bodies and their Metamorphosis or transformation into fiue other equilarer vniforme solides Geometricall, of his owne inuention, hitherto not mentioned by any Geometricians." The cut of a man on the top of a castle taking the distances of two ships at sea: the same as used at chap. 33. "Imprinted—1571." On the back, the arms of Sir Nicholas Bacon, Lord keeper of the great seal of Englande, to whom this book is dedicated; then, a preface to the reader. Hb, in fours; with many cuts and schemes. Licensed. W.H. Quarto.

1571. "The Shippe of assured safetie, wherein wee may sayle without danger towards the land of the Liuing, promised to the true Israelites: contained in four Books. A discourse on Gods Prouidence, &c. by Edw. Cradocke, Dr. and Reader in Diuinity, in Oxford." For W. Norton. Again 1572. W.H. Sixteens.

1571. "A treatise of the right way from danger of synne, and vengeance in this

this wicked worlde, vnto godly wealth, and saluation in Chryst, by Tho. Leuer." For G. Bishop.

" An oration pronounced before the Frenche king at Chentilly, by 1571. the ambassadours of the most noble and excellent princes, the counte of Palatine, the duke of Saxony, and the marlegraue of Brandenborght, electors; Richarde, duke of Bauary; Julius, duke of Bronswyck; Ludowike, duke of Wittenburgh; John Albert, duke of Mecklenborgh; William, landtgraue of Hessen; George Frederike, markgraue of Brandenborgh, and Charles, markgraue of Baden; in the moneth of Januarie, this present yeare 1571." One sheet. Sixteens.

" A most excellent Hystorie, Of the Institution and first beginning of 1571. Christian Princes, and the Originall of Kingdomes: Wherunto is annexed a treatise of Peace & Warre, and another of the dignitie of Mariage.— First written in Latin by Chelidonius Tigurinus, after translated into French by Peter Bouaisteau of Naunts in Brittaine, and now englished by James Chillester Londoner. Seen & allowed, &c.—Printed—in Knight-rider streat,—1571." On the back is the queen's arms, to whom the book is dedicated. Then, a table of contents. Contains besides, 199 pages. W. H. Quarto.

" Quaestionum et responsionum christianarum libellus. Theod. Beza 1571. Vezelio auctore.—Impensis G. Bishop." Sixteens.

" Justini ex Trogi Pompeii historia libri XLIIII. supra plurimorum edi- 1572. tiones doctorum hominum opera castigatissimi. Cum eruditissimi scholi-olis et argumentis, jam primum apte suo loco dispositis et insertis. Adjecimus monarchiarum omnium tabulam, ex fideliss. historiographis diligenter collectam." Again 1577. Octavo.

" Of ghostes and spirites walking by night, and of strange noyses, 1572. crackes, and sundry forewarnynges, whiche commonly happen before the death of menne, great slaughters, & alterations of kyngdomes. One Booke. Written by Lewes Lauaretus of Tigurine, And translated into Englyshe by R. H." In a compartment with Jugge's mark at top. " Printed—for Richard Watkyns, 1572." On the back begins the translator's epistle to the reader. Then, the author's dedication. " To lorde John Steigtrus Consul of Berna.—Tigurine, in the month of January 1570.---A Table of the Chapters &c.---Fauites escaped &c.---An aduertisement" apologising for them. Contains besides, 220 pages. At the end, " Imprinted—by Richard Watkins, 1572." W. H. Quarto.

" The Surueye of the VVorld, or Situation of the Earth, so much as 1572. is inhabited. Compryfing briefely the generall partes thereof, with the names both new & olde, of the principal Countries, &c. First written in Greeke by Dionise Alexandrine, and now englished by Tho. Twine Gentl.

HENRY BYNNEMAN.

Gentl. Jmprinted—1572." On the back is the cut of the Genius of England, as p. 780. Dedicated " To—William Louelace Esq; Serieant at Law.—Leauing thus any farther to trouble you, with my dutiful cōmendations, and daily interceſſion to the Almightie, for the happie eſtate of your worſhip, & the good gentlewoman miſtreſſe Mary Louelace, your louing wyfe. London, 15 May, 1572." Then; " To the frendly Reader." Contains beſides F 3, in eights. On the laſt leaf is his device, and under it " Imprinted—in Knightrider ſtrette, at the ſigne of the Mermayde, 1572. And are to be ſold at his ſhop at the Northweſt dore of Poules Church, at the ſigne of the three Welles." W.H. Octavo.

1572. " The fall and euill ſucceſſe of Rebellion from time to time: Wherein is contained matter, moſte meete for all eſtates to view. Written in old Engliſhe verſe, by Wilfride Holme. Imprinted—in Knightriders ſtreate, —And are to be ſold—at the Northweſt doore of Paules Church. Anno 1572." Contains 34 leaves. Lambeth Library. Quarto.

1572. " Queſtions of Religion caſt abroad in Heluetia by the Aduerſaries of the ſame: and aunſwered by M. H. Bullinger of Zurick, reduced into 17 Common places. Tranſlated into Engliſhe by Iohn Coxe, 1572. Imprinted—for Geo. Byſhop." It is dedicated " To the right Reuerend—William—Biſhop of Exceſter.—Iohn Coxe.---The preface, by Ioſias Simlerus, addreſſed " To—John Iuel Biſhop of Sariſburie," in which he ſtiles Bullinger his father in law.—Zurich, Idib. Auguſt. 1560.---Henrie Bullinger to the Chriſtian Reader."---A table of the Common places. Contains 148 leaves beſides. W.H. Octavo.

1572. " The offer and order giuen forth by ſir Thomas Smyth knight, & Tho. Smyth his ſonne, vnto ſuch as be willing to accompanye the ſayde Tho. Smyth the ſonne, in his voyage for inhabiting ſome partes of the north of Ireland; the firſt payment to begin four years hence, 1576." Signed by Tho. Smyth's own hand. " God ſaue the Queen." See Camden's Elizabeth, p. 190. A broadſide, and ſixteens.

1573. " MARGARITA THEOLOGICA CONTINENS PRÆCIPVOS locos doctrinæ Chriſtianæ per quæſtiones breuiter & ordine explicatos. Omnibus paſtoribus, verbi præconibus, & Eccleſiæ miniſtris neceſſaria. Autore Ioan. Spangebergio, Herdeſiano, apud Northuſianos verbi miniſtro. Londini, Apud Henricum Bynneman,—1573." On the back begins the preface addreſſed " Illuſtriſſimo Principi & domino D. Philippo Duci Brunſuicenſi, &c.—Datum Northuſiæ, in Ferijs natalitijs Chriſti, 1540." Then, " Ad Dominici gregis paſtores Ioannis Spangebergenſis Hexaſticon." L 7, in eights. W.H. Licenſed. Octavo.

1573. " DE FVRORIBVS Gallicis, horrenda & indigna Amirallij Caſtillionei, Nobilium atq; illuſtrium virorum cæde, ſcelerata ac inaudita piorum
ſtrage

ftrage pafsim edita per complures Galliæ ciuitates fine vllo difcrimine generis fexus, ætatis & conditionis hominum. Vera & fimplex Narratio. Ernefto Varamundo Frifio auctore. Vis confilî expers mole ruit fua. Londini, Ex officina—1573." Contains 212 pages. W.H. Octavo.

" The Garden of Pleafure: Contayninge moft pleafante Tales, worthy 1573. deeds and witty fayings of noble Princes & learned Philofophers, Moralized. No leffe delectable than profitable. Done out of Italian into Englifh, by Iames Sanforde Gent. Wherein are alfo fet forth diuers Verfes and Sentences in Italian, with the Englifhe to the fame, for the benefit of ftudents in both tongs. Imprinted—1573." On the back is the Earl of Leicefter's creft gartered, and under it the author's arms, with this motto, ΟΙΣΤΕΟΝ ΚΑΙ ΕΛΠΙΣΤΕΟΝ. On the oppofite page are verfes, " Ad comitem Leceftriæ," in Greek, Latin, Italian, French, and Englifh. On the back thereof begins the dedication " To—Lord Robert Dudley, Earle of Leycefter," &c. Then, a fhort epiftle " To the Reader." Contains 112 leaves befides. On the laft is his device and colophon, "—And are to be fold at his fhop at the Northweft dore of Poules Church." W.H. See it again in 1576. Octavo.

" Certaine Godlie Homilies on Abdias and Ionas, by Rodolph Gualter 1573. of Tigure,—Tranflated by Robert Norton, minifter of the Worde in Suffolke.—1573." See p. 901. Octavo.

" Certaine briefe rules of Geographie, feruing for the vnderftanding 1573. of Maps & Charts, collected by D. P." Octavo.

Articles of Enquiry by Richard Bifhop of Ely at his vifitation, 1573. an. 1573. Quarto.
An edict of the French king for the appeafing of the troubles. Sixteens. 1573.

" THE ARTE of Reafon, rightly termed, Witcraft, teaching a perfect 1573. way to argue and difpute. Made by Raphe Leuer. Seene & allowed" &c. Dedicated " To—Walter, Earle of Effex, Vifcount Hereford," &c. Then, " The Forefpeache;" which begins thus, " To proue, that the arte of Reafoning may be taught in Englifhe, I reafon thus: Firft, we Englifhmen haue wits, as well as men of other nations haue: Wherby we conceyue what ftandeth with reafon, and is well doone, and what feemeth to be fo, and is not.—For Artes are like to Okes, which by litle and litle grow a long time, afore they come to their ful bigneffe. That one mā beginneth, another oft times furthereth and mendeth: and yet more praife to be giuē to the beginner, then to the furtherer or mender, if the firft did finde moe good things, then the folower did adde. Experience teacheth, that eche thing, which is inuented by man, hath a beginning, hath an increafe, and hath alfo in time a full ripeneffe. Nowe although eache worke is moft commendable, when it is brought to his full perfectio: yet,

yet, where the workmen are many, there is oft times more prayfe to be giuen to hym that begynneth a good worke, then to him that endeth it. For if ye confider the bookes, that are now printed, and compare them with the bookes, that were printed at the firft (Lord) what a diuerfity is there, and how much do the laft exceede the firft? Yet if ye wil compare the firft and the laft Printer together, and feke whether deferueth more praife and commendation: ye fhall finde that the firft did farre exceede the laft. For the laft had helpe of manye, and the firfte had helpe of none. So that the firft lighteth the candle of knowledge (as it were) and the feconde doth but fnuffe it."—Farewell from Durefme, 24 Nouember, 1572." Contains befides, 233 pages, and a table " to vnderftand the meaning of newe deuifed termes." * W.H.   Octavo.

1574.   " Abdias the Prophet, Interpreted by T. B. fellovv of Magdalene College in Oxforde. Seene & allowed, &c. Imprinted—for Geo. Bifhop, 1574." Dedicated " To—the Earle of Huntingdon, Prefident of the Queenes Maiefties Counfell eftablifhed in the North partes, &c.— 21 Auguft, 1574." E, in eights. W.H.   Octavo.

1574.   " A Godly Sermon preched before the Queenes Maieftie at Grenewiche the 26 of March laft paft by Dr. Whitgift, Deane of Lincolne: feene & allowed &c. Imprinted—for Humf. Toy, 1574."   Octavo.

1574.   " Comfort for the Sicke, in 2 partes; the firft for fuch as are vifited with ficknes: the fecond to make them willing to die. Tranflated by Ezekias Fogge."   Sixteens.

1574.   " Historia Breuis Thomæ Walfingham, ab Edwardo primo, ad Henricum quintum. Londini Excufum apud Hen. Binneman Typographum, fub infigno Syrenis." In a fcroll compartment with the Mermaid at bottom. In a divifion at top, " Cicero. Hiftoria eft teftis temporum, lux veritatis,—nuncia vetuftatis." In a divifion at bottom, " Anno Domini 1574." On the back are the portraits of the fix kings.—Within C, the firft letter of the preface, are the arms of Archbp. Parker. After the preface is a table of the hiftory from the year 1273 to 1456; and on the back is a note concerning the author. Contains befides, 458 pages, and an index. No mention of his fhop in Paul's Church-yard. W.H.   Folio.

1574.   " The profitable Arte of Gardening, now the thirde time fet forth:" &c. as p. 857. " Wherevnto is added a treatife of the Arte of graffing and planting of trees. Imprinted—1574." Again 1579. Both thefe editions have " the thirde time fet forth," the fame as the fmall edition, 1568. W.H.   Quarto;

1574.   " Theatrum Mundi," &c. as p. 896. In this curious theatre, fays the printer to the reader, " thou maift fee & behold all the vniuerfall world: thou maift firft fee thy felfe what thou art, & what miferies al humaine creatures are fubiect to.—Alfo, how birds hathe taughte men muficke, and how

to

to cure themselues of many infirmities: they haue taught men how to builde, how to preserue & kéepe for time to come, as how to maintaine a good cōmon wealth, how to expulse Vacabōds. Here also shalte thou sée notable examples against sundry vices frequented at this day.—Thus we sée how God hath raised in the latter end of thys worlde, dyuers wayes, his worthye instrumentes to publish his knowledge, for truly knowledge did neuer so abound as it doth now in these our dayes, (God graūte vs to be thankfull for it) neuer worthier preaching or writing than is and hath bene of late time, and yet sinns did neuer more abound than nowe: God graūt this sentence be not applyed vnto vs, He that knoweth his masters will & doth it not shall be beaten with many stripes. God graūte that these & such like worthy workes may take such déepe roote in our hartes, that we may bring foorthe the fruites of repentance & amendmēt of life to Gods glory & the comfort of our soules. Amen." This theatre is divided into four books; the first three treat of the miseries of man in every stage and condition of life, the 4th " Of the excellencie of man;" wherein mention is made of the invention and skill of some who have excelled in arts and sciences; instances of which may be seen ' below.

' Xeuxis, who painted a vine with grapes so subtilly that birds would pick at them. Apelles, who for the space of ten years employed all his skill to paint a Venus, which was so beautiful, that by public edict he was charged to keep it secret. An artificer of Heraclea is mentioned by Pausanias to have made a brazen horse, *or mare*, to which other horses sought to couple. Archimedes, who with one hand drew by a rope a great ship, freighted with merchandise, athwart the market place of Siracuse, by the science of mathematics: as also a glass with which he burnt the ships of his enemies in the sea. Sabor, king of Persia, caused a glass to be made so great, that he was set in a corner of the same as in the sphere, or compass of the earth, seeing under his feet the cloudes, and stars that did rise and set. The statue of Memnon, mentioned by Strabo, and Corn. Tacitus, which spoke every morning at the rising of the sun. Who would not be ravished in admiration at the mention of a dove of wood, composed by Architas, a scholar of Pythagoras, which being made by certain proportions of mathematics, did fly in the air as other birds: see more of this in the Gent. Magazine, April 1784, p. 245. Induced by this, Albertus Magnus made a brazen head, which could speak plainly. A Spaniard, who in the author's time made glasses which represented two faces together; the one alive, the other dead. Ptolemy mentions others, wherein there would appear as many faces as there are hours in the day, &c. But among all the works of our ancestors, nothing may compare with the invention, utility, and dignity of printing: who, therefore, doth not marvel at the times when, as Strabo reciteth, men first wrote on ashes, afterwards on the bark of trees, stones, laurel leaves, parchment, and finally paper; and as they varied in their manner of writing, so they used divers instruments to write with; on ashes with their finger, on stones with iron, on leaves with " pincers," on barks with knives, on parchment with canes, on paper with quills. At first their ink was the liquor of a certain fish, afterwards the juice of mulberries, then soot, and lastly the decoction of galls, gum and copperas. Also, the invention of guns and gun-powder; and yet what is more marvellous, as Brassavolus writeth, an artillery man at that time found out how to make gunpowder that maketh no noise in the discharge. Leave we these roarings, &c. the invention of the Devil for the destruction of mankind, and pursue the invention of our time. An artificer of Italy presented to prince Urbin a ring with a watch that struck every hour. Jerom Cardan, a man worthy credence, witnesseth to have seen a man at Milan publicly wash his hands and face in molten (*melted*) lead, having first washed them with a certain water. And if this seem so wonderful, how he might resist heat, yet it is no more so than what Alexander ab Alexandro and many others have written,

1575. "A TREATISE of the right way frō Danger of Sinne & vengeance in this wicked worlde, vnto godly wealth and faluation in Chrifte. Made by Th. Leuer, and now newly augmented. Seene & allowed" &c. In a neat compartment, with the refurrection of Chrift at top, with this motto about it, "Confidite, ego vici mundum. Io. xvi;" Peter and Paul at the fides; and the queen's arms at bottom. It has two epiftles prefixed, viz. "T. L. vnto Englande wifheth grace, mercy & peace of God in Chrift.—At London, 1571.---Tho. Lever unto Englande &c.—At Geneva, 1556." J 4, in eights. "Printed—for Geo. Byfhop, 1575." W.H.   Octavo.

1575. "A SERMON preached before the Queenes Maieftie at Richmond, the 6 of March laft paft. By the reuerend father in God the Bifhop of Chichefter. Printed—for H. Toy." In the fame compartment as the laft article, Text, Judges 1; 1—13. D 2, in eights. On the laft page is his device of the brazen ferpent. "Imprinted—1575." W.H. Octavo.

1575. "Henrie Cornelius Agrippa of the Vanitie and vncertaintie of Artes and Sciences:" &c. as p. 939. "Imprinted—in Knightryder ftreete—1575." Bbb, in fours. W.H.   Quarto.

(1575.) "A DECLARATION concerning the needfulneffe of peace to be made in Fraunce, and the meane for the fame: exhibited to the moft Chriftian king, Henrie the thirde of that name, King of Fraunce & Polande, vpon two Edictes put forth by his Maieftie; the one the tenth of September, the other the thirtenth of October, Anno 1574. Tranflated out of Frenche by G. H. Efq; Jmprinted—for Raufe Newberie—a little aboue the Conduit." Dedicated "To—his efpeciall good father Sir Pearciuall Hart Knight, one of the Sewers, and Knight Harbinger to hir Maieftie. —This firft of Ianuarie, 1575. Your humble & obedient fonne, Geo. Harte." K 4, in eights. W.H.   Octavo.

1575. The Palace of Pleafure, in 2 volumes.   Quarto.

1575. "The mariners boke, containing godly and neceffary orders and prayers, to be obferued in euery fhip, both for the mariners, and all other

that in their time in Sicily there was a man, named the Fifh Colas, who from his infancy frequented the fea, and there remained with fuch obftinacy, that he became "aquitall," and departed not from thence the moft part of his life, and was fometimes the fpace of five or fix hours hid under water, like a fifh, and would remain eight or ten days on the water and not come out; and would enter into the veffels that he found at fea, and eat with the mariners, then caft himfelf into the fea again; and fometimes he would come on fhore: he lived to be very old; and faid, while he was out of the water he felt great pain in his ftomach. There remaineth now nothing to man, faith our author, but to penetrate the air, and the firmament, to become familiar with them; and yet there was one Leonard Vincius, the which hath fought out the art of flying, and had almoft luckily atchieved his effect. What would this author have faid, had he lived in thefe days of balloon flying?

whatfoeuer

whatfoeuer they be, that fhall trauaile on the fea, for the time of their voyage. By Tho. Mors." Octavo.

" The Supremace of Chriftian Princes" &c. as p. 935.   Quarto.

" The Booke of Faulconrie or Hauking, For the onely delight & 1575. pleafure of all Noblemen & Gentlemen: Collected out of the beft aucthors, afwell Italians as Frenchmen, and fome Englifh practifes withall concernyng Faulconrie, the contentes whereof are to be feene in the next page folowyng. By Geo. Turberuile Gentleman. Nocet empta dolore voluptas." The cut of a nobleman with his hawk, and two attendants in the drefs of the time. " Imprinted at London for Chr. Barker, at the figne of the Grafhoper in Paules Churchyarde, 1575." Dedicated " To—Ambrofe, Earl of Warwicke, Baron Lifle, Maifter of hir Maiefties Ordinance," &c.---The names of authors.---" Ro. Baynes—in behalfe of the Writer.---In cōmendation of Hawking, *in verfe by* Geo. Turberuile." Contains befides, 372 pages, with cuts. In that at p. 81, is introduced the queen on horfeback: the fame again at p. 112. To this is annexed

" THE NOBLE ARTE OF VENERIE OR HVNTING. VVherein is handled 1575. & fet out the Vertues, Nature & Properties of fiuetene fundrie Chaces togither with the order & maner how to Hunte and kill euery one of them. Tranflated & collected for pleafure of all Noblemen & Gentlemen, out of the beft approued Authors, which haue written any thing concerning the fame: And reduced into fuch order and proper termes as are yfed here in the noble Realme of England." A cut of two huntfmen with hounds. " The Contentes vvhereof fhall more playnely appeare in the page next following." Dedicated, " To—Sir Henry Clinton—Maifter of the Hart Houndes," &c. Your honors moft humble C. B." (*Chr. Barker*) At whofe coft this was collected and tranflated.---" The tranflator to the Reader.—16 June 1575.---Geo. Gafcoigne in cōmendation of the noble Arte of Venerie.---T. M. Q. in prayfe of this booke." Contains befides, 252 pages, including " The meafures of blowing." N. B. Leaf N. 6 is cancelled, and two other leaves inferted, with the Otter's oration, in verfe. This part alfo is adorned with feveral neat cuts; in fome of which the queen is introduced, as at p. 90, 95, 133. W.H.   Quarto.

" ALEXANDRI NEVYLLI ANGLI, De furoribus Norfolcienfium Ketto 1575. Duce, Liber vnus. Eiufdem Norvicus. Ex officina—cıɔ.ıɔ.lxxv." This is the firft book i have obferved dated in this manner. On the back are the author's arms, in nine coats. Latin verfes prefixed, " In obitum campliffimi Præfulis Matthæi Parkeri Cantuar. Archicpifcopi eiufdem Alex. Neuylli carmen.---Libri poft defunctum patronum editi, profopopœia. G. A.---Ad librum poft defunctum patronum editum, Carmen confolatorium. G. A.---In Alex. Neuylli Kettum Tho. Drantæ carmen.
---In

--- In hiftoriam Alex. Neuylli, Hendecafyllaba." Then, the dedication, on p. 1. " Reuerendiffimo in Chrifto patri D. Matthæo Cantuarienfi Archiepifcopo," &c. The archbifhop's paternal arms within the firft letter D. This epiftle, for the elegance of the Latin ftile, Mr. Strype thought proper to infert in his appendix to The Life of Archbp. Parker. No. CIV. Within O, the firft letter of the hiftory, are the archbifhop's paternal arms in pale with the arms of his See, having this motto, Mundus tranfit et Concupifcentia cius. This part, of Kett's rebellion, ends on p. 156. Again 1582.' Octavo. W.H.   Quarto.

1575. "ALEXANDRI NEVYLLI ANGLI' NORVICVS." The brazen ferpent. "Ex officina—cɔ.ɔ.lxxv." On the back, the author's arms, as before. The epiftle "Ad lectorem—x. Kal. Decemb. cɔ.ɔ.lxxv." Then, "Alex. Ne. amico fuo ingenuo atq; omni humanitate perpolito Tho. Dranta Archid. Leuicenfis." In hexameters. The hiftory on 207 pages; at the end of which " Nomina prætorum & vicecomitum Norwicenfium, ab anno 1 Hen. IV, ad 16 (17) Eliz." Both thefe treatifes are very neatly printed on Roman and Italic types. W.H.   Quarto.

1575. "The Poefies of George Gafcoigne Efquire. Corrected, perfected, and augmented by the Author, 1575. Tam Marti quàm Mercurio. Imprinted—for Richard Smith. Thefe bookes are to be folde at the North-weft dore of Paules Church." See Reliques of Ancient Englifh Poetry, Vol. 11; p. 138, &c.   Quarto.

1576. " The Steele Glas. A Satyre copiled by Geo. Gafcoigne Efq; Togither with the Complainte of Phylomene, An Elegie deuifed by the fame Author. Tam Marti," &c. Device, Time bringing truth to light, with this motto " Tempora patet occulta Veritas." Underneath, " Printed for Richard Smith." On the back is the author's portrait.
Quarto.

1576. " ALEXANDRI NEVYLLI AD VVALLIÆ PROCERES APOLOGIA." The brazen ferpent, with NVM. xxi. "Ex officina—cɔ.ɔ.lxxvi. Maij xii." On the back are the author's arms, as before. Nine leaves. W.H.
Quarto.

1576. " Hours of recreation or Afterdinners, Which may aptly be called The Garden of Pleafure: Containing moft pleafant Tales, worthy deeds & wittie fayings of noble Princes & learned Philofophers, with their Morals.—by Iames Sandford Gent, and now by him newly perufed, cor-

* This has its fignatures continued from Oeland's Anglorum prælia, and Elizabetha. See p. 909, &c.

' Some copies of this book have a table of the fucceffion of the Saxon kings during the Heptarchy, of the dukes of Normandy, and of the monarchs of England from the conqueft, with a map of the feven kingdoms, engraved by Richard Lyne; alfo, a genealogical table of the Fnglifh monarchs from William the conqueror to queen Elizabeth, with another of the line of France, engraved alfo on copper by Remigius Hogenbergius. Both thefe engravers ftile themfelves feruants of Archbp. Parker. See Gough's Brit. Topography, Vol. 1; p. 508, and Vol. 21; p. 7.

rected

rected & enlarged. Imprinted—1576." The author's arms on the back. Now dedicated " To—Maister Christopher Hatton Esq; Captain of his Maiesties Garde," &c. 223 pages. Hereunto are annexed " Certayne Poemes dedicated to the Queenes moste excellente Maieftie, by Iames Sandforde Gent." Eight leaves. Licenfed. See it in 1573. W.H. Octavo.

" The VVarfare of Chriftians: Concerning the conflict againft the Flefhe, the World, and the Deuill. Tranflated out of Latine by Arth. Golding." The brazen ferpent. " Imprinted—for Iohn Shepparde, 1576." Dedicated " To—fir William Drewrie knight.—16 Jan, 1576." Contains befides, 75 pages. W.H. Octavo. 1576.

Barfton's " Safegarde of Societie. Defcribing the Inftitutions of Lawes & Policies. For John Shepperd." Mr. William White. Octavo. 1576.

" Bellum Grammaticale, a difcourfe of grete war and diffention between two worthy princes, the noun and the verb, contending for the chiefe place or dignity in oration, turned into Englifh by William Haywarde." Octavo. 1576.
The Prince of Orange's Supplication to Philip K. of Spain. Quarto. 1576.
" A Supplication exhibited &c. as p. 920. Octavo. 1577.

" Balthafaris Caftilionis comitis de Curiali, fiue Aulico, libri quatuor, ex Italico fermone in Latinum conuerfi. Barthol. Clerke Auglo Cantabrigienfi interprete. Nouifs. aediti." Octavo. 1577.

The golden Aphroditis, &c. by I. Grange Gent. See it without date. 1577. Quarto.
" Ouid his Inuectiue againft Ibis. Tranflated into Englifh Meter. Wher vnto is added by the Tranflator, a fhort draught of all the Stories & tales contained therein very pleafant to be read. Imprinted—1577." Dedicated " To—fir Thomas Sackuile knight, lord Buckhurft—Tho. Vnderdowne." Then, " The preface." M 4, in eights. Bought of Tho. Eaft. Licenfed. Octavo. 1577.

" The Gardeners Labyrinth: Containing a difcourfe of the Gardeners life, in the yearly traueis to be beftovved on his plot of earth, for the vfe of a Garden: with inftructions for the choife of Seedes, apte times for fowing, fetting, planting and watering: and the veffels & inftruments feruing to that vfe & purpofe: Wherein are fet forth diuers Herbers, Knottes & Mazes, cunningly handled for the beautifying of Gardens. Alfo the Phyfike benefit of eche Herbe &c. Gathered out of the beft approued writers of Gardening, Hufbandrie & Phyficke: by Dydymus Mountaine." A cut of two gardeners twifting vines about an arbour. " Printed—1577." On the back, " Henrici Dethicki ad lectorem Carmen." By whom this book was finifhed, in compliance with a promife 1577.

to

to the author lately deceased, and by him dedicated "To—Sir William Cecill knighte—of the Garter, Lord high Treasurer of England, &c." His crest within the first letter; then, Tables of contents for both parts. The first part on 80 pages, and four leaves of Knots, in various forms.

"The second part of the Gardeners Labyrinth, vttering suche skilfull experiences & worthy secretes, about the particular sowing & remouyng of the most Kitchin Hearbes, with the wittie ordering of other dayntie Hearbes, delectable floures, pleasant fruites, & fyne rootes, as the like hath not heeretofore bin vttered of any. Besides the Phisicke benefites of each Herbe annexed, with the commoditie of waters distilled out of them, ryghte necessarye to be knowen." This title is over a cut of three gentlemen sitting at a collation of fruits and wine in an arbour. The engraver's mark S. A, in one. This part has 180 pages, and a table of the physical properties of the herbs, &c. Licensed. W.H. Again 1578.
Quarto.

1577. "A Godly treatise declaring the benefits, fruits, & comodities of prayer, &c. Written in latin 40 years passt; by an Englishman of great virtue & learning; (*Bishop Fisher*) and lately translated into English. Printed—for Gabriel Cawood, 1577." Also,

"A Spiritual Consolation written by John Fyssher, Bp. of Rochester to hys syster Elizabeth, at such tyme as he was Prisoner in the Tower of London. Very necessary & comodious for all those that mynde to leade a vertuous lyfe: Also to admonishe them, to be at all tymes prepared to dye."
Octavo.

1577. "A most strange example of the judgment of God vpon Symō Penbrok a coniurer, by his sodaine deathe. 17 Jan. 1577; in the parish of S. Marie Ouerie in Southwark." Licensed.

1577. Holinshed's Chronicles appear to have been printed by him, for John Harrison, having repeatedly the same compartment as to Historia Tho. Walsingham, 1574; and his device of the brazen serpent, at the end of the first volume. This first edition is set off with cuts in plenty, most of them repeatedly used on similar occasions. These are omitted in the subsequent edition for matter more important. See it under John Harrison. W.H.
Folio.

1577. "GABRIELIS HARVEII CICERONIANVS, Vel Oratio post reditum, habita Cantabrigiæ ad suos Auditores. Quorum potissimùm causa diuulgata est." The brazen serpent.—"Ex officina—cIɔ.Iɔ.lxxvii." Prefixed are "Gabriel Haruëius S. D. Gulielmo Leuino Doctori Iureconsulto, & Oratori præstantissimo.---Gul. Leuinus Typographo, & eloquentiæ studiosis, Salutem.---Erratorū Elenchus." 67 pages: on the back of the last, the brazen serpent, "Excudebat—M.D.LXXVII. Mense Iunio." Licensed. W.H.
Quarto.

"GABRIELIS

" GABRIELIS HARVEII RHETOR, Vel duorum dierum Oratio, De Na- 1577. tura, Arte, & Exercitatione Rhetorica. Ad suos Auditores." The brazen serpent.——" Ex Officina—1577." Prefixed are, Gabr. Harueius S. D. Bartho. Clerco Doctori Iureconsulto, & Oratori clarissimo.--- B. Clercus G. Harueio Cantabriginsi Rhetoricæ artis Professori eloquentissimo, parem Salutem optat." Q 2, in fours. " Excudebat——Mense Nouembri: rogatu ornatissimi viri B. Clerci, Legum doctoris." Licensed. W.H. I have another copy without this colophon. Quarto.

" GABRIELIS HARVEII VALDINATIS SMITHVS; VEL MVSARVM 1578. LACHRYMÆ; Pro obitu Honoratissimi viri, atque hominis multis nominibus clarissimi, Thomæ Smithi, Equitis Britanni Maiestatisque Regiæ Secretarij.—" The brazen serpent. ", Ex officina—CIƆ.IƆ.LXXVIII." The dedication in hexameters " Honoratissimo, Præstantissimoq; viro Gualtero Mildmaio, Illustrissimæ Reginæ à Consilijs, atq; Regij fisci Cancellario.--- Insignia Honoratissimi viri & amplissimi Equitis Thomæ Smithi, nuper defuncti." The arms. " Emblema. Quâ pote, Lucet.--- Gabriel Harueius ad Charites de Smithæis Insignibus.---Charites ad G. Harueium." The poem is addressed " Ad Ioannem Vuddum, clarissimi equitis, Thomæ Smithi, dum viueret, amanuensem, & sororis filium." At the end are annexed, " In effigiem, duo honoraria Emblemata; Alterum Illustrissimæ Reginæ: alterum Dominæ Russellæ iam tum fortè ad Aulam aduentantis." My copy wants the cut.---" Tumulus clarissimi viri D. Thomæ Smithi, Legum doctoris, Equitis aurati," &c. A cut of the monument. " Emblema Pictoris. Tendimus huc omnes." L 2, in fours. On the last page, under the brazen serpent, " Excudebat— M.D.LXXVII. Calendis Januarijs; rogatu Aulici cuiusdam generosi." Licensed. W.H.+ Quarto.

" Gabrielis Harueij Gratulationum Valdinensium Libri Quatuor. Ad 1578. Illustriss. Augustissimamque Principem, Elizabetham Angliæ, Franciæ, Hiberniæq; Reginam longè serenissimam atq; optatissimam." The queen's arms. " Ex officina—CIƆ.IƆ.LXXVIII. Mense Septembri." On the back, within a letter C, the queen is represented, crowned and robed, sitting on a throne elevated, holding a naked sword in her right hand, and the mund in her left; three nobles uncovered, attending on the right. 28 pages. The second book is dedicated " Ad Nobilissimum Dominum, & Illustrissimum Heröem Comitem Leicestrensem," &c. 12 pages. The third is dedicated " Ad Honoratiss. spectatissimumq'; virum Dom. Burgleium, magnum Angliæ Thesaurarium," &c. 11 pages. The fourth, " Ad Nobiliss. præclarissimumq; Dominum Comitem Oxoniensem, magnum Angliæ Camerarium: Ad Honoratiss. & Ampliss. virum Christophorum Hattonum Equitem auratum, &c. Ad Clariss. Nobilissimumq; Iuuenem Philippum Sidneium, Henrici proregis Hibernici, filium." 25 pages. On the back of the last leaf is a phœnix in flames, looking up to the sun. " Excudebat—M.D.LXXVIII. Calendis Septembribus. Cum Priuilegio." Licensed. W.H. Quarto.

" A TRVE DISCOVRSE of the late voyages of discouerie for the finding of 1578.

HENRY BYNNEMAN.

a paſſage to Cathaya by the Northvveaſt, vnder the conduct of Martin Frobiſher Generall: Deuided into three Bookes. In the firſt whereof is ſhewed, his firſt voyage. Wherein alſo by the way is ſette out a Geographicall deſcription of the Worlde, and what partes haue bin diſcouered by the Nauigations of the Engliſhmen. Alſo there are annexed certayne reaſons to proue all partes of the Worlde habitable, with a generall Mappe adioyned In the ſecond, is ſet out his ſecond voyage, vvith the aduentures and accidents thereof. In the thirde, is declared the ſtrange fortunes which hapned in the third voyage, with a ſeuerall deſcription of the Countrey and the people there inhabiting. VVith a particular Card therevnto adioyned of Meta incognita, ſo farre forth as the ſecretes of the voyage may permit. At London, Imprinted by—ſeruant to the right Honorable Sir Chriſtopher Hatton Vizchamberlain. 1578." This titlepage of my copy is printed all in black; but by the copy of one among Mr. Ames's papers, ſome were printed in red and black. On the back begins an account of "What commodities & inſtructions may be reaped by diligent reading this Diſcourſe." Dedicated by Geo. Beſt, the author, "To—Sir Chriſtopher Hattō, Knight, Capitaine of the Queenes Maieſties Garde, Vizchamberlaine" &c. whoſe arms are on the oppoſite page. Then, The printer's preface. The firſt voyage on 52 pages; the ſecond on 39 pages; and the third on 68 pages. At the end, "Printed—1578, Decembris 10." W.H.   Quarto.

1578.   "A Sermon preached before the Queenes Maieſtie at Hampton Court the 19 of February laſt paſte, on Ezra iv. 1—3. By William James Doctor of Diuinitie, and Deane of Chriſtes Church Oxford." Licenſed. Octavo.

(1578.)   "A Sermon preached at Paules croſs on Barthlemew day, 1578. on Acts x; 1—10. by Iohn Stockwood, ſcholemaiſter of Tunbridge." Dedicated To the Maſter, Wardens & Aſſiſtants of the worſhipful company of Skinners. 174 pages. Dr. Lort.   Octavo.

1578.   "A Deſcription of the Ports, Creekes, Bayes and Hauens of the Weſt Indies: tranſlated out of the Caſtill tongue by Iohn Framptoe." Licenſed.   Quarto.

1578.   "Common Places of Chriſtian Religion" &c. as p. 608.   "Tranſlated out of Latine into Engliſh by Iohn Man of Merton Colledge, in Oxford. —Seene & allowed" &c. The brazen ſerpent. "Imprinted—1578." On the back are the arms of Sir Chr. Hatton, to whom this edition is dedicated by H. B. Then, "An admonition to the Reader.---The Preface of W. Muſculus.---A Table." The Common Places on 1340 pages. The treatiſes on oaths and uſury annexed, on 42 pages. Then a copious table on cccxiiii columns, three on a page. W.H.   Quarto.

1578.   "The Courtly Controuerſie of Cupids Cautels, containing 5 Tragical Hiſtories out of French, by Hen. Wotton." Licenſed.   Quarto.

Iſaiah's

HENRY BYNNEMAN.

Isaiah's Prophecy of Christ's Kingdom, chap. 9. explained by M. Luther. 1578. Octavo.

" THE Pleasant Historie of the Conquest of the VVeast India, now (1578) called new Spayne, Atchieued by the vvorthy Prince Hernando Cortes Marques of the valley of Huaxacac, most delectable to Reade: Translated out of the Spanishe tongue, by T. N. Anno 1578." The brazen serpent. " Imprinted at London by Hen. Bynneman." Dedicated " To—Sir Francis Walsingham.—Tho. Nicholas.---To the Reader.----Stephan Gosson in prayse of the Translator:" in 6 six-lined stanzas.---" In Thomas Nicholai occidentalem Indiam St. Gosson:" in six distichs. Contains besides, 405 pages, and a table. Licensed. W. H. Quarto.

" The Joyfull Receyuing of the Queenes Maiestie into her Highnes 1578. Citie of Norwich: The Things done in the Time of hir Abode there: And the Dolor of the Citie at hir departure, by Sir Rob. Wood, Maior of the same Citie." Bibl. Askeuiana, No. 3528. Licensed. Quarto.

" A DISCOVRSE OF The Queenes Maiesties entertainement in Suffolk (1579) and Norffolk: With a description of many things then presently seene, Deuised by Thomas Churchyarde Gent. With diuers shewes of his own inuention sette out at Norwich: and some rehearsal of hir Highnesse retourne from Progresse. Wherevnto is adioyned a commendation of Sir Humfrey Gilberts ventrous iourney." Sir. Chr. Hatton's crest. " Imprinted by—seruante to the right hon. Sir Chr. Hatton, Vizchamberlayne." Dedicated " To—Maister Gilbert Gerrard,—Attourney Generall." K iij, in fours. Hereunto is annexed, " A welcome home to M. Martin Frobusher, and all those Gentlemen and Souldiers, that haue bene with him this last iourney, in the Countrey called (Meta incognita) whiche welcome was written since this Booke was put to the printing, and ioyned to the same Booke for a true testimony of Churchyardes good will for the furtherance of Mayster Frobushers fame." L, four Leaues. Licensed. George Mason, Esq; Quarto.

" An Arithmeticall Military Treatise, named STRATIOTICOS: Com- 1579. pendiously teaching the Science of Nūbers, as well in Fractions as Integers, and so much of the Rules & Æquations Algebraicall, and Arte of Numbers Cosicall as are requisite for the Profession of a Soldiour. Together with the Moderne Militarie Discipline, Offices, Lawes & Dueties in euery wel governed Campe & Armie to be obserued: Long since attēpted by Leonard Digges Gent. Augmented, digested, and lately finished by Tho. Digges his Sonne. Whereto he hath also adioyned certaine Questions of great Ordinaunce, resolued in his other Treatize of Pyrotechny and great Artillerie, hereafter to bee published. Viuet post funera virtus." The cut of a camp. " Printed—1579." On the back are the arms of the lord Robert Dudley, Earl of Leicester, to whom this book is dedicated. Then, " The Preface to the Reader.---The Contentes.---

tentes.---The Bookes alreadie published by the Authoure.'---Bookes begon by the Author, hereafter to be published." This treatise begins with an explanation of the figures by the numerical letters. Before the third book is the author's armorial ensign. Towards the conclusion, mentioning certain tables of the different ranges of bullets, invented by Tartalea an Italian, he adds as below.* 191 pages: at the bottom is the phœnix, as to G. Harveij Gratulationes. On the back is the mermaid; under it, " Imprinted—dwelling in Thames Street, neere vnto Baynardes Castle, 1579." W.H.   Quarto.

1579. " A Newyeares Gifte, dedicated to the Popes Holinesse, and all Catholikes addicted to the Sea of Rome: preferred the first day of Ianuarie, in the yeare of our Lorde God, after the course & computation of the Romanistes, one thousand, fiue hundreth, seauentie & nine, by B. G. Citizen of London. In recompence of diuers singular & inestimable Reliques, of late sent by the said Popes Holinesse into England, the true figures and representations whereof are heereafter in their places dilated. —Printed—1579." This book is introduced with some Alexandrine verses " Ad Archipapistam." On the back are " The Contentes of the

* These are: A treatise of the art of Navigation. 2 Of nautical Architecture, to build ships for all burthens. 3 Commentaries upon the revolutions of Copernicus. 4 Of dialling. 5 Of great artillery. 6 Of fortification. " All these and other long sithens the Author had finished and published, had not the Infernall Furies, enuying such his Fœlicitie, & happie Societie with his Mathematicall Muses, for many years so tormented him with Lawe Brables, that he hath bene enforced to discontinue those his delectable Studies," &c.

" " The same among other was indeed by my father long practised,—as also by Reflection of Glasses to fire Powder, and discharge Ordinaunce manye miles distaunt. And such was his Fœlicitie & happie successe not only in these Conclusions, but in Optikes & Catoptikes, that he was able by Perspectiue Glasses onely scituate vpon conuenient Angles, in such sorte to discouer euery particularitie in the Countrey rounde aboute wheresoeuer the Sunne beames mighte pearse: As sithence Archimedes (Bakon of Oxforde only excepted) I haue not read of any in Action euer able by meanes natural to performe the like. Which partly grew by the aide he had by one old written booke of the same Bakons Experiments,—that came to his hands, though chieflye by conioyning continual laborious Practise with his Mathematical Studies. The which I thought not amisse to rehearse, to animate such Mathematicians as enioye that quiet & rest, my froward Constellations haue hitherto denyed me, to imploy their studies & trauels for Inuention of these rare seruicable Secretes. But such is my harde Destinie, that as Gods pleasure was to take my Father from me in my yong & tender yeares, and euen at that verie tyme when I began to grow capable of those Secretes, and himselfe (hauing bene long debarred his owne inheritaunce, & natiue Soyle being restoared) ment then immediately to return to his wonted places of Exercise, there to haue deliuered me experimentally the fruites of his long Trauels & Practises. So sithence his death, hauing fostered by studie & conference those Theorical sparkes Mathematicall from infancie by him impressed, after I grewe to some Maturitie of yeares & iudgement, fitte to enter into triall & practise of these Conclusions, by continuall Lawe Brables (torments as repugnant to my nature as the Infernal Furies to Celestial Muses) I haue for manye yeres bene so vexed & turmoyled, that of all those rare Conclusions & Secrets I haue scarsely hitherto had any time of repose or quiet to wade effectually in any one, saue onelye that of great Artillerie."

Booke:"

Booke:" which are tranfcribed below,* as i know not how better to defcribe the book. L, in fours; with a large cut of Bulls, Beads, Agnus Dei, &c. Licenfed: W.H. Quarto.

Theoria Analytics, by Everard Digby, in three books. Licenfed. 1579. Quarto.

" The Arte and Science of preferuing Bodie and Soule in all Health, 1579. Wifdome, and Catholique Religion," &c. as p. 906. Dedicated " To the mofte high, excellent, and renowmed Princeffe, Elizabeth,—Queene of Englande, &c. and in earth, vnder God, of this Church of Englande & Irelande chiefe Gouernoure.—Iohn Iones." The author of The Bathes of Bathes Ayde. S, in fours; but C is omitted. At the end of the table is the brazen ferpent, and " Imprinted at London.—Cum Priuilegio." Licenfed. W.H.+ Quarto.

" A moft hanaus & Traytorlike Fact of Thomas Appletree; with 1579. her Maiefties meffage by Sir Chriftopher Hatton. Licenfed. Quarto.

" Courfe of Chriftianitie, or as touching the dayly reading & medita- 1579. tion of the holy fcriptures; two bookes : written in Latine by Andr. Hipe- rius: tranflated by Iohn Ludham." Licenfed. Quarto.

" Exhortation for the comfort of thofe that are in the Agonie of 1579. Death. Alfo a praier & catechifme for the ficke, by I. S. Licenfed. Octavo.

" Three proper and wittie familiar letters, lately paffed betwene two 1580. vniuerfitie men, touching the earthquake in April laft, and our Englifh refourmd verfifying." With a preface of a well willer to them both. Cum priuilegio, &c. 49 pages. Licenfed. Quarto.

" Two other very commendable letters of the fame mens writing; 1580. both touching the forefaid artificiall verfifying, and certain other particu- lars; more lately deliuered vnto the printer." 69 pages. Quarto.

" The treafurie of the French tong, teaching the waye to varie all 1580. fortes of verbes. Enriched fo plentifully with wordes and phrafes (for the benefit of the ftudious in that language) as the like hath not before

---

* " A preface to the Reader, whiche, fheweth the reafon of the writing of this Bboke. The Argument of a Booke, or Letter fent to Cardinall Poole. A Pre- face to this Booke, made by the Authoure thereof. The Booke—fent to Cardinal Poole. The manner and meanes of the Popes beginning. The proofe thereof. Another touching the charitie of the Pope. The liues of 2 popes; Alexander II, and Gregory VII. Exceptions againft them. A comparifon betwixt Chrift & the pope. The Popes Wares or Merchandife. Leo, a child of noble houfe, and well brought vp, grew wicked by being a Pope. Pope Vr- bans prefent to the Emperour of Grecia. Lennoy of the Authour vpon the fayd 4 popes. The poyfoning of K. John. The- holy Mayde of Kent. The expofition of his miracles. Howe great enimies the Pope and his Legates haue bin to Chriftian Realmes, and how he hath bin expelled. Diuers letters Inuective againft the Pope."

bin publiſhed. Gathered and ſet forth by Cl. Hollyband, with ſpeciall priuilege." In his preface he ſays, " he had ſet forth De pronunciatione linguae gallicae, French Littleton, &c. Printed by Thomas Vautrollier."
                                                                                                                                 Quarto.

1580. "A SHORTE AND briefe narration of the two Nauigations and Diſcoueries to the Northweaſt partes called Newe Fraunce. Firſt tranſlated out of French into Italian by that famous learned man Gio. Bapt. Ramutius, and now turned into Engliſh by Iohn Florio: Worthy the reading of all Venturers, Trauellers & Diſcouerers. Imprinted—in Thames ſtreate, &c. 1580." Dedicated "To—Edmond Bray Eſq; High Sherife within hir Maieſties Countie of Oxenford.—25 June 1580.---To all Gentlemen, Merchants & Pilots." 80 pages. Licenſed. W. H. Quarto.

1580. "The Mirror of Mans life. Plainely deſcribing what weake mould we are made of: What miſeries we are ſubiect vnto: how vncertain this life is: and what ſhall be our ende. Engliſhed by H. Kerton." On the title-page is a neat cut of a gay young man on horſeback, and another meeting him, with this label, " O froath: O vanitie: Why art thou ſo inſolent? O wormes meate." Dedicated "To Anne, Counteſſe of Penbroke, mother vnto—Lord Compton." Who is told that this book was written above 360 years paſt. K, in eights. At the end, "Speculum humanum. Made by Ste. Goſſon." 6 eleven-lined ſtanzas. The mermaid on the laſt page." W. H.+   Octavo.

(1580.) A diſcourſe on the Earthquake which happened, 1580. By Arthur Golding.   Octavo.

1580. "SIVQILA. Too good to be true: OMEN. Though ſo at a vewe, Yet all that I tolde you, Is true, I vpholde you: Nowe ceaſe to aſke why? For I cannot lye. Herein is ſhewed by way of Dialogue, the wonderful maners of the people of Mauqſun,ˣ with other talke not friuolous. Seen & allowed &c. Printed—in Thames ſtreate, neere vnto Baynards Caſtell, 1580." On the back are the arms of Sir Chr. Hatton, to whom this book is dedicated, by Tho. Lupton. Herein we learn that the names eſpecially are to be è conuerſo conſtrued. 178 pages. "Imprinted—in Thamis ſtreet, &c. Cum gratia & Priuilegio Regiae Maieſtatis." W.H. Licenſed.   Quarto.

ˣ This country is another Utopia, and Siuqila a pious vertuous man, who weary with the wickedneſs and enormities of his own country, travelled to find out a country and people agreable to his own affection; and when he had travelled through the world, as he thought, chanced to meet Omen, dwelling in this Mauqſun, wherein, as he ſaid, " be ſuch bleſſed Byſhops, ſuch perfect Preachers, vertuous Miniſters, godly Gouernours, merciful Magiſtrates, iuſt Iudges, worthie Lawes, charitable Lawyers, honeſt Attorneys, pitifull Phyſitions, friendly Surgeons, liberal Lords, lowlie Ladies, gentle Gentlemen, louing Huſbands, obedient Wiues, humble Children, modeſt Maydes, diligent Seruauntes, ſuch good & plaine Dealing, ſuch Hoſpitalitie, Charitie, practizing of Godlineſſe, & ſtriuing to Do wel, that Siuqila did wonder at it, ſaying, It is too good to be true." As we learn from the preface.

" The

HENRY BYNNEMAN.

" The Second part and Knitting vp of the Boke entituled Too good to 1581.
be true. Wherein is continued the difcourfe of the wonderfull Lawes,
commendable cuftomes, & ftrange manners of the people of Mauqfun.
Newly penned & publifhed by Tho. Lupton. Printed—1581. Cum priuilegio. In a compartment with the queen's arms fupported by Fame
and Victory, at top; the Stationers' arms at bottom, beneath which is a
tablet, with "Septembris 6." H on one fide, and B on the other. The
Lion and Dragon in the lower corners. On the back are the arms of Sir
William Cicill, to whom this part is dedicated. D 1, in fours. Li-
cenfed. W.H.   Quarto.

" Lexicon Graeco-Latinum, Joannis Crifpini opera, &c. opera et 1581.
ftudio E. G. (*Edw. Grant.*) Cum gratia et priuilegio regiae maieftatis."
Embellifhed with the earl of Leicefter's arms, &c.   Folio.

" DISERTISSIMI VIRI ROGERI ASCHAMI ANGLI," &c. as p. 920. 1581.
Oo 4, in eights.

" Ariftotelis Ethicorum ad Nicomachum libri decem, in gratiam et 1581.
vfum ftudioforum breuiter et perfpicue per quaeftiones expofiti, per Sam.
Heilandum, Tubingenfis fcholae profefforem ethices."   Quarto.

" PLVTARCHI CHERONEI OPVSCVLVM DE LIBERORUM inftitutione. 1581.
Item: Ifocratis Orationes Tres. 1 Ad Demonicum. 2 Ad Nicolem.
3 Nicolis." The brazen ferpent. "Ex officina Typographica, 1581." D,
in eights. In very neat Long Primer Greek. W.H.   Octavo.

" SEB. VERRONIS FRIBVRGENSIS HELVETII, PHYSICORVM LIBRI X. 1581.
Nunc primum in lucem editi." The brazen ferpent. "Ex officina—Ty-
pographj. CIƆ.IƆ.LXXXI. Cum priuilegio Regiæ Maieftatis." N, in
eights. W.H.   Octavo.

" In P. Rami, regii profefforis clariff. dialecticae libros duos, Lutetiae 1581.
anno M.D.LXXII. poftremo fine praelectionibus aeditos, explicationem
quaeftiones; quae paedagogiae logicae de docenda difcendaque dialectica.
Pars prima. Auctore Fred. Beurhufio Menertzhagenzi, fcholae Tremo-
nianae prorectore."   Quarto.

" A Perfuafion from Papiftrie: VVrytten chiefly to the obftinate, de- 1581.
termined, and difobedient Englifh Papifts, who are herein named &
proued Englifh enimies, and extreme Enimies to Englande. Which perfuafion, all the Queenes Maiefties Subiects, fauoring the Pope or his
religion, will reade or heare aduifedlye & thoroughly, efpecially fuch as
woulde be counted friendes to Englande, that wifhe oure Princes profperitie, the fafegarde of the Nobilitie, the concorde of our Comunalty, and
the continuance of this our happy ftate & tranquilitie. Imprinted—
in Thamis Streete—1581. Cum priuilegio" &c. Dedicated " To the
moft merciful & prudent Princeffe, Elyzabeth, &c.—Tho. Lupton."
                                                                                          The

The queen's arms on the back of the title-page. 316 pages. Licenfed. W.H. Quarto.

1583. "VERBORVM LATINORVM CVM GRAECIS ANGLICISQVE CONIVNCTOrum, locupletiffimi Commentarij: Ad Elaboratum Guilielmi Morelii Parifienfis, Regij in Græcis Typographi Archetypum accuratifsimè excufi, Nouaq; vocum paffim infertarum acceffione adaucti, vt ftellulæ, quæ fingulis lucent paginis indicabunt. Confultis praeter ditiffima aliorum dictionaria, viuis etiam nonnullorum doctorum vocibus, quò Anglica verfio perfpicua magis fit, fructuofiorq; ad communem ftudioforum vfum emânet. Quid vtilitatis in his commentariis contineatur, quæq; confcribendi eos ratio à primo authore inita fit ex ipfius Morelii præfatione ftudiofi facillime percipient." The brazen ferpent, between the date 1583. " In ædibus—per affignationem Richardi Huttoni. Cum priuilegio" &c. The earl of Leicefter's arms on the back, to whom the book is dedicated. Under the arms are eight hexameters infcribed " Ad Honoratiff. Comitem Leiceftriae in Clypeum fuum Gentilitium." After the dedication, are a preface, a lift of authors, and fome Latin verfes, addreffed to the ftudious youth, by Abr. Fleming. Contains 1153[7] pages, double columns. W.H. Folio.

1583. " The firft foure bookes of Virgils Æneis, tranflated into Englifh heroical verfe, by Richard Stanyhurft. With other poetical deuifes thereto annexed." At the end, An epiftle of the printer, relating to the work. Dedicated to his brother, the lord baron of Dunfanye. 105 pages. Licenfed, 1566, and 1582. See Warton's Hift. of Englifh Poetry, Vol. III; p. 399, 401.

1583. " A DEFENCE of the fincere and true Tranflations of the holie Scriptures into the Englifh tong, againft the manifolde cauils, friuolous quarels, and impudent flaunders of Gregorie Martin, one of the readers of Popifh diuinitie in the trayterous Seminarie of Rhemes. By William Fulke, D. D." With anfwers to his oppofers. Dedicated to the queen. It has 532 pages, very neatly printed; with fome Greek and Hebrew letter; and 71 pages of his anfwers, &c. Printed with George Bifhop. Dr. Lort. Octavo.

1583. " An Aftrological [a] Difcourfe vpon the great and notable coniunction of the two fuperiour Planets, Saturne and Iupiter, which fhall happen the

---

[7] Swelled to this extent by printing the conjugations of the verbs moftly at length, as " Cúbito cúbitas cubitáui, cubitátum, cubitáre, prima breui, verbum frequentiuum, à Cubo, cubas." Alfo the genitive cafes of the fubftantives, efpecially thofe of the fourth declenfion, which have the addition of the article hujus, as " Cubâtus, huius cubâtus." Likewife the adjectives thus,

" Cubiculárius, cubiculária, cubiculárium, adiectiuum, prima & tertia bréuibus, fecunda & quarta longis." The illuftration of every phrafe alfo is made a feparate paragraph: feveral of the nouns and fome of the verbs however are abbreviated as in modern ufe.

[a] See the confutation of this by Tho. Heth, M. A. Printed by Rob. Waldegrave.

28 day

28 day of April 1583. With a Declaration of the effectes, which the late Eclipse of the Sunne 1582, is yet heerafter to woorke. Written newly by Richard Haruey: partley to supplie that is wanting in common Prognostications: and partley by prædiction* of mischiefes ensuing, either to breed some endeuour of preuention by foresight, so far as lyeth in vs; or at leastwise, to arme vs with pacience beforehande. Seene and allowed." Dedicated " To—Iohn, Bishop of London.—Here in London, this 23 Jan.[b] 1581." The discourse is addressed to his " verie good & most louing Brother, Master Gabriel Haruey, at his chamber in Trinitie Hall;" and contains 76 pages. Thereunto is added " A Compendious Table of Phlebotomie," &c. which ends on p. 84. " From my fathers in Walden, 6 Dec. 1582. Your louing brother euer at commandement, Richard Haruey." Colophon, " Imprinted—with the assent of R. W." (*Richard Watkins.*) Licensed. W.H. . Octavo.

Another edition with a line more in each page, but contained in the 1583. same number of pages; the quotations being in Latin, and printed on Italics, in the foregoing article, are here translated into English, and printed on Roman types. The dedication to this is thus dated, " Here in London these of Ianuarie, 1583." W.H.4 Octavo.

The New Testament. Lambeth List. Quarto. 1583.

" THESAVRVS LINGVÆ ROMANÆ & BRITANNICÆ," &c. as p. 937, 1584. &c. " Jmpressum Londini. 1584." The Latin Dictionary to 6 V, in sixes; the Historical part from 7 A to 7 M 5, in sixes. " Excusum Londini in ædibus Henrici Bynnemani Typographi. Anno salutis humanæ, CIƆ.IƆ.LXXXIIII. Cum serenissimæ Regiæ maiestatis priuilegio, ad imprimendum solum per annos xxi." W.H. Folio.

" Too good to be true," as p. 986. " At London—dwelling in 1584. Thames street" &c. The late William Bayntun, Esq; Quarto.

The dates of these two last articles seem to contradict what has been asserted, that Binneman died in 1583. The former of them, however, being in my own collection, i can avouch to be here copied as printed; and as both the title-page and colophon have the same date, it is not likely there should be any mistake in that: so that probably being printed about the latter end of 1583, it might be dated 1584, as is not unusual. And, if so, this will evince nearly the time of Binneman's death; which appears for certain to have been before the 8th of Jan. 1583-4, the day when Newbery and Denham, his assignes, delivered up certain copies, " which belonged to Hen. Bynneman deceased," to the Stationers' company, for the benefit of their poor.

Mr. Ames mentioned two books as dated 1587. The former of them " All the famous battles," &c. he printed with Francis Coldock; (p. 922.) but that is only the first part, and without date. Licensed to H. B.

* See Holinshed, p. 1356, &c. Edit. || [b] This date evidently is a misprint for 1587. || 1583.
                                               5 &   1 July,

1 July, 1577. The second part dated indeed 1587, is printed for Gab. Cawood, but does not mention by whom it was printed; and the type is very different from the former part. The other book, Dr. James's sermon, i have entered under the year 1578; after Maunsell.

―――― " A booke of the art and maner howe to plante and grafte all fortes of trees. Englifhed by Leonarde Mafcal." Two editions, licenfed. Quarto.

―――― " Granges Garden: Conteyning as well certaine verfes vpon fundry poyntes, in Metre, as alfo diuerfe Pamphlets in profe: Pleafant to the eare, and delightfull to the Reader, if he abufe not the fcente of the Floures." Dedicated " To—Lord Sturton—John Grange." Alfo, in verfe, " To the Court like Dames, and Ladie like Gentlewomen.— Tam Mineruæ quàm Veneri." Another, " Cuiquam." Then, two commendatory poems by C. G. and W. S. The running title of the former part, " Golden Aphroditis," of the latter, " Granges Garden." S, in fours. Licenfed, 1 July, 1578. Quarto.

―――― " An Hundreth fundrie Flowres bounde vp in one fmall Poefie. Gathered partly (by tranflation) in the fine outlandifh Gardins of Euripides, Ouid, Petrarke, Ariofto, & others: and partly by inuention, out of our owne fruitfull Orchardes in Englande: Yelding fundrie fweete fauours of Tragical, Comical, and Morall Difcourfes, both pleafaunt & profitable to the well fmellyng nofes of learned Readers. Meritum petere graue.— Imprinted for Richarde Smith." On the back, " The contents of this Booke." On the next leaf begins " The Printer to the Reader." Then, " Faultes efcaped. Correction." This collection begins with, " Suppofes: A Comodie, written in the Italian tongue by Ariofto, and Englifhed by Geo. Gafcoygne of Grayes Inne Efquire, and there prefented." Next is " Iocafta: A Tragedie written in Greke by Euripides, tranflated and digefted into Acte by Geo. Gafcoygne, and Fran. Kinwelmerfhe of Grayes Inne, and there by them prefented, 1566." The epilogue " Done by Chr. Yeluerton:" under which, " Printed by Henrie Bynneman for R- Smith." The paging and fignatures are continued from the Suppofes, and end on p. 164. Then, on p. 201, begins " A difcourfe of the aduentures paffed by Mafter F. I." This has the appearance of being printed at another prefs. Afterwards follow " The deuifes of fundrie Gentlemen.—And nowe to recomfort you and to ende this worke receyue the delectable hiftorie of fundrie aduentures paffed by Dan Bartholmew of Bathe, read it, and iudge of it." Ends on p. 445. " Imprinted—at London, for Richard Smith." Among Garrick's Plays; and W.H.+ Quarto.

―――― " An extracte of examples, Apothegmes and Hiftories. Collected out of Lycofthenes, Brufonius and others. Tranflated into Englifhe, and reduced into an Alphabeticall order of common places, by I. P. Imprinted—

Imprinted—for Humfrey Toie." Dedicated " To the ryght worshipfull, and his fingular good mafter, Doctor Humfrie, Vice Chaunceler of Oxforde, Deane of Glocefter, & Prefident of Magdalen College. Iohn Parrincheffe his humble feruant, &c.—Datum Bracklei.—1572.—Io. Parinchef." 232 pages, and a table at the end. W.H.   Octavo.

An anfwer to M. Fecknam's objections to Mr. Gough's fermon, preached in the tower, 15 Jan. 1570. See p. 886. My copy wants the title-page, but has Bynneman's device of the Mermaid at the end. To this anfwer are prefixed Fecknam's affertions and obiections, with a preamble, " To the right worfhipfuls, Sir Frauncis Iobfon Knight, Lieuetenaunt of the Toure, Sir Henrie Neuell Knight, and M. Pellam Lieuetenaunt of the Ordinaunce, geue thefe." Thefe affertions &c. end on fol. 17. On the back thereof, " Here enfueth the anfwere to thefe Affertions & Obiections of M. Fecknam. Made by L. T." Query, Laurence Tomfon. Then, the head title, " An Anfwer to certeine Affertions and Obiections of M. Fecknam fometime Abbot of Weftminfter, which he made of late againft a godly Sermon of M. Iohn Goughs preched in the Toure the xv of Ianuarie, 1570." This is addreffed to the fame perfons as the objections had been; and begins thus: " There was delivered vnto me vpon the 4. of March 1570, by a frend of mine, a little pamflet written by M. Fecknam," &c. The numbering of the leaves continued to fol. 79. W.H.+   Octavo.

" A Treatife wherein dicing, dauncing, vaine plaies or enterludes, with other idle paftimes, &c: Commonly vfed one the fabboth daies, are by the worde of God, and auncient writers, reproued. By John Northbrooke preacher."—for Geo. Bifhop.   Quarto.

" The readie path to the pleafaunt pafture of delitefome, and eternall Paradife, &c." Head title. It begins with an addrefs " To the Reader;" At the end thereof, " God faue our Noble Queene Elyzabeth, and fende continuall peace amongft hir louing fubiectes. I. T." 54 leaves. " And to conclude, prayer is the readie path to the pleafaunt pafture of eternall Paradife." On another leaf are fome verfes by " P. D. In prayfe of the Author." Bought of Mr. John Judfon. Licenfed. W.H.+   Octavo.

" A Letter fent by J. B. Gentleman, vnto his very Frende Mayfter R. C. Efq; wherein is contained a Large Difcourfe of the Peoplirg & Inhabiting the Countrie called the Ardes, & other adiacent, in the North of Ireland; and taken in hand by Sir Thomas Smith, one of the Queenes Maiefties Priuie Counfel, and Tho. Smith Efq; his fon." 31 leaves. See Camden's Elizabeth, An. 1572.   Octavo.

He had licenfes for printing alfo, in 1566, " An almanake & prognoft. of M. Mychell Noftrodamus.—of M. Elis Bomelius. The founde of the trūpe.

trūpe. Rofences Pfalmes in Laten. The reſt of the egloges of Mantuan." See p. 988. In 1567, " An almanack & prog. of Hen. Rochforth. Songes & Sonnetes by Tho. Bryce. The fearefull fantyſes of the florentyne Cowper." In 1568;. " A book in Duche." In 1569, " Epiſtolæ Cantabrigienſes. Wytty fayned Saynges of men, beaſtes & fowles marylized. The wonderous workes of God, and man. The mooſte pleaſaunt hyſtory of Peſiſtratus & Cateanea. The ruffull tragedy of Hemidos & Thelay, by Ryc. Robynſon. The triumphante of the Grene Dragon, made vpon the Erle of Pembroke. An epytaphe of the Erle of Pembroke, made by Mr. Hewſon. A dyſcourſe on the ouermoche crueltye of a wydowe towardes a yonge man. Coloquium Eraſmi, in Latine. Terence, in Lat. (Ult. Marcij 1573, changed this with Mr. Marſhe for Stowes Chronacle, per licence.) Cato, in Lat. (Changed this likewiſe for Hilles booke of Gardeninge per licence.) Sententie pueriles, in Laten. Confabulationes Heſſe." In 1570, " The dyſcourſe of Sypers & Candy. The welcome home of Dr. Storye. An alm. & progn. of Geo. Goſcyne.—Of Mr. Mouſlowe. The petifull hiſtory of ij louyng Jtalyons. The marynors boke. A patrone for parentes. An oration pronounced before the Frynſhe Kynge." See this printed by R. Jugge, p. 720.---
In 1576, " The Touchſtone of Trauailers in Arithmetick & Geometrie." In 1577, " A briefe difcourſe of the accidentes of the Deathe of Mr. Serieant Louelace. The diſcription of a monſtrous childe, named John Tremley. The lamentable hiſtorie of the deathe of ij horſes, ſometime ſeruauntes to Nicholas Sinbor Hackeneyman. The praiſe of Solitarynes. Tarltons Tragicall treatiſes, conteyninge ſundrie difcourſes & conceiptes, bothe in proſe & verſe." In 1578, " The Arte of ſhootinge in great ordinaunce, written by Will. Borne. A requeſte to the kinge of Spaine, and the lordes of the Counſell of the Eſtate, by thinhabitauntes of the Lowe Countrey, 22 June, 1578. A Godlie & pithie exhortation made to the Judge & Juſtiſes of Suſſex at thaſſiſes there by William Ouerton D. D. and one of the Juſtices of peace there." In 1579, " The inſtruction of a Chriſtian man in vertue & honeſtye." In 1580, " A warninge for the wife, a feare for the ſonde, a brydell for the lewd, and a glaſſe for the good: wrytte of the late earthquake in London & other places for the glory of God & benefit of men that warylye can walke & wiſely iudge. Euerardi Digbei de duplici methodo. Lib. primus. An admonitio concerninge thearthquake. A Croniele in Latin verſe, by Mr. Ocland." See p. 909. " The deſcription of a monſtrous childe borne at Fenny Stahton, Hunt'ſhire. A ſhorte treatiſe verie comfortable for thoſe Chriſtians that be troubled & diſquieted in their Conſciences, &c. byTho. Sparke Archdeacon of Lowthe." This probably he printed for, or aſſigned to R. Newbery. See p. 907. " Euerardi Digbei Cantabrigiani admonitio Mildapetti de vnica. P. Rami methode retinenda Reſponſio. Tenne bookes of the Iliades of Homer. A briefe collection of all the notable & materiall thinges conteyned in the hiſt. of Guicciardin, being very neceſſarie

cessarie for parliament, Councell, Treatises & Negociations. The stage of popishe Toyes: written by G. N. Johānis Ravisij Textoris Nivernensis Dialogi aliquot. Eiusdem Epigrāmata aliquot." In 1581, " A true reporte of tharaignemente & execution of the late popish traitor Euerit Haunce. A view of the seditious Bull," &c. See p. 909. In 1582, " Oratio ad pontifices per Nicholaū Grenaldum." In 1583, March, 26. " Aristotelis opera oīa coniūctim & diuisim. Homer, Gr. & Latine. Nouū Testamentū, Grece, ex impressione H. Stephani. Donatus in Politica Aristotelis. Limbinus in Ethica Aristotelis. Alwaies prouided that the said H. B. shall from tyme to tyme chose & accept 5 of this companie to be parteners with him in the imprintinge of these bookes, and that if any of Aristoteles bookes be printed separate, except his Politikes & Ethickes, to pay vnto the company for euery such booke vj.d."

THOMAS PURFOOT, or PURFOOTE,

WAS an original member of the Stationers' company, and seems to have been of the old livery, as he serued collector in the years 1557 and 1558, and was fined with Nic. Boreman his compeer " for that they ded not gyue vp thare accounts at a daye appoynted, iijs. iiijd." In 1563, he was fined xij.d. for binding primers contrary to orders. Also, for keeping open shop on St. Luke's day. In 1566, he with Hu. Singleton were appointed to search after and seize books disorderly printed, and were allowed their charges when they rode abroad. By their means "Anthony Kydson was fined iijl. vjs. viijd. Tho. Marshe, ijl. Rec'd xxs. John Whyte, iiijl. Gerard Dewes, xxs. Abr. Vele, ijl. Humf. Toye, ijl. Rec'd xs." However upon the whole, and by the next entry on the company's book, Purfoot especially appears to have been no better principled than the rest; for he was fined vjl. xiijs. iiijd. and to find one sufficient surety for 100l. " for that he ded sell premers to the haberdaffhers, as was proued in 1565, as that doth more playnly apere by a decre noted amongst the boke of Copyes." In 1591, he seems to have taken his son into partnershippe with him, as 30 March, 1591, A table for the payment of tithes &c. was licensed to them jointly. And Dec. 6, 1591, at a court of Assistants it was agreed and ordered " That vpon the decease of Tho. Purfoot the elder, Stationer and Printer, Tho. Purfoot the yonger shalbe admitted a printer in the rowme of his said father, and shall from thensforthe vse the Art of printinge as his father did in his lyfe time." In 1592, May 6. " Tho. Purfoote the yonger was a suitor for the Clerkship when euer it might fall voyd," and his petition was granted; but in 1599, Dec. 3, the company taking into consideration, that for certain years he
had

had left his former trade, and betaken himself to another vocation, whereby they conceived that he wanted sufficient experience to serve in that place; and also that they found him disorderly and obstinate, in breaking the ordinances of the court, though admonished to the contrary; for these and divers other respects, they judged him an unfit person to have that office at any time. October 29, 1595, T. Purfoot (*the elder*) complained of Edw. Aldee, " for printinge of Breues, aperteyning to him the said Thomas. Yt is orderd by assent of the parties, That from tyme to tyme hereafter whensoeuer any breue or breues of letters patents shalbe offered to the said Edward to be printed, whiche shall apperteine to the said Thomas to print, by force of any entrance made, or to be made, the said Edw. shall refuse & forbeare to deale with the printinge thereof, & shall signifie to the parties trauayling therein, that the imprintinge thereof apperteyneth to the said Thomas. Item, that for all other breues, as for starch or otherwise, whiche shall not apperteine to the said Tho. the said Edw. shall not print any of the same, vnles yt be first alowed & entred vnto him. Item, that the said Tho. shall monethly hereafter enter, by allowance of the wardens, all such breues of letters patentes as he shall prynt; and pay the ordinary duties:" but 1 Dec. following, it was agreed, " that he shall, for all the said entrances, pay xijd. euery quarter to thuse of the house." How long this Tho. Purfoot lived afterwards, or followed businefs, i cannot at present say; as there are books printed with both names in almost a continual succession till 1637, when one of both these names is the third person of the 20, who were allowed, by a decree of the Star-chamber, to print for the whole kingdom. He kept shop in St. Paul's church-yard, at the sign of Lucrece, and dwelt within the new rents in Newgate market, and had a shop over against S. Sepulchres Church. Some books he printed for L. Harrison, which may be seen under his name.

1564. " A politique discourse for appeasing of troubles in the realme of Fraunce, shewing how requisite and necessary it is for the conseruation of the state of the croune."* Again 1569.* Again 1571, for L. Harrison. Octavo.

1566. The popish kingdome, or reigne of antichrist, and 4th monarchy. For Will. Pickering. See it by H. Denham, p. 945. Quarto.

1566. " An Exhortation to the Sicke, to bee rehearsed with a lowde voice, to those that be in the agony of death : with a Catechisme to instruct the Sicke. Translated out of French by Rob. Hesse." Octavo.

1566. " A dialogue between Experience and a Courtier, &c. By Sir David Lindsay." The running title " of the Monarche." See Hist. of English Poetry, Vol. II; p. 305. This, which is the first English edition i have met with, has several small neat cuts. The dialogue ends on fol. 104;
but

but it contains besides, The tragedy of David Beton, Cardinal & Archb. of St. Andrew's; The testament of king James V his popinjay; The dream of Sir D. Lindsay; The deploration of the death of Q. Magdalen; but the running title continues the same throughout. See it in 1575. With W. Pickering. Folios 154. W.H.+ Quarto.

" A Shorte Treatyse of the mysterie of the Euchariste: set furth by 1569. Lewys Euans.—Anno M.D.LXIX." This is introduced with an epistle to the reader, whom he informs, that since the printing his last book, (The castle of Christianity. See p. 944.) the Papists had reviled him as a turn-coat, and reported that he was mad, like as they had of bishop Jewel: " but thankes be to God, his honour is knowen to bee so farre from madnes, as they are from modestie," &c. At the end is " An aunswere to certen rebukes" &c. W.H. Octavo.

" A notable historye of Nastagio & Trauersari, no less pitiesull than 1569. pleasaunt, translated out of Italian into English verse by C. T. (Chr. Tye) Imprinted—in Poules churchyarde,—at the signe of Lucrece. 1569." Licensed. Hist. of Eng Poetry, III; 468.

" The Hatefull Hypocrisie, and rebellion of the Romishe prelacie. By 1570. Lewys Euans.—Anno M.D.LXX." On the back are extracts from Boner's preface and Gardiner's oration, De vera obedientia. In the epistle to the reader are recited the lewd practices of some of the priests and monks just before the monasteries were dissolved. Chap. 1. treats of the Hypocrisie; the 2d of the ignorance, iniquitie, & blasphemies of the Roman prelacy. At the end is annexed " A vewe of certaine rebellions, & of their endes;" but not The four paradoxes, as Ath. Oxon. 1; c. 179. E, in eights. On the last page is a cut of his sign, with " Lucrecia Romana" on a tablet. Under it, " Imprinted—in Paules Churchyearde, at the signe of Lucrece, by Tho. Purfoote." Licensed. W.H. Octavo.

" Stirpivm Adversaria Nova, perfacilis vestigatio luculentaq; ac- 1571. cessio ad Priscorum præsertim Dioscoridis & Recentiorum Materiam Medicam, quibus prope diem accedet altera pars. Qua, Coniectaneorum de Plantis appendix, De succis medicatis, et Metallicis sectio, Antiquæ & nouatæ Medicinæ lectiorum remediorū thesaurus opulentissimus, De succedaneis libellus continentur. Authoribus Petro Pena, & Mathiade Lobel, Medicis." This title on a copper-plate. At top, part of the Ptolemaic System, over the queen's arms: in the upper corners, devices on the authors' names. Under the title, a map of Europe. " Londini, 1570." Prefixed are, a dedication to the queen, the preface, and an index. Neatly printed on Roman types, with wood-cuts of the plants indented. Pages 458. " Londini, 1571. Calendis Ianuarijs, excudebat prelum Thomæ Purfœtij ad Lucretię symbolum. Cum gratia 'Priuilegij." Licensed. W.H. Folio.

" An

THOMAS PURFOOT.

1571. "An almanacke and prognostication for three years, that is to faye, for the yeare of oure Lord 1571, 1572, and 1573, nowe newlye added vnto my late rules of nauigation, that was printed 1111 years paſt. Practiſed at Grauſend for the meridian of London, by William Bourne, ſtudent of the mathematical ſcience." Octavo.

1572. "Remedies for diſeaſes in horſes. Approued and allowed by diuers verie learned mareſhalles." With a cut of a horſe. 12 leaves. Quarto.

1572. "A Shorte Dictionarie (*Vocabulary*) moſt profitable for Yong Beginners. The ſeconde tyme corrected, and augmented with diuerſe Phraſys & other thinges neceſſarie therevnto added: By Lewys Euans. ¶ Cicero, 3. de orat. Suauitate ſcientiæ nihil eſt homini iucundius.—Cum Priuilegio—ſolum. 1572." Dedicated to the earl of Leiceſter. Folios 78. Licenſed, 1564. W.H. Again 1579. Quarto.

1573. "Prælections vpon the Sacred and holy Reuelation of S. John written in latine by William Fulke, Doctor of Diuinitie, and tranſlated into Engliſh by George Gyffard. Ioan. 21, v. 24. This is the Diſciple, &c. Reuel. 22. v. 8. J am John, &c. Imprinted—1573." Dedicated "To—Lorde Ambroſe Dudley, Earle of Warwike," &c. Folios 152. At the end is a different cut of Lucrece, with his name on a tablet. "Imprinted—in Paules Churchyard:" &c. W.H. Quarto.

1574. "An Introduction to the loue of God, accounted among the workes of St. Auguſtine, and ſet forth in his name: very profitable to mooue men to loue God for his benefites receiued. Tranſlated by Edw. (*rather* Edm. Freke, *or* Freake) Biſhop of Rocheſter." Octavo.

1575. "A Theological dialogue, wherein the epiſtle of St. Paul to the Romans is expounded. Gathered & ſet togather out of the readings of Anth. Corranus of Seuilla, profeſſor of Diuinitie." Again 1579. Octavo.

1575. "The Pretie and wittie Hiſtorie of Arnalt and Lucenda: With certaine Rules & Dialogues ſet foorth for the learner of th' Italian tong: And Dedicated vnto the Worſhipfull, Sir Hierom Bowes Knight. By Claudius Holliband, teaching in Paules Churchyarde by the ſigne of the Lucrece. Dum ſpiro ſpero. Imprinted—1575." The novel occupies 305 pages; the Engliſh on one leaf, and the Italian oppoſite. Then follow "Certaine rules for the pronunciation of th' Italian tongue," &c. W.H. Sixteens.

1575. "A Dialogue betweene Experience and a Courtier of the miſerable ſtate of the worlde. Compiled in the Scottiſh tung by Syr Dauid Lindſey Knight, a man of great learning & ſcience: Firſt turned and made perfect Engliſhe: And now the ſeconde time corrected & amended according to the firſt copie. A worke very pleaſant & profitable for all eſtates, but

but chiefly for Gentlemen, & such as are in aucthoritie. Heerevnto also are annexed certein other works inuented by the sayde Knight, as may more at large appeare in a table following. Imprinted—in Paules Churchyards—1575." Folios 140. W.H. Again 1581. Quarto.

"The vvhole Summe of Christian Religion." &c. as p. 928. 1576.

"A Treatise of the Immortalitie of the Soule:—By Iohn Woolton." 1576. At the end, "Imprinted—for Iohn Shepperd." Octavo.

"A New Anatomie of the whole man, as well of his bodie as of his 1576. soule: declaring the condition and constitution of the same, in his first creation, corruption, regeneration, & glorification. Made by John Woolton, minister of the Gospell." Octavo.

"The true Coppy of such Conference as passed (concerning the com- 1577. ming to the Church, hearing of the diuine seruice, and of other controuersies of Religion) betweene W. S. gent. and Lewis Euans." Octavo.

"The Forrest of Fancy, wherein is contained very pretty Apo- 1579. thegmes, and Pleasant Histories, both in meeter and prose, Songes, Sonets, Epigrams, and Epistles, &c.—Finis H. C." See Hist. of Eng. Poetry, Vol. III; p. 386. Quarto.

"A Perfite Looking Glasse for all Estates: Most excellently & elo- 1580. quently set forth by the famous & learned Oratour Isocrates, as contained in three Orations of Morall instructions, written—in the Greeke tongue, of late yeeres: Translated into Latine by that learned Clearke Hieronimus Wolfius. And nowe Englished to the behalfe of the Reader, with sundrie examples of pithy sentences, both of Princes & Philosophers, gathered and collected out of diuers writers, coted in the margent, approbating the Authors intent, no lesse delectable then profitable. Imprinted—in Newgate Market, within the new Rents, at the Signe of the Lucrece. 1580." Dedicated "To—Sir Thomas Bromley Knight, Lord Chaunceller of England, &c.—Tho. Forrest," the translator. There are prefixed also, An epistle to the reader.---"The Authors Enchomion vpon—sir Thomas Bromley.---J. D. in commendation of the Author. --- Jn Praise of the Author.—S. Norreis.---The Booke to the Reader." 46 leaues. Licensed. W.H. Quarto.

"Ranarum & murium pugna, Latino versu donata ex Homero." 1580. By Dr. Chr. Johnson. Licensed. See Hist. of Eng. Poetry, Vol. III; p. 433. Quarto.

"An Introduction to the looue of God. Accoumpted among the 1581. workes of S. Augustine, and translated into English by—Edmund Bishop of Norvvitch, that nowe is, and by him dedicated to the Queenes most excellent

excellent Maieſtie, to the glorie of God, and comfort of his choſen. And newlie turned into Engliſhe Meter, by Rob. Fletcher, 1581. Mat. 22. (37-40) Thou ſhalt looue the Lord thy God alone, &c. Imprinted—and are to be ſolde at his ſhop without Newgate, ouer againſt Saint Sepulchers Church." The prefixes are, The dedication " To—Sir Frauncesse Knowles Knight, Maiſter Treaſurer of the Queenes Maieſties houſholde, &c.—To the Chriſtian Readers." Commendatory verſes by Th. Prat; Tho. Lee, & Jo. Breifilde. G 7, in eights. W.H. Octavo.

1581. " A Commentarie of M. J. Caluine vpon the Epiſtle to the Galathians: And tranſlated into Engliſh by R. V. Pray for the peace of Hieruſalem &c. Pſal. 122, 6. At London, Imprinted—and are to be ſolde at his ſhop ouer againſt S. Sepulchres Church, 1581. In the compartment with David and Moſes, uſed by Tho. Marſhe. Dedicated " To—Iohn my Lorde the Biſhop of London.—1 Aug. 1581. R. V." Then, " The Argument of the Epiſtle to the Galathians by M. Iohn Caluine." 191 pages; and two pages more of " Faultes eſcaped in the Printinge." Licenſed, with conſent of Mr. Harriſon and Mr. Biſhop. W.H. Quarto.

(1581.) " A Commentarie of M. Iohn Caluine vpon the Epiſtle to the Coloſsians: And tranſlated into Engliſh by R. V. Pray &c. Pſ. 122, 6. ſold at his ſhop ouer againſt S. Sepulchers Church." In the compartment formerly Whitchurch's, with the ſun at top. Dedicated " To—maiſter Noel, Deane of Poules, M. Mullins Archdeacon of London, maiſter D. Walker, Archdeacon of Eſſex, & maiſter Towers profeſſor of diuinity, his ſinguler good friends and patrons.—At high Eaſter, 1 of Nouember. —R. V." Then, " The Argument—by maiſter Caluin." 86 pages. W.H. Quarto.

1581. " The Huſbandlye ordring and Gouernmente of Poultrie." Practiſed by the Learnedſte, and ſuch as haue bene knowne ſkilfulleſt in that Arte, and in our tyme. Imprinted—for Garret Dewſe, 1581." Dedicated " To Mrs. Katherine, wife of Maiſter James Woodford Eſq. and cheeſe clarke of the Kitching to Q. Elizabeth." By Leon. Maſcall. Octavo.

1582. " A diſcouerie and batterie of the great Fort of vnwritten Traditions: otherwiſe, An examination of the Counſell of Trent, touching the decree of Traditions." Done by Martinus Chemnitius in Latine, and tranſlated into Engliſhe by R. V. Math. 15. Mar. 7. In vaine doe they worſhip me, teaching doctrines, the commaundements of men. Pſal. 122, 6. Pray for the peace of Hieruſalem, &c. Imprinted at London by him and Will. Pounſonbie." Dedicated, " To—M. James Altham Eſq. one of the Queenes Maieſties Iuſtices of the Peace in the Countie of Eſſex; and

*Herein he mentions having tranſlated all this commentary, and now ioyning the ſame Latin ſermon of Biſhop Jewel's, " which ford with the ſermon he thought it his duty to the breſneſs & ſmalneſs had paſt vndedi-|| dedicate both at once to his Lordſhip." cated hitherto; but now having tranſlated || However my copy has it not.

to

THOMAS PURFOOT

to the vertuous Lady, the Lady Iudde, his Wife.—At H. Easter, 26 Mar. 1582." "To the Courteous Reader," 8 pages. Licensed. W.H. Quarto. 1582.

"A Confutation of vnwritten verities, both by the holye Scriptures, and most auntient authours, and also probable Arguments and pithie reasons, with plaine answeres to all (or at the least) to the most part and strongest arguments, which the Aduersaries of Gods truth, either haue or can bring foorth for the proofe & defence of the same vnwritten Vanities, Verities as they woulde haue them called. Made by Thomas Cranmer, late Archebishop of Canterburie, Martir of God, and burned at Oxenforde for the defence of the true doctrine of our Sauiour Christ. The Contentes whereof thou shalt finde in the next side following. Imprinted at London by Th. Purfoote, and are to be solde at his shop, without Newgate, &c. 1582." The signatures continued from the foregoing article. 79 pages. W.H. Quarto.

"A View of vanitie, and Allarum to England, or retrait from sinne: 1582. In English verse, by Phil. Stubs." Licensed. Octavo.

"The well spring of sciences, which teaches the perfect work and 1583. practice of arithmeticke, &c. by Humphrey Baker, 1562." See it in 1591. Octavo.

"A very proper treatise, wherein is breefely set foorth the art of Lim- 1583. ming,—and how to make sundrye syses or groundes to lay siluer or golde vpon,—and to temper golde & siluer and other mettals, &c. to write or limme withall vpon Velym, Parchment or Paper,—and how to vernishe it when done, &c. verye meete to be knowne to all such as delight in Limming, or tricking of armes in their colours, and therefore to be adioined to the bookes of Armes. Imprinted at London by ——, the assigne of R. Totill, 1583. Cum Priuilegio." At the end, " Finished Anno Dom. 1573." C, in fours. W.H. Quarto.

"A profitable booke, declaring dyuers approoued remedies to take out 1583. spottes & staines in silkes, Veluets, Linnen & Wollen clothes. With diuers colours how to die Velvets, Silkes, &c. Also to dresse Leather, and to colour Felles. How to Gylde, Graue, Spwder, & Vernishe. And to harden & make softe Yron & Steele. Verie necessarie for all men;—with a perfite table: not the like reuealde in English heretofore. Taken out of Dutche, and englished by L. M. Imprinted at London by him and W. Pounsonbie, 1583." L. 3, In fours. The pages haue "Fol." before each number. Licensed. W.H. Again with the art of limming, in 1588, and 1596. Quarto.

"De Christo gratis iustificante, Contra Osorianam iustitiam, cæte- 1583. rosque eiusdem inhærentis iustitiæ patronos, Stan. Hosiü Andrad. Canisiü. Vegam, Tiletanü, Lorichium, contra vniuersa denique Turbam Tridentinam

tinam & Iefuiticam. Amica & modefta defenfio Ioan. Foxij." The flower de luce feeding; with "vbique floret," on the fides. "Londini, Excudebat Thomas Purfutius, impenfis Geor. Byfhop, 1583." It has prefixed "Ad afflictas & perturbatas fidelium in Chrifto confcientias epiftola authoris praefatoria." In three books; pages 436. Hereunto is added "Liber quartus, in quo fubfequitur gravis & erudita concio eximii doct. D. Guliel Fulfii, de duobus Abrahæ filiis, ex D. Paulo, Galat. 4. De lingua populari, in Latinum fermonem reddita per Ioan. Foxium." 47 pages. W.H. Octavo.

1583. The Queen's Injunctions and Articles, by affignment from Chr. Barker. Quarto.

1584. "A Sermon preached at Paules Croffe the ix of Februarie Anno Dom. 1583. By J. Hudfon, Maifter of Arte of Oxon. Imprinted—and are to be folde at his fhop ouer againft S. Sepulchres Church, 1584." Text, Hebr. x; 19. H 2, in eights. Lambeth Libr. Octavo.

1586. "A Sermon made in Latine in Oxonforde, in the Raigne of King Edwarde the fixt by the learned & godly father John Jewell late Bifhop of Sarifburie, tranflated—by R. V." Text, 1 Pet. 4; 11. See p. 998, note'. Octavo.

1586. "A proper newe Ballad, declaring the fubftaunce of all the late pretended Treafons againft the Queens Maieftie, and Eftates of this Realme, by fundry Traytors, who were executed in Lincolnes Inne fielde on the 20 & 21 daies of September 1586. To Wilfons new tune." Their names are at the end, For Edw. White.

1586. The bifhop of Chichefter's Vifitation Articles. Quarto.

1586. A Sermon at the Roles, of repentance & iudgement: on 2 Cor. 5; 10." Octavo.

1586. "A Sermon at Paules Croffe, 29 May 1586, by John Chardon, on Mat. 6; 19, 20, 21." Octavo.

1586. "A Letter or commiffion fent from Hell by Sathanas: to the Pope, Cardinals, Bifhops, Friers, Monks, &c. againft Chrift and his beggerly profeffors." Reprinted by him. Forbidden by the Abp. of Canterbury.

1587. "Antithefis or contrarietie betweene the wicked and the godlie, fet forth in forme of a paire of gloues, fit for every man to wear, though not on his handes, yet on his heart, as neceffarie meditations taken out of the word of God, by John Norden."

1587. "A funerall fermon, both godly, learned, and comfortable, preached at St. Maries in Cambridge, anno 1551. at the buriall of the reuerend
Dr.

Dr. and faithfull paftor of the churche of Chrifte, Martin Bucer. By Matthew Parker, D.D. and fince archbifhop of Canterburye. See p. 727.
Sixteens.

A brief to relieve Hector Walkington of Hayton in the county of 1589. Nottingham, 28 Mar. 31 Eliz. With a cut of the queen's head. This was to be collected only in the counties of York and Lincoln. Broad-fide.

" A letter written by the King of Nauarr, to the three eftates of 1589. Fraunce: Containing a moft liuely defcription of the difcommodities &c. dangers of ciuill warre: and a very forcible perfwafion to obedience, vnitie and peace. Together with a breefe declaration vpon the matters happened in Fraunce fithence 23 Dec. 1588. Tranflated out of French by G. R.—fould ouer againft S. Sepulchres—1589." 18 pages. At the end, " Giuen at Chatteleraud, 4 Mar. 1589." W.H. Quarto.

" The whole and true difcourfe of the enterprifes and fecrete confpi- 1589. racies, that haue been made againft the perfon of Henry de Valois, moft chriftian king of Fraunce and Poland. Whereupon followed his death by the hand of a young Jacobin frier, 1 aug. 1589;" &c. 12 pages. Harl. Mifc. IV; 229. Octavo.

" A double fortreffe of Faith, and of the Sacramentes, by L. Rofcio 1590. of Piemont." Licenfed; in French, and in Englifh. Octavo.

" THE WEL SPRING OF SCIENCES: Which teacheth the perfect vvorke 1591. & practife of Arithmeticke, both in whole Numbers and Fractions: fet foorth by Humfrey Baker Londoner. 1562. And novv once againe perufed augmented & amended in all the three parts, by the fayde Author: wherevnto he hath alfo added certain tables of the agreement of meafures & waights of diuers places of Europe, the one with the other, &c. Imprinted—1591." Dedicated " To the Gouernour, Affiftants, and the reft of the Companie of Marchaunts aduenturers.—2 Sept. 1580.---The prologue to the gentle reader." Ee 7, in eights. At the end is his deuice, Lucrecia Romana. Under it, " Printed—within the new rents in Newgate market, 1591." W.H. Octavo.

" A very golden Collection of comforts diuine & humaine, by which 1592. is fhewed how to ouercome all Aduerfitie: tranflated out of Duch." With R. Watkins. Octavo.

" A Rich Store-houfe or Treafury for the Difeafed. Wherein, are 1596. many approued Medicines for diuers & fundry Difeafes, which haue been long hidden, &c. Now fet foorth for the great benefit & comfort of the poorer fort of peoble, &c. By A. T. At London, Printed for him, and Ralph Blower, 1596." V 2, in fours. W.H. Quarto.

" Aftrolabium Vranicum Generale. A neceffary & Pleafaunt folace & (1596) recreation for Nauigators, Containing the vfe of an Inftrument or generall

rall Aſtrolabe: Newly: for them deuiſed by the Author, to bring them ſkilfully acquainted with the Planets,—and their courſes, &c. Agreeable to the Hipotheſis of N. Copernicus. With all ſuch neceſſary ſupplements for Iudiciall Aſtrology as Aſkabitius & Claudius Dariottus haue deliuered by their Tables. Wherevnto for their further delight he hath anexed another inuention, expreſſing in one face the whole Globe terreſtriall; with the two great engliſh voyages lately performed round the world. Compyled by Iohn Blagraue of Reading Gent. the ſame wellwiller to the Mathematicks. 1596. Printed—for Will. Matts." Dedicated "To— Lord Charles Haward, Baron of Effingham, Lord high Admirall of England, &c." I 2, in fours. W.H.+ Quarto.

1597. The Italian Schoole-maiſter, by Claudius Hollyband. ! . Octavo.

1597. "THE WHOLE COVRSE OF Chirurgerie, wherein is briefly ſet down the Cauſes, Signes, Prognoſtications & Curations of all ſorts of Tumors, Wounds, Vlcers, Fractures, Diſlocations & all other Diſeaſes vſually practiſed by Chirurgions, according to the opinion of all our auncient Doctours in Chirurgerie. Compiled by Peter Lowe, Scotchman, Arellian Doctor in the Facultie of Chirurgerie in Paris, & Chirurgian ordinarie to the—King of Fraunce & Nauarre. Wherevnto is annexed The preſages of Diuine Hippocrates." Dedicated " To—Iames the Sixte,—King of Scotland.—London, 20 April, 1597.---To the friendly Reader.---Omnibus clariſſ. Doctoribus Regii collegij chirurgicorum Pariſienſū, nempe D. Rodolpho Lefort," &c.---An epiſtle from W. Clowes, in commendation of the author. With verſes by G. Baker, Iohn Norden, and Tho. Churchyard. The names of authors. The Chirurgical part contains Ii 4, in fours.

1597. "THE BOOKE OF THE PRESAGES of deuyne Hyppocrates deuyded into three partes. Alſo the proteſtation which Hyppocrates cauſed his ſcholIars to make. The whole collected & tranſlated by Peter Low Arellian Doctor &c. 1597." Dedicated To—" Robert Lord Sempile, Sheriffe of Renfrowe," &c. "D, in fours. W.H." Quarto.

1597. "SYRINX, or a ſeauenfold Hiſtorie, handled with varietie of pleaſant & profitable, both commicall and tragicall, argument. Newly peruſed & amended by the firſt author, W. Warner. 1597." See Hiſt. of Eng. Poetry, III. p. 473. Licenſed in 1584. Quarto.

1598. The ſecond part of the loves of Hero and Leander, by Hen. Petowe. Ibid. p. 434. Quarto.

1598. "Briefe introductions, both natural, pleaſant. & delectable vnto the arte of Chiromancy, or Manuell Diuination, and Phiſiognomy; with the circumſtaces vpon the faces of the Signes. Alſo certaine Canons or rules vpon diſeaſes and Sickneſſes. Whereunto is alſo annexed aſwell the artificiall as naturall Aſtrologie, with the nature of the Planets.

nets. Written in the latine tongue by Iohn Indagine Prieft. And nowe lately tranflated by Fabian Withers." Q 7, in eights, " Imprinted—in St. Nicholas Shambles within the newe Rents, 1598." W.H. Octavo.

" A Briefe and moft eafie Jntroduction to the Aftrologicall Iudgement of the Starres.—Written by the moft famous Phifition Claudius Dariot: and tranflated by F. W. (*Fabian Withers*) Gent. And lately renued, and in fome places augmented & amended by G. C. Alfo hereunto is added a briefe Treatife of Mathematicall Phificke, entreating very exactly of the natures & qualities of all difeafes incident to humane bodies by the natural Influences of the Cœleftiall motions, Neuer before handled in this our natiue language. Written by the fayd G. C. practicioner in Phificke. London, Printed 1598." Dedicated by F. W. "To Maifter Edward Dodge Efq;—To the courteous Reader.—G C." The former tract contains V 3, in fours; the latter K 2, and a folding table. Licenfed in 1582. W.H. Alfo without date. Quarto. 1598.

" The true lamentable Difcourfe of the Burning of Teuerton in Deuon- fhire the Third of April laft paft, about the Hower of one of the Clocke in the Afternpone, being Market Day, 1598. At which Time there was confumed to Afhes about 400 Houfes, with all the Money & Goods that was therein : And fyftie Perfons burnt aliue through the vehemencie of the fame ᵇ Fyer." Quarto. 1598.

" The Fearfull Fancies of the Florentine Goopet. Written in Tufcane, By Iohn Baptift Gelli, one of the free Studie of Florence. And for recre- ation tranflated into Englifh by W. Barker. Penfofo d'altrui, Scene & allowed according to the order appointed. At London Printed—for the Companie of Stationers, 1599." This is introduced by an epiftle To the Reader, by the tranflator, 132 leaves. W.H. Octavo. 1599.

" A funerall oration made the xiiii day of January by John Hopet, the yere of oure faluation, 1549, vpon the texte written in the Reuelatyon of Saincte John Ca. 14." C 4, in eights. " Imprynted—in Paules churche yarde." W.H. Octavo.

" A fermon vpon the xci pfalme." L. Afkell for him. See p. 593. Licenfed, 1563.

" A fermon at Paules Croffe, by John Hudfon, on Heb. x; 19—25." Octavo.

" A Briefe and Compendious Expofition vpon the Pfalme called De profundis," &c. as p. 882.

ᵇ The fire was fo fodden and outrageous that in the fpace of an hour and a half, aboue 400 houfes and other buildings were confumed, and not aboue 20 left ftanding, among which were the church and court- houfe, belonging to the Earl of Devon- fhire; to the damage of 5 or 400,000l. This accident began by a poor woman's frying pancakes over a fire made with ftraw. See Oldys, Catal. of Pamphlets in the Harl. Libr. Number 147.

He had licenses besides for the following, viz. In 1561, "An ABC for chyldren." In 1564, "The Castell of Loue. An almanake & provost. of Hen. Rochforthe." In 1565, "A Cathechifme betwene an olde man & a chylde. The story of Jason, &c. translated out of Laten by Nyc. Whyte. A new yeres gyafte. The vij Masters of Rorhes. &c. An exclamation very profytable, gathered out of the Holy Scriptures, vnto the great cöforth of euery faythfull Sowle. Do all thinges to edefy the congregation." In 1566, "An almanake &c. of Hen. Rothforth. A play of fortune to eche one hyr cödicions & gentle manors, &c. A perpetual almanak, seruing for a memoryall. An almanak &c. of Will. Browne, for iij yeres: with serten rules of Nauigation. A monsterus fyshe taken in the est parts of Holland. A monsterus pygge by Market Rayson in Lynconshyre. A strāge syghte of the sonne, & in the elament at Baffell. Serten verces of Cupydo, by Mr. Fayre." In 1567, "A proper historye of ij Duche louers. The cosmography of Peter Apyan." In 1568, "An exortation, treatinge of the mesery of this worlde. Debytor & Credytor, made by Pele. Kynge Rycharde Cur de lyon. Generydis." In 1569, "The Some of Deuinite: boughte of John Sampson. A lamentable complaynte. A delysious Surupe newly claryfied for yonge scholers ÿ thurste for the swete lycore of Laten speache." In 1570, "A true copy of a letter sente from Ferrara."---In 1579, "Thabridgement of Frosardes Cronicle, written in Latyn by J. Sleidan & translated into English." In 1580, "The voyage of Ferdinando Maganaffes vnto the Molucos. Verses compiled by John Merquaunt to diuerse good purposes. De fide vera Chriftiani." In 1582, "Soe muche of a book in Frenche, intitled Metamorphose Chriftienne, fait par Dialogues par Pierre Viret, as doth not belong to any other of this companie. Soe muche &c. Le monde & L'empire, et le monde demoniacle, fait par dialogues, par P. Viret, as doth not belong &c. The articles betwene Mounsieur of Fraūce, and the states of the lowe cuntreyes." In 1583, "A report of the conquest of the Islandes of Terſora, by the marques of S. Cruz, in the behalf of kinge Phillip. A Sermon preached by John Judson Mr. of Artes, at Paules crosse, vpon Heb. x; 19." In 1585, "The kinge of Portugalles book. A sermon preached before the vj clerkes." In 1586, "The learnynge of Vertue: allowed to him for his copye, which was Agnes Pykeringes; printed 1573. Also, An old book of the xij pairs of of Fraūce. And, of Paris & Vienna. Licensed vnto him, A pleasant newe historye of a man coüerted into a newe kinde of purgatorie. The old booke of Valentine & Orson: prouided that the cūpanie shall haue them at his handes. The mappe or discription of Sutten, with certen verses in latyne thereunto added by Will. Malyn. The blasson of the papistes. Briefes of all Lres patentes under the Great Seale for gathering by reason of casualties &c. by Sea & Land. Articles to be enquired of the churchwardens & swornemen within the Archdeconry of St. Albons.

Epigāmata

THOMAS PURFOOT.

Epigrāmata Johānis Lelandi." In 1587, " Billes for pryſes at fencing, as Mrs. Pryſes & ſchollers pryſe &c. prouided that if there ariſe any trouble by this entrance, then P. to beare the charges." In 1589, " Certen obſeruac'ons for Latyne & Engliſh verſyfyinge by H. B." In 1590, " The abſtracts & copies of all ſuch Briefes & L'res teſtymoniall as the bp. of London ſhall ſubſcribe or ſeale for gathering the deuoſions of Her Maieſties louinge ſubiectes for the relief of anye perſon or perſons that ſhall happen to fall in decaye by loſſes or caſualties." 30 March, 1591. Tho. Purfoote the elder, *and* Tho. Purfoote the yonger. " A table for the paiement of tithes," &c. 23 June, " A true diſcourſe of an Ouerthrowe gyuen to the armye of the Leaguers, in Prouince." 30 Octo. " A brief collection or epitome of all the notable thinges in the hiſtorye of Guiccherdine," &c. 12 May 1592, To T. P. the elder, " A Golden collection of comfortes diuine & humane, by which a xpian maye be truely inſtructed howe to ouercome all kinde of aduerſities." 26. Feb. To both, " The Regiſter booke, or Application of the blanck Almanack to the true & orderlie keeping of the Regiſter booke of weddinges, chriſtenings & burials accordinge to the Queenes iniunctyons." 22 Dec, 1593, " Ceaſars Dialogues," Alſo for ſeveral ballads, at times.

ALEXANDER LACY

WAS made free of the Stationers' Company, 30 April, (1556) which being only five days before their charter was granted, may be the reaſon why his name is not found among thoſe annexed thereto. In 1564 he was fined xijd. for printing ballads, other men's copies: the only time he is found on the Black Liſt. He dwelt in Little Britain, and printed chiefly ballads, petty hiſtories, &c.

" A forme of Meditations, very mete to be daley vſed of houſeholders 1565. in ther Houſes in this dangerus & contagius time. Set fourth according to the Queens iniunctions. An exortation to the Syck," &c. as p. 584, &c. Bagford's Papers. Licenſed, 1563. Octavo.

" A poore mannes beneuolence to the afflicted Church. Actes 3, a. 1566. Gold & ſiluer haue J none &c. Jmprinted—in little Britaine, by Alex. Lacy. The 29 of Januarij, 1566." In the ſame compartment as " The Sycke Mans Salue," printed by John Day, p. 634. Part proſe, and part in the octave ſtanza. At the end, " Finis quoth John Pits." 12 leaves. Licenſed. W.H. Mr. T. Baker mentions an edition in 1562. Sixteens.

" The

ALEXANDER LACY.

—— "The complaint of a finner vexed with paine, defiring the ioye that euer fhall remain." Six ftanzas of 14 lines each. "Printed for Richard Applow, dwelling in Pater nofter row, hard by the Caftle Tauern." Licenfed, 1565. A broad-fide.

He had licences for the following, viz. In 1562, "A Dyolege betwene lyfe & deathe. Dyues pragmaticus, verye pretye for cheldren, &c. The xx orders of Collettes or Drabbys. A comendation of mufeke," &c. In 1565, "A tru certificat fente from Gibralter in Spayne of a wonderfull fyffhe. A new yeres geyfte, by Barnarde Garter. A replye agaynfte that fedicious papiftical written ballet late cafte abrode in the ftreets of London." In 1566, "A tru dyfcription of Enuie: with a book againft popery. A catechefme, teachynge the Chriftian fayth ocupyed in all the domynions of prynce Frederyke palfgraue of the Ryne." With variety of ballads, from 1560 to 1570.

THOMAS EST, ESTE, EAST, or EASTE,

WAS made free of the Stationers' company, 3 Dec. 1565, but did not come on the livery till 1 July 1594. He dwelt at divers places, &c. which induced Mr. Ames to doubt whether thefe were not the names of different perfons; but, as he is entered in the Regifter, when made free, by the name of Efte, or Eafte, i make no doubt of all thefe names indicating the fame perfon. (How many ways did Tottel fpell his name?) He had an affignment from Bird and Talis, to whom the queen had granted a patent for printing ruled paper for mufic: this had T. E. printed at the bottom of the leaves. At the end of fome of his books he printed his own coat of arms: Sable, a cheveron between three nags heads erafed argent, with a crefcent for difference; the creft, a black horfe paffant; the motto, MIEVLX. VAVLT. MOVRIR. EN. VERTV. QVE. VIVRE. EN. HONCTE. He printed fome years after 1600. His widow, or daughter, Lucretia, printed a book of Bird's mufic, in 1610, by the affignment of Will. Barley; and next year Tho. Snodham printed another by affignment of Will. Barley alfo, by which it may be prefumed that the faid Lucretia was married to Snodham: however, he afterwards printed moft of Eaft's copies.

1569. "Ouid his inuectiue againft Ibis," &c. as p. 979. With H. Middleton. W.H. Octavo."
1571. "THE Pfalmes of Dauid and others." Licenfed at the fame time with a ballet of The Myce and the Frogs. Octavo.
"M.D.LXXI.

THOMAS EAST.

"M.D.LXXI. The Most excellent workes of chirurgerie, made 1571. and set foorthe by maister John Vigon head Chirurgien of oure tyme in Italie, translated into english. Wherunto is added an exposition of straunge termes and vnknowen symples, belongyng to the arte. Imprinted at London by Tho. East & Hen. Middelton: (dwelling at London wall by the signe of the Ship:) And finished the 17 day of July." Dedicated —" To the earnest Fauourer of all good & godly learning, master Richard Tracie," by Bart. Traheron, the translator. A table prefixed, and the glossary at the end: contains besides, Fol. CC.lxx. W.H. Again 1580, and 1586. Licensed. Folio.

"OYPANOMAXIA, hoc est, astrologorum* ludus, ad bonarum artium, 1572. et astrologiae in primis studiosorum relaxationem comparatus; nunc primum illustratus, ac in lucem editus, per Guilielmum Fulconem Cantabrigiensem. Abacus & Calculi væneunt apud Guilielmum Iones, in longa officina, ad occidentalem Paulini templi portam. Londini, Per Tho. Eastum & Hen. Middeltonum, impensis Guil. Iones. 1572." Dedicated "Honoratiss. viro Dom. Guilielmo Cecilio—Academiæ Cantab. Cancellario dignissimo." G 2, in fours, except A, which has only two leaves.* W.H. Again 1573. Quarto.

"Common places of Christian Religion, Compendiously written by 1572. M. Henry Bullinger, and translated into English by Iohn Stockwood, Minister. Imprinted—by Tho. East & H. Middleton: for Geo. Byshop, 1572, Ianuarij, 11." Dedicated "To the no lesse vertuous then honorable, Henry Earle of Huntingdon, Lord Hastings, &c.—From Battel, 19 Sept. 1571." The author's preface is addressed "To—Prince William, Landtgraue of Hessen, &c.—At Zurich—1556---The translator to the Reader.—Battel, 7 Octob. 1571.---A Table of the chief points," &c. 252 leaves. W.H. Again 1581. Licensed. Octavo.

"A Discourse, Wherein is plainly proued by the order of time and 1572. place, that Peter was neuer at Rome," &c. as p. 878. Quarto.

"The Benefit of the auncient Bathes of Buckstones, which cureth 1572. most greeuous Sicknesses, neuer before published: Compiled by Iohn Iones, Phisition. At the kings Mede, nigh Darby. Anno salutis, 1572. Seene & allowed, &c. Ienuarij, xviii. Imprinted—by Tho. East &

---

* This game is designed to be played after the manner of chess. The table is in two divisions, each consisting of 360 squares, 12 by 30; the former marked on the margin with the signs of the zodiac, the latter with figures, 1—30. The moveable pieces, seven for each antagonist, have the characters of the seven planets. How much more commendable would it be for academics, &c. to amuse themselves with this, and such like scientific games, than with E. O, or any other insipid game of chance?

H. Myddleton for Will. Iones, And are to be fold at his long fhop at the Weft dore of Paules Church." Dedicated " To—George Earle of Shrewefbury: Lord Talbot:" &c. whofe arms are on the back of the title-page. There are prefixed, four Latin Sapphics by Chr. Carlile, eight Alexandrines by Tho. Lupton, a catalogue of the authors cited, and Jones's preface. 20 leaves. Annexed are " A prayer vfually to be fayd before Bathing."---The table. W.H. Quarto.

1572. " THE BATHES or Bathes Ayde: Wonderfull & moft excellent agaynft very many Sickneffes, approued by authoritie, confirmed by reafon, and dayly tryed by experience: with the antiquitie, cōmoditie, propertie, knowledge, vfe, aphorifmes, diet, medicine, and other thinges ther to be confidered & obferued. Compendioufly compiled by Iohn Iones Phifition. Anno falutis 1572. At Afple Hall befydes Nottingam. Printed—for will. Iones:—13, Maij." Dedicated " To—Henry Earle of Penbrooke, Lord Herbert of Kayerdid, &c." whofe creft is on the back of the title-page. Herein the author afferts thefe Baths to have been in ufe 2460 years, or thereabout; that the founder Blæydin Doyeth, or Bladud, the wife and eloquent Philofopher, who ftudied eleven years at Athens, was the ninth king of Britain after Brute: and afterwards gives his pedigree up to Adam. There are commendatory verfes prefixed by Chr. Carlile, and John Lud, archdeacon of Nottingham, in Latin; others in Englifh by Tho. Churchyard, and Tho. Lupton. Then the preface, and a lift of authors. 35 leaves, including " A Prayer made by the Authour, to be faide of all perfons diffeafed, meekely kneeling vpon their knees, before they enter into the Bathes." Then, A table; at the end of which is a fhort epiftle " To his friends, kinsfolkes, &c. From London, 1572." W.H. Quarto.

1572. " The Poft of the World," &c. See it in 1576. Octavo.

1575. " The breuarie of health" &c. as p. 738. Again 1587, and 1598. Licenfed. Quarto.

1576. " THE POST For diuers partes of the world: to trauaile from one notable Citie vnto an other, with a defcripcion of the antiquitie of diuers famous Cities in Europe. The contents doe farther appeare in the next * leafe folowing. Very neceffary & profitable for Gentlemen, Marchants, Factors, or any other perfons difpofed to trauaile. The like not heretofore in Englifh. Publifhed by Richard Rowlands. Jmprinted at London by Tho. Eaft, 1576." Dedicated " To—Syr Thomas Greafham Knight." Then, A prayer to be faid before any journey be taken in hand. The variety of Miles. After thefe is another title-page, " The Poft of the World. VVherein is contayned the antiquities & originall of the moft famous Cities in Europe, with their trade & trafficke: With their wayes & diftance of myles, from country to country: With the

* Rather on the back of this title-page.

true

true & perfect knowledge of their Coynes, the places of their Mynts: with al their Martes and Fayres. And the Raignes of all the Kinges of England. A booke right neceſſary & profitable for all ſortes of perſons; the like before this tyme not Imprinted. As the Bird is prepared to flye, So Man is ordained to labour & trauaile." The device, Mars, ſword in hand, ſeated on a car, drawn by two horſes: on the two near wheels are repreſented, a bull, and a ſcorpion. " Jmprinted—1576." On the back is an almanac for 17 years; then, a calendar of the 12 months. 112 pages. W.H.   Octavo.

" COMMON places of Scripture," &c. as p. 487. 192 leaves in-   1577. cluding the title, &c. W.H.   Sixteens.

" A Sermon preached before the right honorable Earle of Darbie, and   1577. diuers others aſſembled in his honors Chappell at Newparke in Lankaſhire, the ſecond of Ianuarie, Anno humanæ Salut. 1577. Gala. vi. Dum tempus habemus operemur bonum.—Jmprinted at London by T. Eaſt: the xiiii day of March, 1577." Dedicated " To—Henrie Earle of Darbie,—Lorde of Man and the Iles adioyninge, &c.—At Moberley, 8 Febr.—Iohn Caldwell, Parſon of Winwick." Text, Rom. xiii; 11—14. F 4, in eights. W.H.   Octavo.

" THE PRACTISE of preaching, otherwiſe called The Pathway to the   1577. Pulpet: Conteyning an excellent Method how to frame Diuine Sermons, & to interpret the holy Scriptures according to the capacitie of the vulgar people. Firſt written in Latin by—D. Andr. Hyperius: and now lately —Engliſhed by Iohn Ludham, vicar of Wetherſfeld, 1577. Hereunto is added an Oration, &c. Imprinted—1577." Dedicated " To—D. Iohn Elmar,—Biſhop of London.—28 Maye, 1577.---Andr. Hyperius to the Reader.—Marpurge, 10. Octob. 1552."---The faults corrected; and a table of contents. 181 leaves. " Jmprinted—dwellinge betweene Poules Wharfe: & Baynardes Caſtle."

" An Oration, as touching the lyfe & death of the famous & worthy man D. Andrewe Hyperius, penned & pronounced in a ſolemne aſſemblie of all the States of the Citie of Marpurge, by Wygandus Orthius: (27 Febr. 1564.) And done into Engliſh by Iohn Ludham, 1577. Hominis vita varijs fortunę vicibus aſsidue rotatur." The device of Mars &c. Dedicated " To—M. Alex. Nowell Deane of the Cathedrall Church of Sainct Paul in London." At the end are ſome verſes: " The godlie aliue, to D. Hyperius dead.---The ſhadow of Hyperius to the Reader." Licenſed. W.H.   Quarto.

" A moſt godly and heauenly meditation vpon the 80 pſalm of the pro-   1577. phet David. Written by that zealous martyre of Chriſt, Hieronymus
                              Savanorola,

Savanorola, who, for reproving the corrupt religion, &c. was burnt at Florence 1499. Translated into English. Dr. Lort. Octavo.

1578. "A Christian discourse vpon certaine points of religion, presented to the Prince of Conde: Translated by John Brooke of Ashe nexte Sandwich."* Octavo.

1579. "THE Christian disputations by Master Peter Viret: Deuided into three partes, Dialogue wise: Set out with such grace, that it cannot be but that a man shall take greate pleasure in the reading thereoff: Translated out of French into English by Iohn Brooke of Ashe. Math. vii; d, 21. Not euery one that saith" &c. His crest a black horse with a crescent on his shoulder, on a wreath encompassed with the motto as p. 1006. "Imprinted—1579." Dedicated "To—Edmund—Archbishop of Canterbury.---Iohn Caluyn to the Reader."---Some commendatory verses by Tho. Singleton.---Faults corrected. This work is divided into three ᵉ parts, each with a separate title-page, and table, but the leaves are numbered progressively to folio 304. On the last page is his coat of arms described above: under it, "Imprinted—dwelling by Paules wharfe, 1579." W.H. Quarto.

1579. "GVYDOS QVESTIONS, NEWLY ᶠ CORRECTED. Wherevnto is added the thirde & fourth booke of Galen, with a treatise for the helps of all the outward parts of mans body. And also an excellent Antidotary containing diuers receipts, as well of auncient as latter wryters: faythfully corrected by men skilfull in the sayd Arte. A vvorke both learned & profitable for Chirurgions, the lyke wheroff before this tyme hath not bene Printed." The black horse; his crest. It has prefixed, an epistle by "George Baker, Master in Chirurgery to the Reader," A list of authors. "A prayer necessary to be sayde of all Chirurgiens." 201 leaves. At the end, under the arms, "Imprinted—1579." Licensed. W.H. Also, without date. Quarto.

1579. "Of two Woonderful Popish Monsters, to wyt, Of a Popish Asse which was found in Rome in the riuer Tyber (1496) and of a Moonkish Calfe, calued at Friberge in Misne, (1528.) Which are the very foreshewings & tokens of Gods wrath against blind, obstinate and monstrous Papistes, Witnessed & declared, the one by P. Melancthon, the other

---

ᵉ Each part contains two dialogues: the first is entitled ", The Alchymee of Purgatorie;—bicause the ancient Panims, and our priests after their examples, doe trauayle to extract the fifte essence of the soules, which they melt in their furnaces of Purgatory; also, bicause from those furnaces our priests do drawe out the Philosophicall stone." The second dialogue is called "The office of the dead." The third is called " Anniuersaryies, or yeares mindes." The fourth, " The age of the Masse, and of Purgatory." The fifth is called " The Hels, bicause it declares how many hels the Papisticall doctrine hath forged for vs, and the lodgings & chambers that it hath there builded." The sixth, " The Requiescant in pace of Purgatorye."
ᶠ See p. 373.

by

THOMAS EAST.

by M. Luther: Tranflated out of French into Englifh by Iohn Brooke of Afhh, &c. Thefe bookes are to be fould in Powles Churchyard, at the figne of the Parat.ᵍ" With cuts of the two monſters. Licenfed. Quarto.

" A View of Valyance: Defcribing the famous Feates & Martial Ex- 1580. ploits of Two moſt mightie Nations, the Romans & the Carthaginians for the Conqueſt and Poſſeſſion of Spayne. Tranſlated ʰ out of an auncient Recorde of Antiquitie, written by Rutilius Rufus, a Romaine Gentleman, and a Capitaine of Charge vnder Scipio in the fame Warres. Very delightfull to reade, and neuer before this Tyme publyfhed." Dedicated " To the valiant Sir Henry Lee Knight, Mafter of the Armorie & Leafh." 50 leaves. Licenfed. Octavo.

" An Epitaph vpon the death of the worfhipfull Maifter Benedict (1580) Spinola, merchaunt of Genoa, and free denizen of England, who dyed on Tuefday the 12 of Julie 1580." In 22 four-lined ſtanzas, ſigned R. B. Tolerated to him. Broadfide.

" Orders taken & enacted for Orphans & their Portions." For (1580) Gab. Cawood. Sixteens.
" The Boke named the Gouernour," &c. as p. 455. W.H. Sixteens. 1580.

" A friendlie communication or Dialogue betweene Paule & Demas: 1580. wherein is difputed how we are to vfe the pleafures of this life: By Samuel Byrd, M.A. and fellow, not long fince, of Benet Colledge. Imprinted—for Iohn Harifon the younger,—1580." It is introduced with an epiſtle by S. Byrd to the reader. 80 leaves. W.H. Octavo.

" A SHORT DISCOURS Of the excellent Doctour and Knight, maifter Leo- 1580. nardo Phiorauanti Bolognefe vppon Chirurgerie. With a declaration of many thinges neceſſarie to be knowne, neuer written before in this order: wherunto is added a number of notable fecretes, found out by the faide Author. Tranflated out of Italyan into Englifh by Iohn Hefter, Practicioner in the Arte of Diftillation." The black horfe, &c. as above. Dedicated " To—Edwarde de Vere, Earle of Oxenforde," &c. whofe arms are on the back of the title-page.---" To the gentle Reader.---The Table of contentes.---How that our Phificke & Chyrurgerie is better then that which the Auncients haue comonly vfed." 64 leaves. At the end, " If any be difpofed to haue any of thefe afore-fayd compofitions redy made, for the moſt part he may haue them at Paules Wharfe, by Iohn Hefter practifioner in the Arte of diftillations at the fign of the Furnaifes." His coat of arms, as above. " Jmprinted—1580." See p. 841. Licenfed. W.H. Quarto.

ᵍ And. Maunfell dwelt there at that time.
ʰ By Tho. Newton of Butley, Chefhire. See Oldys's Catal. of Pamphlets. Nᵒ 265.

" CERTAINE

THOMAS EAST.

1580. "CERTAINE GODLY and very proffitable Sermons of Faith, Hope & Charitie. First set foorth by M. Barnardine Occhine of Siena in Italy, and now lately collected, and translated out of the Italian tongue,—by William Phiston of London Student. Published for the profit of such as desire to vnderstand the truth of the Gospell." The black horse, &c. Dedicated "To the famous & most reuerend Father in God, Edmond—Archbishop of Canterbury." This work is divided into 38 sermons, or rather sections; 19 on Faith, 8 on Hope, and 11 on Charity. 100 leaves. "Imprinted—betweene Paules Wharfe & Baynards Castle, 1580." The arms, as above. Licensed. W.H.   Quarto.

1580. "The poore mans Iewell, that is to say, A treatise of the Pestilence. Vnto which is annexed a declaration of the vertues of the hearbes Carduus Benedictus, and Angelica: which are very medicinable, both against the Plague, and also against many other diseases. Gathered out of the bookes of diuers learned Phisitions."—Dedicated by Tho. Brasbrige "To—Sir Thomas Ramsey Knight, Lord Maior of London.---To the Reader." D, in eights. "Imprinted—for Geo. Bishop, 1580." W.H. Octavo.

1581. "Theatrum Mundi.—For John Wyght." See p. 782, and 974. Octavo.

1581. "EVPHVES. THE ANATOMY OF WIT. Verie pleasaunt for all Gentlemen to read, and most necessarie to remember: wherein are contained the delights that Wit followeth in his youth by the pleasantnesse of loue, and the happinesse that he reapeth in age by the perfectnesse of Wisdome. By John Lyly, M.A. Corrected &¹ augmented."—Dedicated "To—Sir William West Knight, Lord De la Warre.---To the Gentlemen Readers.---To—the Gentlemen Schollers of Oxford." 88 leaves. Concludes thus: "I have finished the first part of Euphues, whom I left readie to crosse the Seas to England,—I hope to haue him retourned within one Sumer. In the meane season, I will staie for him in the Countrie, and as soone as hée arriueth you shall know of his cōming." The black horse, &c. "Imprinted—for Gabriel Cawood,—1581."

"Euphues and his England. Containing his voiage & aduentures, mixed with sundrie pretie discourses of honest Loue, the description of the Countrie, the Court & the māners of the Isle. Delightful to be read," &c. Dedicated "To—Edward de Vere Earle of Oxenforde, &c.---To the Ladies & Gentlewomen of England, Iohn Lyly wisheth what

¹ I cannot find when the first edition of this part was printed. Mr. Wood mentions an edition of Euphues and his England in 1580; but he has greatly mistaken these two books, or parts, and led bishop Tanner into the same error; supposing Euphues his England to be divided into two parts, &c. and his Anatomy of Wit to be a separate and subsequent work, whereas in fact it is the first part, and introductory to his England, as is evident by the conclusion of this Anatomy of Wit. See Ath. Oxon, Vol. I. c. 296.

they

they would.---To the Gentlemen Readers." Ll, in fours; but should be only Gg. Kk following signature Ee, by mistake. " Imprinted— for Gab. Cawood," &c. W.H. Quarto.

" A cōmentary on the Ephesians, by Nich. Hemingius: translated by Abr. Fleming." Licensed. Quarto. 1581.

" THE PATHWAIE TO MARTIALL Discipline, deuided into two Bookes, very necessarie for young Souldiers, or for all such as loueth the profession of Armes, latelie set foorth by Tho. Styvvard Gent. The first booke entreateth of the Offices from the highest to the lowest, with the lawes of the Field, arming, mustering, & training of Souldiers, with the imbattailing of such numbers to the greatest force of the like Regiment. The second booke entreateth of sundrie proportions & training of Caleeuers, and how to bring Bowes to a great perfection of seruice, with imbattailing of greater Regiments: also how to march with a Campe Royall: likewise how to encampe the same, with diuers Tables annexed for the present making of your battells, as otherwise to know how manie paces they require in their march & battels from 500 to 10,000. Imprinted— for Myles Jenyngs, in Paules Churchyard at the signe of the Bible, 1581." Dedicated " To—Lord Charles Howard, Baron of Effingham," &c. whose arms are on the back of the title-page.---The preface.---The Author to the Reader," in Alexandrines. 167 pages, with plans, and a table at the end. W.H. Quarto. 1581.

Again, " Now newly imprinted, and deuided into three bookes. Whereunto is added, the order and vse of the Spaniards in their martiall affairs: which Copie was lately found in the Fort in Ireland, where the Spaniards & Italians had fortified themselues.—The third Book comprehendeth the very right order of the Spaniards, how to traine, march & encampe, with diuers Tables therein contained. Londini excudebat T. E. Impensis Milonis Jenyngs." Quarto. 1582.

" Examples drawen out of holy scripture, with their application: also a briefe conference betweene the Pope & his secretarie, wherein is opened his great blasphemous pride. By John Marbecke." Licensed. Octavo. 1582.

" A compendious treatise, entituled, De re militari, containing principall orders to be obserued in martiall affaires. Written in the Spanish tongue, by that worthie and famous captain, Luis Gutierres de la Vega, citizen of Medina del Campo. Translated by Nicholas Lichefild.— 1 Jan."* Licensed. Quarto. 1582.

" The Historie of the Discouerie and Conquest of the East Indias, enterprised by the Portingales, &c. Set forth by Hernan Lopes, englyshed by Nich. Lichefild." 164 leaues. Licensed. W.H.+ Octavo. 1582.

6 C "The

1582. "The xi bookes of the Golden Affe," &c. as p. 938. "Imprinted—for Abr. Veale, 1582." After the dedication, is an epiftle "To the Reader.---The life of L. Apuleius.---The preface of the author to his fone Fauftinus, & vnto the Readers." Cc, in eights. W.H.+   Octavo.

1582. "BATMAN vppon Bartholome, His Booke De Proprietatibus Rerum, Newly corrected, enlarged & amended: with fuch Additions as are requifite, vnto euery feuerall Booke: Taken foorth of the moft approued Authors, the like heretofore not tranflated in Englifh. Profitable for all Eftates, as well for the benefite of the Mind as the Bodie, 1582.—Jmprinted—by Paules wharfe." The prefixes are, "The Prologue of the Tranflator," whofe arms, with fome verfes under them, are on the back of the title-page. "The names of the Authors to this booke, in what time & of what profefsion."---The dedication, "To—Lord Henrie Cary, of the moft noble order of the Garter Knight, Baron of Hunfdon, gouernour of—Berwicke, & Lord Warden of the Eaft Marches of England, anempteft Scotland, & one of the Queenes—Counfaile.—Your Honours Chaplaine at cōmaundment, S. B. M.--To the Reader." A table of contents. Another of the principal matters, and where to find them. A catalogue of the hardeft old Englifh words, &c. 426 leaves. This edition concludes with fome account of the author, as cited in p. 71, 72. The arms, &c. as above, to fill up the laft page. Licenfed. W.H.   Folio.

1583. "The Fountaine of flowing felicitie with the waters of life, &c. gathered by Ric. Powlter."   Sixteens.

1584. "A Treatise of Morall Philofophie," &c. as p. 814. "Firft gathered and partly fet foorth by Will. Baudwin, and now the fourth time fince that inlarged by Tho. Paulfreyman, &c. Imprinted—by Tho. Efte. 1584." Fol. 192. W.H.   Octavo.

1584. "The voiage of the wanderinge Knighte." See it without date.   Quarto.

1584. "Pafquine in a Traunce." as p. 703. Z 2, in fours.   Quarto.

1584. "A Regiment for the Sea," &c. as p. 783.   Quarto.

1585. "A nevv Attractiue: Containing a fhort difcourfe of the Magnes or Loadftone, and amongft other his vertues, of a new difcouered fecret & fubtill propertie, concerning the Declining of the Needle, touched therewith vnder the plaine of the Horizon. Now firft found out by Rob. Norman [k] Hydrographer. Herevnto are annexed certaine neceffarie rules for the Art of Nauigation, by the fame R. N. Newly corrected & amended

[k] This part was printed this year alfo for R. Jones.

THOMAS EAST. 1015

by M. W. B. 1581." A Ship under fail. " Imprinted—for Ric. Ballard, 1585." Dedicated " To—Mafter William Borrough Efq; Comptroller of her Maiefties Nauie.---To the Reader.---The Magnes or Loadftones Challenge:" in verfe. M, in fours, with fchemes and tables. Quarto.

" A Discovrse of the Variation of the Cumpas, or Magneticall Needle: 1585. Wherein is Mathematicallie fhewed the manner of the obferuation, effects & application thereof, made by W. B. And is to be annexed to The New Attractiue of R. N. 1585." The preface is addreffed " To the trauailers, Sea-men & Mariners of England.—At Limehoufe, 26 September, 1581. Will. Borough." H 2, in fours. Imprinted—for R. Ballard, and are to be folde at his fhop, at St. Magnus corner in Thames ftreate, 1585." W.H. Quarto.

" The Treasvry of Health," &c. as p. 360. Aa 4, in eights. 1585. Allowed conditionally. Octavo.

" A Briefe and Neceffarie Treatife touching the Cure of the difeafe 1585. called Morbus Gallicus, or Lues Venerea, by vnctions & other approoued waies of curing: Newly corrected & augmented by William Clowes of London, Maifter in Chirurgerie. Hipprocratis Aphorifmus, 1 Vita breuis, &c. Printed for Tho. Cadman,—1585." On the back, the author's arms; and at the end thofe of T. Eaft. 64 pages. Quarto.

" The Schole of Horfemanfhip, by Chr. Clifford, Gent." Quarto.

" The Golden Booke of Marcus Aurelius," &c. See it in p. 455, 1586. &c. Licenfed.

" The ciuile Conuerfation of M. Stephen Guazzo, written firft in 1586. Italian: diuided into foure bookes; the firft three tranflated out of French by G. pettie. In the firft is contained in generall the fruits that may be reaped by Conuerfation, and teaching hovv to knovv good companie from ill. In the fecond, the manner of Conuerfation meete for all perfons, which fhall come in anie companie, out of their ovvne houfes, &c. In the third—the orders to be obferued in Conuerfation within doores, betweene the hufband & the wife, &c. In the fourth—by an example of a Banquet, made in Caffale, betweene fix Lords & four Ladies: And now tranflated out of Italian by Barth. Young, of the middle Temple, Gent." Dedicated " To—the Ladie Norrice.—6 Feb. 1581. Geo. Pettie.---The preface to the Reader.—From my lodging neere Paules. G. P." 230 leaves. On the laft, under his arms, " Printed—betweene Pauls wharfe & Baynards Caftle, 1586." This was licenfed to R. Watkins in 1579 to be tranflated. W.H. Quarto.

" The vvhole worke of that famous chirurgion Maifter Iohn Vigo: 1586. Newly corrected by men fkilfull in that Arte. Wherevnto are annexed

6 C, 2 certain

certain works compiled & published by Tho. Gale, Maister in Chirurgerie." This edition is introduced with an epistle by Geo. Baker, Gent. who informs the reader, that the former edition being very erroneous, he had corrected some part thereof; but as he could not go through with it, being begun, fast enough for the printer, he committed the rest to one Rob. Norton, who hath shewed his skill and pains herein. 460 leaves: on the last, his arms, and colophon, as to the last article. Quarto.

1586. "CERTAINE VVorkes of Chirurgerie, newlie compiled and published by Tho. Gale, Maister in Chirurgerie." Printed—1586. On the back: "The Contents. 1 An Institution of Chirurgerie. 2 An Enchiridion, the cure of wounds, fractures & dislocations. 3—of wounds made with Gun-shot, &c. 4 An Antidotarie, the principall & secret medicines," &c. Dedicated "To—Lord Robert Dudley, Maister of the—horse, &c.—At my poore house in London, 16. Julie, 1563. Tho. Gale.---W. Cuningham Doctor in Phisicke, vnto his approued friend T. Gale.---Tho. Gale, vnto the friendlie Readers.---T. Gale vnto those that desire the knowledge of Chirurgerie.—20 Maie, 1563." P6, in eights. Quarto.

1586. "CERTAINE VVorkes of Galens, called Methodus Medendi, with a brief Declaration of the worthie Art of Medicine, the Office of a Chirurgeon, and an Epitome of the third book of Galen, of Naturall Faculties: all translated into English, by T. Gale, Maister in Chirurgerie. Printed—1586." Dedicated "To—Sir Henrie Neuell Knight, Lord a Bergauene.—7 Nouemb. 1566.---T. Gale—vnto those young men, which are desirous of knowledge in the Art of Chirurgerie." 138 leaves: on the back of the last, his arms, and colophon. Licensed, W.H. Quarto.

1588. "A rule how to bring vp Children:—wherein is declared, how the Father apposeth his Sonne in the holy Scripture," &c. Dedicated "To—the Earle of Huntingdon, Knight of the Garter, &c.—Iohn Lyster Clarke, late Curate of Leedes, and nowe Vicar of Thorpeach, wisheth &c.---To all Ministers &c.—Thorpearche, within the ancienty or liberties of Yorke, 12 Junne, 1587." This is a dialogue between a father and his son, wherein the Christian and relative duties are illustrated and proved by the scriptures. 139 leaves: on the last, are the printer's arms and colophon. W.H. Octavo.

1588. "A most excellent & compendious method of curing woundes in the head, and in other partes of the body, with other precepts of the same Arte, practised & written by that famous man Franciscus Arceus, Doctor in Phisicke & Chirurgerie; and translated into English by Iohn Read Chirurgeon. Whereynto is added the exact cure of the Caruncle, neuer before set foorth in the English toung; With a treatise of the Fistulae in the fundament, and other places of the body, translated out of Iohānes Ardern: And also the discription of the Emplaister called Dia Chalciteos, with his vse & vertues. With an apt Table, &c. Imprinted—for Tho.

THOMAS EAST.

Tho. Chadman, 1588." Dedicated " To—Iohn Bannifter Gent. Maifter in Chirurgerie, and practitioner in Phificke, William Clowes, & William Pickering, Gentlemen & Maifters in Chirurgerie.----The firſt preface to the friendlie Reader.----Aluarus Nonnius to the renowmed preacher B. Aria Montanus.----A Complaint of the abufe of the noble Arte of Chirurgerie.—Iohn Read:" in metre.----" The forme or figure of an inſtrument for the cure of the diſtort foote, &c. Fol. 66." Kk, in fours. W.H.   Quarto.

" The Arte of Warre," &c. as p. 784.   Quarto. 1588.

" MVSICA TRANSALPINA. SEXTVS. Madrigales tranflated of foure, 1588. fiue, & fixe parts, chofen out of diuers excellent Authors, vvith the firſt & ſecond part of La Verginella, made by Maifter Byrd vpon 2 ſtanz's of Ariofto, and brought to ſpeake Englifh with the reſt. Publifhed by N. Yonge, in fauour of fuch as take pleafure in Mufick of voices. Imprinted by—the afſignè of William Byrd, 1588. Cum Priuilegio Regiæ Maieſtatis." Dedicated " To—Gilbert Lord Talbot, ſonne & heire to —George Earle of Shrowefbury, &c. Earle Marſhal of England, &c. (whoſe arms are on the back of the title-page.)—London, 1 Octob. 1588." My copy contains only the Sextus; but by the table at the end, The Quartus contained 12 ſongs; Quintus No. 13—45. Sextus, 46—57. W.H.   Quarto.

" Pſalmes, Sonnets, and Songes of ſadnes & pietie, made into Muſique of fiue parts by William Birde, one of the Gentlemen of her Maieſties honorable Chappell."   Quarto. 1588.

" Songs of ſundrie natures: ſome of grauitie, and others of mirth, fitte for all companies & voices, compoſed into Muſique of 3, 4, 5, and 6 partes, by W. Birde."   Quarto. 1589.

" CONTRATENOR. Liber primus Sacrarum Cantionum quinque vocum 1589. Authore Guilielmo Byrd Organifta Regio Anglo. Excudebat T. Eſt ex aſſignatione G. Byrd. Cum priuilegio. Londini, 25 Octob. 1589." On the back, " Illuſtriſſimo Proceri, claroq; multis nominibus Edoardo Somerfet, Comiti VVoreeſtriæ, Mœcenati ſuo clementiſſimo.—G. Byrd." At the end, " Londini, apud T. Eaſt, typographum in vico Alderſgate." W.H.   Quarto.

A Catechiſme by Richard Jones, ſchool-maſter of Cardiffe. Licenſed. 1589.   Octavo.

" BASSVS. Of Duos, or Songs for tvvo voices, compoſed & made by 1590. Tho. Whythorne Gent. Of the which ſome be playne & eaſie to be ſung, or played on Muſicall Inſtruments, and be made for yong beginners. And the reſt—for thoſe that be more perfect in ſinging or playing: deuided into 3 parts. Now newly publifhed. Imprinted—by T. Eſte, the aſſignè of W. Byrd, 1590." Dedicated " To—maſter Francis Haſtings, brother to the—Erle of Huntington, (whoſe creſt is on the back of the title-page.)—19 Nouemb. 1590." G 2, in fours. On the laſt

last leaf, "—dwelling in Alderſgate ſtreet, at the ſigne of the black horſe." On the back, is the portrait of " Tho. Whithorne, Anno tætis ſuæ xl;" ornamented with four coats of arms, one at each quarter; his creſt at top, and motto, " Aſpra ma non troppo," at bottom. W.H. Quarto.

1590. " A Booke of the Arte and maner how to Plant & Graffe all ſorts of Trees, how to ſette Stones & ſow Pepins, to make wild trees to graffe on, as alſo remedies & medicines. With diuers other new practiſes, by one of the Abbey of S. Vincent in Fraunce, practiſed with his owne hands : deuided into vij Chapters,—with an addition in the ende, of certaine Dutch practiſes, ſet forth & Engliſhed by Leonard Maſcall." A cut of a man grafting."—Imprinted—for Tho. VVight, 1590." On the back, " The Booke to the Reader:" in metre. Dedicated " To—Sir Ihon Paulet Knight, Lord S. Ihon.—-To the gentle Reader.—-The Table. ---An exhortation to the Planter & Graffer :" with a cut of proper inſtruments. 84 pages, and an alphabetical table. W.H. Again 1592.
Quarto.

1590. " The firſt ſet of Italian Madrigalls engliſhed, not to the ſenſe of the originall dittie, but after the affection of the noate : by Tho. Watſon Gent. Where-vnto are annexed 2 excellent Madrigalls of M. VVill. Byrds, compoſed after the Italian vaine." W.H. Superius. Quarto.

1590. " M. Tullius Ciceroes three books of Duties," &c. as p. 808. Again, 1596; and without date. Sixteens.

1590. " The Eight Bookes of Caius Iulius Cæſar," &c. as p. 697. Contains 128 leaves beſides the prefixes. His arms at the end. " Imprinted—by T. Eeſt, in Alderſgate-ſtreete, 1590." W.H. Quarto.

1591. " Contratenor. Liber Secundus Sacrarum Cantionum, quarum aliæ ad quinque, aliæ verò ad ſex voces æditæ ſunt. Autore Guil. Byrd, organiſta regio, Anglo. Excudebat—ex aſſignatione G. Byrd. Cum priuilegio. 4 Nouemb. 1591." Dedicated, " Illuſtriſsimo Proceri—Domino Lumley," &c. E, in fours. W.H. Quarto.

1591. " A ſermon preached at Pawles croſſe by Geruaſe Babington, now Biſhop of Exeter, on Iohn 6; 37." Licenſed. Octavo.

1591. " Diuers & ſundry waies of 2 parts in one, to the number of 40, vpon one playn ſong. Sometimes placing the ground aboue, and 2 parts beneath; and other while, the ground beneath and 2 parts aboue, &c. performed & publiſhed by John Farmer." 24 leaves. " Printed by him, the aſſigne of W. Bird, & ſold by the author in Broad-ſtreet, near the Royal Exchange." Octavo.

" Guil.

"Guil. Damon, late one of her ma. mufitions, his pfalmes in four 1591.
partes, publifhed by William Swayne, gent." Quarto.

"THE WHOLE BOOKE OF PSALMES: WITH THEIR WONted Tunes, as 1592.
they are fong in Churches, compofed into foure parts: All which are fo
placed that foure may fing, ech one a feueral part in this booke: VVherein
the Church tunes are carefully corrected, and thereunto added other short
tunes vfually fong in London, &c. Compiled by fondry authors, &c.
Imprinted—in Alderfgate ftreete, at the figne of the black Horfe, and
are there to be fold. 1592." Dedicated "To—Sir Iohn Puckering
Knight, Lord keeper of the great Seale of England (whofe arms are
on the back of the title-page)—Tho. Eft.---The preface.—T. E."
268 pages; alfo, "A generall confeffion of finnes, with other praiers,
&c. A Prayer for the Queenes moft excellent Maieftie:" fet to mufic by
I. Douland. A Table. His arms & colophon on the laft page. W.H.
Again 1594; varying much in the Pfalm tunes, &c. Octavo.

"CONZONETS. or little fhort Songes to 3 voyces, by Tho. Morley Ba- 1593.
chiler of Mufique." Quarto.
His firft book of Madrigalls, to 4 voices. Quarto. 1594.

"CONTRATENOR. Songs and Pfalmes compofed into 3, 4, and 5 parts, 1594.
for the vfe & delight of all fuch as either loue or learne Mifficke: By
John Mundy Gent. bachiler of Muficke, and one of the Organeft of hir
Maiefties free Chappell of VVindfor." Dedicated "To—Robert Devo-
rax—Earle of Effex," &c. W.H. Quarto.
Alfo feveral books of mufick by him, Bird, Morley and Watfon.

"The Paffions of the Spirit." Mr. T. Baker's Maunfell. Licenfed. 1594.
Octavo.

"Quintvs. of Tho. Morley." The firft booke of Balletts to 5 voyces, De- 1595.
vice: a fcythe, rake and fork tied together: motto, Sed adhuc mea meffis in
herba eft. "In London by T. Efte. cɪɔ.ɪɔ.xc.v." Dedicated "To—
Sir Robert Cecill Knight.—xij Octob. 1595." W.H.+ Quarto.

"A RECORD of auncient Hiftories, intituled in Latin:' Gefta Roma- 1595.
norum. Difcourfing vpon fundry examples for the aduauncement of vertue, and the abandoning of vice. No leffe pleafant in reading, then
profitable in practife. Now newly perufed & ᵐ corrected by R. Robinfon,

¹ A tranflation of thefe was printed by W. de Worde. See p. 233.
ᵐ In the dedication he fays, "This worke before I toke it in hand, was both of imperfect phrafe in the hiftorie, & of indecent application in the Moralitie, bear-ing the title of the Hiftory of, Gefta Romanorum.—And when I had reformed &c. the feueral hiftories, & corrected the application of the Moralitie in many places, I added a forme of Argument before euery hiftory."

Citizen.

Citizen of London. Imprinted at London by T. Eſt.ⁿ 1595." Dedicated " To—Lady Margaret Duglas, hir grace, Counteſſe of Lineux.— Richard Robinſon." Contains 43 hiſtories, or tales; the applications ſpiritualized: 157 leaves. W.H.+ Alſo without date.   Octavo.

1596. " A Regiment for the Sea," &c. as p. 783. " Newly corrected & amended by Tho. Hood D. in Phiſicke, who hath added a new Regiment & Table of declination. Wherevnto is alſo adioyned the Mariners guide, with a perfect Sea Carde by the ſaid Tho. Hood." The ſhip, &c. but the Lincoln arms taken out. " Imprinted—for Tho. Wight, 1596." Dedicated now " To—George Earle of Cumberland, &c.—T. Hood.— To the Reader.—T. H.—To the Reader.—W. B."---A preface.—— " A table of the reigne of Kings, &c.—The Kalender.—A Table ſhewing the Prime, &c. from 1579 to 1603." Contains beſides, 81 leaves, and a table of contents.   Quarto.

1596. " The Mariners guide Set forth in forme of a dialogue, wherein the vſe of the plaine Sea Card is briefely & plainely deliuered, &c. Written by Tho. Hood." 22 leaves; the laſt has only his arms and colophon. Licenſed. W.H.   Quarto.

1596. " Foure Bookes of Huſbandrie," &c. as p. 783. " Printed by T. Eſte for Tho. Wight. 1596." W.H.   Quarto.

1596. Mary's Meditations. Mr. Tho. Baker's Maunſell.   Octavo.

1597. " Altvs. The firſt ſet of Engliſh Madrigalls, to 4, 5, & 6 voyces. Made and newly publiſhed by Geo. Kirbye. Printed by T. Eſte—in alderſgate ſtreet. 1597." 24 Songs. W.H.   Quarto.

1598. " Altvs. Madrigals to fiue voyces: Collected out of the beſt approued Italian Authors: By Tho. Morley Gentleman of hir Maieſties Royall Chappell." 24 Songs. W.H.   Quarto.

1598. " The firſt Set of Engliſh Madrigals to 3, 4, 5, & 6 voices compoſed by John Wilbie."   Quarto.

1598. " The choiſe of change," &c. by S. R. See it by Roger Ward.   Quarto.

1600. " Tenor. Madrigals to foure voices. Publiſhed by Tho. Morley. Now newly imprinted, with ſome Songs added by the Author. Printed by T. Eſte, the aſſigne of Tho. Morley. 1600." 22 Songs. W.H.   Quarto.

1600. " The Mahumetane or Turkiſh Hiſtorie, containing three Bookes. 1. Of the originall, &c. 2 Of their Conqueſts, &c. 3 Of the warres

ⁿ Mr. Warton mentions an edition of this book in 1577, printed by one Robinſon; alſo that there were ſix editions of it before 1601. There were three printers of the name Robinſon, Robert, George, and Richard; but i have not met with any book printed by either of them before 1585. Query, whether this editor be the ſame perſon with Richard the printer? See Hiſt. of Eng. Poetry, Vol. II; p. 18, 19. Alſo, A Diſſertation on the Geſta Romanorum, prefixed to Vol. III.

&

& feege of Malta. Heerevnto haue I annexed a briefe difcourfe of the warres of Cypres, at what time Selimus the fecond tooke from the Venetians the poffeffion of that Iland, and by reafon thereof I haue adioyned a fmall difcourfe conteining the caufes of the greatneffe of the Turkifh Empire. Tranflated from the French & Italian tongues, by *Raffe* Carr, of the middle Temple in London, Gent. Dedicated to the 3 worthy brothers, Rob. Carr, Will. Carr, & Edw. Carr, in the county of Lincolne, Efquires. Printed by T. Efte, in Alderfgate ftreete 1600." Each book is dedicated to one of thefe brothers feparately; and " The Narration of the warres of Cyprus," to them all jointly. 122 leaves. W.H. Quarto.

" CONTRATENOR. Pfalmes, Sonets, & fongs of fadnes & pietie, made into Muficke of fiue parts: whereof, fome of them going abroad among diuers, in vntrue coppies, are heere truely corrected, and th'other being Songs very rare & newly compofed, are heere publifhed for the recreation of all fuch as delight in Muficke: By W. Byrd, one of the Gent. of the Queenes Maiefties Royall Chappell." Device: A hind ftanding on a wreath, in a compartment fupported by Pallas and Mars, with this motto twined about it, " Cerva chariffima, et gratiffimus Hinnulus. Prov. 5." The creft of Sir Chr. Hatton, to whom this book is dedicated. " Printed by T. Efte, in Alderfgate ftreete, over againft the figne of the George." On the back are " Reafons by th'author to perfwade euery one to learne to fing." 35 pfalms & fongs. W.H. Alfo Baffus, &c. Licenfed conditionally in 1587. Quarto.

" Here beginneth the Kalender of Shepardes: Newly augmented & corrected." See p. 208. Folio.

The Image of both Churches. See p. 674, &c. Octavo.

" The hiftory of Ariodanto and Jeneura, dedicated to the king of Scots, in Englifh verfe. By Peter Bevarlay. Dated from Staple-Inn." Octavo.

" A Prognoftication for euer by Erra Pater," &c. as p. 931. 12 leaves. Sixteens.

" A plain Pathwaie to the Frenche Tongue." Octavo.

" A briefe Introduction to the fkill of Song; Concerning the practife: fet forth by William Bathe, Gent." Octavo.

" The new arriual of the three Gracis into Anglia, lamenting the abufis of this prefent age." By Stephen Batman. Quarto.

" Syr Beuis of Southampton." With cuts. Licenfed. Quarto.

" The ftory of the moft noble & worthy king Arthur: the which was the fyrft of the Worthyes Chryften: and alfo of his noble & valyaunt knyghts

knyghts of the rounde Table: Newly Imprynted & corrected.—betweene Paules wharfe & Baynardes Castell." O o 6, in eights. Licensed. Folio.

" The voyage of the wandering knight, shewing the whole course of mans life, how apt he is to follow vanity, and how hard it is for him to attaine to vertue. Devised by John Cartheny, a Frenchman; and translated out of French into English by W. G. (*Will. Gardyner*) of Southampton, merchant. A worke worthy the reading: and dedicated to the right worshipfull sir Francis Drake, knight." By Rob. Norman. This title is taken from an edition printed by Tho. Snodham, who married East's widow about 1611; my copy wanting the title-page. 128 pages. At the end, " Imprinted at London by T. EaR." Tolerated. W.H.↵
Quarto.

He had licenses also for printing the following, viz. In 1566, " The conqueste of synne, wherin is lamented the synfull estate of this presente age." In 1570, " Newes out of Ferrara." With Hen. Middleton.----In 1576, " Robinson's xpmas recreacons of histories & moralizacons aplied for our solace & consolacons." In 1577, " A Daintie nosegay of divers smelles, containing many pretie Ditties to diverse effectes: compiled in Englishe verse by Will. Tregeo. A damentable confession of M'garet Dorington, wief of Rob. Dorington of Westmynster, who was executed in the pallace at Westm. for murderinge Alice Foxe." In 1578, " The mirror of kings their good & princelie deedes. A straunge & wonderfull birth happened in Italie this presente yere, 1578." In 1579, " An epitaphe on the death of th'erle of Arundell." In 1581, " A swete comfort, willinge to die. The sume of the Scriptures. Jacob & his xij sones. Enchiridion, in Englifh. Oliver of Castell. Erra pater. Sir John Mandevile. The booke of Nurture. Exomedon. Arthur of little Britan. Prester Johns land. Foure Sones of Amon. Two edictes of Monf. the French kinges brother, in the lowe Countries." In 1582, " The 2d parte of The Mirror of Knighthoode, to be translated into Englishe, and so to be printed, conditionally notwithstandinge, that the same be perused, & yf any thinge be amiffe therein to be amended. Godly Seaflowers, comfortable for man's confcience: A safegard & healthe to the sole of all penitent synners " In 1586, " Vocabula M. Stanbrigij. nuper emendata." In 1587, " Articles to be enquired of the Churchwardens," &c. In 1592, " A plefant nosegyt plainelie set out & plainelie presented as a Newe yeres gyfte to the quenes maiestie at Hampton courte, An. Dnī. 1592." In 1593, " An approved medicine againft the decereaed Plague." Also, for fome ballets.

RICHARD

RICHARD WATKINS

WAS presented 13 Oct. 1556, as apprentice with William Powell; and made free of the Stationers' company, 27 April, 1557; came on the Livery, 29 June, 1570; served Renter in 1573 and 1574; Under Warden in 1576, and 1577; Upper Warden in 1580, and 1583; Master in 1589, and 1594. He dwelt in Paul's church-yard, and had a shop adjoining to the Little Conduit, in Cheap. He appears to have been a very orderly member; and though he seems not to have been so successful in trade as he deserved, owing probably to a too credulous disposition in giving credit, yet he appears to have maintained a fair character to the last: his name is not found once on the Black List of persons fined for breaking orders, &c. He had a patent with James Roberts for printing almanacs; and in the year 1583, gave up his interest in the sheet almanacs to the company for the use of their poor, after the example of other printers who had exclusive privileges. In 1593 he was chosen treasurer of Seres and Day's Privileges. On the 27 Feb. 1598-9, I find one Will. Leek mentioned as treasurer; probably on Mr. Watkin's resignation; who, 16 April following appears to have been in a languishing state of health, and rather low in his circumstances, seeing his wife attended at a Court of Assistants, that day, to crave a deduction in his accounts, as treasurer for the said privileges, of some bad debts, and to be allowed time to pay the residue. Whereupon Mr. Bonham Norton, and Mr. Tho. Wight, were appointed to peruse his accounts, and to deduct for all such debts as were justly made, so it were not to petty chapmen in the country; and for the remainder, he or his executors to have respite to pay the same: one half at Christmas next, and the other half the 26 Dec. 1600. He seems to have died soon after, as on the 4th of June, 1599, it was ordered by the Court, that Tho. Grantham, late apprentice with Mr. Watkins, shall serve the residue of his apprenticeship with Mess. Geo. Bishop, Ra. Newbery, and Rob. Barker; and at the end of his time to be made free at their charge.

" Tvvo bookes of Saint Ambrose Bysshoppe of Mylleyne, entytuled: 1561. Of the vocation & callyng of all Nations: Newly translated out of Latin intoEnglyshe, for the edifiyng & comfort of the single mynded & Godly vnlearned in Christes Church, agaynst the late strong secte of the Pelagians, the maynteyners of the free wyll of men, and denyers of the grace of God. By Henry Becher Minister of the Church of God. Allowed, according to the order, &c. Anno Christi, 1561." T 4, in eights. Imprinted—in Powles Church yarde by Rycharde VVatkins." W.H. Octavo.

" Principal

RICHARD WATKINS.

1570. " The Popish Kingdome, or reigne of Antichrist, written in Latine verse by Thomas Naogeorgus, and englyshed by Barnabe Googe. 2 Tim: 3. (8,9) Lyke as Jannes and Jambres withstoode Moyses," &c. Printed by H. Denham for him. See p. 945. Licensed. Quarto.

(1571.) " A Newe Almanack for the yeare of our Lorde God M.D.LXXII. Calculated for the Meridian of Oxenforde, By Tho. Hyll." With small cuts over each month in red and black. W.H.+ Broadside.

(1572.) An almanack for the year 1573, by Tho. Hill of London, student. Octavo.

1572. " Of ghostes and spirites, walking by nyght," as p. 971. Quarto.

1574. " The disclosyng of a late counterfayted possession by the deuyll, in twoo maydens, within the citie of London. Whereunto is annexed, part of a homilie of Chrisostome, and also certaine strange stories and practises, as well in England, as in other countries." Octavo.

1574. " The lyues of holy Sainctes, Prophetes," &c. as p. 947. Quarto.

1575. " A dialogue of Witches, in foretime called Lot-tellers, now comonly called Sorcerers; wherein is briefly declared whatsoeuer may be required touching that matter: translated from Lamb. Danæus." Octavo.

(1575) " A sermon* preached before the Queenes Maiestie, 2 Mar. 1575, on Ps. 131. By Dr. John Younge, bishop of Rochester." Octavo.

1577. " Foure bookes of Husbandry," &c. as p. 783. Frequently printed. Quarto.

(1578.) " An almanacke, and prognostication, made for the year of our Lord God 1579, which yeere is from the beginning of the world 5541. Made and written in Salisburie, by John Securis, maister of art and phisicke." With James Roberts. Octavo.

1579. " THE PROVERBES of the noble & woorthy Souldier Sir James Lopes de Mendoza, marques of Santillana, with the Paraphrase of D. Peter Diaz of Toledo: Wherein is contained whatsoeuer is necessarie to the leading of an honest & virtuous life: translated out of Spanishe by Barnabe Googe." 116 leaves. Licensed. Sixteens.

1579. " PHISICKE againſt Fortune, aswell prosperous as aduerse, conteyned in two Bookes.——Written in Latine by Frauncis Petrarch, a most

---

* This sermon seeming to reflect on some at Court, was printed with this preface. — " For that this sermon may be more aduisedly considered of the Readers, which was not well taken in part of some of the hearers, where it was spoken: It is thought expedient therefore that the preacher thereof should cause it to be put openly in print, & to refer it to be expended by the Learned, and others of ripe iudgment."

famous

famous Poet & Oratour: And now firſt Engliſhed by Tho. Twyne." In the compartment uſed by R. Jugge to The Calender of Scripture, 1575. Dedicated " To—Richard Bertie Eſq;---The preface of F. Petrarch, written unto Azo." 342 leaves, and a table at the end. " Imprinted— in Paules Churchyarde—*1579*." Licenſed in 1577 to be tranſlated. W.H. Quarto.

" An almanacke, and prognoſtication, made for the yeere of our Lord (1580.) God, M.D.LXXXI. Refered to the meridian of the citte of Cheſter. By Alexander Monſlowe." With J. Roberts. Octavo.

" An almanacke for ten years, beginning at the year 1581 : with cer- (1580.) tain neceſſarie rules, by Will. Bourne." With J. Roberts. Octavo.

Mich. Reniger De Pij quinti, &c. See it by T. Dawſon. Octavo. 1582.

" A new almanacke, and prognoſtication, for this yeere of Chriſt his (1583.) incarnation 1584, being leap year. Seruing generally for all England, but eſpecially for the meridian of this honourable citie of London. Gathered and made by Thomas Buckminſter." With J. Roberts. Licenſed to R. W. in 1569, 1570, 1581, 1582. Octavo.

" Leap yeare. A compendious prognoſtication for 1584. Collected (1583.) by John Harvey, &c. Directed to his very good and curtuouſe friende, M. Thomas Meade." With J. Roberts. Octavo.

" An Aſtrological Diſcourſe" &c. as p. 988. See " A manifeſt & ap- 1583. parent confutation" of this, " printed by R. Walde-graue, by the aſſent of R. Watkins." W.H. Octavo.
" An addition to the late Diſcourſe vpon the great coniunction of Saturne & Iupiter. By John Haruey. Wher-vnto is adioyned

THE learned worke of Hermes Triſmegiſtus, Intituled Iatromathe- 1583. matica, that is, his Phiſical Mathematiques, or Mathematical Phiſickes, directed vnto Ammon the Aegyptian. A Booke of eſpeciall great vſe for al Studentes in Aſtrologie & Phiſicke. Lately engliſhed by Iohn Haruey, at the requeſt of M. Charles P." C 4, in eights. Licenſed. W.H. Octavo.

" A Mirror for Mathematiques : A Golden Gem for Geometricians : 1587. A ſure ſafety for Saylers, and an auncient Antiquarie for Aſtronomers and Aſtrologians : Contayning alſo an order howe to make an Aſtronomicall inſtrument, called the Aſtrolab, vvith the vſe thereof : Alſo a playne and moſt eaſie inſtruction for erection of a figure for the 12 houſes of the heauens. A work moſt profitable, &c. By Rob. Tanner Gent. practitioner in Aſtrologie & Phiſick. Imprinted—by J. C. and are to be ſold in Pauls Churchyard, by Ric. Watkins, 1587." On the back are the contents.

tents. Dedicated "To—Charles Lord Howard of Effingham, Knight of the—Garter, Lorde Admirall of England:" &c. 56 leaves, with schemes. W.H. *Quarto.*

1587. "The Vertues & strange vse of a new Terra sigillata, lately found in Germanie; with the order of administring & applying it: written in Latine by Andr. Bertholdus, translated by B. G." For him. Again 1589; *Octavo.*

(1587.) "An almanacke, and prognostication, made for the yeere of our Lord 1588, being leape yeere. Rectified for the altitude and meridian of Dorchester, seruing most aptly for the west partes, and generally for all England. By Walter Gray, gentleman." With J. Roberts. *Octavo.*

1588. "A DISCOURSIVE PROBLEME concerning Prophesies, How far they are to be valued or credited, according to the surest rules & directions in Diuinitie, Philosophie, Astrologie, & other learning: Devised especially in abatement of the terrible threatenings & menaces peremptorily denounced against the kingdoms & states of the world, this present famous yeere, 1588, supposed the Great wonderfull, and Fatall yeere of our Age. By I. H. Printed—by Iohn Iackson for *him*." Dedicated "To—Sir Christopher Hatton Knight Lord high Chancelor of England.—At Kingslynn in Norfolke, 20 Aug. 1587.—Iohn ᵖ Harvey.---To the same ᵠ

ᵖ He appears to have been a younger brother of Gab. and Ric. Harvey of Saffron Walden. This first dedication seems to be the same as to his Annuall Kalender; for herein, after mentioning his patron's favourable acceptance of the labours of several learned men, and among others of his eldest brother, he says, "These, and such like, haue emboldened me, that I haue aduisedly præsumed aboue my simple qualitie to offer vnto your woorthy L. by way of humble & officious Dedication, my Annuall Kalender, or briefe Almanacke, for the famous prædestined yeere following; together with the Astronomicall Diarie, the compendious discourse vpon the Eclipses, and the short Astrologicall Prognostication thereunto appending. A slender Pamphlet, —at first (not fully five yeeres expired)— the late Lord Chancelor, your antecessor, —vouchsafed to accept, &c. May it likewise please your excellent L. to deigne me like fauour, in accepting this Schollerly exercise," &c.

ᵠ Herein he says, "Hauing lately præsumed to direct that slender pamphlet vnto your woorthy L. in modest hope of like hon. fauour: I continue the like boldnes in presenting your L. with a treatise not much vnlike, either for quantitie or qualitie.—No man knoweth better than your L. how notoriously the world hath been abused, from time to time, with supposed prophesies & counterfeit soothsayings, intending at least Comicall sturs, but commonly fostering tragicall commotions. Histories, beside a number of other famous books, & many old smokie paper books are very copious in affording examples of this couenous & imposturall kinde. Among which none more dangerous than those which daily experience eftsoones suggests; vpon euery new occasion reuiued to serue present turnes, & to feed the working humor of tumultuous heads. But of all the residue, What comparable to the terrible pretended prophesie, euen now notoriously in *Esse*, concerning the imagined mightie & wonderfull casualties & hurliburlies of the present yeere 1588? In which respect of so vniuersal fame, I was earnestly mooued & importuned—to vndertake some little trauell in examining the naturall causes & artificiall reasons of the said supposed prophesie."

Right

RICHARD WATKINS.

"Right Hon. very good Lord Sir Chr. Hatton Knight, &c.—At Kings Lin, 14 Jan. 1588." Licensed. W.H. Quarto.

" An Almanacke, or annuall Calender, with a compendious Prog- (1588) nostication thereunto appendyng, feruyng for the yeere of our Lord, 1589. —Referred to the longitude & fublimitie of pole Articke of the citie of London: by John Harvey, Maifter of Artes, & practitioner in Phificke. Long. 19°, 51′; Lat. 51°, 34′." At the end, "God preferue our gratious Queene Elizabeth." With J. Roberts. Octavo.

" An Almanacke & Prognostication, in which you may beholde the (1588) ftate of this yeere of our Lord God, 1589. Made & fet foorth by Ioha Dade, Gentleman, practitioner in Phificke." With J. Roberts. Octavo.

" Titi Liuij Petauini Romanae Hiftoriæ Principis, Libri omnes, 1589. quotquot ad noftram ætatem peruenerunt, poft varias doctorum virorum emendationes, & veterum & recentium exemplarium collatione fumma fide ac diligentia recogniti, & ad publicam vtilitatem denuo editi. Quibus adiuncta eft Chronologia noua, accōmodata ad tabulas Capitolinas Verii Flacci, annotationibus vtiliffimis, varietatem feu diffenfionem Authorum circa Confulum Romanorum nomina demonftrata, illuftrata. Cum Indice cuique parti feparatim adiuncto." Device: Abraham leading Ifaac to facrifice him; with this motto, DEVS PROVIDEBIT. Gene. Cha. 22. Under it, "Impreffum Londini, per Edm. Bollifantum, 1589." Very neatly printed in Brevier Roman, and the Epitomies, &c. on Italic types. At the end, "——impenfis R. Watkins." Kkkk, in eights. W.H. Octavo.

The art of Nauigation by Martin Cortes. See it in p. 719, &c. Li- 1589. cenfed. Quarto.

" Ioannis Twini Bolingdunenfis, Angli, de Rebus Albionicis, Bri- 1590. tannicis atque Anglicis Commentariorum libri duo. Ad Thomam Twinum filium. Cicero de Diuinatione, lib. 1. Quis eft quem moueat clariffimis monumentis teftata, confignatáq; Antiquitas? Londini, Excudebat Edm. Bolifantus pro R. Watkins 1590." The dedication thus inscribed, "Probitate, Literis, Ingenio, Nobilitate illuftriffimo, Roberto Sackvillo, Thomæ Buckhurftii Baronis honoratiffimi maximo natu filio ac hæredi: Tho. Twinus Artium ac Medicinæ Doctor, S. P. D.— Vale, Meridionali traiectu Lenifenfi. --- Authores," &c. 162 pages. Licenfed. W.H. Octavo.

" The Anatomie of the foule, containing godlie praiers vpon all the 1590. Pfalmes, tranflated by R. F." (Rob. Filles) See p. 699. Sixteens.

Florentii Wigornienfis Chronicon. Tho. Dawfon for him. Quarto. 1592.

"A

1592. "A Catechifme containing an inftruction vnto fower principall pointes of Chriftian religion by M. A." For him, and John Wolfe. Octavo.

1592. "A treatife of the temptation of Chrift, on Mat. 4; 1—11. by Ifaack Colfe." For him. Licenfed. Octavo.

1592. "A very golden collection of comforts," &c. With T. Purfoot. Octavo.

(1592.) "A new almanack, and prognoftication for 1593. Compofed according to lawfull and laudable art, and referred fpecially to the meridian and eleuation of the northeren pole of Canterburie, but may ferue vniuerfally, without any great error, for moft parts of England; by Gabriell Frende, practitioner in aftrologie and phifick;" who wrote yearly almanacks, and prognoftications, which were printed by Watkins and Roberts. Licenfed to R. W. in 1584. Octavo.

1593. "Oger Ferrier his aftronomical difcourfe of the Judgment of natiuities. Tranflated by Thomas Kelway, Gent." For him. Octavo.

(1594.) "An Almanacke & Prognoftication made for the yeere of our Lord God 1595.—Calculated according to art for the meridian of Norfolke, &c. by Rob. Wefthawe, Gent. Long. 20°, o'; Lat. 52°, 46'." With J. Roberts. Octavo.

(1594.) "A doble Almanacke, with a Prognoftication for the yeere of our redemption 1595.—Conteyning the olde Julian computation which we vfe; and the new Gregorian reformation vfed in foraigne countreyes: Collected by Gab. Frende, ftudent in Aftronomie." With J. Roberts. Octavo.

1594. "THE RESOLVED ʳ Gentleman. Tranflated out of Spanifhe into Englyfhe by Lewes Lewkenor Efquier. Nel piu bel vedere Cieco. Imprinted—1594." In the compartment ufed by Jugge to The Holy Bible, 1577. Prefixed are commendatory verfes by Maur. Kyffin, and Rob. Dillington. Dedication "To—Lady Anne Counteffe of Warwycke.---To the Reader.---To the—Imperiall Maieftie of great Cæfar. (*Charles*

ʳ This title conveys but a faint idea of the purport of this book, which "conteyneth examples of great pietie, laying open the vaine & deceitfull inftabilitie of this wretched worlde: and finally teacheth how to lyue vertuoufly, and dye bleffedly: Befides it maketh honorable mention of fundry excellent Princes: &c.

"See here laid open to thy fight & fence Th' Error & Terror of this wretched Life, Thy many Foes, the meanes for thy Defence, The glorious End fucceeding all this ftrife."

It was firft written in French by Oliver de la Marche, an ancient knight of Burgundy, in 1483, and fince tranflated into fundry languages; among the reft into Spanifh verfe, by Don Hernando de Acunna, whofe tranflation is here followed, as the original could not be met with; but not in verfe; which the Englifh tranflator allowed, as the fubject was moftly allegory, would have been far better; yet doubted not, but that thofe, who efteeme the fubftance more than the fhadow, would like it never the worfe for being delivered in plain fpeach.

the

RICHARD WATKINS.

the fifth)—Don Hernando de Acuna." On folio 54 begins. " A briefe explanation of some Histories but obscurely touched by the Author in this Treatise,". &c. which ends on fol. 73. Licensed. W.H. Quarto.

" A short introduction for to learn to swimme, gathered out of master 1595. Digbies booke of the art of swimming, and translated into English for the better instruction of those, who vnderstand not the Latin tongue, by Christopher Middleton." With wooden cuts of persons swimming. Quarto.

" A new almanacke, and prognostication, for the yeere of our Lord (1597) MDXCVIII. Gathered according to artificial rules, by Robert Watson, phisition." With Roberts. Cum privilegio. With the principal Fairs in England & Wales: gathered by M. G. O. Octavo.

" The Birth of Mankind, &c. Corrected and augmented." See p. 581. 1598. Also without date. 202 pages. Licensed.* W.H. Quarto.

" A doble Almanack or Kalender drawne for this present yeere, 1600. (1599) —The first Kalender seruing generally for al England, and the other necessarie for such as shall haue occasion of traffique beyond the Seas. Collected by Rob. Watson, practitioner of Phisicke in the towne of Brancktry, in Essex." With J. Roberts.* Octavo.

" The tragecall and pleasaunte history of Ariodanto and Jeneura 1600. daughter vnto the kynge of Scots." In verse. Hist. of Eng. Poetry, Vol. III; p. 479, note, l.

" The Newe Testament of our Sauiour Iesus Christ. Diligently (1600.) ouerseene, and faithfully translated out of the Greeke.
" The pearle which Christ commaunded to be bought:
Is here to be found, not els to be sought.
" Imprinted—Cum priuilegio—solum." It has, prefixed " An Almanacke for xvi yeeres;" giving only the golden number, dominical letter, and Easter day; beginning 1565. This and the title printed in red and black. The text is divided into verses. Notes in the margin.* W.H.+ Quarto.

" Prosper his meditations with his wife. Newly translated into English verse." Octavo.

" A Swoorde against swearyng, conteyning these principal poyntes. 1. That there is a lawful vse of an oth, contrary to the assertion of the Manichees & Anabaptistes. 2. How great a sinne it is to sweare falsly, vainly, rashly or customably. 3 That common or vsual swearing leadeth to periurie. 4. Examples of Gods iust and visible punishment, vpon blasphemers, periurers, and such as haue procured Gods wrath by cursing & banning, whiche we call execration. Imprinted—for Will. Towreolde

RICHARD WATKINS.

Towreolde by the aſſent of R. VVatkins, and are to be ſolde at his ſhoppe adioyning to the lytle Cunduite in Cheape." 48 leaves. Licenſed in 1577. W.H. Sixteens.

"A petite pallace of Pettie his pleaſure. Conteyninge many pretie hyſtories by him ſet forthe in comly colours & moſt delightfully diſcourſed." Licenſed in 1576. See Hiſt. of Eng. Poetry, Vol. III; p. 466.
Quarto.

He had licenſes for printing alſo, In 1569, " A book of Heauenly Recreation, or Comforth to the ſowle by Lady Knowell. A newe yeres gyfte to the rebellious perſons in the North partes of Englonde."---In 1579, To be tranſlated into Engliſh, " La ciuile conuerſation &c. Traduite d'Jtalien du S. Eſtienne Guazzo gentilhome de Eaſal, per Gabriell Chappius Tonrangeoys." Of this he delivered a copy, 27 Feb. 1580-1. See p. 1015. In 1580, 1 Nov. " A ſmall treatiſe of the Crinitall Starre, which appeareth this preſente moneth of October." In 1581, " iiij Almanacks, viz. of Buckmaſter, Twyne, Lloyd, & Kynnet." In 1582, For the ſame again. In 1587, " Certayne Godlye treatyſes cõcerning the end of the world, &c." In 1590, " A map, called A newe deſcription of the worlde, to be printed as well in Duche as in Engliſh. A plain & eaſie laying open of the meaning & vnderſtanding of the rules of conſtruccion in the Engliſh Accidens. A Grãmar with a Dictionary in iij languages, gathered out of diuers good authors, very profitable for the ſtudious of the Spanyſhe tonge, by R. Perciuall." In 1593, " Of Judgementes aſtronomicall vpon Natiuities, &c."

※※※※※※※※※※※※※※※※※※※※※※※※※※※※

JAMES ROBERTS

WAS made free of the Stationers' company 26 June 1564, but the regiſter mentions not by whom. He was joined with Ric. Watkins, in a patent for printing almanacs, about 1573; but he had licenſe to print Rob. Moore's almanac &c. for 1570, and ſundry ballads ſoon after. He came on the livery 25 June, 1596. On 27 Aug. following, he was fined, " for printing a book called Newe tydings, contrary to the known decrees of the Star Chamber, & the laudable ordinãuces of this Court, 6s. 8d. to be preſently paid"; but it was not paid till 11 Apr. 1597. It was ordered by the Court of Aſſiſtants, 1 Sep. 1595. " that he ſhould from henſforth ſurceaſe to deale with the printing of The briefe catechiſme, with the A. B. C, Letany, and other things inſerted, which he hath lately printed, contrary to her Maieſties prohibition, and the order of this Court. Alſo, that he ſhall preſently bring into the Hall all ſuch

leaues

JAMES ROBERTS.

leaues of the said book as he hath printed, to be disposed of as the next Court shall determine: to whose order he submitteth himself." As did Edw. Venge, who printed also some other leaves of the said book: but I find nothing more of this affair in the minutes of the next court, nor at any time after. In a dispute between him and Will. Wood, 26 March, 1599, it was ordered with their consent, " That I. R. shall deliver to W. W. all the books of Markhams Horsemanship, which the said I. R. printeth of this impression, and that the said W. W. shall pay him for the same, viij.s. a Ream for paper & printing; whereof 4 li. to be paid between this and Easter next, and the residue before the end of Easter term next. And that the said copye shall remaine to the said W. W; and, I. R, during his life, shall have the printing thereof, as often as it shall be printed hereafter." 26 March, 1602, he paid 5. l. to be excused serving the Rentership. He printed many books after 1600.

A list of such copies as belonged to James Roberts, from a M.S. of the late accurate and ingenious Mr. Coxeter, of Trinity coll. Oxon, which he lent Mr. Ames, is inserted ' below: it seems to contain his copy-right at

1 An abstract of the history of Cesar and Pompey.
2 The summe of divinitie. †
3 The booke of husbandry. †
4 Ridley's conference.
5 Marcus Aurelius. †
6 The old Algerisme. †
7 The argument of apparell.
8 A penny-worth of wit. †
9 A hundred merry tales. †
10 Adam Bell. †
11 The banishment of Cupid. †
12 Crowley's epigrams.
13 Calvine against the anabaptists.
14 The old governance of vertue, short.
15 Hooper upon the ten commandements.
16 Hooper's homilies.
17 King Pontus.
18 The pollicies of warre.
19 Robin conscience. †
20 A proude wives Pater-noster. †
21 The plowman's Pater-noster.
22 A sack full of newes. †
23 Sir Eglamore.
24 Gowre de confess. Amantis. †
25 The good shepheard & the bad. †
26 The conjectures of the end of the world. †
27 The christian state of matrimony.
28 The poore man's garden. †
29 Northbrooke's confession. †

30 The reward of wickednes.
31 Athenagoras of the resurrection.
32 Corvias postill.
33 The tower of trustines.
34 The castle of knowledge. †
35 The rosary of Christian prayers. †
36 The sweet sobs, & amorous complaints of shepheards & nimphs. †
37 The comfort of a true christian.
38 A replication to frier I. Frances. †
39 Twenty-six sermons of Bullenger, upon the first six chapters of Jeremy.
40 The image of love. †
41 The archdeacons articles for Middlesexe.
42 A mirror for mathematicks. †
43 Prayers collected by the lady Katherine Parre, queene, called, the sweet song of a sinner. †
44 The bills for players. †
45 The articles for the diocesse of Glocester and Bristol.
46 The beginning and ending of popery.
47 A sermon upon the 8th of Job, the 5, 6, and 7 verses. †
48 The history of Palmendos and Primaleon, in seaven bookes. †
49 A compendious forme of domesticall duties.
50 A looking glasse for England and the world.
51 A

JAMES ROBERTS.

at the latter part of his life, but has no date. The Company's Register, ult. Maii, 1594, allows him certain copies, which were John Charlewood's "Saluo Jure cuiuscūq;" they are marked with †. The Register has some others, not in Mr. Coxeter's list, which are added with this mark ‡ before them.

The several almanacks he printed with Richard Watkins, see under his name.

1594. "Five sermons by Geo. Phillips. (1) A recreation for the soule, on Col. 3; 16. (2) The end of vsury, on Habac. 2; 9. (3) The armour of patience, on II. Tim. 2; 3. (4) The mirth of Israel, on Psal. 21. 1, 2, 3. (5) Noah his Arke, on Gen. 8; 6—9." Licensed. Octavo.

1594. The second part of Christian Exercise appertaining to Resolution, &c. by Edm. Bunny. For Simon Waterson. Again, 1598. Sixteens.

1594. "Articles of enquiry in the Dioces of Glocester & Bristol by John, bishop of Glocester. Quarto.

1595. "An Introduction to Algorisme, to learn to reckon with the Pen, or with Counters:—Newly ouerseene and corrected. Whereto is annexed.

51 A catechisme, or pithy summe of matters concerning faith and religion.
52 The treasure of gladnes. †
53 The sermon found hid in a wall. †
54 Dering's sermon before the queene.
55 Dering's prayers for housholders. †
56 Dering's catechisme. †
57 Stockwood upon Dearing's catechisme, at large. †
58 Stockwood upon Dearing, abridged. †
59 The pathway to please God.
60 A booke of Diana.
61 The history of Rouland the amorous.
62 A book of hunting and running horses.
63 Seven godly sermons upon the temptation of Christ.
64 Crowly's answer to the offers of a catholicke papist.
These play books.
The weather. Foure P. Love. Youth. Impatient poverty. Hicke Skorner.
Betweene him and N. Ling.
The English Roman life.
Betweene him and H. Car.
A defensitive against the poyson of supposed prophecies.
Betweene him and S. Waterson.
The second part of the resolution.
"This I have bought of John Waterson for 3l."

‡ Turbervilles songes & sonnetes.
‡ Betweeng him and Richard Jones.
‡ The mery metinge of maydes of London.
‡ Hilles palmestry & physiognomy.
‡ Morall philosophy.
‡ A disputacon betwene twoo Spanishe gentlemen concernyng physyk.
‡ The history of Charlos and Julia.
‡ The history of Palmeryn.
To him likewise belongeth the printing of these bookes following.
The second leafe of the primmer.
The psalter in quarto, on the English letter.
The jewel of health, in 32.
The christian A, B, c. in 16 and 32.
The lady Terrets prayers, in 32.
The complaint of the soule, in 16 and 32.
The queenes meditations.
Prayers of the bible, in 32.
The decads of the West-Indies.
The manuel of prayers.
The hop book for gardiners.
The discovery of witchcraft.
The treatise of prayer.
Monopoly in English.
The conduct of comfort.
The fruites of foes.
A handfull of wholsome hearbes.

certaine

certaine notable & pleasant rules of false positions, not seene before in our Englishe tongue.—At London, Printed by Ia. Roberts, and are to be solde in Paules Churchyarde, at the signe of the Sun, 1595." O 4, in eights. See p. 579. W.H. Octavo.

" A short introduction to learn to swimme, &c. By Christofer Middleton." With wood-cuts. See p. 1029. 1595. Quarto.

" A liuely patterne of true repentance. Preached in a sermon at St. Magnes in London, by Nicholas Coult, minister of the word of God, at Danbery in Essex."* For Tho. Chard. 1595. Octavo.

" A thousand notable thinges of sundrie sortes, some wonderfull; others strange, some pleasant, others profitable." &c. For Edw. White. 1595. Quarto.

" The Seconde parte of the Catalogue of English printed Bookes: —which concerneth the Sciences Mathematicall,—also of Phisick and Surgerie:—Gathered into Alphabet, &c. by Andr. Maunsell Bookeseller." Device, the pelican, as p. 715. " Printed—for A. Maunsell, dwelling in Lothburie, 1595." On the back are the arms of Robert Earl of Essex, in 55 coats, to whom this part is dedicated. W.H. 1595. Folio.

" The most honorable Tragedie of sir Richard Grinuile knight." A poem by Jarvis Markham. For Ri. Smith. Hist. of Eng. Poetry, III; p. 319. 1595. Sixteens.

" Saint Peters Complaint, with other Poems. Printed by I. R. for G. C." (*Gabriel Cawood*) Alex. Dalrymple, Esq; 1595. Quarto.

" The Tragicall Legend of Robert D. of Normandy," &c. For Nich. Ling. 1596. Sixteens.

" Peter Lowe, Scotchman, Arellian, doctor in the facultie of chirurgerie in Paris, chirurgion to Henry IV, king of France and Nauare. His easy, certain, and perfect methode to cure and preuent the Spanish sicknefs, whereby the learned and skillfull may heal many other diseases." 1596. Quarto.

" A right profitable Booke for all diseases, called The Pathway to Health,—by Peter Leuens." &c. For Edw. White. W.H. 1596. Quarto.

" The Poem of Poems, or Syons Muse, Contaynyng the diuine Song of King Salomon, deuided into eight Eclogues, by I. M.—Printed—for Mat. Lownes, and are to be sold at his shop in saint Dunstones churchyarde, 1596." Dedicated " To the sacred virgin, diuine Mistrefs Elizabeth Sidney, sole daughter of the euer admired sir Philip Sydney." Hist. of Eng. Poetry, III; p. 318. 1596. Sixteens.

" A sermon

JAMES ROBERTS.

1596. "A sermon preached before the queenes maieſtie, by maiſter Edw. Dearing, the 25 of Februarie." *   Sixteens.

1596. "How to chuſe, ride, traine, and diet, both hunting horſes, and running horſes; and a diſcourſe on horſmanſhip, and the cure of their diſeaſes. By Jarvis Markham. Dedicated to his father Robert, eſq; of Nottingham." Sheets p. iii.   Quarto.

1597. EVPHVES. The Anatomy of Wit: With his England. See p. 1012.   Quarto.

1597. "A Demonſtration of God in his workes Againſt all ſuch as eyther in word or life deny there is a God. By George More.—For Tho. Charde." With a new title-page, 1598. "By Sir George More, Knight." W.H.   Quarto.

1597. "THE PATHWAY TO PERFECTION. A Sermon preached at Saint Maryes Spittle in London on VVedneſday in Eaſter weeke, 1593, by Tho. Playfere, D. D.—For Andr. Wiſe, 1597." W.H.†   Sixteens.

1598. "Politeuphnia. Wits common wealth.—For Nicholas Ling." Octavo.

1598. Fitzherbert's Huſbandry, with additions. For Edw. White. Quarto.

1598. "THE SINNERS GVYDE. Contayning the whole regiment of a Chriſtian life, deuided into two Bookes. VVherein Sinners are reclaimed from the By-path of Vice & Deſtruction, and brought into the High-way of euerlaſting happineſſe. Compiled in the Spaniſh tongue by— F. Lewes of Granada: Since tranſlated into Latine, Italian & French: And now peruſed & digeſted into Engliſh by Francis Meres, M.A. &c. Printed—for Paule Linley & John Flaſket, and are to be ſold in Paules Church-yard, at the ſigne of the Beare,—1598." Dedicated "To— Sir Thomas Egerton—Lorde Keeper of the Great-Seale," &c. "The Second Booke" appears to have been printed by Ri. Field, for Edw. Blount, at the ſame place, but not till 1614; the pages & ſignatures being in exact progreſſion. 525 pages, and a table. W.H.   Quarto.

1598. "THE MIRROVR OF MIRROVRS, or all the tragedys of the Mirrour of Magiſtrates abbreuiated in breefe hiſtories in proſe.—London, imprinted for James Roberts, in Barbican." Hiſt. of Eng. Poetry, III; p. 281.

1598. "A petite Pallace of Pettie his pleſure." See p. 1030.   Quarto.

1598. "The Metamorphoſis of Pigmalions Image, &c. by John Marſton —For Edm. Matts, and ſold at the hand & plough in Fleetſtreet."   Sixteens.

1598. "Skialetheia, or a ſhadowe of Truth in certaine Epigrams, and Satyres."   Octavo.

"A true

" A true Coppie of the transportation of the Lowe Countries, &c. by the (1598)
King of Spayne for the dowrie of his eldest daughter, giuen in marriage
vnto Card. Albert, duke of Austria. Translated out of Dutch by H. W.
1598. Printed by I. R. for Paule Linley." Quarto.

" Wits Theater of the little World." Device, a ling entangled in 1598.
the stalks of a honeysuckle.—I. R. for N. L. (Nich. Ling) and are to be
sold at the West doore of Paules."* Again 1599. W.H. Sixteens.

" The Scovrge of Villanie. Corrected, with the addition of newe 1599.
Satyres. Three Bookes of Satyres." By Iohn Marston. Sixteens.

Three sermons by M. Hen. Smith. For Nich. Ling. Quarto. 1599.

" A Midsommer Nights Dreame.—Written by Wm. Shakespeare." 1600.
Quarto.

" The Excellent History of the Merchant of Venice, with the Extreme 1600.
Cruelty of Shylocke the Iew toward the saide Merchant, in cutting a just
Pound of his Flesh: And the obtaining of Portia, by the Choyse of Three
Caskets. Written by W. Shakespeare." Running title, " The comicall
history of the Merchant of Venice." Quarto.

The Apology of the Church of England. For Tho. Chard. Twenty-fours. 1600.

The Poore mans Garden: &c. By John Northbrooke. Sixteens. 1600.

" The Spanish Mandeuile of Miracles, Or the Garden of curious 1600.
Flowers. VVherin are handled sundry points of Humanity, Philosophy,
Diuinitie, and Geography, beautified with many strange & pleasant Histories. First written in Spanish by Antonio De Torquemeda, and out of
that tongue translated into English.—It is deuided into sixe Treatises,
composed in manner of a Dialogue, as in the next page shall appeare.—
Printed by I. R. for Edm. Matts.—1600." Dedicated " To—Sir
Thomas Sackuile, Knight, Baron of Buckhurst, Lorde high Treasurer
of Englande, Lieuetenaunt of her Highnes within the County of Sussex,
most worthy Chauncellor of the Uniuersitie of Oxenford, Knight of the
—Garter, &c. 23 Aprill, 1600—Ferdinando Valker," the editor.---
" To—Lewes Lewkenor Esq; one of the hon. band of her Maiesties Gen-
tlemen Pensioners in ordinarie," the translator.—" Ferd. VValker."---
" The Authors Epistle Dedicatorie, to the most hon. and reuerent Præ-
late, Don Diego Sarmento, de Soto Maior, Bishop of Astorga.—The
humble Seruant of your L. which kisseth his most Hon. hands. A. de
Torquemeda.---A table of authors, whose authorities are alleadged."
Fol. 158, and " A table of the principall matters." W.H. Quarto.
" A true

1603 "A true and admirable hiftorie of a mayden at Confolens, in the prouince of Poictiers, that for the fpace of three yeares and more, hath lived, and yet doth, without receiuing either meat or drinke. Of whom his maieftie in perfon hath had the view, and (by his command) his beft and chiefeft phifitians haue tried all means to find, whether this faft or abftinence be by deceipt or no. In this hiftory is alfo difcourfed, whether a man may liue many dayes, monthes, or years, without receiuing any fuftenance. Publifhed by the king's fpecial priuiledge at London, by James Roberts, and are to be fold at his houfe in Barbican." This book is dedicated to the barbers furgeons, by A. M.

He had licenfes alfo for the following, viz. In 1569, "An almanacke & prognoftication by M. Rob. Moore, for 1570. A Chriftenmas carrolle by Chr. Payne." I find nothing elfe licenfed to him except a few ballads till 1594; when, by order of the Court of Affiftants, certain copies which were John Charlewood's were entered as his; provided it be not hurtful to any other's Right. See p. 1032.

✤✤✤✤✤✤✤✤✤✤✤✤✤✤✤✤✤✤✤✤✤✤✤✤✤

WILLIAM HOW, or HOWE,

WAS made free of the Stationers' company, 29 Nov. 1556, came on the Livery about 1574; and ferved Renter in 1581 and 1582, but he is not found ferving any other office; however he appears fitting in the Court of Affiftants in Sept. 1584; not after 1597. He was a very orderly member, not once appearing on the Black Lift. At firft he feems to have printed only ballads, for which he had licenfes in 1565. He printed feveral books for Abr. Veale, which may be feen under his name.

1569. "Prouerbes or Adages, gathered out of the Chiliades of Erafmus, by Richard Tauerner. With new additions, as well of Latin Prouerbes, as of Englifh. Imprinted—in Fleete ftreete, by William How, 1569." See p. 406. Contains 71 leaves. W.H. Sixteens.

1569. "The tragicall and lamentable Hiftorie of two fayhfull mates Ceyx kynge of Thrachyne, and Alcione his wife, drawen into Englifh meeter by William Hubbard, 1569. Imprinted at London, by W. Howe, for R. Johnes." In ftanzas: eight leaves. See Hift. of Eng. Poetry, Vol. III; p. 413.

1570. "The tragical comedie of Damon and Pithias: newly imprinted as the fame was playde before the queenes maieftie by the children of her
graces

graves chapple: Made by mayster Edwards, then being master of the children."; See Hist. of Eng. Poetry, Vol. III; p. 284, and p. 289, note x. Licensed to R. Jones. Quarto.

"The feveral confeffions of of Thomas Norton, and Chriftopher Norton, two of the Northern rebels, who fuffered at Tyburn, and were drawn, hanged, and quartered, for treafon, May 27.—For Richard Johnes." 1570.

"The Ende & Confeffion of Tho. Norton, of Yorkfhire, the Popifh Rebell, and Chr. Norton his nephew; which fuffered at Tiburn for Treafon, the 27. of May." In verfe, by Sampfon Davie. 7 leaves. Octavo. 1570.

Auguft. Marlorat his prayers on the pfalmes, tranflated by Rodolph Warcup. Sixteens. 1571.

"A Declaration of fuch tempeftious & outragious Fluddes," &c. as p. 890. 1571.

"The bookes of the golden afs of Apuleius, tranflated by Will. Addington." Quarto. 1571.

"The caftle of Memoire: wherein is conteyned the reftoring, &c. of memorie, made by Guil. Gratarolus. Englifhed by William Fulford," and by him dedicated to Robert lord Dudley, mafter of the horfe to the queen. Licensed in 1568. 1573.

Alfo without date, "Imprinted—in fleeteftreete by W. How, dwelling at Temple barre." Octavo.

"A new interlude entitled, New Cuftom." Quarto. 1573.

"The French Schoolemaifter, wherin is moft plainlie fhewed the true and moft perfect way of pronouncinge of the Frenche tongue, without any helpe of Maifter or teacher:—Unto the which is annexed a Vocabularie for al fuch woordes as bee vfed in common talkes: by M. Claudius Hollybande, profeffor of the Latin, Frenche & Englifhe tongues.—for Abr. Veale. 1573." Dedicated "To—the towardly yonge Gentilman Maifter Robert Sackuill, fonne & heyre to Lorde Buckhurft.---A VVarning to the Reader." In Englifh and Frenche: wherein he cenfures "A new booke which came out of Anworpe, and now of late printed at London." The author of which appears to be Rob. Fontan; by the following letter in Latin: "Claudius Sancto Vinculo Roberto Fontano S.—Vale Lewfhaniæ Nonis Octobris, 1573." Then, verfes in commendation of this book, &c. "Iehan Henry Maiftre d'efchole, Dizain.---Tho. Twyne, to all ftudents of the Frenche tongue." V, in eights. W.H. Octavo. 1573.

Again without date, Newly corrected.

"Jo. Carr, his ruinous fall of prodigalitie, with examples, &c. Octavo. 1573.

"A Garland of Godly Flowers, bewtifully adorned," &c. Printed in borders. Sixteens. 1573.

"A Com-

1575. "A Commemoration of the moſt proſperous & peaceable Raigne of our gratious & deere Soueraigne Lady Elizabeth, &c. Now newly ſet foorth this 17. Day of Nouember, being the firſt of the Eighteenth Yeere of her Maieſties ſayd Raigne: By Edw. Hake Gent." In verſe; with a poem to the queen's council, and a meditation in proſe. About 3¼ ſheets. Oldys's Catal. No. 5.   Octavo.

1585. "The pſalmes or prayers, taken out of holy ſcriptures; commonly called the kinges pſalmes."   Sixteens.

1588. "The lawers logike, exemplifying the praecepts of logike by the practiſe of the common lawe, by Abraham Fraunce."   Quarto.

1590. The kings pſalmes, and queen's praiers, taken out of the holy ſcripture.   Sixteens.

1590. The Hiſtory of two the moſt noble Captains, &c. as p. 446. Octavo.

—— "The regiment of lyfe; whereunto is added, a treatiſe of the peſtilence, with the booke of chyldren: newly corrected and enlarged by Thomas Phayre."*   Sixteens.

—— "A very mery & pythie commedie called, The longer thou liueſt, the more foole thou art. A myrrour, very neceſſarie for youth, and ſpecially for ſuch, as are like to come to dignitie and promotion.; as it maye well appeare in the matter followynge. Newly compiled by W. Wager." The back of the title, " the players names, prologue, fortune, moros, ignorance, diſcipline, crueltie, pietie, impietie, exercitation, people, idleneſs, God's iudgement, incontinencie, confuſion, wrath. ¶ Four may playe it eaſely. The prologue, crueltie, exercitation, God's iudgement, wrath, for one. Moros, fortune for another. Diſcipline, incontinence, impietie, confuſion, for another. Pietie, idleneſs, ignorance, people, for another." For R. Johnes. Contained in 28 leaves. Quarto.

—— "Sir John Conway, knight, his Godly Meditations & Praiers, gathered out of the ſacred letters, and vertuous writers, diſpoſed in forme of Alphabet on the Queen her Maieſties name." W.H.+   Sixteens.

—— Pyrryes Praiſe and Diſpraiſe of Women. Licenſed in 1568. Sixteens.

—— XIII Bloes at the Popes Bull. Suppoſed by Tho. Norton.

He had licenſe alſo in 1576, for " The moſt excellent, pleaſant & variable hiſtorie of the ſtrange adventures of prince Appollonius, Lucina his wife, & Tharſa his daughter." In 1587, for " A book of Anthony Chalmerius: vpon condycyc'on y̓ he can procure yt orderly."

RICHARD

RICHARD JONES, JHONES, or JOHNES,

WAS admitted a brother of the worshipful company of Stationers, 7 Aug. 1564, for which he paid ij.s. vj.d. He kept shop at the South-west door of St. Paul's church, but dwelt, and had his printing-office, at other places, as mentioned in the several books he printed: many of which were performed jointly with, and sometimes for, others. He appears to have been but a disorderly brother, printing books and ballads without licenfe. In June 1579, he was fined "for printing a ballad without licenfe (the ballad not tollerable) iij.s. iiij.d. whereof *was* giuen to the bedell for his pains vjd." In Aug. following, for that he printed " A brief Inftruction, in maner of a catechifme of Mr. Cobh'edes gathering," without licenfe, he was fined x.s, to bring all the books into the Hall, and never fell, nor reprint them. He paid however only vs. In Jan. 1582-3, he was fined again xs. for printing " a thinge of the fall of the gallaries at Paris Garden, without licenfe, and againft the comandment of the Wardens. And the faid Jones & Will. Barrlet to be comitted to prifon. Bartlet for printing it, and Jones for fuffering it to be printed in his houfe."

The end and confeffion of John Felton, &c. as p. 931. Licenfed. 1570.

Damon and Pythias. See p. 1036. Licenfed.   Quarto. 1571.

6 F, 2   " Palf-

1572. "Palsgraues Catechisine, tranflated by W. Turner Doct. in Phificke." Octavo.

1573. "The treafurie of commodius conceits and hidden fecrets, comonly called, The good bufwiues Clofet of prouifion for the health of her houfhold, by John Par." See it in 1580. Sixteens.

1573. "The Breuiary of Brytayne. As this moft noble, and renowmed Jland was of auncient time deuided into three Kingdomes, England, Scotland, and Wales. Contaynyng a learned difcourfe of the variable ftate, & alteration therof, vnder diuers, as wel natural, as forren princes, & Conquerours. Together with the Geographicall defcription of the fame, fuch as nether by elder, nor later writers, the like hath been fet foorth before. Writen in Latin by Humfrey Lhuyd of Denbigh, a Cambre Britayne, and lately Englifhed by Thomas Twyne, gentleman, 1573." Dedicated "To—Edward Deuiere Lorde Bulbeck, Erle of Oxenford, Lorde great Chamberlayne of England."—The preface; and fundry commendatory verfes. The Breviary on 96 leaves, including ' an explanation of certain Britifh words. "Jmprinted at London by R- Iohnes: and are to be folde at his fhop, ioynyng to the Southweft doore of Paules Church." W.H. Octavo.

1574. "'A new booke intituled, The blafinge of bawdrie, daylie procured by Beldame B. principall broker of all iniquitie. Geuen for a new yeares gyfte, as well to all fuche, in whofe charge the due punifhment therof is committed, as alfo to all other that may reap commodytie, by lothyng their practifes, either by readyng, or hearing of the fame, by R. C. citizen.' It begins with a dialogue between the printer and the author, and is in verfe. Sixteens.

1574. "Middleborow: A brief Reherfall of the Accord & Agreement that the Captaines, Burgifes, and Armie of Middleborow and Armew haue made, in yielding themfelues to the Right High & Excellent Prince, the Lord William Prince of Orange, Countie of Naffau, &c. With a lamentable Difcourfe of the Calamities, great Hunger and extreame Miferies that they fuftayned before they yelded up the faid Townes. Tranflated out of the Dutche Coppie, printed at Dordrecht." 8 leaves. Octavo.

1576. "The Schoolemafter, or Teacher of Table Philofophie. A moft pleafant and merie companion, wel worthy to be welcomed (for a dayly Gheaft) not onely to all mens boorde, to guyde them with moderate & holfome dyet: but alfo into euery mans companie at all tymes, to recreate

' If Mr. Ames was not miftaken, he had a different edition of this fame year; containing 96 leaves, befides preface and explanation of Britifh words.

their

their mindes, with honest mirth and delectable deuises: to sundrie purposes of pleasure and paftyme. Gathered out of diuers, the best approued Authours: and deuided into foure pithy & pleasant Treatifes, as it may appeare by the contentes. Imprinted—dwelling ouragaynft S. Sepulchers Church without Newgate, 1576." This curious treatife is introduced by " The Printers preamble to al eftates, for the friendly entertaynment of this Teacher," in 6 fix-lined ftanzas; and then is infcribed To—M. Alexander Nowell, Deane of the Cathedrall Churche of S. Paules in London." Anonymous; but at the end of the book, T. T. Contains V, in fours. W. H. Quarto.

" A knowledge for Kings, and a warning for fubiects: Conteyning 1576. The mofte excellent and worthy hiftory of the Raellyans peruerted ftate, and *the* gouernment of their common wealth: no leffe rare, then ftrange and wonderfull, and moft meete to be publifhed for a fpeciall example, in thefe perylous and daungerous dayes. Firft written in Latin by Iames Glaucus, a Germaine; and now tranflated into Englifh by William Cleuer, Scholemafter. By fpeciall recorde, this Monument was two thoufande yeares of Antiquity: and fo dufked & forworne with age, that being in a plaine writte letter, could fcarce be read: And for that it was great pitie, that fo precious a Jewell fhould quite fade out of remembrance I with my painefull indeuour haue now renewed it into frefh memorye. Imprinted—1576." Dedicated " To—M. Gryffyn Hamden Effquyer: one of the Queenes Maiefties Iuftices, of the Countie of Buckingham.— From Colfhill, in the Parifhe of Amerfom, 3 Nouemb.—VV. Cleauer, Scholemafter.---James Glawcus his Epiftle tranflated by VV. Cleuer." ---A table of contents. Contains befides, 60 leaves. Licenfed. W. H.
Sixteens.

" The fpoile of Antwarpe faithfullie reported by a true Englifheman, 1576. that was prefent at this piteous mafacre, 25 Nouember: alfo a platfourme of the whole battaile thereunto annexed." Licenfed. Octavo.

" A DISCOVRSE Of a Difcouerie for a new Paffage to Cathaia." Written by Sir Humfrey Gilbert Knight. Quid non? Imprinted—by Henry 1576. Middleton for R- Ihones, Aprilis, 1576." It is prefaced by Geo. Gafcoigne Efq; and contains, befides the prefixes, I, 2. At the end, " Thefe Bokes are to be folde at the fhop of R. Iones, at the Weft end of S. Paules Church, betweene the Brafen Piller and Lollards Tower."
Quarto.

" Of Englifhe Dogges, the diuerfities, the names, the natures, and 1576. the properties. A fhort Treatife written in Latine by Johannes Caius of late memorie, Doctor of Phificke in the Uniuerfitie of Cambridge, And newly drawne into Englifhe by Abr. Fleming Student. Natura etiam in brutis vim oftendit fuam. Seene & allowed. Imprinted—by Ryc. Johnes, and are to be folde ouer againft S. Sepulchres," &c. Quarto.

" A delicate

1576. "A delicate Diet for daintie mouthde Droonkards: Wherein the fowle abuſe of common Carowſing & quaffing with heartie draughtes is honeſtly admoniſhed. By George Gaſcoyne Eſquier. Tam Marti quam Mercurio. Imprinted—Aug. 22. 1576." On three ſheets. George Steevens Eſq; alſo Mr. Edw. Jacob, Feverſham. Octavo.

1577. "The Courte of ciuill Courteſie." See it in 1591. Quarto.

1578. "A View of certain wonderful Effects of late Dayes come to paſſe; and now newly conferred with the Preſignification of the Comete or Blaſing Star, which appeared in the Southweſt, upon the 10 Day of Nouemb. laſt paſt. Written by T. T. This 28 Nov. 1578." On 12 leaves. See Oldys's Catal. of Pamphlets. No. 75. Quarto.

1578. "The Maner to die well, ſhewing the fruitfull remembrance of the fower laſt thinges: gathered out of Pharetra Diuini Amoris. Again 1588. Octavo.

1578. "The right excellent and famous Hiſtorye of Promos and Caſſandra: Diuided into Commical Diſcourſes.—The worke of Geo. Whetſtone Gent." Two Parts." Licenſed.

1578. "Alex. Nowell's leaſt catechiſme, in ᵛ Welch." Octavo.

(1578) "The Way of lyfe. A Chriſtian, and Catholique Inſtitution comprehending principal poincts of Chriſtian Religion, which are neceſſary to bee knowne of all men, to the atteyning of Saluation. Firſt delyuered in the Daniſh Language—by Dr. Nic. Hemmingius,—Profeſſor of Diuinitie for the Kynge of Denmarcke in his Uniuerſitie of Haſnia:—Tranſlated into Latine by Andr. Seuerinus Velleius: And now firſt Engliſhed,—By N. Denham, this yeare of our Redemption, 1578. Imprinted—and are to bee ſould ouer agaynſt S. Sepulchers Churche." On 199 pages beſides the prefixes. Licenſed. W.H. Quarto.

1579. "Cyuile and vncyuile life. A diſcourſe very profitable, pleaſant, and fit to bee read of all Nobilitie and Gentlemen. Where, in forme of a Dialogue is diſputed, what order of lyfe beſt beſeemeth a Gentleman, in all ages and times, as well for educatiõ, as the courſe of his whole life, to make him a parſon fit for the publique ſeruice of his prince and Countrey, and for the quiet, and comlyneſſe of his own priuate eſtate and callinge." Dedicated to ſir Francis Walſingham by the printer. Sheets, N. 3. Licenſed. Quarto.

ᵘ See "Six old Plays on which Shakeſpeare founded his Meaſure for Meaſure," Reprinted by J. Nichols, 1779.
ᵛ It was agreed between him and J. Day, that he having the copy in Welch licenſed to him, he ſhould print it in the name of J. Day, and, as aſſignee unto him; but that the copy was to belong to Jones as amply as if he had printed it in his own name. Stat. Regiſter, Fol. 431, b.

"The

RICHARD JONES.

"The liues of diuers excellent Oratours & Philofophers: written in 1579. Greeke by Enapius of the citie of Sardis in Lydia, and tranflated into Englifhe by W. B." Licenfed. Quarto.

"The firft part of The Eyghth liberal Scyence: Entituled Ars Adu- 1579. landi, The Art of Flatterie, with the Confutation therof; both very pleafaunt and profitable; deuifed & compiled by Ulpian Fulwell. Newly corrected and augmented.

"Who Reads a Booke rafhly, at Random doth runne:
Hee goes on his errand, yet leaues it vndone."

Dedicated to "Lady Mildred Burgley, wife to the Lord Treafurer." Eight dialogues, chiefly in profe: on 37 leaues. See Oldys's Catal. of Pamphlets in the Harleian Libr. No. 386. Licenfed. Quarto.

"Complaint (for reformation) of diuers vaine & wicked abufed ex- 1580. ercifes practifed on the Sabbaoth day, which tend to the hindraunce of the Gofpell, and increafe of many abhominable vices, by Humfry Robartes." Octavo.

"The Treafurie of commodious Conceyts, and hidden Secretes. 1580. Commonly called, The good hufwiues Clofet of prouifion, for the health of her houfhold.—Gathered out of fundry Experimēts lately practifed by men of great knowledge: With neceffarie and new Additions now adioyned at the fecond Jmpreffion. Imprinted—18 Iulij, 1580." On the back are verfes by "The Printer to all that couet the practife of good Hufwiuery, afwell Wiues as Maides," which conclude with,

"And if ye reape commodite, by this my freends deuife,
Then giue him thankes, and thinke not much of foure pence for the price."

Dedicated "To the worfhipfull Maifter Richard Wiftow Gent. one of the Afsiftants of the Company of the Barbours & Surgions." By John Partridge. F, in eights. "Thefe Bookes are to be folde ouer againft S. Sepulchres Church," &c. W.H. Sixteens.

"A fhorte and pithie difcourfe concerning the engendering tokens & (1580) effects of all Earthquakes in generall: Particularly applyed to that 6 April, 1580, by T. T." *Tho. Twine.* Licenfed. Quarto.

"A right rule of Chriftian Chaftitie, profitable to be read of al Godly 1580. & vertuous youthes of both fexes,—an expofition on the 7 Commandement, Exod. 20; 14, by Will. Hergeft. Licenfed. Quarto.

"The difcouery, and conqueft of the prouinces of Peru, and the na- 1581. uigation of the South fea, along that coaft. And alfo of the ritche mines of of Potofi, 6th February. By Aug. Sarat; and tranflated out of the Spanifh, by T. Nicholas." With cuts. Licenfed. Quarto.

(1581) "Two GODLIE AND learned Sermons, appointed, and Preached before the Jefuites, Seminaries, and other aduerfaries to the Gofpell of Chrift, in the Tower of London. In which, were confuted to their faces, the moft principall and cheefe poincts of their Romifh, and VVhoarifh religion: And all fuch Articles as they defend, contrarie to the woord of God, vvere layed open and ripped vp vnto them. In Maye, 7, and 21. Anno 1581. By Iohn Keltridge, Preacher of the vvorde of God in London.—Jmprinted—by R. Ihones, dwelling without Newgate, neere vnto Holburne Bridge." Dedicated "To—Sir Fraunces Walfingham Knight, &c. From my Chamber in Holborne at London, Iune 10." The firft fermon on 28 leaves, the 2d, on 30. Licenfed.* W.H. Quarto.

1581. The nurcerie of gentlewomans names, in verfe. Tolerated. Quarto.

1582. "An heptameron of ciuill difcourfes; containing, the chriftmas exercife of fundrie well courted gentlemen and gentlewomen, &c. The report of George Whetftone, gent. in feuen days exercifes.—3d of February." Not paged. Licenfed.* Quarto.

1582. "Two godlie and learned Sermons, preached at Manchefter—before a great Audience, both of Honor & VVoorfhip. The firft containeth a reproofe of the fubtill practifes of diffembling Neuters, &c. The other, a charge and Inftruction for all vnlearned, negligent, and diffolute Minifters: And an Exhortation to the common people, to feeke their amendment, by prayer, vnto God. By Sim. Harward,—late of Newe Colledge in Oxfoord." With Charlewood. "The Preface to the Chriftian Reader.—VVarrington, 8 Maie, 1582." The firft fermon from Rom. 10; 19, on 46 leaves; the fecond at an ordination, from Luke 10; 2. on 72 leaves: Another at the end, with citations from the Fathers. W.H. Sixteens.

1583. "The feuerall Executions and Confeffions of *John* Slade and *John* Bodye," &c. Octavo.

1583. "The anatomie of abufes: contayning a difcouerie, or briefe fummarie, of fuch notable vices and imperfections, as now raigne in many criftian countreyes of the worlde, but (efpeciallie) in a very famous ilande called Ailgna: together with moft fearefull examples of Gods judge-

---

* One was executed at Winchefter, the other at Andover, Hants. Dr. Bennet attended Slade at his execution, exhorting him to acknowledge Q. Elizabeth for head of the Church. The tract was fuppreffed, and the author punifhed. See Allen's Anfwer to The execution of Juftice in England, p. 5, 6.

mentes,

RICHARD JONES.

mentes, executed vpon the wicked for the fame, as well in Ailgna of late, as in other places elſewhere. Verie Godly to be read of all true chriſtians ewerie where, but moſt needful to be regarded in Englande. Made dialogue wiſe by Phillip Stubbes. Seene and allowed according to order." 1 Maii. * The ſecond part was printed for Will. Wright.   Octavo.

" A declaration of the death of John Lewes, a moſt deteſtable & obſti-   1583. nate Hereticke, at Norwich, 18 Sept. 1583. To the tune of John Care-leſſe." At the end, " Finis Th. Gilbart."   A broadſide.

" The Welſpring of wittie Conceights: tranſlated from the Italian, by   1584. W. Phiſt." See Hiſt. of Eng. Poetry, Vol. III ; p. 308. Licenſed.   Quarto.

" A mirour for Maieſtrates of Cities, by G. Whetſtone." In two parts.   1584. Licenſed. *   Quarto.

" A new Yorkſhyre Song. Intituled :- Yorke, Yorke, for my monie :   1584. Of all the Cities that euer I ſee, For mery paſtime and companie, Except the Citie of London." The merry report of Archery, in 22 ſix-lined ſtanzas. " From Yorke, by W. E. Jmprinted—dwelling neere Holbourne Bridge. 1584." Licenſed.   Broadſide.

" The New Attractiue :" &c. as p. 1014.   Quarto.   1585.

" The vertues of the duke of Bedford, who died 27th of July 1584,   1585. aetat. 78." By G. Whetſtone.   Quarto.

" The ſinfull mans ſolace, moſt ſweete and comfortable for the ſicke   1585. & ſorrowfull ſoule, contriued in 7 daies conference between Chriſt and a careleſſe ſinner." By John Norden. Licenſed.   Octavo.

" A moſt ioyfull Songe, made in the behalfe of all her Maieſties faith-   1586. full & louing Subiects : of the great ioy which was made in London at the taking of the late trayterous Conſpirators, which ſought oportunity to kyll her Maieſty, to ſpoile the Cittie, and by forraign inuaſion to ouerturn the Realm : for the which haynous Treaſons, 14 of them haue ſuffered Death on the 20, and 21 of Sept. Alſo, a deteſtation againſt thoſe Conſpirators, and all their confederates, giuing God the prayſe for the ſafe preſeruation of her Maieſty, and their ſubuerſion. Anno 1586. To the tune of O man in deſperation." 25 four-lined ſtanzas. Heads of the traitors at top ; and between the two columns, their names, and the times of their execution. At the end, " Finis, T. D." Licenſed.   Broadſide.

" Capt. Tho. Sanders his diſcription of a lamentable voyage made to   1587. Tripoly, in 1584; with the barbarous vſage of our men there." Licenſed to him & Edw. White.   Quarto.

" A ſhort admonition or warning, vpon the deteſtable treaſon, where-   1587. with ſir William Standley, and Rowland Yorke, haue betrayed and deliuered for monie, vnto the Spaniards, the towne of Deuenter, and the ſconce of Zutphen." Licenſed. *   Quarto.

6 G   " Chriſtian

1587. "Chriſtian ethickes, or morall Philoſophy, containing the difference & oppoſition of virtue & voluptuouſneſs, by Will. Fulbecke, M. A." Licenſed. Octavo.

1587. "The Cenſure of a loyal ſubiect vpon certaine noted ſpeeches & behauiour of thoſe 14 notable Traitors, (*Ballard, Babington,* &c.) at the place of their execution (*Lincoln's Inn Fields*) the xi (20) and 12 (21) of September laſt paſt. Wherein is handled matter of neceſſary Inſtruction, &c. By Wm. Kemp. Dedicated to Lord Burleigh, by G. W." (*Geo. Whetſtone.*) Licenſed. Alſo without date. Quarto.

1588. "The A. B. C. [x] for children, newly deuiſed with ſyllables, the Lordes praier, our Belief, and the ten Comandements." Two ſheets. Sixteens.

1588. "The good huſwiues handmayde, contayning many principall poyntes of Cookerie," &c. Printed with Anthony [y] Hyll. Octavo.

1588. "The Voyage and Trauaile of M. Cæſar Frederick, Merchant, of Venice, into the Eaſt India, the Indies, and beyond the Indies. Wherein are contained very pleaſant and rare matters, with the cuſtomes & rites of thoſe Countries. Alſo herein are diſcouered the merchandiſes and comodities of thoſe countreyes, aſwell the aboundance of Goulde & Siluer, as Spices, Drugges, Pearles and other Iewels. Written at Sea in the Hercules of London, cōming from Turkie, the 25. of March, 1588.— *Tranſlated* Out of Italian by T. H." (*Tho. Hickock*) Printed with Edw. White, 18 June, 1588. Dedicated "To—Charles Lord Howard, Baron of Effingham, Knight of the—Garter, Lorde high Admirall of England," &c. L. in fours. Licenſed. Alex. Dalrymple Eſq. Quarto.

1588. "A true Report of the inditement, arraignment, conuiction, condemnation, and execution, of John Weldon, William Hartley, and Robert Sutton, who ſuffered for high treaſon in ſeuerall places about London, October 5th. With the ſpeeches which paſſed between a learned. Preacher and them. Faithfully collected, euen in the ſame words as neer as might be remembered, by one of Credit, that was preſent at the ſame." 12 leaves. See Catal. of Harl. Pamphlets, No. 179. Licenſed. Quarto.

[x] This was allowed him again in 1590, on this proviſo, "that there ſhalbe no additions made to the ſame hereafter." But it was cancelled by order of a Court holden 15 May, 1605.

[y] This printer having, in divers inſtances, tranſgreſſed the ordinances of the Court of Star-chamber, had thereby incurred the danger of impriſonment and other forfeitures; to avoid which, he ſigns an order of the Court of Aſſiſtants of the Stationers' company, 12 October, 1586, "That he the ſaid Anthonye ſhall not at any tyme hereafter by himſelf, or any other by his procurement, keepe any printing houſe of hys owne as maſter (except he be admytted thereunto according to the orders of the ſaid decrees) but onlye to worke & lyue as a Journeyman & workman for wages in the trade & facultye of pryntinge." By his being concerned in printing this book, he ſeems to have been now readmitted to the liberty of printing as a maſter; but i have met with no other book of his printing.

"The

RICHARD JONES.

"The Schoole of Vertue, and booke of good nurture." See p. 758. 1588. Licenfed. Octavo.

"The Englifh Ape," &c. Printed by Rob. Robinfon, was fold by him at Holborne Conduit, at the figne of the Rofe and Crowne." Quarto. 1588.

"CERTAIN Difcourfes, written by fir John Smythe, knt. concerning the formes, and effects of diuers fortes of weapons, and other verie important matters Militarie, greatlie miftaken by diuers of our men of warre in thefe daies; and chiefly, of the Mofquet, the Caliuer, and the Longbow; As alfo, of the great fufficiencie, excellencie, and wonderfull effects of Archers: With many notable examples, and other particularities, by him prefented to the Nobilitie of this Realme, and publifhed for the benefite of this his natiue Country of England." T. Orwin's device. "Printed by R- Johnes, at the figne of the Rofe & Crowne neere Holburne Bridge. 1 Maij, 1590." On 50 leaves, befides "—his Proëme Dedicatorie to the Nobilitie of the Realme of England." Licenfed. W.H. Quarto. 1590.

This was anfwered by "A brief difcourfe, concerning the force and effect of all manual weapons of fire, and the difability of the long-bowe, or archery, in refpect of others of greater force now in vfe. Written by Humphrey Barwick, gentleman, fouldier, captaine. Et encore plus oultre." Quarto.

"THE BOOKE OF HONOR and Armes. VVherein is difcourfed the caufes of Quarrell, and the nature of Injuries, with their repulfes. Alfo the meanes of fatisfaction and pacification; with diuers other things, neceffarie to be knowne of all Gentlemen and others, profeffing Armes and Honor." Licenfed. * Quarto. 1590.

"Tamburlaine the great, a Tragedy: in 2 parts." See his licenfes, p. 1054. Octavo. 1590.

"The pathwaye to Readinge, or the newefte fpellinge A. B. C. Conteyninge a mofte fhorte, eafie, & profitable way of teaching to fpell & reade, fette forthe by Tho. Johnfon." Five fheets. Licenfed. (1590)

"Inftructions and orders mylitarie, compofed by fir John Smith knt." (1590) This book was often printed. But there muft be a miftake in this date: fee it in 1594.

"A Sermon vpon the 3 laft verfes of the firft Chapter of Job: tending to the confideration of Gods prouidence, planting of patience, & ap- 1591,

* This copy being controuerted by Rob. Dexter, it was ordered by a Ct. of Affiftants, 3 May, 1591, with confent of the parties, "That R. D. fhall enioye the faid copie to his owne vfe,—And thentrance thereof for R. J. in the Hall booke be croffed out. In confideration whereof R. D. fhall pay to the faid R. J. xx.s. before Mydfomer next: And yf the faid R. J. will deliuer to the faid R. D. the remainder whiche the faid R. J. hath of the faid books (about 300) the faid R. D. fhall accept the fame at vj.s. viij.d. a Reame for paper and printing. Item, that the faid R. J. fhalbe woorkeman to the faid R. D. for thimprintinge thereof at all tymes hereafter, entringe into bond fufficientlie to the faid R. D. to deale truelie with him at all tymes in thimprintinge thereof."

plieng of Confolation."—Device: A wyvera rifing out of a ducal coronet; the creft of Lord Clifford. Printed with Edw. Aggas. Dedicated "To—Lady Margaret, Countefſe of Cumberland: and Lady Anne Counteſſe of Warwick.—1 Jan. 1590.—Henry Peacham." Licenſed. W.H.₄ Sixteens.

1591. "The Triplicitie of Triumphes, Containing The order, folempnitie & pompe, of the Feaſtes, Sacrifices, Vowes, Games & Triumphes vſed vpon the Natiuities of Emperours, Kinges, Princes, Dukes, Popes, and Conſuls, with the cuſtome, order & maners of their Inaugurations, Coronations & annointing. Wherein is alſo mentioned the three moſt happy, ioyfull, & triumphant daies in September, Nouember, and Ianuary, by the name of Triplicia * Feſta. With a briefe rehearſall of the Funerall Solempnities at ſome Emperors, Kings, and Princes burials. By Lodowike Lloyd, Eſquier. Imprinted—by R. Ihones, at the Roſe & Crowne, neere Holborne Bridge. Ianuary, 1591. Liber minimus, labor maximus.". 33 leaves. Licenſed. W.H. Quarto.

1591: "THE COVRT of ciuill Courteſie, Fitlie furniſhed with a pleaſant port of ſtately phraſes and pithie precepts: aſſembled in the behalfe of all young Gentlemen & others that are deſirous to frame their behauiour according to their eſtates, at all times, and in all companies, Therby to purchaſe worthy praiſe of their inferiours, and eſtimation & credite among their betters. Out of the Italian, by S. R. Gent. Imprinted—1591." E, in fours. W.H. Quarto.

1592. "THE English Secretorie: or plaine and direct Method for the enditing of all manner of Epiſtles or Letters, aſwell Familiar as others:— The like whereof hath neuer hitherto beene publiſhed. Studiouſlie now corrected, refined & amended, in far more apt & better ſort then before: according to the Authors true meaning, deliuered in his former edition: Togeather (alſo) with the ſecond part then left out, and long ſince promiſed to be performed. Alſo, a declaration of all ſuch Tropes, Figures or Schemes as either vſually, or for ornament ſake, are in this method required. Finally, the partes & office of a Secretorie, in like maner, amplie diſcourſed. All which for the beſt & eaſieſt direction—for young learners, &c. are now, newlie, wholelie & ioyntly publiſhed, By Angel Day. Imprinted—dwelling at the Roſe & Crowne—1592." Dedicated "To—Edward de Vere, Earle of Oxenford, &c." whoſe arms are on the back of the title-page. The epiſtle to the reader is dated, 24 Jan. 1592. Licenſed by conſent of R. Waldegrave. W.H.₄ Quarto.

* The three triumphant days here intimated, were the days of her majeſty's birth, acceſſion, and coronation, "three moſt happy, ioyful & triumphant daies to England: through the which we triumphed 20295 daies with triumphs of Ouation:" as we read in the author's dedication to the queen: "euery day being a triumphant day, ſithence her Maieſties byrth vnto this preſent time." As in the chapter "Of the happy natiuitie of our gratious Queene Elizabeth."

"Pierce

RICHARD JONES.

" Pierce Penilefse his Supplication to the Diuell. Defcribing the 1592.
ouer-fpreading of Vice, and fupprefsion of Vertue. Pleafantly interlaced
with variable delights: and pathetically intermixt with conceipted re-
proofes. Written by Tho. Nafh, Gent. Imprinted—1592." 42 pages.
Licenfed. Mr. Reed. Quarto.

Sim. Harward his fermon, named " The Sumum bonum, or chief 1592.
Happines of a faithfull Chriftian:" preached at Crowhurft on Pfalm 1; 1."
Licenfed. Octavo.

" A Chriftiall glaffe for Chriftian women, contayning a difcourfe of 1592.
the godly life & chriftian death of miftris Katherine Stubs, who departed
this life at Burton on Trent, 14 Dec." Licenfed. Quarto.

" AVRELIA. The Paragon of pleafure and Princely delights: Con- 1593.
tayning The feuen dayes Solace (in Chriftmas Holy-dayes) of Madona
Aurelia, Queene of the Chriftmas Paftimes, & fundry other well-courted
Gentlemen & Gentlewomen, in a noble Gentlemans Pallace. A worke
moft fweetely intercourfed (in ciuill & friendly difputations) with many
amorous, and pleafant Difcourfes, to delight the Reader: and plentifully garnifhed with Morall Notes, to make it profitable to the Regarder:
By G. W." Device: a fweet-william &c. as in the frontifpiece. X, in
fours. " Printed by R- Iohnes, at the Rofe & Crowne, neere Holburne
Bridge, 1593." W.H. Quarto.

" Solace of Sion, and ioy of Ierufalem, &c. being an expofition on 1594.
the 87 pfalme, by Vrbanus Regius: tranflated by Rich. Robinfon."
Octavo.

" PAN HIS PIPE, conteyninge Three paftorall Egloges in Englyfhe 1594.
hexamiter, with other delightfull verfes." See Hift. of Englifh Poetry,
Vol. III; p. 405, note n.

" The good Hufwifes Handmaide for the Kitchin. Containing Manie 1594.
principall pointes of Cookerie, afwell how to dreffe meates after fundrie
the beft fafhions vfed in England & other Countries, with their apt and
proper fawces, both for flefh and fifh, as alfo the orderly feruing of the
fame to the Table. Hereunto are annexed, fundrie neceffarie Conceits
for the preferuation of health. Verie meete to be adioined to the good
Hufwifes Clofet of prouifion for her Houfhold." See it in 1580. 60
leaves. W.H. Sixteens.

" The Treafurie of commodious Conceits," as p. 1043. " Now 1594.
newly corrected, and inlarged with diuers neceffary phificke helpes," &c.
W.H.+ Alfo without date. Sixteens.

" Inftructions, Obferuations, and orders Militarie: requifit for all 1594.
Chieftaines, Captaines, and higher & lower men of charge, and Officers
to vnderftand, knowe, & obferue. Compofed by fir Iohn Smythe
Knighte, 1591, And now firft Imprinted, 1594." His device. " Imprinted—at the Rofe & Crowne, neer to Saint Andrewes church in Holborne, 1594." Dedicated " To the Knights, &c. of the Englifh Nation

tion, that are honorably delighted in the Art & Science Militarie.—From my house at Badewe in Essex, 1 May, 1594.—Io. Smithe." This epistle on four pages; a table of contents, and of faults, on four pages; and the Instructions, &c. on 220. Licensed. W.H. Quarto.

1595. These were published again the next year, with a new title-page, differing only in the mode, and orthography, except leaving out the word, *first*; changing 1594 to 1595; and, instead of *neer to Saint Andrewes*, has *next aboue S. Andrewes*. The epistle dedicatory now enlarged to 15 pages; but dated as before. W.H. Quarto.

1595. "MODERATVS, The most delectable & famous Historie of the Blacke Knight." Head title. Dedicated "To—Henry Townshend, Esq; one of her Maiesties Iustices of Assise of the countie Pallatine of Chester, &c. —Rob. Parry." Some verses in praise of the author, in Greek and Latin. In this romance, or Fancie, as the author calls it, are introduced several ditties, songs, &c. X 3, in fours. "Imprinted—at the Rose & Crowne, neer to S. Andrewes—1595." Licensed. W.H. 4 Quarto.

1595. "THE GARDEN of Prudence: Wherein is contained, A Patheticall Discourse, and godly Meditation, most brieflie touching the vanities of the world, the calamities of hell, and the felicities of heauen. You shall also find planted in the same, diuers sweet & pleasant Flowers, most necessarie and comfortable both for body & soule." His device, as in the frontispiece. "Printed—at the Rose & Crown next aboue S. Andrewes Church in Holborne, 1595." On the back is the Earl of Leicester's crest, with his motto. Dedicated "To—the most vertuous & renowmed Lady, Anne, Countesse of Warwick.—Bartholmew Chappell.—To the Reader, health," &c. The descriptions "Of the Vanities of the World The Calamities of hell, Of the felicities of heauen," are in verse; to which are added, "A praier to eschue worldly vanities,—to escape the calamities of hell,—to attaine heauenly felicities." The sweet flowers &c. are aphorisms on the virtues and vices alphabetically. F 3, in eights. W.H. Octauo.

1595. "The Arraignment & execution of 3 detestable witches, John Newell, Joane his wife, and Hellen Calles: two executed at Barnett, and one at Braynford, 1 Dec. 1595."

1595. A Dialogue of Constancie, &c. Quarto.

1596. "Corn. Shilander his Chirurgerie: Containing A briefe methode for the curing of Woundes & Ulcers; with an easy manner of drawing Oyle out of Wound-Herbs, Turpentine, Guiacum & Waxe: by Corn. Shilander. Translated out of Latin—by S. Hobbes." For Cutbert Burbie. G, in fours. W.H. Quarto.

1597. "THE WOMAN in the Moone. As it was presented before her Highnesse. By Iohn Lyllie maister of Artes." The prologue is on the back of the title-page. G 2, in fours. For Will. Jones. W.H. Quarto.

1597. Nic. Briton's "Boure of delights: Containing epigrams, pastorals, sonnets," &c. Licensed, in 1591. Quarto.

"THE

"THE ARBOR of amorous Deuices: Wherein young Gentlemen may 1597. reade many pleasant fancies, & fine deuices: And thereon meditate diuers sweete Conceites to court the loue of faire Ladies & Gentlewomen: By N. B. Gent. Imprinted—at the Rose & Crowne, neere S. Andrewes Church." Licensed.

"Gods arithmeticke: Written by Francis Meres, master of arte of 1597. both vniuersities, and student in diuinity." 25 leaves. Octavo.

"Three Sermons vpon some portions of the former lessons appointed 1599. for certaine Sabbaths: The first containing, A displaying of the wilfull de-uises of wicked and vaine vvorldlings: Preached at Tanridge in Surrey, 1 Feb. 1597. The two latter describing the dangers of discontentment and disobedience: Preached, the one at Tanridge, and the other at Crow-hurst, in July then next following: By Sim. Harwarde. Imprinted by Ri. Bradocke for Ri. Iohns, 1599." Dedicated "To—Iohn Lord Arch-bishoppe of Canterbury, primate, &c.—Tanridge, 2 Ian. 1598." D 3, in eights. These three sermons appear to have been printed so as to sell separate, occasionally; the second is as follows; the third i have not seen.

"A Sermon describing the Nature and Horrour of stubberne Disobe- 1599. dience: Preached at Tanridge in Surrey the xvi day of July—1598. And at the same time written to be added as a second part, to the danger of dis-contentment; By Simon Harwarde.—Imprinted by Richard Bradocke for R- Iohns, 1599." Dedicated "To—M. Michaell Murgatrod, Steward in the houshold to the reuerend Father in God, the Lord Archbishop of Canterbury his grace.—Tanridge, 2 Jan. 1599." The preface is dated, 18 Iuly, 1598. D 2, in eights. W.H. Sixteens.

"A most straunge and true Discourse of the wonderfull Judgment of 1600. God, of a Monstrous deformed Infant, begotten by incestuous copula-tion betweene the Brothers sonne, & the Sisters daughter, being both vn-married persons: Which Child was born at Colwall in the county and diocese of Hereford, vpon the sixt day of January last, being the feast of the Epiphany, commonly called Twelfth-Day, 1599. A notable and most terrible example against Incest and Whoredom." By J. R. See Catal. of Pamphlets in the Harl. Libr. No. 313. Printed for him. 9 leaves. Quarto.

"The Schoole of honest & vertuous lyfe: by Tho. Pritchard." Where-unto is added a Discourse of the worthines of honorable Wedlocke, by I R. Licensed, 1569. Quarto.

"A ryght pleasant and merye history of the Mylner of Abington, with his Wife and faire Daughter, and two poore Scholars of Cambridge." See Hist. of Eng. Poetry, Vol. I.; p. 432. Quarto.

"The wyll of the Deuyll; With his ten detestable Cōmaundementes: Directed to his obedient & accursed Chyldren; and the reward promised to all such as obediently will endeuer themselues to fulfil them. Where-unto is adioyned a Dyet for dyuers of the Deuylles dearlings, cōmonly called

called dayly Dronkardes. Very neceſſarie to be read, and well conſidered of all Chriſtians." B 4, in eights. Octavo.

———
"A worthy Myrrour, wherin ye may marke,
    An excellent Diſcourſe of a Breeding Larke."
Compoſed very finely, by Arth. Bour, to ſhew there is but little dependance to be had of friends, or kindred, but that each muſt do for himſelf.
It begins, "A Larke ſometime did breed, within a field of Corne:
    And had increaſe when as the Grayne was ready to be ſhorne."
    Ends, "God ſend her lucke to ſhun both Hauke & Fowlers Gin,
    And me the hap to haue no neede of Freinde, nor yet of Kin.
        Finis. Arthur Bour." Licenſed.   Broadſide.

He had licenſes alſo for the books following, viz. In 1565, "Theſioſus & Arradne. Luſus paſtorales: newly compiled." In 1566, "The naturall Judgement betwene Lyf & Death." In 1567, "A tragedy of Apius & Virginie." In 1568, "A remedy againſt the mutability of Fortune." See p. 1024. "Serten Inſtructions from an houſholder to his children. An Epetaphe of the worthy Lady, my Lady Knowelles." In 1569, "A ſtrange & petyfull nouell—of a noble Lorde & his Lady, with the tregicall ende of them & thayre ij cheldren, executed by a blacke morryon. An Epytaphe of the erle of Pembroke. A Larum to the truehearted ſubiectes of London. The moſte famous hiſtory of ij Spaneſſhe louers." In 1570, "An Epytaphe of my lady mares. Morrall pheloſiphe. A playne pathway to perfect Reſte. The Forreſt: a collection of Hiſtories." See p. 838. "An anſwere to a pologe thrawne abrode in the Couurte to fulkes."—In 1576, "Howe a yonge gentleman may behaue himſelfe in all cūpanies. The lief of the piſmarre moralyzed. Tarltons Toyes: in verſe. A floriſhe vpon Fancie: as gallante a gloſe of ſuche a trifling texte as euer was written: compiled by N. N. gent. To which are annexed manie pretie pamphletes for pleaſaūt heads to paſſe time awaie withall: by the ſame author. The Temptations of the Deuill: with remedies againſt the ſame. A handfull of hidden Secretes—certaine Sonets & other pleaſant deuiſes, pickt out of the cloſet of ſundrie worthie writers, and collected together by R- Will'ms. N. B. This book is entytuled *Delicate Dainties to ſweten buties lips withall*." This obliterated; and inſtead thereof is added, "A gorgious gallery of gallant inuetic'ons." In 1577, "A brief & dolefull Lamentac'on for the loſſe of the liues, aſwell of Sir Rob. Bell knight, lorde chief Baron of the Exchequer, as alſo of diuers other Juſtices, & men of good wurſhip at thaſſiſe laſte holden at Oxford. A merie reioiſinge hiſtorie of the notable feaſtes of Archerye of the highe & mightie prince William Duke of Shoreditche. A true declarac'on of the lamentable burninge, loſſe & deſtructōn of the market towne of the naas, within the countye of Kildare in Jrelande, the 3 of Marche, 1577. The portrature of a truſty ſeruaūt. The maner to dye well

RICHARD JONES.

well. Philemi Sifterno. The flower of Fame. The Booke of Witches. A tragicall memorie of the plagues of adulterye, by a late example of the death of 4 hanous trefpaffers: with the confeffion of Margaret Doryngton, one of the ffoure." In 1578, " A readie remedie againfte the laweles Luft of Loue. A mournefull memorie of the Deathe of Sir Rob't Bell: whereunto is added a fhort epitaphe of f'ieant Louelace. The payne of Pleafure : compiled by N. Britten. Eldertons folace in tyme of his ficknes, cont. fundrie fonets vpon many pithie parables. An Admonic'on to Catholickes, or rather—rebellious & trayterous Papiftes, intituled A glifter to their confiences that truft to coniurde Jmages. A pithie & pleafaūt difcourfe, dialoguewyfe, betwene a welthie citizen & a miferable foldier : brieflye touching the cōmodyties & difcōmodyties of warres & peace, by W. Warren. A merry deuyce concerninge Cardinge, &c. A cōmunicac'on betwene a carefull wyfe & hir comfortable hufband. The chippes of Saluation hewed out of the tymber of Faythe: Alfo, The Regiment of honeft life: by Tho. Pritchard. An epitaphe of my L. Keeper. Certen notable effectes of the Comet. The progreffe of ẏ Plage. Gods warninge fent as a proclamac'on. A pleafant fonet, Of the Joyfull ;receyuinge of the Q. maieftye into Norwyche, with the Dolor of the fame, at hir departure." See p. 983. " The poore knights poefies." In 1579, he with I. Charlewood bought of H. Denham the 15 following copies. " The Arbor of Amitie. Turberuiles Songs & Sonettes. The merrie meetinge of Mindes. Newes from Nynyue. The Caftell of a Chriftian. A ryche ftorehoufe for gentlemen. The greene forrefte. The 4th tragedie of Seneca. Newes out of Paules Churchyard. Palmiftrie. The pityefull ftate of the tyme prefent. Hilles Phifiognomye. The trauelled Pilgrym. A contemplac'on of Mifteries. Morrall Philofophie.---The 3d booke of the painfull pilgrim. A caveat, to beware of falfe cofeners by a late example of a Lancafhire man cozened of v.l. The merrie hiftorye of Steven Broome, howe he becam pope of Rome. The Labirinth of Libertye, written by Aug. Saker, gent. Upon Jones his promife to bring the whole impreff. into the Hall, in cafe it be difliked when it is printed. A Dittie of Mr. Turberuyle murthered, and John Morgan that murthered him: with a letter of the faid Morgan to his mother, and another to his fifter Turberuyle. The fatall fall of m'maduke Glouer. An earneft admonic'on to Repentaūce, vnto England, efpecially to London." In 1580, " A dolefull difcourfe of a Dutche gentlewoman diftraughte of hir wittes: To which is added the hard happe of twoo Norfolke gentlewomen." In 1581, " The exhortac'on of London vnto her children & f'uants to fubmiffion & obedience. Mr. Campion the fedicious Jefuit is welcome to London. The moft traiterous Proteftac'on of Hen. Euerit, al's Euerit Duckette. A doleful difcourfe of a lamentable fpoile done by Fyer in the towne of Faft Durham on Tewfdaie, 18 of Julie, 1581. The picture of two pernicious varletts called Prig Pickthanke & Clem Clawebacke, defcribed by a peeuifh painter. A true & dreadfull teftimonie of Gods wrath fhewed in the parifhe of

6 H   Llandillo

Llandillo in the countie of Merioneth. The manfion of Myrthe, by C. Edwardes." In 1582, " The Edict of tharchbp. & elector of Cullen touchinge the bringinge in of thexercife of Chriftian Religion within his iurifdiction." In 1586, " Mr. Dc'or Bridges praier. A praier for the quenes maieftie & the ftate, which aforetyme had ben printed by Mr. Day. Articles to be enquired of within the archd. of Worcefter. An Epytaphe on the death of Sir Yeuan Lloyd knight, by Hu. Gryffith, pryfoner." In 1587, "A fermon preached by Mr. Dc'or Morgan at the funeral of Sir Yeuan Lloyd knight, to be printed in Welche. A booke of four thinges: tranflated out of Latin." In 1588, " A miraculous & monftrous, but moft true & certen difcourfe of a woman (now to be feen in London) of the age of lx yeres, in the middle of whofe forehead, by the wonderfull woorke of God, there groweth out a crooked horne of 4 ynches longe." Entered to him, with Edw. White, Tho. Orwin and Hen. Carre. " The device of the Pageant borne before the Rt. hon. Martyn Colthorpe, Lorde maior of the Citie of London, 29 Octob. 1588." In 1589, " An Eglogue gratulatorie intitled, To the Rt. hon. and renowned Shepherd of Albions Arcadia Roberte Erle of Effex & Ewe, for his welcome into England from Portugall. The hyftory of Glaucus & Sylla. Tarltons repentance, or his farewell to his frendes in his ficknes, a little before his deathe. Articles of houfhold difcipline, containing Precepts & Prohibitions for a Xpian familye." In 1590, " A comedie of the plefant & ftatelie morral of the three lords of London. The twooe comicall difcourfes of Tomberlein the Cithian Shepparde." See p. 1047. " The life & fortune of Dōn Frederigo di Terra noua, &c. Sir Marten marr-people his Coller of Effes, Or fimple Sym-Sooth-Saier his fcrole of Abufes. An hundreth Godly inftructions. The Triumphes of the Churche, conteyninge the fpirituall Songes, & holie Himnes of godlie men, patriarckes & prophettes. The Shepherdes Starre, &c. dedicated by Tho. Bradfhaw to Therle of Effex." In 1591, " The hunting of Cupid, by Geo. Peele, M. A. of Oxford. A lamentable difcourfe of the deathe of the Rt. Hon. Sir Chr. Hatton, late Lorde Chancellor of England." In 1592, " The lamentac'on of Chr. Tomlinfon, horfe corfer, cōmonly called Kytt with the wry mouth, who killed his wife with a dagger, and was executed for the fame, 4 Dec. 1592, at Tyborne. A plefant fancie, or merrie conceyt, called the paffion at morrys, dauft by a crue of 8 couple of wores, all meere Enimyes to loue." In 1593, " A cōmedie entitled, A knack to knowe a Knaue, newlye fett fourth, as it hath fundrye tymes ben plaid by Ned Allen & his Companie: with Kemps applauded Merymentes of the men of Goteham. Oenōne & Paris, wherein is deciphered the extremitie of Loue/ the effectes of Hate/ the operation of them both. The mery metinge of Maydes of London:" This and five other copies, as p. 1032, between Ja. Roberts and him. In 1594, " Twoo bookes of Conftancie: to be tranflated into Englifh. Newes from Jack begger vnder the bufhe, with the aduife of Gregory Gaddefman his fellow begger, touching the deare prizes of Corne & hardnes of this
prefent

prefent yere. The fifhermans tale, conteyning the ſtorye of Caſſander a Grecian knight. A glaſſe for vaine glorious women: an inuectyue a-gainſt the ffantaſticall deuices in Womens apparell. Helens rape by the Athenian duke Theſeus. Certen queſtions—Merry-pate, the Knaue of Clubbes beinge Aunſwerer." He had licenſes likewiſe, at times, for a great number of ballads.

[b] Particularly 8, Aug. 1586, he had allowed to him 123 ballads, on theſe ij conditions, 1. That he, and none other have a lawfol Right. 2. That they be lawfull and allowable to be printed. For default of either of theſe, the entrance to be void. They were filed up in the Hall, with his name to every ſix; and " mentioned in a bill of R. Iones his own wrytinge filed up with the ballads."

HENRY MIDDLETON, or MIDLETON,

DWELT at the ſign of the Falcon in Fleet-ſtreet, and kept a ſhop in St. Dunſtan's Church-yard. The device he generally uſed, was the ſame as deſcribed in p. 902. He became a member of the Stationers' company 16 Feb. 1566-7; and very probably was the ſon of Will. Middleton, as he was made free without being found an apprentice with any maſter; and ſeems to have taken up his freedom by patrimony rather than by redemption, ſeeing he paid only iij.s. iiij.d. for his admiſſion; the uſual price. And, if ſo, this evinces his father to have been a member of the old company, who died before the charter of the preſent company was granted; and that the immunities of the former company were preſerved and valid in the new one. He was ſworn on the Livery 1 July, 1577, and ſerved Renter in 1582, and 1583. He was choſen Under Warden in July, 1587, but died [c] ſoon after.

He printed in partnerſhip with T. Eaſt from 1569 to 1572, for W. Norton, L. Harriſon, &c. and ſome books alone, for R. Newbery, F. Coldock, H. Denham, and R. Jones, which may be ſeen under their ſeveral names.

" Chriſtian praiers and holy meditations, as well for priuate as publike exerciſe, gathered out of the godly learned in our time by H. B. Now lately augmented, and newly imprinted againe. In the Euening &c. Pſ. 55. Imprinted at London by Henry Middelton—1570." In the calendar prefixed are noted the remarkable days mentioned in the Bible. " An   1570.

[c] Mrs. Middleton appears to have carried on the buſineſs after her huſband's deceaſe, till the 4 March, 1587-8, when ſhe was forbid by the company to print any more " till ſuch time as the Maſter, Wardens, and four of the Court of Aſſiſtants ſhall preſent her name to the High Commiſſioners for cauſes eccleſiaſtical, and that they admit her to be a printer, and governor of a preſſe and printing houſe, according to a decree of the Starr Chamber." Since we find nothing of her printing, ſhe ſeems not to have been approved of.

"An Almanack for xx yeares," beginning 1565. Thefe with the title-page are printed in red and black. 332 pages, and a table at the end. The whole printed with in black lines: the firft i have obferved. W.H.
Sixteens.

1573. "A Supplication to the Kinges Maieftie of Spayne, made by the Prince of Orange, the States of Holland and Zeland, with all other his faithfull fubiectes of the Low Countrie, prefently fuppreffed by the tyranny of the duke of Alba and their Spaniards. By which is declared the originall beginning of all the cōmotions & troubles happened in the faid Low Countrie: To the relief whereof they require his Maiefties fpeedy Redreffe & Remedie. Faithfully tranflated out of Duytfch—by T. W." 23 leaves. Octavo.

1573. "A brief and true rehèrfall of the Noble Victory & Ouerthrow, which, by the Grace of God, the proteftants of the North parts of Holland had againft the duke of Alba his fhips of Amfterdam: with the taking of the earl of Boffu, and their admiral Boffhuyfen, with diuers other Gentlemen, the 12 of October, 1573. Imprinted—in Fleeteftreet, at the fign of the Faucon,—and are to be fold at his fhop in St. Dunftanes Church-yarde." Seven leaves. Octavo

1573. "Erafmi Roterdami de duplici copia verborum ac rerum commentarii duo, multa acceffione nouifque formulis, Iocupletati." Octavo.

1573. "Hiftoriae Brittaniae, defenfio Joanne Prifeo eq." See p. 935. Quarto.

1573. "Saluftii Crifpi coniuratio Catilinae, et bellum Jugurthinum." Again 1583. Octavo.

1574. "Articles conteining the requeft prefented to the French king by the Deputies of the reformed Churches of the country of Languedoc, and other places adioyning, affembled by his Maiefties cōmaundement. Alfo another requeft to him prefented by the perfons of the third eftate of the countrey of Prouence. With his Maiefties anfwere to the faid requefts, &c. Tranflated out of French. For Tho. Cadman." Octavo.

1574. "Chriftian prayers and meditations." Some of them figned J. Bradford. Hereunto are annexed "The Letanie," with fome additional collects, and "Lydleys prayers." A calendar, and an almanac for 17 years, beginning 1572, are prefixed: a table at the end. Kk, in eights. "Imprinted—in Fletftreat, &c. 1574." W.H.+ Again 1587; alfo without date. Sixteens.

1575. "A DICTIONARY in Latine & Englifh, heretofore fet forth by Mafter John Veron, and now newly corrected and enlarged: For the vtilitie & profite of all young ftudents in the Latine tongue, as by further fearch therein they fhall find. By R. W." See p. 912. "Imprinted—for John Harifon, in Paules Churchyarde, at the figne of the white Greyhound, 1575." Sheets, Xx. Quarto.

1576. "A perambulation of Kent," &c. For him. Alfo, for R. Newbery. See p. 903.

"A true

HENRY MIDDLETON.

" A true reporte of the laſt voyage into the weſt and northweſt regions, 1577. &c. worthily atcheiued by captaine Forbiſher, of the ſayd voyage the firſt finder, and generall; with a diſcription of the people there inhabiting, and other circumſtances notable. Written by Dionyſe Settle, one of the company in the ſayde voyage, and ſeruant to the right honourable the earle of Cumberland." Quarto, and Octavo.

" The Sermons of M. Iohn Caluin vpon—Epheſians," &c. as p. 929. 1577. At the end, " Printed by *Him* for L. Harriſon & G. Biſhop." W.H.₊.
Quarto.

" A Commentarie of John Caluine vpon the firſt booke of Moſes, 1578. called Geneſis: Tranſlated out of Latine into Engliſh, by Tho. Tymme, Miniſter." In the magnificent compartment, late L. Harriſon's. See p. 926. Dedicated " To—Ambroſe, Earle of Warwicke," &c. 925 pages. On the back of the laſt, " Printed by H. Middleton, for John Hariſon, & G. Byſhop, 1578." This colophon over his device. W.H.
Quarto.

" D. HESKINS, D. SANDERS, AND M. Raſtel, accounted (among 1579. their faction) three pillars, and Archpatriarches of the Popiſh Synagogue, ouerthrowne and detected of their ſeuerall blaſphemous hereſies: By D. Fulke, Maiſter of Pembroke Hall in Cambridge. —For G. Biſhop, 1579." On the back, " The contentes of the ſeuerall treatiſes," &c. Then, " A catalogue of all ſuch Popiſh Bookes either aunſwered, or to be aunſwered, which haue bene written in the Engliſh tongue from beyond the ſeas, or ſecretly diſperſed here in England haue come to our hands ſince the beginning of the Queenes Maieſties reigne." 803 pages, finely printed on long-primer, Roman and Italic. W.H. Octavo.

" THE Defence of Militarie profeſsion: Wherein is eloquently ſhewed 1579. the due cōmendation of Martiall proweſſe, and plainly prooued how neceſſary the exerciſe of Armes is for this our age. —for Iohn Hariſon. 1579." Dedicated " To—Edward de Vere, Earle of Oxenford, &c.— 23 Dec. 1578,—Geffrey Gates." 63 pages. W.H. Quarto.

" SERMONS of M. John Caluin on the Epiſtles of S. Paule to T*i*mothie 1579. and Titus: Tranſlated out of French—by L. T. (*Leonard Tomſon*) Imprinted for G. Biſhop & T. Woodcoke, 1579." In the ſame compartment as the Commentary on Geneſis. There are 30 ſermons on the two Epiſtles to Timothy, and 17 on the Epiſtle to Tit us; occupying together (excluſive of the preface, and table, prefixed) 1248 pages, in double columns, neatly printed on Pica Roman. At the end is Middleton's device of the Loſt-ſheep, with the elephant. W.H. Quarto.

" The Diſplaying of an horrible Secte of groſſe and wicked Here- 1579. tiques naming themſelues the Family of Loue, with the liues of their Authors, and what doctrine they teach in corners. Newly ſet foorth by I. R. (*John Rogers*) Whereunto is added certein letters ſent from the
ſame

fame Family mainteyning their opinions, which letters are aunfwered by the fame I. R. —for G. Bifhop, 1579." Prefixed are " The Preface of the Author," and " Stephan Bateman to the gentle Reader." Contains befides, O, in eights. W.H.+ Sixteens.

1580. " Teftamenti veteris BIBLIA SACRA, five Libri canonici prifcae Iudeorum Ecclefiae a Deo traditi, Latini recens ex Hebreo facti, brevibúfque Scholiis illuftrati ab Immanuele Tremellio & Francifco Iunio: Accefferunt Libri qui vulgo-dicuntur Apocryphi, Latinè redditi & notis quibufdam aucti a Fr. Junio. Multo omnes quam ante emendatius editi, numeris locifq; citatis omnibus capitum diftinctioni quam hæc editio fequitur exactius refpondentibus: quibus etiam adjunximus novi Teftamenti libros ex fermone Syriaco ab eodem Tremellio in Latinum converfos." His device of the loft fheep; fmaller, and without the elephant. " Excudebat H. Middletonus, impenfis ' J. H. M.D,LXXX." This Bible is divided into fix parts, each with a feparate title-page, viz. The Pentateuch, The Hiftorical, The Poetical, The Prophetical, The Apocryphal: thefe laft four are dated 1579. The New Teftament, having Vautrollier's initials, i fhall place under his name; though it has the device of the loft fheep, as have all the others. The j and v properly ufed throughout. W.H. Again 1581,—" Jmpenfis W. N." but the New Teftament was printed by Vautrollier. W.H4. Again 1585, " impenfis C. B." The New Teftament with Beza's tranflation from the Greek, in one column, and Tremellius's from the Syriac on the other. W.H. Quarto.

1580. " CERTEINE comfortable expofitions of the conftant Martyr of Chrift, M. Iohn Hooper Bifhop of Glocefter & Worcefter, written in the time of his tribulation & imprifonment, vpon the XXIII, LXII, LXXIII, and LXXVII Pfalmes of the Prophet Dauid. Newly recognifed, and neuer before publifhed.—1580." It is introduced with an epiftle to the reader, by A. F. The Argument to the 23d Pfalm is on fol. 9, though only four leaves are before it, including the title. The leaves numbered to 129; the table on another. Licenfed. W.H. Quarto.

1580. " A Godly Sermon preached in Latin at Great S. Maries in Cambridge, in Marche 1580, by Rob. Some: and tranflated by himfelfe into Englifh. —for G- Bifhop, 1580." Dedicated " To—Mafter William Killigrew Efq; one of her Maiefties priuie Chamber, and Myftreffe Margerie Killigrew his wife.—Camb. 20. April, 1580." The text, Hof. 14; 3, 4. B, in eights. W.H. Sixteens.

1580. " The rooting out of the Romifhe Supremacie: Wherein is declared, that the authoritie which the Pope of Rome doth challenge to himfelfe ouer all Chriftian Bifhops and Churches, is vnlawfully vfurped: contrarie

---

d C. B. in Ames, p. 350. Pehaps a joint concern.

e The text is introduced with this fhort prayer, " That all may be done to Gods glorie and our edifying, let us pray to his Maieflie, that his holie worde, which is a-moft precious treafure be not as a SHUT BOOKE, and as a SEALED letter vnto vs. Our Father, &c."

to

to the expresse word and institution of our Sauiour Jesu Christ, who did giue equall power and authoritie to all the Apostles; Bishops and Ministers of his Church, whereof he is the true cornerstone, and only head: Set foorth by Will. Chauncie Esq; —for John Perin, 1580." Dedicated " To—Lorde Robert Dudley, Earle of Leycester, &c.—May 1580—To the Christian Reader." 141 pages; on the back of the last leaf is the author's coat of arms. W.H. Sixteens

" T. STAPLETON and Martiall (two Popish Heretikes) confuted, and of their particular heresies detected: By D. Fulke.—Seene and allowed. —for G. Bishop, 1580." Prefixed is " A catalogue of all such Popish Bookes," &c. as to D. Heskins, above. 217 pages. W.H. Octavo.

" Francisci Maldapetti, Navarreni, ad Everardum Digbeium, Anglum, admonitio de unica P. Rami methodo, rejectis caeteris, retinenda." for him. Sixteens. 1580.

" A GODLY and learned sermon, preached before an honourable auditorie. 26 Feb. 1580.—for Tho. Man. C, in eights. W.H. Sixteens. 1580.

" Psalmi Davidis ex Hebraico in Latinum conuersi, scholiisque pernecessariis illustrati, ab Immanuele Tremellio et Francisco Junio. —Impensis I. H." Sixteens. 1580.

" A Treatise: wherein is declared the sufficiencie of English Medicines, for cure of all diseases, cured with Medicine. —for Tho. Man." Dedicated " To—Lord Zouch.—T. B." 48 pages. W. H. Octavo. 1580.

Hieronymi Osorii Lusitani, De Gloria, Libri v. 1580.

" A Sermon of sure Comfort, preached at the Funerall of Master Rob. Keylwey Esq. at Exton in Rutland, the 18 of Marche, 1580: By Anth. Anderson Preacher and Parson of Medburne in Leicestershire. Printed— in Fleetstreete at the signe of the Falcon, 1581." 87 pages Sixteens. 1581.

" A GODLY sermon preached in the Court at Greenwich the first Sunday after the Epiphanie, Anno Domini 1552. And in the sixt yere of y raigne of king Edward the sixt, the right godly and vertuous king of famous and blessed memory, by B. G. *(Bernard Gilpin)* Imprinted—for Tho. Man, 1581." 76 pages. W.H. Also without date. Sixteens. 1581.

" A reioynder to Bristows replie in defence of Allens scrole of Articles and booke of Purgatorie, by William Fulke, D. D." 792 pages. Octavo. 1581.

" THE FAITH OF THE Church Militant, Moste effectualie described in this exposition of the 84 Psalme, by that reuerend Pastor and publike Professor of Gods word in the famous vniuersitie of Hasnne in Denmarke, Nicholas Hemmingius. A treatife written, as to the instruction of the ignorant in the groundes of religion, so to the confutation of the Jewes, Turkes, 1581.

HENRY MIDDLETON.

Turkes, Atheists, Papists, Hereticks, and al other aduersaries of the truth whatsoeuer. Translated out of Latine into English, &c. by Tho. Rogers. —for Andr. Maunsel, 1581." Dedicated "To the right hon. and virtuous Ladie, the Comitisse of Sussex, &c. 4 Nouember, 1581---To the hon. and for wisdome, godlines and vertue, the renowmed Lorde Peter Oxe, Lord of Gisselfelde, Master of the Palace both of the King, and also of the Kingdome of Denmarke, &c. N. Hemmingius." 534 pages besides ; and 2 tables. W.H.   Octavr.

1581.   " An Exposition of the Symbole of the Apostles, or rather of the Articles of Faith: In which the chiefe points of the euerlasting and free couenant betweene God and the faithfull is briefly and plainly handled. Gathered out of the catechising Sermons of Gasper Oleuian Treuir, and now translated out of the Latine tongue into English for the benefite of Christ his Church: By John Fielde. —for Tho. Man and Tobie Smith, 1581." Dedicated "To—Ambrose Earle of Warwicke, Master of the Queenes Maiesties Ordinance, &c —I. F.----To—Friderike, Countie Palatine by the Rhene, Duke of either Bauaria, Elector, &c.—Heydelberge, 19 Mar. 1576. G. O." 253 pages. W.H. Again 1582, for the same persons. W.H.   Octavo.

1581.   " Jesuitismi pars prima ; siue de praxi Romanæ curiæ contra resp. & principes ; & de noua legatione jesuitarum in Angliam, προθιεάτεια & premonitio ad Anglos. Cui adjuncta est concio ejusdem argumenti, Laur. Humfredo, &c. Edit secunda.—Impensis Geo. Byshop."   Octavo.

158.   " A Faithfvl and familiar exposition vpon the prayer of our Lorde Jesus Christ, and of the things worthie to be considered vpon the same: Written in French, Dialogue wise, by Peter Viret : And translated into english by John Brooke. —for Ric. Sergier, 1582." Dedicated "To— Syr Roger Manwood, Knight, and Lord chiefe Baron of the Queenes Maiesties Excheker.—I.B.---To the Christian Reader,—J. Fielde---To the hon. Lords the Burghmaisters and Counsell of Lausanna.—27 Aug. 1547. P. V." The first dialogue begins on signat. B. numbered 5, though the prefixes occupy 8 leaues. The tenth and last dialogue ends on fol. 196, as figured. An alphabeticall table annexed. WH.   Quarto.

1582.   The Institution of Christian Religion, &c. See p. 880.   Octavo.

1582.   " Of the end of this world, and second coming of Christ, a comfortable and most necessarie discourse, for these miserable and dangerous dayes. 1 Pet. 4. The end of all thinges is at hand: &c. Luke 21. Watch continuallie & praie, &c.—for Andr. Maunsell, 1582." On the back are verses by T. Beza on the new star which appeared—1572 & 1573. Dedicated "To—Edmond,—Archbishop of Canterburie, &c. and John, Bishop of London.—Tho. Rogers----To the vniuersal Church throughout the worlde, &c.—From Emden----The preface." 88 leaues. At the end, " Emden the 29 day of August, An. 1577." W.H. Again 1583.
Octavo.

" Hygieina

" Hygieina, id eft, de fanitate tuenda medicinae pars prima, authore 1583.
Timotheo Brighto Cantabrigienfi, medicinae doctore.—Impenfis T. M."
Alfo without date.   Octavo.
" Andreae Hyperij compendium Phyfices Ariftotelae, cum locuplete 1583.
Rerum & verborum memorabilium Indice. Excudebat—Impenfis Georgii
Bifhop. Anno 1583." From the late Mr. T. Martin's papers.

Babington's Expofition of the Ten Commandments. See it in 1586. 1583.

De officiis, M. T. Ciceronis Libri tres.   Sixteens.   1583.
" M. Tullii Ciceronis confolatio. Liber, quo fe ipfum de filiae 1583.
morte confolatus eft. Nunc primum repertus, et in lucem editus.—
Excud. pro Gul. Ponsonbio."   Octavo.
" Refponfionis ad Decem illas Rationes, quibus fretus Edmundus Cam- 1583.
pianus certamen Ecclefiae Anglicanae miniftris obtulit in caufa fidei, Defenfio contra Confutationem Ioannis Duraei Scoti, Presbyteri, Iefuitae. Authore Guil. Whitakero, &c. Impenfis Tho. Chardi." 887 pages, and
an Index. W.H.   Octavo.
" DE REPVBLICA ANGLORVM. The maner of gouernment, or policie of 1583.
the Realme of England, compiled by the Honorable man Thomas Smyth,
Doctor of the ciuil lawes, Knight, and principall Secretarie vnto the two
moft worthie Princes, king Edwarde the fixt, and Queene Elizabeth.
Seene and allowed. Printed for Gregorie Seton."* W.H. Again 1584.
                                                                                                               Quarto.
" The Sermons of M. Iohn Caluin vpon the fifth booke of Mofes, 1583.
called Deuteronomie : Faithfully gathered word for word as he
preached them in open Pulpet; together with a preface of the Minifters of the Church of Geneua, and an admonifhment made by the
Deacons there : Alfo there are annexed, two profitable Tables, the one
containing the chiefe matters, the other the places of Scripture herein
alledged. Tranflated out of French by Arth. Golding." His largeft
device as p. 992, but without the elephant. " Printed—for John Harifon, 1583." Dedicated " To—Syr Thomas Bromley Knight, Lord
Chancelour of England, &c.—21 Dec. 1582." Here are 200 fermons,
which (exclufive of the prefixes and affixes) occupy 1247 pages, in double columns ; very neatly printed on Long-primer Roman, and the
texts on Pica. The fignatures in fixes. W.H.   Folio.

" A briefe Conference betwixt mans Frailtie and Faith, wherein is 1584.
declared the true vfe and comfort of thofe bleffings pronounced by
Chrift in the fifth of Matthew, that euery Chriftian man and woman
ought to make and take hold of in their feueral tentations and conflicts.
—By Geruafe Babington.—Printed—for Tho. Charde, 1584." Dedicated " To—the Ladie Marie, Counteffe of Penbrooke.—1 Dec. 1583."
The conference and prayers, on 147 pages. W.H.   Octavo.

*By Dyonis Raguenier, borne in Bar, upon the river Seine.

HENRY MIDDLETON.

1584. " De succeſſione eccleſiaſtica, et latente ab Antichriſti tyrannide eccleſia, liber contra Thomae Stapletoni principiorum fidei doctrinalium librum decimum tertium. Authore Gulielmo Fulcone Anglo, aulae Pembrochianae in Cantabrigienſi academia praefecto.—Impenſis George Biſhop." Octavo.

1585. " The Sermons of that reuerend Father in God Edwin, Archbiſhop of Yorke, Primate of England, and Metropolitane." Head-title. 22 ſermons. —for T. Chard. W.H. Quarto.

1586. " The ſtorehouſe of breuitie in workes of arithmeticke, &c. Amended by the firſt author, Dionis Gray, of London, Goldſmith.—Printed for W. Norton and J. Harriſon." Licenſed to them in 1576. Sixteens.

1586. " Doctrinae chriſtianae compendium, ſeu commentarii catechetici, ex ore D. Zachariae Urſini, &c. Impenſis T. Chardi." Octavo.

1586. " A GODLIE Sermon preached the xxj day of June 1586, at Penſehurſt in Kent, at the buriall of the late Right honourable Sir Henrie Sidney, Knight of the—Garter, Lord Preſident of Wales, and of her Maieſties —priuie Councell, By Tho. White profeſſor in Diuinitie. Printed by Henrie Midleton. M.D.LXXXVI." C. 4, in eights. Licenſed. W.H. Octavo.

1586. " A verie Godlie and moſt neceſſarie ſermon ful of ſingular comfort for ſo manie as ſee their ſundry ſinnes, and are inwardly afflicted with a conſcience and feeling thereof. Preached at Ridlington in the Countie of Rutland, and penned at the importunate requeſt of ſome very Godlie affected: by Iohn Deacon Miniſter.—for Andr. Maunſel. M.D.LXXXVI." Text, Rom. 8; 1. On 84 pages. Octavo.

1586. " A very ſruitfull Expoſition of the Commaundements by way of Queſtions & Aunſweres for greater plainneſſe : Together with an application of euery one to the ſoule & conſcience of man, &c. By Geruaſe Babington.—for T. Charde, 1586."—Dedicated " To—Henrie Earle of Penbrooke, Lorde Harbert of Cardiffe, &c.—1 Dec. G. B.---To——Sir Edwarde Manxell, Sir. Edward Stradling, Sir William Harbert, knights, and to M. Will. Mathew, & M. Tho. Lewis Eſquiers, with all other Gentlemen in Glamorganſhire that feare God.—London, 1 Dec. G. B.----To the Godlie readers, and eſpecially to them amongſt whom this author and my ſelfe exerciſe our function.—Abr. Conham." 514 pages beſides; and a table. W.H. Octavo.

1587. " A godlie garden, out of the which moſt comfortable hearbes may be gathered for the health of the wounded conſcience of all penitent ſinners. Peruſed and allowed." Twenty-fours.

—— " Petri Baronis Stempani, ſacræ theologiæ in academia Cantabrig. doctoris ac profeſſoris, de præſtantia et dignitate dluinæ legis libri duo, in quibus varii de lege errores refelluntur, et, quomodo lex gratuitum Dei cum hominibus fœdus, ac Chriſtum etiam ipſum comprehendat, fidemque iuſtificantem a nobis requirat, explicatur; eaque doctrina ſacrarum

crarum literarum authoritate, theologorumque veterum ac recentiorum testimoniis, confirmatur. Adiectus est alius quidam tractatus eiufdem authoris, in quo docet expetitionem oblati a mente boni, et fiduciam ad fidei iuftificantis naturam pertinere.—Impenfis G. B." Octavo.

" A Booke of Epitaphes made vpon the death of the right worfhipfull Sir William Buttes knight, who deceafed 3 Sept. 1583." John Fenn Efq;
Sixteens.

" A Teftimonie of the true Church of God, confirmed as well by the Doctrine as Liues of fundry holy men, both Patriarkes & Prophetes, and alfo by the Apoftles and their true fucceffours: Wherein is manifeftly fhewed how that God hath in all ages rayfed vp fome, yea euen in the moft horrible darkeneffe, which haue beene faithfull Stewards, and true difpencers of his will, with a Catalogue of their names: Tranflated out of French by Will. Phifton.—Printed by H. M. for Tho. Charde, at the figne of the Helmet in Pauls Church yarde." Dedicated "To the worthie & right worfhipfull M. A. Nowell, Deane of Paules." 161 pages, and a table. W.H. Quarto.

" A Golden Chaine" &c. See p. 949. " Printed—by Henrie Middleton,⁵ 1589." Sixteens.

⁵ As Middleton died in 1587, this date muft be a mifprint; either by fetting the 9 the wrong way, for 1586, or putting the 8 for a 7, fuppofing it printed with Denham in 1579.

WILLIAM WILLIAMSON

WAS allowed a Brother of the Stationers' company, 19 Nov. 1562, for which he paid ijs. vjd. April 23, 1570, he was admitted a freeman, at the fame time as Sir Tho. Smyth, privy counfellor, and Sir Humf. Gilbert; each of them paying iijs. iiijd. for their admiffion. He kept Shop in Paul's church yard, at the fign of the White Horfe, in 1571, and afterwards at the Sun. The device he ufed was a triton on the fea blowing a conch within a ferpent, having the tail in its mouth, about which is this motto, " Immortality is gotten by the ftudy of letters," his cipher at bottom, with a fun over it, and 1573 under it: the whole in a very neat compartment.

" —A breefe and pithie fumme of chriftian faith." See it p. 838.   1571;

" An almanack and prognoftication for xxxiiii yeares, very profitable for all men, fpecially for phifitions, chirurgions, men of law, merchants, mariners, hufbandmen, and handycraftes men. Gathered out of Ciprianus Leontius works; with the refolutions at the ende, and all the mofte neceffarye rules, that are needfull to be put into any almanacke. With a fly for young mariners to practife themfelues in, very   1573.
eafie

WILLIAM WILLIAMSON

eafie to vnderftand. Gathered by Philip More, practicioner in phyfick and chyrurgery." Printed—for Anthony Kitfon. Octavo.

1573. " The moft excellent difcourfe of the Chriftian Philofopher Athenagoras, touching the Refurrection of the dead; englifhed from the Greek of Peter Nannius by Richard Porder." Octavo.

1573. " A Spirituall Pofeaye, contayning moft godly and fruictfull confolations, and prayers to be vfed of all men in the time of fickeneffe and mortalitie, as at all times elfe. Gathered out of the facred Scriptures, By N. Nichols, Minifter. Imprinted—in Pavvles Churchyarde—for John Harrifon, 1573." On the back begins the dedication, " To—M. Richard UVarren Efq;—Thus leauing anye further to trouble you, I befeeche God to fende you, and good Miftris VVarren your beddfellowe, long life, continual health, with the increafe of all godlineffe and vertues. From my houfe in Iremongerlane, 16 Octob. 1573. Nicholas Nichols, perfon there.—To the Chriftian Reader." G, in eights. W.H. Octavo.

1573. " De fyllabarum et carminum ratione, libri duo, auctore Rodolpho Gualthero Tygurino." Octavo.

1573. " The Arte of Warre,—An. M:D.LXXIII." &c. as p. 784. Hereunto is annexed " Certaine wayes for the ordering of Souldiours in battelray" &c. as ib. N, in fours. The laft leaf has only this colophon, " Imprinted: —by VV. VVilliamfon: for Ihon Wight Anno falutis M.D.LXXIII,. Menfe Septembris." W.H. Quarto.

1574. " Moft Briefe Tables to knovve redily hovve manye ranckes of footemen—go to the making of a iuft battayle," &c. as p. 784. I 1, in fours. Imprinted—in Paules Churchyarde,—for Iohn.VWight." W.H. Quarto.

1574. " The Paradife of the Soule côtaininig Chriftiã meditations with moft fruitful praiers: containing thefe books, 1 Of the fanctified loue of God, 2 Of Chriftian patience, 3 Of the innocent eftate of man that was loft, 4 Of the conflict of the foule & the flefh, 5 Of death, 6 Of the life of Chrift according to the holy Script. Tranflated by Will. Cleauer. Octavo.

—— " Spiritus eft Vicarius Chrifti in terra [b]. The poore mans Garden, wherein are flowers of the Scriptures & Doctours—: truely collected,— by Iohn Northbrooke, Minifter and Preacher of the woorde of God : And nowe newly corrected and largely augmēted by the former Aucthour. The Song of Songes made by Salomon, Chap. 2, verf. 12. The flowers appeare in the earth," &c. Dedicated " To—VVilliam, by the mercifull prouidence of God thorough Jefus Chrift, Bifhop of Excefter,—To the Chriftian Reader, &c.—Thine in the Lorde Jefu, Tho. Knel. Ju." 271 pages befides, and a table; at the end of which is his device as above mentioned. " Imprinted—in Paules Churche Yarde," &c. W.H.
Octavo.

[b] This motto is fet on the head of all books by this author.

Certayne

Certayne Newes of the whole defcription, ayde & helpe of the Chriftian Princes and nobles, the which for the comfort and deliverance of the poore Chriftians in the Low Countries are gathered together, and are now with their Armies in the Fielde: Drawn and copied out of a Letter fent vnto vs out of the fame campe. Printed at Dodrecht. Imprinted at London by W. Williamfon. Octavo.

THOMAS VAUTROLLIER,

A Frenchman, was a fcholar and printer, as it is faid, from Paris or Rouen, who came into England about the beginning of Q. Elizabeth's reign, and was admitted a brother of the Stationers' company, Oct. 2, 1564, for which he paid ij s. vjd. He fet up his prefs in Blackfriars, where it appears to have continued all his lifetime, notwithftanding his refidence for fome time in Scotland. Mr. T. Baker, in a letter to Mr. Ames, fays, "he was the printer of Jordanus Brunus in the year 1584, for which he fled, and the next year being at Edinburgh in Scotland, he firft taught that nation the way of good printing, and there ftaid until fuch time as by the interceffion of friends he had got his pardon, as appears by a book dedicated to the right worfhipful Mr. Thomas Randolph Efq; where he returns him thanks for his great favour, and for affifting him in his diftrefs: printed in octavo, 1587." He appears however to have been abfent from his bufinefs in July, 1581. See the note in p. 846. That he was in Scotland in 1584 and 1585 is evident by the books of his printing there; and yet the books printed in London the fame years bear his name alfo, as indeed all do till the time of his deceafe, which was before March 4, 1587-8; according to the [1] Order of that day, by the Court of Affiftants. As we find nothing of his printing in Scotland after 1585, 'tis highly probable he returned to London, foon after. He married his daughter Jakin to Richard Field, who fucceeded him in his houfe and bufinefs, 12 Jan. 1588; and buried feveral children in the parifh of Blackfriars, as appears by their church books. He was a moft curious printer, as will be evident to any who look into his books; and commonly ufed the device of an anchor and two fprigs of laurel within a compartment, with this motto, Anchora Spei; fome of them with the addition of a hand from the clouds holding the anchor by its ring, and the compartment more richly ornamented; alfo a vignette of a female head between two cornucopias, and T. V. under it: this he gene-

---

[1] ——, "That Mrs. Vautrolfier, late wife of Tho. Vautrolfier deceafed fhall not —hereafter print anye manner of book or books whatfoever, afwell by reafon that her hufband was noe printer at the tyme of his deceafe, as alfoe for that by the decrees fette downe in the Starre Chamber fhe is debarred from the fame."

rally.

rally used at the end of his books, prefaces, &c. He enjoyed several privileges [k].

1570. "A booke Containing Diuers Sortes of hands, as well the English as French secretary, with the Italian, Chancery & court hands: Also the true and iust proportiō of the Capitall Romaē. Set Forth by John de Beau Chesne, and Mr. John Baildon. Imprinted—dwelling in the blackefrieres, 1570." Dedicated " Illustrissimo comiti domino Arundelio, &c. —Vale, Londini in nostra typographia apud Carmelitas, quarto Kalendas Ianuarias, Anno à partu Virginis, 1569. Thomas Vatrolerus." This set of copies of the various hands usually written at that time is very ingenious and curious, and well deserves the character of the last of them in the book, " Cedo nulli." I apprehend them to have been written by Mr. Beauchesne, a schoolmaster in Black-friars, and cut on wood by Mr. Baildon. Licensed. Alexander Dalrymple, Esq;--- Broad quarto.

1574. Again 1574, when the dedication concludes thus, " Londini apud nostra typographia apud Jacobitas, quarto Kalendas Ianuarias, Anno— M.D.LXXIIII." Licensed. Dr. Lort.

1573. Liber Precum Publicarum in Ecclesia Anglicana. See it in 1574. Quarto.

1573. " De coniunctionibus magnis Insignioribus superiorum Planitarum, Solis defectionibus, & Cometis, in quarta Monarchia, cum eorundem effectuum historica expositione: Auctore Cypriano Leouitico, à Leonicia, Boemo, &c. His accessit, ab anno Domini 1564, in viginti annos sequentes prognosticon. In quo quid planetæ de proximo totius orbis Interitu portendat apertè ostenditur. Item pij cuiusdam viri, de stella quæ citra naturæ ordinem proximo mense Decembri aparuit epigrammate conclusum Iudicium.—ex officina—1573." L, in fours. W.H. Again 1575. Quarto.

1574. " JESV CHRISTI, D. N. NOVVM TESTAMENTVM, Theodoro Beza interprete. Additæ sunt summæ breues doctrinæ in Euangelistas, et acta Apostolorum. Item, methodus apostolicarum epistolarum ab eodem autore, cum breui phraseon et locorum difficiliorum expositione, ex ipsius autoris maioribus annotationibus desumpta, paucis etiam additis ex Joach. Camerarii notationibus in Euangelistas et Acta." Again frequently. Octavo.

[k] One privilege granted to Lodowik Lloyd, 18 Apr. 15 Eliz. 1573. for 8 years, for Plutarchus de vitis imperatorum, which was then translated by the said Lloyd. Another, dated 22 Apr. the same year, granted to him for 10 years. Also for these two books; Gia Silva Horgiardino cosmographico coltivato B. Sylva; and Puræ elegantes linguæ Latinæ phrases, olim ab Aldo Manutio Pauli filio conscriptæ jam vero in ordinem alphabeticam redactæ, & in Anglicum sermonem conversæ. Another privilege dated 19 Junii, 16 Eliz. 1574. granted to him for 10 years, for these following, viz. Novum Testamentum Bezæ, cum annotationibus, & sine annotationibus. Summæ Scripturæ thesaurus, in locos communes digestus. Aug. Marloratus adversus Gul. Fenguercei opera, in tabulas. Biblia Latina ex versione Sancti Pagnini. Synonymorum Sylva Kelegroms, ex Germanica in Latinam conversa. Ovidii Nasonis opera omnia. Ciceronis opera omnia, ex Lambini & doctissimorum virorum annotationibus; simul & separatim. Dialectica Rami illustrati per Makalmaneum. He had liberty also to retain in printing the said books, 6 Frenchmen or Dutchmen, or such like. Stat, Regist. B, fol. 487.

THOMAS VAUTROLLIER.

In the 16 of Eliz. 19 June 1574, a patent, or licenfe, was granted him, which he often put to the end of the new teftament, in thefe words, " Regiae maieftatis priuilegio cautum eft, ne quis hoc Nouum Teftamentum ex verfione, Th. Bezae interpretis infra decennium imprimat, aut alibi extra Angliae regnum impreffum diuendat, praeter illud, quod Thomas Vautrollerius, typographus Londinenfis in clauftro vulgo Blackfriers, commorans, fuis typis excuderit."

" LIBER PRECVM PVBLICARVM, SEV Minifterij Ecclefiafticæ adminiftrationis Sacramentorum, aliorumque rituum et cæremoniarum in Ecclefia Anglicana. Excudebat per Affignationem Francifci Floræ—Cum priuilegio Regiæ Maieftatis." The compartment as in the middle of our frontifpiece. 299 leaves befides the prefixes. A beautiful Pica Roman type. W.H. Again 1585. Octavo. 1574.

" Propheticæ et Apoftolicæ, id est, totius diuinæ ac canonicæ Scripturæ thesaurus,——ordine alphabetico digeftus. Ex Auguftini Marlorati aduersariis a Guiliel. Feuguereio in codicem relatus.—Cum priuilegio." This piece hath the approbation of Matthew Parker, archbifhop of Canterbury. " Lambethi, idibus Januariis," and the queen's privilege, dated at Greenwich, 19 June. Folio. 1574.

A BRIEF INTRODVCTION TO MVSICKE. Collected by P. Delamote, a Frenchman. Licenfed. Octavo. 1574.

" Difcantus cantiones, quæ ab argumento facræ vocantur, quinque et fex partium. Autoribus Thoma Tallifio et Guilielmo Birdo, Anglis, fereniffimæ reginæ maieftati a priuato facello generofis et organiftis.— Cum priuilegio." This piece has a remarkable dedication to the queen, and their patent at the end.* long Quarto. 1575.

The New Teftament, with diverfities of reading, and profitable annotations. An epiftle by J. Calvin prefixed. Printed for Chr. Barker. W.H+. Quarto & Octavo. 1575.

" Confeffio Chriftianae Fidei, et eiufdem collatio cum Papifticis Hærefibus: Per Theo. Bezam Vezelium. Adiecta eft altera breuis eiufdem Bezæ fidei Confeffio." His device. " Excudebat—1575." Dedicated " Meliori Volmario Rufo, præceptori.—Geneuæ, 4to Idus Martii, 1560." X 4, in eights. His vignette at the end. W.H. Again 1581. Octavo. 1575.

" A Commentarie of M. Doctor Martin Luther on the Epiftle of St. Paule to the Galathians." See it in 1577. Quarto. 1576.

" The Lyfe of the moft godly valeant and noble capteine & maintener of the trew Chriftian Religion in Fraunce, Jafper Colignie Shatilion fometyme greate Admirall of Fraunce. Tranflated out of Latin by Arthur Golding." His device. " Imprinted—1576." H 6, in eights.* W.H. Octavo. 1576.

" INSTITVTIO CHRISTIANÆ RELIGIONIS, Joanne Caluino authore. Additi funt nuper duo Indices, hac poftrema editione longè quam anteà caftigatiores, ab Auguftino Marlorato pridem collecti: quorum prior res præcipuas, pofterior expofitos facræ Scripturæ locos continet. Item accefferunt 1576.

cesserunt annotatiunculæ perutiles, de quibus agetur sequenti pagella." His device. "Excudebat—1576." Prefixed are, "Typographus Lectori.—Beza Ioanni Caluino Geneuensi Ecclesiastæ, in publico cœmenterio Geneuensi, nullo sepulchri apparatu, condito. Carmen τρίμικτον.—Iohannes Caluinus lectori,—Gen. Calend. August. Anno M.D.LIX.—Potentiss. illustrissimoq; monarchæ, Francisco Francorum Regi Christianissimo, principi suo, Iohannes Caluinus pacem ac salutem in Christo precatur.— Basileæ, Calend. Augusti, An. M.D.XXXVI.—Præcipua capita &c. Φραγμυκων τοῦ Πόρνου.—Ἐλληκαν τοῦ Ἰαιθέμου.—Eidem Florus Christianus.— Sonnet sur la mort de Iehan Caluin." The Institutions on 743 pages, neatly printed on Brevier Roman. At the end, are, A table of contents, an index by Calvin, and another by Marlorat. W.H. Octavo.

(1576.) "An Edict, or Proclamation, set forthe by the Frenche Kinge vpon the pacifying of the troubles in Fraunce, with the Articles of the same pacification: Read & published in the presence of the sayd King, sitting in his Parliament, the xiiij day of May, 1576. Translated out of Frenche by Arth. Golding." His device. Cum priuilegio. 64 pages, George Mason Esq; Sixteens.

1576. "Commentarius in epistolam Pauli ad Ephesios, scriptus a Nicolao Hemmingio." Quarto.

1576. "A Warning to take heede of Fowlers psalter, giuen by Thomas Sampson. Printed for G. Bishop." Sixteens.

1576. "A Chronographie continevved from the birth of Christ,—vnto the twelf yeare of the raygne of Mauricius the Emperour, &c.—faithfully collected—by M. H." Annexed to,

1577. "The Auncient Ecclesiasticall Histories of the first six hundred yeares after Christ, wrytten in the Greeke tongue by three learned Historiographers, Eusebius, Socrates & Euagrius;—whereunto is annexed Dorotheus, of the liues of the Prophetes, Apostles, and 70 Disciples,—faithfully translated—by Meredith Hanmer, Maister of Arte, &c.—1577." Dedicated "To—the Godly, wise & vertuous Ladie Elizabeth, Countesse of Lyncolne, wife to the right noble Edwarde Earle of Lyncolne, Lorde highe Admirall of England, &c.—London, 1 Sept. 1576.—The translator—touching the translation." Eusebius, 208 pages. Socrates and Evagrius have separate title-pages, dated 1576, as also has Dorotheus, dated 1577, but the paging is continued progressively; Socrates to page 402, and a short note to the reader; Evagrius to p. 512, Dorotheus to p. 521. At the end of the Chronography is an index for the whole. W.H. Again 1585. Licensed. W.H. Folio.

1577. "A Commentarie vpon the fiftene psalmes called Psalmi Graduum; that is, Psalmes of Degrees: faithfully copied out of the lectures of D. Martin Luther,—Translated out of Latine by Henry Bull.—Cum priuilegio. 1577." It has prefixed, an epistle to the reader by John Fox, and the preface of M. Luther. Pages 303. Licensed. W.H. Quarto.

1577. "Commentarius in epistolam Pauli ad Romanos, scriptus a Nicolao Hemmingio.—Impensis George Byshop." Quarto.

"A Com-

THOMAS VAUTROLLIER.

"A Commentarie of M. Doctor Martin Luther vpon the epistle of 1577.
S. Paul to the Galathians first collected and gathered vvord by vvord out
of his preaching, and now out of Latine faithfully translated into English for the vnlearned.—With a table in the ende &c.—Diligently reuised, corrected, and nevvly imprinted againe by — dvvellinge vvithin
the Blackefriers by Ludgate. Cum priuilegio. 1577." On the back
is a commendatory epistle " To the Reader—Aprilis 28, 1575. Edvvinus
London."---An addrefs " To all afflicted confciences, which grone for
Saluation" &c.   Fol. 296.   Frequently printed.   W.H.   Quarto.

"A shorte introduction of Grammar, generally to be vfed: &c. Im- 1577.
printed at London by the assignes of Frauncis Flowar, 1577." In a compartment of the same design as to Liber Precum Publicarum, but larger.
My copy has been stript of the laft leaf of the Accidence, and the
title page of the Grammar, probably for the sake of the cut and the
compartment. Sheets, L, in eights. At the end is a cut of boys gathering
apples; the engraver's initials R. B. Under it, 1577. W.H+ Quarto.

"Commentarius in epistolam Jacobi apostoli, scriptus a Nicolao 1577.
Hemmingio."   Quarto.

"In D. Pauli ad Romanos epistolam Exegema. In quo partium 1577.
dispofitio clariffime oftenditur. Authore Andrea Hyperio." Again
1583.   Octavo.

Tvvo Treatises written againft the Papiftes: the one being An an- 1577.
fwere of the Chriftian Proteftant to the proud challenge of a Popifh Catholicke; the other A Confutation of the Popifh Churches doctrine touching Purgatory & prayers for the dead; by Williame Fulke Doctor in
diuinitie." The firft treatise on 110 pages; the laft, 461 pages. A
table of the arguments of both is at the end. Licenfed. W.H. Octavo.

"QVÆSTIONVM ET RESPONSIONVM CHRISTIAnarum Libellus. In 1577.
quo præcipùa Chriftianæ Religionis capita κατ' ἐπιτομὴν proponuntur
Theodoro Beza Vezelio auctore.—Excudebat—impensis Georgij Bishop, 1577."—Dedicated " Ornatiff. viro, eruditione & virtute præstanti,
D. Io. C. P. Domino & amico plurimùm obferuando.---Summa capita
quæstionum," &c. Pages 139. W.H.   Octavo.

"QVÆSTIONVM ET RESPONSIONVM CHRISTIANARVM PARS ALTERA, 1577.
QVÆ EST de Sacramentis. Theodoro Beza Vezelio, authore. Adiectus
eft index quæstionum copiofus.—Excudebat—impensis Ioan. Harrifon,
1577." Pages 166. W.H.   Octavo.

Beza's tragedy of Abraham's sacrifice, tranflated by A. G. Octavo. 1577.

"A Soueraigne falue for a finfull foule, comprising a neceffarie and 1577.
true meanes wherby a finfull confcience may be vnburdened & reconciled to God: wherin you fhall find all the epithetons or titles of the fon
of God, which for the moft part are found in the fcriptures: by Nath.
Baxter, minifter." Licenfed.   Octavo.

"Special and chofen Sermons of D. Martin Luther, collected out of 1578.
his writings and preachings, &c,. Englifhed by VV. G. (Will. Gace.)
Philip. 3; 8, 9. I think all thinges but loffe &c. Imprinted—Cum
priuilegio, 1578." These 34 sermons are dedicated " To—Syr Thomas
Heneage

6 K

THOMAS VAUTROLLIER.

Heneage Knight, Treafurer of her Maieftie spriuie chamber." Then, "An admonition—by J. Foxe," Pages 481. He was fined for printing this book without licenfe, xs. W.H. Again 1581. Quarto.

1578. "An Excellent Treatife of Chriftian Righteoufnes, written firft in the French tongue by M. I. de l'Efpine, & tranflated into Englifh by I. Feilde, for the comfort of afflicted confciences, very neceffarie & profitable to be read of all Chriftians, as well for eftablifhing them in the true doctrine of Iuftification, as alfo for enabling them to confute the falfe doctrine of all Papiftes and Hereticks in that point." On the back are fome verfes "To the Reader." Dedicated "To the vertuous and my very deare Frend the Lady Elizabeth Terwhit, I. F. wifheth encreafe of godlines, &c.—Fare you well moft humbly from my poore houfe in Grubftreat this 2d of Nouember, 1577." Pages 128. Licenfed. W.H. Alfo without date. Octavo.

1578. "A Right Comfortable Treatife containing fourteene pointes of confolation for them that labor & are laden: Written by D. Martin Luther to Prince Friderik Duke of Saxonie, he being fore ficke, thereby to comfort him in the time of his great diftreffe: Englifhed by W. Gace." Dedicated "To the worfhipfull M. Henrie Dale, Citizen and Marchant of the citie of London—W. Gace.---To the famous Prince and Lord, the Lord Friderik, Duke of Saxonie, &c.—Martin Luther.---The preface." The Confolations on 84 pages. Licenfed. W.H. Again 1580. Octavo.

1578. METROMAXIA, fiue Ludus Geometricus, Auctore Gulielmo Fulcone Anglo.—Excudebat—1578." This very fcarce tract was unknown to bifhop Tanner, and is not found in the Bodleian or Harleian catalogues. See his Ouranomachia, p. 1007. Dedicated "Illuftriff. heroi ac domino, D. Roberto Dudleio comiti Leceftrenfi," &c. 51 pages, and a whole fheet fcheme. W.H. Alfo without date. Quarto.

1578. Lentulo's Italian Grammar, put into Englifh by Henry Grantham. See it in 1587. Octavo.

1578. The French Littleton. See p. 777.

1579. "The Historie of Guicciardin, conteining the warres of Italie and other partes, continued for many yeares vnder fundry Kings and Princes, together with the variations & accidents of the fame, deuided into twenty bookes:—Reduced into Englifh by Geffray Fenton. Mon heur viendra." His device, Anchora Spei. M. Ames had a copy with W. Norton's rebus. Dedicated "To the Queenes moft excellent Maieftie,—Elizabeth —Queene of England, &c.---At my lodging neare the Tower of London, vij Januarij, 1578." Then, a table of contents. Pages 1184, and an index. Licenfed to W. Norton and him. See p. 879. W.H. Folio.

1579. Ciceronis orationes. Lib. III. Printed frequently. Octavo & Sixteens.
1579. Phrafes linguæ Latinæ ab Manutio. Octavo.
1579. "De Rep. Anglorum inftauranda libri decem, Authore Thoma Chalonero Equite, Anglo. Huc acceffit in laudem Henrici Octaui Regis quondam Angliæ præftantifs. carmen Panegyricum. Item, De illuftrium quorundam encomiis mifcellanea, cum epigrammatis, ac Epitaphijs,

phijs nonnullis, eodem authore.—Excudebat—1579." On the back is the author's portrait, with the queen's arms at top, and five other coats of arms about the head, indicated in six Latin verses by G. M. "The prefixes are "Guliel. D. Burghleius, Thesaurarius Angliæ, in memoriam Thomæ Chaloneri Equitis aurati ab Hispanica Legatione redeuntis, & ad cælestes beatorum oras proficiscentis." In 32 Latin hexameters.---" Illustriss. viro—D. Gulielmo Cæcilio, aureæ periscelidis Equiti aurato, &c. —E schola Paulina, pridiè Idus Augusti. 1579.—Guliel. Malim.---Gul. Malim candido lectori."---Commendatory Latin verses, by L. Humfrey; W. Fleetwood, recorder of London; Chr. Carlile, and Edw. Webb of Rochester.---A Latin epistle from Bart. Clerk, dean of the court of Arches, to W. Malim.—vij Calendas Augusti, 1579.---Guliel. Malimi in singulorum decem de Repub. librorum argumenta, Ογδοάισ]χον." Contains besides, 379 pages; but the panegyric, and miscellany have separate title-pages. W.H. Quarto.

"A learned and very profitable exposition made vpon the cxi Psalme." 1579. Dedicated "To the right worshipfull the Master, Felowes, Schollers, and other Students of the Colledge of the holy & vndiuided Trinity in Cambridge.—By him which in the reuerende good will he beareth to you hath taken an obligation of himselfe dayly to pray for you.---The verses of M. Portus vpon thautor." The exposition on 112 pages, neat Long Primer Roman.* For printing this contrary to Orders he was fined xs. W.H. Sixteens.

Plutarch's Lives, from the French of Amyott, by Sir. Tho. North. Licensed to him and J. Wight. 1579. Folio.

Nich. Hemingius on the 25 Psalm, translated by Rich. Robinson. 1580. Octavo.

"Psalmorum Dauidis et aliorum Prophetarum, Libri quinque Argumentis & Latina Paraphrasi illustrati, ac etiam vario carminum genere latinè expressi. Nunc postremum recogniti & a variis mendis repurgati. Theo. Beza Vezelio Auctore.—Typis Tho. Vautrollerij & impensis Herculis Francisci, 1580." Dedicated "Illustrissimæ nobilitatis Domino, D. Henrico Huntingtoniæ comiti, &c.—Geneuæ, xvi Maii, anno vltimi temporis 1579."---Commendatory verses in Hebrew, Greek and Latin. 690 pages, neatly printed with Roman and Italic types of various sizes. W.H. 1580. Sixteens.

"Paraprasis psalmorum Davidis poetica, multo quam antehac castigatior; authore Georgio Buchanano, Scoto, poetarum nostri saeculi facilè principe. Eiusdem Buchanani trogoedia, quae inscribitur Jephthes— Impensis H. Denham." 1580. Sixteens.

"Jesv Christi D. N. Novvm Testamentvm, e lingua Syriaca Latino sermone redditum. Interprete Immanuele Tremellio, theologiae doctore et professore. Excudebat T. V. Typographus, impensis, J. H." But has H. Middleton's device, and is annexed to his edition of the Old Testament of this date. See p. 1085. W.H. 1580. Quarto.

"A Retentive to stay good Christians in the true faith & religion, against the motiues of Rich. Bristow. Also, A Discouerie of the daunger- 1580.

THOMAS VAUTROLLIER.

one Rocke of the popish Church commended by Nich. Sander D. of Diuinitie. Done by VVill. Fulke, D. D. &c.—for Geo. Bishop, 1580." Prefixed are, A catalogue of popish books either answered, or to be answered; and Dr. Fulke's request to all learned Papists, viz. to conclude the controversies of religion that are between us in the strict form of Logicall arguments. 316 pages. W.H. Octavo.

1580. " Claudii a Sancto Vinculo de pronuntiatione linguæ Gallicæ libri duo: Ad illustrissimam simulq; Doctissimam Elizabetham Anglorum Reginam. Dum spiro spero.—Excudebat—1580." Pages 150. To which is subjoyned, " Claudii a Sancto Vinculo de resurrectione Domini, ad consulem, et consulares vrbanos, ceterosque oncives Londinenses, oratio." Latin on one side, and French on the other. Together 199 pages. Licensed. W.H. Octavo.

1580. " A Treatise for declining of verbs, which may be called the second chiefest worke of the French tongue: Set forth by Claudius Hollyband. Pages 151. Octavo.

1581. " The Frenche Littelton, a most easie, perfect, and absolute way to learne the Frenche tongue: Set forth by Claudius Hollyband." 192 pages. Octavo.

1581. Apologia Eccl. Anglicanæ, authore Johanne Juello. Sixteens.

1581. " Psalterium Davidis carmine redditum per (*Helium*) Eobanum Hessum, cum annotationibus Vti Theodori Norimbergensis, que comentarij vice esse possunt: Cui accessit Ecclesiastes Solomonis, eodem genere Carminis redditus.—Excudebat, 1581." W.H. 4. Sixteens.

1581. " Ad rationes decem Edmundi Campiani Jesuitae, quibus fretus certamen Anglicanae ecclesiae ministris obtulit in causa fidei, responsio Guilielmi Whitakeri, theologiae in academia Cantabrigiensi professoris regii." Reprinted the same year with an index. Octavo.

1581. " A Briefe Discourse of Royall Monarchie, as of the best Common Weale: VVherein the subiect may beholde the Sacred Maiestie of tho Princes most Royall Estate. VVritten by Charles Merbury, Gent: in duetifull Reuerence of her Maiesties most Princely Highnesse. Wherunto is added by the same Gen. A Collection of Italian Prouerbes," &c. Inscribed " To his especiall good fauourers and Friends"---A commendatory addresse by " Henry Vnton to the vertuous Reader." 52 pages, and on another leaf, " M. Tho. Norton counsailler, and solliciter vnto the citie of London, hauing by th'appointment of the L. Bishop of London reade this Treatice, & diligently perused the same, maketh this reporte therof vnto the Reader.—verie commendable, and safely to be reade" &c. This book appears to have been printed only to be presented by the author to his friends. W.H. Quarto.

(1581.) " Marci Tullij Ciceronis Epistolæ ad Familiares a D. Lambino." See the note in p. 846.

1581. " Positions, wherin those primitiue circumstances be examined, which are necessarie for the training vp of children, either for skill in there booke, or health in their bodie. VVritten by Richard Mulcaster, master of the schoole, erected in London *anno* 1561, in the parish of saint Laurence

rence Pouuntneie, by the vvorshipfull companie of the merchaunt tailers, of the said citie." Dedicated "To the most vertuous Ladie, his most deare, & soueraine princesse, Elizabeth,—Queene of England, &c.—Author ipse ad librum suum," in 18 distichs.—The arguments of the 45 chapters. Pages 303, and the Errata on another. W.H. Againe 1587. Quarto.

"The First Part of the Elementarie which entreateth cheefelie of the 1582. right writing of our English tung, set furth by Richard Mulcaster." Dedicated "To—L. Robert Dudlie Earl of Leicester, &c.—Autoris ipsius ad librum suum," in 36 distichs.—"The titles handled in this book." Pages 272. I do not find the author ever prosecuted this subject, by printing a second part. W.H. Quarto.

"P. Ouidii Nasonis Metamorphoseon libri xv. ab Andrea Naugerio 1582. castigati, et vict. Giselini scholiis illustrati.—Cum priv." Sixteens.

"P. Ouidii Nasonis Fastorum libri vi. Tristium libri v. De Ponto lib. 1583. iiii. In Ibin. Ad Liviam.—Cum priv." Sixteens.

"P. Ouidii Nasonis Heroidum epistolae. Amorum libri iii. De arte 1583. amandi libri iii. De remedio amoris libri ii, &c. Cum priu." These are in a small neat type, not at all inferior to Gryphius's classics. Sixteens.

"Alexandri Dicsoni Arelij De vmbra rationis & Iudicij, Siue de me- 1583. moriae virtute Prosopopæia. Ad illustrissimum D. D. Robertum Dudlaeum comitem Licestaraeum," &c. On the sides of his device, "Sub vmbra illius, quam desideraueram Sedi." Octavo.

"An Answere to a Supplicatorie Epistle of G. T. for the pretended 1583. Catholiques: written to the right Hon. Lords of her Maiesties priuy Councell, By Walter Travers, Minister of the worde of God." Pages 389. Octavo.

"A learned and very profitable exposition made vpon the Cxi Psalme, 1583. for the comfort of the child en of God, by R.T." (Rob.Travers.) Sixteens.

"Campo di Fior, or else, The flourie field of four languages of M. 1583. Claudius Desainliens, alias Holiband; for the furtherance of the learners of the Latine, French, English, but chieflie of the Italian tongue." 387 pages. Sixteens.

"Institutionis christianae religionis, a Joanne Caluino conscriptae, 1583. epitome. Per Guil. Launeum, in Eccl. Gallicana ministrum." Again 1584. Octavo.

"Ad Nich. Sanderi demonstrationes quadraginta, in octauo libro vi- 1583. sibilis monarchiae positas, quibus Romanum pontificem non esse Antichristum docere instituit, responsio Guilielmi Whitakeri, theologiae in acad. Cantab. professoris regii," &c. Octavo.

"JVSTITIA BRITANNICA: Per quam liquet perspicue aliquot in eo 1584. regno perditos ciues, Seditionis & armorum ciuilium authores, regniq; hostium propugnatores acerrimos, vt communi Ecclesiæ Reiq; publicæ paci, cautius prospiceretur, morte mulctatos esse: propter Religionem vero; aut cæremonias Romanas, neminem in capitis discrimen vocarum: licet ab aduersarijs secus multò, & admodum malitiose publicetur. Perscriptum primo in nostrate lingua, Deinde versum in Latinam. Londini, Excudebat

THOMAS VAUTROLLIER.

Excudebat—Anno M.D.LXXXIIII, Anno Regni Reginæ Elizabethæ xxvi, Menſe Martio." 72 pages. Hereunto is annexed, with the ſignatures continued, "De Summa eorum Clementia, qui habendis quæſtionibus præfuerant contra proditores quoſdam, deq; tormentis quæ in eoſdem, ob Proditionem, non ob Religionem, exprompta ſunt." 13 pages. W.H.
Octavo.

1584. "The Cauteles, Canon, and Ceremonies of the moſt blaſphemous, abominable, and monſtrous Popiſh Maſſe: Togither *with* The Maſſe, intituled Of the Bodie of Jeſus Chriſt. Fully and wholy ſet downe both in Latine and Engliſhe.—With annotations for the vnderſtanding of the text, ſet forth by—Peter Viret, and tranſlated out of French—by Tho. Sto. Gent. Jmprinted—for Andr. Maunſell,—in Paules Churchyarde, at—the braſen Serpent. 1584." Dedicated "To—Syr William Cycell—Lorde High Treaſurer of England, &c.—London, 6 Feb. 1584.—Tho. Stocker.—-A Preface," &c. 256 pages. W.H. Octavo.

1584. "The hiſtory of the reformation of the church of Scotland, by John Knox." See Mackenzie's Writers of the Scots Nation, Vol. III; p. 136. Alſo, Bancroft's Survey, p. 48. Quarto.

1584. "Spaccio della Beſtia[1] Trionfante, propoſto da gioue, effettuato dal conſeglio, reuelato da Mercurio, recitato da Sophia, Vdito da Saultino, regiſtrato dal Nolano. Diviſo in tre dialogi, ſubdiuiſi in tre parti. Conſecrato al molto illuſtre et excellente caualliero Sig. Philippo Sidneo. Stampato in Parigi. M.D LXXXIIII."* Pages 261. The epiſtle dedicatory to Sir Phillip Sidney is on 15 leaves. Sixteens.

1585. "Giordano Bruno Nolano, degli Heroici Furori, al molto illuſtre & excellente Caualliero, Signor Philippo Sidneo.—Parigi appreſſo Antonio Baio." Octavo.

1585. Savanorola, Meditations Chretiennes, &c. Octavo.

1585. "Albercei gentilis de legationibus libri tres" Dedicated to ſir Phillip Sidney. Octavo.

1585. "D. Nathanaelis Baxteri Colceſtrenſis quaeſtiones, et reſponſa in Petri Rami dialecticam." 91 pages. Octavo.

1586. "A Treatiſe of Melancholie: Containing the Cauſes thereof, and reaſons of the ſtrange effects, it worketh in our minds and bodies; with the phiſicke cure, and ſpirituall conſolation for ſuch as haue thereto adioined an afflicted conſcience. The difference betwixt it & melancholie, with diuerſe philoſophical diſcourſes, &c. By T. Bright, Doctor of Phiſicke." Dedicated "To—M. Peter Oſbourne—from litle S. Bartlemews,—23 May, 1586.---To his melancholicke friend, M.---The contentes," &c. Pages 284, and on another the faults corrected. W.H. Octavo.

1587. "Novvm Testamentvm ad editionem H. Steph. impreſſum, et nunc cum vltima doctiſſimi Theodori Bezae editione diligenter collatum, cum quoad exemplar ipſum, tum etiam quoad obſcuriorum vocum, et loquendi generum interpretationes, margini aſcriptas." Sixteens;
"La

---

[1] This little atheiſtical book is ſaid to have been ſold for £50. Mr. Toland imagined his copy the only one extant. At the ſale of the late Ch. Bernard, eſq; in 1711 it ſold for £28. to Walter Clavel eſq; At that of James Weſt eſq; in 1773, this, with the following piece of Bruno's, ſold together for £9. 15s.

THOMAS VAUTROLLIER.

"La Grammatica di M. Scipio Lentulo Napolitano da lui in latina 1587. lingua Scritta & hora nella Italiana & Inglefe tradotta da H. G. An Italian Grammer, written in Latin by Scipio Lentulo a Neapolitane; And turned into Englifhe by Hen. Granthan." Dedicated " To the right vertuous Myftres Mary, and Myftres Francys Berkley, daughters to—Henry Lord Berkley.—4 Dec. 1574." Pages 259. W.H. Sixteens.

" Of Apoftacie: directed againft the Apoftates of the Church of 1587. Fraunce: By John de l'Efpine." Tranflated by *anonymous*. Octavo.

" M. Tullii Ciceronis orationum volumen primum a Joan. Michaele 1587. Bruto emendatum.—Impenfis J. Harifon. Cum priuilegio." Sixteens.

" The copie of a letter fent out of England to Don Bernardin Men- 1588. doza, ambaffadour in France for the king of Spaine declaring the ftate of England, contrary to the opinion of Don Bernardin, & of all his partizans, Spaniardes & others. This Letter, although it was fent to Don Bernardin Mendoza, yet by good hap the copies thereof as well in Englifh as French were found in the chamber of Ric. Leigh, a feminarie prieft, who was lately executed for high treafon, cōmitted in the time that the Spanifh Armada was on the feas. Whereunto are adioyned

" Certaine Aduertifments out of Ireland concerning the loffes & dif- 1588. treffes happening to the Spanifh Nauie vpon the VVeft coaftes of Ireland, in their voyage intended from the Northerne Ifles beyond Scotland towards Spaine. Pf. 118; (26.) This is the Lords doing, &c. For Ri. Field." Two editions. W.H. Quarto.

He had licenfes alfo for the following, viz. In 1581, "Paraphrafes aliquot pfalmorum Dauidis, carmine heroico: Scipio Gentili aucthore. Le Jardin de vertu, & bonnes meurs. per J. B. gent. Cadomois." In 1584, " Sūma veterum interpretum in vniuersā dialecticā Ariftot. Authore Joh'e Cafe, Oxon. Olim collegij Diui Joh'ls percurforis focio." In 1586 " La main Chriftienne aux tomber."

CHRISTOPHER BARKER Efq; Queen's printer,

CAME of an ancient family, being defcended, as Mr. Ames was informed, from Sir Chr. Barker, knight, king at arms. Edward Barker, who by a will dated 31 Dec. 1549, was appointed heir to one Will. Barker his coufin, might probably be father to Chr. Barker the printer. Q. Elizabeth granted him a very extenfive patent, in confideration of his great improvement of the art of printing, dated 28 Sept. 19 Eliz. 1577. He dwelt at firft in Pater-nofter-row at the fign of the Tiger's Head, and kept a fhop in St. Pauls churchyard at the fign of the Graffhopper; but he feems afterwards to have dwelt wholly in St. Paul's churchyard, at the * Tiger's

---

* He fet up the fign of the Tiger's Head, and ufed the fame in various devices to his books; whereby he feems to announce Sir Francis Walfingham for his patron, whofe creft it appears to have been, being placed over againft the Walfingham arms on the fides of a compartment ufed to feveral of his octavo volumes.

Tiger's Head. He had also a printing office in Bacon house, near Foster-lane, in which he printed Acts of Parliament, &c. Originally he was a member of the worshipful company of Drapers, but translated to that of the Stationers, 4 June 1578, and taken into their livery on the 25th following. For a dispensation from serving the Renter's office he paid a fine of £3; but seems to have been excused, through favour, serving Under Warden. The upper wardenship he served in 1582, and 1585. From the year 1588 he seems to have carried on the printing business wholly by deputies. He died 29 Nov. 1599, in the 70th year of his age, and was buried at Datchet near Windsor in Bucks; where against the North wall of the chancel of St. Mary's church was erected an handsome altar monument of white and black marble, supported by two pillars of red-veined marble: at the top on the dexterside are these arms, Or, on a fesse dancetty azure, three fleur de lis of the first, being "Barker's: On the sinister side, are the same arms impaled with these, Party per chevron or and azure three mullets counterchanged, being, as is supposed, the arms of William Day, bishop of Winchester, whose daughter was wife to Robert Barker, the son and successor of our Christopher. On a black marble tablet was the inscription below °.

Mr. Ames says the Barkers continued printing till 1599, when Christopher the father died. I do not find that they either printed, or assigned jointly; but the business was carried on in the father's name only, till about 1588, and from thence forward by his assignment to deputies. I have therefore alloted each a separate place.

I do not find any book of his printing before 1574, but it appears that he had licenfe for printing in 1569.

1574. "Lady Eliz.ᵖ Tirwits Morning and Euening praiers, with diuers Psalmes, Himnes & meditations. Printed for Christopher Barker." Sixteens.

1575. The Bible. Lambeth list. N. Testam. See p. 1067. Octavo.
1575. "The Glasse of Gouernment, a tragicall comedie, so entituled, because therein are handled as well reward for vertues, as also the punishment for vices. Done by Geo. Gascoigne esq: Blessed are they that fear the Lord, their Children," &c. The Dramatis personæ, on the back. Then, "The Argument.---The Prologue.---This vvorke is compiled vpon these sentences—set downe by mee C. B.---In Comœdiam Gascoigni carmen

○. From a pedigree in the Herald's office, entered by his grandson, Matth. Barker, 1634, it appears that Chr. Barker the printer bore Or, on a fess dancette vert, three fleurs de lis argent.

* Piæ Memoriæ
Christopheri Barker armigeri,
Serenissimæ Reginæ Elizabethæ typographi,
Qui typographiam Anglicanam
Lateritiam invenit, Marmoream reliquit:
Parentis optimi,
Et
Rachaeli Barker dilectissimæ conjugis,

Quæ duodena prole suscepta,
Undena superstite,
Quotquot illam noscere contigit,
Triste suum desiderium reliquit
Posui.
Obiit Julii 13, Anno Domini 1607,
Ætatis suæ 35.

ᵖ Q. Elizabeth had a copy of this book bound in solid gold, which is now in the possession of Mrs. Ashby, of Barrow, Suffolk. It was reprinted in The Monument of Matrons, by H. Denham.

CHRISTOPHER BARKER. 1077

then B. C." in 5 diftichs. Under them is his rebus, fomewhat like that in our frontifpiece; over it, " A Barker if ye will, In name, but not in fkill." Dedicated " To—Sir Owen Hopton knight, hir Maiefties Lieutenant in hir tower of London.—26 Apr. 1575." The comedy¹ N 1, in fours. At the end, " Imprinted at London by H. M. for C. Barker, at the figne of the graffehopper in Paules churchyarde." W.H. Quarto.
Gafcoign's poems. Printed for him. See p. 978, & 990. Quarto. 1575.
¶ The Booke of Faulconrie or Hauking," &c. as p. 977. At the end 1575. of the book of Hunting, " Jmprinted by H. Bynneman for Chr. Barker." Quarto.

The Bible, &c. as p. 800. " Imprinted—by Chr. Barker, dwelling 1576. in Povvles Churchyard at the figne of the Tygers head, 1576. Cum priuilegio." On the back of the title-page of the New Teftament is a device which he frequently ufed afterwards, viz. A tiger's head with a ducal coronet about his neck, erafed out of a mural crown, with a crefcent for difference, having this legend on a riband twirling down to a lamb ftanding on a hill, " Tigre: reo animale del Adam vecchio figliuolo, merce L'Euangelio fatto n'eftat agnolo." The whole in a compartment. The fame again at the end of the Revelation. This Bible is very neatly printed on Long primer Roman, with the arguments in Italic. The Old Teft. 365 leaves, the Apocrypha 84 leaves, the N. Teft. 115 leaves. W.H. Again 1577. Short folio.

The New Teftament after Beza, tranflated by L. Tomfon.' Again 1576. 1577, and frequently. Octavo.

" A treatife of the Excellencie of a Chriftian man, and how he may be 1576. knowne: Written in French by M. Peter de la Place, one of the kings councel, and chief prefident of his court of Aides in Paris. Whereunto is added the life and death of the faid author. Tranflated by L. Tomfon." Dedicated to Mrs. Urfula Walfingham.* Again 1577, and 1585. Octavo.

A CHRISTIAN CONFESSION of the late mofte noble and mightie 1577. Prince, Frederich of that name the third, Count Palatine by ŷ Rhein, &c. wherein conftantlie and merkelie he departed out of this world the 26 of October—1576. Taken word for word out of his laft will & teftament. Whereunto is added the Lantgraue his anfwere to the French King." Device, a fmall compartment with 1576 in it, and his creft at top. " Imprinted—by Chr. Barkar, 1577." This book is introduced with a preface by John Cafimire, count Palatine, &c. F 7, in eights. W.H. Sixteens.

" DE OLIVA EVANGELICA Concio in Baptifmo Iudæi habita Londini, 1578. primo menf. April. Cùm enarratione capitis vndecimi D. Pauli ad Romanos. In qua, de principijs & fundamentis Chriftianæ fidei, de vera & fyncera ecclefia, de Chrifto Meffia, eiufq; regni æterna amplitudine, atq; infinita gloria, Difputatio cum Iudæis ex Propheticæ fcripturæ cer,

---
¹ This is omitted by Abel Jeffes in what he entitles " The whole woorkes of Geo. Gafcoigne, efq;"
' See Lewis, p. 264, &c.
* Under-Secretary to Sir Fran. Walfingham.

6 L tiffimis

tissimis testimonijs instituitur. Per Ioan. Foxium. Londini, Ex officina Chr. Barkeri, Reginæ Maiestati Typographi, 1578." Dedicated "Magnifico cum primis & inter patricios summates Landavissimo viro D. Francisco Walsinghamo Reginæ Elizabethæ consiliario, illiq; à secretis comentarijs intimo.—Londin. Die Natalis Dom. Tuus in Christo, Ioan. Foxius.---Præfatio parænetica." The sermon on 77 pages. To this is added, with the signatures continued, " De Christo triumphante eiusdem autoris," on 23 pages. Then, " Gemitus piorum ad Christum pro maturo redditu," in 30 distichs. The whole neatly printed on Roman and Italic types. W.H. Octavo.

1578. " A Sermon preached at the Christening of a certaine Iew; at London, by John Foxe, conteining an exposition of the xi Chap. of S. Paul to the Romanes. Translated out of Latine by Iames Bell. Imprinted—by Chr. Barker, Printer to the Queenes Maiestie, at the signe of the Tygres head in Pater Noster Rowe." In a neat compartment with the queen's arms at top; on the dexter side, a tiger's head, with a crescent for difference erased out of a mural crown, but without the ducal coronet about its neck; on the sinister side is the coat armour of Sir Fr. Walsingham; a tablet at bottom, in which is " Anno 1578." Dedicated "To—Sir Francis Walsingham Knight, principall Secretarie to her Maiestie.— Lond. Anno 1578.—Iohn Foxe.---The Preface" &c. N 2, in eights. On the last page " Imprinted &c. 1578," over the device of the tiger's head and lamb, as to the Bible 1576. Licensed. W.H. Sixteens.

1578. " A verie godlie & necessary Sermon preached before the yong Countesse of Comberland in the North, 24. Nouember 1577. By Chr. Shutt. 1 Pet. 4; 7 The end of all thinges is at hād, &c. Eccles. xii; 13, 14. Let vs heare the ende of all: feare God," &c. In the same compartment as to the translation of Fox's Sermon: on the tablet at bottom, " Imprinted—by Chr. Barker, Printer to the Queenes Maiestie." Dedicated " To—the Margaret Countesse of Comberland." G 4, in eights. On the last leaf, " Imprinted—1758;" over the device as to the Bible 1576. Licensed. W.H. Octavo.

1578. " The Bible' Translated according to the Ebrew and Greeke" &c. In a new compartment with the queen's arms at top, which are seated by Religion and Iustice; at the bottom, a tablet supported by the lion and and dragon, with C. B. under it: the tablet is blank, but has a label pasted on it, with " Imprinted at London by Chr. Barker, Printer to the Queenes Maiestie: Cum gratia & priuilegio Regiæ Maiestatis" The title-page to the New Testament, and the colophon, are dated, 1578. W.H. Again frequently, in various sizes. Folio.

1578. " A Most Excellent and comfortable Treatise, for all such as are any maner of way either troubled in mynde or afflicted in bodie, Made

---

¹ This edition has the former version of the psalms, as to the Great Bible, placed beside those of this Geneva translation: this seems to be the reason why they are not annexed to the Common Prayer prefixed; and which has only references to find the epistles and gospels. Mr. Lewis calls this a middling folio, but my copy is as large as any of the Great Bibles. Perhaps he printed two sizes. See Hist. of Eng. Transl. of the Bible, p. 271, &c.

by Andr. Kingsmyl Gent. sometime fellowe of Alsolne Colledge in Oxforde. Whereunto is adioyned a verie godlie and learned Exhortation to suffer patiently all afflictions for the Gospel of Christ Iesus: And also a conference betwixt a godly learned Christian & an afflicted conscience: &c. Imprinted—Cum priuilegio Regiæ Maiestatis, 1578." This is introduced by a short addrefs "To the Christian Reader, by. f. myllea the publisher, and friend of the author, deceased. R, in eights. W.H. Again 1585. Sixteens.

"ALLARME To England, foreshewing what perilles are procured, 1578. where the people liue without regard of Martiall lawe: with a short discourse conteyning the decay of warlike discipline, conuenient to be perufed by Gentlemen, such as are desirous by seruice to seeke their owne deserued prayse, and the preservation of their countrey: Newly deuised and written by Barnabe Riche, Gent.—Peruſed and allowed, 1578." Dedicated "To—Syr Christopher Hatton Knight, Captaine of her Maiesties Garde, Vicechamberlaine &c.---To the gentle and friendly Reader.--- To the valiant Captaines &c.---To my very louing friend Captaine Barnabe Riche—Barnabe Googe."---Then follow commendatory verses by Lod. Flood, Tho. Churchyard, S. Stronge, and Tho. Lupton; others by the author, "Why he tooke in hand to write this booke." K 2, in fours. Licenſed. W.H.   Quarto.

"A Fourme of Prayer with Thankſgiuing to be vſed &c. euery year, 1578. the 17th of Nouember, the day of her Highneſs entrie to her Kingdome."ᵃ Licenſed. Quarto.

"The true vnderſtanding of theſe words: This is my bodie, &c. By 1578. Tho. Eraſtus. Alſo Bezas treatiſe of the Sacraments in generall, Tranſlated by John Shutt." Licenſed. Sixteens.

"AN EXCELLENT and godly ſermon, moſt needefull for this time, (1578.) wherein we liue in all ſecuritie & ſinne, &c. Preached at Paules Croſſe 26 Octob. 1578, By Laurence Chaderton, B. D." In the ſame compartment as to Fox's ſermon abovementioned. Dedicated "To M. Richard Martin, Warden of the Queenes Maieſties Mints, & Alderman of—London.---To the Chriſtian Reader." H 2, in eights. Licenſed. W.H. Alſo 1580. Sixteens.

"Two right profitable and fruitfull Concordances, or large & ample 1578. Tables Alphabeticall.—And will ſerue as well for the tranſlation of the Bible called Geneva as for the other authorized to be read in Churches: By Rob. F. Herrey. Frequently printed. W.H. Quarto.

"A Concordance or table containing the principal woordes & mat- 1579. ters which are comprehended in the N. Teſtament, By T. W." Octavo.
The New Teſtament: The Biſhops' tranſlation. Sixteens. 1579.
"An abſtract of all the penal ſtatutes, which be general, in force and 1579. vſe, wherein is conteyned the effect of all thoſe ſtatutes, which do threaten to the offenders thereof the loſſe of life, member, lands, goods, or any other puniſhment, or forfaiture whatſoeuer. Collected by Ferdinando

---

ᵃ In the entry, "A pſalme or ſong of ‖ the 17 day of Nov. for the Queues Mapmiſe and thankſgiuing to be ſonge on ‖ ieſtie."

Pulton,

CHRISTOPHER BARKER.

Pulton, of Lincolnes-Inne gent. Deut. 17. 8. &c.—20th October." See p. 374. Printed again 1581, 1586. Quarto.

1579. Articles—in the Convocation, 1562. Put forth by the queen's authority. These, with the Queen's Injunctions, were frequently printed; in a compartment with the queen's arms at top, Faith and Humility on the sides, the tiger's head at bottom, and the ensigns of the 4 Evangelists in the corners. Quarto.

1579. "Of the principall pointes which are at this day in controuersie concerning the holy Supper of Iesus Christ, and of the Masse of the Romaine Church, by Pet. Viret; translated by Iohn Shut." Licensed. Octavo.

1579. "A Notable Treatise of the Church, in which are handled all the principall questions that haue bene moued in our time concerning this matter, by Philip of Mornay, Lord of Plessis Marlyn, &c. Translated by Io. Fielde." Dedicated "To—Robert Dudley, Earle of Leycester, &c. ---To the most excellent Prince Henry King of Nauarre, &c. To the Reader." 384 pages. Licensed. W.H. Again 1580; and 1581, without the word Notable. Octavo.

1579. Will. Rastall's Collection of the Statutes, alphabetically. Again frequently. Folio.

1580. "Sweet Consolac'on for all such as are afflicted & oppressed with the weight and burden of their sinnes: wherunto is adioyned a treatise against the fear of Death; and of the happynes of the faithfull after this life: Gathered out of the writings of sundry learned and Godly Fathers, by Jo. S." Licensed. Octavo.

1580. "The order of Prayer and other Exercises to auert & turn Gods wrath from us, threatned by the late terrible Earthquake, &c. With a prayer to be vsed of Housholders euery Euening; a godly admonition; and a Report of the Earthquake, which happened on Easter Wednesday, 6 Apr. 1580, somewhat before six of the clock in the afternoon." Licensed. Quarto.

1580. "A Preparation to a due consideration and reuerent coming to the holy Communion of the Body & Blood of our Lorde.—Printer to the Queenes most excellent Maiestie." F, in eights. 1580. Licensed W.H. Again 1585. Also without date. Thirty-two's.

1580. "An answere to a seditious pamphlet lately cast abroade by a Iesuite, with a discouerie of that blasphemous sect: By Will. Charke. 1 Reg. 20; 11.—Imprinted &c. 1580. Decembris 17." G, in eights. Licensed. W.H. Again 1581. Octavo.

1580. "A Conference or Dialogue discouering the sect of Jesuites: written in Latin by Christian Francken, and translated by W. C." Octavo.

1580. A Proclamation against New Buildings & Inmates. 7 July. 22 Eliz.

1580. — against the sectaries of the Family of Loue. Given at our Manor of Richmond 3 Octob.

1580. Articles to be enquired—in the province of Canterbury. Quarto.

1580. — in the Diocess of Chester. Dr. Lort. Quarto.

1581. "A declaration of the recantation of Iohn Nichols (for the space almost of two yeeres the Popes Scholer in the English Seminarie or Colledge

ledge at Rome) which defireth to be reconciled & receiued as a member into the true Church of Chrift in England.—Imprinted—Anno 1581, Februarii 14." Dedicated " Illuftri Equiti, clarifsimoq; Regiæ Maieftatis Londini Caftrorum præfidi, cuftodiq; infigni, præftanti, ac nobilitate veræ fapientiæ, & virtutis infignibus decorato, Domino Odoeno Hopton, ---The preface to the gentle Reader." N 2, in eights. W.H. Sixteens.

" A REPLIE to a Cenfure written againft the two anfwers to a Iefuites 1581. feditious Pamphlet: By Will. Chaike. 1 Reg. 20; 11.—Imprinted &c. 1581." Q 7, in eights. Licenfed. W.H. Octavo.

" EIRENARCHA:" &c. as p. 823. N. B. The arms and creft on the 1581. fides of the compartment are Sir Fr. Walfingham's; not Barker's. Octavo.

" An aduertifement and defence for Trueth againft her *Majefty's* 1581. backbiters, and fpecially againft the whifpring Fauourers, and Colourers of Campions, and the reft of his confederats treafons." One fheet. At the end, " God faue the Queene, long to reign to his honour." W.H.
Quarto & Octavo.

" Lectures of I. B. vpon the xii Articles of our Chriftian faith, briefely 1581. fet forth for the comfort of the godly, and the better inftruction of the fimple & ignorant. Alfo hereunto is annexed a Confeffion of the Chriftian faith, conteining an hundreth articles, according to the order of the Creede of the Apoftles: Written by that learned & godly martyr J. H. fometime Bifhop of Glocefter in his life time." The former treatife feems to have been printed alone, and dedicated " To Sir Francis Walfingham knight principal Secretarie" &c. in 1579, by Iohn Baker. The latter to have been annexed on account of the fimilarity of its fubject; yet, having a particular title-page, might fell feparate, occafionally; the fignatures however are continued in it to H h 2, in eights. At the end, " Imprinted—1581." Licenfed. W.H. Again 1583, 1584. Octavo.

" A true report of the arraignement and execution of the late popifh 1581. traitor, Euerarde Haunce; executed at Tyborne." Octavo.

Acts of Parliament: " Anno xxiii Reginæ Elizabethæ—holden at 1581. Weftminfter the xvj day of Ianuarie—and there continued vntil the xviii day of March following." &c. 63 leaves. Colophon, " Imprinted—in Bacon houfe, neere Fofter lane." W.H. Folio.

" Chriftian Meditations vpon Eight Pfalmes of the Prophet Da- 1582. uid, &c. by Theodore Beza: Tranflated out of Frenche by I. S.—Imprinted in Bacon Houfe" Licenfed. Again 1589. Sixteens.

A treatife on the facraments generally; and efpecially of Baptifme and 1582. the Supper: by John Prime. Licenfed. Octavo.

" EIPHNAPXIA. De peccatiffimo Angliæ ftatu, imperante Elizabetha: 1582. Compendiofa oratio." Licenfed to T. Woodcock and him. Quarto.

" Variæ meditationes et preces piæ, variis defignatæ vfibus, Latino, 1582. Italico, Gallico & Anglico fermone confcriptæ. Excudit Chr. Barkerus Juffu." Licenfed. Quarto & Octavo.

" A particular declaration or teftimony of the vndutifull & traiter- 1582. ous affection borne againft her Maieftie by Edmond Campion Jefuite,
and

CHRISTOPHER BARKER.

and other condemned Prieftes, witneffed by their own confeffions: in reproofe of thofe flaunderous bookes and libels deliuered out to the contrary by fuch as are malitioufly affected towards her Maieftie & the ftate. Publifhed by authoritie"—An. Do. 1582." 14 leaves. W.H.
Quarto.

1583. " A true report of the Difputation, or rather priuate Conference, had in the Tower of London, with Edm. Campion, Iefuite, the laft of Auguft 1581. Set downe by the Reuerend learned men themfelues, that dealt therein. Whereunto is ioyned, alfo a true report of the other three dayes conferences, had there with the fame Iefuite. Which nowe are thought meete to be publifhed in print by authoritie.—Januarii 1, 1583." Licenfed.* W.H.
Quarto.

1583. Certaine Sermons preached before the Queenes Maieftie, and at Paules croffe, by the reuerend father Iohn Jewel late Bifhop of Salisburie. Whereunto is added a fhort Treatife of the Sacraments, gathered out of other his fermons made vpon that matter in his cathedrall Church at Salifburie.—1583." Dedicated " To—Sir William Cicil Knight, &c. and to—L. Robert Dudley, Earle of Leicefter, two moft worthie Chauncelours of the Vniuerfities, Oxforde and Cambridge.—Iohn Garbrande." Z, in eights. Licenfed to him & Fr. Coldock, W.H. Octavo.

1583. " Articles to be enquired in the vifitation, in the firft yeere of the raigne of our mofte dread Soueraigne Lady Elizabeth," &c. as p. 717. "Anno Domini 1583." In the fame compartment as the Injunctions, and the Articles of Religion. Cum priuilegio, &c. 7 leaves. W.H. Again 1591.
Quarto.

1583. " A Declaration of the fauourable dealing of her Maiefties Commiffioners appointed for the Examination of certaine Traitours, and of tortures vniuftly reported to be done vpon them for matters of religion. 1583." 4 leaves. W.H.
Quarto.

1583. " The Execution of Iuftice in England for maintenaunce of publique and Chriftian peace, againft certeine ftirrers of fedition, and adherents to the traytors and enemies of the Realme, without any Perfecution of them for queftions of Religion, as is falfely reported and publifhed by the fautors and fofterers of their treafons. xvii Decemb. Imprinted at London, 1583." Reprinted with the following alteration in the title-page, " Secondly Imprinted at London menfe Ian. 1583, An. Reg. Eliz. 26. With fome fmall alterations of thinges miftaken or omitted in the tranfcript of the firft Originall." Each, 20 leaves. W.H. both. Quarto.

1584. " A difcouerie of the treafons practifed, and attempted againft the Queenes Maieftie, and the Realme, by Francis Throckemorton, who was for the fame arraigned and condemned in Guyld Hall, in the citie of London, the 21 day of May laft paft." 14 leaves. W.H. Quarto.

1584. " A true and plaine declaration of the horrible Treafons, practifed by Will. Parry the Traitor, againft the Queenes Maieftie. The

* This book is very fcarce, and reckoned the beft vindication of the proceedings againft Edm. Campion.

maner

maner of his Arraignment, Conuiction, and execution, together with the copies of fundry letters, of his and others, tending to diuers purpofes, for the proofes of his Treafons. Alfo an addition not impertinent thereunto, containing a collection of his birth, education, and courfe of life." &c. 53 pages. W.H. Alfo without date. Quarto.

" Articuli per archiepifcopum, epifcopos, et reliquum clerum Cantu- (1584.) arienfis prouinciæ, in fynodo inchoata Londini, 24to menfis Nouembris, anno Domini 1584, Regniq; fereniffimæ in Chrifto principis, Dominæ Elizabethæ, Dei gratia Angliæ, Franciæ, & Hyberniæ reginæ, fidei defenforis, &c. 27mo, ftabiliti, et regia auctoritate approbati & confirmati. Londini, in ædibus C. B."* Quarto.

" A true and fummarie reporte of the declaration of fome part of the Earle of Northumberlands Treafons, deliuered publiquelie in the Court at the Starrechamber by the Lord Chancellour, and others of her Maiefties moft Hon. Priuie Counfell, and Councell learned, by her Maiefties fpecial cōmandement, together with the examinations & depofitions of fundrie perfons touching the maner of his moft wicked & violent murder cōmitted vpon himfelfe with his owne hand, in the Towre of London, the 20 day of Iune, 1585. In ædibus C. Barker." 22 pages befides the preface. W.H. Quarto. 1585.

" A Declaration of the Caufes moouing the Queene of England to giue aide to the Defence of the people afflicted, and oppreffed in the Lowe Countries." * W.H. 14 leaves. (Alfo in Latin.) Octavo. 1585.

" Certain prayers, and other godly exercifes, for the 17th of Nouember, wherein we folemnize the bleffed reigne of our gracious Souereigne lady Elizabeth. By Edmund Bunny, fubdeacon of York." Quarto. Acts. An. 27 Eliz. "Imprinted—in Baconhoufe, neere Fofter Lane." Folio. 1585.

" An Order* of Prayer and Thankfgiuing for the preferuation of her Maieftie and the Realme from the traiterous and bloody practifes of the Pope, and his adherents: to be vfed at tymes appointed in the preface." Licenfed. Quarto. 1586.

" Orders deuifed by the efpeciall commandment of the queenes maieftie, for the reliefe and ftaie of the prefent dearth of graine within the realm." Quarto. 1586.

" The Copie of a Letter to the right hon. the Earle of Leycefter, Lieutenant generall of all her Maiefties forces in the vnited Prouinces of the lowe Countreys, written before, but deliuered at his return from thence: With a report of certeine petitions &declarations made to the Queenes Maieftie at two feueral times from all the Lordes and Commons lately affembled in Parliament: And her Maiefties anfweres thereunto by her felfe deliuered, though not expreffed by the reporter, with fuch grace and life as the fame were vttered by her Maieftie. Imprinted—Printer to the Queenes moft excellent Maieftie, 1586." The queen's arms, very neatly cut on wood, facing the title-page. 32 pages. W.H. See it in Holinfhed, p. 1580, &c. As alfo, Quarto. 1586.

* This, with feveral other forms of prayer, are in one volume, at Emanuel College, Cambridge: given by Abp. Sancroft.

A Pro-

1586. A Proclamation concerning the fentence againſt Mary queen of Scots. Richmond, 4 Dec. 1586.
— Concerning corn. Greenwich, 2 Jan.
1587. Acts of Parliament. 29 Eliz. W.H. Alſo, Folio.
1587. The whole volume of ſtatutes at large, from magna charta, to 29 Elizabeth, With marginal notes, and a table. Again 1588. Folio.
1588. " A Fourme of prayer neceſſarie for the preſent time & ſtate:" with a preface ſuited to the occaſion. Printed by his deputies. Quarto.
1588. " Y BEIBL CYS-SEGR-LAN. SEF YR HEN DESTAMENT, A'R NEWYDD, 2 Tim. iii. 14, 15.—TESTAMENT NEWYDD EIN Harglwydd Jeſu GRIST, Rom. i. 16." Dedicated by William Morgan to the queen. 155 leaves.[y] Folio.
1588. " Pſalmau Dafydd or' vn cyfieithiad ar Beibl. cyffredin. Jaco. 5; 13." Quarto.
(1588.) " A Proclamation againſt certaine ſeditious and ſchiſmatical[a] bookes, libels, &c. 13 Feb." By his deputies. Broadſide.
1588. " A godlie treatiſe, containing and deciding certaine queſtions, mou'd of late in London and other places, touching the miniſtrie, ſacraments, and church; written by Rob. Some, D. D." Preface dated 6 Maij, 1588. 36 pages. Quarto.
1588. " A Defence of ſuch Points in R. Somes laſt treatiſe as M. Penry hath dealt againſt: And a refutation of many Anabaptiſtical, blaſphemous & Popiſh abſurdities touching Magiſtracie, Miniſtrie, Church, Scripture and Baptiſme, &c. conteined in M. Penryes treatiſe, &c. By R. Some, D. D." The preface dated 19 Sept. 1588. The pages and ſignatures are continued from the former treatiſe, to p 200. Both printed by G. B. (Geo. Biſhop) his deputy. W.H.+ Quarto.
1588. " A packe of Spaniſh lyes, ſent abroad in the world: firſt printed in Spaine in the Spaniſh tongue, and tranſlated out of the originall. Now ripped vp, vnfolded, and by iuſt examination condemned, as conteyning falſe, corrupt, and deteſtable wares, worthy to be damned and burned." 8 leaves. By his deputies. W.H. Quarto.
1589. " An Admonition to the People of England: VVherein are anſvvered, not only the ſlaunderous Vntruethes, reprochfully vttered by Martin the Libeller, but alſo many other Crimes by ſome of his broode, obiected generally againſt all Biſhops, and the chief of the Cleargie, purpoſely to deface and diſcredite the preſent ſtate of the church. Detractor et libens auditor vterque Diabolum portat in lingua. Seene and allowed by authoritie." The epiſtle " To the Reader" is ſigned[a] T. C.----" The Con-

---

[y] Weſtm. Library. M.S. on a ſpare leaf, " Gulielmus Morgan, ſacraeTheologiaeprofeſſor, hanc Britannicam ſacrorum bibliorum tranſlationem Weſtmonaſterienſi bibliothecae dono dedit, viceſimo die menſis Novembris, anno reſtaurationis humani generis milleſimo quingenteſimo octogeſimo octavo." Then ſome Hebrew. " Teſte Jaſpar Gryffy." W. Jones, eſq;
[p] Shewing that they were " ſlanderous to the ſtate, and to the eccleſiaſtical gouernment eſtabliſhed by law.—That they ſhould immediately be brought in and deſtroyed; and that no author, printer, or diſperſer, ſhould dare to offend herein vnder pain of her Maieſty's diſpleaſure, and being procecuted with ſeverity."
[a] Tho. Cooper, at firſt biſhop of Lincoln, and tranſlated to Wincheſter in 1584.

ten s

tents of this treatife." 252 pages. Deputies. Another edition in 245 pages. W.H. Quarto.

" Declaratio cauſarum quibus ſereniſſimæ Maieſtatis Angliæ Claſsiarij 1589. adducti, in expeditione ſua Luſitancaſi, quaſdam naues frumento, alióq; apparatu bellico ad vſus Hiſpaniarum Regis, in vicinis Baltici maris regionibus comparato, dum ab ijs in Vliſsiponam tenditur, atq; in ipſis faucibus Vliſsiponæ, ceperunt. 30 Junij An. Domini 1589, ac regni Maieſtatis ſuæ 31.—Excudebant Deputati Chr. Barkeri, Sereniſsimæ Reginæ Angliæ Maieſtati Typographi, 1589." 10 leaves. W.H. Alſo in Engliſh. Quarto.

" The Text of the New Teſtament of Ieſus Chriſt, tranſlated out of 1589. the vulgar Latine by the Papiſts of the traiterous Seminarie at Rhemes. With Arguments of Bookes, Chapters & Annotations, pretending to diſcouer the corruptions of diuers tranſlations, and to clear the controuerſies of theſe dayes. Whereunto is added the tranſlation out of the original Greeke, commonly vſed in the church of England, With a confutation of all ſuch Arguments, Gloſſes & Annotations as contein maniſeſt impietie, of hereſie, treaſon & ſlander againſt the Catholike Church of God, and the true teachers thereof, or the tranſlations vſed in the Church of England: Both by aucthoritie of the holy Scriptures, and by the teſtimonie of the ancient fathers. By William Fulke, D. D. Imprinted—by the Deputies—1589." Dedicated to Q. Elizabeth: with a copious preface, &c. 496 folios, and a table of controverſies. W.H. Folio.

" A Godly Treatiſe, wherein are examined & confuted many execra- 1589. ble fancies, giuen out & holden, partly by Hen. Barrow and Iohn Greenwood: partly by other of the Anabaptiſtical order: Written by Rob. Some D.D." Dedicated To—Sir Chr. Hatton knight, L. Chancellour—, and Sir William Cecill knight, Baron of Burleygh, L. High Treaſurer, &c. and Chancellours of the Vniverſities of Cambridge & Oxford.—At my Lordes Grace of Canterburie his houſe in Lambeth, 12 Maij, 1589." Pages 40. Printed by G. B. his deputy. W.H. Quarto.

" A Forme of Prayer, thought fit to be dayly vſed in the Engliſh 1589. Armie in France." Deputies. Quarto.

" The principall Nauigations &c. of the Engliſh nation." See p. 914. 1589.

Acts of parliament, " Anno xxxj Reginæ Elizabethæ." Deputies. 1589. W.H. Folio.

" The Communion booke, or Booke of Common prayers and admi- 1590. niſtration of the ſacraments, &c. Alſo the Act of Parliament for vniformitie, alſo the tracte of ceremonies why ſome be aboliſhed & ſome retained.—1590." Deputies. Folio.

" The third part of the bible (after ſome diuiſion) conteyning fiue 1591. excellent bookes, moſt commodious for all Chriſtians: faithfully tranſlated out of the Ebrewe, and expounded with moſt profitable annotations vpon the harder places." Deputies. Sixteens.

" A Declaration of great troubles pretended againſt the Realme by a 1591. number of Seminarie Prieſts and Ieſuits, ſent, and very ſecretly diſperſed in in the fame, to workɇ great Treafons. vnder a falfɇ. pretence of Religion: With a prouifion very neceffarie for a remedie thereof. Publifhed by this her Maiefties Proclamation." Deputies. W.H. Quarto.

1592. " Η ΚΑΙΝΗ ΔΙΑΘΗΚΗ. Nouum Jefu Chrifti D. N. Teftamentum. Cum obfcuriorum vocum & quorundam loquendi generum accuratis interpretationibus margini adfcriptis. Londini, Excudebant Reg. Typog. Anno falutis humanæ, cɪɔɪɔxcɪɪ." Matthew 1 ; 11.. is read thus, Ἰωσίας δὲ ἐγέννησε τὸν Ἰακεὶμ. Ἰακεὶμ δὲ ἐγέννησε τὸν Ἰεχονίαν καὶ τοὺς ἀδελφοὺς αὐτῶ, &c. Jofias autem genuit Jacheim: Jacheim autem genuit Jechoniam, et fratres ejus, &c. in exceeding neat Greek letter, 463 leaves. Sixteens.

1592. " Conspiracie, for Pretended Reformation, viz. Prefbyteriall Difcipline. A Treatife difcouering the late defignments and courfes held for aduancement thereof, by William Hacket Yoeman, Edmund Coppinger, and Henry Athington, Gent. out of others depofitions and their owne letters, writings, & confeffions vpon examination: Together with fome part of the life and conditions, and the two Inditements, Arraignment, and Execution of the fayd Hacket. Alfo, An anfwere to the calumniations of fuch as affirme they were mad men: and a refemblance of this action vnto the like, happened heretofore in Germanie. 20 Sept. 1591." 100 pages.* In my copy " Vltimo Septembris, 1591.—Publifhed now by authoritie.—1592." Pages 102, befides an epiftle " To the Reader," and " The preface," prefixed. Deputies. W.H. Quarto.

1592. " An Anfwere to a certaine Libel Supplicatorie, or rather Diffamatory, and alfo to certaine Calumnious Articles, and Interrogatories, both printed and fcattered in fecreet corners, to the flaunder of the Ecclefiafticall ftate, and put vnder the name and title of a Petition directed to her Maieftie: Wherein not onely the friuolous difcourfe of the pititioner is refuted, but alfo the accufation againft the Difciplinarians his clyents iuftified, and the flanderous cauils at the prefent gouernement difciphered. By Mathew Sutcliffe. Dedicated " To—Sir Edm. Anderfon Lord Chief Iuftice of her Maiefties court of Common Pleas.—20 Decemb.--- The Preface." 208 pages, and " An Aduertifement," &c. Deputies. W.H. Quarto.

1592. " Orders thought meete by her Maieftie, and her priuie Councell to be executed throughout—this Realme, in fuch Townes, &c. as are or may be hereafter infected with the plague, for the ftay of further increafe of the fame. Alfo, an aduife fet downe vpon her Maiefties expreffe cōmaundement, by the beft learned in Phyficke within this Realme, containing fundry good rules & eafie medicines, without charge to the meaner fort of people, afwell for the preferuation of her good fubiects— before infection, as for curing & ordering them after—infected." D 2, in fours. W.H. Alfo without date. Quarto.

1593. " De imperandi Authoritate, et Chriftiana Obedientia, Authore Hadriano Saravia. Excudebant Reg. Typogr." Quarto.

1593. " Certaine Praiers collected out of a fourme of godly Meditations fet foorth by her Maiefties authoritie in the great Mortalitie, in the fift yeere of her Highneffe raigne, and moft neceffarie to be vfed at this time

in the like prefent vifitation of Gods heauie hand for our manifolde finnes. July 1593." See p. 721. Deputies. 8 leaves. W.H. Quarto.
Acts of parliament, " Anno xxxv Reg. Elizabethæ." Deputies. 1593. W.H. Folio.
" The perpetual gouernment of Chriftes Church.. Wherein are 1593. handled; The fatherly fuperiority which God firft eftablifhed in the Patriarkes for the guiding of his Church, and after continued in the Tribe of Leui and the Prophetes; and laftlie confirmed in the New Teftament to the Apoftles & their fucceffors: As alfo the points in queftion at this day; Touching the Iewifh Synedrion: the true kingdome of Chrift: the Apoftles cōmiffion: the Laie Prefbiterie: the diftinction of Bifhops from Prefbyters, and their fucceffion from the Apoftles times and hands: the calling & moderating of Pruinciall Synodes by Primates and Metropolitanes: the allotting of Diœcefes, and the Popular electing of fuch as muft feed & watch the flocke: And diuers other points concerning the Paftorall regiment of the houfe of God; By Tho. Bilfon Warden of Winchefter Colledge. Perufed and allowed by publike authoritie." Deputies. An epiftle to the reader. 414 pages. W.H. Quarto.

" The Practice, proceedings & Lawes of armes, defcribed out of the 1593. doings of moft valiant & expert Captaines, and confirmed both by ancient & modern examples, and præcedents, By Matthew Sutcliffe. Luke 14; 31. —1593." Deputies. Dedicated—" To the Earle of Effex.---The Preface." 342 pages. W.H. Quarto.

" An Apologie for fundrie proceedings by Iurifdiction Ecclefiafticall, 1593. of late times by fome chalenged, and alfo diuerfly by them impugned: By which—all the Reafons & Allegations fet down as well in a Treatife,[b] as in certaine Notes (that go from hand to hand) both againft proceeding ex officio, and againft Oaths miniftred to parties in caufes criminall are alfo examined and anfwered: Vpon that occafion lately[c] reuiewed, & much enlarged aboue the firft priuate proiect, and now publifhed, diuided into three parts: &c. Refpectiuelie fubmitted to—the reuerend Iudges & other Sages of the Common lawe: &c. Whereunto—'I haue prefumed to adioine that right excellent & found determination (concerning Oaths) which was made by M. Lancelot Andrews, D. D. in the Vniuerfitie of Cambridge in July 1591. Lex iuftitiæ; Iuftitia reipub. bafis.— 1593." Deputies. Each part has a feparate title-page. W.H. Quarto.

" The renevving of certaine Orders deuifed by the fpeciall cōmande- 1594. ment of the Queenes Maieftie, for the reliefe & ftay of the prefent dearth of Graine within the Realme: in the yeere of our Lord 1586. Nowe to be again executed this prefent yere 1594 vpon like occafions as were feene the former yere: With an addition of fome other particular orders

[b] Entitled " A briefe of Oathes exacted by Ordinaries &c. wherein is proued that the fame are vnlawfull." Some account of which is given in the epiftle to the reader prefixed to this Apology.
[c] Of the former edition, confifting of but two parts, about forty copies only were printed, without any purpofe of further publifhing. Ibid.
[d] The author was Dr. Richard Cofens, dean of the Arches. Heylin's Hift. of the Prefb. p. 318.

CHRISTOPHER BARKER.

for Reformation of the great abuses in Ale-houses & such like." Deputies. 12 leaves. W.H. Quarto.

1594. "An Order for Prayer and Thankes-giuing (necessary to be vsed in these dangerous times) for the safetie & preseruation of her Maiestie and this realme. Set forth by Authoritie." D 3, in fours. Deputies. W.H. Quarto.

1594. "An answer vnto a certain calumnious Letter published by M. Job Throkmorton, entitled A Defence against the slanders of M. Sutcliffe, wherein the vanity both of the defence of himself & the accusation of others is manifestly declared, By Matth. Sutcliffe, D. D." Deputies. Again 1595. Licensed. Quarto.

1594. "The recantation of Tho. Clarke (sometime a seminarie priest of the Colledge of Rheimes) made at Pauls Crosse, 1 July 1593: Whereunto is annext a former Recantation made by him in a public Assembly, on Easter day, 15 Apr. 1593." Deputies. Octavo.

1596. "A Declaration of the Causes mouing the Queenes Maiestie of England to prepare & send a Nauy to the Seas, for the defence of her Realmes against the King of Spaines Forces, to bee published by the Generals of the saide Nauy, to the intent that it shall appeare to the World, that her Maiestie armeth her Nauy onely to defend her selfe and to offend her enemies, and not to offend any other that shall forbeare to strengthen her enemie, but to vse them with all lawfull fauours.—1596." Signed " R. Essex. C. Howard." Deputies. 3 pages. W.H. Printed also in French, Italian, Dutch, and Spanish. Quarto.

1596. "A Prayer of Thanksgiuing for continuance of good Successe to his Maiesties forces." Deputies. Broad sheet.

"A Declaration of the iust causes moouing the Queenes Maiestie to send a Nauie and Armie to the Seas, and towardes Spaine." Head title, my copy wanting the title-page. 6 pages. W.H. Quarto.

1596. "The Examination of M. Tho. Cartvvrights late apologie, Wherein his vaine and vniust challenge concerning certaine supposed slanders pretended to haue bene published in print against him, is answered, and refuted, By Matth. Sutcliffe.—1596." Hereunto are prefixed, an epistle superscribed " To M. Tho. Cartwright Master of the Hospitall at Warwike, giue these in the Ile of Gernesey, or els where he shalbe then resiant. ---To the Reader." 58 leaves. Deputies. W.H. Quarto.

1597. "Certain Prayers sett foorth by Authoritie, to be vsed for the prosperous successe of her Maiesties Forces & Nauy." Deputies. Quarto.

1597. "Capitula, siue Constitutiones ecclesiasticae, per archiepiscopum, episcopos, et reliquum clerum Cantuariensis prouinciae, in synodo inchoata. Londini, vicesimo quinto die mensis Octobris, anno Domini 1597, regnique serenissimae in Christo principis, Dominae Elizabethae, &c. tricesimo nono, congregatos tractatae, ac postea per ipsam regiam maiestatem approbatae et confirmatae, et utrique prouinciae, tam Cantuariensi quam Eboracensi, ut diligentius obseruentur, eadem regia authoritate sub magno sigillo Angliae promulgatae." Deputies.* Again 1599. Quarto.

* They are eight in number; the first remarkable for the style, composed by the Queen.

" A Pro-

CHRISTOPHER BARKER.

"A Proclamation againſt the inordinate exceſſe of Apparel." &c. 1597.
Deputies. Brit. Muſeum. 4 Broad ſides.
Acts of Parliament, "Anno xxxix Reg. Elizabethæ." Deputies. 1597.
W.H. Folio.
" Vita et Obitus ornatiſsimi celeberrimiq; viri Richardi Coſin Legum 1598.
Doctoris Decani Curiæ de Arcubus, Cancellarij ſeu Vicarij generalis Reue-
rendiſſimi patris Ioannis Archiepiſcopi Cantuarienſis, &c. per Guil. Bar-
lowum Sacræ Theologiæ Baccalaureum, amoris ſui & officii ergo edita.
—Excudebant Deputati—1598." Pages 47. W.H. Quarto.
" CARMINA FVNEBRIA, In eiuſdem Venerandi Doctoris triſte fatum, 1598.
a quibuſdam Cantabrigienſibus, illius amicis, multo mœrore fuſa magis,
quàm condita. Ille dolet verè, qui ſine teſte dolet. Marti. Anno ſalutis
humanæ, 1598." The pages continued to 80. W.H. Quarto.
" LIVER GWEDDI GYFFREDIN, a Gweni dogaeth y ſacramentau, ac 1598.
erraill gynneddfau a ceremoniau yn eglwys loegr." By his deputies.
Quarto.
" A Prayer for the good ſucceſſe of her Maieſties Forces in Ireland." 1599.
Deputies. A broadſide.
" A Letter written out of England to an Engliſh Gentleman remain- 1599.
ing at Padua, containing a true report of a ſtrange conſpiracie contriued
between Edw. Squire lately executed for the ſame Treaſon as Actor, and
Ric. Walpoole a Jeſuite, as Deuiſer & Suborner, againſt the perſon of
the Queenes Maieſtie." Deputies. 8 leaves. Quarto.
" LAVVES and Ordinances ſet downe by Robert Earle of Leyceſter, the
Queenes Maieſties Lieutenant & Captaine General of her armie & forces
in the Lowe Countries: &c. Jmprinted by Chr. Barker, Printer," &c.
W.H. Quarto.
" The Booke of Common praier, and adminiſtration of the Sacra-
ments, and other Rites & Ceremonies in the Church of England. Im-
printed—by the Deputies of Chr. Barker, Printer, &c. Cum gratia &
priuilegio Regiæ Maieſtatis." On the back is " An Almanacke for xxvj
yeeres," beginning 1578. Quarto.
" The Pſalter or Pſalmes of Dauid, after the tranſlation of the great
Bible, pointed as it ſhalbe ſung or ſaid in Churches. Imprinted," &c.
Each in the ſame compartment as his quarto Bibles. W.H. Quarto.
I have an edition in octavo, wanting the title-page, with an almanac
beginning alſo 1578; but the Pſalter is dated 1595.
" A Catechiſme with praiers annexed, meet for chriſtian families:
written to the congregation of Dedham, by E. C." Octavo.
" A ſhort Catechiſme to bee learned before the admiſſion to the Lordes
Supper. By Ric. Saintbarb." Octavo.
He had licenſes alſo for printing the following, viz. In 1569, " Serten
prayers of M. Bullion."---In 1576, " Carminū prouerbialiū totius
humanæ vitæ Loci cōmunes." In 1578, " An antheme or ſonge, begyn-
ninge, Lord ſaue & bleſſe with good encreaſe the churche, our quene,
& Jreland in Peace." In 1580, " An order of praier & other exerciſes
to turn away Gods wrathe: To be vſed in the prouince of York. The
kinges

CHRISTOPHER BARKER.

kinges maiefties acte of Proclamation of Scotland, together with certen actes of proclamac'on made againfte the aduerfaries of Chriftes gofpelL A fermon preached by[f]—*before* a Jefuit in the Tower on Sondaie the fiueth daie of Feb. 1580." In 1582, " A catechifme & certen praiers of M. Dr. Chapman's doinge, which hath bene printed before and hath nowe bene fhewed againe to the wardens. An introduction to the Bible, or entrance into the writinges of the Prophetes & Apoftles." In 1583, For a book by comandment touching the State. In 1584, " Remedies for a Flixe, where the phifition is not prefent. A prayer for the queene vfed in her Maiefties chappell." In 1585, " An anfwere to the excommunication lately denounced and publifhed by Sixtus quintus pope of Rome fo called, againft the twoo Xpian prynces, Henry K. of Nauarra & the prynce of Conde: made by the faid prynces, and fent to Rome." In 1586, " A true copie of a letter fent from her Maiefte to the Lord mayor and Cytizens of London, withe a fpeache of Mr. Dattons to that effecte."

ROBERT BARKER, Efq, queen's printer,

THE only fon of Chriftopher Barker beforementioned, was made free of the Stationers' company by patrimony, 25 June 1589; and called to the livery 1 July 1592. After the deceafe of his father, 3 Jan. 1599-600, he exhibited to the Court of Affiftants his letters patent granted him by her majefty for the term of his life, bearing date the 8 Aug. in the 31 year of her reign, comprehending very extenfive privileges, and appointing him her highnefs's printer. He ferved Upper Warden in 1601, having been affeffed 20l. to be excufed ferving Under Warden. In 1604 he paid 20l. for not ferving the fecond Upper Wardenfhip. About the year 1593 he appears to have commenced trade, which, while his father lived, he carried on in conjunction with Meff. Bifhop and Newbery, as his father's deputies. None of his title-pages or colophons fhew where he dwelt, or had his printing office; but moft probably he had the fame that was his father's. Mr. Ames ftyles him Robert Barker of Southley, or Southlee, in the county of Bucks efq; and fays, he married two wives; Rachael, daughter of Richard Day, bifhop of Winchefter, by whom he had feveral children; and Ann, relict of Nicholas Cage of London. Alfo, that others, befides his fons, were concerned with him in the bufinefs of printing. The 19th of July, 1603, a fpecial licenfe was granted him for printing all the ftatutes during his life.[g] King James the firft, in confideration of the fum of three hundred, and an annual rent of 20l. demifed to him the manor of Upton, to hold for two and twenty years, tefte 28 Sept. the rent two years after was raifed to 40l. per Annum. William Ball, efq; fays, Robert Barker had paid for amending,[h]

---

[f] This feems to be the fame as afcribed by bifhop Tanner to Dr. Will. Fulke "Apud Turrim in Joh. xvii, 17. M D LXXX."
[g] See Dugdale's Orig. Jurifd. p. 24.

[h] In his treatife on printing 1651. But this muft rather have been towards the expences of making the new tranflation.

or

ROBERT BARKER.

or correcting the tranflation of the bible 3500l. &c. therefore his heirs had the right of printing it. This family enjoyed the office of King's printer, and had exclufive privileges many years; however they had their changes in fortune, for Robert lay in prifon above ten years, as appears from a certificate communicated to Mr. Ames, in thefe words: Thefe are to certify whom it may concern, that Robert Barker, efq; was committed a prifoner to the cuftody of the marfhal of the king's bench, the 27th of November 1635, and died in the prifon of the king's bench the 10th of January 1645, *per* Tho. Wigg, clerk of the papers to the marfhal of the king's bench, 16 Jan. 1679. He was buried the 12th Jan.[i]

Biblia[k] Sacra. Printed for G. B. R. N. and R. B. *(Geo. Bifhop, Ralph Newbery, and Rob. Barker.)* See p. 881. Again 1597. Folio.   1593.

"The Second part of the French Academie: Wherein as it were by a naturall hiftorie of the bodie & foule of man, the creation, matter, compofition, form, nature profite & vfe of all the partes of the frame of man are handled, with the naturall caufes of all affections, virtues & vices, and chiefly the nature, powers, workes, and immortalitie of the Soule: By Peter de la Primaudaye Efquire, Lord of the fame place & of Barre. And tranflated out of the fecond Edition, which was reuiewed and augmented by the author. At London, printed for G. B. R. N. and R. B. 1594." Dedicated "To—Sir Iohn Puckering knight, Lorde Keeper" &c. by T. B.[l] the tranflator. Then, a copious preface, and a table of contents. 600 pages. W.H. Quarto.   1594.

"Rerum Anglicarum Scriptores poft Bedam præcipui, ex vetuftiffimis codicibus manufcriptis, nunc primum in lucem editi. Willielmi Monachi Malmesburienfis de geftis rerum Anglorum, lib. v. Eiufdem Hiftoriæ Nouellæ, lib. 11. Eiufdem de geftis Pontificum Angl. lib. 1111. Henrici Archidiaconi Huntindonienfis Hiftoriarum lib. v111. Rogeri Houedeni Annalium, pars prior & pofterior. Cronicorum Ethelwerdi lib. 1111. Ingulphi Abbatis Croylandenfis hiftoriarum lib. 1. Adiecta ad finem Chronologia." The device; an open book, on which rays of light are darted from a circle of glory in the clouds; the legend, enclofing thefe, "Dat effe manus: fupereffe Minerva." Without are Mercury fupporting the book on one fide, and Pallas on the other pointing to the glory above. At the upper corners are angels holding feftoons of printing materials. "Londini. Excudebant G Bifhop, R. Nuberie, & R. Barker Typographi Regij Deputati. Anno ab incarnatione, cIɔ Iɔxcvi." Each of thefe authors works has a feparate title, in the compartment defcribed in p. 881; but the paging of them is continued to p. 525. "Fafti regum & epifcorum Angliæ vfq; ad Willielmum feniorem," 30 leaves. W.H. Folio.   1596.

"THE NAVIGATORS SVPPLY." &c. as p. 916, &c. Quarto.   1597.

[i] Smith's Obituary, by Peck.
[k] I take this to be the fame that was printed for W. Norton (who died this year) and probably by them, as the deputies of Chr. Barker, that book being printed with feveral of the faid Chriftopher's vignettes. My copy wants a leaf at the end, which might contain a colophon confirming this.
[l] He tranflated alfo the firft part, printed for Geo. Bifhop, in which he fubfcribes with the initials T. B. C.

" THE

ROBERT BARKER.

1599. " THE HAVEN FINDING ART, Or the way to find any Hauen or place at Sea by the Latitude & variation: Lately publiſhed in the Dutch, French & Latine tongues, by cōmandement of—Count Mauritz of Naſſau, Lord High Admiral of the vnited Prouinces of the Low countries, enioyning all Seamen that take charge of ſhips vnder his iurifdiction, to make diligent obſeruation in all their voyages according to the directions preſcribed herein: And now tranſlated into Engliſh for the common benefite of the Seamen of England. Imprinted—by G. B. R. N. and R. B. 1599." Dedicated " To—Charles Earle of Notingham, Baron howard of Effingham, knight of the—Garter, Lord high Admiral &c. And her Maieſties Lieutenant and Captaine general ouer all her ſubiects leuied in the South parts of this Realme, &c.—23 Aug. 1599.—E. Wright." 27 pages. Alexander Dalrymple Eſq; Quarto.

1599. " The principal Navigations," &c. as p. 917. The 1ſt & 2d Vol. Folio.

1600. " The third and laſt volume of the voyages," &c. Folio.
1600. Articles of Religion: and the queen's injunctions. Quarto.
1600. " BELLVM PAPALE, ſiue Concordia Diſcors Sixti Quinti & Clementis octaui, circa Hieronymianam editionem. Preterea, in quibuſdam locis grauioribus habetur comparatio vtriúſq; editionis, cum poſtrema & vltima Louanienſium; vbi mirifica induſtria Clementis & Cardinalium ſuper caſtigatione Bibliorum deputatorum, notas duntaxat marginales Louanienſium in textum aſſumendo, claré demonſtratur. Auctore Thoma Iames, Noui Collegij in alma Academia Oxonienſi ſocio, & vtriúſq; Academæ Artibus Magiſtro Excudebant Georgius Biſhop, Radulphus Newberie, & Robertus Barker. Anno 1600." Dedicated " Reuerendiſſimo in Deo patri Ioanni Archiepiſcopo Cantuarienſi, &c.--- Sixtus Epiſcopus ſeruus feruorum De ad perpetuam rei momoriam.— Dat. Romæ apud ſanctam Mariam Maiorem, Anno incarnationis Dominicæ 1589, Kalendis Martij, Pontificatus noſtri anno Quinto.---Index Locorum Manu ipſius Sixti quinti, vel aliorum poſt impreſſionem correctorum.---Biblia Sacra vulgatæ editionis Sixti quinti Pont. Max. juſſu recognita atq; edita. Accipe & deuora. Apoc. 10. Romae Ex Typographica Apoſtolica Vaticana, 1592, Præfatio ad lectorem.---Clemens P. octavus ad perpetuam rei memoriam. Dat. Romæ, apud S. Petrum ſub Annulo Piſcatoris, die 9 Nouembris, 1592. Pontif. noſtri Anno 1. M. Veſtrius Barbianus.---Abreuiationes" &c. K 3, in fours. W.H. Quarto.
1600. A Proclamation to give heed to the converſation of perſons, and their ſlanderous words & rumours againſt Government. " Imprinted by Rob. Barker printer to the queenes Maieſtie, 1600." Brit. Muſeum.

"—Concerning the Earl of Eſſex, &c. who had layed plots with the Traitour Tirone." The title cut off. Ibid.

1601. " A DECLARATION of the practiſes & Treaſons—by Robert late Earle of Eſſex & his Complices, againſt her Maieſtie and her Kingdoms; and of the proceedings, &c. Together with the Confeſſions & other parts of the Euidences—taken out of the Originals. Imprinted at London,

London, by Rob. Barker, Printer to the Queenes most excellent Maiestie, 1601," Q, in fours. W.H. Quarto.
" An admonition to the people of England;" &c. as p. 1084. Octavo.
" THE SEDVCTION of Arthington* by Hacket, especiallie with some token of his vnfained repentance & Submission: Written by the said Henrie Arthington the third person of that wofull Tragedie. Printed by R. B. for Tho. Man." Richard Gough Esq; Quarto.

* See Conspiracie for pretended Reformation, &c. p. 1086.

JOHN CHARLEWOOD, CHARLWOOD, or CHARLEWOODE, stationer,

SEEMS to have printed so early as queen Mary's reign, in a temporary partnership with John Tysdale at the Saracen's Head near Holbourn conduit; how long this lasted is uncertain, as nothing of their printing with a date, has (as yet at least) reached these times. In 1560 Tysdale is found resident in Knight-rider street. Charlewood at first printed only ballads and small things, for which he had several licenses. Aug. 18, 1578. He was fined " ijs. vjd. for printing a booke concerninge foraine reports, and not alowed." Again, " Sep. 20. for printing Newes out of the Low Cuntrey, xijd." Again, " Dec. 1. for printinge Fourbyshers voiage, without Lycence, vs." Again, " 26 March, 1580, for printinge a lewde book, and that without Lycence, xs." For which 3s. 5d. was received in full payment, 3 Octo. following. From this time he appears to have behaved very orderly. A few of his books inform us of his abode in Barbican, at the sign of the Half-Eagle and Key, (the Geneva arms) and he used the same with this motto, " Post tenebras lux," to such books, as a device or cognizance; but most of his books are without any direction, and several of them have only the initials of his name. M. Marprelate, in his first epistle, p. 23, mentions him by his initials, " I. C. the earle of Arundel's man, hauing a press and letter in a place called Charterhouse, in London." See the extract at large in the account of printing in Wales. This must have been before the year 1589. He died early in the year 1593.

" This Booke is called the Treasure of Gladnesse," &c. as 943. 1563. Licensed. Again 1575, and 1590. Sixteens.
" Of the signes and tokens of the later day: by Rob. Crowley." For him, 1567. Octavo.
" The BVKOLIKES, of P. Virgilius Maro, with alphabeticall Annotations,—Drawne into plaine & familiar Englishe verse by Abr. Fleming student," &c. Hist. of Eng. Poetry, III; p. 402. 1575.
" A philosophicall discourse, Entituled, The Anatomie of the minde: Nevylie made & set forth by T. R. Imprinted at London by I. C. for Andr. 1576.

Andr. Maunfell,—1576." Dedicated "To—M. Chr. Hatton Efq; capitane of the Queenes Maiefties Guard, &c. Tho. Rogers.---The Preface" &c.---Commendatory verfes, viz. " Iofua Hutten to the Booke.---Iuftiniani Baldwini, carmen ad Lectorem.---Abr. Fowlers needeles Hædera.---Epigrāma Guil. Camd. in Anatomiam a Tho. Rogerio, elaboratam." 206 leaves, and a table. W.H. Octavo.

1576. " THE CHRISTIAN Manuell, or the life and maners of true Chriftians. A Treatife, wherein is plentifully declared how needefull it is for the feruaunts of God to manifeft & declare to the world: their faith by their deedes, their words by their works, and their profeffion by their conuerfation. VVritten by Ihon VVoolton Minifter of the Gofpel, in the cathedral church of Excetor. Imprinted at London by I. C. for Tho. Sturruppe—in Paules Church yarde, at the—George 1576." Dedicated " To—Sir William Cordell knight, Maifter of the Rolles.—At Whymple 20 Nouember 1576." N 7, in eights. W.H. Octavo.

1576. " The Caftell of Chriftians, and Fortreffe of the Faithfull, befeged & defended, now almoft fixe thowfand yeares. Written by John Wolton, one of the Cathedrall Church in Exon." In the compartment with The refurrection at top &c. as p. 976. It has a table of contents before the dedication "To right honorable Sir Frauncys VValfyngham Secretary to the Queenes moft excellent Maieftie, &c.—From Exetor the laft day of May 1577." N 7, in eights. " Imprinted by I. C. for Tho. Sturrup: —1577." W.H. Octavo.

1577. " A Tragedie or Enterlude, manifefting the chiefe promifes of God vnto man, by all ages in the olde lawe, from the fall of Adam to the incarnation of the lord Jefus Chrifte. Compyled by John Bale, anno Domini 1538. And now fyrft imprinted 1577. John i. Interlocutores: Pater caeleftis, Juftus Noah, Moifes Sanctus, Efais Propheta, Adamus primus homo, Abraham fidelis, Dauid rex pius, Johannes Baptifta. Printed for Stephen Peele, in Roode-lane." 16 leaves. See Hift. Eng. Poetry III; p. 78 Quarto.

1578. " WHARTONS Dreame: Conteyninge an inuective agaynft certaine abhominable Caterpillers as Vferers, Extorcioners, Leafmongers and fuch others, confounding their diuellyfh fectes by the aucthority of holy fcripture, felected & gathered by Iohn VVarton Scholemaifter. Seneca. In iuuentute cogitaui bene facere, vt in feneetute bene moriar. Sapiens, 3. (1.) The foules of the righteous" &c. Imprinted—by Iohn Charlevvod, for Paull Conyngton, & are to be fold at his fhoppe in Chauncery lane, at the—blacke Beare, 1578." On the back, " Perufed and thought well of,—by thefe following: John Fox. Rob. Crowley. W. Wager. Tho. Buckmafter, and others." Dedicated " To—Maifter Alexander Nowell, Deane of the Cathedrial Church of S. Paules in London.---To the Reader.---Commendatory verfes by Tho. Buckmaifter, Tho. Smith. Will. Vallans, and Geo. Rogers." E, in fours. W.H. Quarto.

1578. " A Commemoration of the right noble & vertuous Ladye Margrit Duglafis good Grace, Countis of Lenox, Daughter to the renowned & moft

JOHN CHARLEWOOD.

most excellent Princesse, Margarit Queene of Scotland espowsed to King James the fourth, of that name. In the daies of her most puissaunt & magnificent father Henry the seaventh, of England, Fraunce & Ireland King. Imprinted—in Barbycan at the halfe Eagle & Key, 1578." John Fenn Esq; Quarto.

"The Popes pittiful Lamentation for the death of his deere darling (1578.) Don Ioan of Auſtria: and Deaths aſwer to the ſame. With an Epitaphe vpon the death of the ſaid Don Ioan. Tranſlated after the French printed copy by H.C." In verſe. Under the epitaph, "The fyrſt of October, 1578.—Imprinted by I. C." 4 leaves. Licenſed. W.H. Sixteens.

"A dolorus diſcourſe of a—bloudy Battel fought in Barbarie, 4 Aug. (1578.) 1578." With Tho. Man. Licenſed. Sixteens.

A ſermon on Luke xvi, 2. Found hid in a wall. See it in 1584. (1579.) Sixteens.

"Cohelet, ſeu concio Salomonis de ſumo hominis bono, paraphraſi (1579.) explanata ex prælectionibus Anth. Corani Hiſpalenſis: cum noua verſione marginibus addita eodem Cor. interprete. Lond. 25 Martij 1579." This book was licenſed to John Wolf, but has in the margin, "printed by Jhon Charlwood."

"Newes Newes contayning A ſhorte reherſall of the late enterpriſe (1579.) of certaine fugytiue Rebelles: fyrſt pretended by Capt. Stukeley, and ſithence continued, & put in practiſe by Mac Morice (his Lieutenant) vpon the Countrey of Ireland in the monthe of July laſt: who ſince (among others) was diſcomfited & beheaded: as by the diſcourſe followinge more plainely maye appeare. Tranſlated out of Dutch into Engliſh, 6 Octo. 1579. Imprinted by I. C." 8 leaves, Great Primer, black. Licenſed. W.H. Octavo.

"A ſermon preached before the queenes Maieſtie, by maiſter Edward (1580.) Dering, the 25 Feb. 1569." On Pſ. 78; 70, 71, 72. W.H. Alſo without date. Sixteens.

"Newes from Antvverp, 10 Aug, 1580: Contayning a ſpeciall view (1580.) of the preſent affayres of the lowe Countreyes, Reuealed & brought to lyght by ſundrie late intercepted Letters of certaine vizarded & counterfeyt Countreymen of the ſame Countreyes. Tranſlated—partly out of French, and partly out of Lattin; According to the originall Copie printed at Antwerp by Wm. Riuiere a ſworne Printer & Bookſeller." E 2, in fours. Licenſed. W.H. Quarto.

"The Oration and Sermon made at Rome by cōmaundement of the 1581. foure Cardinalles & the Dominican Jnquiſitour, vpon paine of death. By Iohn Nichols, latelie the Popes Scholler. Which—was preſented before the Pope & his Cardinalles in his Conſiſtorie, 27 Maie, 1578, and remaineth there regiſtred: Now by him brought into the Engliſh tongue, &c. Heerin alſo is aunſwered, an infamous Libell, maliciouſlie written & caſt abroad, againſt the ſaide Iohn Nichols, with a ſufficient diſcharge of himſelfe from all the Papiſts lying reports, and his owne life both largelie and amplie diſcouered.—Imprinted—and are to be ſold at the little North dore of S. Paules church, at the—Gunne, by Edw. White, 1581."

JOHN CHARLEWOOD.

1581." On the back are the queen's arms, to whom there follows a pompous dedication in Latin, signed " Ioannes Nicholaus Camberbritannus.---To the courteous---Reader---To the worshipfull companie of Merchant Aduenturers at Embden & at Antwerp," &c. P 4, in eights. Licensed. W.H. Also without date. Sixteens.

1581. An Introduction to Algorisme, &c. as p. 1033. Octavo.

(1581.) " The Araignement and Execution of a wilfull & obstinate Traitour, named Euералде Ducket, alias Hauns: Condemned at the Sessions house, for high Treason, on Friday, beeing the 28 of Iuly, and executed at Tiborne, on Monday after,—1581. Gathered by M. S. Imprinted —by him, and Edw. VVhite." 8 leaves. W.H. Sixteens.

1581. " A learned and a Godly sermon, to be read of all men, but especially for all marryners, captaynes, & passengers, which trauell the seas, preached by Richard Madoxe, M. A. and fellow of All Soules Oxforde, at Waymouth, & Melcombe Regis, a porte in the countie of Dorsett, 3 Oct. 1581." On Matt. 8; 23. Dedicated by Thomas Martin to the mayor, bayliffs, and aldermen of the same town. Octavo.

1581. Instructions for Christians, containing a fruitfull & Godly exercise, aswell in fruitfull praiers, as in reuerent discerning of Gods holy comandements and Sacramentes, by Richard Jones Scholemaster of Cardiffe; translated by Doroth. Martin. Octavo.

1581. " Two NOTABLE Sermons, Made by—Maister Iohn Bradford;—now newlie Imprinted. Perused & allowed," See p. 782. W.H. Octavo.

1581. " An Aunswer to sixe Reasons, that Thomas Pownde, gentleman, and prisoner in the Marshalsey, at the commaundement of her Maiesties Commissioners, for causes ecclesiastical, required to be aunswered.— Written by Robert Crowley." Licensed. Quarto.

1581. " A Briefe discourse concerning those fower vsual notes whereby Christes Catholike Church is knowen." &c. Quarto.

1582. " Two godlie & learned Sermons preached at Manchester," &c. as p. 1044. Licensed.

1582. " THE ENGLISH Romayne Lyfe: Discouering The liues of the Englishmen at Roome, the orders of the English Seminarie, the dissention Betweene the Englishmen & the VVelshmen, the banishing of the Englishmen out of Roome, the Popes sending for them againe, a reporte of many of the paltrie Reliques in Roome, their Vautes vnder the grounde, their holy Pilgrimages: &c. There vnto is added the cruell tirranny vsed on an Englishman (Rd. Atkins) at Roome, his Christian suffering & notable Martirdome, for the Gospell of Jesu Christ, in Anno 1581. VVritten by A. M. (Anth. Munday) sometime the Popes Scholler in the Seminarie among them. Honos alit Artes. Seene & allovved. Imprinted by him for Nich. Ling—in Paules Churchyarde, at the—Maremaide, 1582." Dedicated " To—Sir Thomas Bromley Knight, Lord Chaunceller, &c. William Lorde Burleigh, and Lord Treasurer; Robert Earle of Leicester, with all the rest of her Maiesties—priuie Councell.— To the—Reader." 75 pages; and a cut of the sufferings of Mr. Atkins, in 4 divisions. Licensed. W.H. Again 1590. Quarto.

" Ip.

JOHN CHARLEWOOD.

" Io. Northbrooke, his briefe and pithie fumme of the Chriftian faith, 1582. made in forme of a confeffiõ, with a confutation of popifh objections." The addrefs to the reader is dated " from Redcliffe in Briftoll." O O, in eights. On the laft page is his fign, and under it, "—Printed by *him* at the figne of the Haulfe Eagle & Keye in Barbican." W.H+ Octavo.

" *The* wifh of a poore wifher, wifhing health & faluation to all men. 1582. By John Pitt minifter." Octavo.

A fermon, in 1388, on Luke 16; 2. See it in 1584. Octavo. 1582.

" A defenfatiue againft the poyfon of fuppofed prophefies: Not 1583. hitherto confuted by the penne of any man, which being grounded, eyther vppon the warrant and authority of olde paynted bookes, expofitions of Dreames, Oracles, Reuelations, Inuocations of damned fpirits, Judicialles of Aftrologie, or any other kinde of pretended knowledge whatfoeuer, de futuris contingentibus: haue beene caufes of great diforder in the common wealth, and cheefely among the fimple and vnlearned people: very needefull to be publifhed at this time, confidering the late offence, which grew by moft palpable & groffe errours in Aftrology.— Printed by *him*, Printer to the right Hon. Earle of Arundell, 1583." Dedicated " To—Sir FrauncisWalfingham, principall Secretarie, &c.— From Howarde houfe, this 6 of Iune—Henrie Howarde.---To the Reader." R r, in fours, but the firft alphabet ends with Y. My copy is not paged. Licenfed. W.H. See Gent. Mag. LIV ;. p. 254.. Quarto.

" A Booke of Prefidents" &c. as p. 521. " Printed by *him*, the affigne 1583. of Rychard Tottle." W.H. Sixteens.

" A lecture, or expofition vpon a part of the fifth chapter of the 1583. epiftle to the Hebrues." &c. See p. 887; alfo his xxvii lectures, p. 928. Sixteens.

" A TRVE REPORTE Of the late difcoueries, and poffefsion, taken in 1583. the right of the Crowne of Englande of the Newfound Landes: By that valiaunt & worthye Gentleman, Sir Humfrey Gilbert Knight. Wherein is alfo breefely fette downe her highneffe lawfull Tytle therevnto, &c. Seene & allowed. At London, Printed by I. C. for Iohn Hinde,—in Paules Church-yarde, at—the golden Hinde, 1583." Dedicated " To —Sir Fraucis Walfingham Knight, &c.—From my lodging in Oxforde, 12 Nouember. Your Honours poore Scholler, &c. G. P." *(Sir Geo. Peckham.)* Then follow commendatory verfes by Sir Wm. Pelham, Sir Fran. Drake, M. Iohn Hawkins, Mr. Capt. Chefter, Mat. Roydon M. A. Mr. Anth. Parkhurft, Arth. Hawkins, Iohn Achelley.---The table of contents. The report, which may be feen in Hackluyt's collection of voyages, &c. is on 25 leaves. At the end are annexed The articles between the principal affignees of Sir H. Gilbert and the four forts of adventurers with them in this voyage, viz. Affociates, or fuch as adventured £100; Affiftants,—£50; Adventurers in the firft degree,—£25; in the fecond degree,—£12. 10. W.H+ Quarto.

" The great Cicle of Eafter: Containing A fhort Rule To knowe 1583. vppon what day of the month Eafter day will fall, &c. With other neceffarie necessarie Tables to learne out the course of the yeere, By Io. P. 1583.—Imprinted—by I. C. for Tho. Butter." Dedicated "To—Maister Wm. Roe, Alderman of London.—Seuenoek in Kent—Iohn Pett.—To the Reader." 24 leaves. W.H. Sixteens.

1583. "An addition to the proues of Scripture noted in maſter Deeringes Catechiſme, and printed with the Catechiſme. By John Stockwood, miniſter, & preacher at Tunbridge." Octavo.

1583. "The Pathway to pleaſe God, which way whoſoeuer walketh (eſpecially in the time of any worldly affliction) ſhall be defended againſt all tentation, with ſundry Godly exerciſes: Alſo, a comfortable perſwaſion to patience. By T. Wallis." Sixteens.

1583. "The Roſarie of Chriſtian praiers & meditations for diuers purpoſes, and alſo at diuers times, as well of the day as of the night, by Phill. Stubbes." Licenſed. Sixteens.

1584. "A godlie and famous Sermon, preached in the yeere of our Lord 1388," at Paules Croſſe, on the Sunday of Quinquageſima, by R. VVimbeldon, and found out hyd in a vvall." Head title, on A, iij. The text, "Luke 16, (2) Redde rationem villicationis tuæ. Come, giue a reckoning of thy Bayliwicke." D, in eights. "Imprinted at London by him, 1584," over his device. W.H+ Again, 1588. Dr. Lort. Sixteens.[n]

1585. "A diſcourſe of Engliſh poetrie," as p. 735. Licenſed. Again, 1586. Quarto.

1585. "The declaration of the king of Nauarre, touching the ſlanders publiſhed againſt him in the proteſtations of thoſe of the league, that are ryſen vp in armes in this realme of Fraunce." Tranſlated from French. Licenſed. Sixteens.

1585. "A letter lately written from Rome, by an Italian gentleman, to a freende of his in Lyons in Fraunce. Wherein is declared, the ſtate of Rome; the ſuddiane death of pope Gregory the thirteenth; the election of the newe pope, and the race of life this newe pope ranne, before he was advanced, &c. Tranſlated from Italian, by J. F." Sixteens.

1585. "The Preparation of a Chriſtian to the ſupper of the Lord, taught in the Church of Epping in Eſſex, by R. H.—Printed by I. C." Octavo.

1585. "The true trauaile of all faithfull Chriſtians, how to eſcape the daungers of this wicked world; alſo a Chriſtian exerciſe for priuate houſholders, by William Chub." Octavo.

---

[n] Caſley, in his catalogue of MSS. in the Royal Library, p. 273, mentions "Tho. Wimbleton's Two Sermons, at Paul's-Croſs, on Luke xvi. 2: preached A. D. 1388." In the catalogue of MSS. in Sidney-Suſſex college, is "A Sermon preached at Pauls Croſs An. 1389. on Quinquageſima Sunday, by Tho. Wymbledon. Ioc. Redde rationem &c. Luc. 16." In the catalogue of MSS. in Gonvil and Caius college, is "R. Wimbledon Concio; extat quoq; Anglice, 1593. 8vo." In the Bodleian is "R. Wimbleton's Sermon at Paul's Croſs on Quinquageſima Sunday, An. 1388." Bale nor Pits have not either of the names. Bp. Tanner mentions the printed edition of 1588, and the M. S. in Sidney-Suſſex coll. under the name of Tho. Wimbledon. This ſermon was firſt printed by John Kynge, without date; who printed between 1550 and 1561. Fox has inſerted it in the ſecond edition of his Martyrology, 1570, p. 653; and mentions two MS. copies with R. Wimbeldon's name.

"Short

JOHN CHARLEWOOD.

"Short queftions betweene the father & the fonne, by Wm. Cotts." 1585.
Octavo.

"Two fruitfull and godly fermons preached at Dorchefter in Dorfet- 1585.
fhyre, the one touching the building of Gods temple; the other what the
temple is. Dedicated by William Chub, minifter from Froome Zel-
wood, 12th December 1585, to George Trenchard, efq;" Again, 1586.
Sixteens.

"An oration, or funerall fermon, vttered at Rome, at the burial of 1586.
the holy father Gregory the 13th, who departed in Jefus Chrift the 11th
of April," &c. Sixteens.

"A comfortable Sermon of Fayth in temptations and afflictions;" 1586.
&c. as p. 887. "Imprinted by him, dwelling in Barbican, at—the
halfe Eagle & the Key, 1586." Dedicated "To the right worfhipfull & godlye Mrs. Marie HarrysWyddowe.—I.Y." The text, Mat. 15;
21—28. E 4, in eights: D, omitted. W.H. Octavo.

"A Replication to that lewd aunfweare which Frier Iohn Francis (of 1586.
the Minimies order in Nigeon neere Paris) hath made to a letter that
his mother caufed to bee fent to him out of England: By Rob. Crowley."
Licenfed. Quarto.

"An excellent and pleafant worke" &c. as. p. 897. 1587.

"A fhort declaration of the end of Traytors & falfe Confpirators 1587.
againft the ftate, and of the dutie of Subiectes to theyr foueraigne Gouernour: and wythall, howe neceffarie Lawes & execution of Iuftice are
for the preferuation of the Prince & Common wealth. Wherein are alfo
breefely touched fundry offences of S. Queene comitted againft the crowne
of this land, and the manner of the honorable proceding for her conuiction thereof, and alfo the reafons & caufes alledged & allowed in Parliament, why it was thought dangerous to the ftate, if fhe fhould haue
liued. Publifhed by Ric. Crompton, an Apprentice of the common
Lawes. Séene & allowed. Eccl. 10; (20) Wifh no euill to the King
&c. Printed by him, for Tho. Gubbin, & Tho. Newman, 15°7." Dedicated "To—John—Archbyfhoppe of Canterbury, &c. From the
Myddle Temple, the xij of Feb. 1587." F 4, in fours. W.H. Quarto.

The Sick-man's Salve. By Tho. Becon. See p. 634. Octavo. 1587.

"Amorous Fiammetta. Wherein is fette downe a catalogue of all & 1587.
finguler pafsions of Loue & iealofie incident to an enamored yong Gentlewoman, with a notable caueat for all women to efchewe deceitfull &
wicked Loue, by an apparant example of a Neapolitan Lady, her approued & long miferies, and wyth many founde dehortations from the fame.
Firft wrytten in Italian by Mafter Iohn Boccace, the learned Florentine,
and Poet Laureat. And now done into Englifh by B. Giouano del M.
Temp. (Bart. Young of the Mid. Temple.) With notes in the Margine,
and with a Table in the ende of the cheefeft matters, &c. Printed by
I C. for T. Gubbin, & T. Newman, 1587." Dedicated "To—Sir
VVilliam Hatton Knight.—Thomas Newman.—To the noble & gallant.
Dames of the Cittie of Caftale in Mon Ferrato: Gabriel Giolito ---The
Authour his Prologue. Fiammetta fpeaketh." 123 leaves. W.H. Quarto.

"A Mirror

1587. "A Mirror for the Mathematiques:" as p. 1025. Licensed.
1587. "A Mirrovr of Monsters;" &c. as p. 898.
1587. "A Path Way to Military practise." as p. 735.
1587. "The Secrets & Wonders of the Worlde." &c, as p. 898.
1587. "A short catechism for House holders, by John Stockwoode Schole-maister of Tunbridge." Sixreens.
1588. "The Recantation made at Paules Crosse by Wm. Tedder Seminarie Priest, 1 Decemb. 1588. Pf. 118; (13.) VVhereunto is adioyned: The recantation or abiuration of Ant. Tyrell (sometime Prieste of the English Colledge in Rome) pronounced by himselfe at Paules Crosse the next Sunday following in the same yeere. Seene & allowed &c. Printed by him, and VVylliam Broome,—1588." Tyrrell's recantation has a separate title-page, wherein after the word Rome is added "but nowe by the great mercie of God conuerted & become a true professor of his word) Pronounced by himselfe at Paules Crosse, after the Sermon made by M. Pownoll Preacher: 8 Dec. 1588. Ier. 31; 19." The paging and signatures however are continued to p. 47. Then, on a single leaf is their joint dedication "To the most noble vertuous & gracious Princesse Elizabeth—Queene of England, &c. defendresse of the fayth," &c. W.H.
Quarto.

1588. "A Deliberat answere made to a rash offer, which a popish Antichristian Catholique made to a learned protestant (as he saieth) and caused to be publyshed in printe: Anno Do. 1575. Wherein the Protestant hath plainly & substantially prooued, that the papists that doo nowe call themselues Catholiques are in deed Antichristian schismatiks: and that the religious popish protestants are in deed the right Catholiques: VVritten by Rob. Crowley.:—1587. London, Printed by him, and are to bee sold at—the black Beare in Paules Church yarde, 1588." An epistle "To all and singular Recusants.—16 Jan. 1587.—Robert Crovvley." The head-title, "The deliberate aunswer that Ro Crowley (a Protestant Christian) hath made to that rash offer, that a popish Antichristian Catholique made to a learned Protestant: Imprinted in Dovvaie (as he saith.)—1575." The running-title, "First trie, and then trust." 88 leaves. Licensed. W.H. Quarto.

1588. "A most necessarie treatise, declaring the beginning and ending of all poperie, or the popish kingdome. Drawne out of certaine old prophecies aboue 300 yea.s since; and nowe newly set forth with the auncient pictures thereto belonging, in K. Edw. vi daies." See p. 752. Licensed. Quarto.

1588. "A Catechisme, with a preface prefixed, by M. Edw. Dering." Octavo.

1589. "A Sermon at the Tower, on the Gospell, John 6; 34. By M. Edw. Dering." Octavo.

1589. "A Briefe Answer vnto those idle & friuolous quarrels of R. P. (Ro. Parsons) against the late edition of the Resolution: By Edm. Bunny. Whereunto are præfixed the booke of the Resolution, and the treatise of Pacification, perused & noted in the margent on all such places as are misliked of R. P. shewing in what Section of this Answer following, those places are handled. Pf. 120; 7. I labour for peace: &c. Printed by him,

him, 1589." The preface dated, " From Bolton-Percy in the auncienty of Yorke, 1589." Pages 166. W.H. Octavo.

" A cōpendious forme for domesticall dutyes: Also, Our truft againſt trouble: By Charles Gibbon." Licenſed. Octavo. 1589.

A ſermon entitled, " The ſpectacle of Gods mercie carued out of the 8 Chap. of Iob, verſ. 5, 6 & 7. By William Joliph." Octavo. 1589.

" A SERMON PREACHED at Torceter, in the Countie of Northampton the 8 of Iune,—1588, at the viſitation of the right reuerend Father in God, the Biſhop of Peeterborow, By Iohn Beatniffe Preacher of the woord of God in Brackley. 2 Cor. 22; (11) 28.—Printed by *him* for Roger VVard, 1590." Dedicated " To. the moſt reuerend Father in God, my Lords grace, Archbiſhop of Cāterbury: Sir Chr. Hatton knight, Lord high Chancellor of England: Sir Wm. Scicil Knight—Lord high Treaſurer of England: Sir Chr. Wray Knight, Lord chiefe Iuſtice of England: and to al other which ſincerely fauour the word of God." W.H. Sixteens. 1590.

" The Rare and Singuler worke of Pomponius Mela," &c. as p. 898. Quarto. 1590.

" Priuate praiers for houſholders to meditate vpō, & to ſay in their families: By Edw. Deringe." Octavo. 1590.

" A homelie or ſermon of the good & euill Angell: on the 18 Mar. ver. 10. By Vrbanus Regius: tranſlated by Rich. Robinſon." Licenſed in 1582. Octavo. 1590.

" A Treatiſe of the cure of the French Poxe, with all other diſeaſes ariſing & growing thereof, by Th. Paracelſus: tranſlated by Io Heſter." Quarto. 1590.

" THE Second part of the Booke of Chriſtian exerciſe, appertayning to Reſolution: Or a Chriſtian directorie, guiding all men vnto theyr ſaluation Written by the former Authour, R. P. Pſ. 27; 4.—Printed by *him* for Simon Waterſon, in S. Paules Church-yarde, at Cheape-gate, 1591." Dedicated " To—Sir Thomas Heneage, Knight, Chaūcellour of the Dutchie of Lankaſter, vize Chamberlaine to her excellent Maieſtie, Treaſurer of her royall Chamber, and one of her Highneſſe—priuie Councell.---To the Reader.---The table." Y, in twelves, or 527 pages. Licenſed. W.H. Again 1592. Twenty fours. 1591.

" The Widowes treaſure, plentifully furniſhed with ſundrie precious approued ſecretes in Phiſicke & Chirurgerie.—for Edw. White." Octavo. 1591.

" A preparatiue to mariage: The ſumme whereof was ſpoken at a contract, and enlarged after. Whereunto is annexed a treatiſe of the Lord's ſupper, and another of vſurie. By Henrie Smith. Newly corrected, and augmented by the author.—For Thomas Man." * Sixteens.

" A brief Treatiſe contayning many proper Tables, &c. Augmented by W. W.—For Tho. Adams." See p. 819. Octavo. 1591.

" A Catechiſme with prayers, by Bartimeus Andrewes, preacher at Yermouth." Octavo. 1591.

" The wonderfull Combat (for God's Glory & Man's Saluation) betweene Chriſt & Sathan; opened in 7 moſt excellent & zealous ſermons, &c.—for Rich. Smith." Aſcribed to Dr. Lancelot Andrewes, who died Biſhop of Wincheſter. Licenſed. Octavo. 1592.

6 O " Archaipo

JOHN CHARLEWOOD.

1592. "Archaioplutos: Or the Riches of Elder ages. Proouing by many good & learned Authors, that the auncient Emperors & Kings were more rich & magnificent then such as liue in these daies. Hereto is annexed the honours of the braue Romaine Souldiours, with the seauen Wonders of the Worlde. Written in French by Guil. Thelin, Lord of Gurmont & Morillonuilliers, and truely translated into English. Patere aut Abstine. Printed by I. C. for Rich. Smith,—1592." Dedicated "To—Gilbert, Lord Talbot, the honorable Earle of Shrewesburie, Knight of—the Garter, &c.—°An. Monday, One of the Messengers of her Maiesties Chamber." O 2, in fours. Licensed, W.H. Quarto.

1592. "Gallathea. As it was playde before the Queenes Maiestie at Greenewiche, on Newyeeres day at Night: By the Chyldren of Paules.— Printed by him for the VVidow. Broome, 1592." H 2, in foura. W.H. Quarto.

1592. "THE MASQVE OR THE League and the Spanyard discouered." Wherein, 1. The League is painted forth in all her collours. 2. Is showen, that it is not lawfull for a Subiect to Arme himselfe against his King, for what pretence so euer it be. 3. That but few Noblemen take part with the enemy: an Aduertisement to them concerning their dutie. To my lord, the cardinall of Bunbon. Faythfully translated out of the French coppie: Printed at Toures by Iamet Mettayer, ordinarie Printer to the King. Patere aut abstine—For Richard Smyth,* 1592." 72 leaues. At the end, A. M. Licensed, W.H. Quarto.

1593. "A DISCOVRSE OF Horsmanshippe: Wherein the breeding & ryding of Horses for seruice, in a breefe manner is more methodically sette downe then hath been heeretofore, &c. Also the manner to chuse, mayne, ryde & dyet, both Hunting-horses & Running-horses: with all the secretes thereto belonging discouered. An arte neuer heeretofore written, by any Author. Bramo assai, poco spero, nulla chieggio." Smith's device, Time bringing Truth to light. "At London, Printed by I. C. for Rich. Smith,—1593." Dedicated "To the Right worshipfull, & his singuler good Father, Ma. Rob. Markham of Cotham, in the County of Nottingham, Esq;—Jeruis Markham.—To the Gentlemen Readers." L, in fours. Licensed, 29 January 1592-3. W.H. Quarto.

"Remember man both night and day,
Thou must neede die, there is no nay."

In sixteen four-lined stanzas, each with the said two lines for a chorus. "Imprinted at London at Holburne Conduite by John Tysdale, and John Charlewood." A broadside.

"An Epitaph vpon the death of King Edward." Begins
"Adewe pleasure, Gone is our treasure, Morning mai be our mirth, For Edward our king, That rose did spring, is vaded & lyeth in earth.

---

* On the back of this dedication, "If they—may passe wyth your wonted kind acceptance: expect a worke from the Presse, very shortly, more aunswerable to your humours; namely, the sweet conceited Historie of Orlando Amoroso. Though farre inferiour to that already extant of Orlando Furioso, doone by so rare a Scholler of the Muses: yet what wants in cunning, good wyll shall supply to compleat Orlandos whole Historie together. A. M."

\² Published again 1605, with a new title.

JOHN CHARLEWOOD.

"Imprinted at London in Holburne next the Conduite at the ſign of the Sarſins head by John Charlewood, and John Tyſdale." A ſheet.
"A Spectacle for a blind papiſt. Made by J. S." With E. White. Octavo.
"Spiritus eſt Vicarius Chriſti in terra. The poore mans Garden," &c. as p. 1064. At the end, under his device, "Imprinted—by *him*, dwelling in Barbican, at—the halfe Eagle & Key." W.H. Octavo.
"A comfortable Treatiſe ſent to all thoſe that haue a longing deſire for their ſaluation, and yet knowe not how to attaine thereto by reaſon of the miſchieuous ſubtiltie of Sathan the Archenemy of mankinde, by Tho. Burlz." Licenſed. Octavo.
"The Heroycall Epiſtles of the learned Poet P. Ouidius Naſo," &c. as p. 943. &c. X, in eights. George Maſon Eſq; Octavo.
"The Sickmans meditation, by William Wels." Octavo.
"A cōmemoracōn of the life of Peter Kempe, late of Staunford, deceaſed: Alſo, the tragicall diſcourſe of him and his wyfe, deceaſinge bothe within the ſpace of v. houres, reported by Tho. Clarke as witnes of the ſame." Licenſed 1577. Octavo.
"Greens viſion. A penitent paſſion for the folly of his pen. Serò ſed ſeriò." Quarto.

He had alſo licenſes for the following, viz. In 1563, "The tēne cōmandementes of almyghty God/ other ſhorte treatis in tyme of ſheughes. The pathe waye into the holy ſcriptures, with a compendious introduction or preface vnto the epiſtle to the Romans. A book of ſerten Godly prayers of Lady Fanes." In 1564, "A Dyaloge of too Landelordes, &c." In 1565, "The opening of the wordes of the prophet Joell in his 2d and 3d Chap. Reherſed by Chriſte in Mat. (24; 29.) Mark, (13; 24.) Luke (21; 25.) Actes (2; 17.)" In 1566, "An interlude of the repentance of Mary Magdalen. An interlude named, The colledge of canonycall clerkes." In 1569, "An epytaph of the death of Lady Jane Gryffen."---In 1576, "The deſcription of the glaſſe of comforte." In 1577, "The defence of pouertie againſte the deſire of worldlie richeſ: Dialogue wiſe, collected by Anth. Mundaie. A ſonet, neceſſarye for this tyme of God's viſitac'on, aſwell againſte the exceſſiue pride & abuſe of apparell, as other vices, to warne the citie & countrey to retu͡rn to the Lorde. A brief & clere confeſſion of the xpian faith, cont. 100 articles of thapoſtles, made & declared by Jhon Gardener." In 1578, "An epitaphe of the Lady Lomley. An epitaphe vpon the death of Sir Andr. Corbet. Cotes compariſon of Hollie & Iuye with true Religion & Superſtition, deſcribing both their nature & qualities: whereunto is anexed 12 ſentences, called preſeruatiues for the ryche agaynſt ỹ day of vengeance: and alſo the talke of ỹ wycked atheiſtes, &c. taken out of the book of Wiſdom, Chap. 2; and Ecclus, Ch. 5. A ſhrowing for Ladies. A dialoge betwene a Ladie called Lyſtria & a pilgrim cōcerninge ỹ gouernment & cōen weale of the great prouince of Crangulor." In 1579, "A true declarac'on of ỹ greate valiancye of the noble towne of Maſtricht. A book concerning ỹ Welles found at Newnam in Warc'ſhire." Allowed

JOHN CHARLEWOOD.

Allowed to R. Jones and him certain copies which they bought of H. Denham. See p. 1053. " A ballad made by Ant. Monday, Of thencoragement of an Englifhe foldior to his fellowe mates. A dilectable dialoge—betwene ij Spanifhe gent. concerninge phifick & phificians, &c. tranflated, by T. W. A thinge feene in thayre." In 1580, " Certen newes of the Turke." With Edw. White, " An exhortac'on to amendmente of life by fignes & tokens feene in ÿ ayre, and of the lafte blafinge ftarre, 8 Octo. 1580. The true & naturall proportion of a monftrous childe born in Chieri in Piemonte. The hift. of Charles & Julia twoe Brittaine Louers. A difcourfe in cōmendac'on of the valiante & vertuous mynded gent. Mr. Fr. Drake. The A. B. C. or inftruction for Chriftians,—newelie tranflated out of Frenche, by D. M. which he bought of John Arnold. The voice of the lafte trumpett, blowen by ÿ 2d Angell: as Apoc. xj. which he bought of Wm. Bartlett. The hift. of Palmerin of Englande: *conditionally*. A friendly well wifhinge to fuch as endure, &c. by Nich. Bourmā." In 1581, " A true report of the late horrible murder cōmitted by Wm. Sherwood, prifoner in ÿ Q. Benche." With Ed. White, " The conquefte atchiued by Capt. Norrice, generall Colonell in ÿ Campe in Friefland, 9 Julie, 1581. A brief difcourfe concerninge thofe 4 vfuall notes whereby Chriftes Catholike Church is knowen: Written by Ro. Crowley, clerke. The wrath of God in ÿ punifhmēte of twoo drūkardes at Nekershofen, ih Almayne. An Abftracte of ÿ hift. of Cefar & Pompeius. Theis copies 'undermentioned: alwaies prouided, That yf—anie other hath righte to printe anie of them, then this Lycence as touchinge euerie fuch of thofe copies—fhalbe void, &c. A wafpes neft found aboue the ground." In 1583, " The fweete fobbes & amorous complaints of Shepardes & Nymphes—by Ant. Munday." In 1584, " The comfort of a trewe Chriftian." In 1586, " xxvj fermons of Mr. H. Bullinger vpon the firft fermon of the prophet Jeremye, contained in ÿ firft vj chap. of his pro-

"Copies which were Sampfon Awdeleys:	Copies which were Wm. Williamfons.
The Some of dyuintie.	The poore mans Garden.
The booke of Hufbandrye.	Northbrookes Confeffion.
Ridleis Conference.	The rewarde of wickednes.
Marcus Aurelius.	King Pontus.
Tholde Algorifme.	The policies of warre.
Thargumente of Apparell.	Robin Confcience.
A pennyworth of witte.	A proude wyues Pater nofter.
A hundred merry tales.	The plowmans Pater nofter.
Adam Bell.	A fackfull of newes.
The banifhment of Cupid.	Sir Eylamore.
Crowleys Epigrams.	Gowre De confeffio. amantis.
Caluin ag. thanabaptifts.	The good fheppard & the bad.
A Foxe Tale.	Coniectures of the end of the world.
Tholde gov. of vertue, fhort.	Athanagoras of the Refurrecc'on.
Hooper on the x Comands.	Coruius poftyll.
Hoopers Homilies	The tower of truftines.
A Morninge praier.	The Caftell of knowledge.
Againfte praife of womens Bewtye.	Plaia bookes.
Chriftian ftate of matrimonie.	The Weather. iiij P. Loue. Youth.
	Ympacient pou'tie. Hyke Skorner."

phecie

JOHN CHARLEWOOD.

phecie. The Ymage of Loue. Articles to be enquired of ẏ church-wardens & fwornemen within the Archdeconry of Middlefex. The iufte Judgemēt of God vpon a myferable hard harted fermor." In 1587, " Praiers or meditations collected out of certen holie workes by the mofte vertuous pryncefſe Katheryne Parre Q. of England, &c. An. 1545, and nowe newlye imprinted at ẏ requeſt of Mrs. Eliz. Rous, and is intytuled The ſweete ſonge of a ſynner, 1587. A table: Foure elemētes, 4 feaſons, 4 humors, 4 vertues. The onlie ymprinting of all manner of Bills for players: Prouided ẏ if any trouble ariſe, then C. to beare ẏ charges. A Prayer & thankſgiuing vnto God for ẏ proſperous eſtate & long contynuance of ẏ Q. maieſtie, to be ſonge on the xvij of Nouemb. 1587. Articles to be enquired of within the Dyoces of Gloceſter & Bryſtoll. An epitaphe vpon the life & death of ẏ counteſſe of Oxon." In 1588, "An epitaphe of Mr. Wm. Lynakers death. The newes of a moſt certen victorie ag. the Spanyards by ẏ helpe of God 19 Nouem. 1588, by ẏ captaines of Holland. The hon. hiſtories of Palmendos & Primaleon of Grece, ſōnes to the famous emperor Palmerin d'Oliuia of Conſtantinople: in vij partes." In 1589, " An epitaphe vpon ẏ death of the Erle of Leiceſter, by Hen Robertes. A Looking glaſſe for England & the whole world. A catechiſme, or pythie ſome of principall matters conc'ning faith & religion." In 1590, " A moſte ſhorte & profitable introduction to learne to reade wrytten & prynted hand within a monethes ſpace, by Tho. Fowler. The order of martiall diſcipline. A particuler account of ẏ yielding vp of the towne of Zulphen and ẏ belegering of Deuener: (*See p. 1043.*) with the hon. enterpriſe of Sir Rog. Williams performed vpon 1200 of ẏ enemies foldyers lyeing at Cinque Saume 9 leages from Deepe," In 1591, " Hiſtoire de Roland L'amoureux, cōprenant les cheualeureux faictes d'armes & d'amours deuiſee en trois liures: to be tranflated into Engliſh." In 1592, " A ſecond procedinge in ẏ harmony of K. Dauids harpe." Likewiſe, at times, for ſeveral ballads.

ALICE CHARLEWOOD,

His widow, carried on the buſineſs a ſhort time after her huſband's deceaſe, and printed

A brief treatiſe containing tables, &c. as by J. C. in 1591. Octavo. 1593.
" The Honour of the Garter: Diſplaied in a Poeme Gratulatorie. Entitled to the worthie & renowned Earle of Northumberland, created Knight of that Order, and inſtalled at Windſore, Anno Regni Elizabethæ 35, Die Junii 26. By Geo. Peele, Maiſter of Artes in Oxenforde. At London, printed by the Widowe Charlewood," &c. 11 leaves. See Oldys's Catal. of pamplets &c. N°. 224.   Quarto.

She had alſo licenſes to print the two following books, viz. 23 Apr. 1593, " Geruis Mackwin his Thyrſys & Daphne." 8 Sept. 1593, " Chriſtes teares oᵥer Jeruſalem."

THOMAS

THOMAS WOODCOCK, stationer,

WAS bound apprentice to Francis Coldock for 9 years from Midsummer, 1564; and made free of the Stationers' company, 9 July 1570; came on the Livery 7 May, 1582; serued Renter in the years 1589, and 1590; was chosen Under Warden in July 1593, but died 22 April 1594, and Tho. Styrrup serued the remainder of his year. He married Isabel, second daughter of John Cawood esq; and dwelt at the sign of the Black Bear, in S. Paul's church-yard.

1575. Virgil's eclogues translated into English verse (rythmical) by Abr. Fleming. Printed for him. Probably by John Charlewood. See p. 1093. In the Bodl. Libr. Again 1589, with the Georgicks.

1576. "A REGISTRE of Hystories, conteining Martiall exploites of worthy warriours, Politique practises of Ciuil Magistrates, wise Sentences of famous Philosophers, &c. Written in Greeke by Ælianus a Romane, and deliuered in English (as well, according to the truth of the Greeke text, as of the Latine) by Abr. Fleming. Seene & allowed.—Imprinted —for him:—Anno a Messia nato. 1576." The printer's name not mentioned. Dedicated "Ornatissimo viro,—Doctori Goodmanno Westmonasteriensis ecclesiæ Cathedralis Collegijque celeberrimi dignissimo Decano."---Several commendatory verses in Greek, Latin and English. 178 leaves. W.H. Quarto.

1577. "A LEARNED AND fruitefull Commentarie vpon the Epistle of Iames the Apostle, vvherein are diligently & profitably entreated all such matters—of Religion as are touched in the same Epistle: Written in Latine by the learned Clerke Nich. Heminge professour of Diuinitie &c. and nevvly translated into English by VV. G. (Wm. Gace) Imprinted by him, and Gregorie Seton, & are to be solde at—the blacke Beare in Paules Churchyarde,—1577." Dedicated "To—M. Alex. Novvell, Deane of the Cathedrall Church of S. Paule in London." 90 leaves. Licensed. W.H. Quarto.

1577. "Of all blasing starrs in generall, as well supernaturall as naturall, to what countrie or people so euer they appeare, &c. The iudgement of the right reuerend Frederike Nause, Bishop of Vienna. Written & dedicated to the high & puissaunt Emperour Ferdinand. Translated out of Latine—by Abr. Fleming." A blazing star. "Imprinted at London for him, 1577." Dedicated "To—Sir William Cordell Knight, Maister of her Maiesties Rolles.---To the Christian Reader.---The contents.---A Prognostication of blasing starres, after the opinion of the Poet Pontanus," in 3 seuen-lined stanzas. E, in eights. Licensed. W.H. Sixteens.

(1578.) "Inuentions or Deuises. Very necessary for all Generalles & Captaines, or Leaders of men, as wel by Sea as by Land: Written by Wm. Bourne. An. 1578." For him. In a compartment used by T. Marshe. Dedicated "To—the Lorde Charles Howard, Baron of Effingham, Knight of—the Garter.---The Preface.---Table of the contents." 99 pages. Alex. Dalrymple Esq; and W.H. Quarto.

"A booke

THOMAS WOODCOCK.

"A booke called the Treasure for trauellers, deuided into fiue Bookes, 1578. or partes, contaynyng very necessary matters for all sortes of trauailers, eyther by Sea or by Lande, written by William Bourne." With schemes. For him, 1578. Dedicated " To---syr Wm. Winter Knight, Maister of the Q: Maiesties Ordinance by Sea, Suruaior of her highnesse marine causes," &c. whose arms are on the back of the title-page.---The preface. ---" A briefe note, taken out of M. Dees Mathematical Preface, before Euclides Elementes." Each part has its leaues numbered separately, with its proper argument, and table of contents. At the end of the last part is a table of " Faultes escaped in printing." Licensed. Alex. Dalrymple Esq; and W.H. Quarto.

"XXVII Lectures, or readings," &c. as p. 928. In the same com- 1578. partment. For him, 1578. Again 1584, and 1590. Quarto.

" FLORIO His firste Fruites: which yeelde familiar speech, merie (1578.) Prouerbes, wittie Sentences, and golden sayings. Also a perfect Intro- duction to the Italian, and English toagues,---The like heretofore, neuer by any man published." In the same compartment the 2 foregoing ar- ticles. A dedication, in Italian, to the earl of Leicester, whose crest gar- tered is on the back of the title-page; another to the same in English. Also an address in Italian to such English Gentlemen as delight in the Italian Language; another, in English, " Vnto the friendly, curteous, and indifferent Reader," another, in Italian, to such Italian Gent. & Merchants as delight in the English language. Then; sundry commen- datory verses in Italian, English, and French. Tables of the contents, and faults. 163 leaues. " Di Londra à di 10 Agosto, 1578.—Imprinted —by Tho. Dawson, for T. Woodcocke." Licensed. W.H. Quarto.

" THE LECTVRES of John Knewstub," &c. as p. 929. In the same 1579. compartment. For him, 1579. Again 1584. W.H. Quarto.

" The schoole of abuse, conteining a pleasant Jnuective againft Poetes, 1579. Pipers, Plaiers, Jesters, & such like caterpillers of a Cōmonwelth. A Discourse as pleasaunt for Gentlemen that fauour Learning, as profitable for all that wyll follow vertue. By Ste. Gosson, Stud. Oxon.—1579." See p. 950. Dedicated " To---M. Philip Sidney Esquier." Licensed. Again 1587. Quarto.

" Sermons of M. John Caluin on the Epistles of S. Paule to Timothie." 1579. &c. as p. 1057. Quarto.

The 2d part of Beza's Questions & Answers, translated by John Field. 1580. For him. Octavo.

" The benefite that Christians receiue by Jesus Christ crucified." See 1580. it in p. 929. " Jmprinted—by T. Dawson for G. Bishop and him, 1580." W.H. Octavo.

" A view of Mans estate:" &c. as p. 928. George Bishop for him. 1580. Octavo.

" THE Treasure of Truth, touching the grounde worke of man his 1581. saluation, & chiefest pointes of Christian Religion: with a briefe summe of the comfortable doctrine of God his prouidence, comprised in 38. short Aphorismes. Written in Latine by Theod. Beza, and newly turned into

into English by Iohn Stockwood. Whereunto are added, these Godly Treatises: One of the learned & Godly Father, Maister I. Foxe:—The other of M. Anth. Gylbie." Both on election, &c. "All newly ouerseene & corrected. Seene & allowed," &c. For him, 1581. 8, in eights. W.H. Octavo.

1581. "The Badge of both Churches, shewing how the Papists differ from the Protestantes, in euery article of the Creed, in euery precept of the morall lawe, and in the whole doctrine of the Sacraments: by Io. Iacob Grineus. Translated by T. M. Phisition." Octavo.

1582. "A Booke of Bertram the Priest, concerning the body & blood of Christ, written in Latine to Charles the great, being Emperour about 700 yeares agoe, and first translated & printed in English, 1546, and nowe newly reuiued, corrected & published by T. W. 1581." Printed by T. Dawson for him, 1582." John Fenn Esq; Octavo.

1582. "THE Christian against the Iesuite: Wherein the secrete—writer of a pernitious booke, entituled A Discouerie of I. Nicols Minister &c. priuily printed, couertly cast abrod, & secretely solde, is not only iustly reproued: But also a booke, dedicated to the Q. Maiestie, called A persuasion from papistrie, therein derided & falsified, is defended by Tho. Lupton the authour thereof. Reade with aduisement, & iudge vprightly, & be affectioned only to truth. Ps. 7; (15) Seene & allowed." Dedicated "To—Sir Francis Walsingham, kn"." &c.—An alphabetical table, or Index. 100 leaves: on the last, "Imprinted—by T. Dawson for him, 1582." See p. 987. Licensed. W.H. Quarto.

1582. "DIVERS voyages touching the discoverie of America, and the Ilands adiacent vnto the same, made first of all by our Englishmen, and afterward by the Frenchmen & Britons: And certaine notes of aduertisements for obseruations, necessarie for such as shall hereafter make the like attempt: With two mappes annexed—for the plainer vnderstanding of the whole matter." For him. Prefixed are "The names of certaine late writers of Geographie.---The names of—trauaylers.---Probalitie of a passage, by the Northwest.---To the—most vertuous Gentleman master Phillip Sydney Esq;—R. H;" (Richard Hakluyt.) After signature D, in fours, begins another alphabet, to R, in fours. At the end, "Imprinted —by T. Dawson, 1582." Licensed. W.H. Quarto.

1582. "A Summarie of the principles of Christian religion, selected in manner of Comon-places out of the writings of some of the best Diuines of our age, by Swithurne Butterfield." For him. Octavo.

1583. "A Godly, zealous & learned Sermon vpon the 18, 19, 20, 21 verses of the 10. Chap. to the Romaines: Wherein is set foorth vnto vs the greate mercy of God in the calling ot the Gentiles, and his iust iudgement in the reiecting of the vnbeleeuing Iewes & vs also, if we with like obstinacie contemne his profered mercies. By Frauncis Tayler, Preacher of Gods word." C, in eights. "—T. Dawson for him 1583." W.H. Again 1593. Sixteens.

1584. "SERMONS of Maister Iohn Caluin, vpon the Booke of Iob, Translated out of French by Arth. Golding." In the magnificent compartment

described

THOMAS WOODCOCK.

described in p. 881. Dedicated " To—Robert Earle of Leicester, &c.
—31 Dec. 1573."---M. Calvin's preface.—" At Geneva, 1 June, 1563.
---A Table of the principall matters." 159 sermons, on 751 pages, double columns; very neatly printed on Small Pica Roman; the words properly and regularly spaced, close, yet distinct. " Imprinted—by T. Dawson for G. Bishop, and *him*. 1584." W.H. Folio.

" Two Sermons preached, The one at Paules Crosse, the eight of Ianuarie, 1580, the other at Christes Church in London, the same day in the after-noone; by Iames Bisse, Maister of Arte & Fellowe of Magdalene Colledge in Oxenford,—Robert waldegraue for *him*—1585." Dedicated " To—Sir Iohn Horner, and sir George Rogers, knightes.—19 Jan. 1580." J 6, in eights. Again 1585; also without date. W.H. Sixteens. 1584.

" A Treatise of the sinne against the holy Ghost: Written by M. Austen Marlorate. Translated out of French for the consolation of all such as repent, and the astonying of those that mocke & despise the Gospell. James 5; 19, 20; Brethren, if any of you have erred from the truth," &c. C, in eights. " Printed by R. Walde-graue for *him*, 1585." W.H. Sixteens. 1585.

Holinshed's Chronicles. See p. 961, &c. Folio. 1587.

" Thirteene most pleasaunt and delectable questions, entituled, Philocopo, or A disport of diuers noble personages, composed in Italian by M. Iohn Bocace, Florentine, and Poet Laureat: And nowe turned into English by H. G." Dedicated " To—M. Wm. Rice, Esq;" L, in eights. " Imprinted—by Abell Jeffes, and are to be sold in Paules churchyard by *him*—1587." 1587.

" The Arte of shooting in great Ordinaunce: Contayning very necessary matters for all sortes of Seruitoures eyther by Sea or Lande: Written by Wm. Bourne." His rebus, a cock, crowing upon a pile of wood, with this motto on a scroll about it, CANTABO IEHOVÆ QVIA BENEFECIT MIHI. Dedicated " To—Ambrose Dudley, Earle of Warwick," &c.---The Preface." 94 pages, and a table of contents. With schemes. T. Dawson for him. W.H. Quarto. 1587.

" A Declaration of the x holy commaundements of Almightie God," &c. as p. 714. " Imprinted—for *him*, 1588." 93 leaves. Licensed. W.H. Octavo. 1588.

" A view of the marginal Notes of the Popish Testament, translated—(1588.) by the English fugitiue Papists resiant at Rhemes in France. By Geo. Wither.—Printed by Edm. Bollifant for *him*." Dedicated " To—Iohn Archbishop of Canterburie, &c.—At Dunburie, 12 Apr. 1588." Pages 316. W.H. Quarto.

¶ I have an edition of this book in quarto, having a MS. title-page with this colophon, " Printed by A. I. for Tho. Woodcock, 1567." This date i should apprehend a mistake, either in the printing or copying, for 1576, but that the dedication is dated " 6 Martij 1566." If 1567 be the true date of printing, it will be difficult to reconcile it with either of these parties, seeing neither of them were then out of their apprenticeship. And though Mr. Ames had a book printed by A. Jeffes, dated 1561; that very probably was a misprint for 1591; i have two editions with this date.

" Ephemeris

1589. "Ephemeris expeditionis Norreyfi & Draki in Lufitaniam.—Impenfis Thomæ V.Voodcocke, apud fignum Vrfi nigri, 1589." An epiftle prefixed "Michaeli Abiffelt Amoris fortio. O. H. S. D. P." 34 pages. Licenfed. W.H. Quarto.

1589. "A Trve Coppie of a Difcourfe written by a Gentleman employed in the late voyage of Spaine & Portingale: Sent to his particular friend, & by him publifhed for the better fatisfaction of all fuch as hauing been feduced by particular report, have entered into conceipts tending to the difcredit of the enterprife, & Actors of the fame." For him, 1589. Licenfed. Quarto.

1589. "Ioannis Brunfuerdi, Maclesfeldenfis Gymnafiarchæ Progymnafimata quædam Poetica. Sparfim collecta & in lucem edita, ftudio & induftria Thomæ Newtoni Ceftrefhyrij.—Impenfis Tho. VVoodcoki,—ad interfignium Vrfi nigri, 1589." Dedicated "Viris, pietate, fapientia, eruditione, honore, præftantibus, munificifq; bonarum literarum Mecœnatibus, D. Thomæ Henneagio & D. Ioanni Fortefcuto, fereniffimæ Reg. Elizabetæ a fanctioribus Concilijs, &c." On the back are fome commendatory verfes. 56 pages. At the end, ' Diftichon Epitaphicum, æri inclfum, & tumulo D. Ioannis Brunfuerdi, Præceptoris optimi, in templo Maclesfeldenfi apud Ceftrefhyrios fepulti & repuluerafcentis, affixum: qui placidifsimè in Domino obdormiuit, 15 Aprilis, 1589.

Alpha Poetarum, Coryphæus Grammaticorum,
Pœdonomún Phænix, hac fepelitur humo. Tho. Newtonus."
W.H. Quarto.

1589. "The Bucoliks of Publius Virgilius Maro, Prince of all Latine Poets; otherwife called his Paftoralls, or fhepherds meetings: Together with his Georgiks or Ruralls, otherwife called his hufbandrie, conteyning foure books. All newly tranflated into Englifh verfe by A.F." (Abr. Fleming) His rebus, as above. "Imprinted—by T.O. (Tho. Orwin) for him, 1589." Dedicated "To—Iohn, Archbifhop of Canterburie, &c.—The maine Argument of Virgils Bukoliks." 32 pages. W.H. Quarto.

1589. "The Georgiks—Otherwife called the Italian Hufbandrie, &c. Grāmaticallie tranflated into Englifh meter, in fo plaine & familiar fort, as a learner may be taught thereby to his profit & contentment." &c. On the back, "To the Reader whatfoeuer." 77 pages. W.H. Quarto.

1590. A catechifm by William Horne. Octavo.

(1591.) "The True Hiftory of the Ciuill VVarres of France, betweene King Henry the 4, and the Leaguers. Gathered from the yere of our Lord 1585, vntill this prefent October, 1591. By Ant. Colynet. Rom. 13. (2)" In the magnificent compartment as p. 926. Dedicated "Reuerendiffimo —D. Iohanni Cantuarienfi Archiepifcopo, &c.—20 Octob. 1591.—A, C. ---To the Chriftian Reader." 549 pages. Tho. Orwin for him. Licenfed. W.H. Quarto.

1591. A catechifm by Rob. Linaker. Octavo.

1591. "Martin Marfixtus, a fecond reply againft the defenfory, and apologie of Sixtus the fifth, late Pope of Rome, defending the execrable fact of the Iacobine Frier, vpon the perfon of Henry the 3, late Kinge of
Fraunce,

Fraunce, &c. Wherein ỹ said Apologie is faithfully translated, directly answered, and fully satisfied. By R. W." For him. Licensed. Quarto.

"FLORIOS SECOND FRVTES, To be gathered of twelue Trees of diuers but delightsome tastes to the tongues of Italians & Englishmen. To which is annexed his Garden of Recreation yeelding six thousand Italian Prouerbs.—Printed for *him*,—1591." In the grand compartment, as p. 926." Dedicated " To—the kinde entertainer of vertue, & mirrour of a good minde Master Nich. Saunder of Ewel Esq; &c.---To the Reader.— 30 Apr. 1591.---Phæton to his friend Florio," in verse. 205 pages, and tables of the contents, in Italian, & English. The Proverbs, printed separately, on 217 pages; on the back of the last is a table of the number of proverbs under each letter of the alphabet, in the whole 6150. Also, under T. Orwin's device, " Finito di stampare in Londra, apresso Thomaso Woodcock l'vltimo di Aprile, 1591." W.H. Quarto. 1591.

"THE Third part of the Countesse of Pembrokes Yuychurch: Entituled: Amintas Dale. Wherein are the most conceited tales of the Pagan Gods in English Hexameters: together with their auncient descriptions & Philosophicall explications: By Abr. Fraunce. ἴκας, ἴκας, ἑσλις ἄλιηρος.—Printed for *him*,—1592." In a compartment, with Moses and David on the sides, late T. Marshe's. 60 leaves. The first and second parts were printed for Wm. Ponsonby. Licensed. W.H. Quarto. 1592.

" A Dictionarie French and English :—Gathered & set forth by Claudius Hollyband.—Dum spiro, spero. Imprinted—by T. O. for *him*, 1593." In the grand compartment, as p. 926. Dedicated " A Tresnoble, Honorable, & vertueux Seigneur Monseigneur Edward Zouche, ferme base, pillier et vray Mecœnas de toutes mes estudes.—De Londres ce dixiesme jour de Avril, 1593.—Desanliens.---Cl. Holliband to the students of the French tongue." Sheets K k 2, in eights. W.H. Quarto. 1593.

" A Treatise concerning the right vse & ordering of Bees: Newlie made & set forth, according to the authors own experience : (which by any heretofore hath not been done) By Edm. Southerne Gent.—" T. Orwin's device, who printed it for him, 1593. Dedicated " To—Mrs. Margaret Astley, wife to John Astley Esq; Master & Treasurer of her Maiesties Iewels, and Plate, &c.---To the Reader." E 1, in fours. Licensed. W.H. Quarto. 1593.

" A Mirrour of Popish Subtilties: Discouering sundry wretched & miserable euasions & shifts which a secret cauilling Papist in the behalfe of one Paul Spence Priest, yet liuing, and lately prisoner in the Castle of Worcester, hath gathered out of Sanders, Bellarmine, &c. for the auoyding & discrediting of sundrie allegations of Scriptures & Fathers against the doctrine of the Church of Rome, concerning Sacraments, the sacrifice of the Masse, Transubstantiation, Iustification, &c. Written by Rob. Abbot, Minister of the word of God in the Citie of Worcester.—Perused & allowed." Creed's device. " Printed by Tho. Creede for. *him*,— 1594." Dedicated " To—the L. Archbishop of Canterbury his Grace Primate &c. and to—the L. Bishop of Worcester.—From Worcest. Janu. 7. 1593. 1594.

THOMAS WOODCOCK.

7. 1593.---The preface.---The speciall matters that are discussed in this Treatise.---Mr. Spences answer to 2 places of Chrysostome & Gelasius, with a Reply," &c. The running-title of the book, " A defence of the authorities alledged in the Reply." 226 pages. Licensed. W.H.
Quarto.

1594. " The Tragedie of Dido Queene of Carthage: Played by the Children of her Maiesties Chappell. Written by Chr. Marlowe, and Tho. Nash, Gent. Actors: Iupiter, Ganimed, Venus, Cupid, Iuno, Mercurie or Hermes, Æneas, Ascanius, Dido, Anna, Acates, Ilioneus, Iarbas, Cloanthus, Sergestus. At London, Printed by the Widdowe Orwin, for *him*, and are to be solde at his Shop in Paules Church-yeard, at the signe of the blacke Beare, 1594." Pages 52. The only copy of this play, that we haue heard of is in the possession of Mr. Reed of Staples-Inn, London.
Quarto.

―― " A Postill, Or Exposition of the Gospels," &c. as p. 927. In the grand compartment. For G. Bishop, and him. W.H. Quarto.

―― M. Luther's preface on the epistle to the Romans, translated by W. W.
Octavo.

―― " The ground of Christianitie, containing all the principall pointes of our Saluation in Christ, by Alex. Gee." For him. Octavo.

―― A catechism, entitled " The way to euerlasting Saluation, by R. L." For him. Octavo.

He had also licenses for the following, viz. In 1576, " The figure of ij monstrous children borne at Wemme in Shropshire." In 1578, " Their copies'---sold by Mrs. Harrison, wief vnto Mr. Luke Harrison deceased, & which apperteined to *him* in his lief-tyme. Also the part of their copies, assigned & sold by *the said* Mrs. Harrison. A briefe." In 1580, " Manilia: A lookinge Glasse for the ladies of England." In 1581, " Eiphnapxia, siue Elizabetha." In 1584, " A true discourse *of what* happened by an Earthquake, 1 Mar. 1584, in the places adioyninge to the lake of Geneua." In 1588, " A godlie prayer for the preseruac'on of ỹ Q. Maiestie, and for her armyes, bothe by Sea & Land: aganste ỹ enymies of the Churche, & this realme of England: with condic'on &c. The Entraūce of Christianytie. A letter from ỹ Fr. king vnto his court

" Mr. Deeringes lectures.
Mr. Knewstubes lectures.
Caluins last will & testament.
Bezas 2d part of questions.
Dictionarie, French & English.
Confess. of the reformd churches.
Agreement of the scriptures.
The preparat. to the L. supper.
Fulkes S. on 4th to the Gal.
Andersons sermons.
The synne of the H. Ghoste.
A dialoge betw. cap & the head.
Nortons book to the rebells.
Cope on the 1. Pro°rbes,"

" Caluins serm. on Deut.
――――――― Job.
――――――― the Psalmes.
――――――― the Galathians.
――――――― the Ephesians.
――――――― Tim. & Titus.
Marlorat on the Reuelat.
Holinsheds Cronicles.
Hemingius on the Gospels.
Chitreus on the Epistles.
Tomson ag. Fecknam.
Philosophie of the Court.
The view of mans estate.
The benefite of Christe.
Sume of Xpian religion."

THOMAS WOODCOCK.

of Parliament of Burdeux, cōcernyng ỹ death of the duke of Guife." In 1589, "A method of meafuring & furueying of land: publifhed by J. F. practic'oner in phifik." In 1592, "Idea: The fhepherds garland, fafhioned in x ecloges. A fermon of hearinge, or Iuell of the cave."

WILLIAM HOSKINS, ftationer,

WAS bound apprentice to Richard Tottel for feven years, from Michaelmas 1560, and probably made free fometime between July 1570 and July 1576, the regifter for which interval is miffing. He appears to have been but a diforderly member. " Sept. 3, 1582. This day Wm. Hofkins was awarded to prifon for iij daies & to pay xs. for keeping a prentice above vij yeres vnprefented, contrary to all order: and yt is ordered ỹ the faid aprent. fhall neuer be admitted to be free of this company." The fine was not paid till 2 Sept. 1583, when he was alfo fined xxs. for keeping another apprentice unprefented above 7 years; whereof was then paid xs., and promifed to pay the other xs. on the next quarter day. In June 1593, he prefented, or bound, another apprentice in the regular way. At firft he dwelt in Fleet-ftreet, at the Temple-gate; but latterly he appears to have been connected with John Danter, and to have lived in Fetter-lane.

" THE FLOWER OF FAME: containing the bright renowne, and mofte 1575. fortunate reigne of king HENRY VIII, wherein is mention of matters, by the reft of our cronographers, ouerpaffed. Compyled by Vlpian Fulwell. Hereunto is annexed (by the author) a fhort treatife of III noble and vertuoufe queeenes: and a difcourfe of the worthy feruice, that was done at Haddington in Scotlande, the fecond yere of the reigne of king Edward the fixt." In profe and verfe. Dedicated to the Lord Treafurer Burleigh. Quarto.

" A FRVITFVLL SERMON, Vpon part of the 5 Chap. of the firft Epiftle 1591. of S. Paul to the Theffalonians, by Henrie Smith: Which Sermon being taken in Chara&terie is now publifhed for the benefite of the faithfull." C 7, in eights. " Imprinted At London, by Will. Hofkins, Henrie Chettle,[1] & Iohn Danter, for Nich. Ling, and are to be fold at his fhop at the Weft end of Paules, 1591." W.H. Sixteens.

" THE AFFINITIE of the Faithfull: Being a verie Godlie & fruitfull 1591. Sermon made vpon part of the 8 Chap. of the Gofpel of S. Luke, By Henrie Smith.—Printed by Will. Hofkins & Iohn Danter, for Nich. Ling & Iohn Bufbie, 1591." C 3, in eights. W.H. Sixteens.

[1] By the Court of Affiftants, 3 Aug. 1591. " Yt is thought good that Wm. Hofkins maye accept to be partners with him in printinge Hen. Chettle & John Danter; prouided alwayes that there fhalbe no alien- ac'on, or tranfportinge made by hym to them, or either of them, or to any other, of his Rowm or place of a mayfter printer, without the confent of the Mafter, Wardens & Affiftants for the tyme beinge."

" THE

1592. —" THE ENGLISH Phlebotomy: Or, Method & way of healing by letting of blood: Very profitable in this spring time for the preseruatiue intention, and most needful al the whole yeare beside, for the curatiue intention of Phisick. Collected out of good & approued authors at times of leasure from his other studies, and compiled in that order that it is: By N. G. Pro. 30; 15.—Ch. 27; 9.—Imprinted—for Andr. Mansell, and are to be solde at his shop in the Royall Exchange, 1592." Dedicated " To—Master Reginald Scot, Esq;—Nich. Gyer, minister of the word." 297 pages, and a table of contents; another of " The names of the authors whose help is chiefly vsed in this collection." On the back in a small compartment, " London, Printed by Wm. Hoskins & Iohn Danter, dwelling in Fetter-Lane. 1592." W.H. Octavo.

JOHN SHEPHERDE, SHEPPERDE, or SHEPPARDE, stationer,

WAS apprentice with Reynold Wolfe eight years from Lady-day 1566, and seems to have taken up his freedom of the Stationers' company in the interval, whereof the Clerk's book is missing. He dwelt in S. Pauls church-yard, at the sign of the brazen serpent, and used the same for his device.

1576. " AN ARMOVRE of Proufe: Very profitable, as well for Princes, noble men, and gentlemen, as all other in authoritie, shewing the firme fortresse of defence, and hauen of rest in these troublesome times & perilous dayes. Made by Iohn VVolton, Minister of the Gospell." His device, the same as was used by his Master, and sometimes by Bynneman. " Imprinted—by Iohn Shepperde, Anno 1576." Dedicated " To—Sir William Cicil Knight, &c.—Exceter, the last of Febru. 1576." Sir William's crest is on the back of the title-page. 48 leaves. W.H.
Sixteens.

1576. " The VVarfare of Christians:" &c. asp. 979. At the end is a cut of the sun as p. 543; under it " Imprinted—by Hen. Binneman for Iohn Shepherde."

1576. " A TREATISE of the Immortalitie of the Soule: Wherein is declared the Origine, Nature & Powers of the same, together with the state & condition thereof, both as it is conioyned, and dissolued from the body. Made by Iohn Woolton Minister of the Gospell. Imprinted—in Paules Churchyarde, at the—Brasen Serpent by Iohn Shepperd,—1576." On the back is the cut of the Holy Trinity covenanting for the redemption of mankind. Dedicated " To the right hon. and vertuous Lady, Bryget, Countesse of Bedforde.—Exceter, 2 March, 1576." Fol. 96. " Imprinted at London by Thomas Purfoote for Iohn Shepperd." W.H.
Sixteens.

1576. " The Safegarde of Societie: Describing the Institution of Lawes, and

JOHN SHEPHERDE.

and Policies to preserue euery felowſhip of People by degrees of ciuil gouernmente: Gathered of the Moralls & Policies of Philoſophie by John Barſton." At the end, "—I. Shepherd for Hen. Benyman." Sixteens.

"Flowres of Epigrammes out of ſundrie the moſt ſingular authors, 1577. ſelected &c. by Timothie Kendall, late of the vniuerſitie of Oxford, now of Staple Inn. London, 1577." Licenſed. Sixteens.

THOMAS DAWSON, ſtationer,

WAS bound apprentice to R. Jugge for eight years from S. Bart. 1559, and made free 18 Feb. 1567-8; Came on the Livery 6 May, 1585; ſerved Renter in 1592, and 1593; Under Warden in 1595, and 1596; Upper Warden in 1600, and 1603. He was fined for printing a book for Tho. Butter, before the Wardens hands were to it, as alſo was T. Butter; ij.s and vj.d. each. For keeping apprentices unpreſented, he was fined thrice: for the firſt two x. s. each, and for the laſt vijs. vjd. "but in reſpect of his ſtate, & for that he took in thap'ntiſe for charities ſake, being a poor child, the court reſtored him vs. thereof." He dwelt at the ſign of the three cranes in the Vintry, and uſed the ſame for his device. I find one Wm. Dawſon made free 21 Oct. 1561; but nothing more of him.

"The Workes of a young wyt truſt' vp with a Fardell of prettie 1577. fancies, profitable to young Poetes, preiudicial to no man, and pleaſaunt to euery man to paſſe away idle tyme withall: Whereunto is ioined an odde kinde of wooing with a Banquet of Comfettes to make an end withall. Done by N. B. Gent. Imprinted at London nigh vnto the three Cranes in the Vintree by Tho. Dawſon, and Tho. Gardyner, 1577." N.B. This is a curious little book, and gives us ſeveral pictureſque deſcriptions of the manners of the time, &c. John Fenn Eſq; Quarto.

"The prayſe of Follie. Moriæ Encomium, a book made in Latine by 1577. that great clerke Eraſmus Roterodame: Engliſhed by Sir Tho. Chaloner knight. Imprinted—by Tho. Dawſon, & Tho. Gardiner, 1577." P 2, in eights. Licenſed." W.H. Sixteens.

"THE TESTIMONIE of a true Fayth: Conteyned in a ſhorte Cate- 1577. chiſme, neceſſarie to all Families, &c. Gathered & written for the benefite of Gods well diſpoſed children, by C. S. *(Chr. Shutte)* Preacher. Imprinted, &c. 1577. March 10." C. 2, in eights. On the laſt leaf, a colophon, as before, over their joint rebus; on one ſide a gardener with a pair of ſheers trimming a flower-pot, on the other a Daw, picking its feathers under the ſun. T. D. and T. G. ſet in a ſquare in the middle. Licenſed. W.H. Again 1584. Sixteens.

* On condition not to print above 1500; and "that any of the company may laie on with him reaſonablie at euery impreſſion, as they think good, & that he ſhall gyue reaſonable knowledge before to them, as often as he ſhall print it."

"An

1577. "An expoſition of the 4 Chap. of St. Johns Reuelation made by Bar. Traheron in ſundry Readynges before his countrymen in Germanie: Wherein the prouidence of God is treated, with an aunſwer made to the obiections of a gentle aduerſary." Again 1583. Sixteens.

1577. "Soueraigne approued Medicines and remidies, as well for ſundry diſeaſes within the bodie, as for all ſores, wounds, goutes, and other griefes whatſoeuer." By him and Thomas Gardiner. Licenſed. Octavo.

1577. "Of the ende of this world, and ſecond coming of Chriſt," &c. as p. 1060. 48 leaves. "Imprinted—by Tho. Gardyner, & Tho. Dawſon, for A. Maunſell," &c. W.H. This treatiſe had great effect on peoples minds in general, ſo that in the courſe of the next year it was printed with this addition to the title,

1578. ———"Now the thirde tyme corrected, and wyth a learned Epiſtle ſent by the author out of Frieſelande to the tranſlator, (*Emden*, 29 *Aug.* 1577.) with verſes of one Sibil Erithrea, and with other things, hitherto not put in, augmented.—Imprinted—by T. Dawſon (*alone*) for A. Maunſell,—1578." On the back of the title-page are "Verſes written by Th. Beze vpon the new ſtarre which appeared in—1572 and 1573." Under them "Apoc. 6. (10) Howe long tarryeſt thou Lorde," &c. 60 leaves including the title-page. W.H. Quarto.

(1578.) "FLORIO his firſte Fruites:" &c. as p. 1107.

1578. "A Commentarie of M. Iohn Caluine, vpon the Booke of Ioſue, finiſhed a little before his death: Tranſlated out of Latine by W. F. Whereunto is added a table, &c.—for Geo. Biſhop, 1578." Fol. 105, beſides The argument prefixed, and a table at the end. W.H. Quarto.

1578. "A true report of M. Martin Frobiſher his third & laſt voyage: written by Tho. Ellis, ſailer & one of the company." See p. 981, &c.

1578. "A caſtle for the ſoule, conteining many godly prayers, and diuine meditations, tending to the comfort and conſolation of all faythful chriſtians, againſt the wicked aſſaults of Satan. Dedicated to the right honorable lord Ambroſe, earle of Warwicke, with an alphabet upon his name. Sene and allowed." Printed within a pretty border for Robert Walgrave.* Sixteens.

1578. "The third part of the Secretes of M. Alexis," &c. See p. 781.

1579. "AN Aunſweare vnto certaine aſſertions, tending to maintaine the Churche of Rome to bee the true & Catholique church. By I. Knewſtub." Dedicated "To thoſe gentlemen in Suffolke whom the true worſhipping of God hath made worſhipfull." 62 leaves: on the laſt, over the ſame rebus as to The Teſtimonie &c. 1577, with T. D. only, in the ſquare. "Imprinted—at the 3 Cranes in the Vintree, by *him*, for Richard Sergier, 1579." W.H. Quarto.

1579. "A Confutation of monſtrous and horrible hereſies taught by H. N. (*Hen. Nichols of Leyden*) and embraced by a number, who call themſelues the Familie of Loue, by I. Knewſtub. Ephe. 4; 14, 15. Seene & allowed, &c.—for R. Sergier, 1579." Dedicated "To—Ambroſe, Earle of Warwick, maiſter of her Maieſties Ordinance, &c.—To the Reader.— The iudgement of a godly learned man touching this matter, &c.— W. C."

THOMAS DAWSON.

"W. C." 94 leaves. Hereunto is annexed, "A Sermon preached at Paules Crosse the Fryday before Easter,—1576, by I. Knewstub." Text, Titus 2; 11—15. The signatures continued to S 4. W.H. Quarto.

" A veey fruiteful Sermon preched at Paules Crosse, the tenth of May last, being the first Sunday in Easter Terme: in which are conteined very necessary & profitable lessons & instructions for this time: By Iohn Stockevvood Schoolemaister of Tunbrydge. Ps. 119; 104. By thy precepts" &c. Dedicated " To—my very good Lorde & Maister Henry Earle of Huntingdon, Lord President &c.—30 Sept. 1579." 70 leaves. " —for G. Bishop, 1579." W.H. Sixteens. 1579.

" Spiritus est vicarius Christi in terra. A Treatise wherein Dicing, Dauncing," &c. as p. 991. " Made Dialoguewise by Iohn Northbrooke Minister &c. Cicero lib. 1. de officiis. VVe are not to this end born,— for play & pastime: &c.—for G. Bishoppe, 1579." Dedicated "To—Sir Iohn Yong.—At Bristowe.---To the Reader.—From Henburie.---An admonition," in verse. 72 leaves. W.H. Quarto. 1759.

" Thirteene Sermons of Maister Ioha Caluine, Entreating of the Free Election of God in Iacob, and of reprobation in Esau,—firste published in the French toung, & now translated—by Iohn Fielde, For the comfort of all Christians. Rom. 11. 33.—Imprinted—for Tho. Man & Tobie Cooke. 1579." Dedicated " To—the Earle of Bedford,—and to the Hon. godly & vertuous Ladie the Countesse his wife.—25 Octob. 1579." At the end of the sermons is " An Answere to a Libel against Predestination." 176 leaves. W.H. Quarto. 1579.

" Foure Sermons of M. Iohn Caluin, Entreating of matters very profitable for our time, as may be seene in the Preface.: With a briefe exposition of the LXXXVII Psalme: Translated out of Frenche—by Iohn Fielde.—for Tho. Man, 1579." Dedicated " To—Henry Earle of Huntington, Lord Hastings, Hungerford, Botreaux, Mullens & Moyles, of the most Hon. order of the Garter Knt. and Lorde President of the—Councell established in the North partes.---I. Caluin to all true Christians —Geneua, 20 Sep. 1552." W.H. Quarto. 1579.

" Sermons (16) vpon the 10 Commandements of the law of God, geuen by Moises, taken out of the sermons on Deuteronomy: By John. Caluin; translated by John Harmar." For G. Bishop. Again 1581. Quarto. 1579.

" A SERMON Preached at Yorke before the right Hon. Henrie Earle of Huntington, Lorde President of her Maiesties Councell established in the North, and other noblemen & gentlemen, at a general Communion there, 23 of Sept. in the eightienth yeare of her Maiesties raigne: By Mathewe Hutton, Deane of Yorke.—for Rich. Sergier, 1579." With a Preface by W. C. (Wm. Cecil) W.H. Sixteens. 1579.

" A most godly and learned Discourse of the woorthynesse, authoritie & sufficiencie of the holy Scripture: &c. Wherein is discussed this famous question: Whether the Canonical Scriptures haue authoritie from the Church, or rather the Church receiue authoritie from the Scriptures: By occasion wherof are touched the dignities & duties of the Church, 1579.

6 Q touching

touching traditions, with aunfwers to all obiections. Tranflated out of Latine—by Iohn Tomkys:" &c. Dedicated " To—Sir Richard Pipe, Knight, Lord Maior of—London.—From mine houfe in Bilftoin, 10 Feb. 1579.---Henrie Bullinger, to the gentle Reader,—Zurich, Aug. 1571. ---The table." 119 leaves. " —for Wm. Pounfonby, 1579." W.H.
Sixteens.

1579. " The ordinance and edic, vppon the fact of the execution of both the religions, ftatuted by the bailiefes, fchepens of both the benches, and both the wardens of the citie of Ghaunt, by aduifement of my lord the prince of Orange, &c. 27 December, 1579. Tranflated out of Dutch."*
Sixteens.

1579. " The Ephemerides of Phialo, deuided into three Bookes. 1 A method which he ought to follow that defireth to rebuke his freend, when he feeth him fwarue: without kindling his choler, or hurting himfelfe. 2 A Canuazado to Courtiers in foure pointes. 3 The defence of a Curtezan ouerthrowen: And a fhorte Apologie of the Schoole of Abufe againft Poets, Pipers, Players, & their Excufers: By Step. Goffon, Stud. Oxon.—1579." Dedicated " To the right noble Gentleman, M. Philipp Sydney Efq;---Literarum Studiofis in Oxonienfi Academia Steph. Gofson Sal.—Londini, 5 Kal. Nouemb. 1579.---To the Reader." 92 leaves. Licenfed. W.H. Again 1585. Sixteens.

1579. " The fume of chriftianitie reduced to eight propofitions, briefly and firmly confirmed by the holy worde of God." 24 leaves.* Sixteens.

1580. " A Fort for the afflicted. Wherin are miniftred many notable and excellent remedies againft the ftormes of tribulation. Written chiefly for the comforte of Chriftes little flocke, which is the fmal number of the Faithfull, by Iohn Knoxe. Iohn 16; 23."* This is an expofition vpon the 6th Pfalm. It has prefixed, an epiftle " To the Religious Reader by Abr. Flemming.---To his beloued mother I. K. fendeth greeting in the Lorde." At the end of the expofition is " A comfortable epiftle fent to the afflicted churche of Chrift, exhorting them to bear his croffe with patience &c. Written at Deepe 31 May 1554." F 4, in eights. Licenfed. W.H. Sixteens.

1580. " Three Propofitions or Speeches, which that excellent man M. Iohn Caluin, one of the Paftors of the church of God in Geneua had there. To which alfo is added, an expofition vpon that part of the catechifme, which is appointed for the three and fortieth Sunday in number. Tranflated into Englifh by T. VV."* Dedicated " To the right worfhipfull Syr Richard Knightley Knight, and the Right Hon. the Ladie Elizabeth his wyfe.—London, 20 Dec. 1579." L 3, in eights. For G. Bifhop, 1580. W.H. Octavo.

1580. The inftitution of Chriftian Religion by I. Caluin. See p. 880. Octavo.
1580. Caluin on Iob. See p. 1108. Folio.
1580. " Two and twentie Sermons of Maifter Iohn Caluin: In which Sermons is moft religioufly handled the 119 Pfalme of Dauid, by 8 verfes aparte according to the Hebrewe Alphabet. Tranflated out of French—by T. S." Dedicated " To—Sir Robert Iermyn, and to—his godly & vertuous

tuous wife the Lady Iudith Iermyn.—Mildenhall, 4 Nouemb. 1579. Tho. Stocker.---To all faithfull Readers." 190 leaves. "—for Iohn Harifon & Tho. Man, 1580." W.H. Quarto.

"The Bee hiue of the Romifhe Churche. A worke of al good Catho- 1580. likes too bee read, and moft neceffary to be vnderftood. Wherin both the Catholike Religion is fubftantially confirmed, and the Heretikes finely fetcht ouer the coales. Tranflated out of Dutch into Englifh by Geo. Gilpin the Elder. 1 Thef. 5; 21. Newly imprinted with a table thereunto annexed, 1580. Thefe bookes are to be folde in Paules Church yarde, at the figne of the Parret." Dedicated "To the—wife, and vertuous Gentleman, Maifter Philip Sidney Efq;—Iohn Stell.---To the Reader.—interpretation of the Epiftle of M. Gentian Haruet, lately fet forth in Frenche & Dutche: &c.—Made by Ifaac Rabbotenu of Louen, Licentiate in the Popifh lawes.---To Doctor, and Magifter nofter, Maifter Francifcus Sonnius,—Bifhop of Shertoghenbofch, *(afterward Bifhop of Antwerp)*—Datum in our Mufeo, the v. of Ianuary, the Eeuen of the 3 kings of Collen, at which time all good Catholikes make merry, & crie, The king drinkes, in anno 1569. Ifaac Rabbotenu.---The argument; &c.---The firft Table conteyning Authours names. Gathered by Abr. Fleming.---The 2d. *Table conteyning all fuch doctrines as are expreffed in this Beehiue. Gathered by A. F. Impenfis Andreæ Maunfel Bibliopole." This ironical treatife concludes on fol. 350. "Now followeth further the expofition & declaration of the Bee Hiue, and the defcription of the Bees, the Honie & Honie combe," &c. At the end, on fol. 365, is "The locke of this Booke." On the back, "Imprinted at Lōdon, at the 3 Cranes in the Vinetree, by Tho. Dawfon for Iohn Stell, 1580." There are two very droll cuts belonging to this book, though rarely found in it. Each of them reprefent the Beehive by the pope's triple crown, and the bees flying about it in different pofitions and attitudes, fome with cardinals hats on, others with mitres, the reft tonfured. W.H. Again 1598, for himfelf. W.H. Sixteens.

Ioyfull Newes out of the New-found World. See p. 879. Quarto. 1580.

"A Godly and learned Affertion in defence of the true Church of 1580. God, and of his woorde: written in Latine by—D. Philip Melancthon, after the Conuention of Ratifbona,—1541. Tranflated by R. R. Seene, perufed & allowed. Efay 66; (10)." Dedicated "To Edwarde Earle of Rutlande, Lorde Roos of Hamelake, of Beauuoyr, and of Trufbuz, &c. Rd. Robinfon." 67 leaves. " Imprinted, &c. 1580." W.H. Octavo.

"A Sermon at S. Peters in Exeter, on Luke 21; 25. Of the ende of 1580; the worlde, by Iohn Chardon." Octavo.

---

* This 2d. Table feems to have been added after the firft publication, and to be the table intimated in the title-page. The book was licenfed 21 Jan. 1576, to John Skell, or as he is afterwards called, Hans Skell, a Dutchman, at whofe expence it was tranflated and printed; however it does not appear to have been publifhed before April 1579, when he prefented a copy to the company, accordibg to the cuftom of that time.

"A Godly

1580. "A Godly and learned Expofition vppon the Prouerbes of Solomon: Written in French by Maifter Michael Cope, Minifter of the woorde of God in Geneua: And tranflated—by M. O." In the elegant compartment, as p. 926. Dedicated "To—Sir William Cecil, Knight, &c.—Marcelline Outred.---An ample Index.—Gathered by Abr. Fleming." 639 leaves. "—for G. Byfhop, 1580." W.H. Quarto.

1580. "Papa confutatus. Sanctæ & Apoftolicæ Ecclefiæ in confutationem Papæ Actio prima. Londini in ædibus Thomæ Dawfon, impenfis Richardi Sergier, 1580." Iohn Fenn Efq; Quarto.

1580. "THE POPE CONFVTED. The holy & Apoftolique Church confuting the Pope. The firft Action. Tranflated out of Latine—by Iames Bell. Pfal. 27. Dominus illuminatio mea," &c. It is introduced by two epiftles; "The Tranflator to the Reader.---The Athour to the freendly Reader." The head title, "A Defence of the holy and Apoftolyke Church, againft the Bifhop of Rome." The fecond action begins on fol. 48, with this head title "The holy and Apoftolical Church confuting the Errours of the Popes doctrine." Ends on fol. 121. On another leaf, "Imprinted &c. for Rd. Sergier, 1580." W.H. Quarto.

1580. "The tryal of Truth: wherein are difcouered 3 great enemies to mankind, as Pride, Priuate Grudge, & Priuate gain: which corruptions are the difturbers of houfes, cities, comon weals, & people: By Edw. Knight." Licenfed to him, and T. Butter. Octavo.

1580. "The Nofegay of Moral Philofophy, &c. Englifhed by Tho. Crewe." Queftions and Anfwers. Sixteens.

1580. "The benefite that Chriftians receiue by Jefus Chrift crucified." See p. 1107. Alfo without date. See p. 929.

(1580.) "A treatife of the Plague:—1 Whether it be infectious or no? 2 Whether, and how farre, it may be of Chriftians fhunned by going afide: Written in Latin by Th. Beza, and turned into Englifh by Iohn Stockwood—of Tunbridge." Dedicated to Sir Henry Sidney. For Geo. Bifhop Octavo.

1581. "The vnfolding of fundry vntruths and abfurde propofitions, lately propounded by one I. B. a great fauourer of the horrible herefie of the libertines," Printed for T. Man. Quarto.

1581. "A fourme of catechifing in true religion, confifting in queftions and anfwers, with obferuations thereon, for the further declaration and vfe of the fame: By Wm. Woord." Sixteens.

1581. "A Checke or reproofe of M. Howlets vntimely fhreeching in her Maiefties eares, with an anfweare to the reafons alleadged in a difcourfe therunto annexed, why Catholikes (as they are called) refufe to goe to church: Wherein (among other things) the Papifts traiterous & treacherous doctrine, and demeanour towardes our Soueraigne & the State is—vnfolded: their diuelifh pretended confcience alfo examined, &c. And laftly fhevved that it is the duety of all true Chriftians & fubiectes to haunt publike Church affemblies. 1 Cor. 4; 3.—Imprinted—for Toby Smyth, 1581." Z z z, in fours. W.H. Quarto.

1581. "A briefe Confutation of a Popifh Difcourfe, Lately fet forth, & pre-

presumptuously dedicated to the Q. most excellent Maiestie, by Iohn Howlet, or some other Birde of the night under that name: Contayning certaine Reasons, why Papistes refuse to come to Church, which—are here inserted—at large, with their seuerall answeres: By D. Fulke,—. Seene & allowed.—for G. Bishop, 1581." On 58 leaves. W.H. Quarto.

"A Sermon at the Tower, on Iohn 17; 17, by Wm. Fulke, D. D." 1581. For Geo. Bishop. Sixteens.

"The Iesuites Banner: Displaying their original and successe: their 1581. vowe & othe: their hypocrisie & superstition: their doctrine & positions: with à confutation of a late Pamphlet secretly imprinted, and intituled: A Briefe Censure vpon two bookes written in answeare to M. Campions offer of disputation, &c. Compiled by Meredith Hanmer, M. A. and Student in Diuinity. Imprinted by *him* & Rd. Vernon, and are to be solde in Paules Churchyard at the Brazen Serpent, 1581." Dedicated "To—Sir Thomas Bromley knight, Lorde Chanceller of England: William Lorde Burleigh, and Lorde Treforer: Robert, Earle of Leicester: Edward, Earle of Lyncolne: with the rest of her Maiesties most Hon. counsell:—London, 3 Mar. 1580.---Iesuitis, Seminarijs, sacrificulis, alijsq; omnibus, qui Pontificij erroris caligine obducti tenentur: M. Hanmerus, Anglo Brytannus, salutem," &c. K, in fours. W.H. Quarto.

"Iohn Niccols Pilgrimage, wherin is displaied the liues of the 1581. proude Popes, ambitious cardinals, lecherous Bishops, fat bellied Monkes, & hypocriticall Iesuites. Apoc. 18; (2, and 24) Imprinted—for Tho. Butter, & Isaac Godfrey, 1581." Dedicated, "Illustrissimæ serenissimæq; Principi Angliæ,—Reginæ, Elizabethæ, fidei Catholicæ defensori, &c. ---To the indifferent Reader.---To the Reader." This is in Italian, and concludes with "Essendo il prencipe giusto, &c. The prince being iust, the Cleargie holy, the Church fauoured well, the comon weale amended, and all the Realme peaceable, that Prince, that Cleargie, that Church, that Comon Weale, & that ealme shalbe blessed of God. R, in eights. W.H. Sixteens.

"St. Augustin's Ladder to paradice; translated by T. W. Dedicated 1581. "To—Lady Fane.---To the reader." E 3, in fours; half sheets."—for Tobie Cooke, 1581." W.H. Octavo.

"A verie true Report of the apprehension & taking of that Arche- 1581. Papiste Edmond Campion, the Pope his Right Hand, with 3 other lewd Iesuite Priests, & diuers other Laie People, most seditious persons of like sort: Containing also a Controulment of a most vntrue former book, set out by A. M. (*Ant. Munday*) concerning the same; as is to be proued & iustified by Geo. Ellyot, one of the ordinary Yeomen of her Maiesties Chamber; Author of this Booke, and chiefest cause of the finding of the sayd lewde & seditious people;" &c. 13 leaves. See Oldys's Catal. of Pamphlets, N°. 143. Octavo.

"Foure Sermons vppon the 7 chiefe vertues or principal effects of 1581. faith, & the doctrine of election: &c. Preached at Malden in Essex by Maister Geo. Gifforde, penned from his mouth & corrected; and geuen to the Countesse of Sussex for a New yeeres gift. Iames 2; 18. Shevve me me thy fayth by thy vvorkes." Dedicated " To—the Lady Frauncis Counteſſe of Suſſex.—Rd. Ioſua Senior.---To the godly Reader.---Rd. Ioſua Iunior." Text, 2 Pet. 1; 1—11. G 4, in eights. "—for Tobie Cooke, 1581." W.H. Octavo.

1581. Two ſermons—By James Biſſe, M.A. &c. See p. 1109. Sixteens.
1581. " The Chriſtian mans Cloſet, Wherein is conteined a large diſcourſe of Godly training vp of children; as alſo of thoſe duties that children owe to their parents: made Dialogue wiſe.—Collected in Latine by Bart. Batty, of Aloſtenſis: And nowe Engliſhed by Wm. Lowth. Eſt adoleſcentis ætas ſuſpectior, ætas lubrica, delitijs ebria, Legis egens." Dedicated " To—M. Tho. Darcie, and M. Brian Darcie Eſquiers.—Malden, 31 May 1581.---To the Reader." It conſiſts of two books. 1. The duty of parents towards their children. 2. The duty of children to their parents. 101 leaves. " Imprinted—at the 3 Cranes in the Vintree, by him & Gregorie Seton, 1581." W.H. Again 1582. Quarto.

1581. " The Arte of Nauigation, wherein is contained all the rules, declarations, ſecretes & aduiſes, which for good Nauigation are neceſſarie & ought to be knowen & practiſed:—made by (maſter Peter de Medina) directed to the right excellent & renowned Lord don Philippe, prince of Spaine, &. of both Siciles. And nowe newly tranſlated out of Spaniſh —by John Frampton." Dedicated " To—maſter Edwarde Dier, Eſq;— London, 4 Aug. 1581. Wherein we learn that this treatiſe was compiled by M. P. de Medina in 1545; and was allowed by the Chief Pilot and Coſmographers of the Contractation houſe of Seuil.---To Don Phillip, Prince of Spain, &c.---A preamble, vpon the Arte of Nauigation, &c. ---The table." 83 leaves, in double columns, with ſchemes & tables. " Imprinted—at the 3 Cranes in the Vinetree,—and are there to be ſolde, 1581." Licenſed. Alexander Dalrymple, Eſq;- Again in 4to. 1595. Folio.

1582. " A Booke of Bertram the Prieſt," &c. as p. 1108. Octavo.
1582. " To the king of Fraunce, Francis the firſt: The relation of JohnVezarianus, a Florentine, of the land by him diſcouered in the name of his maieſtie. Written in Dieppe 1524. The true diſcouery by Captain John Ribault, in the yeare 1563. Tranſlated into Engliſhe by one Thomas Hackitt." With maps, &c. This is only part of Divers voyages touching the Diſcovery of America &c. as p. 1108.

1582. " A Godly and ſhorte Treatiſe of the Sacraments: VVritten by Robert Some. By one ſpirit we are all baptiſed into one body, &c. 1 Cor. 12; 13. Imprinted—for Geo. Biſh, 1582." Dedicated " To—Lord Robert Dudley, Earle of Leiceſter, &c.—Cambridge, 15 May, 1582." F 6, in eights. W.H. Sixteens.

1582. The Urinal of Phyſick—By Rob. Record. See p. 599. W.H+ Octavo.
1582. " The Seuenth Lampe of virginitie," &c. See p. 956.
1582. Card. Granville's Letters touching the ſtate of Flanders, & Portugal. Octavo.

1582. " A Dialogue againſt light, lewde, & laſciuious Dauncing, wherein are refuted all thoſe reaſons which the cōmon people bring in defence thereof, by Chr. Fetherſtone." Licenſed. Octavo.

" An

THOMAS DAWSON.

" An Epiftle to the Faithfull, neceffary for all the children of God: 1582.
efpecially in thefe dangerous dayes: Written by Maifter Peter Viret in
French, and englifhed by F. H. Efq; Imprinted—for Tobie Smith,—
1582." No direction to " The Epiftle Dedicatory." H 3, in eights.
W.H. Sixteens.

" THE Chriftian againft the Jefuite:" &c. as p. 1108. 1582.

" The teftimonies of Scripture quoted in Pagets Catechifme,—by Rob. 1582.
Openfhaw." Octavo.

" The examination & confeffion of the Witches taken at St. Ofes in 1582.
Effex, whereof fome were executed." Licenfed.

" Mich. Reniger, De Pij Quinti & Gregorij xiij Romanorum Pontifi- 1582.
cum furoribus contra potent.—principem Elizabethem Reginam. Londini, excudebat—pro R. Watkins." Octavo.

" A Briefe difcourfe of certaine points of the religion, which is among 1582.
the commō fort of Chriftians, which may bee termed the Countrie Diuinitie: With a manifeft confutation of the fame, after the order of a Dialogue. Compiled by Geo. Gifforde. Imprinted—for Toby Cook,—
1582." Dedicated " To—Ambrofe Earle of Warwick," &c. 84 leaves.
W.H+. Octavo.

" A Catechifme, conteining the fumme of Chriftian Religion; giuing a 1583.
moft excellent light to all thofe that feeke to enter the path-way to faluation: Newlie fet foorth by G. G. Preacher of Gods worde at Malden
in Effex. Pf. 19; 8.—" An Epiftle " To the Reader.—Geo. Gyffard."
L 4, in eights. W.H. Octavo.

" AN Anfweare for the time, vnto that foule, & wicked Defence of the 1583.
Cenfure that was giuen vpon M. Charkes Booke, and Meredith Hanmers:
Contayning a maintenance of the credite & perfons of all thofe woorthie
men:—M. Luther, Caluin, Bucer, Beza, &c. whom he, with a fhame-
leffe penne moft flanderoufly hath fought to deface: finifhed fometime
fithence: And now publifhed for the ftay of the Chriftian Reader till
Maifter Charkes Booke come foorth. Exod. 20; 16.—1 Tim. 5; 19.
Imprinted—by *him* & Tobie Smith, 1583." Addreffed " To the Chriftian
Reader." 107 leaves. W.H. Quarto.

" A Perfwafion to Godlie purpofes: Written to a certaine Gentlewo- 1583.
man, correcting fuch vices as remayned in her: By W. N. Delight
thou in the Lord, &c. Pf. 37; 4, 5.—for Wm. Ponfonby, 1583." An
addrefs " To the Reader." by the author. " The Preface—A. H." the
author's kinfman, who publifhed the book without the author's knowledge, for the common good. On a feparate leaf is the argument, or
the author's* title to his MS. D 4, in eights. Licenfed in 1580.
W.H+. Octavo.

Two

---

* " A Booke intituled, and rightlie named: An Exhortation to the diligent & continual reading of Gods word. Wherein is declared, 1 A Diffwafion from Pride & hawtineffe of minde. 2 Remedies againft the fame. 3 Howe wee are knowen to haue Gods loue in vs; and how farre our loue fhould extend one to another. 4 What true wifedome is, with her fruites. 5 What companie to keepe, and what to efchew. 6 What

1583. Two treatises: one, Of the Church; the other, againſt oppreſſion: written by Rob. Some. The dedication is dated, "At Queens colledge in Cambridge, 1 Nouemb. 1582." H 4, in eights. "—for Geo. Biſhop. 1583." W.H+ Sixteens.

1583. "Ad P. Rami dialecticam, variorum et maxime illuſtrium exemplorum, Naturali artis progreſſu, inductio: &c. Quæ Pædagogiæ Logicæ pars tertia, ad artis vtilitatem demonſtrandam. A uctore Fred. Beurhuſio Menertzhagenſi, Scholæ Tremonianæ Rectore. Editio ſecunda." Dedicated "—Dominis Conſulibus, &c.—Cal. Martij, 1583.---F. B. diſcipulis S.---Prooemium de artium vſu," &c. 438 pages. "Ex Officina,— Impenſis Georgij Biſh. 1583." W.H. Alſo without date. Octavo.

1583. Fr. Tayler's ſermon, of the calling of the Gentiles. See p. 1108.

1583. "OF The duetie of a faithfull & Wiſe Magiſtrate, in preſeruing & deliuering of the cōmon wealth from infection, in the time of the Plague or Peſtilence: Two Bookes. Written in Latine by Iohn Ewich, ordinary Phiſition of the woorthie cōmon wealth of Breame, & newlie turned into Engliſh by Iohn Stockwood, Schoolemaiſter of Tunbridge. A VVorke verie neceſſarie for our time, and countrie," &c. Dedicated "To—Sir Iohn (Thomas) Blanke, Knight, L. Maior of the moſt renowmed Citie of London: and to—M. Wm. Fleetwood, Sergeant of the Law & Recorder of the ſaide Citie: and alſo to—the Sheriffes and Aldermen, with the whole ſtate of the ſame citie.—Tunbridge, 19 May, 1583.---A ſhort admonition, &c.---The Preface of the Author.---To—the L. Conſuls, and to—the Senators of the Common wealth of Bream, Stadeen, & Boxtehud.—Iohn Ewich, D. of Phiſicke." 117 leaves. Concludes with a Prayer in verſe. W.H. Octavo.

1583. "THE WORLDE poſſeſſed with Deuils, conteyning three Dialogues. 1. Of the Deuill let looſe. 2. Of Blacke Deuils. 3. Of VVhite Deuils. And of the comminge of Jeſus Chriſt to iudgment: a very neceſſarie and comfortable diſcourſe for theſe miſerable and daungerous dayes. Luke 21 (36) Watch ye therefore at all times, and praye, &c. Imprinted— for Iohn Perin,—1583." By the difference of orthography between this and Mr. Ames's there ſeem to have been two editions of this book the ſame year; and by the ſignatures in my copy it appears to have been printed at different preſſes. The firſt two dialogues, and the contents of the third, occupy ſignature H 2, in eights. The third dialogue begins with a freſh ſet, and ends on e 3, in eights: probably there might be another leaf with a colophon; which, if ſo, my copy wants. W.H. Again 1588. Octavo.

1583. "The ſecond part of the demoniacke worlde, or worlde poſſeſſed with diuels, containing three dialogues. 1. Of familiar diuels. 2. Of lunatick diuels. 3. Of the coniuring diuels. Tranſlated out of French into Engliſh, by T. S. (Tho. Stocker) gent. Tranſlated from Viret, for Iohn Perin."* Octavo.

What talke to vſe. 7 What maketh men riche, and what poore. 8 How to obtaine a wife or huſband, and what perſons they ought to be. With diuers other godlie inſtructions and heauenlie commodities thereby comming as in this littleTreatiſe, though briefly, yet pithilie, both to Gods glorie, and the Readers great profite."

Spaniſh

Spanish cruelties in West India. Quarto. 1583.
" A Commentarie vpon the Epistle of St. Paul to the Romanes, writ- 1583.
ten in Latine by M. Iohn Caluin, and newely translated—by Chr. Rosdell, preacher.—Col. 3; 16. Let the worde of Christ dwel in you" &c. Dedicated " To—Sir Edwarde Seimer knight, Baron Beauchampe, and Earle of Hertforde.—London, 11 Jan. 1583.---Iohn Caluine vnto Gryney a man very worthie to bee honoured.—At Argentine, the 15 of the Calend. of Nouember 1539.---The Argument" &c. 202 leaves, and a table or index. " Imprinted—for John Harrison, & Geo. Bishop, 1583."
W.H. Quarto.
" Aduertisements partely for due order in the publique administration 1584.
of Common prayers," &c. See p. 609. B 3, in fours. Dr. Lort.
Quarto.
" A HARMONIE vpon the three Euangelists, Matthew, Mark, and 1584.
Luke, with the Commentarie of M. Iohn Caluine: Faithfully translated out of Latine—by E. P. Whereunto is also added a Comentarie on the Euangelist of S. Iohn by the same authour.—Impensis Geor. Bishop, 1584." Dedicated " To—Fraunces Earle of Bedford, &c.—From Kiltehampton in Cornewall, 28 Jan. 1584.—Eusebius Paget.---To the renowmed Pieres & noble Lordes, the Consuls, and the whole senate of the famous Citie of Frankeford. I. Caluine—Geneua, 1 Aug. 1555.---A Table shewing the Chap. verse & Fol. of all the principall matters, &c.---A Table of those thinges which are expounded, &c. The argument." 806 pages. W.H. Quarto.
" THE holy Gospel of Iesus Christ, according to Iohn, with the Com- 1584.
mentary of M. Iohn Caluine: Faithfully translated out of Latine—by Chr. Fetherstone, student in Diuinitie." Dedicated " To—Lord Robert Dudley, Earle of Leycester, &c.---To the Reader." 464 pages, and a table, or index. " —for Geo. Byshop, 1584." W.H. Quarto.
" Sermons of M. Iohn Caluin vpon the Booke of Iob" &c. See p. 1108. 1584.
" The Catechisme of C. W." Octavo. 1584.
" A Sermon of the destruction of the Idumeans. Obadiah 5, 6. By 1584.
Iohn Rainolds, Doct. of Diuinitie." Octavo.
" A Sermon on the destruction of Ierusalem, by Iohn Stockwood; on 1584.
Luke 19; 41—44." Octavo.
" The Mathematical Iewel, &c. By Iohn Blagraue." For Walter 1585.
Venge. Folio.
" Balthasaris Castilionis comitis de Curiali siue Aulico libri quatuor, 1585.
ex Italico sermone, Latinum conuersi. Bartholomaeo Clerke in Anglo Cantabrigiensi interprete. Novissime edit." Licensed. See Strype's Abp. Parker, p. 384, &c. Octavo.
" THE Nauigations, peregrinations & voyages, made into Turkie by 1585.
Nicholas Nicholay Daulphinois, Lord of Arseuile, Chamberlaine & Geographer ordinarie to the King of Fraunce: conteining sundry singularities which the Author hath there seene & obserued: Deuided into foure Bookes, with threescore figures, &c. With diuers faire & memorable histories, happened in our time. Translated out of the French by T.
Washington

Wafhington the younger. Imprinted—1585." Dedicated "To—the Right Hon. Sir Henrie Sidney, Knight of—the garter, &c. And to the Right Worfhipfull Sir Phillip Sidney Knight.—Iohn Stell," at whofe coft this book was publifhed. 163 leaves, and a table. Licenfed. W.H. Quarto.

1585. "The Commentaries of M. Iohn Caluin upon the Actes of the Apoftles, Faithfully tranflated out of Latine—by Chr. Fetherftone ftudent in Diuinitie." Dedicated " To—Henrie, Earle of Huntington, Lord Haftings, &c.—From Maighfield in Suffex, 12 Octo. 1585.---The Epiftle to the Reader.---To—L. Nicolas Radziwill, duke in Olika, Countie Palatine of Vilna, chief Marfhall & head Chauncellar of the great Dukedome Lethuania, &c.—At Geneua, 1 Aug. 1560.---The argument." 598 (printed 298) pages, and a table or index. Very neatly printed in white letter, " for Geo. Bifhop, 1585." W.H. Quarto.

1586. Demofthenis Oratio in Midiam. Græcé. Quarto.

1587. " De Arte Natandi libri duo, quorum prior regulas ipfius artis, pofterior vero praxin demonftrationemque continet. Authore Everardo Dygbeio, Anglo, in artibus magiftro." Licenfed. Quarto.

1587. " The Arte of fhooting in great Ordinaunce :" &c. as p. 1109.

1587. " A NOTABLE HISTORIE, containing foure voyages made by certayne French Captaynes vnto Florida: Wherein the great riches & fruitfulnes of the countrey with the maners of the people hitherto concealed are brought to light, written all, fauing the laft, by Monf. Laudonnier, who remained there himfelfe, as the French Kings Lieutenant, a yere & a quarter. Newly tranflated out of French.—by R. H." *(Rd. Hackluyt)* Device, the three cranes in a vine, within a compartment. Dedicated " To the right worthie & Honorable Gentleman, Sir Walter Ralegh knight, feneical of the Duchies of Cornewall," &c. Whofe arms are on the back of the title-page.---The preface. 64 leaves, and a table. " Imprinted—1587." Alexander Dalrymple Efq; Quarto.

1588. " Three Bookes of Colloquies concerning the Arte of fhooting in great & fmall peeces of Artillerie, variable randges, meafure & waight of leaden, yron, & marble ftone pellets, minerall faltepeeter, gunpowder of diuers fortes, and the caufe why fome fortes of Gunpowder are corned, and fome—not corned: Written in Italian & dedicated by Nich. Tartaglia vnto—Henrie the eighth, late King of England &c. And now tranflated—by Cyprian Lucar, Gent. who hath alfo augmented the volume—with the contents of euery Colloquie, and with all the Corollaries & Tables. Alfo the faid C. Lucar hath annexed—a treatife named Lucar Appendix collected by him—to fhew—the properties, office & dutie of a Gunner, and to teach him to make & refine artificial faltpeeter, to fublime brimftone for gunpowder,—to fhoote well at any marke—to mount morter peeces,—to make—fire workes, mynes," &c. A cannon lying level on a carriage, with a Gunner's quadrant in its mouth. " Imprinted—for Iohn Harrifon, 1588." Dedicated " To—Robert Earle of Leicefter,—Lord Steward of her Maiefties houfhold, Chiefe Juftice in Oyer of all her Maiefties Forrefts, &c. by South Trent, and Knight of the

the moſt hon. orders of the Garter, and St. Michael in France, &c. (whoſe arms gartered and collared are on the back of the title-page.)—Wherfore licenſed to diſpoſe, as I will, of this Engliſh worke made by M. C. Lucar with a dutifull zeale to benefite his natiue ſoyle, I thought it my duetie to offer it to your noble patronage, as a preſent moſt fit for your honour, and moſt profitable for theſe times.—Your Honors moſt humble and duetifull orator, Iohn Harriſon, Stationer.---To the moſt puiſant & mercifull prince Henrie VIII.—Nich. Tartaglia.---Ad lectorem. *In* 12 *diſtichs.* G. B. Cantab." 80 pages, and at the end Lucar's arms quartered. The Appendix had a ſeparate title-page, with Dawſon's ſign, or device, of the three cranes in the vine, and Sir Walter Ralegh's arms on the back; but that ſeems to have been cancelled for one with a cut of a gunner firing a mortar, throwing a bomb into a caſtle, on the front, and the Earl of Leiceſter's arms on the back. " The names of Authors out of whoſe Bookes the greateſt parte of this—hath been collected." 120 pages. Though the paging is begun anew, the ſignatures, in ſixes, are continued from the former part. " Printed by *blm*, for Iohn Harriſon the elder,—1588." With cuts, ſchemes, &c. W.H. Folio.

" CARMINVM PROVERBIALIVM Totius humanæ vitæ ſtatum breuiter 1588. deliniantium, necnon vtilem de moribus doctrinam iucundè proponentium. Loci communes, in gratiam iuuentibus ſelecti.
Si Chriſtum diſcis, ſatis eſt, ſi cætera neſcis:
Si Chriſtum neſcis, nihil eſt ſi cætera diſcis."
The three cranes in the vine. " Impreſſum Londini, 1588." This book is introduced with a page of the ſame ſort of rhymed Latin verſes as the motto, by " S. A. I. Ad emptorem." Indeed moſt of the proverbial maxims throughout the book are of this kind. 216 pages. " Excudebat—ex aſſignatione Chr. Barkeri." W.H. Again 1595. Sixteens.

" Certaine ſermons preached of late at Ciceter, in the countie of Gloceſter, vpon a portion of the firſt chapter of the epiſtle of James; wherein the two ſeueral ſtates of the riche and poore man are compared, &c. by Philip Jones, preacher of the word of God, in the ſame towne." Dedicated " To—John biſhop of Gloceſter, and Comendatory of Briſtow." For Tho. Butter. Octavo. 1588.

Jerom of Ferrara his mediations on the 51, and 31 pſalms; tranſlated 1588. and augmented by Abr. Fleming. Alſo without date. Licenſed, 1578.
Sixteens.

" A Briefe deſcription of Ireland: Made in this year 1589, by Robert 1589. Payne, vnto xxv of his partners, for whome he is vndertaker there. Truely publiſhed verbatim, according to his letters, by Nich. Gorſan, of *Trowell, Nottinghamſhire*, one of the ſayd partners, for that he would his countreymen ſhould be partakers of the many good notes therein contayned." The three cranes, &c. Sixteens.

The hiſtory of the Eaſt Indies, &c. See it in p. 1013. Quarto. 1589.

" Short Queſtions and anſweares, conteining the ſumme of Chriſtian 1590. Religion: Newly enlarged with the Teſtimonies of Scripture. Rom. 8; 38,

38, 39.—Imprinted—1590." Dedicated " To the Worſhipful Maiſter Mayor, the Bayliffes, the Aldermen, Burgeſſes, &c.—28 Jan. 1584. Your Paſtor in the Lorde much bounden & to be cōmanded, R. O.---A Prayer —before Catechiſing." At the end, " A Prayer—after catechiſing." E 7, in eights. " Fins quoth Robert OPenſhawe, Paſtor of the Church of Waymouth & Milcombe Regis, in the countie of Dorſet." Licenſed in 1586. W.H. Sixteens.

1591. A catechiſm by Euſebius Paget. Octavo.

1591. " A Godly letter written by Sir Henry Sidney knight of the honor. Garter, and Lorde Preſident of Wales, to Phillip his ſonne. With An epitaph vpon the life & death of Sir Henry Sidney." Licenſed. Octavo.

1591. " OF THE RVSSE Common Wealth, or Maner of Gouernement by the Ruſſe Emperour (commonly called the Emperour of Moſkouia) with the manners, & faſhions of the people of that Countrey.—Printed by T. D. for Tho. Charde, 1591." Dedicated to the queen by G. Fletcher. 116 leaves. W.H. Octavo.

1591. " The Vineyarde of Vertue, digeſted into a tripartite order, containing 32 moſt excellent plants of fruitfull vertue: ſhewing the definition, teſtimony & example of each vertue." Dedicated " To—M. Hen. Vvedale alias Vdale of More Criche, in Dorcetſhier Eſq; &c.—London, 2 July, 1591.—Rd. Robinſon, Citizen." 132 leaves. Licenſed, 1579. W.H₊ Sixteens.

1591. " The grounds of the longitude, with an admonition to all thoſe, that are incredulous, and beleeue not the truth of the ſame. Written by Simon Forman, ſtudent in aſtronomie and phiſique." Licenſed. Quarto.

1592. " CHRONICON ex Chronicis, ab initio mundi vſque ad annum Domini 1118, deductum, Auctore Florentio Wigornienſi monacho. Acceſſit etiam continuatio vſq; ad annum Chriſti 1141, per quendam eiuſdem cœnobij eruditum. Nunquam antehac in lucem editum.—Excudebat— pro Ricardo Watkins, 1592." Dedicated "—D. Gulielmo Cecilio,—ſummo Angliæ Theſauro: &c.—Gul. Howardus.---Candido Lectori.—Calend. Aug. 1592." 584 pages. W.H. Quarto.

1592. Of the profit and neceſſity of catechiſing, by Rob. Cawdray. Octavo.

1592. " The vſe of both the Globes, Cœleſtiall & Terreſtiall, moſt plainely deliuered in forme of a Dialogue: containing moſt pleaſant and profitable concluſions for the Mariners, &c. Written by Tho. Hood, mathematicall lecturer in the citie of London, ſometime fellow of Trinitie college in Cambridge."* Octavo.

1593. " A ſermon preached at Newport paignell, in the countye of Bucks, by Roger Hacket, on 2 Cor. 5 ;. 20, 21." Licenſed. Octavo.

1595. " The ſeamens ſecrets, deuided into two parts; wherein is taught the three kinds of ſailing; horizontall, paradoxall, and ſailing vpon a great circle, &c. by John Dauis of Sandrudge near Dartmouth, gent." Licenſed. Octavo.

1595. " The Worldes Hydrographical Diſcription: Wherein is proued—that the worlde in all his Zones, Clymates & places is habitable & inhabited, and the Seas likewiſe vniuerſally nauigable,—whereby it appears that there

THOMAS DAWSON.

there is a short & speedie passage into the South Seas, to China, &c. by Northerly Nauigation, to the renowne, honour & benifit of her Maiesties state and Comunality. Published by I. Dauis of Sandrudg." Dedicated "To the—Lordes of her Maiesties—priuie Counsayle.—27 May, 1595" C, in eights. Licensed. Alexander Dalrymple Esq; Octavo.

"A profitable and necessarie Booke of Obseruations, for all those that 1596. are burned with the flame of Gun powder, &c. and also for curing of wounds made with Musket & Caliuer shot, &c. Last of all is adioined a short Treatise for the cure of Lues Venerea, by vnctions, &c. now againe newly corrected & augmented in—1596." See p. 667. "The Epistle Dedicatorie. To all the true professors of Chirurgerie," &c. Though the Lues Venerea has a separate title-page, the paging is continued to p. 229. "Edm Bollifant for *him*." W.H. Quarto.

"The life and death of Thomas Wolsey, cardinall; diuided into 1599. three partes; his aspiring, triumph, and death. By Thomas Storer, student of Christchurch in Oxford." In verse.* See Oldys's Catal. N°. 114. Quarto.

"A fruitfull exhortation giuen to all godly & faithfull Christians: Wherin they are instructed to cloathe themselues with the true & spirituall Adam Christ Jesus, to detest sinne, and to forsake the vaine inticinge pleasures of this wicked world, to vanquish the straying & rebellious lusts of the flesh, and to bringe foorth the sweet smelling fruites of vnfained repentance: published by Iohn Phillips. Mat. 3. (10) The Axe is layd to the roote of the tree," &c. Dedicated "To—Lady Lettis, the noble Countesse of Leicester." F 4, in eights. Sixteens.

"The good huswiues Iewell,—for conceits in Cookerie." Octavo.

A Treatise of the way of Life. In 3 partes. By B. A. See p. 1042.
Octavo.

"A Perpetuall Kallender." At bottom,

"Take this as a Remembrance, and for no other Cause, Esteeme not the gifte, but the good will of Dawse." Broadside.

He had also licenses for the following, viz. In 1578, "A handfull of honniesuccles gyuen for a neweyeres gift vnto ladies & gentlewomen of the priuiechamber" In 1579, "An excellent discourse of an exployt of Iohn Fox, an inglishman, who had been prisoner 14 yeres vnder the Turks, and killing the goaler deliuered 266 Christians, that were also prisoners," &c. To him & Steph. Peele. "Children his garden, wherein you maie find many a pleasant flower to delight a Xpian mynde. The region of Tartaria, and of the lawes & power of the Tartares. Of Scithia, and the man's of the Scithians, Of the other side of the Ganges, Of Cataia, & the region of Sina, And of the m'uailous wonders that haue ben seene in those parts. ij thinges, of newes in the Lowe cūtreyes; and in Frizelande." In 1580, "A short & fruitfull treatise of the profit & necessitie of cate hising vong. & ignorant persons in the principalla & groundes of the Christian religion." In 1581, "A godly & short treatise of the Sacramentes." In 1582, "A prognosticall Judgement of the great coniunction which shall happen ȳ 28. Aprill, by Rob. Tanner."

In

THOMAS DAWSON.

In 1586, " Demosthenes in Greeke: wholie, or any parte of yt." In 1594, " A true difcourfe of a moſt cruell & barbarous murther, comitted by Tho. Merrey on ẏ perſons of Tho Beeche & Tho. Wincheſter his feruant, on fridaie night, 23 of Aug. With his arraynment & execution. Item, Beche his ghoſte coplayninge on ẏ wofull murther, &c.

CHARLES YETSWEIRT, Efq;

WAS the fon of Nicafius Yetfweirt Efq; clerk of the Privy Seal, and fecretarie to her Maieſty for the French tongue;[y] who alſo had a licenfe granted him for printing all manner of books concerning the common laws of this realm for thirty years, dated 18 Nov. 1577, 20 *Eliz. Rd. Tottel however ſeems to have demurred againſt it, and to have ſet it by, having had a prior patent, 1 Eliz. for the ſame purpoſe, during his natural life: accordingly no book is found with Nicafius's name, whereas Tottel printed the Law Books to 1593, about which time he deceaſed; and then Cha. Yetfweirt began to print both law and other books. His father appears to have died about the ſame time, ſeeing his own patent was dated 36 Eliz. (From 17 Nov. 1593 to—1594.) In the Herald's office is found a memorandum concerning this family, which is inſerted below.[a]

He dwelt in Fleet-ſtreet, near the Middle Temple gate; probably the ſame houſe that Tottel lived in; had a patent 36 Eliz. for printing all books concerning the laws, for thirty[b] years, and ſucceeded to his father's places at Court, but did not long ſurvive him, as appears by the note[c] below.

" The

[y] College of arms. Rymer's Fœd. Vol. xvi. p. 20.

[a] Dugdale's Orig. Jurid. p. 61. edit. 1671.

[a] " That Mrs. Mary Yetfweirt, the daughter of James Bouchier efq; late wife of Mr. Nicafius Yetfweirt, deceaſed at her huſband's houſe in the pariſh of Sunburie in the county of Middleſex the 29th of September anno Domini 1568; and was buried in the ſame pariſh church the 4th of October. The ſaid Nicafius Yetfweirt had iſſue by the ſaid Mary a ſon and four daughters, viz. Charles Yetfweirt, ſon & heir, at this preſent of the age of 21 years; Mary of the age of 25 years; Anne of the age of 23 years; Frances of the age of 13 years, and Sufanna of the age of 10 years. The ſaid Mary and Frances were mourners at the ſaid interment. The aforeſaid Charles bare the pynnon of arms, and at the ſaid funeral ſerved Nichas Dethick, alias Bluemantle, purſivant of arms." Sufanna was afterward married to Daniel Rogers efq; clerk of the council to Q. Elizabeth,' who on 16 Feb. 1590, was buried in the church of Sunbury near his father in law Nicaſ. Yetfweirt. See Ath. Oxon. Vol. 1. col. 246, &c.

[b] Dugd. Orig. Jur. p. 61. Ed. 1671.

[c] " Charles Yetfweirt efq; her majeſty's fecretary for the French tongue, and one of the clerks of the ſignet, died at his houſe at Sunburie the 25th day of April, anno 1595, and was buried in the Church of the ſame pariſh, the 5th day of May next following. He married Jane Elkin, and had iſſue Frances, who died the 2d of February 1594. This funeral was ſolemnized by York Herald (deputy for Clarencieux king of arms) and Portcullis officer of arms. *Subſcribed by* Shotbolte, *and* Francis Galle." Coll. of arms, 6 funeral book. Thomas Edmonds ſucceeded him as ſecretary for the French tongue, with the ſalary of 66l. 13s. 4d. beſides

CHARLES YETSWEIRT.

"The arrainement of the whole focietie of Jefuits in Fraunce; holden 1594.
in the honorable court of parlement in Paris, the 12th and 13th of July
1594." Cum priv. By—Arnauld. Quarto.
"L'authoritie et jurifdiction des courts de la maieftie de la roygne. 1594.
Nouelment collect & compofe, per R. Crompton del milieu Temple,
Efquire, Apprentice del Ley.—In ædibus Caroli Yetfweirti Armig. Cum
priuilegio Regiæ Maieftatis. 1594." 232 leaves. "Imprinted—in
Fleeteftreete, by Charles Yetfweirt Efq; and are to be fold at his houfe
within Temple Barre, neare to the Middle Temple gate." W.H. Quarto.
"Littletons tenures in Englifh, lately perufed and amended.—Sold 1594.
at his houfe within Temple Barre, neere to the Middle Temple." Sixteens.
Alfo in French. W.H. Twenty-fours. 1594.
"A true report of the horrible confpiraces of late time, detected to 1594.
haue (by barbarous murders) taken away the life of the queenes moft
excellent maieftie, whom almighty God hath miraculoufly conferued
againft the treacheries of her rebelles, and the violences of her moft
puiffant enemies. Nouember. Printed by Charles Yetfweirt efq;" 31
pages. Quarto.
Weft's Symbolæography. See it p. 826. Quarto. 1594.
"Of the interchangeable courfe, or variety of things in the whole 1594.
world; and the concurrence of Armes and Learning, thorough the firft
and famoufeft Nations, from the beginning of ciuility, and Memory of
man, to this Prefent. Moreover, whether it be true or no, that there
can be nothing fayd, which hath not bin faid heretofore: and that we
ought by our owne Inuentions to augment the doctrine of the Auncients;
not contenting our felues with Tranflations, Expofitions, Corrections,
and Abridgments of their writings. Written in French by Loys le
Roy, called Regius; and tranflated into Englifh by R. A." *(Robert Afkley)*.
Dedicated to the lord keeper Puckering. Cum privilegio, &c. 130
leaves. W.H. Folio.
"Les Commentaries, ou Reportes de Edmund Plowden," &c. See 1595.
p. 822.
"An Expofition of certaine difficult & obfcure words and termes of 1595.
the Lawes of this Realme, newly fet forth & augmented, both in French
& Englifh, for the help of fuch yong Students as are defirous to attaine
to the knowledge of the fame. At London, Printed by th'Affignee of
Charles Yetfweirt Efq; deceafed. Cum priuilegio—1595." A table
prefixed. 192 leaves. See p. 826. W.H. Octavo.

JANE YETSWEIRT,

The widow of Charles Yetfweirt efq; continued exercifing the art of
printing and felling, fome time after the deceafe of her hufband, but met
with a great deal of trouble from the Stationers' company, as may appear
to fuch as will enquire into her cafe. For Mr. Ames found among the
fides all perquifites enjoyed by Nicafius & or any other. Dated at Weftm. 17th of
Charles Yetfweirt, Iohn Mafon, Ste. Tuke, May, 1596. Rymer's Fœd. Vol. xvi. p. 290.

late.

late lord Oxford's MSS. thefe letters, complaining of her hard ufage, viz. A letter of the earl of Effex to the lord keeper Puckering, in favour of Mrs. Jane Yetfweirt widow, againft the Stationers, or printers of London, relating to the printing of law, dated 27 March, 1595. From the earl of Effex, to the lord keeper, in favour of Mrs. Yetfweirt, dated 7 May, 1595. Another from Mrs. Yetfweirt to lord treafurer Cecil, dated 7 May 1595. Another from Mrs. Yetfweirt to lord keeper Puckering, dated from Sunburie, 1595.

1595. "REGISTRVM OMNIVM BREVIVM, tam Originalium, quam Iudicialium, correctum & emendatum ad vetus exemplar manufcriptum, cuius beneficio, à multis erroribus purgatum, ad vfus, quibus inferuit, redditur accommodatius.—In ædibus Ianae Yetfweirt, relictæ Caroli Yetfweirt Ar. nuper defuncti. Cum priuilegio Regiæ Maieftatis." In the compartment ufed by H. Bynneman to Hiftoria Brevis, &c. p. 974; but the mermaid taken off. Within the firft letter, a blooming E, both of the original and judicial briefs, is reprefented Q. Elizabeth, royally apparelled, fitting on a throne. The original briefs on 321 leaves; the judicial on 85. See p. 808. W.H. Folio.

1596. A collection of entries &c. by Wm. Raftell. 704 leaves. See p. 815. Folio.

1596. An abftract of all the penal ftatutes—in force, &c. See p. 813. Quarto.

1597. "Anni regum Edwardi quinti, Richardi tertii, Henrici feptimi, et Henrici octaui, omnes qui antea impreffi fuerunt, iam recens poft priores editiones emendati & repurgati, numeris interlinearibus notatis tum in margine, tum in capite, priorum editionum foliis refpondentibus, quorum principiis figna ifta ❀❀ in textu præponuntur. Acceffit nunc primum annalium regis Henrici octaui, in honorandi illius iudicis, domini Roberti Brooki, equitis, epitomes titulis digeftorum, in margine annotatio." Folio.

1597. "Symbolaeography, which may be termed the art, defcription, or image of inftruments. Or, the paterne of Praefidents. Or, the Notarie, or Scriuener. The firft part of inftruments extraiudiciall, the fourth time Corrected by William Weft of the Inner Temple, Efquire, firft author thereof." W.H. Probably fhe printed the 2d part alfo. Quarto.

1597. "Annalium tam regum Edwardi quinti," &c. See p. 822, &c. Sixteens.

1597. "A profitable booke of Mafter Iohn Perkins, Fellow of the Inner-Temple: Treating of the Lawes of England.—In ædibus."—Cum priv. 168 leaves. W.H. Twenty-fours.

1597. Littleton's tenures. Alfo without date. Sixteens.

1597. "Afcuns nouel cafes de les ans et temps le roy Hen. 8. Edw. 6. et la royne Mary, efcrie per fir Roberti Brooke chiualer," &c. See p. 821. Octavo.

—— "Abridgment des touts les cafes reportez, alarge per Monfieur Plowden, &c. compoffee et digeft per T. A." 92 leaves. Octavo.

HUGH

HUGH JACKSON, or JAXON, stationer.

ONE Heugh Jackson was bound apprentice to Wm. Powell for 10 years from the feast of Pentecost, 1562. I find also Hewghe Jackson, son of Wm. Jackson of London, plumber, bound to Hen. Denham. Perhaps both entries mean the same person, bound first to Powell, and afterward to Denham; for the register at that time makes no distinction between apprentices bound, or turned over, except from some other company; mentioning only their being presented. Reckoning from either time, the apprenticeship expired in 1572; so that we cannot ascertain when he took up his freedom, or who made him free, for want of that part of the company's register, from 1570 to 1576, which is missing. However one Hugh Jackson bound, or presented, an apprentice in 1577. I do not find any person of these names on the livery. He, by the name of Hugh Jaxon, as it is frequently written in the register, was fined, 6 Oct. 1577, for not coming at the quarter day, vjd. Be these as they may, the printer whose works are now to be declared, succeeded T. Colwell, at the sign of St. John the Evangelist in Fleet-street, a little beneath the conduit.

" Of the Conscience.—by Iohn Woolton." See p. 936.   Octavo.   1576.
" Andr. Bourd his Dietarie of health." See p. 739.   Octavo.   1576.
" Foure straunge, and lamentable tragicall Histories, translated out of French—by Ro. Smythe." Licensed.   Quarto.   1577.
" The renowned historie of Cleomenes & Juliet."   Octavo.   1577.
" The garden of eloquence, conteyning the figures of grammar and rhetorik, &c. by Henry Peacham, minister."   Quarto.   1577.
" The boke of Nurture, or Schoole of good Maners: for Men Seruants & Children: with Stans puer ad Mensam: &c. Imprinted—in Fleteftrete, beneath the Conduite, at the signe of S. Iohn Euangelist, by H. Jackson, 1577."   Octavo.   1577.
" A Briefe & most excellent Exposition of the xij Articles of our Fayth, comonly called the Apostles Creede, &c. Written in Italian by—D. Peter Martyr Vermilius, and lately translated into English by T. E." Psal. 118. 17. I wyll not dye, &c. R 4, in eights. " Imprinted—1578," over a cut of his sign. Licensed. W. H. Also without date.   Sixteens.   1578.
" The perfite way to paradice." This small piece is prettily printed within a border, and contains prayers and meditations; one is, " for a woman that is great wyth child, to be repeated by her hartely before her trauell, which wil be marueylous comfort unto her." In verse, 44 lines. Again 1588, and 1590.   Sixteens.   1580.
" THE SHIELD of our Safetie: Set foorth by the Faythfull Preacher of Gods holye Worde, A. Anderson, vpon Symeons sight in his Nunc dimittis. Seene & allowed. Micha 2; 11." Dedicated " To—Father in God, by Gods appoyntment, the Byshop of London.—Medborne in Lecesterfhire, 12 Dec. 1580." X, in fours. " Imprinted, &c. 1581." Licensed. W. H.   Quarto.   1581.

6 S   " A won-

HUGH JACKSON.

1583. "A wonderfull Judgment of God vpon two Adulterers in S. Brides parish, 3 Feb. 1583. Set forth by Sam. Saxey, Diuine." Octavo.
1584. "Ioh. Phillips his Sūmon to Repentance." Again 1590. Octavo.
1585. "The prayse of Nothinge, by Edw. Da." Licensed. Quarto.
1585. "The knowledge of thinges unknowen," &c. as p. 382. Again 1588; also, without date. Sixteens.
1590. "Baculum Familliare, Catholicon siue Generale. A Booke of the making and vse of a Staffe newly inuented by the Author, called the Familiar Staffe: As well for that it may be made—familiarlie to walke with, as for that it performeth the Geometrical mensurations, &c. Newlie compiled, and at this time published for the speciall helpe of shooting in great Ordinance, &c. and may as well be imployed—for measuring of land, &c. By Iohn Blagraue of Reading Gent.—Printed &c. 1590." Dedicated " To—Sir Frances Knolles."—The table 70 pages, with schemes, &c. Licensed. W.H. Also without date.
  Quarto.
——— "The Mirrovr or Glasse of Health." &c. as p. 375. W.H.
  Sixteens.
——— "The Husbandmans practise, or prognostication for euer, as teacheth Alberte, Alkin, Haly & Ptholome." Mr. T. Martin's papers. Sixteens.
——— "The wife lapped in Morels skyn: or the taming of a Shrew."
  Quarto.

He had also licenses for the following, viz. In 1576, "An Enterlude. The tide tarieth for no man. The description & figure of a monstrous childe borne at Taunton, 8 Nouemb. 1576." In 1577, "A caueat or warning vnto maysters & seruants, instructing them howe they oughte to leade their lyues in that vocac'on. Precious perles of perfect Godlines, &c. begun by lady Frances Aburgauenny, & finished by John Phillip." In 1580, "An exposic'on of the ap'les Creede, translated & taken out of ŷ Catecheticall sermons of Oleuianus." In 1584, "Arbasco: the Anatomie of Fortune." In 1586, "Bullinger's Decades, in Latin; with R. Newbery." Likewise for some ballads, in the early part of his time.

✿✿✿✿✿✿✿✿✿✿✿✿✿✿✿✿✿✿✿✿✿✿✿✿✿✿✿✿

ANDREW MAUNSELL, Draper,

WAS rather a bookseller than a printer, yet had many books licensed to him, and printed for him; particularly his catalogue of books printed in the English language to his own time, which will be described in its place, and for which he ought to be remembered with honour. Mr. Ames thought he began trade about 1570: possibly he might, as the Stationers' register from that time to 1576 is missing; however there appears no certainty of it. I find nothing printed for him, nor mentioned in his catalogue, before 1576. He dwelt sometime at the Parrot in St. Paul's church yard; afterwards at the Brasen Serpent there; but latterly dwelt in Lothbury, and had a shop in the Royal Exchange.

" A phi-

" A philosophicall discourse,—The Anatomie of the minde." See 1576.
p. 1093.
" A free Pardon with many Graces therein contained, granted to all 1576.
Christians by our most holy & Reuerente Father God Almighty, the
principal high priest & Bish. here in earth. First written in Spa-
nish, & translated by Iohn Daniel." A similar book, if not the same, was
dedicated to the lord mayor, &c. Octavo.
" Cuthb. Mutton his confutation of the—sect of Anabaptistes: 1576.
Wherein you may behold the perfecte humanity of Christ." Octavo.
" " The Moste Profitable, aud commendable science of Surueying of 1577.
Landes, &c. Drawen & Collected by the Industrie of Valentyne Leigh.
Wherunto is annexed, by the same author, a right necessarie Treatise of
the measuryng of all kyndes of Lande," &c. The preface dated 27 Oct.
1562. Q, in fours. Iohn Baynes, Esq; Licensed. Again 1578. WH.
Octavo.
" Of the ende of this world," &c. as p. 1116. Again 1578. Licensed. 1577.
Quarto.
" Tho. Rogers, translation of Phil. Cæsars generall discourse against 1578.
the damnable sect of Vsurers, grounded vpon the worde of God, and
confirmed by the Authoritie of Doctors, old & new: also is added a
treatise of the lawfull vse of riches, taken out of Heming on Iames."
Licensed. Quarto.
" The wonderfull woorkmanship of the World: wherin is conteined 1578.
an excellent discourse of Christian naturall Philosophie, concernyng the
fourme, knowledge, and vse of all thinges created: specially gathered
out of the Fountaines of holy Scripture, by Lambertus Danæus, and now
Englished by T. T." (Thomas Twyne.) Dedicated " To—Syr Francis
Walsingham.---To—the Lord Friderike of Nachod, &c.—Written the
Calendes of Decemb. 1575." 87 leaves, and a table." Licensed. W.H.
Quarto.
" The profession of the true Church and Poperie compared." Trans- 1578.
lated by Tho. Rogers, M. A. and preacher of the Word. Licensed.
Octavo.
" A prayse, and reporte of maister Martyne Forboishers voyage to Meta 1578.
incognita (a name giuen by a mightie and most great personage) in which
praise and reporte is written diuers discourses, neuer published by any
man as yet. Now spoken by Thomas Churchyarde, gent. and dedicated
to M. Secretarie Wilson, one of the queenes maiesties most honourable
councell." Licensed. Quarto.
" The English Creede, wherein is contained in tables, an exposition 1579.
on the articles which euery man is to subscribe vnto. Where the article
is expounded by scriptures & confessions of all the reformed Churches,
and heresies displaid; by Tho. Rogers." Licensed. Folio.
" Of two wonderful Popish Monsters, &c." as p. 1010. 1579.
" The Bee-hiue of the Romishe Churche," &c. See p. 1119. 1580.
" The Faith of the Church militant. An exposition of the 84 Psalm, 1581.
by N. Hemingius, translated by Tho. Rogers." Licensed. Octavo.

" Christian

1581. "Christian Instruction, containing a collection of sundry places of scripture, seruing for an expofition on the Lords prair, Creede & ten Comandements, and the Sacraments: tranflated out of French by Geo. Capelin, Gent." Licenfed. Again 1585, in 32. Sixteens.

1581. "Generall Sefsion. A Difcourfe Apologeticall of God his generall Judgment, by Tho. Rogers." Licenfed. Octavo.

1582. "Of the end of this world," &c. as p. 1060. Again 1589.

1583. "The Common places of—D. Peter Martyr," &c. as p. 957. Licenfed.

1583. "Of the foolifhnes of men in putting off the amendement of their wicked liues from daie to daie; by Iohn Riuius: tranflated by Tho. Rogers." Licenfed. Again 1586. Octavo.

1584. "Cato conftrued, Or a familiar & eafie interpretation vpon Catos morall verfes. Firft doen in Laten & French by Maturinus Corderius, and now newly englifhed, to the comfort of all young Schollers. Imprinted at London, For *him*, dwelling in Paules Church-yard, at—the Brafen Serpent, 1584." K 2, in eights. W.H. Octavo.

1584. "The Cauteles, Canon &c. of the—Popifh Maffe:" &c. as p. 1074. Licenfed.

1585. "The Englifh Creede, confenting with the true aunciernt catholique, and apoftolique Church in al the points & articles of Religion, which euerie Chriftian is to knowe & beleeue that would be faued. The firft parte, in moft loyal maner to the glorie of God, credit of our Church, and difplaieng of al hærifies & errors, both olde & newe, contrarie to the faith, fubfcribed vnto by Tho. Rogers. Allowed by aucthoritie," &c. Windet's device of Time, &c. "Imprinted by Iohn Windet for *him* at the brafen Serpent in Pauls church yard, 1585." Dedicated to Edmund bifhop of Norwich. The preface dated, 6 Feb. 1585. The creed on 79 pages, printed only 63. W.H. Folio.

1586. "A verie Godlie—fermon—by Iohn Deacon." See p. 1062. Licenfed.

1587. "The Englifh Creede,—The fecond part," &c. as the firft part. "Printed by Rob. Walde-graue for *him*, at the Brafen Serpent,—1587." Dedicated to Sir Chr. Hatton. W.H. Folio.

1589. "An Hiftoricall Dialogue touching Antichrift & Poperie, drawen & publifhed for the comfort of our Church in thefe dangerous dayes, againft the defperate attempts of the vowed aduerfaries of Iefus Chrift his Gofpell, and this flourifhing ftate. By T. Rogers." Octavo.

1589. "The feconde founde of the Trumpet vnto Iudgment: wherein is proued that all the tokens of the later day are not only come, but well neere finifhed. By Anth. Marten, gent." Quarto.

1592. "The Englifh Phlebotomy:" &c. as p. 1114.

1592. "Soliloquium Animæ of Tho. de Kempis, entituled, The fourth booke of the Imitation of Chrift, tranflated & corrected by T. Rogers." Sixteens.

1592. "Sermons on the hiftorie of Melchifedech: of Abrahams refcuing his nephew Lot:—his faith & iuftification:—and of his offering his fon Ifaack. &c. By Iohn Caluin, tranflated by Tho. Stocker," Octavo.

"Generall

" Generall Calenders, or—Aftronomicall tables, containing afwel the 1594.
Names, Natures, Magnitudes, Latitudes, Longitudes, Afpects, Declinations & right Afcentions of all the notableft fixed ftarres, vniuerfally feruing all Countries: as alfo their mediatiō of heauen: alfo their fcituatiō in the 12 houfes of the cœleftial figure, indifferently fitting all the middle of the 8 climate, but very precifely the Lat. 51 deg. 42 min. of the Pole-articke: Alfo certaine perpetuall tables for the exact placing of the Planets, &c. Moreouer, a Callender of the Cofmicall & Acronicall rifing & fetting of all the fayd Starres. By Geo. Hartgill, Gent. and Preacher of the word." Folio.

"The First Part Of the Catalogue of Englifh printed Bookes: 1595, Which concerneth fuch matters of Diuinitie as haue bin either written in our owne Tongue, or tranflated out of anie other language: And haue bin publifhed, to the glory of God, and edification of the Church of Chrift in England. Gathered into Alphabet, and fuch Method as it is, by Andrew Maunfell, Bookfeller. Vnumquodque propter quid." Jugge's device of the pelican. "London, Printed by Iohn VVindet for him, dwelling in Lothburie, 1595" Dedicated "To the queens moft facred Maieftie---To the Reuerend Diuines, &c.---To the Worfhipfull the Mafter, Wardens & Affiftants of the Companie of Stationers, and to all other Printers & Bookfellers in generall." 123 pages. W.H. Folio.

"The Seconde parte of the Catalogue of Englifh printed Bookes: 1595. Eyther written in our owne tongue, or tranflated—: which concerneth the Sciences Mathematicall, as Arithmetick, Geometrie, Aftronomie, Aftrologie, Mufick, the Arte of VVare, & Nauigation: And alfo of Phyfick & Surgerie: which haue beene publifhed to the glorie of God, and the benefit of the Common-weale of England. Gathered into Alphabet, &c. Printed by Iames Roberts, for him,—1595." Dedicated "To—Robert Earle of Effex," &c. whofe arms are on the back of the title-page.---" To—the Profeffors of the Sciences Mathematicall, &c.---To—the Maifter, Wardens &c. of the Company of Stationers:" &c. 27 pages. W.H. Folio.

"Three Sermons, or Homelies to moue compaffion towards the poor 1596. & needy in thefe times." On Heb. 13; 16. and Luke 14; 13, 14. By authority. Quarto.

"A Table of the lawfull vfe of an oath, and the curfed ftate of vaine fwearers: by Tho. Rogers."

"Peter Droet his newe counfell againft the Plague, tranflated by Tho. Twine." Octavo.

---

d " Apr. 19. 1596. Whereas Andr. Maunfell hath taken paines in collecting & printing A catalogue of bookes, whiche he hath dedicated to the Company: Hauing alfo ben a petic'oner to them for fome confiderac'on towards his paines & charges. Be yt remembred that thereupon the Comp. of their meere beneuolence haue beftowed vpon him in money & books the fome of —— for which he yieldeth thankes, holding himfelfe fully contented without any further matter, or benefit, &c. The particulars of whiche money & bookes appere in a Book thereof made, conteyning the names of the perfons that contributed to the fame." Stat. Reg. B. 461.

He

He had alfo licenfes for the following, viz. In 1576, "An alphabet & plaine pathwaye to ẏ facultie of Readinge: With ij other copies, fo that he haue them prynted by a free prynter, being a freeman of the myfterie of Stationers, whereof one is of furueying lands by V. Leigh, the other a difcourfe of Hufbandrye." In 1578, "Of the miferie of Flaunders, the calamitie of Fraunce, misfortune of Portugale, vnquietnes of Ireland, troubles of Scotland, and ẏ bleffed ftate of England. The 2d part of Daneus Chriftian Philofophy: in Englifh. The olde Faithe by M. Couerdale." Apr. 3. 1587, "Forafmuch as he dothe bind his prentifes to fremen of this cumpanie to th'entent to make them free of this cumpanie; and for ẏ he hathe & dothe promife to fhewe himfelfe conformable to thordinance here: yt is thought conuenient to allowe vnto him for his copie, the 2d parte of the Englifh Crede."

HENRY BAMFORD.

1577. "A Profitable treatife of the anatomie of mans body: compyled by that excellent chirurgion, M. Thomas Vicary efquire, feriaunt chirurgion to king Edward the vi. to queene Mary, and to our moft gracious foueraigne lady, queene Elizabeth, & alfo cheefe chirurgion of St. Bartholomews hofpital. Which work is newly reuyued, corrected & publifhed by the chirurgions of the fame hofpital now beeing (1577.) Imprinted at London." Licenfed. Octavo.

RICHARD WEBSTER, or WEBBER.

1578. "THE Seconde part of the Mirrour of Magiftrates, conteining the falles of the infortunate Princes of this Lande: From the Conqueft of Cæfar, vnto the commyng of Duke William the Conquerour. Imprinted by Richard Webfter,—1578." In a neat architective compartment; and on the fell, "Goe ftraight and feare not." It is introduced with an epiftle from "The Printer to the friendly Reader." Then "The Authour's Epiftle vnto his friende." Which concludes with "keepe thefe trifles from the view of all men, and as you promyfed, let them not raunge out of your priuate Study.—15 Maye, 1577.—Tho. Blener Haffet." 66 pages. See the firft and third parts printed by T. Marfhe. Licenfed. W.H. Quarto.

He had alfo licenfed to him on 6 Apr. 1579, "A Difcourfe of ẏ lowe countries fince Don Jhons Deathe, with ẏ eftate & particularities of ẏ laft yere there, with A briefe Declaration of the comynge of Duke Cafimyr hither: And his honorable enterteynment in England."

ROBERT WALDE-GRAVE, or WALGRAVE,

SON of Richard Walgrave, of Blaoklay, Worcefterſhire, yeoman, deceaſed, was preſented to the company of Stationers by Wm. Greffeth, to be apprentice for eight years from Midſummer, 1568. As the time of his apprenticeſhip expired during the chaſm in the company's regiſter, it is not known when he was made free. He was fined 1 Dec. 1578, " for that he tooke a prentice, bound & inroled him, without licence, and preſented him not, contr. to the ordinances of this cūpany." In Jan. 1580, he preſented one orderly. At the cloſing of the Wardens account for the year from 10 July 1582, to 10 July 1583, in an inventory of Stock, &c. delivered over to the new Wardens is " an obligation of Rob. Walgraue of 40l. to the Company, concerning the not printing of any thing in Mr. Seres priuilege." The next year the Company " lent Rob. Waldegraue on bond with Geo. Bryar his Surety, to pay 24 June 1584, 5l." At the time of payment, " Item, in full for 5l. lent to Ro. Waldegraue, 4l." At firſt he printed in the Strand, near Somerſet-houſe, then in Foſter-lane, afterward at the White Horſe in Cannon-lane, according to an addition in the margin by Mr. Ames; but no book with ſuch a direction has yet come to my ſight: however, he appears to have dwelt again without Temple-bar in 1586. He met with great hardſhips on account of his printing books for the Puritans, particularly that entitled, "The ſtate of the Church of Englande laide open," printed by Chr. Barker, 1589, as may be ſeen in the note on that book, a contract to which you will find in An admonition to the people, &c. in the account of printing in Wales. However at length getting over his troubles by the aſſiſtance of friends, he was appointed printer* to K. James VI. of Scotland in the year 1580. His patent, and the titles of ſuch books as he printed there, will be found in our account of printing in Scotland. He ſeems to have returned to England, when K. James came to this crown, ſeeing Arnold Hatfield printed the Dæmonologie for him in 1603. He uſed for his device a Swan ſtanding on a wreath, with GOD IS MY HELPER about it in an oval compartment: ſometimes GOD IS MY DEFENDER up and down the ſides thereof.

" A caſtle for the ſoule," &c. as p. 1116. Licenſed. Again 1586. 1578.
" Ane ſhorte and generall confeſſion of the trewe chriſtiane faytn and (1580) religion, according to God's word and actis of our parliamentis, ſubſcriued to the kingis maieſtie and his houſholde, with ſindrie vtheris. To the glorye of God, and good example of all men. At Edinburgh the 28 January 1580, and the 14 of his maieſties regne. Imprinted *alſo* at London by Robert Waldegraue, without Temple bar, near vnto Somerſet-houſe." Alſo printed in Roman letter at London, for Thomas Man, dwelling in Pater-noſter-row, at the ſign of the Talbot, 20 June, 1580.
Sixteens.

* In 1600 Rob. Charteris ſtiles himſelf Typographus regius; but Waldegrave retains the ſame title in 1602.

1580.	" The order of Matrimony by R. W." Maunfell, p. 77.	Octavo.
1581.	" A dialogue between a chriftian vnlearned, and a catholike valearned," &c. Licenfed.	Octavo.
1581.	" A dialogue betweene a Gentleman, and a popifh prieft; very pleafant and profitable:" &c.	Octavo.
1581.	" The wonderfull worke of God fhewed vpon a chylde, whofe name is William Withers, being in the towne of Walfam, within the countie of Suffolke, who being eleuen years of age, laye in a traunce the fpace of tenne dayes, without taking any manner of fuftenaunce, and at this prefent lyeth, and neuer fpeaketh, but once in twelue, or foure and twentie houres, and when he cometh to himfelf, he declareth moft ftraung and rare thinges, which are to come, and hath continued the fpace of three weeks." Licenfed.	Sixteens.
1581.	" A fermon of—M. Iohn Caluine,—conteining an exhortation to fuffer perfecution for following Iefus Chrifte & his Gofpell, vpon Heb. 13; 13.—Tranflated out of French.—Imprinted—for Edw. White, 1581." At the end " An anfwere to the flanders of the Papiftes againft Chrifts fyllie flock, &c. Finis quod I. P." In verfe. C 6, in eights. W.H.	Octavo.
(1581.)	" The confeffion of the true and chriftian fayth according to Gods word, and acts of parliament, holden at Edenborgh, the 28 day of January 1581." At the end it is faid, " From Edenburghe the 17 Auguft 1560, thefe acts and articles were red in the face of the parliament, and ratified by the thre eftates." The king's charge is dated 1581. Licenfed.	Sixteens.
1581.	" Singuler & fruitfull manner of prayer vfed by D. M. Luther, paraphraftically written on the Lordes praier, beliefe, and the commandements." Licenfed.	Sixteens.
1582.	" A comfortable Treatife on 1 Pet. 4; 12—19. By O. Pigge.—Seene & allowed.—for Iohn Harifon the yonger, & Tho. Man. 1582." W.H.	Sixteens.
1582.	" A difcourfe of the true & Vifible Marks of the Catholick Church, By Th. Beza; tranflated by Tho. Wilcox." Alfo without date.	Sixteens.
1583.	" A manifeft & apparent confutation of an Aftrological difcourfe, lately publifhed to the difcomfort (without caufe) of the weake & fimple fort, as will by the fequel of that which followeth, euidently appeare. With a briefe Prognoftication, or Aftronomicall prediction of the coniunction of the two fuperior planets, Saturn & Iupiter: which fhalbe in the year of our Lord God 1583, the 29 of Aprill, at three of the clocke in the morning. Written the 25 of March by Tho. Heth, M. A." The fwan, &c. " Printed by *him*, dwelling in Fofter Lane, ouer againft Gold-fmiths Hal, at the figne of the George. By the affent of Rd. VVatkins." Addreffed " To—George Carey knight and Knight Marfhall of her Majefties—houfholde." F 2, in eights. See p. 988. W.H.	Sixteens.
1583.	" Certaine verie worthie, godly & profitable fermons vpon the 5th Chap. of the Songs of Solomon:—By Bartimeus Andreas."—for T. Man.	Sixteens.

" A notable

ROBERT WALDEGRAVE.

"A notable & comfortable expofition of M. Iohn Knoxes, vpon the (1583.) fourth of Mathew, concerning the tentations of Chrift: &c. Printed—for T. Man." Dedicated "To—Mres. Anne Prouze of Exeter by I. Field.— London, this firft day, of the firft moneth, in the yeare 1583." W.H. Sixteens.

"Anotations on Rom. 13. By John Hooper, Bifhop of Glocefter." (1583.) Octavo.

"A Fruitful Sermon, preached at Occham in the County of Rut- (1583.) land, 2 Nouemb. 1583, by Tho. Gybfon. Prov. 29; 8.—VVhere there is no Vifion, the people decay." The fwan, &c. "Printed by him, dwelling without Temple Barre." Dedicated "To the—Earle of Bedford, —with Sir Walter Mildmay, chauncellour of the exchequer." The text introduced with "Hearken with feare & reuerence to the worde of the Lorde, written by the holy Apoftle S. Paule, in his firft Epiftle to the Corinthians. Chap. 9, verfe 16. Woe be to me, if I preach not the Gofpell." E, in eights. Licenfed, 1586. W.H. Again 1584: Alfo without date. Sixteens.

"Andr. Hiperius his Catechifme, tranflated by I. H." Sixteens. 1583.
"In Canticum Canticorum, &c. Authore Gul. Thomfon." Octavo. 1583.
"Gods Judgment fhewed at Paris Garden, 13 Jan. 1583, being the 1583. Sabath day, at Beare bayting, at the meeting of aboue 1000 perfons, whereof diuers were flayne, moft maymed & hurt: Set out with an exhortation for the better obferuation of the Sabaoth, by John Field, minifter." With Hen. Carr. Again 1588, for him. Octavo.

"A fruitfull Sermon vpon the 3, 4, 5, 6, 7 & 8 verfes of the 12 Cha- 1584. piter of the Epiftle of S. Paule to the Romanes: Very neceffarie for thefe times to be read of all men," &c. The Swan, &c. with "God is my Helper," up and down the fides. "At London; Printed by him, 1584." Prefixed are "The chiefe heades of the Sermon;" and, an analitical table, or "The parts & order of the Sermon." Thefe on a neat Long Primer Roman, No. 2. As in Mr. Ames's copy thefe were on Italic types, there muft have been two editions: befides his copy contained 92 pages, mine only 80. Licenfed. W.H. Sixteens.

"Amendment of life; three fermons on Acts 2; 37, 38; by John 1584. Vdall, preacher of the word of God at Kingfton vpon Thames." For Tho. Man, W. B. and N. L. Alfo two fermons on obedience to the gofpel, by the fame author; and printed for the fame perfons. Octavo.

"A Briefe and plaine declaration, concerning the defires of all thofe 1584. faithfull Minifters, that haue & do feeke for the Difcipline & reformation of the Church of Englande: Which may ferue for a iuft Apologie againft the falfe accufations & flaunders of their aduerfaries." The fwan, &c. "At London;—1584." It is introduced by "A præface to the Chriftian Reader." The running title, "A learned Difcourfe of Ecclefiafticall Gouernment." 148 pages. W.H. Sixteens.

"Libellus de memoria, veriffimaque bene recordandi fcientia. Au- 1584. thore G. P. Cantabrigienfe. Huc acceffit eiufdem admonitiuncula ad A. Difconum, de artificiofae memoriae, quamquam publicè profitetur, vanitate." E, fheets. Sixteens.

"Two fermons—by James Biffe, M. A." &c. with a prayer at the 1584. end

6 T

end by Nich. Hemming. Again 1585; alſo without date. W.H. As neither of my copies have the ſaid prayer annexed, perhaps it was only accidental in Mr. Ames's. Sixteens.

1584. "Io. SLEIDANI de quatuor ſummis imperiis, Babylonico, Perſico, Graeco, & Romano, libri tres." Octavo.

1584. "A DIALOGVE, CONCERNING the ſtrife of our Churche: Wherein are aunſwered diuerſe of thoſe vniuſt accuſations, wherewith the godly preachers and profeſſors of the goſpell are falſly charged; with a briefe declaration of ſome ſuch monſtrous abuſes, as our Byſhops haue not bene aſhamed to foſter." The ſwan, &c. 136 pages. W.H. Sixteens.

(1584.) "A COVNTER-POYSON, modeſtly written for the time, to make aunſwere to the obiections and reproches, wherewith the aunſwerer to the Abſtract would diſgrace the holy Diſcipline of Chriſt. Luke 19; 40. I tell you," &c. It has two epiſtles prefixed, one by " The Authour to the Reader." the other from " A Faithfull Brother to the Chriſtian Reader." 195 pages. W.H. See p. 958. Quarto.

1585. " The Latine Grammar of P. Ramus: Tranſlated into Engliſh. Seene & allowed." The queen's arms on the title-page. Rev. Mr. Samuel Aſcough. Octavo.

1585. " An A. B. C. For Layemen, othervviſe called, The Lay-mans Letters. An Alphabet for Lay-men, deliuering vnto them ſuch Leſſons as the holy Ghoſt teacheth them in the worde, by thinges ſenſible, very neceſſary to be diligently conſidered. Printed—for Tho. Man, & Wm. Brome, 1585." Dedicated by Geo. Wilkes the author " To—Syr William Cicill,—and to his hon. Patrone, Syr Walter Mildmay, knight, &c.—Danbury 29 Jan. 1585." Pages 173. W.H. Octavo.

1585. " A Comfortable Treatiſe for all ſuch as as are ouerladen with the burden of their ſinnes &c. being a treatiſe againſt deſperation." Octavo.

1585. " A Treatiſe of the ſinne againſt the holy Ghoſt:" &c. as p. 1109.

1585. " A briefe and Pithy ſumme of the Chriſtian fayth, made in forme of a confeſſion—of al ſuch ſuperſtitious errors, as are contrary therevnto. Made by Theod. De Beza. Tranſlated out of French by R. F.—Printed by *him*,—without Temple-barre neere Somerſet-houſe, 1585." See p. 964. Pages 445. W.H. Sixteens.

1585. " An exerciſe for a Chriſtian Familie; Contayning a ſhort ſum of certayne poyntes of Chriſtian religion, with—godly Prayers, &c. Very neceſſary to be vſed in euery Chriſtian Familie. By R. M." 238 pages. W.H. Twenty-fours.

1585. " —The Inſtruction of a Chriſtian Woman." See p. 937. "—without Temple Bar, near to Somerſet houſe." Rev. Dr. Lort. Octavo.

1585. " The good Angell and miniſtring Spirite, whiche am the Meſſenger of God, ſent to miniſter for their ſakes which ſhalbe heyres of ſaluation, the tidings bringer of great comfort & ioy, ſendeth greeting vnto al my fellow-ſeruants & brethren, which are the elect children of God." The ſwan, &c. D 3, in eights. W.H. Sixteens.

1585. " The ſchoole of beaſtes, intituled, the good houſholder, or the oeconomickes. Made dialogue wiſe, by M. Peter Viret, and tranſlated out of French into Engliſh by J. B." * Sixteens.

" The

"The Shielde of Saluation, containing very Chriſtian & heauenly   1586.
meditations." Sixteens.
A ſermon preached in the Tower before Sir Henry Hopton, Lieute-   1586.
nant of the Tower, on Pſ. 2; 1. Sept. 25, 1586. Sixteens.
"THE ENGLISH Secretorie. Wherein is contained, A perfect method   1586.
for the inditing all manner of Epiſtles & familiar Letters, &c. Nowe
firſt deuized & newly publiſhed by Angel Daye.—Printed by him, and
are to be ſold by Rd. Jones—1586." Dedicated "To—Edward de
Vere, Earle of Oxenford," &c. whoſe arms are on the back of the title-
page.---To the Reader.---The table. 251 pages. At the end is the
device deſcribed in p. 646. The 2d. part licenſed to him. W.H.
Quarto.
"THE BOOK of Ruth expounded in 28 Sermons by Levves Lauaterus   1586.
of Tigurine, and by hym publiſhed in Latine, and now tranſlated—by
Ephraim Pagett, a childe of eleuen yeares of age. Printed—without
Temple-bar, 1586." Dedicated "To the worthy Matrons & Mirrours
of vertue, the gratious Lady Anne Dutches of Somerſet,—Brigitt Coun-
teſſe of Bedford,—Miſtreſſe Philip Prideaux, Miſtreſſe Luce Cotten, and
Miſtreſſe Mary Watts." 163 leaues. W.H. Octavo.
"The profitable arte of gardening," &c. as p. 974. Quarto. 1586.
"The Baptizing of a Turke. A Sermon preached at the Hoſpitall of(1586.)
St. Katherin—2 Octo. 1586, at the Baptizing of one Chinano a Turke,
borne at Nigropontus: by Meredith Hanmer, D. D. Printed—without
Temple-barre." Dedicated "To—Raphe Rokeby Eſq; Maiſter—of S.
Katherine, &c.—Shordich, 12 Oct. 1586." F 4, in eights. W.H.
Sixteens.
"A Confutation Of the tenne great plagues, Prognoſticated by Iohn(1586.)
Doleta from the Country of Calabria to happen in the yeare of our Lorde,
1587.—Printed—without Temple barre." 8 leaves. Licenſed. W.H.
Sixteens.
"The true remedie againſt Famine & Warres. Fiue ſermons vpon(1586.)
the firſt chap. of the propheſie of Joel; preached by John Vdall—1586.
Printed—for T. Man & T. Gubbins." Sixteens.

¶ "After the Sermon ended the Turke confeſſed in the Spaniſh tongue before the face of the congregation, the Preacher out of the Pulpit propounding the queſtions, and receiuing the anſwers by ſkilfull Inter-pretors, in ſumme as followeth. 1 That he was verie forie for the ſinful life which he had lead in times paſt—and hoped to ob-teine pardon in Ieſus Chriſt. 2 Hee re-nounced Mahumet the falſe Prophet of the Moores, &c. and bleſſed God which had opened his eyes to beholde the truth in Ieſus Chriſt. 3 Hee—beleeued the Trinitie of perſons,—and the ſame to bee one God in vnitie,—bleſſed for euer. 4 That hee be-lieued verily, that Ieſus Chriſt was & is the ſonne of God, and God from Euerlaſting, the onely true Meſſias, & ſauiour of the worlde, that he ſuffered for the ſinnes of al that beleeue in him, and that there is no way to be ſaued but onely by the merits of the death & paſſion of Ieſus Chriſt. 5 He de-ſired hee might be receiued as one of the faithfull Chriſtians, & bee baptized in the faith of the bleſſed Trinitie, promiſing from henceforth newnes of life, &c.
In the middeſt of the congregation there was a comely Table ſet, couered with a faire lin-nen cloth, and thereon a Baſen with water. After the congregation had bleſſed God for his great mercies, after ſundry godly Praiers & Collects, according vnto the reuerend order of holye Church, ſuche as broght him thither, deſired his name might be William, ſo was he baptized: In the name," &c.

1587.	Meditations on eternal judgment, by Rob. Openshaw.	Sixteens.
1587.	" Combat betweene the flesh & the spirit: a sermon on Colloss. 2; 6, 7, by Wilfrid Ros."	Sixteens.
1587.	" A Christian & learned exposition on Rom. 8; 18—23. Written long agoe by T. W." (Tho. Wilcocks)	Octavo.
1587.	" The English Creede—The second part," as p. 1136.	Folio.
1587.	" The xv Bookes of Ouidius Naso,—Metamorphosis." &c. 200 leaves, including the preface. See p. 698. W.H.	Quarto.
1588.	" Amendment of life." &c. as in 1584.	Octavo.

1588. "Two very lerned Sermons of M. Beza, togither with a short sum of the sacrament of the Lordes supper: Whereunto is added a treatife of the subftance of the Lords Supper, &c. By T. W. 2 Cor. 13; 5.— Printed—for T. Man & T. Gubbins, 1588." Dedicated "To—Brigit countesse of Bedford: and to—sir Charles Morisine, and ladie Dorothie —his wife.—London, 6 Feb. 1587." Pages 250, and two prayers at the end. W.H.   Octavo.

1588. "A Catechism by Henry Gray."   Octavo.

—— "A Declaration of the x holie Commaundements," as p. 1109. For Tho. Woodcock.

—— "A Booke of the forme of common prayers, adminiftration of the Sacraments: &c. agreeable to Gods Worde, and the vse of the reformed Churches.—1 Cor. 1; 11.—At London; Printed by Rob. Waldegraue." 77 pages, numbered over the inner margins. See p. 547. W.H.   Octavo.

—— A poem on the life & death of Sir Philip Sydney. Licensed.   Quarto.

—— The examinations of Anne Askew. See our General History, 1546.

—— "THE COMBATE betwixt Christ and the Deuill. Foure Sermons vpon the temptations of Christ in the wildernes by Sathan, &c. By Iohn Vdal Preacher &c. Printed—for T. Man & Wm. Brome." Dedicated "To—Henry Earle of Huntyngdon," &c. K 6, in eights. At the end J. Day's device, as p. 646, with R and W on the sides. W.H.   Octavo.

—— "A WORTHY TREATISE of the eyes; contayning the knowledge and cure of one hundreth & thirtene difeases incident unto them: First gathered & written in French by Iacques Guillemeau Chyrurgion to the French King, and nowt ranflated into Englifh; togeather with a profitable treatise of the Scorbie; and another of the Cancer, by A. H. Also next to the treatise of the eies is adioyned a work touching the preseruation of the sight, set forth by VV. Bailey D. of Physick. Printed—for T. Man & Wm. Broome." M 10, in twelves. Licensed, 1586. W.H. Twenty-fours.

—— " A Perfuasion vnto patient receiuing & embracing of all such afflictions as God shall bring vpon vs for the testimony of the Gospell: by Wilfrid Ros."

—— "Fower great liers striuing who shal win the siluer whetstone. Also a resolution to the countreyman, prouing it vtterly vnlawful to buy or vse our yearely prognostications: by W. P."   Octavo.

A sermon

ROBERT WALDEGRAVE.

A sermon on Matth. 4; 1—11. by Tho. Beatham. Octavo.
" A treatise of the lawfull boundes of all buying & selling, entituled, Nobodie is my name, that beareth euery bodies blame." Octavo.
" P. Melangton his treatise of praier, translated by John Bradford." Maunsell p. 86.
" An Answer to a letter, whether it be lawfull to be present at the Popish Masse, and other superstitious Church seruice: By John Bradford." Sixteens.
The Anker of Faith, or Christian exercises: containing godly meditations, meete for all Christians thirsting after godlines &c.
" A fruitful treatise of Baptisme and the Lords Supper: of the vse & effect of them: of the worthie & vnworthie receiuers of the same Supper." Octavo.
A sermon of Election, on Gen. 25; 23. *By Euseb. Pagett.* For T. Man. Octavo.
" A Caveat for Parsons Howlet, concerning his vntimelye flighte, and schriching in the cleare daylighte of the Gospell, necessarie for him & all the rest of that darke brood & vncleane cage of Papistes, who with their vntimely bookes seeke the discredite of the trueth, and the disquiet of this Church of England. VVritten by Iohn Fielde, Student in Diuinitie. Reuel. 18; 6.—Imprinted—for T. Man, & Tobie Smith." Dedicated " To—Lord Robert Dudley, Earle of Leicester, &c.—30 Aug. 1581." H 4, in eights. W.H. Octavo.
" A sermon on the parable of the sower, taken out of the xiii Matthew, by M. G. Gifford." Printed for Tobie Cooke.* Sixteens.
" Io. Dauison[e] his shorte Christian instruction." Maunsell, p. 29.
" The state of the Church of Englande,[b] laide open in a conference betweene Diotrephes a Byshopp, Tertullus a Papiste, Demetrius an vsurer, Pandochus an Inne-keeper, and Paule a preacher of the worde of God. Psal. 122; 6. Pray for the peace of Hierusalem, &c. Reuel. 14, 9, 10.

---

[a] " There was the last sommer, a little catechisme, made by M. Dauison, and printed by Walde-graue; but before he coulde print it, it must be authorized by the Bb. either Cante. or London. He went to Cant. to haue it licensed, his grace committed it to doctor Neuerbegood (Wood) he read it over in halfe a yeare, the booke is a great one of two sheets of paper. In one place of the booke the meanes of saluation was attributed to the worde preached: and what did he thinke you? he blotted out the word (preached) and would not haue that word printed, so ascribing the way to work mens saluation to the worde read." M. Marprelate's first Epistle, p. 34; of which see our Gen. Hist. 1588.

[b] Though this book is without date, name of the author, or printer, the following minute from the Company's Register will justify me in ascribing it to Walde-graue, and shew nearly the time of printing.

" May 13, 1588. Whereas Mr. Caldock, warden, Tho. Woodcock, Oliver Wilkes, & John Wolf, on the 16 of April last, vpon search of Rob. Walgraues house, did seise of his & bring to Stationers' hall. according to the late decrees of the Starrechamber, and by vertue thereof A presse with twoo paire of cases, with certaine Pica Romme, & Pica Italian letters, with diuers books entituled: The state of the Churche of England laid open &c. For that the said Walgraue without aucthority, and contrary to the said Decrees had printed the said book. Yt is now in full Court—ordered & agreed by force of the said decrees & according to the same, That the said books shall be burnte, and the said presse, letters & printing stuffe defaced & made vnseruiceable."

9, 10, And the third Angel, &c." 1 2, in eights; but D, E, F, and H are omitted, and yet the book appears perfect. W.H. Octavo.
He had also licensed to him as follows. In 1579, "A fighte in thelement in Cornewall." In 1580, "A dialogue betwene Baldwin & vi Sailors. The Auchthor of faithe, grounded vppon the Sacred Scriptures. Flatteries displaie. A cōfeſſion & declarac'on of praiers, made by John Knox. The cōfeſſion of the faithe & doctrine believed by the proteſtants of Scotland. A sermon by Luther vpon y̆ xx of John. The groundes of Chriſtianytie. The tryumphe ſhewed before the queene & y̆ Frenche Embaſſadors." In 1581, " A diſcourſe of the Eſtate of Cambray duringe y̆ ſiege." In 1585, "The lamentations of the prophet Jeremye, with a paraphraſe vpō the ſame: publiſhed by Dan. Toutſaintz, & engliſhed by F. S. Vpon condic'on he procure it to be auchthoriſed accordinge to her maieſties Jniunctions." In 1587, "A tragedie of the vniuſt vſurped primacie of the bp. of Rome, by Barn. Occhine, in ix dialogues: tranſlated by Dr. Ponet." May 13, 1588, " A copie, whereof he is to bring the title."

GEORGE BISHOP, ſtationer,

WAS preſented to the company by Mrs. Toy, widow of Rob. Toy, as her apprentice, 13 Octo. 1556; which was ſoon after they had received their charter. Probably he was bound to her huſband in her huſband's lifetime, for he was made free 16 April, 1562; and was taken into the livery in 1568; ſerved the Renterſhip in 1571 and 1572; Under Warden in 1578 and 1579; Upper Warden in 1584 and 1586; Maſter in 1590 and 1592, as alſo the remainder of the year 1592, after the deceaſe of Wm. Norton; again in 1600 and 1602. He was concerned with, and employed others, in ſeveral large works; and was one of the deputies to Chr. Barker eſq; printer to the queen's majeſty. In the year 1589, He bought ſo many of the books that were ſeized as he paid 3l. for; but " 40 of them, being Harmonies of the Churches, rated at ij s. le peece, were had from him by warrant of my L. of Canterb. and remain at Lambeth with M. Dr. Coſen; and the reſidue remain in the Hall to the vſe of Yarrette James." He married Mary, eldeſt daughter of John Cawood eſq; queen's printer.[1] Afterwards he became an alderman of London; and, among other legacies, bequeathed 6l. yearly to his company; 10 l. yearly for ever towards maintaining preachers at Paul's croſs; and 6l. yearly to Chriſt's hoſpital. Stow's London, B. 1. p. 267. He died in 1610, according to the Table of benefactors hung up in St. Faith's, 1630; on which was entered, " 1610. Geo. Biſhop, Stationer, Alderman, gave 10l. to be diſtributed." Query, whether the ſame as mentioned above? Alſo, " 1613. Mary Biſhop, the wife of Geo. Biſhop, gave 10l. to be diſtributed." Stow, ib. B. 3. p. 147.

[1] Dugdale's St. Paul's, p. 125. Edit. 1658.

" A Poſtill

GEORGE BISHOP.

" A Poſtill or Expoſition of the Goſpels that are vſually red in the 1569. churches of God, vpon the Sundayes and Feaſt dayes of Sainctes. Written by Nich. Hemlnge, a Dane, a Preacher of the Goſpell in the Vniuerſitie of Hafnie: and tranſlated by Arth. Golding.—Imprinted by H. Bynneman for L. Harriſon & *him*." See p. 927.   W.H.   Quarto.
" A Postil—of certeine Epiſtles" &c. as p. 926. 1570.
" The Pſalmes of Dauid, &c. With—Caluins Commentaries." &c. as ib. 1571.
" Common places of Chriſtian Religion," as p. 1007. 1572.
" Queſtions of Religion—aunſwered by M. H. Bullinger," &c. as p. 1572.
972.
" The fall of Hugh Surean from the Goſpell," &c. as p. 927. 1573.
" Sermons of M.—Caluin vpon—Job." &c. as ib. Again 1580 & 1574.
1584.
" ———— vpon the Epiſtle—to the Galathians." &c. as ib. 1574.
" A Catholike expoſition vpon the Reuelation" &c. as ib. 1574.
" Abdias the Prophet, Interpreted by T. B." as p. 974. Again 1577. 1574.
" A Treatise of the right way frō Danger of Sinne" &c. as p. 976. 1575.
" The Philoſopher of the Court," as p. 927. 1575.
" A Viewe of mans eſtate" &c. as p. 928. Again 1580. 1576.
A warning to take heede of Fowlers pſalter. as p. 1068. Again 1578. 1576.
" The vvhole Summe of Chriſtian Religion," &c. as ib. 1576.
" The Sermons of M. Iohn Caluin vpon—the Epheſians." &c. as 1577.
p. 1057.
" A ſermon at Alphages on Gal. 4; 21—31. by W. Fulke, D. D." 1577.
See p. 929.
" A Commentarie vpon S. Paules Epiſtles to the Corinthians. Writ- 1577. ten by M. Iohn Caluin, and tranſlated—by Tho. Tymme Miniſter. Imprinted at London for Iohn Harriſon & *him*, 1577." Dedicated " To —Edmund—Archebiſhop of Canterburie.---The Argument." My copy imperfect at the end. The blooming letters being the ſame as to Calvin's ſermons vpon the Epiſtle to the Epheſians, conclude this was printed alſo by H. Middleton. W.H.   Quarto.
" Quæſtionum & Reſponſionum Chriſtianarum Libellus." &c. as 1577.
p. 1069. Licenſed to him in 1570.
" Commentarius in epiſtolam—ad Romanos." &c. as p. 1068. 1577.
" A Commentarie of J. Caluine vpon—Geneſis :" &c. as p. 1057. 1578. Licenſed to him, with John Harriſon, ſenior.
" ———— vpon the Booke of Ioſue," &c. as p. 1116. Licenſed. 1578.
" The poore mans Iewell." &c. as p. 1012. Again 1580, 1592. Li- 1578. cenſed.
" A ſermon at Paules Croſſe on Acts 3 ; 12, 13. by John Walfall." 1578.
For him.
" The diſplaying of a ſect of hereticks, naming themſelues the family 1578. of loue." Licenſed. Again 1579. See p. 1057.   Octavo.
" A ſermon at Paules Croſſe," &c. as p. 982. For him. Licenſed. W.H. (1578.)
" ———— on the 10 May—on Math. 9; 35." Licenſed.   Octavo, 1579.
" A Sum-

1579.	" A Summarie, and short meditations, touching certaine points of Christian Religion, by Tho. Wilcoxe." Licensed. Octavo.
1579.	A treatise of dicing, dancing, &c. See it in p. 1117. Also without date. Licensed 1577, 1578.
1579.	" Sermons (16) vpon the x Comandements," &c. as p. 1117.
1579.	Heskins, Sanders & Rastel detected, &c. by Dr. Fulke. See it in p. 1057.
1579.	Calvin's sermons on the epistles to Timothy & Titus. &c. as ib.
1579.	" Responsio ad Thomæ Stapletoni cauillationes calumnias in sua principiorum doctrinalium demonstratione," &c. Guil. Fulke. Octavo.
1580.	" A Godly—Exposition vppon the Prouerbes" &c. as p. 1120. Licensed.
1580.	" A Godly Sermon—by Rob. Some;" &c. as p. 1058.
1580.	" A Retentive to stay good Christians in the true faith," &c. as p. 1071.
1580.	" The benefit that Christians receiue by Iesus Christ" &c. as p. 929. T. Dawson for him and T. Woodcock.
1580.	" Three propositions, or speeches by M. Caluin, at Geneua." See p. 1118.
1580.	Stapleton and Martiall confuted by Dr. Fulke. See p. 1059. Licensed.
1580.	" A treatife of the Plague:—by Th. Beza," &c. as p. 1120. Licensed.
1581.	" A briefe Confutation of a Popish Discourse," &c. as ib.
1581.	" A sermon, at the Tower, on Iohn 17; 17," &c. as ib.
1581.	" A Reioynder to Bristowes replie in defence of Allens scroule of articles, and booke of Purgatory: also the cauiles of Nich. Saunders, about the Supper of our Lord, &c. confuted by Wm. Fulke, D.D." Licensed. Octavo.
1581.	" Sermons (27) concerning the Diuinitie, humanitie, &c. of Christ: By M. John Caluin, translated by Tho. Stocker." Licensed. Octavo.
1581.	" A forme of catechising: consisting in questions & answeres, with obseruations thereon, &c. by W. Wood." Licensed. Octavo.
1581.	" Of the Church & certaine ceremonies thereof: By Tho. Sampson."
1582.	" A Godly—Treatife of the Sacraments:" &c. as p. 1122.
1582.	" Jesuitismi pars prima;" as p. 1060. See Strype's Annals, Vol. x; 473.
1582.	" A Treatife of reformation in religion, diuided into 7 sermons, preached in Oxford, on Math. 21; 12, 13. Also two sermons touching the supper of the Lorde, on 1 Cor. 11; 28, 29, and on Math. 26; 26, 27, 28. By Harbert Westphaling, D.D." Licensed. Quarto.
1583.	" Two treatises:—Of the Church;—Against oppression:" See p. 1124.
1583.	" A Commentarie vpon the Epistle—to the Romanes" &c. as p. 1126.
1583.	" A Defence of the—Translations of the holie Scriptures" &c. as p. 988.
1583.	" A Treatife on the Lordes Praier, twelue Articles of Faith, and ten Commandements: by Rob. Some, DD." W.H+. Licensed. Octavo.
1583.	" Of nature and grace: with answers to the enemies of grace vpon incident occasions, offered by the late Jesuits notes vpon the New Testament. By John Prime, Fellow of New College, Ox." Octavo

" Andr.

GEORGE BISHOP.

"Andr. Hyperij compendium Phyſices Ariſtotelæ," &c. as p. 1061.   1583
Licenſed.
"Ad P. Rami dialecticam, &c." as p. 1124. Licenſed.   1583.
"A Harmonie vpon the three Euangeliſts Matthew," &c. as p. 1125.   1584.
"The ſumme of the Conference betwene Iohn Rainoldes & Iohn   1584.
Hart, touching the Head and Faith of the Church.—Penned by Iohn
Rainoldes, according to the notes ſet down in writing by them both:
peruſed by Iohn Hart, &c. Whereto is annexed a Treatiſe entituled,
Six concluſions touching the Holie Scripture, and the Church, written
by Iohn Rainoldes. With a defenſe of ſuch thinges as T. Stapleton & Gr.
Martin haue carped at therein. 1 Ioh. 4; 1.—" Dedicated "To—Ro-
bert Dudley, Earle of Leiceſter, &c.—At London, 18 July, 1584. I. R.
---Iohn Hart to the indifferent Reader.—From the Tower, 7 July.---
I. R. to the Students of the Engliſh Seminaries at Rome & Rhemes.---
The contents." 750 pages, including title, &c. "Printed by Iohn
VVolfe for *him*, 1584." W.H. Again 1588, and 1598, printed by *him*.
W.H.   Quarto.
"The Commentaries of M.—Caluin vpon the Actes," &c. as p. 1126.   1585.
"De Sacris Eccleſiæ miniſteriis ac beneficiis, Libri viii.—Item, Pro   1585.
libertate Eccleſiæ Gallicæ aduerſus Romanam aulam Defenſio Pariſienſis
curiæ, Ludouico xi Gallorum Regi quondam oblata. Authore Franciſco
Duareno Iureconſulto &c. Opus ab authore denuo auctum ac emenda-
tum." On 113 leaves beſides the prefixes. "His inſuper Petri Rebuffi
Iuris Vtriuſq; doctoris tractatum de decimis viſum eſt annectere.—Im-
penſis G. Biſhop, 1585." The tract De decimis has a ſeparate title-
page, by which it appears that John Jackſon was the printer. 68 leaves.
Licenſed. W.H.   Octavo.
"The Brutiſh Thunderbolt:—of Pope Sixtus the fift," &c. as p. 913.   1586.
"Lipſii Epiſt. ſelectarum centuria prima." Secunda, 1590. Octavo.   1586.
"Eicasmi, ſeu meditationes in ſacram Apocalypſin. Authore Jo. Foxo,   1587.
Anglo." Dedicated by his ſon Sam. Fox to Archb. Whitgift.   Folio.
Holinſhed's Chronicles. See p. 914, 961.   Folio.   1587.
"Edwardi Livelei, Hebraearum literarum in academia Cantabrigenſi   1587.
profeſſoris, annotationes in quinq; priores ex minoribus prophetis, cum
Latina eorum interpretatione, ejuſdem opera ac ſtudio ad normam He-
braicae veritatis diligenter examinata." Rev. Dr. Lort.   Octavo.
"Sacrorum parallelorum libri tres; id eſt, comparatio locorum ſcrip-   1588.
turae ſacrae, qui ex teſtamento vetere in nouo adducuntur, &c. Franciſci Junii Biturigis. Editio ſecunda. Impenſis G. Biſhop." Licenſed.
Alſo without date. The title in Chr. Barker's compartment, as p.
1078, with Sir Fr. Walſingham's arms. 374 pages, beſides the preface,
&c. Very neatly printed on Brevier Roman and Italic. W.H. Octavo.
"Joannis vel Curicnis commentarium, libri iiii. in Vniuerſam Ariſ-   1588.
totelis phyſicen; nunc recens ſumma fide exactaque diligentia caſtigati
et excuſi. Rerum praeterea et vocum memorabilium pleniſſimus index
praefixus. Impenſis Geo. Biſhop."   Sixteens.
"Enchiridion locorum communium theologicorum, ex Malorati the-   1588.
ſauro," &c.   Octavo.
"M. Annaei Lucani Pharſalia."   Octavo. 1588.

6 U   "A Godlie

GEORGE BISHOP.

1588. " A godlie treatife, &c. by Rob. Some, D. D." See p. 1084. Licenfed.
1588. " The Portraiture of Dalilah: the bridle of luft: Seale of fecrets, expounding Judges 16; 16, 17. Alfo, a meditation on Luke 10; 41, 42, containing the profit of reproof; together with the neceffitie & excellencie of Gods word: By E. R." Licenfed. Octavo.
1588. " An Anfwere to ten friuolous & foolifh reafons fet downe by the Rhemifh Iefuits & Papifts in their Preface before the New Teftament by them lately tranflated into Englifh, which haue mooued them to forfake the originall fountain of the Greeke, wherein the Spirit of God did indite the Gofpell, and the holie Apoftles did write it, to follow the ftreame of the Latin tranflation, tranflated we know not when, nor by whom. With a difcouerie of many great Corruptions & faults in the faid Englifh tranflation—. By E. B.—Impenfis—1588." Dedicated " To—Sir Francis Walfingham Knight, &c.—At Woodhull, 9 Apr. 1588. Edw. Bulkeley." 103 pages. Licenfed. W. H. Quarto.
1589. " De Ecclefia Dei ab Antichrifto per eius Excidium liberanda, eaque ex Dei promiffis beatiffimè reparanda Tractatus: Cui addita eft ad calcem veriffima certiffimaq; ratio conciliandi diffedii de Cœna Domini.—Impenfis—1589." Licenfed. Lamb. Libr. Quarto.
1589. " P. Rami regii profefforis dialecticæ libri duo. Defenfio eiufdem dialecticæ per fcholafticas quarundam interpretationum, animaduerfionum, triumphorum, & emendationum, difquifitiones: Authore Fred. Beurhufio.—1589. Impenfis G. Bifhop." In C. Barker's compartment. 286 pages, befides the prefixes. Licenfed. W. H. Octavo.
1589. " Antimartinus, fiue Monitio—contra—M. Marprelat" See p. 914.
1589. " The Principall Nauigations, Voiages," &c. as ib.
1589. " THE FRENCH ACADEMIE,[k] wherin is difcourfed the inftitution of Maners, & whatfoever els concerneth the good & happie life of all eftates & callings, by precepts of doctrine, and examples of the lives of ancient Sages & famous men: By Peter de la Primaudaye Efquire, Lord of the faid place, and of Barree, one of the ordinarie Gentlemen of the Kings Chamber: dedicated To the moft Chriftian King Henrie the third, and newly tranflated into Englifh by T. B. *(Tho. Bowes.)* The fecond Edition. Impenfis—1589." Dedicated " To—M. Iohn Barne Efq;—17 Octo.[l] T. B. C." The author's dedication is dated, " At Barre, in the moneth of Februarie, 1577.---The[m] Author to the Reader.---The Con-

[k] N.B. This is the firft Englifh book, noticed by me, in which the letters j and i, v and u, are ufed with propriety. See the fame remark on a Latin book, in p. 681, note o.

[l] " ——— as manie as are defirous to be bettered by the reading of this booke, they muft thinke ferioufly upon that end unto which this Author had regard when he penned it, which was the fame that Ariftotle had in writing his Ethicks, or booke of Manners: namely, *The practife of vertue in life, & not the bare knowledge & contemplation thereof in the braine.*"

[m] " We fhall fee heere the order & eftablifhment of Policies & Superiorities; what is the dutie of the Heads of them,—as alfo what the dutie of their Subjects is. Briefly, both great & fmall may drawe out from hence the doctrine & knowledge of thofe thinges which are moft neceffarie for the government of a Houfe, and of a Commonwealth; with inftruction how to frame their life & maners in the mould of holie vertue, and how they may run the race of their daies in joy and tranquillitie of Spirit, and that in the midedft of the greateft adverfities, which the uncertaintie & change of humane things may bring upon them.—And this thou maieft do, if thou takeft paines to read well, to underftand better, and (which is beft of all) to followe the precepts, inftructions, & examples which thou fhalt find heer.—Spe certa quid melius?"

tents."

GEORGE BISHOP.

tents." 757 pages, and an index. Licenfed to him and R. Newbery. W.H. See the 2d part in 1594. Quarto.

" Iobvs, Theodori Bezæ partim Commentariis partim Paraphrafi 1589. illuftratus. Cui etiam additus eft Eccleſiaſtes, Solomonis Concio de ſummo bono, ab eodem Th. B. paraphraſticè explicata.—Typis Georgij Biſhop, 1589." In C. Barker's compartment. Dedicated " Potentiſſimæ et Sereniſſ. Dei gratia Angliæ, Franciæ, Hiberniæ & circumiacentium inſularum Reginæ Elizabethæ, Gallorum, Belgarum & Italorum pro Chriſti nomine exulantium nutriciæ, & Chriſtianæ totius veræ religionis inuictæ propugnatrici.—Geneuæ à Sabaudo Principe circumſeſſe xii Auguſti, anno vltimæ Dei patientiæ, cıɔ ıɔ xxcıx." Job is concluded on p. 290; a blank leaf, then the paging continued to p. 344. Licenſed. W.H. Octavo.

" A ſermon—at Pawles,—by Wm. James, D. D." See p. 916. 1590.

" Ivsti Lipsi Politicorvm, ſiue Ciuilis Doctrinæ libri ſex. Qui ad Principatum maximè ſpectant. Editio altera, quam Auctor pro germanâ & fidâ agnoſcit.—Typis ejus, cıɔ ıɔ xc." 231 pages beſides the prefixes. Licenſed. W.H. Octavo.

" Ivsti Lipsi ad Libros Politicorum breues notæ.—Typis &c." 47 pages.

" ΑΠΟΛΙΝΑΡΙΟΥ ΜΕΤΑΦΡΑΣΙΣ ΤΟΥ ΨΑΛΤΗρος διὰ ςίχων ήρωικῶν. 1590. Apolinarij interpretatio Pſalmorum, verſibus Heroicis : Ex Bibliotheca Regia. Londini, Excudebat—Regiæ Maieſtatis Typographi Deputatus. cıc ıɔ xc." In C. Barker's compartment. Elegantly printed on Pica Greek. 202 pages, and Errata. W.H. Octavo.

" D. Io. Chryſoſtomi—Homiliæ," &c. as in p. 915. 1590.
" A Remonſtrance, or plaine detection," &c. as ib. 1590.
" The Conſent of Time," &c. By Lod. Lloid. &c. as ib. Alſo with- 1590. out date.

" Academiarum—enumeratio breuis." as p. 916. 1590.
" Analyſis Logica epiſtolarum Pauli ad Romanos, Corinthios, Gala- 1590. tas, Epheſios, Philippenſes, Coloſſenſes, Theſſalonicenſes Vnà cum ſcholijs, &c. Autore M. Iohan. Piſcatore,—Impenſis 1590." In C. Barker's compartment. 476 pages, excluſive of the preface. Licenſed. W.H. Again 1591, 1594. Octavo.

" Britannia," &c. as p. 913. " Nunc tertio recognita & magna 1590. acceſſione adaucta." The queen's arms. "—impenſis—.Cum gratia &c. 1590." England on 662 pages; Scotland continued to p. 671: Ireland, having a ſeparate title-page, with the device of Abraham and Iſaac, is continued to p. 719; Inſulæ Britannicæ to p. 762. At the end are tables and an index. W.H. Octavo.

" That the pope is that Antichriſt; and an anſwer to the obiections of 1590. ſecretaries, which condemne this church of England. Two notably learned, and profitable treariſes or ſermons vpon the 19 verſe of the 19 chapter of the Reuelations; the firſt whereof was preached at Paules croſſe, in Eaſter terme laſt; the other purpoſed alſo to haue bene there preached. By Lawrence Deios, bachelor in diuinitie, and miniſter of God's holy word." 184 pages. With R. Newbery. Sixteens.

" A Treatiſe of Eccleſiaſticall Diſcipline :" &c. as p. 916. 1590.
" ΟΜΗΡΟΥ ΙΛΙΑΣ. Homeri Ilias, id eſt, de Rebus ad Troiam geſtis.— 1591.

6 U 2 Excu-

GEORGE BISHOP.

Excudebat—Regiæ Maieſtatis Typographi Deputatus,—cɪɔ ɪɔ xcɪ." In C. Barker's compartment. Elegantly printed on Pica Greek; the ſame as S. John Chryſoſtom's homelies. 545 pages. Licenſed. W.H. Octavo.

1592. " Η ΚΑΙΝΗ ΔΙΑΘΗΚΗ." N. Teſtamentum, as p. 1086.
1593. " Teſtamenti Veteris Biblia Sacra," &c. as p. 881. See p. 1091, note k. The letters j and i, v and u, are herein applied throughout, as in modern uſe.
1594. " BRITANNIA," &c. as in p. 913. " Nunc quarto recognita,—poſt Germanicam æditionem adaucta." The queen's arms. "—impenſis— 1594." The whole on 717 pages, beſides prefixes and affixes. W.H. Quarto.
1594. " The Second Part of the French Academie," &c. as p. 916, and 1091.
1594. " An anſwer vnto a certain calumnious Letter," &c. as p. 1088.
1595. " The Amendment of Life, compriſed in ſower Bookes: Faithfully tranſlated according to the French Coppie. Written by M. John Taffin Miniſter of the word of God at Amſterdam. Math. 3; 2, and 4; 7.— Impenſis—1595." In a compartment with the ſtory of Diana and Actaeon. Dedicated by the author " To the chiefe Magiſtrates—of the Towne of Amſterdam.—June, 1594."---The contents. 549 pages. W.H. Quarto.
1595. " Articles to be enquired within the dioceſſe of Briſtoll, in the viſitation of the—Archbiſhop of Canterburies grace.—The epiſcopal ſee of Briſtoll being void." Quarto.
1596. " Rerum Anglicarum Scriptores poſt Bedam," &c. as p. 1091.
1597. " THE NAVIGATORS SVPPLY," &c. as p. 916, &c.
1598. " THE Workes of our Antient and lerned Engliſh Poet, Geffrey Chaucer, newly Printed. In this Impreſſion you ſhall find theſe Additions; 1 His Portraiture & Progenie ſhewed. 2 His Life collected. 3 Arguments to euery Booke. 4 Old & obſcure words explained. 5 Authors by him cited, declared. 6 Difficulties opened. 7 Two Books of his neuer before printed:" theſe are his Dream, and The flower & the leaf." " ImpenſisGeor.Biſhop,—1598." In the magnificent compartment deſcribed in p. 881. On 402 leaves, beginning with The knight's tale. Probably it was printed by Adam Iſlip, who printed it alſo for Tho. Wight; changing the name when a certain quantity was printed. W.H. Folio.
1598. " Praxis Medicinæ vniuerſalis; Or A generall Practiſe of Phyſicke:— The like whereof as yet in engliſh hath not beene publiſhed. Compiled & written by the moſt famous & learned Dr. Chr. Wirtzung, in the Germane tongue, and now Tranſlated into Engliſh; in diuers places corrected, and with many additions illuſtrated & augmented, by Iacob Moſan Germane, Doctor in the ſame facultie." A. Hatfield's Device, a caduceus between two cornucopias. " —Impenſis Georg. Biſhop, 1598." Contains 790 pages, and three indexes, beſides the prefixes. W.H. The late Mr. Bayntun had a copy printed by or for Edm. Bollifant this year. Folio.
1599. " The principal Navigations," &c. as p. 917.

" The

GEORGE BISHOP.

" The Hauen finding Art," &c. as p. 1092.   1599.
" Plutarchi Chaeronei opufculum de liberorum inftitutione. Item, 1599.
Ifocratis orationes tres. I. Ad Demonicum. II. Ad Nicoclem. III.
Nicoclis. Typis Georgii Bifhop, anno falutis humanae."   Octavo.

" A Geographical Hiftorie of Africa, Written in Arabicke and Italian 1600.
by Iohn Leo a More, borne in Granada, and brought vp in Barbarie,
—Before which, out of the beft ancient & moderne writers, is prefixed a
generall defcription of Africa, and alfo a particular treatife of all the
maine lands & Iflands vndefcribed by Iohn Leo. And after the fame is
annexed a relation of the great Princes, and the manifold regions in that
part of the world. Tranflated & collected by Iohn Pory, lately of Gone-
uill & Caius Coll. in Cambridge." Hatfield's device. " Impenfis—
1600." Dedicated " To—fir Robert Cecil Knight, &c.—At London
this three and fortieth moft ioifull Coronation-day of her facred Maieftie,
1600.---To the Reader." Some particulars of the Hiftory and the
author. 420 pages. W.H.   Folio.

" BELLVM PAPALE." &c. as p. 1092.   1600.
" Britannia." &c. as p. 913. " Nunc poftremò recognita," &c. as in 1600.
1594. This edition has before the title-page a neat frontifpiece from
copper plate. In a tablet at top, is Britannia fitting on a rock ; in the
center a map of Great Britain, and part of Ireland, fupported by Neptune
and Ceres ; at the bottom a view of Bath, with Stonehenge in the diftance ; a cathedral in one corner, and a fhip under fail in the other. W.
Rogers fculpfit. Alfo, a dedication to the queen, which concludes with
" De noftris annis tibi Chriftus adaugeat annos." There are alfo added
in their proper places two maps of England, as under the Romans, and
the Saxons ; 7 plates of Roman coins ; and a view of Stonehenge : all
from copper plates. 831 pages ; and at the end of the tables, an epiftle
" Ad lectorem," 30 pages. W.H.   Quarto.

" Ecloga Oxonio-Cantabrigienfis, tributa in libros duos ; quorum Prior 1600.
continet Catalogum confufum Librorum Manufcriptorum in—Bibliothe-
cis,—Oxon. & Cantab. Pofterior, Catalogum eorundem diftinctum &
difpofitum fecundum quatuor facultates, &c.—opera & ftudio T. I.
(Tho. James)—Impenfis G. Bifhop, & Io. Norton, 1600." Dedicated
" Rev. Archiepifcopis, Epifcopis & Præfulibus dignifl. Eccl. Anglicanæ." The firft book on 144 pages. " Liber Secundus," &c. has a
feparate title-page, with the device of Time bringing Truth to light,
" Excudebat Arnoldus Hatfield, 1600." Pages 132. W.H.   Quarto.

" A fhorte Difcourfe of the Ciuill warres—in Fraunce," &c. as p. 929.
The hiftory of Gefta Romanorum. See p. 1019.

He had alfo licenfes for the following, viz. In 1576, " Colloquiorū
fcholafticorum Libri tres, Authore Maturino Cordelio ab ipfo aucti &
recogniti :" With Jhon Harrifon the elder ; as alfo, " The cōmentaries
of Jhon Caluin vppon the prophet Efaie, to be tranfl. into Englifhe, & fo
to be prynted." In 1577, " Hollinfheds Chronicle with John Harrifon.
A warning to take hede of Fowlers Pfalter, gyuen by T. Sāpfon." In 1588,
" Ad ep'lam Sanifiei Hofij de exp'ffo Dei verbo Guil. Foulconi Refponfio,
&c.

GEORGE BISHOP.

&c. Dialogues en quattre langues; Flamen, Francoys, Espagnol, & Italien, with the Englishe to be added. The whole Bible in Latin, with the annotations of Tremellius & Junius:" with John Harrison senior, Wm. Norton, & Chr. Barker. " Crispins Lexicon, Gr. & Lat. quarto. Apophthegmatū—Loci cōmunes denuo aucti & recogniti: Cum Parabolis siue similitudinibus olim ex grauissimis auctoribus collectis, aūc vero per Conrad. Lycosthenem in Locos cōmunes digestis:" with J. Harrison senior, and Wm. Norton. " An answere made by Oliver Carter, to certen popishe quest. & demands. Rocci Pilorcij Marsianensis Ciuis Perusini, de scribendi rescribendiq; epistolas ratione liber. Daneus of Christian Amitie." In 1579, " Psalmorum Dauidis," &c. as p. 1071. " Nota; Vautrolier had printed this without licence before this was entered to Mr. Bishop Sex Theses de S. Scriptura & ecclesia—à J. Raynoldo: Item, in Englishe." In 1580, " In P. Rami Dialecticam animadversiones Joānis Piscatoris Argentinens. exemplis sacrarum literarū passim illustratæ:" with J. Harrison, senior. In 1582, " Joannis Velcuriuius cōmentariorum Libr. iiij. in Aristotelis phisicen. Aristotelis problemata. Dialectica Hunei. De Xpo gratis iustificante, per J. Fox. Gul. Templei philosophi Cantabr. ep'la de dialectica Rami ad Joh. Piscatorē Argentinens:" with J. Harrison, senior. In 1583, " Triumphus Logicæ Ramæ:" with J. Harrison, senior. " Secunda pars Jesuitismi Laur. Hüfrido authore." In 1584, " Hemyngius on the epistle to the Romans:" In 1585, " Lipsius de Constantia. Erasmi in sacrā Apocalipsin per J. F. The same book in Englishe." In 1586, " A treatise of Melancholie by Mr. Dr. Bryght:" with John Winder. " Nota, Mr. Dr. Bright hath promised not to meddle with augmenting, or altering the said book vntill thimpression which is printed by the said J. W. be sold." See p. 1074. In 1587, " Annotac'ons in Minores Prophesias. Enchiridion Locorū cōmuniū theologicorū ex Marlorati thesauro, et Obenhinij promptuario." In 1588, " Mr. Dr. Fulkes Answere against the Remishe Testament. Touching the quietnes & contentac'on of mynd, transl. from the French of L'estrine of Angers. A godlie treatise wherein are examined & confuted many execrable fancies gyuen out by H. Barrowe & John Grenewood, partly by other of the Anabaptistical order." In 1589, " Lucianus poeta. A Letter of Sir Fr. Walsinghams. De diuersis ministrorū Euangelij gradibus," &c. with R. Newbery. In 1590, " Dr. Sucklif, de Presbyteriū &c. Esopes fables in Greeke." With R. Newbery. " Apolonarius, in Greeke. Orat. Isocrat. Grece." In 1592, " De imperandi authoritate, & Christiana obediencia." In 1593, " The thirde part of the French Academie." As the entrance money for this does not appear to have been paid; query whether it was printed.

JOHN

[1155]

JOHN HARRISON,[a] the elder, stationer,

WAS made free 19 Aug. 1556; came on the livery in the year 1564; served Renter in 1568 and 1569. Under Warden in 1573, and Upper Warden in 1578 and 1579; Master in 1583, 1588, and 1596. He dwelt in St. Paul's Church-yard at the sign of the White Greyhound; afterward in Pater noster-row. At first setting out in life, he seems to have been in but a low way. In 1558 he was fined 4d. for printing a ballad without licence. In 1560, for opening his shop on the Sunday, 6d. In 1564, with others, for keeping his shop open on St. Luke's day; also for stitching books, contrary to Orders. Latterly, by industry &c. he appears to have moved in a higher sphere of life; was often concerned in copies with Geo. Bishop; and was an orderly and respectable member of his company. He used the same rebus as Rd. Harrison, and probably was related to him.[b] I have not found any book of his publishing earlier than

"Frederick Count Palatine his Catechisme, translated out of Latine &c. 1570. Dutch.—for him." Maunsell p. 29.

"A Spirituall Poseaye," &c. as p. 1064. 1573.

"A plaine description of the Auncient Petigree of dame Slaunder 1573. together with her Coheirs, and Fellowe Members, Lying, Flattering, Backebyting, (being the Diuels deare Darlinges. &c. Imprinted at London by John Harrison, 1573."

"Dominici Mancini de quatuor virtutibus, &c. Libellus. Excusum 1574. apud Ioan. Harison, Anno Domini, 1574." This, and the former article from the late Mr. T. Martin's papers. Sixteens.

"The pathewaie to knowledge," &c. as p. 601. The titles to both the 1574. first and second books are set in the compartment with Moses and David on the sides. At the end, "Imprinted at London by Jhon Harrison, 1574." W.H. Quarto.

"A Catholike and Ecclesiasticall exposition of—S. John" &c. Maun- 1575. sell, p. 70. See p. 864.

"A DICTIONARY in Latine & English," &c. as p. 1056. 1575.

"THE Firste volume of the Chronicles of England, Scotlande, and 1577. Ireland: Conteyning, The description & Chronicles of England from the first

---

[a] So named to distinguish him from John Harrison, or Haryson, the younger, of whom some account will be given by and by. I cannot satisfactorily determine their relation to each other. In a memorandum before the Stationers' Register, B, is this note, "28 Nov. 1583, John Harrison the younger by con. from ΗIS BROTHER nowe our master hath entred Isagoge" &c. which was licensed to him accordingly the 2 Dec. following. As this was only a mem. taken perhaps in a hurry, it may the easier be sup- posed a mistake; for we can hardly think it likely for a younger son to have been named John, while there was an elder son of that name living. In the minutes of the said Register, 4 Feb. 1599 60, "John Haryson the younger" is styled "son of Mr. John Haryson the elder." This has a much greater appearance of probability; however, he was his apprentice.

[b] See Strype's Life of Archbp. Parker, p. 451.

JOHN HARRISON, the elder.

first inhabiting vnto the conquest.—of Scotland,—till the year of our Lorde 1571.—of Yreland,—vntill the yeare 1547. Faithfully gathered and set forth by Raphaell Holinshed. At London, Imprinted for John Harrison. See p. 980. The date, 1577, at top; and "God saue the Queene," at bottom. Each kingdom has a separate title-page, paging, &c. England is in two parts; before, and after the conquest; the former of these, with Scotland, and Ireland, make the first volume; the latter makes the second. The whole is dedicated "To —Sir William Cecill,—Lord high Treasourer of England." See the subsequent edition, p. 961. Licensed to him and G. B. W.H. Folio.

1577. Quaestionum & Resp. Christianarum, Pars altera, &c. as p. 1069.

1577. "A COMMENtarie vpon S. Paules Epistles to the Corinthians. Written by M. Iohn Caluin," &c. as p. 1147.

1578. "Quinti Horatii," &c. as p. 879. Again 1585, with Juvenal & Persius.

1578. "A Commentarie of John Caluin vpon—Genesis:" &c. as p. 1057.

1579. "The Defence of Militarie profession:" &c. as ib. Licensed.

1579. "The grounde of artes;" &c. as p. 605. "and now of late diligently ouerseene & augmented with new & necessarie additions, by Mr. John Dee." Printed again 1582, and 1590. "Augmented by John Mellis of Southwarke, schoolmaster." Octavo.

1580. Psalmi Dauidis ex Hebraico, &c. as p. 1059.

1580. Testamenti veteris Biblia Sacra, &c. as p. 1058. Also N. Testam. as p. 1071.

1582. "A Sermon of Repentance—preached at Lee in Essex by Arth. Dent,—7 Mar. 1581." Text Luke 13; 5. Again 1583, and 1590, "Imprinted for *him*, and are to bee solde at the Grey-hound in Paules Churchyard, 1560." D 7, in eights. W.H. Licensed. Also in 1600. Sixteens.

1582. "Christian questions & answeres, wherein is set forth the chiefe pointes of Christian religion.—for *him*, 1582." Licensed 1578. Octavo.

1583. "A Commentarie vpon the Epistle—to the Romanes," &c. as p. 1125.

1583. "Praiers vsed by M. John Caluin at the end of his readinge on the prophet Hoseah; translated by John Field. Printed for *him* and Hen. Carre, 1583." Sixteens.

1583. "A treatise of the word of God; written by Anth. Sadeell against the traditions of men; wherein also is set down a methode to dispute diuinely & schoolelike: translated by Iohn Coxe.—for *him*, 1583." Octavo.

1583. "An Answere vnto the Confutation of Iohn Nichols his Recantation, &c. By Dudley Fenner, Minister of Gods word." The flower de luce; on the sides VBIQVE FLORET. "Jmprinted by John Wolfe for *him* & T. Manne, in Pater noster rowe, and are there to be solde, 1583." Dedicated "To—Lord Robert Dudley, Earle of Leicester, &c.---To the Reader." About 100 leaves. W.H. Quarto.

1583. "The Sermons of M. Caluin vpon—Deuteronomie," &c. as p. 1061.

1586. "Dionis Gray, his Storehouse of Breuitie," &c. as p. 880.

1587. Holinshed's Chronicles. See p. 961.

" M. T.

JOHN HARRISON, the elder.

" M. T. Ciceronis orationum volumen primum," &c. as p. 1075.   1587.
" A Treatife named Lucar Appendix," &c. See it p. 1126. Licenfed. 1588.
" A Treatife named Lucarfolace deuided into fovver bookes, which in   1590.
part are collected out of diuerfe authors—and in part deuifed by Cyprian
Lucar Gent." Field's device; the fame as was ufed by T. Vautrollier.
" Imprinted by R. Field for him, and are to be fold at his fhop in Paules
church yard, at the—greyhound, 1590." Dedicated " To—his brother
in law Maifter Wm. Roe Efq; and Alderman of the hon. Citie of London:" whofe arms are on the back of the title-page.---The contents. 167
pages, with fchemes. Licenfed. W.H.   Quarto.
" Sebaft. Verronis Fribugenfis Helvetii, phyficorum libri x. Impen- 1590.
fis J. Harrifon."   Octavo.
" The Spaniards Monarchie, & Leaguers Olygarchie." For him. 1592.
   Quarto.
" An admonition to all fuch as intend hereafter to enter the ftate of 1594.
Matrimonie godly & agreeably to the Laws." Peter Short for him.
See Strype's Life of Abp. Parker, p. 87, &c. Licenfed in 1578.
   Broad fide.
The Rape of Lucrece, a Poem, by Wm. Shakefpeare. R. Field for 1594.
him, 1594. (Again 1598, Peter Short for him, 8vo.) Licenfed. Quarto.
" Phrafes linguae Latinae, ab Aldo Manutio, P. F. confcriptae; nunc 1595.
primum in ordinem abecedarium adductae, et in Anglicum fermonem
conuerfae. Accefsit huc index dictionum Anglicarum, cujus ope quilibet hoc libello quam commodifsime vti poterit.—Impenfis J. Harrifon."
Licenfed.   Octavo.
" Two treatifes: The firft, of the liues of the Popes & their Doc- 1600.
trine; The fecond, of the Maffe :—collected of that which the Doctors,
& ancient Councels, and the Sacred Scripture do teach. Alfo, A Swarme
of falfe Miracles, wherewith Marie de la Vifitacion, Prioreffe de la Annuntiada of Lifbon, deceiued very many; and how fhe was difcouered,
and condemned. Reuel. 17; 1, and 15. The fecond edition in Spanifh
augmented by the Author himfelfe, M. Cyprian Valera, & tranflated—by
Iohn Golburne, 1600. Printed—by *him*, and—fold at the Greyhound in
Pater nofter row, 1600." Dedicated " To—Sir Thomas Egerton, Lord
Keeper &c. Chamberlaine of the Countie Palatine of Chefter; &c.—24
Octo. 1600."---to the Reader.---The author's epiftle.—25 June 1588.
Pages 446. W.H.   Quarto.
" The Beliefe of Hen. Bullinger, conteyning his iudgement on the
Lordes Supper, with an expofition on the 6 Article of the Chriftian
Faith; tranflated by Fr. Shakelton." For him. Licenfed 1578. Octavo.
" Hiftorie of the 12 men—fent to fpie out the land of Canaan, written in the book of Numeri, 13 & 14 Chap. that we being admonifhed by
their example, may the better know how to behaue ourfelues in our iourney, euery man in his calling, fpecially fuch as are purpofely fent: By Iohn
(Phil.) Nicols." See p. 756.   Octavo.
He had alfo licenfes for the following, viz. In 1576, " Colloquiorū
libri tres," as p. 1153. " A notable example of Gods vengeance vpon a
faitheles kinge, quene & hir children. The cōmentaries of J. Caluin
vppon—Efaie," &c. as p. 1153. In 1577, " Adoniari Talei" &c. as p. 882.

6 W   In

JOHN HARRISON, the elder.

In 1578, "The whole Bible in Latin," &c. as p. 1154. "Crispin's Lexicon," as ib. In 1580, "Osorij libri quinq; de Gloria : et Eiusdem, de nobilitate libri v. Methodus cōscribendis epistolis à Geo. Macropedio secundum veram artis rationem tradita. In P. Rami dialecticam," as p. 1154. In 1582, "Gul. Templi—ep'la" &c. as p. 1154. "Johan. Fridgij. pedagogus. Audomari Talei Rhetoricæ, cum cōment. per Cl. Minoem. P. Rami Regij professoris Dialectica; Lib. duo." In 1583, "Triumphus Logicæ Ramæ." In 1585, "A cōpendious Chyrurgerye, by John Canyster:" with T. Man. In 1586, "Ramus Logike, in Latin, & Talleus Rhetorike:" with Wm. Norton. In 1588, "The Psalmes of Dauid in meter," &c. as p. 912. In 1589, " A notable example of Gods Judgement vppon John Chambers gent. in these last daies, teaching vs to walk as becometh the Gospell of God, soundly without hipocrisye. Isocrates three oratyons, in Greeke : ſoe ẏ yt doe not belong to any of the Cumpanye." In 1590, "The Spanish scoolemaster, conteyning 7 dialogues, according to euerie daie in the weeke, & what is necessarie euerie daie to be done, &c. whereunto are ānexed most fine prouerbes & sentences ; as also the Lordes prayer, &c. by Wm. Stepney. M. Tullij. Ciceronis Tuscul. questionū—Libri v." In 1591, "Simonis Veropæi de epistolis Latine conscribendis Lib. v. 'Analisis Logica quinq; postremarū epistolarū Pauli,—vnacum Scholijs &a auctohore Joh. Piscatore : set down to be his copie when it shall come ouer from ẏ marte, vnles it shalbe found to belong to any other person by aucthoritie of ẏ lordes of her Maiesties Counsell." In 1593, "Orcheftra, or a poëm on daūſinge. Venus & Adonis : assigned him from R. Field." In 1594, "Coniuratione Catilinæ, ac bello Jugurthino."

JOHN HARRISON, the younger, stationer,

PROBABLY the son of John Harrison the elder, to whom he was bound apprentice for eight years from Christmas 1561 ; and made free 16 Octo. 1569 ; came on the Livery 6 May 1585 ; chosen Renter 2 Apr. 1599, but paid 5l. for a dispensation. He dwelt at the Golden Anchor in Pater noster Row. See p. 1155, note m.

1580. "Sermons (22) of Maister Iohn Caluin," &c. as p.1118. Licensed.
1580. "A friendly communication," &c. as p. 1011. At the end, "Imprinted—by Tho. East, for Iohn Harison the younger, dwelling in Pater noster Roe, at the signe of the Anker, and are there to be solde, 1580." Licensed.
1581. "Praiers and meditations for the use of priuate families, and sundry other persons according to their diuers states & occasions, by John Field." For him. Licensed. Sixteens.

In 1582 he was fined xijd. for that he and Tho. Man began to print An answer to a confutation concerning Nycolls. It was however licensed to them afterwards.

" A com-

JOHN HARRISON, the younger. 1559

"A comfortable Treatife on 1 Pet. 4; 12—19." &c. as p. 1140. 1582.
"The Compaffe of a Chriftian, directing them that bee toffed in the 1582.
waues of this world vnto Chrift Jefus, by A. P." Octavo.
"Preparation to the way of Life, with a direction into the right vfe of 1583.
the Lords Supper." For him and T. Man. Octavo.
"A verie profitable & neceffarie difcourfe concerning the obferuation 1584.
& keeping of the Sabboth day, &c. Written in Latine by Zach. Vrfinus,—and very newly turned into Englifh by John Stockvvood." Dedicated "To—Ladie Pellam of Laughton, Suffex.—Tunbridge, 20
Oct. 1584." E 2, in eights. W.H. Sixteens.
"An expofition on Hefter, by Iohn Brentius; tranflated by Iohn 1584.
Stockwood." Dedicated, "To—Sir Fr. Walfingham. J. Wolfe for
him. Octavo.
"A Godlie and learned Commentarie vpon the excellent book of 1585.
Solomon, commonly called Ecclefiaftes, or the preacher.—on the which,
yet there hath neuer bin fet forth any expofition in the Englifh tong, in
fuch large & profitable manner. VVritten in Latin by Iohn Serranus, and newly turned into Englifh by Iohn Stockwood, Printed by John
VVindet for *him*, 1585." Dedicated "To—M. Tho. Vane of Bufton,
Efq;—and M. Hen. Vane of Hadlow, Efq;—From the Schoole of Tunbridge. 24 June, 1585."---A recommendatory epiftle "Vnto the godly
Chriftian Reader.—W. Fulke.---The Præface of Iohn Serranus,—conteining the Argument & Contents." E e, in eights. W.H. Octavo.
"An expofition of the 51 Pfalme, by Wolph. Mufculus, tranflated by 1586.
Iohn Stockwood." For him. Sixteens.
"HAGGEVS the Prophet. Where vnto is added a moft plentifull 1586.
commentarie gathered out of the publique Lectures of D. Iohn Iames
Gryneus, &c. Faithfully tranflated out of Latin—by Chr. Fetherftone
ftudent in Diuinitie." The flower de luce. "Printed by Iohn Wolfe
for *him* in Paternofter-row, at the—Golden Anchor, 1586." Dedicated
"To---Iohn Lord Saint Iohn, Baron of Bletfoe.—From my chamber in
Southwarke this 21 July, 1586.---To the Reader---To—the lord Huldrich Fugger, lord of Richberge & Weiffenhorne, the patron of the
learned.—Bafill, 1 Mar. 1581. I. I. Gryneus." 294 pages; on another
leaf, the tranflator's arms. Licenfed. W.H. Octavo.
"THE KINGS MAIESTIE OF SCOTLAND, James the 6. his fruitfull me- 1589.
ditation, containing an expofition or laying open of Reuel. 20; 7—10.
Firft printed in Scottifh, at Edenburgh, by Hen. Chartis, 1588. Since,
printed at London for *him*, 1589." Licenfed. Octavo.
"Barthelmew fairing, fhewing that children ought not to mary with- 1589.
out confent of their parents: wherein is fufficiently proued what in this
point is the office of Fathers, & the dutie of all obedient children. By
John Stockwood, minifter," &c. For him. Licenfed. Octavo.
"THE Shepheards Calender: Conteining twelue Aeglogues propor- 1591.
tionable to the 12 Monethes. Entituled, To the noble & vertuous Gentleman moft worthie of all titles both of learning & chivalry Maifter

6 W 2 Philip

Philip Sidney. Printed by I. Windet for *him*,—in Pater noſter Roe at the ſign of the Anger, 1591." On the back, the author " To his booke." Dedicated " To the moſt excellent & learned, both Orator & Poet, Mr. Gab. Haruey.—E. K. 10 Apr. 1579.---The Argument." 51 leaves, with cuts to every month. W.H.   Quarto.

——   " The Commentaries of M. Jhon. Caluin vpon the firſt Epiſtle of Sainct John, and vpon the Epiſtle of Jude :" &c. as p. 842.

——   " Chriſtian meditations on the 6. 25. & 32 pſalmes, by Phil. Morney: Moreouer a meditation on the 127 pſalme, by P. Pileſſon P. Both tranſlated by John Field." For him.   Sixteens.

He had alſo licenſes for the following, viz. In 1579, " The treatiſe of the faſt." with T. Man. In 1580, " The Teſtimony of a true Church." In 1583, " Iſagoge ad Chriſtianorū theologorū locos cōmunes libri duo, Lamb. Daneo authore; cum prefatione Theo. Beze: provided it be lawfully allowed." In 1586, " Queſt. & Reſp. grāmaticales," ' In 1587, " A book in French, intytuled, Hiſtoire du grand Rouaume de la Chine: to be tranſlated."

※※※※※※※※※※※※※※※※:※※※※※※※※※※※※※※※※

ABEL JEFFES, ſtationer,

WAS apprentice to Hen. Bynneman, and made free 26 Mar. 1580; but ſeems to have been of too deſpicable[p] a character to be taken on the Livery. Mr. Ames has placed a book to him as printed in 1561, which

[p] I find the following minutes concerning him in the Stationers' Regiſter, viz. 7 Aug. 1592, " Whereas Abel Jeffes in July laſt did reſiſt the ſerche whiche Mr. Sterrop, warden, T. Dawſon, & T. Man, renters, were appointed to make,—according to thordonances & decrees; and for that he contemptuouſly proceeded in printing a book without authority, contrary to our Maſter his commaundement, and for that he refuſed to deliuer the barre of his preſſe, neither would deliuer any of the books to be brought to the hall, according to the decrees; and alſo for that he vſed violence to the officer in the ſerch: Yt is now therefore ordered, by a full court, that for his ſaid offence he ſhall be committed to ward, according to the Ordinances," &c. He ſeems to have continued in hold from this time till " 18 Dec. following, when " In full court—Abell Jeffes, according to the direction of the lord Abp. of Cant. his grace, appered and humbly acknowledged his former offences & vndutyfulnes, crauing pardon & favour for the ſame, and promyſing hereafter to lyue as becometh an honeſt man; and to ſhew himſelf obedient & dutifull in the company, & to the ordonances thereof, Abel Jeffes." The ſame day " Yt is ordered, that if the book of Dr. Fauſtus ſhall not be found in the Hall Book ente ed to Rd. Oliff before Abell Jeffes claymed the ſame, which was about May laſt, That then the ſaid copie ſhall remayne to the ſaid Abell his proper copie from the tyme of his firſt clayme." Alſo, " Whereas Edw. White & Abell Jeffes haue each of them offended, viz. E. W. in hauing printed The Spaniſh tragedie belonginge to A. J. And A. J. in having printed The tragedie of Arden of Kent, belonginge to E. W. Yt is agreed that all the books of each impreſſion ſhalbe confiſcated & forfayted, according to thordonances, to thuſe of the poore of the company. Item, yt is agreed that either of them ſhall pay for a fine—10s a pece, preſently or betweene this & our Lady day next. [In the margin, " Solut. xs. per E. White, in May 1593."] And as touching their impriſonment for the ſaid offence, yt is referred ouer to ſome other conuenient tyme, at the diſcrec'on of the Mr. Wardens & Aſſiſtants.

ABEL JEFFES.

which one might suppose to have been taken out of some catalogue, had he not marked it as in his own possession, and rightly queried whether the printer was the same person who printed the subsequent articles; there being a chasm of twenty-three years between it and them: therefore it must have been a mistake[q] of the printer's; probably by setting the figure 6 the wrong way.

"An Introduction to the true arte of Musicke, wherein are set downe exacte & easie rules, &c. whereby any by his owne industrie may shortly, easily & regularly attaine to all such things as to this Arte doe belong, by Wm. Bathe, Student in Oxforde." Quarto. 1584?

"A godlie Dittie to be song for the preseruation of the Queenes most excellent Maiesties raigne.—by him, in the Fore-streete, without Cripplegate." With the tune set in score. Broad side. 1586.

"The Vvhole woorkes of George Gascoigne Esquyre: Newlye compyled into one Volume, That is to say: His Flowers, Hearbes, Weedes, the Fruites of warre, the Comedie called Supposes, the Tragedie of Iocasta, the Steele glasse, the Complaint of Phylomene, the Storie of Ferdinando Ieronimi, and the pleasure at Kenelworth Castle.'—Jmprinted by—dwelling in the Fore Streéte, without Creéple-gate, neere vnto Grubstreét, 1587." See p. 978, and 990. At the end of the princely pleasures, are " Certaine notes of instruction concerning the making of verse or rime in English." W.H. Quarto. 1587.

"Tragical tales by (Geo.) Turberuille, in time of his troubles, out of sundrie Italians, with the argument and lenuoye to eche tale." Mr. Wood supposed this to be the same as his Epitaphs, Epigrams, &c. with additions. See p. 945. Quarto. 1587.

"Thirteene most pleasaunt—questions," &c. as p. 1109. 1587.

"Toxophilvs: The Schoole, or partititions of Shooting contayned in 1589.

Assistants. Item, Abell hath promised to pay the vjd in the li. to those of the poore which he owth for Quintus Curtius." Notwithstanding what had passed, "3 Dec. 1595. Abell Jeffes hath disorderly—printed a lewde booke called The most strange prophecie of Dr. Ciprianus, &c. and diuers other lewde ballads & thinges very offensiue, Yt is therefore ordered—that his presses & letters & other printing stuffe which were seized & brought to the Hall for the said offences, viz. One presse, xij paire of cases, and certen fourmes of letters shall be defaced & made vnseruiceable for printinge."

q I conclude so, not only from my having two editions of the same book dated 1591; but because the author appears to have been only fifteen years of age in 1575, when he was admitted a scholar in Linc. Coll. Oxf. He was lecturer at St. Clement Danes, and much followed, being styled the Silvertongued Smith. Being so popular, the petty printers were eager to get copies of his sermons taken in short hand; but as they were hastily so they were incorrectly printed; of which the author in his preface to this book in Edit. 1591 complains: "Hearing howe fast this book hath been vttered, and yet how miserablye it hath bin abused in Printing, as it were with whole lims cut off at once & cleane left out, I haue taken a little paines (as my sicknesse gaue me leaue) both to perfit the matter, & to correct the print."

r I know not of any other edition of this princely entertainment, made by Robert Earl of Leicester for Queen Elizabeth, on a summer excursion, 1575. Indeed, a letter, giving an account of some part of it, was published that same year, by R. Laneham, of which some account will be given in our Gen. History; besides which, not the least notice appears to have been taken of it by any of our historians till this publication, though Holinshed has given so particular an account of her majesty's summer progress in 1579, to Suffolk and Norwich.

ABEL JEFFES.

in two bookes, Written by Roger Afcham, And now newly perufed,—by the confent of H. Marfh.—" See the former edition, 1545, printed by E. Whitchurch. 64 leaves. " Printed by *him*, dwelling in Phillip Lane, at the—Bell—1589." On the laft leaf is his device of the Bell, with this motto about it, WITH HARPE AND SONGE PRASE THE LORDE. W.H. Quarto.

1589. Afcham's Schoolmafter. See p. 650. 60 leaves. W.H. Quarto.
1589. " Eft natura hominum nouitatis auida. The Scottifh Queens Buriall at Peterborough, vpon Tuefday beeing Lammas day, 1587. London, Printed by A. I. for Edw. Venge, and are to be fold at his fhop without Bifhops-gate." 4 leaves. W.H. Sixteens.
1590. " The Sacke of Roome, Exfequuted by the Emperour Charles armie euen at the Natiuitie of this Spanifh Kinge Philip. Notablie defcribed in a Spanifh Dialogue, with all the Horrible accidents of this Sacke, &c. Tranflated latelie into the English tongue, neuer fitter to bee read, nor deeplier confidered, then euen now at this prefent time. London, Printed by *him* for Roger Ward, 1590." K 3, in fours. Licenfed. W.H. Quarto.
1590. " The knowledge or appearance of the Church, gathered out of the Holy Scriptures, declaring & plainly fhevving, both the Church that cannot but erre, and alfo the Church that cannot erre. &c. VVritten by R. Phinch, and now publifhed in this yeare 1590; for the benefite of all fuch as defire the truth concerning the church. London, Printed by *him*, for Roger Ward,—at the—Purffe in the Little Old-baily, 1590." G 1, in fours. W.H. Quarto.
1590. " The Benefite of Contentation by H. Smith, Taken by Characterie, and examined after. London, Printed by *him*, for R. Warde, 1590." 16 leaves. W.H. Sixteens.
1591. " The Benefite of Contentation : Newly examined & corrected by the Author." His device in a compartment divided into four parts; 1 The queen's arms; 2 The city arms; 3 The Stationers' arms; 4 His rebus, the Bell, with A. I. " London Printed by *him*, 1591." This edition is introduced with a fhort epiftle by the author, complaining of the former edition, &c. W.H. This is the book of which Mr. Ames had a copy, dated 1561. See note q, p. 1109. Another edition, with a colophon, " Printed at London by *him*, dwelling in Paules church yard, at the great North doore of Paules, 1591." Sixteens.
1591. " The Wedding Garment. Rom. 13; 14. Put yee on the Lorde Iefus Chrift. By H. Smith. At London printed 1591." This has not Jeffes's name, but has the fame ornamental pieces as the former article. On the back is a fhort epiftle, " To controll thofe falfe copies," &c. W.H. Sixteens.
1591. " A Geometrical Practical Treatize, named Pantometria," &c. as p. 970. " Lately reuiewed by the author himfelfe, and augmented with fundrie Additions, Diffinitions, &c. to prepare a way to the vnderftanding of his Treatize of Martiall Pyrotechnie, and great Artillerie hereafter to be publifhed." The cut of a man taking the diftance of a fhip at fea,

ABEL JEFFES

sea, from a castle; the same is at Chap. 21. " At London Printed by *him*, Anno 1591." The additions begin at p. 175 and are continued to p. 195. On the back of which is the author's coat armour. W.H. Folio.

The history of Quintus Curtius. See p. 813. Quarto. 1592.

" EVPHVES SHADOW, THE Battaile of the Sences. Wherein youthfull 1592. folly is set downe in his right figure, and vaine fancies are prooued to produce many offences. Hereunto is annexed the Deafe mans Dialogue, contayning Philamis Athanatos: fit for all sortes to peruse, and the better sort to practise. By T. L. Gent." Device, an anchor held by a hand\ from the clouds: behind the anchor are a kind of brackets, in the form of cornucopiæ, crossed; on the top of one is a child's head, on the other a skull; and two sprigs of laurel crossed also with them. " London Printed by *him* for Iohn Busbie, and are to be sould at his shop in Paules Churchyard, neere to the West doore of Paules, 1592." Dedicated " To—Robert Ratcliffe, Viscount Fitzwaters.—Rob. Greene, Norfolciensis.---To the Gentlemen Readers.---Philautus to his sonnes liuing at the Courte." N, in fours. Some verses interspersed. W.H. Quarto.

" Pierce Penilesse his supplication to the diuell. Barbaria grandis 1592. habere nihil. Written by Tho. Nash, gent." See another edition, p. 1049. Quarto.

" THE Famous Chronicle of king Edward the first, sirnamed Long- 1593. shankes, with his returne from the holy land: Also the Life of LLeuellen, rebell in Wales: Lastly, the sinking of Queene Elinor, who sunck at Charing crosse, and rose again at Potters-hith, now named Queenehith." The same device as to Euphues Shadow. " London Printed by *him*, and are to be solde by Wm. Barley, at his shop in Gratious streete, 1593." I. in fours. At the end, " Yours, George Peele, Maister of Artes in Oxenford." Licensed. George Mason, Esq; Quarto.

" A most strange and wonderfull prophisie vpon this troublesome 1595. world. Calculated by the famous doctor in astrologie, maister John Cypriano, conferred with the judgements of James Marchecelsus and sinnior Guivardo," &c. See note p, p. 1161. Quarto.

" Articuli per archiepiscopos, episcopos, et reliquum clericum," &c. as under C. Barker.* Quarto.

He had also licenses for the following, viz. In 1584, " A book concerninge the choise of frendes. A psalme to be songe as a thankesgyuinge on the xvij of Nouember 1584, for the Queenes happie reigne, &c." In 1586, " A book of pretie conceiptes, &c. to be printed for the Companie. A sackfull of newes: being an old copie, which Edw. White is ordered to have printed by Abell Jeffes." In 1589, " By consent of Rog. Ward, The Chaos of Histories. Newes from Nynnynghem, of Skynkes farewell to England." In 1591, " Newes from Lisbone, &c. The hon. accions of y most worthy gent. Edw. Gleinham of Benhall in Suff. Esq; with his most valiant conquestes ag. the Spaniards." In 1591, " The treasons of Geo. Bysley, als'. Parsey, & Mountford, seminarie prestes, whoe suffered in Flete streete 1 Julye, 1591." In 1592, " The second part

part of The Defiance to Fortune. The 1, 3 and 4th partes of Gerillion:" T. Scarlet had licenfe for the 2d part. Chaucers workes: to print for the company. The Spanifhe tragedie of Don Horatio & Bellmipera." In 1595, " Sainct Peter's teares." Alfo, for feveral ballads.

✤✤✤✤✤✤✤✤✤✤✤✤✤✤✤✤✤✤✤✤✤✤✤✤✤✤✤

THOMAS SCARLET, ftationer,

THE fon of Wm. Scarlet, of Wardon, Heref. Yeoman, was bound to T. Eaft for eight years, from L. Day 1577; made free, 12 Octob. 1586. For keeping an apprentice feven years, and never prefenting him at the hall, nor binding by order of the company, to the manifeft breach of divers of the ordinances, 12 Dec. 1593, he was fined 32s. 8d. He ufed for a device, a kite, or fome bird of prey, with a fmall bird in his talons, flying over a caftle on a hill; the fun in full blaze; in a compartment with this motto, SIC CREDE.

1590. " A Plaine Confutation of a treatife of Brovvnifme, publifhed by fome of that Faction, Entituled: A defcription of the vifible Church.— Whereunto is annexed an anfwere vnto two other Pamphlets, by the faid Factioners latelie difperfed of certaine conferences had with fome of them in prifon.—There is alfo added a fhort anfvvere vnto fuch argumentes as they haue vfed to proue the Church of England not to be the Church of God. London, Printed by Tho. Scarlet for Wm. Wright, 1590." Dedicated " To—Sir Thomas Henedge Knight, &c.—R. Alifon.---To the Reader." 129 pages. Licenfed. W.H.   Quarto.

1591. " THE Pride of King Nabuchadnezzer. Dan. 4; 26, 27. By Henrie Smith." His device. " Printed by *him*, 1591." Without Wm. Wright's name. Another edition, with running titles. " Printed by *him* and are to be fold by Wm. Wright, 1591." This edition has an epiftle prefixed by W. W. acquainting the reader that he had caufed thefe fermons to be examined by the beft copies, & to be corrected accordingly. W.H. Sixteens.

1591. " THE FALL OF King Nabuchadnezzer. Dan. 4; 28, 29, 30. By Henrie Smith." His device. " Printed by *him*, 1591." W.H.   Sixteens.

1591. " THE Reftitution of King Nabuchadnezzer. Dan. 4; 31—34. By Henrie Smith." His device. " Printed by *him*, 1591." At the end, " —fold by Wm. Wright." Licenfed. W.H. 2 editions. Sixteens.

1591. " THE CHRIftians Sacrifice. Seene & allowed. At London Printed* for Tho. Man,—1591." This fermon is introduced with a pathetic and affectionate 'addrefs " To my late auditors, the Congregation of Clement Danes, all the good I can wifh." W.H.—H. Smith.   Sixteens.

---

* This has no printer's name, but the title-page has one of Scarlet's head-pieces.
† The pious author appears to have been now unable to preach, by ficknefs; probably his laft, as this fame year we find fome more of his fermons printed by R. Field, " Perufed by the author before his death." Confequently he departed this life about the 32d year of his age. See note q, p. 1161.

" NEVVES

THOMAS SCARLET.

"NEVVES from France: VVhere Monſieur de Signiers in the Kings (1591.) behalfe, moſt brauely diſcomfited the Armie of the King of Spaine & the Pope,—18 Sept. 1591. With ſome notes & newes from Deruente in Holland.—Printed by *him*, for Wm. Wright." 5 pages. W.H. Quarto.

" Thirteene Sermons—Containing neceſſarie & profitable doctrine, 1592. as well for the reformation of our liues, as for comfort, &c. By Henrie Smith." His device. " Printed for T. Man,—1592." Dedicated " To —Sir Wm. Cicell Knight, &c.—T. M.;—To the Reader." 166 leaves. W.H. Octavo.

" Satans compaſſing the Earth. By Henrie Smith.—1592." W.H. 1592. Octavo.

" A Sermon preached vpon 1 Cor. 10; 12. By H. Smith.—1592." 1592. W.H. Octavo.

" MARIES CHOISE. By H. Smith: With prayers written by the ſame 1592. author." His device; larger. " Printed by *him*, for Cuthbert Burby, 1592." Octavo.

" The ſinfull mans ſearch: or ſeeking of God. Preached by H. 1592. Smith, and publiſhed according to a true corrected Copie, ſent by the author to an Hon. Ladie." His larger device. " Printed for C. Burby, 1592." W.H. Octavo.

" A CONFVTATION of the Popiſh Tranſubſtantiation: Together with 1592. a narration how that the Maſſe was at ſundrie times patched & peeced by ſundrie popes.—Tranſlated out of French—by Pet. Allibond.— Printed at London by *him*, for T. Man, 1592." C 5, in eights. W.H. Sixteens.

" MIDAS. Plaied before the Queenes Maieſtie vpon Tvvelfe Day at 1592. night. By the Children of Paules." His ſmall device. " Printed by *him*, for I. B. and are to be ſold in Paules Churchyard at the—Bible, 1592." W.H. Quarto.

" Tancred and Giſmund, a tragedy acted before the queen at the Inner 1592. Temple, in 1568. Newlie reuiued and poliſhed according to the decorum of theſe daies, by R. W." *(Rob. Wilmot)* See Hiſt. Eng: Poetry, III; p. 376, note g. Quarto.

" A notable Diſcouery of Cooſenage. Now daily practiſed by ſundry 1592. lewd perſons called Connie-Catchers, and Croſſe-byters: Plainely laying open thoſe pernitious ſleights that hath brought many ignorant men to confuſion. Written for the general Benefit of all Gentlemen, Citizens, Aprentices, Countrey Farmers & Yeomen that may hap to fall into the Company of ſuch cooſening companions. With a delightfull diſcourſe of Colliers. Naſcimur pro patria. By R. Green, Maiſter of Arts. London Printed by *him*, for Tho. Nelſon, 1592." 4 ſheets. Geo. Steevens Eſq; See the 2d part by J. Wolfe. Quarto.

" Fruitfull Leſſons vpon the Paſſion, Buriall, Reſurrection, Aſcenſion, 1593. and the ſending of the holy Ghoſt: Gathered out of the foure Euangeliſts: with a plain expoſition of the ſame. By Miles Couerdall. John 14; 6. I am the waie," &c. His large device. " London Printed by *him*, 1593." Q q 3, in fours, beſides the preface. Licenſed. W.H. Mr.

6 X

Mr. Ames has given an edition in 1576, but this printer was not then bound apprentice. Quarto.

1593. "A Penitent mans Prayers. By P. R." Sixteens.

1594. "The vnfortunate traueller: or the life of Jack Wilson. By Thomas Nash. Printed—for G. *(query John)* Busby." Quarto.

1595. "The Estate of English Fugitiues vnder the king of Spaine & his ministers." &c. His small device. "Printed for Iohn Drawater, and are to be solde at his shop in Canon lane neere Powles, 1595." The author's preface "To the Reader." and "The copie of a Letter," &c. that was sent with the narrative. S 3, in fours. W.H. Again 1596, pages 136. Quarto.

1596. "A Christian familiar Comfort and Incouragement vnto all English Subiects not to dismaie at the Spanish threats.—an admonition to all English Papists, who openly or couertly couet a change.—With requisite praiers to almightie God for the preseruation of our Queene & Countrie. By the most vnworthie I. N." His large device. "Printed at London for I. B. 1596." An epistle to the queen by Iohn Norden; another to his "Christian countrymen." 70 pages. W.H. Quarto.

1596. "A Preparatiue to platting of Landes & Tenements for surueigh.—Patched vp as plainly together, as boldly offered to the curteous view & regard of all worthie Gentlemen, louers of skill, And published instead of his flying papers, which cannot abide the pasting to poastes. London, Printed by *him*, 1596." At the end, "Radolph Agas," the author. Above this is the following direction, "—— at the Flower de Luce, ouer against the *ca*stle without Fleetbridge." On 20 pages. W.H. Quarto.

He had also licenses to print the following, viz. In 1590, "Les devises Heroiques de M. Claude Paradin Chanoi ne de Beauieu, &c. Le Theatre des Engins, &c. *Both* to be printed in English. A newe Copie booke conteyning theis handes,—Eng. and Fr. Secretaire, with the Italian, Roman, Chancerie & Courtehandes; Spanishe, Jerman & Dutch handes. Sōmarie discqurs au vray de ce qui est aduenneu L'armee du Roy tres christiē, depuis que le duc de Parma sest ioynt a celle des enemies Jusques au 17 du Sept. enuoyez par sa maiestie au M. de Beauar. In English." In 1591, "The thirde & last parte of Cōnye Catchinge, with the newe deuysed knayshe Arte of foole takinge." In 1592, "Le second Liure de la plaisante & delectable historie de Gerileon Angleterr: To be translated into English. The myrrour of Alkamye of Rog. Bacon. Also, Of the wonderfull power of Art & Nature; wherein is treated of the Philosophers stone: Both to be translated out of French, into English."

* I cannot tell whether this is the residence of the author, or the printer; part of the last two leaves of my copy being burnt, by which accident the former part of this direction is lost. What is extraordinary, we do not find the printer's place of abode to any of his performances. Mr. Ames calls the author a relation of Edw. Aggas, the printer; he appears however to have been a famous topographer, who made plans of Oxford, Cambridge, London, &c. of which see Brit. Topog. Vol. I; p. 209, 744. Vol. II. p. 96, 141, 249.

EDWARD

EDWARD AGGAS, or AGAS, stationer,

THE son of Rob. Agas of Stoke-naylonde, Suffolk, yeoman, was apprentice to Humf. Toy for nine years from Easter 1564; and probably made free in the time of the chasm in the Stationers' Register. He is said to have been a scholar: we find no other instances of it than, if the initials E. A. may be applied to him, his translating some French books. He was rather a bookseller than a printer, and dwelt at the sign of the red dragon, the West end of St. Paul's church-yard. The device he used was a wyvern rising out of a ducal coronet, the crest of George earl of Cumberland, baron Clifford, &c. Probably he was related to Ralph Agas, of whom see p. 1166.

"The Defence of Death." &c. as p. 891. Licensed. 1576.

"Politiqye discourses, treating of the differences and inequalities of Vocations, as well Publique as Priuate: with the scopes or endes wherevnto they are directed. Translated out of French, by Ægremont Ratcliffe Esquire. Suas habet Respublica ligaturas." His device. " Imprinted —for him, 1578." Dedicated " To—Sir Francis Walsingham Knight, &c.---To the—most Christian King of Fraunce, Charles the ninth," &c. The whole 81 leaves. Licensed. W.H. Quarto. 1578.

"A sermon at Pauls (cross) by Wm. Fisher, on Math. ix. 11—13, Printed for Tho. Chard & him, 1580." Licensed. Octavo. 1580.

"A Letter intercepted from R. H.* one of Brownes faction discouering in part his great dislikings of the said Brownes schismatical practises. —1583." Octavo. 1583.

"A Declaration & Exhortation to Princes. Remonstrance & Exhortation aux Princes. Viz. Aux Princes Christiens a donner secours a l'Eglise de Dieu & Royalme de France. In French and English, par Pierre Erondell natife de Normandie. A Londres chez Edoard Aggas, 1586." Octavo. 1586.

"Three Letters written by the King of Nauarre, first Prince of the bloud, and chiefe Peere of France, to the States of the Cleargie, Nobilitie, & third Estate of France, &c. All faithfully translated out of French. London, Edw. Aggas, 1586." Octavo. 1586.

"A true Report of the afflictions of Margaret Harrison, in Norfolk, by an euil spirit. And the conference between Mr. Robinson, minister & the spirit.—1586." Sixteens. 1586.

"The Politicke and Militarie Discourses of the Lord De la Novve. VVhereunto are adioined certaine obseruations—of things happened during the three late ciuill warres of France. &c. All faithfully translated out of French by E. A. Printed for T. C. and E. A. by Tho. Orwin, 1587." Dedicated " To—his verie good Lorde, George Earle of Cumberland, &c. E. A.---To the King of Nauarre.—Lausanna, 1 Aprill, 1587.—De 1587.

---

* In " A parte of a register," &c. A collection of Puritan tracts, is " Master R. H. his letter to the B. of Norwich, 1576."

Fresnes."

EDWARD AGGAS.

Frefnes."—The Table of contents. 458 pages. His device. "Imprinted—for Tho. Cadman and Edw. Aggas. 1588." W.H. Quarto.

1587. "A treatife againft the feare of death; and of the refurrection; written P.M. I.D. L.E. &c. tranflated from the French by Lifle Caury, Gent.—1587." Licenfed. Alfo without date. Sixteens.

1588. "The hiftorye of Aurelio and Ifabella, daughter of the king of Scotts. In Italian, Spanifh, French & Englifh, 1588." Licenfed. See Hift. of Eng. Poetry. III; 477, &c.

1588. "Analifis on a parte of the fecond Chapt. of Saint James, from verfe 14 to the end; with a brief confutation of the Khemiftes annotations thereupon written. By John Morgan—1588." Octavo.

1588. "Patr. Galoway his Catechifme.—for him & John Bowen." Octavo.

1588. "A Difcourfe of a greate and furious Battaile fought neere to Cracouia in Pologne, the 25 of December laft, betweene Maxamilian, Archduke of Auftrich, the Emperours Brother, and Sigifmund, fonne to the King of Sweden, each pretending to be the elect King of Pologne. Tranflated out of the French." His device. "Printed by Tho. Orwin for him, 1588." Licenfed. Sixteens.

1588. "A difcourfe vpon the prefent eftate of France. Togither with a copie of the kings letters patents, declaring his mind after his departure out of Paris. Whereunto is added, the copie of the two letters, written by the Duke of Guize. Tranflated out of French into Englifh: And now newly reprinted, and corrected by E. Aggas." 67 pages. W.H. Quarto.

1589. "A fermon, Preached at Reyfham,—Norff. 22 Sept. 1588. And eftfoones at requeft publifhed by R. H. minifter of Gods worde.—John Wolfe for him, 1589." Dedicated "To—Edmund,—Byfhop of Norwich. —Rob. Humfton." Text Hab. 3; 3. On 28 leaves. Brit. Mufeum. Octavo.

1591. "A Sermon vpon the three laft verfes of the firft Chapter of Iob: By Henry Peacham," &c. as p. 1047.

1591. "A fermon preached at Pauls croffe, 17 Nouember 1590,—commonly called the Queenes day, by John Duport, D. D." Text Pf. 118; 24. John Wolf for him. Licenfed. Octavo.

1592. "A fermon at Paules (crofs) by Wm. Fifher, on Malach. 3; 16, 17." Octavo.

1592. "A fermon by Rob. Temple on 1 Cor. 14; 1." Octavo.

1594. "The Jefuite difplayed, containing the original and proceedings of the Jefuites, together with the fruites of their doctrine, openly difcourfed in an oration made in the Parliament houfe of Paris, by one Maifter Pafquier, in that action aduocate for the Vniuerfitie againft the Jefuites plantifes in that Court. Faithfully tranflated out of French by E. A." Licenfed. Quarto.

—— "An oration conteyning an expoftulation, as well with the queenes highneffe faithfull fubiects for their want of due confideration of Gods bleffings, enioyed by means of her maieftie, as alfo with the vnnatural Englifh for their difloyaltie, and vnkindnefs towards the fame foueraygne.

At

EDWARD AGGAS.

At the first pronounced vpon the queens maiesties birthday, in the Guildhall of the burrowe of New Windsore, by Edw. Hake of Grayes Inne gent. then mayor of the same burrowe, and now newly imprinted 17 Nov. xxx of her reign. Spoke 10th Aug. 1586." Licensed. Sixteens.

" The true report of such occurrences, as fell out at Marseiles the 8, 9, 10, dayes of April."* Sixteens.

" A declaration and protestation, published by the king of Nauarre, prince of Conde, and the lord duke of Montmorency, concerning the peace concluded with the house of Lorrayne, to the preiudice of France, &c. 10 Aug. 1585. The King of Nauarres Letters to the Parliament, & Doctors of the Sorbonne, and Lord Mornay's epistle to the French King. Translated out of French."* Twelves.

" The Ladder to Paradise, to be read of euery Christian that is willing to tread the steps which leade to heauen." See page 1121. Octavo.

" A catholick Apologie against the Libels, Declarations, Aduices and Consultations, made, written & published by those of the League perturbers of the Estate of France. By E. D. L. I. C." Octavo.

" Fiue diuine branches, springing in the Garden of vertue. By Ioseph Caldwell." Octavo.

He had also licenses to print the following, viz. 1577, " A remembraunce of the well employed life & godlie ende of Geo. Gascoign esq; who deceased at Stamford in Lincolneshire, the vij day of Octob. 1577, reported by Geo. Whetstons gent." In 1579, " Certen briefe anotations of the lowe countrey affairs from the yere 1566, till the yere 1579. Certen l'res of the prince of Orang: with the protestac'on of the faythfull of Antwerpe, dedycated to the confession of Ausburghe." In 1580, " Bucanani de Iure Regni: in Eng. & Lat." In 1581, " The French kinges edict touchinge y Pacificae'on of the Troubles of his Realmes: published in the court of Parliament at Rouan, 3 Feb. 1581." In 1584, " Vrsins catechism. The consolac'on of y foule vppon y assurance of y forgiuenes of synnes, &c." In 1586, " La Main Chrestienne aux Tomber: to be translated into Eng. An oration latelie pronounced by y Ambassadors of the protestant Princes of Germanye vnto y Frenche kynge: with y kinges answere &c. Translated out of Frenche. The voyage whiche Ant. D'Espeio made in y yere 1583 for y dyscouerye of newe Mexico. Responce a la profession de foy, publiée contre ceux de l'eglise reformée: Auec la refutation, &c. A declaration exhibited to the Fr. king by his court of Parliament. Grene his farewell to follie. Penelopes webbe." In 1587, " The tokens of the children of God: and comfortes in their afflictions. A description of y pope, both by the papistes & protestantes. Phidamore his figure of fancye. Lettre d'un Gentilhome Catholique Fransois a Messieurs de la Sorbonne de Paris. An ease for a diseased man. A viewe of vnytie. De L'authorite du Roy, &c. A declarac'on of y Fr. Kinges pleasure. A discourse of y ouerthrow gyuen to the

---

* This ladder consists of four steps; reading, meditation, prayer and contemplation.
" Seeke by reading, and ye shall find, by meditation; knock by prayer, and it shall be opened by contemplation."

popes catholikes at Doiezy & Jametz, by y̌ pater of y̌ dutcheſſe of Bouillon." In 1588, " Ramſies farewell to his late lorde & maſter therle of Leiceſter, whiche departed this worlde at Cor'burye the 4 Sept. 1588." In 1589, " La Courligue a certaines lettres enuoyes a Reynes per vn tigneue ſe diſant Seigneur de la valee du maine, &c. Thaccurrences of ſixe monethes, by way of diſcourſe, concernyng French matters. La manifeſte de la France : to be printed in Eng. The 6 bookes of Politiques by Juſtus Lipſius, to be printed in Eng." In 1590, " Tho. Pearſton his catechiſme: With John Wolfe. Certen tragicall caſes, conteyning Lv hiſtories, with their ſeu'rall declamac'ons, &c. written in French by Alex. Vandenbuſhe; tranſlated by R. A. With John Wolfe. xxviij Chriſtian diſcourſes touching y̌ eſtate of the world, and of the Churche of God. Mr. Beza his ſermon vpon y̌ hiſt. of y̌ Paſſion & Buriall of our Lord Jeſus Chriſt, wrytten by y̌ iiij Euangeliſts: to be tranſlated & prynted in Eng." In 1592, " De la Saincte Philoſophie Liuret, auec pluſieurs autres petites traictez de pietie: to be tranſlated into Eng." In 1593, " Le Threſor et Manuel des Princes: to be tranſlated into Eng." In 1594, " A diſcourſe of eſtate vpon y̌ late hurt of y̌ Fr. kinge. A godly & learned treatiſe, declaring y̌ waye & meanes to attayne vnto y̌ very ſtaye & perfect quietnes of y̌ mynde, &c."

JOHN WOLF, or WOLFE, city printer,

WAS apprentice with John Day for ten years from Lady-day 1562; and if he was made free of the Stationers' company, it muſt have been during the time of the chaſm in their regiſter. Mr. Strype in his edition of Stow's Survey calls him " a Fiſhmonger uſing printing;" but this (with ſubmiſſion) ſeems to be a miſtake, according to the Stationers' regiſter.[y] It is ſaid alſo of Wolf, that in a conteſt between the patentees and the Sationers' company, " taking upon him as a captain in this cauſe, was content with no agreement, but generally affirmed, that he might & would print any lawfull book, notwithſtanding any commandment of the queen," &c. See B. v. Ch. xiv. But however Wolf might at firſt behave, nothing appears in the Stationers' regiſter, in its preſent ſtate, but what evinces him to have been a very orderly member,[z] whatever the

part

---

[y] " 9 Apr. 1587. As Tim. Rider the Bedell by reaſon of his infirmities cannot execute the buſines of his office, John Wolf, a brother & freeman of this company, is choſen & admitted to the ſaid office—untill further order to the contrary.—And on the avoidance of the ſaid office,—the ſaid John, being found of good behaviour, &c. ſhall have the preferment thereof before any other." And, 23 July following he was choſen beadle of the company inſtead of Tim. Rider. In 1591, he had his ſalary raiſed from 6 to £10. a year.

[z] His name is not once found on the black liſt, that remains, for either printing or behaving diſorderly. By the quotation given in note y, he appears to have been in ſpecial eſtimation with the Court of Aſſiſtants, of which moſt of the privileged printers were members. His being choſen beadle ſuppoſes him to have been diligent in finding out, and giving intelligence of books diſorderly

part that is miffing might exhibit. He fucceeded Hugh Singleton as printer to the honourable city of London, and John Windet fucceeded him. He dwelt in feveral places, as mentioned on the title-pages of his books; and ufed for his device a flower de luce of different fizes and forms, as defcribed with the books to which they are ufed; but chiefly a fmall plain one, not noticed. Sometimes he ufed the device of other printers; probably when printed for them.

"Sapientifsimi Regis Salomonis Concio De fummo hominis bono, 1579. quam Hæbrei Cohelet, Græci & Latini Ecclefiaften vocant. In Latinam Linguam ab Antonio Corrano Hifpalenfi verfa, & ex eiufdem prælectionibus illuftrata. Accefserunt & notæ; &c.—Londini, per Johannem Wolfium, expenfis ipfius Authoris, 1579." Dedicated to Lord Chancellor Bromley. 381 pages. Octavo.

A fhort introduction to Arithmetick, Again 1590. Octavo. 1581.

"A Defence of the old & true profefsion of Chriftianitie againft the 1581. new counterfeite fect of Iefuites, by Peter Boquine, tranflated by T. G." Octavo.

Vita di Carlo Magno Imperatore. Licenfed. Quarto. 1581.

"Two common places taken out of Andreas Hyperius, a learned 1581. diuine, whereof in the one he fheweth the force, that the fonne, moone, and ftarres haue ouer men, &c. In the other, whether the deuils haue bene the fhewers of magicall artes, &c. Tranflated into Englifh by R. V." (R. Vaux.) See Ath. Oxon. Vol. 1. col. 171. Octavo.

"A fermon at S. Iames, by Bart. Chamberlain, on Heb. 9; 28." 1581. Octavo.

"A fermon at Paules croffe, by Rob. Palmer, on Mal. 1; 2, 3." 1581. Octavo.

"Part of the harmony of king Dauids harp. Conteyning the firft xxi 1582. pfalmes of king Dauid, briefly and learnedly expounded by the reuerend D. Victorinus Strigelius, profeffor in diuinitie in the vniverfity of Lipfia, in Germany. Newly tranflated into Englifh by Rich. Robinfon." Dedicated "To—Ambrofe Dudley, earl of Warwick," &c. In MS. on the title page, "Stringelius wrote two vols. on the book of pfalmes; his chief aim is to apply the pfalms to the reformation, which he wrote as a general directory to the church."* His 2d part, containing 23 pfalms more, tranflated by R. Robinfon, was printed for Abr. Kitfon, in quarto, 1593. Quarto.

"An Anfvveare vnto certaine crabbed queftions pretending a reall 1582. prefence of Chrift in the Sacramente: Latelie propounded by fome fecret Papift, &c. Together with a Difcouerie of the Iefuiticall opinion of Juftification, guilefully vttered by Sherwyne at the time of his execution. Gathered & fet foorth by Peter Whyte, very neceffary &c. Seene & allowed," &c. His device. "Imprinted by him & Hen. Kirkham and

derly printed, &c. He was at the fearch of Waldegrave's houfe for fuch, and doubtlefs very active therein; infomuch that M. Mar- | prelate calls him "John Woolfe (alias Machivill) — moft tormenting executioner of Waldegrave's goods."

are

are to be fold at his fhop at the little north doore of S. Paule." Dedicated "To—Ambrofe Dudley Earle of Warwicke," &c. 44 leaves. The Difcovery has a feparate title-page, with a like colophon as the Anfweare: 39 leaves. At the end, "London Jmprinted by John Wolfe, 1582." W.H. Octavo.

1582. "A Learned and True Affertion of the original, Life, Actes, and Death of the moft Noble, Valiant, and Renoumed Prince Arthure, King of Great Brittaine.—Collected and written of late years in Lattin, by the learned Englifh Antiquarie of worthy memory, John Leyland. Newly tranflated into Englifh by Richard Robinfon citizen of London, 1582." His device: "Jmprinted by *him*, dwelling in Diftaffe Lane, ouer againft the fign of the Caftell, 1582." 47 leaves. George Steevens, Efq; Quarto.

1583. "The Auncient Order, Society, and Unitie Laudable of Prince Arthure, and his knightly Armory of the Round Table: with a Threefold Affertion friendly in fauour & furtherance of Englifh Archery at this day. Tranflated & Collected by R. R. (*Rd. Robinfon*) Pfal. 133; 1, and 4. Imprinted by *him*, &c. 1583." In verfe. M, in fours. George Steevens Efq; Quarto.

1583. "An Anfwere vnto the confutation of Iohn Nichols," &c. as p. 1156.

1583. "A Fruitfull treatife, and full of heauenly confolation; againft the fear of Death: Wherunto are annexed certaine fweete meditations of the Kingdom of Chrift, of life euerlafting, and of the bleffed ftate & felicity of the fame. Gathered by that holy Martyr of God, John Bradford." Twenty-fours.

1583. "A declaration made by the Archbifhop of Collen, vpon the deede of his marriage, fent to the States of his Archbifhoprike. VVith the letter of Pope Gregorie the 13. againft the celebration of the fame marriage, and the Bifhops aunfwer thereunto. According to the coppie Jmprinted at Collen, 1583. London Printed by Iohn VVoolfe, 1583." Dedicated "To—Iohn Bifhoppe of London.—Tho. Delone." C 4, in eights. W.H. Octavo.

1584. "Scipii Gentilis Solymeidos libri duo priores de Torquati Taffi Italicis expreffi. Scipii gentilis in xx Davidis pfalmos Epicae paraphrafes." Quarto.

1584. "Scipii Gentilis Nereus five de Natali Elizabethae illuftris. Philippi Sydnaei filiae." Quarto.

1584. "The fumme of the Conference betwene Rainolds and Hart." See p. 1149.

1584. "A Sermon on the Parable of the Sower, &c. as p. 1145.—for Toby Cooke, 1584." C 4, in eights. W.H. Sixteens.

1585. "The Pfalmes in Mufick of 5 and 6 partes, made by John Cofyn." Octavo.

1585. "La vita di Giulio Agricola fcritta finceriffimamente, Da Cornelio Tacito Suo Genero, et Mefsa in volgare da Giouan Maria Manelli." The arms of the lord Robert Sidney, to whom this book is dedicated. "Londra Nella Stamperia di Gouanni Wolfio, 1585." Pages 48. W.H. Quarto.

Admonition

JOHN WOLF.

"Admonition directed to the Frenchmen. See it in 1587. Octavo. 1585.
"Haggevs the Prophet," &c. as p. 1159. 1586.
"A sermon preached 30 Jan. at Bletsoe, &c. Concerning the doctrine 1586.
of the Sacrament of Christ's body & blood. By Edw. Bulkley, D. D."
Dedicated to lord & lady St. John. Octavo.
"S. Avgvstines Manuell," &c. as p. 682. 1586.
"The Gardeners Labyrinth:" as p. 979, &c. 1586.
"The Shepheards Calender." &c. See it in p. 1159. 1586.
"Tichborns elegie, written with his owne hand in the Tower, before 1586.
his execution." In three six-lined stanzas. See Holinshed, Castrations
p. 1570.
"A sermō against bad spirits of malignitie, malice, & vnmerciful- 1586.
nesse. (*Preached* 15 *Nov.* 1571.) by Th. Banks, on Luke 6; 37, 38."
"The pilgrimage of Princes,* penned out of sundry Greeke and 1586.
Latine aucthours, by Lodowicke Lloid Gent." Quarto.
"Esamine di varii Giudicii de i politici: e della dottrina e de i fatti 1587.
de i protestanti veri, et de i cattolici Romani. Libri quattro. Per 1587.
Gio. Battista Aurellio." Quarto.
"A Christian and wholesom Admonition, Directed to the French- 1587.
men, which are reuolted from true religion, and haue polluted themselues
with the superstition and Idolatrie of Poperie.—in Distaffe Lane, neare
the signe of the Castle." Dedicated "To—M. Henry Neuill Esq; and
Maistresse Anne Neuill his wife—At Maighfield in Suffex, 24 Iune,
1587.—Chr. Fetherstone." The translator; being the first fruits of his
knowledge in the French tongue. There are two sonnets, in six-lined
stanzas, prefixed. I 6, in eights. W.H. Octavo. 1587.
"The Lamentations of Ieremie, in prose and meeter, with apt notes 1587.
to singe them withall: Also Tremelius annotations, by Chr. Fetherstone."
Octavo.
"The Complaint of England." Running title, my copy wanting the 1587.
title-page. Dedicated "To—Sir George Barne Knight, Lord Maior,
&c.—Wm. Lightfoote.---Ad pontificios Apostrophe." J 2, in fours.
W.H+ Octavo.
"The blessednefs of Brytaine, or a celebration of the queenes holy- 1588.
day. By Mr. Maurice Kyffin. Published with authoritie." In verse.
This M. Kyffin translated Andria, the first comedy in Terence; printed
1588, probably by him. Quarto.
"The Holy Bull, And Crusado of Rome: First published by the Holy 1588.
father Gregory the xiii. and afterwards renewed and ratified by Sixtus
the fift: for all those which desire full pardon and indulgence of their
sinnes: and that for a litle money, to weete, for two Spanish Realls, viz.
thirteene pence. Very plainely set forth, and compared with the testimony of the holy scriptures, to the greate benefite and profite of all good

---

*This book was printed also by Wm. Jones, without date; dedicated "To—Maister Christofor Hatton Esquier, Capitaine of the Queenes Maiesties Garde, and Gentleman of her highnesse priuie Chamber;" and had several commendatory verses prefixed. Kkk, in fours. W.H. Quarto.

6 Y   Christians.

JOHN WOLF.

Chriftians. 2 Pet. 2; 18.—Together with a briefe declaration—found in the Armado of Spaine, of the prowde prefumption of the Spaniard: which through the inftigation of the aforefaide Bulle, hath taken in hand the fetting forth the inuincible Army (as they terme it) out of Portingale, towards Englande & the Lowe countries,—Which Armado is come to confufion through the hand of the Almighty. Pf. 2; 15.—Jmprinted firft By Rd. Schilders Printer to the States of Sealand: with confent of the States, Giuen at Middleborrowe, the xii of September, 1588.—And reprinted—by Iohn Wolfe, dwelling in the Stationers Hall, 1588." The whole on 59 pages. W.H. Quarto.

Mr. Ames mentions the fame in Octavo.

1588. "The Hiftorie of the great and mightie kingdome of China, and the fituation thereof: Togither with the great riches, huge Citties, politike gouernement, and rare inuentions in the fame. Tranflated out of Spanifh by R. Parke.—for Edw. White, and are to be fold at the little North doore of Paules, at the—Gun. 1588." Dedicated "To—M. Tho. Candifh Efq;—1 Jan. 1589." Pages 410. W.H. Quarto.

1588. A briefe difcoverie of Dr. Allen's feditious drifts, &c. See p. 921.

1588. "A Caueat for France vpon the prefent Euils that it now fuffereth. Together with the Remedies neceffarie for the fame. Tranflated out of French—by E. Aggas." 29 pages. See Oldys's Catal. of Pamphlets, Nº. 394. Quarto.

1588. "Perimedes the Blacke Smith: A Golden Methode how to vfe the Minde in pleafant and profitable Exercife. Wherein is contained fpeciall Principles fit for the Higheft to Imitate, and the meaneft to put in Practife; how beft to fpend the wearie Winter Nights, or the longeft Summer Euenings, in honeft & delightfull Recreation. Wherein we may learn to auoid Idlenefs and wanton Scurrilitie, which diuers appoint as the End of their Paftime. Herein are interlaced 3 Merrie & Neceffarie Difcourfes fit for our Time. With certain pleafant Hiftories & tragicall Tales, which may breed Delight to all, and offence to none. Omne tulit punctum, qui mifcuit Utile Dulci." Dedicated to Gervis Clifton Efq; by the author Ro. Green. 31 leaves. See Oldys, ibid. Nº. 389. Quarto.

1588. "A Difcourfe vpon the prefent ftate of France." Has no printer's name; only his device, the flower-de-luce feeding, with Vsique Floret on it. "Imprinted at London. 1588." A different tranflation from that by E. Aggas. 98 pages. W.H. Licenfed in Fr. & Eng. Quarto.

1588. "A briefe Treatife, Difcoueringe in fubftance the offence & vngodly practifes of the late 14 Traitors condemned the 26 of Auguft, 1588. With the maner of the execution of 8 of them—on the 28 of Aug. following. Seene & allowed.—for Hen. Carre." Octavo.

1588. "Prepofitus his Practife. A Worke very neceffary to be vfed for the better preferuation of the Health of Man. Wherein are—moft excellent and approued Medicines, &c. With a Table—of euery the Difeafes, and the Remedies for the fame. Tranflated out of Latin—by L. M.—for Edw. White,—1588." O, in fours. Licenfed. W.H. Quarto.

"A true

JOHN WOLF.

"A true Difcourfe of the Armie, which the King of Spaine caufed 1588.
to bee affembled in the Hauen of Lifbon, in the Kingdome of Portugall,
in the yeare 1588, againft England. The which began to go out of
the faid Hauen on the 29 and 30 of May. Tranflated out of French,—
by Daniel Archdeacon. Whereunto is added the verfes, that were printed
in the firft page of the Dutch copy printed at Colen, with anfweres to
them, and to Don Bernardin de Mendozza." The device, a fhip. 70 pages,
befides one leaf of the verfes aforefaid, in Latin. See Oldys's Catal. Nº.
77. Licenfed. W.H.   Octavo.

"The Courtier of Count Baldeffar Caftilio, deuided into foure Bookes." 1588.
See p. 694, 948. In three columns, Englifh, French, and Italian. "Print-
ed for the Cūpany, & vjd in the li. alowed to thufe of ỹ poore out of yt."
Sheets P p. Licenfed. W.H.   Quarto.

"Certaine briefe and fpeciall Inftructions for Gentlemen, merchants, 1589.
ftudents, fouldiers, marriners, &c. Employed in feruices abrode, or anie
way occafioned to conuerfe in the kingdomes & gouernementes of forren
Princes." His device, with UBIQVE FLORET, on it. Dedicated "To the
moft valiant & renowned Knight, Sir Francis Drake, the ornament of his
Country, the terror of the enimie, the Achilles of this age.—24 Ianuarie,
1589—Philip Jones." There are prefixed fome Latin verfes, Lectori;
ad—D. Fr. Dracum; P. J. 16 leaves. Licenfed. W.H. Quarto.

"Directions from the King, to the gouernors of the Prouinces, con- 1589.
cerning the death of the Duke of Guyfe. Togither with the kings letter
to the Lord of Taian. Tranflated out of French—by E. A." 4 leaves.
Licenfed. W.H.   Quarto.

"An Apologie, or Defence of our dayes, againft the vaine murmur- 1589.
ings & complaints of manie: Wherein is plainly proued, that our dayes
are more happie & bleffed than the dayes of our forefathers." Dedicated
by Fr. Trigge "To —Sir Anth. Thorrold Knight, and the vertuous
Ladie Anne Thorrold his wife.—Welborne 15 Nouember, 1589."
Pages 42. W.H.   Quarto.

"A Politike Difcourfe moft excellent for this time prefent: Com- 1589.
pofed by a French Gentleman, againft thofe of the League, which went about
to perfwade the King to breake the Allyance with England, and to con-
firm it with Spaine." Prefixed is an epiftle to the reader by Francis
Marquino, the tranflator. 41 pages, printed 27. W.H. Quarto.

"A LETTER, vvritten by a french Catholicke gentleman. Conteyn- 1589.
ing a briefe aunfwere to the flaunders of a certaine pretended Englifhman." 70 pages. W.H.   Quarto.

"An Admonition giuen by one of the Duke of Sauoyes Councel to 1589.
his Highneffe, Tending to diffwade him from enterprifing againft France.
Tranflated out of French, by E. A." 10 leaves. Licenfed. W.H.
                                                             Quarto.

"THE COVNSELLER, a Treatife of Counfels and Counfellers of Princes, 1589.
written in Spanifh by Bartholomew Phillip, Doctor of the Ciuill and
Cannon lawe. Englifhed by I. T. (*John Thorius*) Graduate in Oxford."
Dedicated "To—M. Iohn Fortefcue Efq; Maifter of her Maiefties great
Garderobe," &c. 191 pages. W.H.   Quarto.

JOHN WOLF.

1589. "A fermon Preached at Reyfham," &c. as p. 1168.
1589. "A book of praiers entituled, Bewtiful Baybufh to fhrowd vs from the fharp fhowers of fin." Licenfed in 1586.
1589. "A BAITE FOR MOMVS, So called Vpon occafion of a Sermon at Bedford iniurioufly traduced by the factious. Now not altered, but augmented. With a briefe Patrocinie of the lawfull vfe of Philofophie in the more ferious & facred ftudie of diuinitie. By Tobie Bland Chaplaine to— Iohn lord Saint Iohn, Baron of Bletfoe," &c. Device, a tun floating on the fea, with a fprig of lillies iffuing from the bung-hole, and a viper twifted about the ftalk, within a compartment having this motto, INVIDIA SIBI ALIAS VENENVM. At the bottom a coat of arms, a bend between fix martlets. Gregory Seaton's rebus. "London Printed by Iohn Wolfe, 1589." On the back is a cut of St. George. An epiftle "To the Reader.—London, Junij 14." Pages 38. Licenfed. W.H.
Quarto.
1589. "A fermon at Paules croffe, 27 April, 1589, by Barth. Chamberlain, D.D. on Amos 3; 6." Licenfed. Octavo.
1589. "The Contre-League and anfwere to certain letters fent to the Maifters of Renes, by one of the League who termeth himfelfe Lord of the valley of Mayne, & gentleman of the late Duke of Guizes traine. Faithfully tranflated—by E. A." 78 pages. Licenfed. Mr. Reed. Quarto.
1589. "The Contre-Guyfe: VVherein is deciphered the pretended title of the Guyfes, and the firft entrie of the faide family into Fraunce, with their ambitious afpiring & pernitious practifes for the obtaining of the French Crowne." M 3, in fours. W.H. Quarto.
1589. "The Declaration of the Lord de la Noüe vpon his taking Armes for the iuft defence of the Townes of Sedan, and Jametz, &c. Truely tranflated,—by A. M." 12 leaves. See Oldys's Catal. N°. 520. Licenfed. Quarto.
1589. "A difcourfe vpon the declaration publifhed by lord de la Noüe." Licenfed, in French and Englifh. Quarto.
1589. "A true difcourfe of the moft happy victories obtained by the French king againft the rebells and enemies and his maiefty, of the taken the fuburbs of Paris by the king." With a map, &c. Tranflated by Luke Wealfh. Dedicated to Robert lord Effex. For him, and Edw. White. Licenfed. Quarto.
1589. "The Letter of Henry the IIII. King of France & Nauarre. Conteyning the death of the late king deceafed, the declaration of his intent, afwell concerning the eftate of Religion, as Politikes. Together with the oth, afwell of Princes of the bloud, as of the Nobilitie & Officers of the Crowne. Both in French & Englifh." A 7, in eights. Sixteens.
1589. A difcourfe of the ancient faith of England, by Chr. Rofdell. Octavo.
1589. "A difplay of Dutie, deckt with fage fayings, pithie fentences, & proper fimilies, by L. Wright. Licenfed. (Quarto.)
1589. "The Hunting of Antichrift, with a caueat to the contentious. By Leonard Wright." 28 pages. Licenfed. Mr. Read. Quarto.
1590. "A friendly Admonition to Martine Marprelate & his mates. By Leonard Wright." Licenfed. Quarto.

"De

" De Furtivis literarum notis vulgo. De Ziferis libri IIII. Joan. 1590.
Baptifta Porta Neopolitano autore." Again 1591. W.H. Licenfed.
Quarto.
" A Myrrour for Martinifts, and all other Schifmatiques which in 1590.
thefe dangerous daies doe breake the godlie vnitie, and difturb the Chrif-
tian peace of the Church. Publifhed by T. T." 34 pages. Quarto.
" Miles Chriftianus, or defence of all neceffarie writings & writers, 1590.
written agaynft an Epiftle prefixed before a Catechifme, made by Miles
Mofse." Licenfed. Quarto.
" A Catechifme, briefely noting our profeffion, exercife & obedience, 1590.
required of vs in this life. By Tho. Pearfton." Licenfed, with Edw.
Aggas. Octavo.
" Sermo de Morte Henrici III." Bibl. Bodl. II; p. 516. c. 2. Quarto. 1590.
" Anti Sixtus, An oration of Pope Sixtus v/ vpon the death of the late 1590.
French King Henrie III. with a confutation, &c. Tranflated out of
Latin by A. P." Licenfed. Quarto.
" The Coppie of the Anti-Spaniard, made at Paris by a Frenchman, 1590.
a Catholique. Wherein is directly proued how the Spanifh King is the
onely caufe of all the troubles in France. Tranflated out of French."
41 pages. Licenfed. W.H. Quarto.
" Newnams Nightcrowe. A bird that breedeth braules in many 1590.
Families & Houfholdes." &c. His device, with VBIQVE FLORET. H, in
fours. Licenfed. W.H. Quarto.
" A fpeciall treatife of Gods Prouidence, and of comforts againft all 1590.
kinde of croffes, &c. With an expofition on the 107 Pfalm. Heerunto
is added an appendix of certaine Sermons, and Queftions, — as they were
vttered and difputed ad Clerum in Cambridge. By P. Baro, D. in Diui.
Englifhed by I. L. *(John Ludham)* Vicar of Wetherffielde." 541 pages.
Alfo without date. Licenfed in 1588. W.H. Octavo.
" A Treatife of confeffion of finnes vnto God. By Iohn de l'Efpine; 1590.
tranflated by Nicholas Becket." Licenfed. Octavo.
" The Sickmans comfort againft death & the Deuill, the lawe & Sinne, 1590.
the wrath & iudgment of God, Tranflated out of French by I. E." Octavo.
" The Treafure of the Soule, wherein we are taught, how in dying to 1590.
finne, we may obtaine the perfect loue of God, our neighbour, &c. trans-
lated out of Spanifh by Adrian Pointz." Licenfed. Sixteens.
" Pepetuall Prognoftication of the—weather—by I. F." Licenfed. 1590.
Octavo.
" The Sergeant Maior, or a Dialogue of the office of a Sergeant Maior, 1590.
written in Spanifhe by the Maifter of the Campe, Francifco Valdes, tranflated by Iohn Thorius." Quarto.
" Green's mourning Garment given him by Repentance at the Fune- 1590.
rals of Love." Licenfed. Quarto.
A Spanifh Grammar, by Anth. de Corro. Licenfed. Quarto. 1590.
" An extract of the Prophecys, entitled Sericum mundi filum, Of 1590.
the ouerthrow of the abominable Idolatry of the Pope, &c." Quarto.
" THE GEOMANCIE of Maifter Chr. Cattan Gent. A Booke no leffe 1591.
pleafant & recreatiue, then a wittie inuention, to knowe all thinges, paft,
prefent,

present, and to come. Whereunto is annexed the wheele of Pythagoras. Translated out of French." 244 pages besides the prefixes. "Printed by *him*, and are to be solde by Edw. White,—1591." Licensed. W.H.
Quarto.

1591. "Il pastor Fido, tragicomedia pastorale di Battista Guarini, al serenes. D. Carlo Emanuele, duca di Sauoia, &c. Dedicata Nelle Reali Nozze di S. A. con la sereniss. Infanta D. Caterina d'Austria. Londra per Giouanni Volfeo, a spese di Giacopo Casteluetri." 298 pages in Italic letter.
Sixteens.

1591. "The Path-way to Penitence. VVith sundry deuout prayers, &c. Luke 13: (24) Striue to enter by the narrow gate," &c. 316 pages. Licensed. W.H.
Twenty fours.

1591. "Praiers for the time of warre, by Ant. Anderson." Sixteens.

1591. "Discourses of Warre and single Combat, Translated out of French by I. Eliot." His device, with VBIQVE FLORET. "Printed by *him*, and are to be solde at his shop right over against the great South doore of Paules, 1591." Dedicated "To—Robart, Earle of Essex, &c.---The Authors Epistle—to the—King of France &c.—From Castel-geloux, 5 Octo. 1590.—B. de Loque." 68 pages. W.H.
Quarto.

1591. "The Ship of Saluation, containing certaine praiers fit to be vsed by all sorts of men, women & children." Licensed.
Sixteens.

1591. "D. SARAVIA. 1. Of the diuerse degrees of the Ministers of the Gospell. 2. Of the honor—due vnto the Priestes & Prelates of the Church. 3. Of Sacrilege, and the punishment thereof.—Iob 8; 8, 9, 10.—Printed by *him*, and are to be sold by Iohn Perin, at the—Angell in Paules Church-yard, 1591." Prefixed are "The contents.---To the Reader,—the translator.---To—John—Archbishop of Canterbury, &c. Sir Chr. Hatton, Knight of the—Garter, &c. As also, Sir Wm. Cecill,—high Treasurer of England, &c.—London 4 Kal. Aprill, 1590. Hadrian Sarauia.--- To the—Ministers of the Church of Christ, &c.---To the Reader.---The Preamble." 240 pages. W.H.
Quarto.

1591. "Le vite delle Donne illustri. Del Regno D'Inghilterra, & del Regno di Scotia, & di quelle, che d'altri paesi ne i due detti Regni sono stato maritate. Doue si contengono tutte le cose degne di memoria da esse, ò da altri per i rispetti loro state operate, tanto di fuori, quanto di dentro de i due Regni. Scritte in lingua Italiana da Petruccio Vbaldino Cittadin Florentino. Londra Appresso Giouanni Volfio, 1591." Dedicated "Alla Ser. et prudentissima Elisabetta, &c.---Proemio.---Aggiunta al Lettore." 117 pages, and a table. Licensed. W.H. Quarto.

1591. "The Pilgrimage to Paradise. Compiled for the direction, comfort, and resolution of Gods poore distressed children, in passing through this irksome wildernesse of temptation & tryall. By Leonard Wright.—Seene & allowed. Printed by *him*, and are to be solde at his shoppe against the broad South doore of Paules, 1591." Dedicated "To— Lord Saint-Iohn, Baron of Bletso.---To the Reader." 58 pages, and a table of the chapters. "Chap. III. Of the miserie of Adams brats, & vanitie of the world." Licensed. W.H.
Quarto.

"A Dis-

"A Difcouery of the great fubtiltie and wonderful wifedome of the 1591.
Italians, whereby they beare fway ouer the moft part of Chriftendome, and
cunninglie behaue themfelues to fetch the Quintefcence out of the peoples
purfes. Difcourfing at large the meanes, howe they profecute & continue the fame: and laft of all, conuenient remedies to preuent all their
pollicies herein." Dedicated " To the moft—inuincible Henrie the
fourth, King of France & Nauarre, and to all the princes of the bloud
royall, and all others Ecclefiafticall perfons, both noble & others affift-
aunt in the affemblie of States: the royall French-man wifheth grace,"
&c. 94 pages and a table of contents. Licenfed. W.H.   Quarto.

" A fermon—at Pauls croffe,—by John Duport, D.D. See it p. 1168. 1591.

" New and fingular patternes and workes of linnen, feruing for paternes 1591.
to make all fortes of lace, edginges, and cut-workes. Newly invented
for the profite and contentment of ladies, gentilwomen, & others, that
are defireous of this art." With great variety of wooden patterns.
Printed with Edward White.   Quarto.

" The Hon. Entertainment gieuen to the Queenes Maieftie in Pro- 1591.
greffe at Eluetham, in Hampfhire, by the—Earle of Hertford, 1591."
Gent. Mag. Vol. 49; p. 81, and 121. Licenfed.   Quarto.

" A fermon preached at Farington by Bart. Chamberlain."   Octavo. 1591.

" The Comforter, wherein are contained many reafons to affure the 1591.
forgiuenes of finnes to the confcience that is troubled with the feeling
thereof, by Iohn Freeman."   Sixteens.

" Short inftructions for Gardening, and Graffing, with diuers Plotts 1591.
& Knots for Gardiners." Licenfed.   Quarto.

" A Fig for the Spaniard or Spanifh Spirits: Wherein are liuelie por- 1591.
traited the damnable Deeds, miferable Murders, & monftrous Maffacres
of the curfed Spaniard. With a true Reherfal of the late Troubles &
troublefome Eftate of Aragon, Catalonia, Valencia & Portugal. Whereunto are annexed Matters of much Marueile, & Caufes of no lefs confequence." Oldys's Catal. N°. 180. Again 1592. See our Gen. Hift.
   Quarto.

" A true declaration of the ftreight fiedge laide to the cytty of Steen- 1592.
wich, and of the fkirmifhes and battailes, which happened on both fides,
very ftrange and aduenturous. Tranflated from the Dutch by I. T." *
   Quarto.

" A Quip for an vpftart Courtier: Or A quaint Difpute between Vel- 1592.
uet breeches and cloth-breeches. Wherein is plainely fet downe the Dif-
orders in all Eftates and Trades.—Imprinted by *him*, and are to bee fold
at his Shop at Poules Chayne, 1592." Dedicated to—Tho. Burnabie
Efq; by Rob. Greene. 48 pages. Licenfed Harl. Mifcel. v; 371. Quarto.

" The furuey, or topographical defcription of France; with a new 1592.
mappe, helping greatly for the furueying of euery particular country,
cittye, fortreffe, river, mountaine, and forreft therein. Traflated from
the French."   Quarto.

" An Excellent difcourfe of the now prefent eftate of France: tranf- 1592.
lated out of French by E. A." Licenfed. The rev. Dr. Lort. Quarto.
   " A Cate-

JOHN WOLF.

1592. "A Catechifme—by M. A." See p. 1028.
1592. "Analyfis Logica epiftolarum Pauli," &c. See p. 1151. Another vol. containing the epiftles ad " Timotheum, Titum, Philemonem, Hebræos," &c. 220 pages. Licenfed. Octavo.
1592. " The fecond and laft part of Conny-catching, With new additions, containing many merry tales of all lawes worth the reading, becaufe they are worthy to be remembred: Difcouering ftrange cunning in Coofnage, which if you reade without laughing Ile giue you my cap for a noble. Mallem non effe quam non prodeffe patriæ. R. G." With a print. For Wm. Wright. 5 fheets. Licenfed. Geo. Steevens Efq; See the firft part, p. 1165. Quarto.
1592. " A patterne of pietie, meete for houfholders, for the better education of their families in the feare of God, by Iohn Parker." Licenfed. Octavo.
1592. " The Poore-mans ftaffe, wherein is declared, how a Chriftian ought to make his laft will & teftament: how to aunfwere the Deuill: how to make anfwere at the iudgment feat of God: By what meanes to come to heauen. By R. B." Licenfed. Octavo.
1592. " The poore-mans teares, a fermon by H. Smith, on Mat. x; 42." Octavo.
1593. " A new letter with notable contents; with a ftraunge fonet intituled. Gordon, or the wonderfull yeare." Quarto.
1593. " CHVRCHYARDS challenge." The device, a flower de luce feeding, within a compartment, with I. W. at the bottom of it. Dedicated to Sir Iohn Wolley, Secretary for the Latin tongue, &c." This collection confifts of 21 pieces. Licenfed. George Mafon Efq; Quarto.
1593. " PHILADELPHVS, or A Defence of Brutes, and the Brutans Hiftory. Written by R. H." Device, a flourifhing palm tree, with ferpents and toads about the root, having this motto, IL VOSTRO MALIGNARE NON GIOVA NVLLA. This was ufed by A. Iflip. " Imprinted—by *him*, 1593." Dedicated to the Earl of Effex, by Richard Harvey. 107 pages. Licenfed. W.H. Quarto.
1593. " The coppy of a Letter written by the Lord of Themines, &c. to the Lord Marfhall Matignon, the Kinges Lieuetenant Generall in Guyenne, concerning the battaile at Villemure, &c. Alfo, A Decree of the Court of Parliament fitting at Chaalons, againft a refcript in forme of a Bull, directed to the Cardinall of Plaifance, & publifhed by the Rebels in Paris, in October laft. Faithfully done into Englifh by E. A. Heereunto are adioyned, the reportes of certaine letters of Newes out of France, and Sauoya." 8 leaves. W.H. Quarto.
1593. " ORTHO-EPIA-GALLICA. Eliots Fruits for the French: Enterlaced vvith a double nevv Inuention, vvhich teacheth to fpeake truely, fpeedily & volubly the French-tongue. Pend for the practife &c. of all Englifh Gentlemen, who will endeuour by their owne paine, ftudie & diligence, to attaine the naturall Accent, the true Pronounciation, the fwift & glib Grace of this Noble, Famous & Courtly Language. Naturâ & Arte." Device. with I. W. Dedicated to Lord Rob. Dudley, in Italian.---To the learned profeffors of the French tongue, in—London.

175 pages. Hereunto are annexed three other dialogues, on 60 pages. Licenfed. W.H. Quarto.

"Beza's "Cordial for a ficke confcience, tranflated by H. Aires." Octavo. 1593.

"Two epiftles by Theo. Beza, tranflated by R. Vaux." Octavo. 1593.

"A fermon at S. Maries, Cambridge, by M. Butler, on Luke 22; 26." 1593. Quarto.

"Moriemini, a fermon preached at Court, by H. B. on Pfal. 82; 6, 7." 1593. Quarto.

"Pierces Supererogation, or a new prayfe of the Old Affe. A pre- 1593. paratiue to certaine larger Difcourfes, intituled Nafhes S. Fame. Gabriell Haruey." Device, a palm-tree, &c. as to Philadelphus. "Imprinted by *him*, 1593." Addreffed "To my very gentle & liberall frendes, M. Barnabe Barnes, M. Iohn Thorius, M. Antony Chewt, and euery fauorable Reader.—London, 16 July, 1593." With fome fonnets, &c. 10 leaves Licenfed. W.H. Quarto.

"A SVRVAY of the pretended Holy Difcipline. Contayning the 1593. beginninges, fucceffe, parts, proceedings, authority, and doctrine of it: with fome of the manifold & materiall repugnances, varieties & vncertainties in that behalfe. Faithfully gathered, by way of hiftoricall narration, out of the bookes & writinges of principall fauourers of that platforme Anno 1593.—1 Tim. 1; 7.—" His device with I. W. "Imprinted —1593." It has prefixed an epiftle "To the Reader," and "The contents," &c. 464 pages, and the errata. Licenfed. W.H. Quarto.

"Refurgendum. A notable fermon eoncerning the Refurrection, 1593. preached not long fince at the Court, by L. S. on Phil. 3; 20, 21." Quarto.

"Daungerous Pofitions and Proceedings, publifhed & practifed within 1593. this Iland of Brytaine, vnder pretence of Reformation, and for the Prefbiteriall Difcipline. Δεινὰ τὰ τῶν τυραννῶν λήμματα. My fonne feare the Lord & the King: &c. Prov. 24; 21. They defpife gouernment, & fpeake euill of them that are in authority, Iude." (8.) Device, a flower de luce, with this motto IN DOMINO CONFIDO, fupported by two cupids kneeling, and I W. on another fcroll. "Imprinted—1593." Prefixed are "An aduertifement to the Reader.---The Contents." Then a collection of fentences from the fcriptures, and the fathers. 183 pages. W.H. Quarto.

"The Aunfwere Of the Lords the Eftates generall of the vnited Pro-(1594.) uinces of the Lowe-Countries, to the Letter of the Archduke of Auftria, heereafter inferted. Together VVith the propofition done in the name of the fayde Arch-duke to the forementioned States, by Otto Hartius, and Jeronimus Coomans, learned in the Lawes. Alfo, the extract of certaine ᵇLetters, written out of the Campe before Groning. Printed

---

ᵇ Thefe letters contain "A true declaration, howe the ftrong Cittie of Groning was forced to yeelde it felfe, through the mightie fiege of the Lords, the generall Eftates of the vnited Prouinces of the Lowe-Countries, vnder the conduct of the excel- lent Prince Morrice, prince of Orrenge, Earle of Naffow, &c. Generall Captaine. Doone the xxii of July, 1594. Firft printed in Dutch at Amfterdam by Ellert de Veere, dwelling vpon the newe Bridge. London, Printed by *him*, 1594."

firft

first at Middleburgh, by Rd. Schilders, Printer to the Estates of the Land and Earledome of Zealand, 1594. London, Printed by John Wolfe." W.H. Quarto.

1595. "VINCENTIO SAVIOLO his Practise. In two Bookes. The first intreating of the vse of the Rapier and Dagger. The second, of Honor and honorable Quarrels." His device, with I. W. "London Printed by *him*, 1595." Dedicated "To—Robert Earle of Essex, &c.---To the Reader." The first book, with cuts, ends on N 4. The second book has a separate title-page, dated 1594. The signatures are continued to M m 3, in fours. W.H. Quarto.

1595. "The Reward of the mercifull, by C. Hailes." Sixteens.

1595. "St. Peters Complaint, with other Poems." See p. 1033. Quarto.

1595. "The Lawves of the Market." The Stationers' Arms. "Imprinted by *him*, Printer to the honorable Citie of London, 1595." On 11 leaves. W.H. Sixteens.

1595. "The History of the Warres betweene the Turkes & the Persians. Written in Italian by Iohn-Thomas Minadoi; and translated—by Abr. Hartwell. Containing the description of all such matters, as pertaine to the Religion, to the Forces, to the Gouernment, and to the Countries of the kingdome of the Persians. Together with—a new Geographicall Mappe of all those Territories, &c. And last of all—is discoursed, what Cittie it was in the old time, which is now called Tauris," &c. His device, with I. W. "Imprinted—by *him*, 1595." Dedicated "To—Iohn—Archbishop of Canterbury, &c.—At Lambehith, this New-yearesday, 1595.---The Authors Epistle to the Reader." 500 pages, and a table explaining names, &c. My copy wants the map. Licensed in 1588. W.H. Quarto.

1596. "In obitum ornatissimi viri Guilielmi Witakeri, Doctoris in Theologie, in Academia Cantabrigiensi, professoris Regii, & in eadem Collegii Sancti Iohannis praefecti, Carmen funebre, Caroli Horni.—Excudebat— 1596." On 33 leaves. Lambeth Library. Quarto.

1596. "The first book of cattel; wherein is shewed the gouernment of oxen, kine, calues, and how to vse bulls, and other cattle to the yoake, and fell; with remidies. The second booke treateth of the gouernment of horses, gathered by L. M. *(Leonard Mascall.)* The third booke intreateth of the ordering of sheep and goates, hogs and dogs; with such remidies to help most diseases, as may chaunce vnto them. Taken forth of learned authors, &c. And are to be sold by John Harison the elder, at the White Grayhound in Pater noster row." Quarto.

1596. "THE DECREE For Tythes to be payed in London. Printed by *him*, Printer to the Hon. Cittie of London, 1596." On 10 leaves. W.H. Octavo.

1596. "THE TREASVRE OF the Soule. Wherin we are taught how in dying to Sin, we may attayne to the perfect loue of God, & our neighbour; and consequently vnto true blessednes & Saluation. Many yeares since written in the Spanish tonge, & now newly translated—By A. P. 1 Tim. 1. (5.)—Gal. 5. (24.)—Printed—1596." Dedicated "To—his good

good Vncle, Maifter Richard Saltonſtall, Alderman of the Citty of London, and his wife Miſtres Suſan Saltonſtall, his louing Aunt—Adrian Pointz.---The Preface of the Authour." 302 pages, in borders. W.H.
Octavo.

"A Spirituall Wedding of the higheſt Kings meſſage. Written firſt 1597. in the high Dutch tongue, and dedicated to the Empreſſe Leonora, now newly tranſlated into Engliſh. Imprinted—1597." Dedicated " To— the lord Byſhop of Durham—John Thorius." 176 pages, in borders. W.H.
Twenty-fours.

"The Charter of Romney Marſh." In Latin and Engliſh. W.H. 1597.
Sixteens.

" A REPORT OF THE KINGDOME of Congo, a Region of Africa, &c. 1597. Drawn out of the writinges & diſcourſes of Odoardo Lopez, a Portingall, by Philippo Pigafetta. Tranſlated out of Italian by Abr. Hartwell" Device, a ſmall flower-de-luce, in a compartment, with this motto, IN DOMINO CONFIDO. " Printed—1597." Dedicated " To—Iohn—Lord Archbiſhop of Canterbury, &c.—From your Graces houſe in Lambehith, 1 Ianuarie, 1597.---To the Reader." After theſe is another title-page, differing only in the device, which is a plain flower-de-luce. 217 pages, and a table; with wood-cuts of habits, &c. and two large copper-plate maps, engraved by Wm. Rogers, and printed by Wolf. W.H. Quarto.

" THE DESCRIPTION of a voyage made by certaine Ships of Holland 1598. into the Eaſt Indies.—Who ſet forth on the ſecond of Aprill 1595, and returned on the 14 of Auguſt, 1597. Tranſlated out of Dutch— by W. P." A ſhip. " Imprinted—1598." Dedicated " To—Sir James Scudamore.—16 Ianuarie 1597.—W. Phillip.—To the Bayliefes, Burghemaiſters & Counſell of—Middleborgh in Zeelande.—19 Octo. 1597.—Barnardt Langhenez." 40 leaves, with maps, and the money of Java, from wood-blocks, neatly cut. W.H.
Quarto.

" IOHN HVIGHEN VAN LINSCHOTEN his Diſcours of Voyages into ẏ 1598. Eaſte & Weſt Indies. Deuided into foure Bookes.—Printer to the Hon. Cittie of London." In a ſuitable compartment; the whole deſigned and engraved on copper-plate by Wm. Rogers, " ciuis Londinenſis." Dedicated " To—Iulius Caeſar, Doctor of the Lawes, Iudge of the High Court of Admiralty, Maſter of the Requeſts—and Maſter of Saint Katherines.—Iohn VVolfe.---To the Reader." The 2d, 3d, and 4th books ſeparate title-pages, " Tranſlated out of Dutch by W. P." The paging continued throughout to p. 462. Illuſtrated with maps, plans, and views, from copper-plates, ſome engraved by the ſaid W. R. and four by Rob. Beckiby, two by R. E. (*Raynold Eſtrak*) Copied from the Dutch. Some, the originals, by Bapt. à Doetechum. My copy has alſo thirty cuts engraved by Ioan. à Doet; but theſe ſeem to have been ſupplied from the Dutch editions. W.H.
Folio.

" A SVRVAY OF LONDON. Containing the Originall, Antiquity, 1598. Increaſe, Modern eſtate, and deſcription of that Citie, written in the yeare 1598, by Iohn Stow Citizen of London. Alſo an Apologie (or defence) againſt the opinion of ſome men, concerning that Citie, the greatneſſe thereof.

JOHN WOLF.

thereof. With an Appendix, containing in Latine, Libellum de situ & nobilitate Londini: Written by Wm. Fitzstephen, in the raigne of Henry the second." His device, with I. W. " Imprinted by *him,* Printer to the hon. Citie of London: And are to be sold at his shop within the Popes head Alley in Lombard street, 1598." Dedicated " To —the Lord Mayor of the Citie of London, to the communaltie, and Citizens of the same.---A Table of the Chapters." 484 pages. W.H. Quarto.

1598. " A true Coppy of the Admonitions sent by the subdued Prouinces to the States of Holland; and the Hollanders answere to the same. Together with the Articles of Peace concluded betweene the high & mighty Princes, Philip—King of Spaine, &c. and Henry the fourth—the most Christian King of France,—1598. First translated out of French into Dutch, and now into English by H.W." This title over a cut applicable to the times. " Imprinted—1598." D, in fours. W.H. Quarto.

1599. " A womans woorth defended against all the men in the world; prouing them to be more perfect, excellent, and absolute in all vertuous actions, than any man of what quality soeuer. Written by one, that hath heard much, seen much, but knows a great deal more." * Quarto.

1599. " A discouery of the fraudulent practises of Iohn Darrel, B. A. in his proceedings concerning The pretended possession and dispossession of Wm. Somers at Nottingham: of Tho. Darling, the boy of Burton at Caldwall: and of Katherine Wright at Mansfield & Whittington: and of his dealings with one Mary Couper at Nottingham, detecting in some sort the deceitfull trade in these latter dayes of casting out Deuils." &c. His device, as to Dangerous Positions. " Imprinted—1599." Introduced with " The Epistle to the Reader.—S. H." (*Sam. Harsonet*) 324 pages, and a table of contents. W.H. Quarto.

1599. " The first part of the life & raigne of King Henrie the IIII. Extending to the end of the first yeare of his raigne. Written by I. H." Device, with I. W. " Imprinted—by *him,* and are to be solde at his shop in Popes head Alley, neere to the Exchange, 1599." Dedicated " Illustrissimo—Roberto Comiti Essexiæ, &c.—I. Haywarde.---A. P. to the Reader." 149 pages. W.H. Quarto.

1599 " Certaine Articles concerning the Statute lately made for the reliefe of the poor, to be executed in London, by the Churchwardens and Ouerseers of euery parish, according to the effect of the same Statute." The Stationers' arms. "—Printer to the Hon. Citie of London, 1599." On eight leaves. W.H. Sixteens.

1599. " The happy entraunce of the high borne Queene of Spaine, the Lady Margarit of Austria, in the renovvned Citty of Ferrara. With feastiuall ceremonies vsed by Pope Clement the eight, in the holy Mariage of their Maiesties. As also in that of the high borne Archduke Albertus of Austria, with the Infanta Isabella Clara eugenia, Sister to the Catholique King of Spaine, Phillip the third. First translated out of Italian after the Coppy printed at Ferrara, allowed by the Magistrates." His device, with I. W. " Imprinted by Iohn Woolfe, and are to be solde—in Popes head Alley,—1599." On four leaves. W.H. Quarto

" The

JOHN WOLF.

" The Speeche which the French King made to the Lords of the Parliament on the fifth of Ianuarie 1599. Faithfully tranflated out of French by H. W.—folde—in Popes head Alley,—1599." On one fheet. Mr. Reed.   Quarto. 1599.

" A briefe relation, of what is hapned fince the laft of Auguft 1598. by comming of the Spanifh campe into the Dukedome of Cleue: and the bordering free Countries, which with moft odious & barbarous crueltie they take as enemies, for the feruice of God, and the King of Spaine (as they fay.) &c. Together with a defcription of the VVhale—which came on fhoare at Berckhey in Holland, 3 Feb. 1598. with anotation therupon. &c.—folde—in Popes head alley,—1599." On 28 pages, with a map of Cleve and Munfter, &c· W.H.   Quarto. 1599.

" The Letter of the Emperour of Germanie to the Admirant of Arragon, Generall for the Archeduke Albertus in—Cleue and Munfter, &c. With the—anfwere. The confpiracie of the 3 Bishops. The death of the Earle of Brooke. &c. Faithfully tranflated out of the Dutch copie at Roterdam." His device, with I. W. "—folde—in Popes head Alley,—1599." Annexed to the former article. 13 pages. W.H. Quarto. 1599.

" A Pageant of Spanifh Humours.—out of Dutch by H. W." Quarto. 1599.

" MASTER BROVGHTONS LETTERS, Efpecially his laft Pamphlet to and againft the Lord Archbifhop of Canterbury, about Sheol and Hades, for the defcent into Hell, anfwered in their kind. Pf. 85. (75; 5. Bifhops Tranflation) I fayd vnto the fooles, Deale not fo madly. &c. Imprinted by him, 1600." Introduced with an epiftle " To the Reader." 45 pages. W.H.   Quarto. 1600.

" Difce mori. Learne to Die. A Religious Difcourfe, moouing euery Chriftian man to enter into a ferious remembrance of his ende.— Efay 38; 1.—Printed by him, 1600." A wood-cut frontifpiece of Adam and Eve. Dedicated " To—Lady Elizabeth Southwell,—Chr. Sutton." A wood cut of the laft judgment.---The Preface. Another cut of a dying man, with this label to a youth, in armour, " As thou art, I once was. As I am, thou fhalt be." Then, " A Copie of a letter fent from Oxford to the Authour.—From L. Col. the 6. of Auguft, 1600. R. K.---The Contents." Printed in lines. Q 6, in twelves. W.H. Twenty-fours. 1600.

" THE Comforter: OR A comfortable Treatife, wherein are contained many reafons taken out of the word, to affure the forgiuenefle of finnes to the confcience that is troubled with the feeling thereof. &c. Written by John Freeman, fometime Minifter of the word, in Lewes in Suffex.--- Printed by him, and are to bee fold by Edw. White, at the little North dore of Paules, at—the Gun, 1600." Dedicated " To—the whole congregation of Lewis.---The Epiftle to the Reader.---The heads of the Arguments drawne into a Table." (Analytical) Half a fheet, folded. I, in twelves. Licenfed in 1590. W.H.   Twenty-fours. 1600.

" A Booke of Fifhing with Hooke & Line, and all other inftruments thereunto belonging. Another of fundrie Engines and Traps to take Polcats, Buzzards, Rats, Mice, and all other kinds of Vermine, and Beafts whatfoever, moft profitable for all Warriners, and fuch as delight in this kind 16:0.

kind of fport & paftime. Made by L. M." *(Leonard Mafcall.)* A woodcut adapted to both fubjects, and under it, " Thefe Treatifes have many wood-cuts, efpecially the latter.—" Printed by *him*, and are to be fold by Edw. White,—1600." The book of fifhing, 50 pages; the other has a feparate title-page, but the pages are continued to p. 92. Licenfed in 1587. Sir John Hawkins. Quarto.
—— " The ΕΚΑΤΟΜΠΑΘΙΑ, or paffionate Centarie of Love." Quarto.
—— " A little bcoke of Secreats, to make Inke and diuers Colours, alfo to write Gold, and Siluer. &c." Licenfed in 1590. Octavo.
—— " Beawtie difhonored, written vnder the title of Shores wife." In fixlined ftanzas. Licenfed in 1592. Quarto.
—— " Alberici Gentilis I. C. profefforis regii de iure belli commentatio prima." Licenfed in 1588.
He had alfo licenfes for the following, viz. In 1580, " Youthes witte, or the witte of greene youthe, by Nic. Atkinfonne. The lyfe of Charles the greate Emperour of Rome, tranflated out of Italian." In 1581, " The Caftle of Curtefy, the holde of Humility, & the Chariot of Chaftity." In 1583, " A difcourfe of the buriall of—Thomas Erle of Suffex." In 1584, " Alberici Gentilis De lege Civili." In 1586, " Another Debora with ỹ king of Navarra." In 1587, " The Maryn's flie. The mourning mufes of Lod. Bryfkett vpon the death of the moft noble Sir Phillip Sydney knight, &c. A difcourfe of Ingratitude, in 2 parts: englifhed by W. F. Conditionū liber; Albericke Gentil author. afwell in Eng. as Fr. Singulares pourtraicts et ouurages de L'ingerie per Fred. de Vinciolo Venitien. Il decamerone di Boccafio: both in Italian & Englifh. Hiftorio de Nicolo Machiauelli Cittadino et Secretario Florentino. The horne A, B, C. Difcours fommaire, de la miraculeufe victoire obtenue par le Roy de Navarre contre ceux de la Ligue le 20 Octo. 1587: in Fr. and Eng. The defcripc'on of Scotland by Petruccio: Ital. & Eng. The genealogie of the kings of England from Will'm the conqueror, in a table with pictures. The oration of Neptarne to Jupiter; in ỹ praife of Q. Elizabeth. A newe yeres guifte cōprehending ap'parition againft ỹ prognofticated daungers of the yere 1588, cōpyled by Tho. Tymme minifter: vpon condic'on, &c. Newes out of Italye. De authoritate Sacre Scripture: vpon Cond. &c. The moft cruel & tyranous murther, cōmitted by a mother in lawe vpon a child of 7 yeres of age, in Weftm. in this yere: vpon cond. &c. An Apologie for Chriftian Soldyors. A praier dayly vfed in Stepney parifhe." In 1588, " A ioyfull fōnet of ỹ redines of the fhires & nobilitie of England to her maiefties fervice. The quenes vifiting the cāpe at Tilberye, & her enterteynment there the 8 and 9 of Aug. 1588. with cōdition. Pfalmes of Invocation vpon God to preferve her Maieftie & the people of this land from ỹ power of our enemies, gathered by Xpofer Stile. A treatife teaching everye Chriftian howe to trye & examine himfelf. Aufcuns Articles propofez per les chefs de la Ligue en l'affemblee de Nancye en Januier 1588 pour eftree anōfez en la generalle de Mars prochain. The martiall fhewes of horfmen before her maieftie at St. James. L'afine d'oro dy Nicolo Macchavelli. Recuel, contenant les chofes memorablos
advennes

JOHN WOLF.

advennes foubs la Ligue, Qui s'eſt faicte et eleuee contra la Religiõ Reformee pour L'abolir: in Eng. The popes bull in Dutche, with ẏ anſwere thereto; to be tranſlated: and ẏ no perſon ſhall print any parte or parcell thereof to his hindrance. Dialogo di Pietro Aretino vel quale ſi parla del graco con moranta Piaceuole. Ragionamento nel quale M. Pietro Aretino figura quattro ſuoi amici che ſanellano delle corti del mondo, e di quella del cielo. Lettre di Pietro Aretino. A letter to Don Bernardin di Mandozza/, with ẏ advertiſements out of Ireland, in Italian:" See p. 1075. " La Harrangue faicte par le Roy Henrye Trois de Fraūce, 16 Octo. 1588. A diſcourſe of ẏ Spaniſh navie by theæaminac'on of Don Diego Prementelli: in Dutch & Eng'. A ſonge wherein is cōteyned ẏ treacherie of ẏ wicked, & is made to be ſong on ẏ coronc'on daye, or at any other time. Roberts his welcome of good will to Capt. Candiſhe. Lettre del Bettye; in Italian. Alcida Grenes metamophoſis. A myrror to all ẏ love to followe ẏ waves. The beſieging of Berghen vppon Zome. A diſcourſe touching ẏ firſt planting of the Xptian faith in this land, &c. A declarac'on made by ẏ Duke of Moopenſier for reuniting the Clergy & Nobylytye of Normandy vnto ẏ kings obedience. A diſcourſe againſt ſeditious preachers. A copie of a letter touching Moūtpenſiers procedings in his gov'nment of Normandy, & thov'throwe of certen Rebells. Newes ſent vnto ẏ ladye Princeſſe of Orange. Tho. Eraſtus de excōmunicatione." In 1589, " Two ſev'ral places of Fr. Guychdyne which have bene lefte out of his old true copies, to be printed in all languages. The drunkards maſſe. A letter ſent to the good townes ẏ are in obedience to the king, &c. in Fr. & Eng. Examen pacifique de la doctrine des Huguenots provant contre les Catholiques rigoureux, &c. in Fr. & Eng. Juſti Lipſii Politiorū libri ſex: in Lat. & Eng. A treatiſe of the ſhortnes of māns life: Salus iure cuiuſcūq; The true garment of a Chriſtian. Refutatio cuiuſdā libelli ſine aucthore cui titulus eſt, De Jure magiſtratū in ſubiectos, & officio ſubditorū erga magiſtratus, aucthore Johāne Boccaraci. Een Schadt der Sielen, &c. Die gheſtelijk Brijloſt, &c. Een profitelick en twoſtelijck Boecpken vanden Gheloone ende hoope wat dat oprechte ghelone is, &c. in Eng. The French kings letter to M. de Verune touching ẏ victorye againſt ẏ rebelles & leagers 14 Mar. 1590. The decree of ẏ courte of p'liment of Normandie for ẏ ſeaſing of ẏ rebelles goods. ij canticles touching ẏ ſaid victorye. Certen prophecies of Pawle Grebner: in Lat. Fr. Dutch & Eng. Reports from Fraūce & Flaūders. Vergiſſ' Nitmein, &c. Sterbens Kunſt, &c. in Eng. Alcoranus Franciſcanorū : in Lat. Fr. & Eng. De quatuor hominis noviſſimis. Englands mourning gowne. The rendring of St. Denys, with ẏ taking of iij traytors in the p'ſence chamber there: the taking of Marcyllia & Grenoble for ẏ kinge." In 1590, " Le procez criminel des Pariſiens Liguers, autheurs de tous les troubles, guerres & calamitez de la France : Et a preſent qu'ils ſe ſentent preſſeq; pris & ſubiugex, demandans cōpoſition : in Fr. & Eng. A letter from ẏ colledge of Sorbonne in Paris to ẏ pope, with the gen'all cōfeſſiou of ẏ pillers of ẏ holly vnion, and certen epigrams. Eſſame de gli Jugegous:

in

JOHN WOLF.

in Ital. & Eng. Certen tragicall cafes" &c. as p. 1170. "Difcours veritable des horribles meurtres & maffacres comis & perpetrez de fang froid par les troupes du duc de Savoy conduicte par dom. Amidee baftard de la dicte Savoy, fur les paures paifans du Bailliage de Gex. &c. The tables & mappes of ÿ Spaniars p'tédid Invafion by fea, together with ÿ defcription thereof by booke &c. in all languages. A Spanifhe grāmer &c. by Tho. D'oyley. Jufti Lipfij adverfus dialogiftā, Liber de vna religione, &c. Les Rodomantades du Capitaine Viques, and a letter from a French gent. to my lady Jaquet Clement. An addic'on to thappendix of ÿ Familiar ftaffe. Les Lauriers du Roy, contre les foudres pratiques par l'Efpagnol: in Fr. & Eng. The fucceffe which fell out in ÿ purfute of ÿ prince of Parma, with a letter of ÿ kings to ÿ m'fhall of Byron, &c. Copie de la Refponce de Meffieurs les eftats generales fur la propoficion a eux faites per Meff. les Ambaffadeurs des Circles d'Allemagne. A true defcripc'on of ÿ fearfull yearthquake—at Vienna in Auftria, 15 Sept. 1590. Bibliotheca Hifpanica, cōtayning a grāmer with a Dictionary, in iij languages, gathered &c. by R. Percivall. Admirandū & ineftimabile opus Steganographiæ à Johāne Tritēno, Abbate Spanhymenfi, &c. La celeftina comedia: in Spanifh. The Judgment of Vrine. An Aūfwere to certen of Mr. Barrows affertions, &c. Newes out of Fraūce, Savoy, Dauphine, Aleppo, Soria & Tripole. A difcourfe vppon a queftion of eftate of this tyme. The method curative of the Venerious difeafe, &c. The phifionimye of dreames. The arte of p'fumitorye. Cookerye for all māner of Dutch vyctuall. To p'ferve wyne from marryng. To meafure all māner of veffels. Epulario, ÿ dreffing—flefhe, byrdes, & fifh, & to make all fortes of fawce, tarts, pafties, &c. Chr. Meffifbugo his booke of Cookerye. The right making Aqua vitæ. Of planting & graffing. Howe to brewe all forts of beire. Alkamye. iij little bnoks of fifhing: to be tranflated from ÿ Dutch. A cōtrov'fie betwene ÿ flees & women. Articles of agreement vpon ÿ yielding of Grenoble, & adv'tifements out of Prouince to ÿ Fr. King. Refponce a la fupplicac'on contre Celux lequel faifāt femblant de donner advis au Roy de fe faire Catholiciveult exciter fes bons fubiects a rebellion. Newes from ÿ Englifhe armye out of Britaine. A Journall of ÿ late fervice in Britaine by ÿ Pr. de Dombes, affifted with her maiefties forces vnder—Sir Jo. Norreis." In 1591, "Godlye prayers—by Anth. Anderfon. A letter cōcerning—ÿ Eng. forces nowe in Fraūce vnder the E. of Effex. Defcription of ÿ armye levyed this yere by ÿ princes of Germanye for Fraūce, &c. A memorial fyt to obferve in mans life. The lamētac'on of ÿ pr. of Parma. Late adv'tifements out of Brytayne, in Sept. 1591. The popes armye cōducted by ÿ Erle Hercules difcōfited, the takinge of St. Efpryte, &c. The difcov'rye of x Eng. Lepers. True reporte of a greate Galie ÿ was broughte to Rochell. A difcours of fuch matter as hath fallen out vpon ÿ ov'throw of ÿ vicoūt Guyerche, &c. The poore mans tears, A harmony from Heaven, A memento for magiftrates: 3 fermons by Hen. Smith. A prophecie for 8 yeres to come. Gargātua his prophefie. The 2d, 3d, 4th, & 5th bookes of Amadis de Gaule: in Eng. A catechifme

chifme—by Mard. Alden. A true relac'on of ỹ Fr. kings good fucceffe againſt ỹ duke of Parma, A certen mountain borninge in ỹ Iſle of Palme 5 or 6 weekes. A Journall of ỹ doinges of both armies from ỹ cominge of ỹ duke of Parma into Fraūce vntill 18 May, 1592. &c. An inſtruction for yong gentlewomen. Adryan Pointz his colleƈtions of Alkamye. A Diƈtionary Geographicall, Aſtronomicall, & Poetical. Philomela ỹ ladye Fitzwalters nightingale, by Ro. Greene." In 1592, "A cōmemoration of ỹ moſt valiant & worthie knight Sir Wm. Sackvill ſlayn in ỹ warres of Fraūce.' Naturall & morall queſtions & aūſwers. A defence of ſhorte haire, &c. The theater of ỹ earth. The garden of Goodwill. Palma Chriſtiana. La fleur de lice, &c. A ſhort anſwere to ỹ reaſons which ỹ popiſh recuſants alege why they will not come to our churches, by Fr. Bonny. Analyſis logica, 7 epiſtolarū apoſtolicarū que Catholice appellari ſolent, Authore Johāne Piſcatore. Parthenophil & Parthenope. A ſhort dialogue concerning ỹ arraignment of certen Caterpillers." In 1593, " The certentye of our Salvac'on. The vnfortunate travellor. Lucans firſt booke of the Ciuill warr betwixt Pompey & Ceſar, engliſhed by Chr. Marlow. Hero & Leander, deviſed by Chr. Marlow. A book ſhewinge—howe muche 1 pound, 2, 3, 4, 5—to 50000 li. after ỹ rate of 6 per Cent. &c. will yield daily,—or yeerely. A letter from Dr. Harvey to J. Wolf. The firſt part of ỹ Mount Calvarie. Procris & Cephalus. Newes of twoo angells ỹ came before ỹ cytie of Droppa, in Sleſia. The newe founde arte of catching of Connyecatchers, or a trap to take a knave. A letter ſent by Amorathe ỹ greate Turke to Chriſtendome. The table of ỹ Tenne Cōmaūdements, with the pictures of Moyſes & Aaron. A bill for ỹ cure of ỹ Fiſtola. Theſoro politico cioe Relatione Jnſtructione Tratato diſcorſi varij damloa ſciadori Nell' Academia Italiana di Colonia l'anno, 1589: in Eng. A letter ſent by ỹ Fr. kinge to Monſ. de Vielliers, vpon ỹ overthrowe gyven to Countie Manſſeild, &c. Michall Renichōn his araynment & execuc'on." In 1594, " The Articles of gyuenge over of Grōing. &c. Thamendment of life,—by John Taffyn. The Sheppherds prattle. The reward of ỹ mercyfull. The eſtate of Chriſtians—vnder ỹ ſubieƈtion of ỹ Turke. Lycanthropia, or Cupids phrenſie." He had licenſes alſo for ſeveral ballads, eſpecially about the time of the Spaniſh invaſion.

※※※※※※※※※※※※※※※※※※※※※※※※※※※

ROGER WARD, ſtationer,

THE ſon of Humphrey Ward, of Drayton, Salop, was apprentice to Tho. Marſh for nine years from Lady-day 1566; and ſeems to have been made free in the time of the chaſm in the company's Regiſter; but he does not appear to have printed any thing, unleſs ballads, &c. for himſelf before 1582. He dwelt then at the Talbot, near Holbourn conduit; in 1589 on Lambeth-hill, near Old-fiſh-ſtreet; in 1590, at the Purſe in the little Old-bailey; and laſtly, in 1595, in Saliſbury Court, at the figne of the Castle In Strype's edition of Stow's Survey of London, p. 223, he is said to have been a contumacious printer, as well as J. Wolfe, and indeed with more propriety, as appears by some extracts from the Stationers' books, inserted below.* He used for his device a pheasant

" In the Wardens account of difburfements for the year from July 1584 to July 1585, is charged, Paid " Hyde for carrying Ward to the counter, 1 s."

On Monday 17 Octo. 1586, " The Wardens, vpon serche of Roger Wards houte dyd find there in printing, a book in verse intytled Englands Albion, beinge in English & not aucthorised to be printed, which he had ben forbidden to prynte, aswell by the L. Archbp. of Canterburye, as also by the Wardens at his own house. Item, they found there in printing The grammar in 8vo. belonging to the privilege of Mr. Fr. Flower. Item certen former readie set of the Catechisme belonginge to Rd Dayes priviledge. And of the prymers belonginge to Wm. Seres privilege, by her maiesties l'res patents. Item, Psalter & callender ready sett. And certen other formes ready sett of other mens copies. And for asmuche as all this he hath done contrary to the late decrees of the hon. court of Starre chamb.' The said Wardens seised iij heapes of the said Englands Albyon; the first leaf of the grammer, in 8°; iij presses & divers other parcells of pryntinge stuffe, by vertue of the said decrees; and accordingly brought the same to Stationers Halle. Whereupon it is now concluded & ordered, according to the said decree, That the said presses & pryntinge stuffe shalbe made vnserviceable, defaced & vsed in all points accordinge to the tenor of the decrees aforesaid."

In 1587—1588, " To Cole an officer of the Abp. of Cant. for Roger Ward, about the presse that was conveyed out of Lothbury, and Southwark spittle, 10 s." Also, " To John Wolf, who rid to Croydon for a warrant for Rog. Ward, 4 s."

In 1589—1590, " About Rog. Ward, and defacing his press & letters, 1 L. 10s. 4d."

July 4, 1590, " Whereas vpon serche lately made by thappointment of the Wardens, it was found that Roger Warde had, contrary to thorder of the Company & decrees of the Starrechamb.' printed The Christian sacrifice, being forbidden by my lords grace of Canterbury, Burtons sermon, & A treatise of a reformed church; and had a form standinge redy to goe to the presse of the iiijth leaf of the Grammer in viij°. And did also kepe & conceale a presse & other printinge stuff in a Taylors house neere adioyninge to his owne; and did hyde his letters in a henhouse neare St. Sepulchres churche, exp'ssely against the decrees of the Starchamb.' All the which stuffe—were brought to Stac'hall, according to the said decrees. And yt is nowe therfore concluded, that all his presses & printinge instruments shalbe defaced & made vnserviceable for printing, according to the said decrees."

" 3 M'cij. 1590-1. Whereas yt is manifestly proved by the testimonie of John Leigh & Tho. Street, that Roger Ward about half a yere past did contrary to the decrees of the Starrechamb.' erect & conceale a printinge presse with other printinge furniture in the house of one Ofield a Tannor by the bankside in Southv'k, & therewith printed The sermon of Repentance, & the grammer in 8, contrary to the decrees. The leaves of which books were fetched wett from the printing house by Mr. Platt, & the said Leigh fetched the sermons so printed from Platts house to Roger Wards house. And the said Leigh confesseth that Anth. Hill & Hen. Jefferson wrought vpon the grammar, and that they did ij daies work vpon a heape, & wrought vpon it about a moneth, & wrought about 8 leaves. And the said Street confesseth that they printed about xj or xij leaves of the grammar, and that all the Latin parte thereof was fully synished, and that they had a newe Vineyat to the 1st leaf. And whereas the said presse with about iij formes of l'res in div'ers sortes, & iij cases with other printing stuff were, 2 M'cij 1590, found at Hammersmyth by Mr. Warden Cawood, Mr. Harrison the elder, & Mr. Watkins, and by them seised & brought to Stac'oners hall, by vertue of the said decrees. Yt is now ordered & decreed in full Court—by force of the said decrees, That the said presse, letters, cases & printing stuff shall p'sently be defaced & made vnf'viceable for printing, according to the decrees aforesaid."

In 1590—1591, " M. Bedells man for the copies of 6 men exam. about Rog. Wards presse taken at Hamersmith; and the copy of Rog. Wards Bond, 6 sh."

February 9, 1595-6, " Forasmuch as Roger Ward hath of late erected twoo presses, in sev'rall obscure places, for printing pheafant couchant on a wreath, in a compartment with R. and W. on the fides.

"A Discovrse, Concerning two diuine Pofitions. The firft effec- 1582. tually concluding, that the foules of the faithfull fathers deceafed before Chrift, went immediately to heauen. The second fufficientlye fetting foorth vnto vs Chriftians, what we are to conceiue, touching the defcenfion of our Sauiour Chrift into Hell: Publiquely difputed at a Commencement in Cambridge, Anno Domini 1552. Purpofely written at the firft by way of a confutation, againft a Booke of R. Smith of Oxford, D. D. entituled a Refutation, imprinted 1562, & publifhed againft John Caluin & C. Carlile. And now firft publifhed by Chr. Carlile, 1582.— Imprinted—by Roger Ward, dwelling by Holborne conduit, at the figne of the Talbot, Anno 1582." Dedicated "To—Henry Earle of Huntington, &c.—13 May 1582.---To the Reader." Then fome verfes by Sir John Cheek, &c. At the end is a translation in verfe of "The beliefe of a Chriftian, called The Comon Credi, as it is repeated in Auguftine ad Petrum Diaconum," &c. 174 leaves, and a table. W.H. Sixteens.

"A Difcourfe of Peters Lyfe, Peregrination and Death. Wherein is 1582. plainly proued by the order of time and place, that Peter was neuer at Rome. With a confutation of fuch coniectures as are alledged to the contrary. Furthermore," &c. as p. 878. In the compartment to Cranmer's Defence of the Sacrament, p. 600. Dedicated "To—his fingular good Lorde, Sir Thomas Wentworth knight, Lorde Wentworth.—by whom I have bene liberally fuftained thefe xxx yeares.—C. Carlile.—To the Reader." 104 pages; then "A Defcription of the Pope by H. I. The Table. Imprinted—by Holburne Conduit—1582." W.H. Alfo without date. Quarto.

"A Discovrse vpon vfurie by waie of Dialogue and oracions, for the 1582. better varietie, & more delight of all thofe, that fhall read this treatife. By Tho. Wilfon," doctor of the Ciuil lavves, one of the maifters of hir maiefties hon. courte of requefts. Seene & allowed &c. Imprinted— 1582." Dedicated "To—Robert Dudley, Earle of Leicefter, &c. From the Queenes maiefties Hofpitall at St. Katreynes, 20 July 1569." See p. 819. Again 1584. W.H. Octavo.

"The Duties of Conftables, Borfholders, Tythingmen, and fuch 1582. other Lowe Minifters of the Peace. Whereunto be alfo adioyned the feuerall offices of Church Wardens: of Surveiors for amending the highwaies; of diftributors of the Provifion for Noifome Foule & Vermine: of the Collectors, Overfeers, and Gouernors of the poore: and of the Wardens & Collectors for the houfes of Correction. Collected & Penned by

ing contrary to the decrees of the Starchamber, & hath employed the fame with other printing ftuffe in printinge the prymers & other thinges contrary to her maiefties priuilege & the faid decrees; and did alfo fet a forme at the temple of part of a booke not alowed, & hath alfo otherwife offended with the faid ftuff. Yt is therefore ordered

by vertue of the faid decrees, that, according to the fame, the faid preffes with the reft of the faid printing ftuffe fhalbe defaced & made vnf'viceable for printinge; and the ftuff thereof fo defaced to be redeliv'ed to the faid Roger."

[d] See Bibl. Topogr. Brit. N°. V. Append. p. 84.

Wm. Lambard, of Lincolns Jnne Gent. 1582." Pages 83. The late Mr. T. Martin's papers. Sixteens. Feb. 24, 1582-3, he was fined xſ. for printing a ſermon of Repentance; a copy of Mr. John Harriſon's, the elder.

1583. " Philotimus. The warre betwixt Nature and Fortune: compiled by Brian Malbancke, Student in Graies Inne." Quarto.

1583. A ſermon on predeſtination. Sixteens.

1584. " A briefe diſcourſe of the moſt renowned acts, and the right valiant conqueſts of thoſe puiſant princes, called, The nine worthies, of their ſeveral proportions, and what armes euery one gaue, &c. compiled by Richard Lloyd gent." With wood-cut portraits. Quarto.

1585. " THE GLORIOUS and beautifull Garland of Mans Glorification. Containing the Godlye Miſterie of heauenly Ieruſalem, the helmet of our Saluation. The comming of CHRIST in the fleſhe for our glorie and his glorious cōming in the end of the world, to crowne men with crownes of eternall glorie. Beeing an heauenly Adamant to drawe thee to CHRIST, and a ſpirituall Rod to mortifie thy Life. Made and ſet. foorth by Fr. Kett, Dr. of Phyſick: Phillip. 3. (21) Chriſte is our glorie, &c.—printed—1585." Fronting this title-page, are the queen's arms, and over them, VIVAT ELIZABETHA REGINA. Again at the head of the dedication, " VIVAT SERENISſima Regina. To the moſt Mightie, Imperial, and vertuous Princeſſe, The Lordes Anointed, Queene Elizabeth," &c. Q, in fours. W.H. Quarto.

1585. " THE Choiſe of Change: Containing the Triplicitie of Diuinitie, Philoſophie, and Poetrie. Short for memorie, Profitable for Knowledge, and Neceſſarie for Maners: Whereby the Learned may be confirmed, the Ignorant inſtructed, and all men generally recreated. Newly ſet foorth by S. R. Gent. and Student in the Vniuerſitie of Cambridge. Tria ſunt omnia.—Printed &c. 1585." Dedicated " To—Sir Henry Herbert, Knight of the—Garter, Lord of Cardiffe mannor, and S. Quintin, & Earle of Pembrocke;—Sir Philip Sidney Kaight, with—M. Robert Sidney Eſquier.----To the Reader. 1 He that knoweth not that he ought to know, is a brute beaſt among men. 2 He that knoweth no more than he hath need of, is a man among brute beaſts. 3 He that knoweth all that may bee knowen, is a God among men. 1 Read willingly. 2 Correct friendly. 3 Judge indifferently." N 2, in fours. W.H. Quarto.

1589. " A Briefe Deſcription of vniuerſal Mappes and Cardes, and of their vſe; and alſo the vſe of Ptholemey his Tables. Neceſſarie for thoſe that delight in reading of Hiſtories: and alſo for Traueilers by Land or Sea. Newly ſet foorth by Tho. Blundeuille of Newton Flotman, in the Countie of Norffolke, Gent." His device. "—for Tho. Cadman—1589." Dedicated " To—M. Francis Windam, one of the Iudges of—Common Pleas.—From my poore Swans neſt, 17 Decem. 1588.----To the Reader." F 2, in fours, with The Mariners Quadrant, to fold. W.H. Quarto.

1589. " A ſūmarie of Sir Francis Drakes Weſt Indian voyage." Quarto.

1589. " The Fountaine and VVelſpring of all Variance, Sedition, and deadlie

ROGER WARD.

deadlie Hate. Wherein is declared at large, the opinion of the famous diuine Hiperius, &c. that Rome in Italie is signified and noted by the name of Babylon,—in the 14. 17. and 18 Chapters of the Reuelation of S. Iohn.—Printed by *him*, dwelling vpon Lambard hil, neere vnto olde Fifhftreet, 1589." Dedicated " To—the Earls of Huntington and Warwicke.—Chr. O." 39 pages. W.H. Quarto.

" Robert Greens Spanifh Mafquerado." St. John's Coll. Camb. Quarto. 1589.

" The French Kinges Declaration vpon the Riot, Felonie, & Rebellion 1589. of the Duke of Mayenne, & the Duke and Knight of Aumalle, and all their affiftantes. Whereunto is adioyned Another Declaration of the fame King, againft the tovvnes of Paris, Orleance, Amyens, and Abbe- uille and their adherentes. Faithfully tranflated out of the French." His device. " Printed—for Tho. Cadman, 1589." Pages 18. W.H. Quarto.

" The Confeffion of Faith, with a confutation of fuperftitious errours 1589. contrarie thereunto, by Theod. Beza; tranflated out of French by Rob. Fills." Octavo.

" A fermon preached at Torceter," &c. as p. 1101.   1590.
" The Benefite of Contentation," &c. as p. 1162.   1590.
" The knowledge or appearance of the Church," &c. as ib.   1590.
" The Sclopotarie of Iofephus Quercetanus, Phifition. Or His booke 1590. containing the cure of wounds receiued by fhot of Gunne or fuch like Engines of warre. Whereunto is added his fpagericke antidotary of medicines againft the aforefayd woundes. Publifhed into Englifh by Iohn Hefter, practitioner in the faid fpagericall Arte." His device. "—Printed—for Iohn Sheldrake, 1590." Dedicated " To—Robert Deuorax, Earle of Effex, &c.---To the Reader." Then, fome commen- datory verfes on the author, Monf. du Chefne, Lord of Morence & Li- ferable, a phyfician, and Philofopher. 95 pages, and a table. W.H. Quarto.

" The Flower of Phificke, comprehending a methode for mans affured 1590. health, with 3 Bookes of Philofophie, for the due temperature of mans Lyfe, &c. written by Will. Cleuer." Quarto.

" The Sacke of Rome," &c. as p. 1162.   1590.
" A brief refolution of a right religion, touching the controuerfies, 1590. that are now in England, written by C. S. for John Proctor. Dedicated to the right worfhipfull man, Francis Flower, efq; juftice of the peace and quorum."

" A TALE OF TWO SWANNES. Wherein is comprehended the original 1590. and encreafe of the riuer Lee, commonly called Ware riuer; together with the antiquitie of fundrie places and townes feated vpon the fame. Pleafant to be read, and not vnprofitable to be vnderftood, by W. Val- lans." This tale is related in a poem of 266 verfes: in which, writing of Waltham crofs, the author ftiles it " The ftately croffe of Elnor, Henries wife," inftead of Edward's. To the tale is annexed " A com- mentarie or expofition of certain proper names vfed in this Tale." Wherein mention is made of a paper mill at Hartford, of which fee p. 200.

200. In 12 leaves. For John Sheldrake.* Reprinted in Leland's Itinerary, Vol. v. Quarto.

1590. "The First part of the Diall of Daies, Containing 320. Romane triumphes, besides the triumphant Obelisks & Pyramydes of Aegyptians, the Pillers, Arches, & Trophies triumphant, of the Græcians, and the Persians, with their Pompe & Magnificence: Of Feastes & Sacrifices both of the Iewes and of the Gentils, with the stately games & plaies belonging to these Feastes & Sacrifices, with the birthes & funeral Pomps of Kinges & Emperours, as you shall finde more at large in the 2. part, wherein all kind of triumphs are enlarged. By Lodowick Lloid Esquire. Prou. 20. (27)—London, Printed for *him*, dwelling at—the Purse in the little Old Bailie, 1590." Dedicated "To—Sir Chr. Hatton, Knight, &c.—How and when all Nations begin in their yeares, and daies," &c. 197 pages. W.H. Quarto.

1591. "The Arte of VVarre. Beeing the onely rare booke of My Militarie profession: drawne out of all our late & forraine seruices, by Will. Garrard Gent. who serued the King of Spayne in his warres 14 yeeres, and died—1587. Which may be called, the true steppes of warre, the perfect path of knowledge, and the playne plot of warlike exercises: &c. Corrected & finished by Captaine Hichcock,—1591—for *him*,—at the—Purse in the Olde-balie—M.D.XCI." Dedicated "To—Robert Deuorax, Earle of Essex, &c.—Tho. Garrard.—Capt. Tho. Hichcock, his commendations of this Booke: &c." 368 pages, with schemes, whereof six are on folding sheets. W.H. Quarto.

1593. "A briefe discouery of the Damages that happen to this Realme by disordered and vnlawfull Diet: The Benefites and Commodities that otherwise might ensue. With a perswasion of the People, for a better Maintenance of the Nauie. Briefly Compiled by Edw. Jeninges." 30 pages. Licensed. Oldys's Catal. N°. 112. Quarto.

1595. "A briefe note of the benefits that growe to this Realme, by the obseruation of Fish-daies: with a reason & cause wherefore the lawe in that behalfe made is ordained. Very necessarie to be placed in the houses of all men, specially common Victualers. Seen & allowed by the most hon. Priuie Counsell,—1595. the 20 of March. At London, printed by *him*, dwelling in Salisburie Court at the signe of the Castle." Licensed. A sheet.

He had licenses also for the following, viz. In 1557, "An enterlude, All for money." Also, for several ballads.

✦✦✦✦✦✦✦✦✦✦✦✦✦✦✦✦✦✦✦✦✦✦✦✦✦✦✦

THOMAS CHARD, or CHARDE, stationer,

SON of Tho. Chard of Dartford, Kent, was apprentice for ten years, from Christmas 1565, to Humphrey Toy, and seems to have taken up his freedom within the time of the chasm in the Company's register. In 1581, he was fined "for non appearance on the quarter day, xijd."

Also,

Alfo, "for keeping an apprentice iij. qu. of a yere, vij f. vid." I find him no more on the black lift. In 1587, he dwelt at the Helmet in St. Paul's church-yard: In 1600 he dwelt in Bifhopfgate church-yard. He feems rather to have been a bookfeller than a printer, as i have not yet met with any book printed by him.

"Ierome Sauanarola martyr, his Meditations on the 80 pfalme. Tranflated." Printed for him, 1577.   Octavo.   1577.

" A fermon at Paules (crofs) by Wm. Fifher," &c. as p. 1167.   1580.

" A true difcourfe of the affault committed vpon the perfon of the moft noble prince, Williams prince of Orange, countie of Naffau, marqueffe de la Vere, &c. by John Iauregui, Spaniarde. With the true copies of the writings, examinations, depofitions; and letters of fundrie offenders in that vile and diuelifh attempt. Faithfullye tranflated out of the French copie, printed at Antwerp by Chriftopher Plantine. Imprinted—for Thomas Charde, and W. Broome." 47 leaves. See Oldys's Catal. Nº. 76. Licenfed.   1582.

" Refponfionis ad Decem illas Rationes,—Defenfio" &c. as p. 1061.   1583.

" Ad' Nicholai Sanderi demonftrationes quadragiata, &c. refponfio Guil. Whitakeri, &c. London."   Octavo.   1583.

" The Common Places of—D. Peter Martyr," &c. as p. 957.   1583.

" A briefe Conference betwixt mans Frailtie and Faith," &c. as p. 1061. Again 1590.   1584.

" An Answer To the two firft and principall Treatifes of a certeine factious libell," &c. as p. 958.   1584.

" The Sermons of—Edwin, Archbifhop of Yorke," &c. as p. 1062. Prefixed is an epiftle to The Chriftian Readers, and a table of contents. W.H.   Quarto.   1585.

Whitaker's anfwer to Rainolds's Refutation. There are two other editions of it, this yeare; one by Tho. Thomas of Cambridge, the other without the printer's name.   Octavo.   1585.

" Doctrinae Chriftianae compendium," as p. 1062.   1586.

" A very fruitfull Expofition of the Commaundements," &c. as ib. Again 1590. Tho. Orwin for him. W.H.   1586.

" Salomon's fong tranflated into Englifh verfe, with annotations, by Robert Fletcher." Alfo without date.   Octavo.   1586.

" Fruitfull inftructions and neceffarie doctrine, to edifie in the feare of God, by Iohn Frewen." See Maunfell, p. 51, 52.   1587.

" Explicationum catecheticarum, quae tractationem locorum Theologicorum κατ' ἐπιτομὴν complectuntur," &c. Printed by Thomas Thomas, of Cambridge, for him.   Octavo.   1587.

" A Defence of the Gouernment eftablifhed in the Church of Englande for Ecclefiafticall Matters. Contayning an aunfwere vnto a Treatife—intituled, A briefe and plaine declaration," &c. as p. 1141. "Aunfwering alfo to the argumentes of Caluine, Beza, &c. by Iohn Bridges Deane of Sarum. Come & fer.—Take it vp & Read.—Iohn Windet, for *him*, 1587." This is the book pointed at by M. Marprelate in his Epitomes, "Oh read ouer D. John Bridges, for it is a worthy worke:"—   1587.

of which see our General History. 1401 pages, besides the preface. W.H. Quarto.

1587. " —— Militarie Discourses of Lord de la Novve." &c. as p. 1167.

1588. " A profitable Exposition of the Lords Prayer, by way of Questions and Answers for more playnnes; Together with many fruitfull applications to the life and Soule, aswell for the terror of the dull & dead, as for the sweet comfort of the tender harted. By Geruase Babington.— Tho. Orwin for *him*, 1588." Dedicated " To—Henry Earle of Penbrooke, Lord Harbert of Cardiff, &c. and to the—vertuous Ladie the Countesse his wife.—VVilton, 11 May, 1588." Pages 582, and a table, Licensed. W.H. Again 1596, quarto. R. Robinson for him. Sixteens.

1589. " A Sermon at Pauls Crosse, 17 Nouemb. 1583, on Tit. 3; 1, 2." Octavo.

1589. " A Sermon of Obedience, on Luke 20; 25. by Edm. Suckling, at Norwich, 22 Feb. For him and Wm. Lownes." Octavo.

1591. " An Ansvvere to a great number of blasphemous cauillations written by an Anabaptist, and aduersarie to Gods eternall Predestination. And confuted by Iohn Knox, Minister of Gods word in Scotland. &c. Prov. xxx." (12) Field's device. " Imprinted for T. Charde, 1591." Pages 443, including the preface. W.H. Octavo.

1591. " APOLOGIA ECCLESIÆ ANGLICANÆ. Authore Iohanne Iuello, olim episcopo Sarisburiensi. Rom. 1. (16) Non enim me pudet Euangelij Christi:" &c. Device the lost sheep. " Impensis—1591." Prefixed is Peter Martyr's epistle to bishop Jewel. 197 pages. W.H. Twenty-fours.

1591. " OF THE RVSSE Common Wealth," &c. as p. 1128.

(1591.) A Sermon at Court, 24 May, 1591, by G. Babington, on 2 Kings 15; 13—16. Octavo.

1592. " Hugonis Latimeri Anglicani Pontificis Oratio, apud totum Ecclesiasticorū conuentū, antequam consultatio publica iniretur, de Regni statu per Euanglium reformando, regni Inuictiss. Regis Henrici VIII. Anno vigesimo octauo habita. Impensis—1592." It is introduced with an epistle by Sim. Grinæus to the reader. 53 pages. W.H. Sixteens.

1592. " Certaine plaine, briefe, and comfortable notes, vpon euerie chapter of Genesis. Gathered and laid downe for the good of them, that are not able to vse better helpes, and yet carefull to reade the worde, and right heartilie desirous to taste the sweete of it. By G. Babington, bishop of Landaph.—for *him*." Licensed. Again 1596. Quarto.

1593. " THE Description of the Low countreys &c. gathered into an Epitome out of the Historie of Lodouico Guicchardini. Imprinted—by P. Short for *him*, 1593." Dedicated " To—Lord Burghley, high Treasorer &c. Tho. Danett ---A rable." 122 leaves. W.H. Octavo.

1595. " A liuely patterne of true repentance." as p. 1033.

1595. A funeral sermon, on 2 Sam. 10; 1,—4, by G. Babington. Octavo.

1597. " A Demonstration of God in his workes," as p. 1034.

1600. " THE Apologie of the Church of England. With a briefe & plaine declaration of the true Religion professed and vsed in the same. Published by the most reuerend Father in God, Iohn Iuell, Bishop of Sarisbury." Translated by lady Anne Bacon: see p. 608. P. Martyr's epistle is prefixed. 323 pages. J. Roberts for him. W.H. Twenty-fours.

The

The hiftory of the civil wars of France, by T. Churchyard. Quarto. 1600.
" A Continuation of the Hiftorie of France, from the death of Charles 1600.
the eight, where Comines endeth, till the death of Henry the fecond.
Collected By Tho. Danett Gent. London, Printed by Tho. Eaft for *him*,
1600." Dedicated "To—Lord Buckhurft, Lord high Treaforer of England," &c. 148 pages. W.H. Quarto.
Articles miniftred by Arthur, Bifhop of Chichefter to the Churchwar- 1600.
dens, &c. W.H. Quarto.
" A Teftimonie of the true Church of God," &c. as p. 1063.
The Lives of holy men. Quarto.
He had alfo licenfes for the following, viz. In 1578, " The tranflation of Mr. Juels replye againft Hardinge, done into Lattin by Mr. Whit-
takers." In 1580, " Pofitions/ whereupon the trayninge vp of children,
& fo confequentlie ỹ wholle courfe of learninge, ys grounded." See p.
1072. In 1582, " Letters intercepted of ỹ Cardinall of Grandvelle. A
newe order for bankeruptes." In 1584, " A difcourfe touchinge ỹ
meanes to p'ferve the ftate & religion in ỹ Lowe Cuntreyes."

EDWARD WHITE, or WHYTE, ftationer,

SON of John Whyte of Bury St. Edmond's, Suffolk, was apprentice for
feven years, from Michaelmas, 1565, to Wm. Loble, who feems to
have been related to Mich. Lobley, however he was made free by him in
Auguft, 1577. White feems to have taken up his freedom in the time of
the chafm in the Company's regifter. He was admitted on the livery
6 May 1588, ferved Renter in 1596 and 1597, and Under Warden in
1600. He appears to have been but a diforderly[e] member, being often
found on the Black Lift. He dwelt at the little north door of St. Paul's
church, at the fign of the Gun. Very few books appear to have been
printed by him; thofe that were are particularly noticed.

" Chr. Iohnfon his Councell againft the Plague, or any other infec- 1577.
tious difeafe; Alfo a queftion, whether a man for preferuation may be
purged in Dogge-dayes or no.—for *him*." Licenfed. Octavo.

[e] In 1579, he was fined three times for printing ballads contrary to Orders. In 1588,
for keeping an apprentice unprefented, one
year. In Octob. 1589, on a matter of controverfy with Meff. Newbery and Bifhop,
which he had requefted might be determined
by the Court of Affiftants, it was ordered
and agreed by affent of the parties, That
he fhould pay unto them iiij f. for ten books
of D. Fulke's anfwer to the Rhemifh Teftament, which he had diforderly bought of
one of their workmen; and fhall alfo prefently pay x f. for a fine, to the ufe of the
poor. " Solut. 3 Augufti, 1591." Alfo
in 1592 for fundries, as in p. 1160, note p.
In 1594, for printing a ballad without licenfe. Again in June, 1600, " Yt is ordered touching a diforderly ballad of the
wife of Bathe, printed by Edw. Aldee, &
Wm. White, and fold by Edw. White, That
all the fame ballads fhalbe brought in &
burnt; and that either of the printers fhall
pay v f. a pece, and that Mr. White for
felling it fhall pay x f."

EDWARD WHITE.

1579. "An Hospitall for the diseased. Wherein are to bee founde—approoued Medicines, aswell Emplasters of speciall vertue, as also notable Potions, &c. bothe for the Restitution & the Preseruation of bodily healthe. Very necessary for this tyme of common Plague and immortalitie, &c. With a newe addition. Gathered by T. C. Ecclesias. 38; 4.—Imprinted —for him, at the little Northdoore of Paules Churche, &c. 1579." Prefixed is an epistle "To all suche Readers as have care of their bodylie healthe." Licensed. W.H.₊ Also, without date. Quarto.

1579. "A detection of the damnable drifts practized by 3 witches arraigned at Chelmisford in Essex, and executed in Aprill 1579."

1579. "The horrible acts of Eliz. Style," &c. as p. 891.

1580. "The Lectures or daily Sermons of that Reuerend Diuine, M. Ihon. Caluine,—vppon the Prophet Ionas. Whereunto is annexed an excellente exposition of the two last Epistles of sainct Jhon, doen in Latine by that worthie Doctor August. Marlorate, and Englished by N. B. And newly corrected & amended. Matth. 12; 39.—Imprinted—by him, dwellyng at the little Northdoore of Paules, at the—Gun, 1580." Dedicated "To—Sir Fraunces VValsyngham Knight, &c. And to—Sir Ihon Broket, of Broket Hall, and Sir Henrie Cocke, Knightes.—From my house in Ridborn, 22, Januarie, 1577.—Nath. Baxterus." 70 leaves. The epistles of St. John have a separate title-page, and are dedicated "To—the Ladie Ursula VValsingham, wife to—sir Frances.—Ridborne, 26 Marche, 1578.—N. B." 16 leaves. Licensed. W.H. Quarto.

(1581.) "The Araignement &c. of—Eueralde Ducket," &c. as p. 1096.

1581. "A sermon—of M. Iohn Caluine," &c. as p. 1140.

1581. "The Oration and Sermon made at Rome," &c. as p. 1095.

1582. "A Discouerie of Edm. Campion, and his confederates, their most horrible & traiterous practises, againft her Maiefties moft royall person, and the Realme, &c. Whereto is added, the Execution of Edm. Campion, Raphe Sherwin, & Alex. Brian,—at Tiborne the 1 of December. Published by A. M. sometime the Popes Scholler, &c. Seene & allowed. Imprinted—for him,—29 of Janua. 1582." Dedicated "To—Sir Tho. Bromeley Knight, Lord Chanceller,—with the reft of her Maiefties Councell.—A. Munday."—To the Reader. G 7, in eights. Licensed. W.H. See Oldys's Catal. N°. 144. Sixteens.

1583. "A wonderfull and ftraunge newes, which happened in the countye of Suffolke, and Essex, the firft of February, beeing Fryday, where it rayned Wheat, the space of vi or vii miles compas; a notable example to put vs in remembraunce of the iudgements of God, and a preparatiue, sent to moue vs to speedy-repentance. Written by William Auerell, ftudent in diuinitie." At the end, "witneffe hereunto these men, whose names doe followe: maifter Willyam Geffreyes, dwelling in Ipfwhich ; John Bull, feruant to Juftice Germaye ; Richard Boothe of Ipfwhich, dwelling with Olyuer Boothe ; Richard Kaye, with diuers others." Printed for him. 14 leaves. Licensed. See Oldys's Catal. N°. 8.
Octavo.

1586, "The Old-mans Dietary, Neceffary for the preferuation of old perfons
in

EDWARD WHITE.

in perfect health & soundness; translated by Tho. Newton. Printed for *him*, 1586." Licensed. Octavo.
"Historie de Aurelio & Isabella fille du Roy d'Escoce." French, 1586. Italian, and English. Licensed. See Hist. of Eng. Poetry, III; 477.
A thousand notable things, &c. as p. 1033. Licensed. Also without date. 1586. Quarto.
"A proper new Ballad," &c. as p. 1000. 1586.
"A short Discourse: Expressing the substaunce of all the late intended (1586.) Treasons against the Queenes Maiestie, and Estates of this Realme, by sondrie Traytors; who were executed for the same on the 20 and 21 daies of September last past, 1586. Whereunto is adioyned a Godly prayer for the safetie of her Highnesse person, her hon. Counsaile, and all other her obedient subiects. Seene & alowed. Imprinted by Geo. Robinson for *him*." 4 leaves. Lambeth Libr. Quarto.
"Capt. Tho. Sanders his discription of a—voyage," &c. as p. 1045. 1587.
"The Pathway to health, wherein are most excellent & aproued 1587. Medicines—the distilling of diuers waters, making of Oyles, and other comfortable receytes, by Peter Leuins.—for *him*." Licensed. Quarto.
"Io. Carpenter, his 2 sermons on Luke 17; 32.—for *him*." Octavo. 1588.
"Prepositas his Practise," &c. as p. 1174. 1588.
"The Historie of the—kingdome of China," &c. as p. 1174. 1588.
"The Voyage and Trauaile of M. Cæsar Frederick," &c. as p. 1046. 1588.
"Two letters written ouer into England: the one to a godly Ladie, 1589. wherein the Anabaptistes errours are confuted, & the sin against the holy Ghost plainely declared; the other an answer to a godly merchants letter, written for his comfort, being greeued with the heauy burden of his sinnes, wherein is declared the true confession of sinnes. By T. C.—for *him*." Sixteens.
A true discourse of the victories—by the Fr. King. &c. as p. 1176. 1589.
"—prognostications of the weather." &c. as p. 1177. Again 1598. 1590.
"The Good huswifes Iewell,—for conceits in Cookerie.". In 2 parts. 1590. Tho. Dawson for him. Octavo.
"The Geomancie of M. Chr. Cattan Gent." &c. as p. 1177. 1591.
"De neutralibus et mediis,—Jack of both sides." &c. as p. 883, &c. 1591.
"The Vineyard of Deuotion, comprehending sundry godly Praiers, 1591. Psalmes & meditations, meet to comfort the wounded consiences of all penitent sinners.—for *him*, 1591." Licensed. Sixteens.
"The Widowes treasure," &c. as p. 1101. Licensed in 1583. See 1591. p. 683.
"New and singular patternes," &c. as p. 1179. 1591.
"The lamentable and true Tragedie of M. Arden, of Feversham in 1592. Kent, who was Most wickedlye murdered, by the Meanes of his disloyall & wanton Wyfe, who for the Loue she bare to one Mosbie, hyred two desperat Ruffins, Blackwill and Shagbag, to kill him. Wherin is shewed, the great Malice & Discimulation of a wicked Woman, the vnsatiable desire of filthie lust, and the shamefull End of all Murderers.— Printed for *him*, 1592." Licensed. Mr. Edw. Jacob, of Feversham. Reprinted 1770. See Holinshed, Anno 1551. Quarto.

7 A 2   A sermon

1593. A sermon of Faith and Repentance, by Hen. Summer. Mark 1; 15. Octavo.
1595. " A short introduction to learn to swimme," &c. as p. 1033.
1595. " The Mount of Caluarie. COMPYLED BY THE reuerend—Lord Anthonie de Gueuara, Bishop of Mondonnedo, &c. Wherin is handled all the mysteries of the mount of Caluarie, from the time that Christ was condemned by Pilat, vntil he was put into the sepulcher, by Ioseph and Nichodemus." Device the palm-tree, &c. " Printed by A. Islip for him,—1595." A prologue prefixed, 409 pages, and a table of the chapters. W.H. Quarto.
1596. " Mary the Mother of Christ her Tears. Set forth by L. T." Octavo.
1596. " ——— The Pathway to Health," &c. as p. 1033.
1597. " Mount Caluarie, THE SECOND PART:—In this Booke the Author treateth of the Seuen Words which Christ our Redeemer spake hanging vpon the Crosse. Translated out of Spanish into English." The palm-tree, &c. " Printed by Adam Islip for him,—1597." A table of the chapters prefixed, and 502 pages. W.H. Quarto.
1597. The Comforter: &c. by John Freeman. See p. 1179.
1598. Fitzherbert's Husbandry, with additions. J. Roberts for him. Quarto.
1599. " The Key to vnknowne Knowledge, Or a shop of fiue Windowes, Which if you doe open, to cheapen and copen, You will be vnwilling, for many a shilling, To part with the profit, that you shall haue of it. Consisting of 5 necessarie Treatises: Namely, The iudgement of Vrines: Iudiciall rules of Physicke. Questions of Oyles. Opinions of curing Harquebush-shot. A discourse of human Nature.—Printed by Adam Islip for him,—1599." L, in fours. W.H. Quarto.
1599. " A breefe Treatise Of the vertue of the Crosse: And the true manner hoyv to honour it. Translated out of French into English. &c. Printed for him,—1599." Dedicated " To—Sir Stephen Soame, Knight, Lord Maior of—London: And to the vertuous Ladie his wife." F, in eights. W.H. Sixteens.
1600. " The Comforter:" &c. as p. 1185.
1600. " A Booke of Fishing with Hooke & Line," &c. as ib.
——— " A Spectacle for a blinde papist. Made by J. S." See p. 1103.
——— " Pretty conceytes taken out of Latine, Italion, French, Dutch, and English. Reprinted for him." In 1586, This was allowed to Ab. Jeffes, to print for the Company. Octavo.
——— Churchyards Choice. Licensed in 1579. Perhaps the Choice Mirrour of Honour, &c. as Ath. Oxon. Vol. 1. col. 319. Quarto.

He had also licentes for the following, viz. In 1576, " The lamentation of a Xpian, servinge as a target againste Temptations. A cruell murder done in Kent." In 1577, " A faithfull relac'on of a most horrible murder comitted by Alphonse Diazius, a Spaniard,—on the bodie of his brother Jhon Diazius." In 1578, " The confession of certen witches at Abington. The myrror of modestie, meete for all mothers, matrons & maydes. Thexaminac'on of certen wytches in Essex, as mother Stanton," &c. In 1579, " Certen pointes of foraine husbandrie,—about

Cattell,

EDWARD WHITE.

Cattell, &c. An epitaphe of ẏ lady Anne Lodge. An epitaph of Sir Wm. Drury. The footepathe leadinge ẏ highewaye to heauen:ᶠ Turned over to him from Wm. Hoſkins. The 2d earthquake in Kent." In 1580, " A paper of ẏ artis of ſhameles ſhiftenge. A pleaſant & ſpeedy pathe for ẏ bringing vp of yonge children. The deſcription of great wonders ſeene ẏ xiij of Jan'ry 1580, and fearfull wyndes & earthquakes at Rome. The true Report of ẏ proſperous ſucces which God gave vnto our Engliſh ſouldiors againſte ẏ forraine bādes of our Romaine Enemyes latelie arrived but ſoon inoughe to their coſte in Jreland, in the yere 1580." In 1581, " The conqueſt atchiued by Capt. Norrice generall Colonell in ẏ campe in Friſeland, ẏ ix of Julie, 1581. A replie vnto twoo ſeditious papers printed, thone in French, thother in Engliſh, contayning a defence of Edm. Campion, &c. Their horrible treaſons & practiſes againſt her maieſtie & her Realm." In 1583, " The Paradice of Daintie Devices: Put over to him from Tim. Rider." In 1584, " A book of Cookerye:" Conditionally. In 1586, " A pleaſant baite, or recreation for wayfaringe men, compiled by ẏ poore pylgrim. Morando: The tritameron of love. A ſackfull of newes, an old copie, which be is ordered to have printed by Abell Jeffes." In 1587, " Euphues his cenſure to Philautus. Perymides ẏ blackſmith:" See p. 1174. In 1588, " A miraculous—diſcourſe," &c. as p. 1054. In 1589, " Greens Orphariọn." In 1590, " The Parle betwene ẏ Fr. king & the Pariſians, 6 Aug. 1590,—with what happened in his camp aboute ẏ ſame tyme, &c. Deſcription veritable dez battailles, victoires, & trophies de.duc de Parme: In Eng. and Fr. The arraynement, & cōdemnac'on of Arnalt Coſby for murdering ẏ lord Burghe." (14 Jan'ry 1590.) In 1591, " A treatiſe of meaſuring land," &c. Conditionally. The arte of Conye katchinge: with Tho. Nelſon." In 1592, " The tragedye of Salamon & Perceda. The ſermon preached at Bromley before Sir Gilb. Gerrard Mr. of the Roles, 1 Octo. 1592." In 1593, " The footepath to faſtinge. The Hiſtorye of Fryer Bacon & Fryer Boungaye. Tho moſt famous chronicle hiſtorye of Leire kinge of England, & his 3 daughters. The famous hiſtorye of John a-Gaunte ſōne of kinge Edwarde the third, with his conqueſt of Spayne, & marriage of his 2 daughters to the kings of Caſtile & Portugale. The booke of David & Bethſheba. A paſtoral pleaſant comedie of Robin Hood & Little John. A Wynters nightes paſtime. A newe propheſie ſeene by ẏ Viceere ſinau Baſſa at his cominge into Hungarie." Alſo at various times for many ballads. He printed after 1600.

ᶠ This ſeems to be the ſame with The Footpath of Faith, which Hen. Denham had printed in his Diamond of Devoticn, and for which he was ordered by the Court of Aſſiſtants, 9 Jan. 1581-2, to pay Edw. White, whoſe copy it was, iij l. vj f. viij d. for the injury done him.

WILLIAM

WILLIAM BARTLET, or BERTHLET,

As he has spelt his name in the two following books:

1578. "THE true reporte of the skirmish fought betwene the states of Flaunders, and Don Joan, duke of Austria, with the number of all them, that were slayne on both sides, which battel was fought 1 August, being Lammas day." Sixteens.

1583. "Two examples of Gods judgment: upon a wicked swearing woman, and of one Strangman, who gave himself to the deuill." Octavo.

He had also licenses for a few ballads.

WILLIAM BROME, or BROOME, stationer,

PROBABLY was made free in the interval of the chasm in the Stationers' register, either by purchase or translation from some other company; for there is not found one of that name on the Company's books before he presents John Drawater as his apprentice, 6 Dec. 1585. He died in 1591, and his widow followed the business from that time to 1596, inclusive.

1576. "A Treatise of the peace made and concluded betwene the states of the lowe countries, assembled within the city of Bruxels, and the prince of Orenge, the states of Holland and Zealand, with the associates, published the VIII day of November 1576. with the agreement and confirmation of the kings maiestie, as followeth. Translated out of a Dutch copy printed in Bruxels, by the kings printer, with said kings priuiledge." Sixteens.

1577. "Hugo Cardinalis, exposition on certain words of S. Paule to the Romans, entitled A treatise of the workes of three daies, on Rom. 1; 20, the inuisible things of God. Also another treatise of the truth of Christes natural body: translated by Rich. Curtis, Bishop of Chichester.—for *him*, 1577." Octavo.

1582. "A true discourse of the assault committed vpon the person of—William prince of Orange," &c. as p. 1195. Licensed.

1583. "The Spainish colonie, or brief chronicle of the acts & gestes of the Spaniards in the West-Indies, called the new world, by Bart. de las Casas, or Casaus. Englished from the Spainish, by M. M. S." For *him*. Quarto.

1583. "The Common Places of—D. Peter Martyr," as p. 957.

1584. "The discouerie of witchcraft, Wherein the lewde dealing of witches & witchmongers is notablie detected, the knauerie of coniurors, the impietie of inchantors, the follie of soothsaiers, the impudent falshood of cousenors, the infidelitie of atheists, the pestilent practises of Pythonists, the curiositie of figurecasters, the vanitie of dreamers, the beggerlie art of Alcumystrie, The abhomination of idolatrie, the horrible art of poison-

ing

WILLIAM BROOME.

ing, the vertue & power of naturall magike, and all the conueiances of Legierdemaine & iuggling are deciphered: &c. Heerevnto is added a treatife vpon the nature & fubftance of fpirits & diuels, &c: all latelie written by Reginald Scot Efq: 1 Iohn 4; 1. Beleeue not euerie fpirit, &c. 1584." Dedicated " To—Sir Roger Manwood,—Lord cheefe Baron &c.---To—Sir Thomas Scot.---To—his louing friends, Maifter Dr. Coldwell Deane of Rochefter, and Mr. Dr. Readman Archdeacon of Canturburie, &c.---To the Readers.---Forren & Englifh authors vfed in this Booke." 560 pages, and a table of contents. W.H. Quarto.

" Amendment of life;" &c. as p. 1141. Octavo. 1584.

" Two Sermons preached by—Richard Bifhop of Chichefter, the firft at Paules Croffe (Apoc. 12; 1—9:) The second at Weftminfter before the Queenes Maieftie. (Acts 20; 28—31) At London, Printed by T. Man, & W. Brome, 1584." W.H. Octavo. 1584.

" An A. B. C. for Layemen," &c. as p. 1142. Licenfed. 1585.

" A COMPENDIOVS CHYRVRGERIE: Gathered, & tranflated (efpecially) out of Wecker,—Publifhed for the benefite of all his countreymen, by Iohn Banifter Maifter in Chyrurgerie.—Imprinted by Iohn Windet, for T. Man, and him, 1585." Pages 530, befides prefixes and affixes. W.H. Twenty-fours. 1585.

" The Recantation made at Paules Croffe," &c. as p. 1100. Licenfed. 1588.

" Campafpe, Played beefore the Queenes maieftie on twelfe day at night, by her Maiefties Children, and the Children of Paules. Imprinted—by Tho. Orwin for Wm. Broome, 1591." W.H. Quarto. 1591.

" Sapho and Phao, Played beefore the Queenes maieftie on Shroue tewfday, by her Maiefties Children, and the Boyes of Paules.—Tho. Orwin for him, 1591." This and the foregoing are two of the 6 Court comedies written by John Lylie. W.H. Quarto. 1591.

" The Combate betwixt Chrift and the Deuill," &c. as p. 1144.

" A Worthy Treatife of the eyes," &c. as ib.

He had alfo licences for the following, viz. In 1580, " Articles for a Bifhop." In 1583, " Logica ad P. Rami Dialecticam conformata, &c. Authore Gofwino Wafferleider Mulhernio. P. Rami Profefforis Regis Gramatica, ab eo demum recognita, &c. P. Rami—Grāmatica græca," &c. In 1584, " M. Wilcox vpon the Cantycles." In 1585, " Seneca's tragedies, in Latin. Chilias fecunda—a Johāne Rodio: In Englifh. The cōmentarie of the Pfalmes, by T. Wilcoxe." The 7 laft articles with T. Man.

JOAN BROME, or BROOME, his widow.

" A prooued practife for all young Chirurgians concerning burnings with Gunpowder, & woundes made with Gunfhot, Sword, Halbard, Pike, Launce, &c. Hereto is adioyned a Treatife of the French or Spanifh Pocks,—by Iohn Almenar, a Spanifh Phifition. Alfo a collection of Aphorifmes,—Englifh & Latine. Publifhed—by Wm. Clowes, Maifter in Chirurgery. Newly corrected & augmented. Seene, and allowed, &c. Printed by T. Orwyn for Wydow Broome, 1591." Pages 200, as numbered. 1591.

numbered; but besides the prefixes, and table at the end, there are 51 leaves inserted before p. 97, with cuts of instruments, &c. W.H. Quarto.

1592. "Gallathea." &c. as p. 1102. By John Lylie. Licensed. W.H. Quarto.
1596. "A reforming glass, precious & profitable for all persons to the right disposing of their thoughts, words & actions, to God their neighbour & themselues, by meditations & prayers. Compiled by John Norden. Printed for Joane Brome widow." Sixteens.

She had also licenfe in 1591 for "Three comedies plaied before her maieftie by the Children of Paules, thone called Endimion, thother Galathea, and thother Midas." All written by John Lylie.

WILLIAM CARTER,

SON of John Carter of London, Draper, was put apprentice to John Cawood for ten years from the Purification 1562-3; there is no further mention of him in the Stationers' register, now exifting We have not met with any book bearing his name as the printer: indeed as he appears to have been an emiffary for the Popifh party, and to have printed fuch books only as ferved their turn [e] they were doubtlefs printed incog, without name or place, or elfe with fictitious ones: And even of thefe we have particular intelligence of only two.

A book in French, afferting the innocence of the Scotch queen.[h] See Strype's Life of bp. Aylmer, p. 45, &c.

Reafons that catholicks ought in any wife to abftain from heretical conventicles; faid to be printed at Doway, but really at London, 1580, in octavo, under the name of John Howlet and dedicated to Q. Elizabeth. The running title, A treatife of fchifm.[i]

[e] Strype fays he was divers times put in prifon for printing lewd pamphlets, popifh, and others, againft the government.

[h] Perhaps a tranflation into that language of bifhop Lefley's Defence of Q. Mary's honour, privately printed in 1569, and reprinted with fome alterations at Liege in 1571. See Anderfon's Collections, Vol. 1.

[i] For printing this book he was tried and condemned at the Old Baily, 10 Jan. 1584; and the next day hanged, bowelled, and quartered at Tyburn. See Ath. Oxon. Vol. 1. col. 357. Stow's Annals, 1605, p. 1176.

I have an edition, 1578, of which an account will be given in our General Hiftory, from which Carter's feems to have been printed, with the addition of a dedication to the queen, under the name of Howlet. (*Parfons.*) Of this edition, with Gregory Martin's name as the author, our hiftorians appear hitherto to have been ignorant. See Camden's Elizabeth, p. 295. Fuller's Ch. Hift. B. ix. p. 169. § 11.

In an anfwer, faid to be written by Card. Allen, to The execution of Juftice in England, p. 10, &c. it is faid, " Cartar, a poore innocent artifan: who was made away onelie for printing a catholique booke, De Schifmate.—The faid young man Cartar of whofe martyrdome we laft treated was examined vpon the racke, vpon what Gentlemen, or catholique Ladies he had beftowed, or intended to beftowe certaine bookes of prayers and fpiritual exercifes, and meditations, which he had in his cuftodie.".

HENRY MARSHE, or MARSH,

THE fon of Tho. Marſhe, was made free of the Stationers' company by his father's copy, 3 Feb. 1583-4. Soon after, his father appears to have aſſigned over certain of his copies to him and Gerard Dewes. Henry ſeems not to have ſurvived his father long however, as in June 1591, one Edw. Marſhe aſſigns the copies of Tho. Marſhe, late deceaſed, to Tho. Orwin. See. p. 847.

" Baptiſtae Mantuani—Adoleſcentia," &c. as p. 941.   1584.
" A Catholicke—Expoſition on S. Jude," &c. as ib.   1584.
" Dialectica Joannis Setoni," as p. 942.   1584.
" The Araygnement of Paris. A Paſtoral. Preſented before the 1584. Queenes Maieſtie, by the Children of her Chappell. Imprinted at London by Henrie Marſhe,—1584." See Preface to Johnſon and Steevens Shakſpeare, p. 240. E, in fours. Garrick's Plays.   Quarto.
" An Epiſtle to diuers papiſtes in England prouing the Pope to be the 1585. beaſt in the 13 Reuelat. and to be the man exalted in the Temple of God, as God, (2 Theſſ. 2; 4.) Whereby the true Church of Chriſt is knowen from the euill. By Franc. Kett."   Octavo.
" THE Mirour for Magiſtrates, wherein may bee ſeene by examples 1587. paſſed in this Realme, with how greeuous plagues vices are puniſhed in great Princes & Magiſtrates, and how fraile & vnſtable worldly proſperity is found, where Fortune ſeemeth moſt highly to fauour: Newly imprinted, and with the addition of diuers Tragedies enlarged." 272 leaves. " Imprinted—by Henry Marſh, being the aſſigne of Tho. Marſh, neare to Saint Dunſtanes Churche in Fleeteſtreete, 1587." Of this and ſome other editions of this book, and the authors contributing thereto, ſee Hiſt. of Eng. Poetry, Vol. III; p. 209—282.   Quarto.

He printed alſo divers almanacs and prognoſtications written by Hen. Lowe, M. D. See Maunſell, Part II; p. 15.

RICHARD YARDLEY,[k] ſtationer,

THE ſon of Tho. Yardley, of Morton, Warwickſhire, Gent. was apprentice to Rd. Jugge for 7 years from Lady-day 1569. Probably he took up his freedom in the interval of the chaſm in the company's

---

[k] Mr. Ames joined him in partnerſhip with Peter Short; but it does not appear that they were any otherwiſe ſo than Day and Seres, Jugge and Cawood, who printed certain books together as concerned in the ſame privilege, &c. So Yardley and Short printed together ſuch books chiefly, if not ſolely, as were for the aſſignees, or partners in Day's, or Seres's privilege; each ſeeming to have had books printed alſo in his own name, within the ſame time that they printed jointly. They appear indeed to have dwelt together in the ſame houſe, on Bread-ſtreet hill, having no other direction

ny's register. He appears to have died in the year 1593, Short printing alone, for others, in that year.

1584. "Hugonis Platti armig. Manuale sententias aliquot divinas et morales complectens; partim è sacris patribus partim è Petrarcha philosopho et poeta celeberrimo, decerptas." Twenty-fours.

1584. "A Dialogue betweene youth and olde age, wherein is declared the persecution of Christes religion since the fall of Adam hitherto. By John Marbeck.—for Rd. Yardley, 1584." Octavo.

1589. "The testamentes of the xij patriarches, &c. Printed for the assignes of Rd. Day." See page 682, Again 1590. Octavo.

1590. "A Booke of Christian Prayers," &c. as p. 683.

1590. "The Psalmes of Dauid," &c. as p. 952. With Denham's device.

1590. "A Paraphrastical explanation—of 14 holie PSALMES chosen out of the booke of the old and new Testament: and may verie aptlie be ioined with Dauids Psalter.—Latelie written in Latin by—Theo. Beza, and now newlie englished by Ant. Gilbie. Imprinted by—R. Yardley and P. Short, dwelling on Bredstreete hill, at—the Starre. Cum priuilegio —1590." Pages 77, and a table. W.H. Twenty-fours.

1590. "Directions declaring the discharge of our dutie towards God and man, &c. Printed by P. Short & R. Yardley, Twenty-fours.

1591. "Seuen sobs of a sorrowfull foule for sinne," &c. as p. 960. "Prin. by Rd. Yardeley, 1591." See Maunsell, p. 61.

1591, "O VTINAM. 1 For Queene Elizabeths securitie, 2 For hir Subiects prosperitie, 3 For a generall conformitie, 4 And for Englands tranquilitie. Printed—by R. Yardley & P. Short, for Iohn Pennie, dwelling in Pater noster row, at—the Grey hound, 1591." Inscribed "To the most noble, famous, renowmed, &c. Realme of England:— Iohn Dauies." G 4, in eights. W.H. Sixteens.

1591. "S. Augustines Praiers:" &c. as p. 953. "Printed for R. Yardley."
1591. "———— Manuel." &c. as ib. "—by R. Yardley & P. Short."
1591. "The enemie of Securitie, &c. as p. 951. With Short. Again 1593.
1591. "The enemie to Atheisme," &c. as p. 962.
1591. "The Pooremans Pater noster, with a preparatiue to praier, wherunto are annexed diuers godly Psalmes & meditations by Tho. Timme printed by Peter Short, 1591." Maunsell, p. 86. Licensed to him with R. Yardley. Sixteens.

1592. "Of the Imitation of Christ, &c. as p. 959, &c. With Short, "for the assignes of Wm. Seres, 1592." On the back is a concise Latin dedication to Sir Thomas Bromley, lord chancellor; then, an epistle "to the faithful Imitators of our Sauior Christ in England.—30 of Julie,

tion to the books printed with both their names; but that is not entirely clear, since the printing might be done, at Short's house, who lived there to the last, while Yardley kept shop elsewhere, and yet be concerned in the printing office:. thus Yardley and Co. might print for R. Yardley, or for P. Short, separately. I have not seen any book with Yardley's name alone, which would have enabled me to speak more positively. Maunsell mentions three; one in 1584, for him, two in 1591, by him; but he is not always exact. I find the title of another in 1593, among Mr. Ames's papers. I apprehend they were all printed jointly, though each had their separate copies, from the time of Short's being free, in 1588, to the decease of Yardley.

1580.

RICHARD YARDLEY.

1580. T. Rogers.—A second Epistle concerning the translation & correction.—A godly preface made by him, whoever he was, that translated this booke out of the Latine toong into French.—A. G." A neat woodcut of king Dauid on his knees. 277 pages, and a table. Denham's device over the colophon at the end. W.H. Twenty-fours.

"Soliloquium Animæ," &c. as p. 1136. "Neuer before published. 1592. At London printed. And are to be solde in the Royall Exchange at the shop of Andr. Maunsell, 1592." Prefixed are, an epistle by the translator, and the author's preface. 223 pages, and a table. Licensed to Yardley and Short. W.H4. Twenty-fours.

"Axiomata philosophica Venerabilis Bedae, &c. studio M. Joannis 1592. Kroeselii. Impensis Richardi Oliff." Octavo.

" A discouerie of ten English Leapers verie noisome & hurtfull to the 1592. Church & common wealth. 1 A Schismatike, 2 a Church robber, 3 A Simoniacke, 4 an hypocrite, 5 A proude man, 6 A glutton, 7 An adulterer or fornicator, 8 a couetous man, 9 A murtherer, 10 a murmurer. By T. Timme. Printed by P. Short, 1592." Maunsell p. 114. Licensed also to Short alone. Quarto.

" The Life and death of Wm. Longbeard the most famous & witty 1593. English Traitor, borne in the City of London: accompanied with many other most prettie histories, by T. L. (Tho. Lodge) of Lincolns Inne Gent. Yardley." Mr. Ames's papers. Quarto.

"Thomas Tussers 500 points of Good husbandrie." Yardley. Quarto. 1593.

PETER SHORT, stationer,

Was made free by redemption, 1 March 1588-9, for which he paid vi s. and came on the livery 1 July 1598. He dwelt at the sign of the star on Breadstreet hill, where R. Yardley was concerned with him in printing till 1593. See p. 1206. &c. He used the device of an open Bible encompassed with light, hanging on a chaplet of laurel with a pair of wings, held by a hand in the clouds; the whole enclosed in a compartment with this motto, ET VSQVE AD NVBES VERITAS TVA. Sometimes that of the brazen serpent, used by Binneman.

" The Description of the Low countreys," &c. as p. 1196. 1593.
News from Brest, or a diurnal of Sir John Norris, &c. Quarto. 1594.
" An admonition to such as intend Matrimonie," &c. as p. 1157. 1594.
" THE Jewell House of Art and Nature. Conteining diuers rare & 1594. profitable Inuentions, together with sundry new experimentes in the Art of Husbandry, Distillation & Moulding.—by Hugh Platte, of Lincolnes Inne Gent.'—Printed by him—on Breadstreat hill, at the—Star, and are to be solde in Paules Churchyard, 1594." On the back are the armes

---

[1] This Jewel house consists of five apartments, or books, each with a separate title-page, &c. so as to sell single, occasionally; but have the same running title, 'The Jewel-house of Art and Nature. It is so at least with the first three books, viz. Divers new experiments, 2 Divers conceits of Husbandry, 3 Chimical concl sions concerning Distillation. 4 Of moulding, casting, &c. 5 An offer of certain new inventions, which the author proposes to disclose upon reasonable considerations, &c.

of Robert earl of Essex, to whom this book is dedicated. Licensed. W.H. Quarto.

1594. "An ould facioned Loue: translated from Watson's Amintas by I. T. Gent." Quarto.

1594. "The Lamentation of Troy for the death of Hector, &c. by I. O." Quarto.

1594. "THE PRESENT state of Spaine. Translated out of French." His device. "Imprinted—by P. S. for Rd. Serger, 1594." W.H. Again 1596. Quarto.

1595. "The Copie of a Letter from the French King to the People of Artoys & Henault, requesting them to remoue the forces gathered by the King of Spaine, from the Borders of France, otherwise denouncing open warre. Also, a Declaration of the French Kings proclaiming open warre against the King of Spaine & his adherents, and the causes mouing thereto.—for Tho. Millington, and are to be sold at his shop vnder S. Peters in Cornhill, 1595." Quarto.

595. "THE FIRST FOWRE Bookes of the ciuile wars between the two houses of Lancaster and Yorke. By Sam. Daniel.—Printed—for Simon Waterson, 1595." In the octave stanza. The 4th book ends on folio 88. A 5th is added; the folios continued to fol. 108. W.H. Quarto.

1595. "A Briefe Description of Hierusalem and of the Suburbs therof, as it florished in the time of Christ. Wherto is annexed a short Commentarie concerning those places which were made famous by the Passion of Christ, &c. confirmed by—Histories of Antiquity.—Hereunto also is appertaining a liuely & beawtifull mappe of Hierusalem, with Arithmeticall directions correspondent to the numbers of this Booke. Translated out of Latin—by Tho. Tymme, Minister. Printed—for Tho. Wight, 1595." Dedicated "To—Sir John Puckering—Lorde Keeper," &c.—The preface.—Some verses by Tho. Newton. 112 pages, and a table. W.H. Quarto.

1595. "The Secrets of the reuerend Maister Alexis of Piemont, containing excellent remedies against diuers diseases," &c. Four parts; each with a separate title-page, but the signatures and paging continued progressiuely. "Imprinted for Tho. Wight, 1595." See the several parts described under John Wight. W.H. Quarto.

1595. "The Tragedie of Antonie. Doone into English by the Countesse of Pembroke." His device. "Imprinted at London for Wm. Ponsonby, 1595." Borders of metal flowers at the top and bottom of every page. G 7, in eights. At the end, "At Ramsbury, 26 Nouember, 1590." On the back of the last leaf "Printed—by P. S. for Wm. Ponsonby, 1595." W.H. Sixteens.

1596. "Sundrie nevv and Artificiall remedies against Famine. Written by H. P. (*Hugh Plat*) Esq. vppon thoeccasion of this present Dearth. Non est quo fugias à Deo irato nisi ad Deum placatum. Aug." His device. "Printed by P. S.—on Breadstreet hill, &c. 1596." C, in fours. W.H. Quarto.

1596. "ACTES And Monuments of matters—happening in the Church, &c. Now

PETER SHORT.

Now againe, as it was recognised,—by the Author Maister John Foxe, the fift time newly imprinted, Anno 1596. Menf. Iun.—Printed by *him*, dwelling on Breadsteete hill at the—Starre, by the assigne^m of R. Day." 728 pages besides the prefixes. W.H. Demy Folio.
" The second volume of the ecclesiasticall historie," &c. as p. 648. 1597.
" New recognised" &c. Folio.
" A Plaine and Easie Introduction to practicall Musicke, Set downe 1597. in forme of a dialogue: Deuided into three partes, The first teacheth to sing, with all things necessary for the knowledge of pricktsong. The second treateth of discante, and to sing two parts in one vpon a plain song or ground, &c. The third & last part entreateth of composition of 3, 4, 5 or more parts, with many profitable rules to that effect. With new songs of 2, 3, 4, and 5 parts. By Tho. Morley, Batcheler of musiek, & one of the gent. of hir Maiesties Royall Chappell. Imprinted &c. 1597." In the same compartment as The Cosmographical Glasse, p. 630, and 631. Dedicated " To the most excellent Musician Maister William Birde, one of the gentlemen of her Maiesties chappell." Commendatory verses by Ant. Holborne, A. B. and I. W.---To the Reader. B b 6, in fours; then " Annotations necessarie for the vnderstanding of the Booke," &c. 12 leaves more. W.H. Folio.
" CANZONETS. Or little short Songs to foure voyces: collected out 1597. of the best & approued Italian Authors. By Tho. Morley,ⁿ Gent. of her Maiesties Chappell. CANTVS. Imprinted—by *him*, dwelling on Bredstreet hill at the—Star & are there to be sold. 1597." C, in fours. W.H. Also, the TENOR, & BASSVS. Quarto.
" Seuen sobs of a sorrowfull soule for synne," &c. as p. 1260. 1597.
Blundevill's treatises of horsemanship. See p. 950. W.H+ 1597.
Carie's farewell to Physick; with the hammer for the Stone, as p. 957. 1597.
Part of Du Bartas's Divine Weeks, translated by Josuah Sylvester. 1598. The whole was probably printed. I have, The Eden, The Deceipt, and The Babilon, printed by him without intimating for any other person; The Furies, The Handy-crafts, and The Ark, printed by him, " for Wm. Wood,—at the West ende of Paules, 1598." Each with a separate title-page, dedication, &c. Sixteens.
The Imitation of Christ; with the Soliloquium Animæ. As p. 1206, &c. 1598.
" William Alabasters seuen motiues, Remoued and confuted by John 1598. Racster." Deuice the brazen serpent. " Printed by *him* for Andrew

---

^m The partners in this impression, with their shares, were as follow, viz. Mr. Harrison, 100; Mr. Bishop, 100; Mr. Watkins, 200; Mr. Wight, 200; Mr. Newbery, 100; Mr. Coldock, 100; Mr. Norton, 100; Mr. Ponsonby, 100; Mr. Dewce, 100; and Mr. Woodcock, 100. " At a Court holden at Stationers' Hall, 7 April, 1595. Yt is agreed that P. Short shall finish the impression of the B. of Martyrs from the place where Mr. Denham left,—For which he is to have after the rate of xvij s. vj d. for a booke,—for paper & printinge.—The paper shalbe rated at vij s. the ream."

ⁿ He had a patent, dated 28 Sept. 1598, for 21 years, for the sole printing of Set Songs in parts, &c. in any tongue, serving for either Church or chamber; and also for the ruling of any paper by impression, to serve for the pricking or printing of any song, &c. to sung or plaid in church, &c.

Wife

PETER SHORT.

Wife dwelling in Paules Church-yard at the—Angell, 1598." Dedicated " To—Robert Earle of Essex, &c.—Ad Academicos & suos salutem longam, Epistolam breuem mittit. I. R.—Ad lectorem Epigramma Authoris." 43 leaves. W.H. It was printed also with this title, " The seuen Planets, or wandring motiues of Will. Alabafter's wit, retrograded and removed by John Raciter. 1598." Fasti Oxon. Vol. 1; col. 148. Quarto.

1598. " The dailie exercise of a Christian," &c. as p. 962.
1598. " Palladis Tamia, Wits Treasurie, being the second part of Wits Commonwealth. By Francis Meres, maister of artes of both vniuersities. For Cuthbert Burbie." Quarto.
1598. " Henry IV. First part, 1598. P. S. for Andrew Wise.
1599. Cleopatra a tragedy, with other poems, by Sam. Daniel. Quarto.
1599. " The English Secretarie," &c. as p. 1048. " Now newly reuised, and In many parts corrected, &c. Printed by P. S. for C. Burbie, and are to be sold at his shop at the Royall Exchange, 1599." W.H. Quarto.
1599. " The effect of certaine Sermons touching the full Redemption of mankind by the death and bloud of Christ Jesus: &c. Preached at Paules Crosse and else where in London by—Tho. Bilson Bishop of Winchester. With a conclusion—for the cleering of certaine obiections made against the said doctrine.—Imprinted—for Walter Burre, and are to be sold in Paules Churchyard at the—Flower de luce, 1599." Introduced with an epistle To the Christian Reader. 420 pages. Quarto.
1599. " The practise of the new and old phisicke, wherein is contained the most excellent Secrets of Phisicke," &c. as p. 948. W.H. Quarto.
1599. " Foure Sermons preached by Master Henry Smith. And published by a more perfect copie than heretofore." His device. " Printed by P. S. for Cutbert Burby, 1599." On the back, " The contents, 1 The trumpet of the soule. 2 The sinfull mans search. 3 Maries choyce. 4 Noahs drunkennes." 30 leaves. W.H. Quarto.
1599. " Two Sermons preached by master Henrie Smith, with 3 prayers thereunto adioyned. And published by a more perfect copy then heretofore." His device. " Printed—for Wm. Leake, 1599." On the back, " The contents, 1 The sinners conuersion. 2 The sinners confession." &c. 22 leaves. W.H. Quarto.
1599. " A touchstone, whereby may easily be discerned, which is the true catholicke faith, of all them that profess the name of catholiques in the church of England, that they be not deceiued." Quarto.
1600. " Guilielmi Gilberti Colcestrensis, Medici Londinensis, De Magnete, magnetisq; corporibus, et de magno magnete tellure;" &c. The brazen serpent. " Londini, excudebat—MDC." On the back are the author's arms. It has prefixed, a preface; an epistle from Edw. Wright to the author; and an interpretation of certain words. 240 pages, with diuers schemes. W.H. Folio.
1600. " The nevv and admirable Arte of setting of Corne : With all the necessarie Tooles," &c. The device, A spade lying horizontally, and a scroll twisted about it, with this inscription, ADAMS TOOL REVIVED. Under it is a cluster of stems of wheat, full eared, growing out of the ground,

ground, as may be presumed, from one grain, with this motto under it *Magnus Deus in minimis.* Imprinted—by *him*,—at the—Starre on Bred-street hill, 1600." D, in fours. W.H. Quarto.

"Obseruations vpon the fiue first bookes of Cæsars commentaries, 1600. setting foorth the practise of the art military, in the time of the Roman empire. &c. By Clement Edmunds." The brazen serpent. "Printed by *him*, 1600." Dedicated "To—Sir Francis Vere, chiefe commander of her Maiesties forces in the seruice of the states, in the vnited Prouinces," &c. 199 pages, with a copper-plate plan of Cæsar's battle with the Helvetians. W.H. Folio.

He had also licenses for the following, viz. In 1593, "A pleasant conceyted hystorie called y Tamyng of a Shrowe." In 1594, "Hugonis Pratti," &c. as p. 1206. "The prentizs indentures:—for y companie of Merchant Tailors. A reporte of a monster borne at Ottingham in Holdernes. Also of a strange & hughe fishe driven on y Sands at Outhorne in Holdernes." The two last articles with Tho. Myllington. He printed also after 1600.

ARNOLD HATFIELD, stationer,

SEEMS to have been bound apprentice to Henry Denham in the interval of the chasm of the company's register. However he appears to have bought out his time, by the following minute. "Jan. 8, 1579. Yt is agreed y A. Hatfield shall depart from Mr. Denham, at a moneth from this day, paying iij li. whereof y Hall to have x sh. and Mr. Denham 1 sh." This must be understood for his being made free of the company also; for he binds apprentices in it to himself afterwards. He appears by an order of the Court of Assistants, 2 May, 1586, which may be seen under John Jackson, to have been concerned in a partnership with Ninian Newton, Edm. Bollifant, and the said Jackson. I have not however met with any book printed with all their names, nor any one of them, with an indication of a partnership annexed; only some with the names of N. Newton, and A. Hatfield, jointly; and others with Edm. Bollifant and John Jackson together. He dwelt in Lothbury, and kept shop at the Brazen Serpent in St. Paul's Church-yard. He used the device of a caduceus between two cornucopias.

"The Scepter of Iudah:" as p. 782. With Nin. Newton. 1584.
"A Booke of Christian exercise," &c. as p. 733. With N. Newton. 1584.
"C. Iulii Cæsaris Commentarii; Novis emendationibus, &c. Apud 1585. Arnoldum Hatfield & Nin. Newtonum, 1585." Pages 600, with woodcuts and maps. Italic types. W.H. By A. Hatfield alone, 1590. Sixteens.
"Iunii Iuuenalis et A. Persii Satyræ. Annotatiunculis, quæ breuis 1585. commentarij vice esse possint, illustratæ." See p. 880.
"THE HAVEN OF HOPE; Containing godlie Praiers & meditations 1585. for diuers purposes. Gathered by R, A."—The whole printed in borders.

O, in eights. At the end, Bollifant's device, and under it " Imprinted —by Arn. Hatfield & Nin. Newton, for Yareth James, dwelling in Newgate market ouer againſt Chriſt-church gate." Lambeth Libr. Sixteens.

1585. " The Brutiſh Thunderbolt :" &c. as p. 913. With N. Newton.

1586. " THE LORD MARQVES IDLENES : Conteining manifold matters of acceptable deuiſe; as ſage ſentences, prudent precepts, morall examples, ſweet ſimilitudes, proper compariſons, and other remembrances of ſpeciall choiſe. No leſſe pleaſant to peruſe, than profitable to practiſe: compiled by the right honorable L. William, Marques of Winchester, that now is.—Imprinted at London by A. Hatfield, 1586." On the back are the queen's arms upheld by religion and juſtice, over three diſtichs of Latin verſes. Dedicated " To—the Queenes moſt excellent Maieſtie.--- To the friendly Readers.—Baſing this viij of Nouember.—Wincheſter. ---In laudem operis hexaſticon G. Ch. The table." 94 pages.* Licenſed to the four partners W.H. Quarto.

1587. " A ſermon at Paules, on Iohn 16; 33, by Wm. Grauet B. D." Vicar of S. Sepulchers. Maunſell, p. 99. Licenſed to J. Jackſon. Octavo.

1587. " A briefe report of the militarie ſeruices done in the Low Countries, by the erle of Leiceſter. Written by one, that ſerued in good place there, in a letter to a friend of his." With Nin. Newton. For G. Seton. Quarto.

1588. " An Exhortation, To ſtirre vp the mindes of all her Maieſties faithfull Subiects to defend their Countrey in this dangerous time, from the inuaſion of Enemies.—by Anth. Marten, Sewer of her Maieſties moſt honorable chamber. Meliora ſpero." Device, the pelican, as p. 715, but without ſupporters. " Imprinted—by John Windet, and are to be ſold in Paules churchyard, at the Braſen Serpent, 1588." F, in fours. At the end are, " His (2) Prayers to this purpoſe, pronounced in her Maieſties Chappell and elſwhere." W.H. Quarto.

1589. " Repueſtas contra los Falſedades impreſas en Eſpana en biturperio de la Armada Ingleſa y Sennor Don Carlos Conde de Howarde, Grande Almirante de Inglaterra. Londres, en Caſa de Arn. Hatfielde, por Tho. Cadmana, 1589." Alſo, without date. Quarto.

1589. " Ioannis Strumii &c. Epiſtolæ ad R. Aſchamum," &c. as p. 921.

1590. " Diff. viri R. Aſchami Epiſtolarum libri III." See p. 920.

1590. " A Diſcourſe concerning the Spaniſh fleet in 1588, ouerthrown by the Queens Nauy, the Lord Charles Howard Lord high admiral of England. Written in Italian by Petruccio Ubaldina, citizen of Florence, and tranſlated for A. Ryther (a little from Leaden hall) next to the ſign of the Tower." 27 pages, with cuts of the ſeveral exploits and conflicts had with the ſaid fleet, graued by Ryther. Bagford's papers. Quarto.

1594. The Mirror and Manners of men. A poem: Written, he ſays, 50 years before, by Tho. Churchyard. Quarto.

1594. " Ayme for Finſburie archers, or an alphabeticall table of the names of euery marke within the ſame fields, with ther true diſtances, both by the map, and dimenſuration with the line. Publiſhed for the eaſe of the ſkilfull, and behoofe of the yoonge beginhers in the famous exerciſe of archerie; by I. I. and E. B. To be ſold at the ſign of the Swan in Grubſtreet by F. Sergeant." Hatfield alone. Sixteens.

" The

ARNOLD HATFIELD.

"The Historie of Philip de Comines, Knight, &c.—for John Nor- 1596.
ton." Folio.
" The Honor of the Lawe. Written by Tho. Chorchyard Gent. 1596.
Imprinted—by him for Wm. Holme, 1596." Quarto.
" A Sermon preached at Paules Crosse, 4 Dec. 1597. Wherein is 1597.
discoursed that all buying & selling of Spirituall promotion is vnlawfull.
By Iohn Howson, Student of Christes-Church in Oxford. Imprinted—
for Tho. Adams, 1597;" 50 pages. W.H. Quarto.
" A Second Sermon,—21 May 1598, vpon Math. 21, 12, 13. Con- 1598.
cluding a former Sermon—vpon the same Text: By Iohn Howson, &c.
—for Tho. Adams, dwelling in Paules Churchyard, at—the White Lion
1598." Pages 52. W.H. Quarto.
" Praxis Medicinæ vniuersalis;" &c. as p. 1152. 1598.
" Balfami, Opobalsami, Carpobalsami, & Xylobalsami, cum suo Cor- 1598.
tice explanatio. Interprete & Auctore Matthia de L'Obel, Medico
Insulano, Galla-Belga. Ad Clariss. & Honoratiss. Dominum D. Georgium
Careyum, Baronem de Hunsdon, Nobiliss. Ordinis Perifcelidis Equitem
auratum, Sacri Cubiculi Præfectum, Regiæ Maiestati à fanctioribus
Consilijs, & Vectis Insulæ Administratorem." Whose arms on 20 coats,
gartered & crested, are set beneath. " Londini excudebat Arn. Hat-
field, impensis Ioannis Norton, 1598." Dedicated " Clariss. & Honori-
ficentiss. Domino, &c.—Londini ipsis calendis Decembris—1597."—
Some Latin verses by Wm. Mount, Master of the Savoy, and Fran. He-
ring M.D. 40 pages, with cuts. W.H. Quarto.
" A Worlde of Wordes, Or Most copious, and exact Dictionarie in 1598.
Italian & Engliſh by Iohn Florio.—for Edw. Blount, 1598." In the
elegant compartment by C. T. as p. 926. Dedicated " To the
Right Hon. Patrons of vertue, Patterns of Honor, Roger Earle
of Rutland, Henrie Earle of Southampton, Lucie Countesse of Bed-
ford.—To the Reader."—A sonnet to each patron, and A friendes gra-
tulation to the author. 462 pages. At the end, " Imprinted—for Edw.
Blunt: and are to be sold at his shop ouer against the great North dore
of Paules Church." W.H. Folio.
" The Annales of Cornelius Tacitus. The Description of Germanie. 1598.
M.D.XCVIII." Dedicated " To—Robert Earle of Essex &c.—Richard
Grenewey.—To the Reader." 271 pages. " Printed—for Bonham &
John Norton." W.H. Folio.
" Apologia ecclesiae Anglicanae. Authore J. Juello." Sixteens. 1599.
" The Institution of Christian Religion," &c. as p. 612. With mar- 1599.
ginal notes, " conteining in briefe the substance of the matter handled
in each section." His device. " Printed—for Bonham Norton, 1599."
W.H. Quarto.
" A Briefe Replie to a certaine odious & slanderous libel, lately 1600.
published by a seditious Jesuite, calling himselfe N. D. in defence both
of publike enemies, & disloyall subiects, and entitled A temperate ward-
word, to Sir Francis Hastings turbulent watchword. Wherein not only
the honest & religious intention, and zeale of that good Knight is
defended,

7 C

defended, but alſo the cauſe of true catholike religion, and the iuſticeo her Maieſties proceedings againſt Popiſh malcontents & traitors, from diuers malitious imputations & ſlanders cleered, and our aduerſaries glorious declamation anſwered & refuted by O. E. defendant in the Challenge, & encounters of N. D. Hereunto is alſo added a certaine new Challenge made to N. D. in 5 encounters, concerning the fundamentall pointes of his former whole diſcourſe. And that ſuch as haue died in the popes quarrel, were rather falſe traitors than Chriſtian martyrs. Together with a briefe refutation of a certaine calumnious relation of the conference of Monſ. Pleſſis and Monſ. d'Eureux before the French king, lately ſent from Rome into England; and an anſwer to the fond collections & demands of the relator.—Imprinted—1600." Inſcribed "To N. D. alias Noddie, that lately tooke vpon him to pleade for P. P. (*Father Parſons*) and for all the Popiſh faction & hereſie.—The Preface to the Reader." Each of theſe three tracts has a ſeparate title-page, but the ſignatures are progreſſive to Ee 3, in eights. See p. 855, note w. W.H. Quarto.

1600. " Ecloga Oxonio-Cantabrigienſis," &c. as p. 1153.
1600. " The Hiſtorie of the vniting of the kingdom of Portugall to the Crowne of Caſtill: Containing the laſt warres of the Portugals againſt the Moores of Africke, the end of the houſe of Portugall, and change of that Gouernment. The deſcription of Portugall, &c. Of the Eaſt Indies, the Iſles of Terceres," &c. His device. " Jmprinted—for Edw. Blount, 1600." Dedicated " To the moſt noble & aboundant preſident both of Honor & vertue, Henry Earle of Southampton.—Edw. Blount. ---The Authors Apologie vnto the Reader.---The genealogie of the kings of Portugall." 324 pages, and a table. W.H. Folio.

1600. " GODFREY of Bulloigne, or The Recouerie of Ieruſalem. Done into Engliſh Heroicall verſe, by Edw. Fairefax Gent. Imprinted—by *him*, for I. Iaggard & M. Lownes, 1600." Dedicated " To her High Maieſtie." In verſe.---" The Allegorie of the Poem." 362 pages. W.H. Folio.

He had licenſes alſo as follows, viz. In 1586, " A ſermon treating of Affliction." With E. B. and J. J. In 1587, " Tratado del papa y de ſua autoridad, In Spaniſh." With E.B. & J. J. In 1589, " Inſtitutionum dialecticarum libri quatuor, a Johāne Sanderſono Lancaſtrenſi &c." With E. B. & J. J. He printed after 1600.

NINIAN NEWTON, ſtationer,

Son of Tho. Newton, of Upſall, Yorkſhire, Gent. was bound apprentice to Wm. Seres, for ten years from Michaelmas, 1569, and made free 8 Octo. 1579, then ſerving with H. Denham. He was concerned in a partnerſhip with Arn. Hatfield, Edw. Bollifant, and John Jackſon. See their ſeveral memoirs; particularly Hatfield's, where you will find ſeveral books printed in both their names, though one of them in particular is licenſed to all the four partners. I don't find his name joined with any other beſides Hatfield's. The two following books have no printer's

NINIAN NEWTON.

er's name but his own, but Hatfield especially seems to have been concerned in the first.
" Quincti Horatii Flacci Venusini,—poemata," &c. as p. 880. 1585.
" A New Herball, or Historie of Plants:" &c. as p. 941. " Imprinted at London by Ninian Newton, 1586." Pages 916, besides the prefixes and affixes. W.H. Quarto. 1586.

EDMUND BOLLIFANT, alias CARPENTER, stationer, was made free by Hen. Denham, 10 Apr. 1583; dwelt in Eliot's court in the Little Old Bailey; and was concerned in a partnership with Arn. Hatfield, Nin. Newton, and John Jackson, but in what manner is not easy to say. See their several memoirs. I do not find his name printed with any of the others, except Jackson's. 1 Dec. 1585, he and his partners, having printed Cato disorderly, were fined xx s. and all the book forfeited to the use of the company, according to their ordinances, and Bollifant committed to ward. However, 9 Feb. following ijs. vj d. was accepted in full payment. He used the device of Abraham and Isaac, with this motto, DEVS PROVIDEBIT.

" Pontici Virunnii—Britannicæ Historiæ Libri sex, &c. Itinerarium 1585. Cambriæ:" &c. as p. 912. " Cambriæ Descriptio: Auctore Sil. Giraldo Cambrense. Cum Annotationibus Dauidis Poueli," &c. His device. " Londini Apud Edm. Bollifantum impensis H. Denhami & R. Nuberij, 1585." These three treatises have separate title-pages, but the signatures and paging are continued, to p. 284; then an index to the whole. This last tract was included in the number of pages to the Itinerarium Cambriæ, in R. Newbery's Account; the title-page being then overlooked.

" M. T. Ciceronis Opera omnia quæ extant, a Dionysio Lambino 1585. Monstrolienfi ex codicibus manuscriptis emendata & aucta:—Eiusdem D. Lambini Annotationes, seu emendationum rationes singulis tomis distinctæ, &c. Fragmenta omnia quæ extant, &c. Londini, Per Ioh. I. & Edm. C. *(John Jackson & Edm. Carpenter)* 1585." In 9 tomes. W.H. I have also the 2d Vol. of the Orations, of this date and size, with " Ioh. Iacsonum & Edm. Bollifantum." I have not seen any other book with his initial C. or with the name of Carpenter. Other books in the names of Jackson & Bollifant, see under John Jackson's name.
Octavo.

" Aesops Fables in tru ortography, with grammer notz. Her-unto 1585. ar also cooined the shorte sentencez of the wyz Cato, imprinted with lyke form and order: both of which authorz ar translated out of Latin intoo English, by William Bulloker.
Geu God the praiz
That teacheth al waiz.
When truth trieth
Erroor flieth."
The Tables on 319 pages, and a table; the Cato 31 pages. W.H.
Octavo.

" W. Bullokars abbreuiation of hiz Grammar for English, extracted 1586. out

EDMUND BOLLIFANT.

out of his Grammar at larg for the spedi. parcing of English speach, and the eazier coming to the knowledge of grammar for other langages." 68 pages. See Hist. of Eng. Printing. III; 347.

1586. "The French Academie," &c. See it p. 1150. Quarto.
1587. "An Herbal for the Bible. Containing a plaine & familiar exposition of such Similitudes, Parables & Metaphors,—as are borrowed & taken from Herbs, Plants, &c. by observation of their Vertues—and effects, &c. Drawen into English *(from Levinus Lemnius)* by Tho. Newton. *(Cestrisirius)* Imprinted at London by Edm. Bollifant, 1587." Dedicated "To—Robert Earle of Essex, &c.—From my poore house at Little Ilford in Essex, 26 May, 1587." Pages 287, and a table. W.H. Octavo.
1587. "Analysis typica omnium, cum Veteris tum Novi Testamenti, librorum historicorum ad intelligendam rerum seriem, et memoriam iuuandam, accommodata. Autore Mose Pflachero, s.t.d." Quarto.
1588. "The Touch-stone of Wittes, by Edw. Hake. See Hist. of Eng. Poetry, Vol. III; p. 275.
(1588.) "A View of the marginal Notes of the Popish Testament," &c. as p. 1109.
1589. "Tri Livij Patauini Rom. Historiæ Principis," &c. as p. 1027. This book rectifies a mistake of John Albert, who says it was printed at Lyons. Tom. 1. p. 198. Ven. 1728.
1590. "Ioannis Twini—de Rebus Albionicis," &c. as p. 1027.
1591. "Amandi Polani a Polansdorf partitiones theologicæ, iuxta naturalis methodi leges conformatæ, duobus libris, quorum primus est de fide, alter de bonis operibus, &c. Edit. secunda." Licensed. Octavo.
1594. "A Booke of Christian Exercise appertaining to Resolution, that is, shewing how that we should resolve ourselves to become Christians indeed, by R. P. *(Rob. Parsons)* Perused by Edm. Bunny. Heb. 13; 8.—Imprinted by him, for T. Wight, and are to be sold at the great North doore of Paules, 1594." Dedicated "To—Edwin—Archbishop of Yorke," &c. W.H. Twenty-foure.
1595. "A New Herball or Historie of Plants." &c. See p. 1215. Licensed to him and partners.
1596. Clowes's Observations for curing Gun-shot, &c. as p. 1129.
1596. "A Perambulation of Kent:" &c. as p. 902. Dedicated "To—M. Thomas Watton, Esquier.—Seintcleres, 31 Jan. 1570." Pages 588, and a table. W.H. Quarto.
1597. "The Herball, or Generall Historie of Plantes. Gathered by John Gerarde of London, Master in Chirurgerie. Imprinted—by Iohn Norton. 1597." In a suitable compartment, designed and neatly engraved on copper-plate by William Rogers. Dedicated "To—Sir William Cecil—Lord High Treasurer of England," whose arms are on the back of the title-page. There are prefixed several gratulatory epistles and verses to the author, and his portrait, aged 53, anno 1598. The engravers mark WR, in one. 1092 pages, with wood-cuts of the plants, &c. and six indexes and tables. Colophon, "Imprinted—by Edm. Bollifant, for Bonham, and Iohn Norton, M.D.XCVII." W.H. Folio.

"Praxis

EDMUND BOLLIFANT.

" Praxis Medicini vniuerfalis ;"—&c. as p. 1152.   1598.
" The Ende of Nero and Beginning of Galba. Fower bookes of the   1598.
hiftories of Cornelius Tacitus. The life of Agricola. The fecond Edi-
tion, M.D.XCVIII." Dedicated " To her moft facred Maieftie.—Henry
Savile.—A. B. To the Reader." 227 pages. " Printed by *him*, for Bon-
ham & Iohn Norton." W.H.   Folio.
" Diana of George of Montemayor: Translated out of Spanifh—by   1598.
Bart. Yong of the Middle Temple Gent.—Printed by *him*, Impenfis
G. B." (*George Bifhop*) Dedicated "To—Lady Rich.—High Ongery, Effex,
28 Nouemb. 1598." Pages 496. W.H. See Hift. of Eng. Poetry,
in ; 344, note a.   Folio.
" A SPANISH GRAMMAR, firft collected & publifhed by Ric. Perciuale   1599.
Gent. Now augmented &c. by Iohn Minfheu Profeffor of Languages
in London. Hereunto—are annexed Speeches, Phrafes," &c. A Hat-
field's device. " Imprinted—by Edm. Bollifant, 1599." Dedicated
" To—the Gentlemen Students of Grayes Inne.---To the Reader.---
In opus M. Minfhei edendum Ioh. Keperi generofi ἐξαστίχον.---Soneto
de un capitano Hifpanol. del Autór. The Ploeme." 84 pages. W.H.
  Folio.
" A Dictionarie in Spanifh and Englifh, firft publifhed into the Englifh   1599.
tongue by Ric. Perciuale Gent. Now enlarged & amplified, &c. by
Iohn Minfheu. Hereunto—is annexed an ample Englifh Dictionarie,"
&c. Hatfield's device. " Imprinted—by Edm. Bollifant, 1599." Dedicated " To—Sir Iohn Scot, fir Henry Bromley, fir Edward Greuel
Knights, and Mafter William Fortefcue Efq;---To the Reader.---Direc-
tions," &c. 391 pages. W.H.   Folio.
" Pleafant & Delightfull Dialogues in Spanifh and Englifh, profitable   1599.
to the learner, and not vnpleafant to any other Reader. By Iohn Minfheu—Virefcit vulnere Virtus." Device, Time flying in the air, con-
ducting Truth into the Light ; a Demon pulling her back by her hair,
and fcourging her with fnakes. The motto, VERITAS FILIA TEMPORIS.
" Imprinted—by Edm. Bollifant, 1599." Dedicated " Al muy illuftre
Se'nor, Don Eduardo Hobby :—Juan Minfheu." 68 pages. W.H. Folio.
" A briefe declaration of the Sicknefs, laft Words, and Death of Philip   1599.
the fecond, king of Spain."   Quarto.
" TIMES LAMENTATION : or An expofition on the prophet Joel, in   1599.
fundry Sermons or Meditations. Ierem. 13 ; 17.—Printed by *him*, for
Geo. Potter, 1599." Dedicated " To—Sir Charles Blunt, Lord Mount-
ioy, Knight of the—Garter, &c.—Edw. Topfell." 444 pages. W.H.
  Quarto.
" De vera Chrifti ecclefia, contra Bellarminum liber."   1600.
" Reges, reginæ, Nobiles, et alij in Ecclefia collegiata B. Petri Weft-   1600.
monafterij fepulti, Ufque ad annum reparatæ falutis.* Excudebat—
unc." K 6, in fours. Richard Gough Efq;   Quarto.
" Maifon Ruftique, or The Countrie Farme. Compiled in the French   1600.
tongue by Charles Steuens and Iohn Liebault Doctors of Phyficke. And
tranflated into Englifh by Ric. Surflet, Practitioner in Phyficke. Alfo a
fhort collection of the hunting of the Hart, wilde Bore, Hare, Foxe, Gray
                             Conie ;

EDMUND BOLLIFANT.

"Conie; of Birds & Faulcoprie. The Contents—to be seene in the page following." Device, Paul planting, Apollos watering, over all &c. in an illuminated cloud. " Printed by *him*, for Bonham Norton, 1600." Dedicated " To—Sir Peregrine Bertie knight, *&c*.---To the Reader.---To—Jaques of Crufoll, Duke of Vzez, *&c*. Paris, Octob. 1582.—Io. Liebault." Some verfes. Liebault's preface. A caveat to the reader. With A table, when to fow divers feeds. 901 pages, and an Index, &c. W.H. Quarto.

1600. " The Hofpitall of incurable Fooles: Erected in Englifh, as neer the firft Jtalian modell—as the vnskilfull hand of an ignorant Architect could deuife. I pazzi, e li prudenti, fanno giuftifsima bilancia.—for Edw. Blount, 1600." Thus infcribed, " To the good old Gentlewoman, & her fpecial Benefactreffe, Madam Fortune, Dame Folly, (Matron of the Hofpitall) makes curtefie, *&c*.---Prologue of the Author to the beholders.---To my moft neere & capriccious Neighbor, ycleped Iohn Hodgfon, alias Iohn Hatter, or (as fome will) Iohn of Paules Churchyard, (Cum multis alijs, quæ nunc imprimere longum eft) Edw. Blount wifheth profperous fucceffe in his Monomachie, with the French & Spaniard. ---Not to the wife Reader.—Il pazziffimo." 158 pages. W.H. Quarto.

He had alfo licenfes for the following, viz. In 1587, " Le Theatre du Monde: in French only." In 1588, " An Eitaphe, or epigram, or elegies by Mr. Morfet." In 1590, " Analyfis Logica Evangelij, fecundum Johannem, Authore Joh. Pifcatore." Thefe all licenfed to him alone. He printed Warner's poetry 1602, and other books afterwards.

JOHN JACKSON, or JACSON, grocer, Was connected in a partnerfhip with Hatfield, Newton and Bollifant, the three laft preceding printers. This appears only by an order of the Court of Affiftants of the Stationers' company, 2 May 1586, " John Jackfon grocer, vfing the trade of printing, fhall have the fervice of this apprentice (*Ric. Browne, bound to Rd. Collins the company's clerk*) to be brought vp in the trade of printing, fynding him all neceffaries, &c. and the faid John Jackfon, in the name of himfelf, Nynyan Newton, Arn. Hatfield, & Edm. Bollyfant his partners in printing doth agree y as long as they or any three of them continue partners, they or any of them fhall not have reteine, or kepe but only this apprentice & one other which heretofore hath ben allowed vnto them, and bound to Arn. Hatfield. By me John Jakfon." I do not find him concerned in printing, either alone or with others before.

1585. " M. T. Ciceronis opera omnia," &c. as p. 1215.
1585. " ——— Epiftolæ ad familiares. A. D. Lambino—ex codicibus manufcriptis emendatæ. *&c*. Item P. Manutii annotationes breuiff. in margine adfcriptæ.—Excudebant Ioh. Iacfonus & Edm. Bollifantus, 1585." Perhaps the partners printed all Tully's works alfo of this fize. Sixteens.
1585. " Petri Rebuffi—Tractatus de Decimis." &c. Device, Abraham and Ifaac, &c. as p. 1215. " Londini Apud Iohan. Iacfonum, pro Geo. Bifhop, 1585." This is annexed to the treatife De facris Ecclefiæ minifteriis, &c. as p. 1149. W.H. Octavo.

" T he

JOHN JACKSON.

"The true difference betweene chriftian fubjection and vnchriftian 1586. rebellion; wherein the princes lawful power to command for truth, and indepriueable right to beare the fword, are defended againft the Popes cenfures, and the Iefuits fophifmes, vttered in their Apologie and Defence of Englifh Catholikes: With a demonftration, that the things reformed in the Church of England, by the lawes of this Realme, are truly Catholike, notwithftanding the vaine fhewe made to the contrarie, in their late Rhemifh Teftament: by Thomas Bilfon, Warden of Winchefter. Perufed and allowed by publike authoritie. Printed by *him* and Edm. Bollifant, 1586."* The queen's arms on the back, to whom the book is dedicated. ---The contents.---To the Chriftian Reader. Part 1 & 2, pages 430; 3 & 4, 686, and a table. W.H. Octavo.

"A Booke of Chriftian Exercife," &c. as p. 783. 1586.

"Pfeudographia G. Goffenii Natione Brabanti anno 1586. In Evan- 1587. gelicæ veritatis maximum præiudicium edita; A. I. Æengelrammo vero in defenfionum eiufdem veritatis & chriftianæ ipfius profeffionis ac famæ, quæ per eiufdem pfeudographi erroneos, falfos ac famofos libellos nimis villicatæ funt, detecta ac refutata. Iftiusmodi pfeudoapoftoli, operarij dolofi funt. 2 Cor. 11; 13. Londini Apud Joh. Jackfonum, 1587." Licenfed to the four partners. Lamb. Libr. Octavo.

"W. Chub his fermon of afflictions, on Rom. 8; 18. Octavo. 1587.

"A Difcourfive Probleme concerning Prophefies," &c. as p. 1026. 1588.

"Epithetorum Joann. Ravifii Textoris epitome, ex Hadr. Junii medi- 1588. ci recognitione. Accefferunt ejufdem Ravifii Synonima poetica, multo quam prius locupletiora." Printed for John Harrifon. I cannot fay which of them, having never feen the book. The printer's Latin preface to the Reader is dated, Kal. October 1564. Octavo.

"An Anfwer to the vntruthes publifhed & printed in Spaine, in glorie 1589. of their fuppofed victorie atchieued againft our Englifh Nauie, and the Right Hon. Charles Lord Howard, Lord high Admiral of England, &c. Sir Francis Drake, &c. Firft written and publifhed in Spanifh By a Spanifh Gentleman who came hither out of the Lowe Countries from the fervice of the prince of Parma, &c. Faithfully tranflated by I. L. Printed by *him*, for Tho. Cadman, 1589." On the back, "England to hir Queene, S. D." a ftanza of eight lines. Alfo, "England to hir Admirall," in feven lines, by I. Lea. Dedicated "To the---Lord high Admirall.---James Lea.---To the Queenes moft excellent Maiefty, &c.--- D. F. R. de M.---The author to hir Majeftie, S. D." an octave ftanza. "The tranflator." Three five-lined ftanzas. 55 pages, with fonnets, fongs, &c. W.H. Quarto.

"BIBLIOTHECA HISPANICA. Containing a Grammar, with a Dic- 1591. tionarie in Spanifh, Englifh, & Latine, gathered out of diuers good Authors:---By Ric. Percyuall Gent. The Dictionarie being inlarged with the Latine by the aduife &.conference of Mafter Tho. Doyley Doctor in Phyficke. Imprinted by *him*, for Ric. Watkins, 1591." On the back are the arms of the earl of Effex, to whom the book is dedicated.---To the reader.-- Some verfes, Latin and Englifh.---An analytical table. The grammar

grammar ends on F 1, in fours. The Dictionary has a separate title-page, and signatures, with a preface in Spanish and English. Z, in fours. W.H. Again 1592. Quarto.

1592. "The true vfe of Armorie, Shewed by Hiſtorie, and plainly proued by example: the neceſſitie therof alſo difcouered: with the maner of differings in ancient time, the lawfulnes of honorable funerals and moniments: with other matters of Antiquitie, incident to the aduancing of Banners, Enſignes, and marks of nobleneſſe, and cheualrie. By William Wyrley." Imprinted by *him* for Gabriel Cawood. Addreſſed "To—the lords, and others, the profeſſors of martiall difcipline." The real author of the profe part of this treatiſe, which ends on p. 28, was Mr. Sampſon Erdſwicke. See Dugdale's Ancient uſage of Arms, p. 4. Thereunto are annexed two poems in ſeven-lined ſtanzas, each of them ſigned Wm. Wyrley. The one entitled Lord Chandos; the other, Capitall de Buz; two original knights of the Garter. The ſignatures and paging are continued to p. 139. W.H. Quarto.

1593. "THE PHOENIX NEST. Built vp with the moſt rare and refined workes of Noblemen, woorthy Knightes, gallant Gentlemen, Maſters of Arts, and braue Schollers. Full of varietie, excellent inuention, and ſingular delight. Neuer before this time publiſhed. Set foorth by R. S. of the Inner Temple Gent." See Hiſt. of Eng. Poetry, III, p. 401, note b. Licenſed to him and partners. W.H. Quarto.

1594. "Liber Precum Publicarum," &c. as p. 1067. "Excufum Londini, per aſſignationem Franciſci Floræ. Cum priuilegio Regiæ Maieſtatis, 1594." To which is annexed.

1594. "Liber pſalmorum Dauidis prophetæ et regis, ad Hebraicam veritatem a Sebaſtiano Munſterio quam diligentiſſime verſus." With Hatfield's device. At the end, "Excudebat Ioan. Iackſonus." W.H. Octavo.

1594. "GIACOMO Di Graſſi his true Arte of Defence," &c. as p. 826.

(1595.) "Certaine very proper, and moſt profitable Similies, wherein ſundrie—foule vices, &c. are ſo plainly laid open,—that the Chriſtian Reader,—will be very fearfull, euen in loue that he beareth to God, to pollute & to defile his hart, his mind, his mouth or hands with any ſuch forbidden things. And alſo manie very notable vertues,—ſo liuely & truly expreſſed, according to the holy word, that the godly Reader,—will be mightily inflamed with a loue vnto them. Collected by Anth. Fletcher, miniſter of the word of God, &c. This preſent yeere of our happines, 1595. Pſal. 128; (1)—Printed by *him*, for Iſaac Bing." Dedicated "To—Gilbert Taulbot, Earle of Shrewſburie, &c.—22 May, 1595.---To the Chriſtian Reader." On the back of this epiſtle is a cut of an evil tree, of which Covetouſneſs is the root, and Self-love the top, the trunk is an union of the ſeven deadly ſins, the branches conſiſt of every enormity, infidelity, diſobedience, hypocriſy, &c. Juſtice has faſtened a cord to the top of the tree, in order to pull it down, while Truth is hacking at the root with an axe, the word of God. At the end of the Similies is a kind of epilogue in 14 ſhort ſtanzas, at the head of which is a cut of the author, ſitting in contemplation at a table, with a lighted candle burnt down to the ſocket,

having

having this motto, ALIIS INSERVIENDO CONSVMOR, on a scroll twisted about it. 160 pages, and a table of the chief matters. W.H. Quarto.

WALTER VENGE,

SEEMS to have been rather a bookseller, or stationer, by trade, than a printer. I find nothing more concerning him than what is gathered from the following book, viz.

"The mathematical Jewel, Shewing the making, and most excellent vse of a singuler Jnstrument so called: in that it performeth with wonderfull dexteritie, whatsoeuer is to be done, either by Quadrant, Ship, Circle, Cylinder, Ring, Dyall, Horoscope, Astrolabe, Sphere, Globe, or any such like heretofore deuised: yea, or by most Tables commonly extant: and that generally to all places from Pole to Pole. The vse of which Iewel is so aboundant and ample, that it leadeth any man practifing thereon, the direct pathway (from the first steppe to the last) through the whole Artes of Astronomy, Cosmography, Geography, Topography, Nauigation, Longitudes of Regions, Dyalling, Sphericall triangles, Setting figures, and briefely of whatsoeuer concerneth the Globe or Sphere: with great and incredible speede, plainenesse, facillitie, and pleasure: The most part newly founde out by the Author, Compiled and published for the furtherance, as well of Gentlemen and others, desirous of speculatiue knowledge, and priuate practise: as also for the furnishing of such worthy mindes, Nauigators, and traueylers, that pretend long voyages, or new discoueries. By Iohn Blagrave,[o] of Reading, Gentleman, and well willer to the Mathematickes, who hath cut all the prints, or pictures, of the whole worke with his owne hands. 1585." The device an armillary sphere, with I. BLAG. SCVLP. "Imprinted at London by Walter Venge, dwelling in Fleetelane ouer against the Maidenhead." Dedicated "To—Sir William Cycill—Lord high Tresurer of Englande, &c. ---The Author, to the curteous Reader.—20 Ian. 1584."---Some verses by "The Authour in his owne defence." Others entitled "The Authours dumpe."---A table of contents; another explaining terms. 124 pages. "Imprinted—by Tho. Dawson for Walter Venge." Licensed.* W.H.   Folio.   1585.

He had also a licence for a ballad, in 1584, "Of a strange example of a mayden child borne vpon Sonday 3 Jan. 1584, in the Mynoryes, without Aldgate of London."

[o] In an edit. of this book, in the Ashmolean museum, is writ concerning Blagrave the wood cutter as follows:
"Here stands Mr. Gray master of this house, And his poor catt, playing with a mouse.
John Blagrave married this Grayes widdow, (she was a Hungerford) This John was symple, had yssue by this widdowe. 1. Anthony, who marryed Jane Borlass. 2. John, the author of this booke. 3. Alexander, the excellent chefs player in England. Anthony had sir John Blagrave, knight, who caused his teeth to be all drawn out, and after had a sett of ivory teeth in agayne."

EDWARD

[1222]

EDWARD VENGE, stationer,

SON of Edward Venge of Reading Berks, painter, was bound apprentice to Henry Carre for nine years, from Christmas, 1578, and made free 3 July, 1588. He was concerned with James Roberts in printing The brief catechism, &c. See p. 1030.

1589. " —The Scottish Queens Buriall at Peterborough," &c. as p. 1162.
1590. " A moste true discourse declaring the damnable lyfe & deathe of one Stubbe Peter, a high Jermayne borne, a Sorcerer, who in the likeness of a Wolfe comited many murders 25 yeres together: and for the same was executed in the cytye of Bedbur, near Coleyn, 31 March, 1590." Licensed.

SIMON WATERSON, stationer,

SON of Ric. Waterson, was made free by his father's copy, 14 Aug. 1583; and came on the livery 1 July, 1592. He dwelt at the sign of the crown in St. Paul's Churchyard, at Cheap gate, and sold books for John Legat of Cambridge, and Joseph Barnes of Oxford. He had many books printed for him without the printer's name. He put up a memorial tablet for his father, in St. Faith's under St. Paul's, 1 Jan. 1599. He lived to be twice master of his company, was chosen a commoncouncilman 1608, and the year following a governor of two of the royal hospitals. He married Frances daughter of Tho. Legat, Esq; in the county of Essex, by whom he had 7 daughters and 3 sons. He died 16 March 1634, aged 72, as appears by his monumental inscription, set up in St. Faith's by John, his youngest surviving son, and executor. Both his father's and his may be seen in Dugdale's St. Paul's p. 126. This son succeeded him in his house and business.

1585. " The Worthy tract of Paulus Iouius, contayning a Discourse of rare inuentions, both Militarie and Amorous, called Imprese; VVhereunto is added, a Preface contayning the Arte of composing them, with many other notable deuises. By Samuel Daniell, late Student in Oxenforde." Supposed to be the first piece of Daniel's. Wood says, " He was 23 years old when he translated this book." Dedicated " To—Sir Edw. Dimmock, Champion to her Maiestie.---To his good frend S. Daniel,—N. W.---To the friendly Reader.—S. D." Of the properties of a perfect Impresa. H, in eights. For him. Licensed. W.H. Octavo.
1591. " Giles Clayton of the approved order of Martiall discipline, with euery particular officers—dutye: also a second booke for the true ordering & Imbatteling of any number soeuer," &c. For him. Quarto.
1591. " The second part of—Christian exercise," &c. as p. 1101. Again 1592.
1593. " The defence of Contraries. Paradoxes against common opinion, debated

SIMON WATERSON.

debated in forme of declamations in place of publia cenfure: only to exercife yong wittes in difficult matters. Wherein is no offence to Gods honour, the eftate of Princes, or priuate mens honeft actions; but pleafant recreation to beguile the iniquity of time. Tranflated out of French by A. M. (*A. Munday*.) one of the meffengers of her Majefties chamber. Parere & abftine. Imprinted—by John Windet for him, 1593." Pages 99: Quarto.

"The fecond part of —Chriftian Exercife," &c. as p. 1932. 1594.
"—of the ciuile wars between—Lancafter & Yorke," &c. as p. 1208. 1595.
"The confolation of the foule, being an affurance of the forgiuenes of finnes, with the moft notable promifes of God, contained in the holy fcriptures, briefly applied with certaine examples touching the great mercy of God towards miferable finners. Made by Jo. Chaffanion. Tranflated by H. S. of Grey's Inne, gent." For him. Octavo.

He had alfo licenfes for the following, viz. In 1584. " Of the vfe of the celeftial Globe, by Geo. Turnebull,". In 1589, " A Chriftian Directory guidinge all men to theire Salvation: with J. Charlewood." In 1591, " Delia, conteyning divers fonets, with the Complainte of Rofamon. In 1593. " The tragedye of Cleopatra."

JOHN WINDET, ftationer,

WAS made free by John Alde, 13 April, 1579. At firft fetting out in bufinefs he was concerned in partnerfhip* with Tho. Judfon. See p. 966, note p. He came on the livery 4 July, 1586. May 12, 1589, he was fined xl f. for printing books contrary to the ordinances, which he promifed to pay; but paid only vij f. in full, the laft of September. He was fined again 12 Jan. 1591-2, for keeping an apprentice unprefented, ij f. vj d. In 1594 and 1595 he ferved Renter. In 1597 he was fined for binding and enrolling two apprentices without prefenting them to the Mafter and Wardens, contrary to the ufual and laudable ordinances of this company, xl f. which was paid 7 Aug. 1598. He ferved Under Warden in 1599. He printed the affignees of R. Day, and fucceeded John Wolf as printer to the Hon. city of London, about 1603. He dwelt at the fign of the White Bear, in Adling ftreet nigh Bernard's Caftle, and in the year 1589, at the Crofs Keys, near Paul's Wharf. He ufed a pretty device of Time mowing down a fheaf of corn behind a clafped book with *Verbum Dei manet in æternum*, the whole in a compartment having the queen's arms at top, the City's on the right fide, and the Stationers' on the left, his fign at the bottom with I. W. over it: the

*" Jan. 15. 1583-4. Yt is ordered that John Windet & Tho. Judfon fhall enter into bond one to another to be partners in pointing for v yeres, and that duringe that time, the' faid W. fhall be accompted the maifter printer. And yf after thexpirac'on of the faid v yeres they breake from partnerfhip by W's meanes, without confent of the Mafter, Wardens & Affiftants, or the more part of them: That then from thensforth Judfon fhall enioy the place of a maifter printer, according to thelection that hath ben already made, wherein the choice hath fallen vpon him." Stat. Reg. B. fol. 436. b.

motto,

JOHN WINDET.

motto, NON SOLO PANE VIVET HOMO. Luke 4. Sometimes he used other devices, perhaps those of his employers.

1581. " Tim. Bright—his abridgement of the booke of Martirs. printed by John Windet, 1581." Maunsell, p. 23. See it in 1589. Quarto.

1584. " PETERS FALL. Two Sermons vpon the Historie of Peters denying Christ. Wherin we may see the causes of mans falling from God, and the manner how, both of the wicked thorough incredulitie, and of the godly by infirmitie: and also the way that God hath set downe in his word to rise againe. By John Vdall, Preacher of the word of God at Kingston vpon Temmes. Prou. 24; 16.—Printed at London by Iohn Windet & Tho. Iudson, for Nich. Lyng, 1584." Dedicated " To—Frauncis Earle of Bedford, &c.---To 'the Godly & well disposed Reader.---The Text out of St. Mathew & St. Luke conferred both togither." G 4, in eights. W.H. Octavo.

1584. Ovid's Metamorphoses. See p. 698. With T. Judson. Quarto.

1584. " Foure Sermons vppon the seuen chiefe vertues or principall effectes of faith, and the doctrine of election: &c. Preached at Malden in Essex by Maister Geo. Gifford, penned from his mouth and corrected & giuen to the Countesse of Suffex, for a Newyeeres gift. Iames 2; 18.—Printed by Tobie Cooke—1584." Dedicated " To—Frauncis Coütesse of Suffex.—Ric. Iosua Senior.---To the godly Reader.—Ric. Iosua Iunior." The text, 2 Pet. 1; 1—11. G 3, in eights. " Imprinted—by *him* & T. Iudson for Toby Cooke,—1584." W.H. Octavo.

1585. " A Compendious Chyrurgerie:—by Ihon Banister." See p. 1203.
1585. " A Godlie—Commentarie vpon—Ecclesiastes," &c. as p. 1159.
1585. " The English Creede," as p. 1136.
1585. " Songes and sonets, written by the right honourable lord Henry Howard, late earle of Surry, and others." See p. 812.
1585. " Compendium Grammaticæ Græcæ Jacobi Ceporini, ex postremaï authoris editione, nunc primum opera Joannis Frisii Tigurini castigatum et auctum, Quæ priori editioni accessere sequens pagella indicabit. Londini Excudebat." At the end, under his device, " Impressum—per Johannem Windetum."

1586. " LLIVER GWEDDI GYFFREDIN a gwenidogaeth sacramentæ, ac eraill grynnedde seu, a ceremoniæ yn eccles. Loecr." The book of Common Prayer, in Welsh. " Vewed, perused, and allowed by the bishops, according to the act stablished for the translation of the bible, and this booke into the British tongue; at the costes and charges of Thomas Chard." Quarto.

1586. " The Droomme of Doomes Day. Wherein the frailties & miseries of man life are liuely portrayed & learnedly set forth. Deuided as appeareth in the Page next following. Translated & collected by Geo. Gascoigne Esquyer. Tam Marti, quam Mercurio.—Imprinted by *him*, for Gabriell Cawood: dwelling in Paules Churchyard, at the signe of the Holy Ghost, 1586." On the back, " This worke is deuided into three partes,—The view of worldly Vanities.—The shame of sinne.—The Needels Eye.—Hereunto is added a priuate Letter,—against the bitternesse of Death."

Death." By I. P. to his familiar friend G. P. Dedicated "To—his singuler good Lord & Maister, the Earle of Bedforde.—From my lodging where I finished this trauaile in weake plight for health as your good Lordshippe well knoweth, this 2 of Maye, 1576." Pages 264. W.H. Quarto.

" An Historicall Discourse, or rather tragicall Historie of the citie of 1586. Antwerpe, since the departure of king Phillip king of Spaine out of Netherland, till this present yeare, 1586." The arms of Antwerp. " Printed by *him*, dwelling in Adling street, at the signe of the White Beare, neere Baynards castle." 27 leaves. At the end, under his device of Time, &c. " Printed—1586." W.H. Quarto.

" A Methode vnto Mortification: Called heretofore, The contempt 1586. of the world, and the vanitie thereof. Written at the first in the Spanish, afterward translated into the Italian, English, and Latine tongues: now last of al perused &c. by Tho. Rogers. Allowed by authoritie. 1 John 2; 15.—Imprinted—1586." Dedicated " To—his good friendes, M. H. Blagge, and T. Pooley, Esquires, Iustices for the maintenance of the peace within the countie of Suff. S." Wherein we learn that the author was F. Diego de Stella of the order of S. Fr. and that it had been translated into English by G. C.—" From Horningsheath, 1 Octo. 1586." Pages 499, and three indexes. W.H. Twenty-fours.

" A Mirror for The Multitude, or Glasse, Wherein maie be seene the 1586. violence, the error, the weakneffe & rash consent of the multitude, and the daungerous resolution of such, as without regard of the truth, endeuour to runne & ioyne themselues with the multitude: With a necessary conclusion, that it is not the name or title of a Protestant, Christian or Catholike, but the true imitation of Christ, that maketh a Christian. By I. N." Device, A pair of wings on a shield, parted by a pale, with the word SVRSVM on it, and this motto up and down the sides, NON VI, SED VERITATE. " Caueamus. Virtutem incolumem odimus, sublatam oculis sero quærimus inuidi.—Printed by *him*, 1586." On the back are the queens arms, to whom the book is dedicated, by John Norden.---To the Christian Reader. 116 pages. Licensed. W H. Octavo.

" A Treatise of Melancholy." &c. as p. 1074.—" Imprinted by *him*, 1586. 1586." Pages 276. Licensed. W.H. A subsequent edition to Vautrollier's. Octavo.

" The English Myrror. A Regard Wherein al estates may behold 1586. the Conquests of Enuy: Containing ruine of common weales, murther of Princes, cause of heresies, and in all ages, spoile of deuine & humane blessings, vnto which is adioyned Enuy conquered by vertues. Publishing the peaceable victories obtained by the Queenes most excellent Maiesty against this mortall enimie of publike peace & prosperitie, and lastly A Fortris against Enuy, Builded vpon the counsels of sacred Scripture, Lawes of sage Philosophers, and pollicies of well grounded common weales: wherein euery estate may see the dignities, the true office & cause of disgrace of his vocation. A worke safely, and necessarie to be read of euerie good subiect. By Geo. Whetstones Gent. Malgre. Seene & allowed.—Printed by *him*, for G. Seton, and are to be sold at his shop vnder

vnder Alderſgate, 1586." On the back are the queen's arms over an acroſtic on Elizabetha regina; to whom the book is dedicated.—" To the moſt Honourable the Nobilitie of this floriſhing Realme of Englande. —R. B. to the Reader:" in verſe. 249 pages. W.H. Quarto.

1586. " The Blazon of Gentrie: Deuided into two parts. The Glorie of Generoſicie.—Lacyes Nobilitie. Comprehending diſcourſes of Armes and of Gentry.—Compiled by Iohn Ferne Gent.—for Toby Cooke, 1586," Dedicated " To Edmund Lorde Sheffilde—13 Sept. 1586.—To the honorable aſſemblyes of the Innes of Court," &c. The Glory of Generoſity on 341 pages; Lacy's Nobility 130. W.H. Quarto.

1586. " Sir William Harbert Knight his Letter to a Romen pretended Catholike, wherein vpon occaſion of controuerſie touching the Church the 12, 13 & 14 chapters of the Reuelation are expounded." Alſo his " Sidney, or Baripenthes, briefely ſhadowing out the rare & neuer ending laudes of that moſt honorable & praiſe-worthy geat. Sir Phillip Sidney knight. 1586." Licenſed. Quarto.

1587. " A Second Sermon vpon the ix Chapter of the holy Goſpel of Ieſus Chriſt according to Saint Iohn. Preached at S. Maries in Oxford, the 11 of December 1586. By Iohn Chardon, D. D. Seene, peruſed & allowed. Eccleſ. xj. vj.—Imprinted for Tobie Cooke, 1587." Dedicated to Ambroſe Dudley earl of Warwick. C, in eights. W.H. Octavo.

1587. " A Sermon Faithfullie and truelie publiſhed: According as it was preached at the Courte at Greenewviche, the Tweſday in Eaſter weeke, before the Right hon. and diligent Auditory. By M. Peter Wentworth, Parſon of Much-Bromelie in Eſſex, and chaplaine to—the L. Darcy.—Printed by him, for Tho. Gubbin, and John Winnington, 1587." The text, Pſal. 2; 10, 11. C 2, in eights. W.H Octavo.

1587. " Academiæ Cantabrigienſis Lachrymæ, tumulo nobiliſſimi equitis, D. Philippi Sidneij ſacratæ, per Alexandrum Nevillum.—Ex officina —impenſis Thomæ Chardi,—cɪɔ. ɪɔ. lxxxvij. Febr. xvj."

1587. " The true Triall of a mans owne ſelfe, which is a ſeuerall examination of a mans conſcience, vpon euery commandements wherein euery Chriſtian may behold his ſpirituall deformitie by nature deſcribed, &c. Written in Latine by And. Hyperius, tranſlated by Tho. Newton." Printed in a very neat Roman letter, having before it an almanac with ſeveral hiſtorical remarks. Twenty-fours.

1587. A Defence of the Execution of Mary queen of Scots. " The Bleſſednes of Brytaine, Or a Celebration of the Queenes Holyday, &c. by Maurice Kyffin. 1587." In verſe. Licenſed. Quarto.

1588. " A Defence of the Government eſtabliſhed in the Church of Englande for Eccleſiaſticall Matters." &c. as p. 1195.

1588. " Characterv. An Arte of ſhorte, ſwift, and ſecrete writing by Character. Inuented by Timothe Bright, Doctor of Phiſike. Imprinted —by him, the Aſſigne of Tim. Bright, 1588. Cum priuilegio Regiæ Maieſtatis. Forbidding all other to print the ſame." Dedicated " To the moſt high & mightie Prince Elizabeth, of England, &c. Queen."

JOHN WINDET.

Queen." &c. Wherein he claims the invention. K 2, in twelves. W.H.
Twenty-fours.

"An Exhortation, To ſtirre vp the mindes of all her Maieſties 1584.
faithfull Subiects, to defend their Countrey," &c. as p. 1212. Licenſed.
Quarto.

On the given up Deventer in Overiſſel unto ſir William Stanley.* See 1588.
p. 921 and 1045. Quarto.

"The raſing of the foundations of Browniſme." &c. Quarto. 1588.

"A Briefe Inſtruction—to keep bookes of Accompts," &c. as p. 743. 1588.
Imprinted by him, at the ſigne of the White Beare, nigh Baynards
Caſtle, 1588." In his epiſtle to the reader he ſays, "And knowe ye for
certaine, that I preſume ne vſurpe not to ſet forth this worke of mine
owne labour and induſtrie, for truely I am but the reneuer and reuiuer of
an auncient old copie, printed here in London the 14 of Auguſt 1543.
Then collected, publiſhed, made, and ſet forth by one Hugh Oldcaſtle,
Scholemaſter, who, as appeareth by his treatiſe, then taught Arithmetike,
and this booke in Saint Olhaues pariſh, in Marke-lane." To this is
added, "A Short and Plaine Treatiſe of Arithmeticke, in whole numbers, compriſed into a briefer method than hetherto hath bin publiſhed.
By John Mellis." The ſame year.* Licenſed. W.H. Octavo.

"Of the end of this world," &c. See p. 1135, and 1136. Quarto. 1589.

"Eight Sermons vpon the firſt foure Chapters & part of the fift, of 1589.
Eccleſiaſtes, Preached at Mauldon, by G. Giffard.—for Toby Cooke.—
1589." Dedicated "To—the Lady Anne Counteſſe of Warwike." 140
leaves. W.H. Sixteens.

"An Abridgement of the Booke of Acts and Monumentes of the 1589.
Church: Written by that Reuerend Father, Maiſter Iohn Fox: and now
abridged by Timothe Bright, Doctor of Phiſicke, for ſuch as either
thorough want of leyſure, or abilitie, haue not the vſe of ſo neceſſary an
hiſtory." A cut of the pope ſitting on a throne in his pontificals flaying
a lamb, held by a friar: martyrs ſeen burning at a ſtake in the offing.
Over it, "All day long are we counted as ſheepe for the ſlaughter. Pſal.
44." Under it, "How long Lord, holy and true? Apoc. 6, 10. Imprinted at London by I. Windet, at the aſſignment of Maſter Tim. Bright,
and are to be ſold at Pauls wharf, at the ſighe of the Croſſe-keyes, 1589.
Cum Gratia, & Priuilegio Regiæ Maieſtatis." Dedicated "To—Sir
Francis Walſingham, Knight, &c.—To the Chriſtian Reader." The
firſt vol. 504 pages; the ſecond 288, and a table. W.H. Quarto.

* He begins thus: "Cicero did account it worthie his labor, and no les profitable to the Roman common weale (Moſt gratious Souéraigné,) to inuent a ſpeedie kinde of wryting by Character, as Plutarch reporteth in the life of Cato the yonger. This inuention was increaſed afterward by Seneca, that the number of Characters grue to 7000. Whether through iniurie of time, or that men gaue it ouer for tedioufnes of learning, nothing remaineth extant of Ciceroes inuention at this day." But query, whether the numerous contractions in ancient MSS. and early printed books be not what are here meant. He goes on, "Vpon conſideration of the great vſe of ſuch a kinde of writing, J haue inuented the like: of fewe Characters, ſhort & eaſie, euery Character anſwering a word: My inuention meere Engliſh, without precept, or imitation of any. The vſes are diuers?" &c.

"The

1589. "The Common-vvelth of England, and maner of gouernment thereof. Compiled by the hon. Sir Thomas Smith, Knight," &c. as p. 1061. "With new additions of the cheefe Courts in England, the offices thereof, & their feuerall functions, by the fayd Author: Neuer before publifhed.—" Seaton's rebus: a tun floating on the fea, with a lilly iffuing out of the bung-hole, and a ferpent twifted about it: in a compartment with his arms at the bottom, and this motto, INVIDIA SIBI ET ALIIS VENENVM. "—Imprinted by *him*, for Gregorie Seton,—1589." An epiftle, "To the Reader." 148 pages. Quarto.

1589. "Ant. Tirrell his fermon on Matth. 12; 43,—45." Sixteens.

1589. "An Hiftoricall Dialogue touching Antichrift," &c. as p. 1136.

1590. "The vfe of the Celeftial Globe in plano, fet foorth in two hemifpheres: Wherein are placed all the moft notable Starres of heauen, according to their longitude, latitude, magnitude & conftellation:—their names, both Latin, Greeke, Arabian or Chaldee: Alfo their nature, and the Poetical reafon of each feuerall Conftellation. Moreouer,—the declination—right afcenfion, &c. Set foorth by Tho. Hood, Mathematicall Lecturer in the citie of London, fometimes Fellow of Trinitie Colledge in Cambridge. The Hemifpheres are to be fold in Abchurch-lane, at the houfe of Th. Hood." Device, a pair of wings, &c. "Imprinted—for Tobie Cooke, 1590." Dedicated "To—Iohn Lumley Knight, Lord Lumley.---To—mafter Tho. Smith, and the reft of the friendly auditours of the Mathematicall Lecturer." 43 leaves. "Imprinted—by *him*, for Tobie Cooke." W.H. Quarto.

1590. "Of The markes of the children of God, and of their comforts in afflictions.—By Iohn Taffin. Ouerfeene againe & augmented by the Author, and tranflated out of French by Anne Prowfe. Rom. 8; 16.— Printed—for T. Man, 1590." Dedicated "To—the Counteffe of Warwick.---To the faithfull of the Low Countrie.—Harlam, 15 Sept. 1586.— Iohn Taffin Minifter of the holye Gofpell in the French Church at Harlam."---The contents. 124 leaves, and fome verfes on "The necefsitie & benefite of affliction." W.H. Octavo.

1590. "CHRONOGRAPHIA. A Defcription of time from the beginning of the world, vnto the yeare of our Lord 137.—Collected out of fundry Authors, but for the greateft part—out of Laurentius Codomannus his Annales facræ fcripturæ. Deut. 4; 32.—Printed—for Rob. Dexter,— 1590." An epiftle "To the Reader.—12 Feb. 1590." Commendatory verfes. 56 pages. Alexander Dalrymple Efq; Octavo.

1590. "A Sermon vpon the 6, 7, and 8 verfes of the 12 Chapter of S. Pauls Epiftle vnto the Romanes; Made to the Confutation of fo much of another Sermon, entitled, A Frutful Sermon &c. as concerneth both the depriuation of the præfent gouerment, and the perpetual & vniforme regiment of our Church By certaine of their defcribed Officers to be in euerie particular Parifh through-out al her Maiefties Dominions; More fullie penned, than could by mouth be expreffed, the tyme limitted to the fpeaker being verie fhort. Publifhed at the requeft of certaine frendes by Tho. Rogers. Allowed by Auctoritie.—Printed—13 April, 1590." Pages 62. Licenfed. Lambeth Library. Quarto.

"A Re-

JOHN WINDET.

" A Reconciliation of all the Paſtors and Cleargy of this Church of 1590. England. By Anth. Marten, Sewer &c. Meliora ſpero. Be of one minde &c. 2 Cor. 13." (11) Device, The Pelican. " Printed—1590." Dedicated "To the Queenes moſt excellent Maieſtie.---The Preface to the Reader." 108 leaves. Licenſed. W.H. Quarto.

" A ſhort treatiſe againſt the Donatiſts of England, whome we call 1590. Browniſts; wherein, by the anſwers vnto certayne writings of theyrs, diuers of their hereſies are noted, with ſundry fantaſticall opinions. By George Giffard, miniſter at Malden." For Toby Cooke. 110 pages. Quarto.

" A briefe treatiſe of teſtaments, and laſt willes, very profitable to be 1590. vnderſtoode of all the Subiects of this Realme of England, &c. By the 1591. Induſtrie of Henrie Swinburn, Bachelar of the Ciuill Lawe. Printed— 1590." Dedicated " To—John—Archbiſhop of Yorke, &c.---To the Reader.---The principall parts."---Analytical tables of each part. 293 leaves, an Epilogue, and an alphabetical table. At the end, " Printed —1591." Licenſed. W.H. Quarto.

" The Shepheards Calender." &c. as p. 1159. 1591.

" An expoſition vpon the canonicall epiſtle of ſaint Iames: with the 1591. Tables, Analyſis & Reſolution, both of the whole Epiſtle & euery Chapter thereof: with the particular reſolution of euerie ſingular place. Diuided into 28 lectures or ſermons, made by Ric. Turnbull ſometimes fellow of Corpus Chriſti Colledge in Oxford, now preacher & miniſter of the word of God & the holy Sacraments, in—London." Device, The reſurrection of Jeſus Chriſt, as p. 668. " Imprinted—1591." Dedicated " To—my Lord his grace, Archbiſhop of Canterburie, &c.—May 10, 1591.---To the Chriſtian Reader." 326 leaves, and a page of faults corrected. Licenſed. W.H. Octavo.

Again, " Whereunto is annexed the expoſition of the ſame Author 1592. vpon the Canonicall Epiſtle of Sainte Iude, with foure Sermons made vpon the fifteenth Pſalme. All lately corrected, enlarged," &c. Device, A reverend old man ſtanding with his arms held up to receive a book and a wheat-ſheaf from two hands extended out of the clouds; at his feet are two doves, holding labels in their bills, the one with PEACE, the other with PLENTIE: the whole encompaſſed with this motto THOV SHALT LABOR FOR i. e. peace and plenty. " Printed by him, dwelling by Paules VVharfe, at the—Croſſe Keyes, 1592." The expoſition on Jude is divided into ten ſermons, and has a ſeparate title-page, with the device of the pelican. The xvth pſalm into four ſermons; the device is the ſame deſign as to St. James, but larger. W.H. Octavo.

" The Principles of Geometrie, Aſtronomie & Geographie:—gathered 1591. out of the tables of Aſtronomicall inſtitutions of Geo. Heniſchius, by Fran. Cooke. Appointed publiquelye to be read in the Staplers chappell at Leadenhall, by the worſhipfull Tho. Hood, mathematicall lecturer of the citie of London." Alſo without date. Octavo.

" QVERIMONIA Eccleſiæ Michææ. 7;" (8,—10.) Device, Peace and 1592. Plenty. For R. Watkins 245 pages. WH. Quarto.

" The Aſſiſe of Bread, newly corrected to the raiſinge the price of 1592. wheat," &c. Licenſed. Again 1597. Quarto.

" The

JOHN WINDET.

1592. "The moste profitable and commendable Science of Suruoying of Landt, Tenementes & Hereditamentes: drawen & collected by—Valentin Ligh. Whereunto is also annexed—of the measuring of all kindes of Landes,—and that afwell by certaine cafie—Rules, as alfo by an exact—Table, &c. Newly Imprinted & corrected.—for Robart Dexter,—1592." Q, in fours, and two folding tables. W.H. Quarto.

1593. "A Dialogve concerning V.Vitches and Witchcraftes. In which is laide open how craftely the Diuell deceiueth not onely the Witches, but many other, and so leadeth them awrie into many great errours. By George Giffard, Minifter of Gods word in Maldon. Printed by him for Tobie Cooke & Mihil Hart,—1593." Dedicated "To—Maifter Robert Clarke, one of her Maiefties Barons of her Highnes Court of Efchequer." M, in fours. W.H. Quarto.

1593. "The defence of Contraries," &c. as p. 1222.

1593, "A Defensative againft the Plague:—shewing the meanes how to preferue vs from the dangerous contagion thereof: how to cure those that are infected therewith. Whereunto is annexed a short treatife of the small Poxe:—Published for the loue & benefit of his Countrie by Simon Kellwaye Gent.—Printed by him,—neere Powles Wharfe, &c. and are there to be foulde, 1593." On the back are the arms of the earl of Effex, to whom this book is dedicated.—To the friendly Reader.—25 Mar. 1592.—Geo. Baker, in commendation of the author.—The author to the Reader; explaining the characters. 48 leaues, and an index. Licensed. W.H. Quarto.

1593. "Ric. Hookers Lawes of Ecclefiafticall politie, written in defence of the present gouernement eftablished, againft the new defired difcipline. Printed—1593." Maunfell, p. 59. Licensed. Folio.

1593. "A Glaffe of vaineglory tranflated out of Auguftine, entit. Speculum peccatoris, by W. Prid. Doct. of the Lawes: with certaine Praiers added thereto." Licenfed. Twenty-fours.

1593. The pfalms in metre, by Sternhold, &c. For R. Day's affignees. Frequently printed, and of various fizes.

1594. "The order of ceremonies obferued in the annointing and coronation of the moft chriftian king of France and Navarre, Henry the IIII. of that name, celebrated in our lady church, in the cittie of Chartres, vppon Sonday the 27th of February. Tranflated by E. A." With the printer's epiftle. Quarto.

1594. "Willobie his Auifa, or the true picture of a modeft maid, and a chaft and conftant wife." Licenfed. Quarto.

1594. "M. Blvndevile His Exercices, containing fixe Treatifes, the titles whereof are fet down in the next printed page; which treatifes are verie neceffarie to be read and learned of all yoong Gentlemen, that have not bene exercifed in fuch difciplines, and yet are defirous to haue knowledge, as well in Cofmographie, Aftronomie, and Geographie, as alfo in the Arte of Navigation, in which Arte it is impoffible to profit without the helpe of thefe, or fuch like inftructions.—Dedicated to the young Gentlemen of this Realme." Each tract has a feparate title-page; but the folios are continued to 350, with tables and projections.* W.H. Quarto.

The

"The refolued gentleman. See p. 1028. 1594.
" The Praife of a good name. The reproch of an ill Name.—With 1594.
certaine pithy Apothegues, &c. by C. G. (Char. Gibbon)—Imprinted
for Tho. Gosson, 1594." Dedicated " To some of the beft, and moft
ciuill fort of the inhabitants of St. Edmonds Bury.—To the Reader."
55 pages. W.H. Quarto.
" Generall Calenders," &c. p. 1137. Licenfed. 1594.
The French king's edict, upon the reducing the citie of Paris under his 1594.
obedience, publifhed the 28th of March 1594. Whereunto is adioyned
the faid king's letters patents for the re-eftablifhment of the court of parliament at Paris, &c. Licenfed Quarto.
" Godfrey of Bulloighe, or the recouerie of Hierufalem. An heroicall 1594.
poeme, written in Italian by feig. Torquato Taffo. Tranflated by R. C.
efq; Printed for Chriftopher Hunt of Exeter." 235 pages. Quarto.
The firft part of Maunfell's catalogue. See p. 1137. Licenfed. 1595.
" Bart. Barnes his Diuine Century of Spiritual Sonnets." Quarto. 1595.
" A PROGRESSE of Pietie, or The harbour of Heauenly harts ease, to 1596.
recreate the afflicted Soules of all fuch, as are fhut vp in anye inward, or
outward affliction. By John Norden.—Printed—for J. Oxenbridge."
Dedicated " To the moft famous chriftian Queene Elizabeth." &c. 100
leaves. W.H. Twenty-fours.
" A Comfort againft the Spaniard. Deut. 20; (2—4.)—Printed for 1596.
I. O. 1596." In a neat compartment of Diana and Actaeon. Dedicated
" To—William Lord Bifhoppe of Norwich.—Tho. Nun, minifter of the
word, at Wefton." 11 leaves. W.H. Quarto.
" A Libell of Spanifh Lies: Found at the Sacke of Cales, difcourfing 1596.
the fight in the Weft Indies, twixt the Englifh Nauie being 14 Ships &
Pinaffes, and a fleete of 20 faile of the king of Spaines; and of the death
of Sir Francis Drake. With an anfwere—by Hen. Sauile Efq; &c.
Alfo an Approbation of this difcourfe, by Sir Thomas Baskeruile, then
Generall of the Englifh fleete in that feruice: Auowing the maintenance
thereof, perfonally in Armes againft Don Bernaldino, if he fhall take exceptions to that which is here fet downe, Touching the fight—or iuftifie that
which he hath moft falfely reported in his vaine printed letter." 47 pages.
W.H. Quarto.
" St. Peters chain, confifting of eight golden linkes, moft fit to adorn 1596.
the necks of the greateft ftates. The Linkes are Faith, Vertue, Know-
ledge, Temperance, Patience, Godlinefs, Brotherly kindnefs, Loue. By
Ra. Manerick." Octavo.
" The Jaylors conuerfion, a fermon on Acts 16; 30. By Hugh Dow- 1596.
rich B. D. Wherein is liuely reprefented the true Image of a foul rightly
touched & conuerted by the Spirit of God." Octavo.
" Three Sermons or Homelies," &c. as p. 1137. 1596.
" The reward of Religion: Deliured in fundry lectures on the Book of 1596.
Ruth." Octavo.
Blundeuile's Exercifes, &c. as in 1594, with the addition of two more 1597.
treatifes, viz. " A Briefe Defcription of vniuerfal Maps & Cards," &c. as
p. 1192. Licenfed. 392 leaves. W.H. Quarto.
7 E, 2 " How

JOHN WINDET.

1597: "How to chuſe, ride, traine, and diet both hunting horſes, and running horſes. With all the ſecrets thereunto belonging diſcovered; an art never heretofore written by any author," &c.* Quarto.

1597. "Chariſma ſiue Donumſanationis: ſeu Explicatio totius quæſtionis de mirabilium Sanitatum Gratia, &c. Auctore Guil. Tookæro S. T. D.— Excudebat—1597." On the back are the queen's arms, to whom the book is dedicated. 124 pages, and "Oratio omnibus horis dicenda pro Elizabetha ſerenifs. Angliæ Reginæ." W.H. Quarto.

1597. "Eſſaies, Religious meditations, Places of perſwaſion, and diſſwaſion. Seen & allowed. Printed for Humf. Hooper, and are to be ſold at the black Beare in Chancery Lane." The firſt edition of lord Bacons Eſſays, containing only theſe ten; "Of Studie, Diſcourſe, Ceremonies & reſpects, Followers & friends, Sutors, Expence, Regiment of health, Honor & reputation, Faction, Negotiating." 13 leaves. To which are adjoined, "Meditationes ſacræ." Alſo, "Of the colours of good & euill; a fragment." Theſe were all printed again next year. 49 leaves. W.H. Twenty-fours.

1598. "The Making and vſe of the Geometricall Inſtrument called a Sector.—Written by Tho. Hood, Doctor in Phyſicke, 1598. The Inſtrument is made by Charles Whitwell,—againſt S. Clements Church. Printed by him, and are to be ſolde at the great North dore of Paules Church by Sam. Shorter." Dedicated "To—Charles Blunt knight, Lord Montioye, Knight of the—Garter, and Captaine of her Maieſties Forte of Porteſmouth.---The contents." 50 leaves, with many projections, and a copper-plate print of the Sector, &c. Quarto.

1599. "The Art of Logike, Plainely taught in the Engliſh tongue, by M. Blundeuile of Newton Flotman in Norfolke, aſwell according to the doctrine of Ariſtotle, as of all other moderne & beſt accounted Authors thereof &c. Imprinted—and are to be ſold at Paules Wharfe,—1599." Preface "To the Reader.---A Poſtſcript.---The Contents." 170 pages. W.H. Quarto.

1599. "The Common-wealth and Gouernment of Venice, Written by the Cardinall Gaſper Contareno, and tranſlated out of Italian by Lewes Lewkenor Eſq;—VVith ſundry other Collections annexed by the Tranſlator.—With a ſhort Chronicle—of the Venetian Dukes, &c. Imprinted—for Edm. Mattes,—at the ſigne of the Hand & Plow in Fleetſtreet, 1599." Dedicated "To—the Lady Anne Counteſſe of Warwicke.—Selſey, 13 Aug. 1598." Verſes by Edw. Spencer, I. Aſhley, Maur. Kiffen, Hen. Elmes, and John Harington.---To the Reader. 230 pages. W.H. Quarto.

1600. "Fowre bookes of the inſtitutions, vſe and doctrine of the holy ſacrament of the Euchariſt in the old church: As likewiſe how, when, And by what Degrees the Maſſe is brought in, in place thereof. By my Lord Philip of Mornai, &c. The ſecond edition, reuiewed by the Author.—Printed—for I. B. T. M. and W. P. 1600." Dedicated "To the—Lords and others of her Maieſties moſt Hon. Priuie Councell.—R. S." the tranſlator.---"The Author his preface, &c.---The end and drift of the Author, &c. The Contentes." 484 pages. W.H. Folio.

A diſ-

A difcourfe of cities with privileged univerfities. Quarto. 1600.

"Lachrimae, or teares, figured in feuen pafsionate pauans, with diuers other pauans, galiards, and almands, fet forth for the lute, viols, or violins, in fiue partes, by John Dowland, bacheler of muficke, and lutenift to the moft royal and magnificent Chriftian the fourth, king of Denmark," &c. Folio.

He had licences alfo for the following, viz. In 1586, "A difplaie of the vanities of ỹ world. The defcription of ỹ Netherland, &c. An Analogie, or refemblance betweene Johane Q. of Naples, & Marye Q. of Scotland." In 1587, "To the right worfhipfull grave & prudent Senators, mafter John Periam maior of ỹ citie of Excifter, and to thaldermen, & his brethren of ỹ cōmon counfell of ỹ fame. Concio ad Academicos Oxonienfes; Tit. 13;. 1." In 1588, "The mifery of Mergiddo. Twelue rules & weapons concerning ỹ fpirituall battel," &c. See p. 223. In 1589, "An ordinarie lecture preached at ỹ Black friers, by Mr. Egerton. A frutefull fermon. of ỹ. nature & vilenes of fynne." In 1590, "Corderius dialoges: in Fr. & Eng." In 1593, "The order of ỹ coronation of the frenche king." He printed till 1610, and perhaps after.

ROBERT ROBINSON, ftationer,

WAS made free 27 June, 1580, by Mrs Seres, late wife of William Seres the elder. By the entries below,' he appears to have been but a diforderly member. He dwelt in "Pewter, or Feter-lane neer Holborne." I find no other place of abode mentioned in any of his books in my poffeffion:

"Les Ans du roy Richard le fecond, collect' enfembl' hors de les Abridgments de Stratham, Fizherbert et Brooke per Richard Bellewe de Lincolns Inne, 1585. Quefq; vn table a c' annexe.—Imprinted by Rob. Robinfon dwelling in Fewter lane neere Holborne." Dedicated "To the Students of the Common Lawes of this Realme, and efpecially to the graue & learned Societie of Benchers Vtterbarrefters & Students of Lincolnes Inne.—10 Jan. 1585." The epiftle in Englifh; but the Reports, &c. all in French. 326 pages, and a table. W.H. Octavo. 1585.

"The Summe of Chriftianitie, containing 8 propofitions." Octavo. 1585.

"A proper new fonet, declaring the lamentation of Beckles a market town in Suffolke, which was in the great winde vpon S. Andrewes eve laft paft, moft pittifully burned with fire to the loffe by eftimation 20,000l. and vpwarde, and the number of fourefcore dwelling houfes. To Wil- 1586.

' "May 12, 1589, for difobedience & other diforder he was fined v f. and committed to ward. June 2, 1590,—for printing a brief diforderly, ij f. Febr. 1, 1593-4, —For buying & difperfing Pfalmes diforder- ly printed—iij l. which he promifeth to pay ten dayes after Efter next. July 15, 1594, paid xiij f. iiij d. Dec. 6, 1596,—for printing the Merchanttailors prentice Indentures, v f."

fons

ROBERT ROBINSON.

'fom tune." In 14 octave ſtanzas. "For Nich. Colm of Norwich, dwelling in St. Andrews."* *An half ſheet.*

1586. "The Mirrour of mans life, deſcribing what weak mould we are made of, what miſery wee are ſubject vnto, how vncertaine this life is, and what ſhalbe our end. Tranſlated by Hen. Kirton." Octavo.

1586. "The Gouernance of vertue," &c. as p. 641. Licenſed to the uſe of the company, being an old copy printed by J. Day.

1586. "Certain deuiſes and ſhewes, preſented to her maieſtie by the gentlemen of Grayes-Inne, at her highneſſe court in Greenwich, the 28 day of February, in the thirtieth yeare of her maieſties moſt happy reigne." Quarto.

1587. "The tragicall Hiſtorie of Romeus and Juliet: Containing in it a rare Example of true Conſtancie: with the ſubtill Counſels and Practiſes of an old Fryer, and their Euent. Imprinted by R. Robinſon." See Mr. Steevens's note 1, on Romeo and Juliet.

1587. "A treatiſe of Morall Philoſophie," &c. as p. 814.

1588. "The Zodiake of life," &c. as p. 904. Licenſed.

1588. "The Engliſh Ape, The Italian imitation, the Footeſteppes of Fraunce. Wherein is explained the wilfull blindneſſe of ſubtill miſchiefe, the ſtriuing for Starres, the catching of Mooneſhine: and the ſecrete found of many hollow hearts; by W. R." Dedicated "To—Syr Chriſtopher Hatton Knight," &c. 24 pages. See p. 1047. W.H. Quarto.

1589. "The Portraiture of Hypocriſie, liuely & pithilie pictured in her colours: wherein you may view the vglieſt & moſt prodigious monſter that England hath bredde.—Imprinted—for Iohn Dalderne, 1589." Dedicated "To—Sir Anthonie Therold.—Iohn Bate.—To the Chriſtian Reader." Running-title "A dialogue betweene a Chriſtian and an Atheiſt." 192 pages. W.H. Sixteens.

1589. Perkins's Treatiſe, whether a man be in a ſtate of Damnation or a ſtate of Grace. Again 1590, for John Porter & Tho. Gubbin. Octavo.

1590. "M. T. Ciceronis epiſtolæ familiares.—A Lambino," &c. Octavo.

1590. "Meliboeus Thomæ Watſoni, ſiue Eclogia in obitum honoratiſſ. viri Dom. F. Walſingham Equitis aurati." Octavo.

1591. "A PENSIVE Mans practiſe. Verie profitable for all perſons wherein are conteined verie deuout—praiers for ſundrie godly pūrpoſes, &c. Written by J. Norden." Singleton's rebus. "Printed by him, 1591." It has prefixed a calendar, in which the ſcripture hiſtories are particularly applied; an almanac for ten years; and a preface. 258 pages, an acroſtic on the author's name, and a table. W.H. Twenty-fours.

1592. Terentii Comœdiæ. Quarto.

1592. "A WORKE Concerning the Trewneſſe of Chriſtian Religion, written in French: Againſt Atheiſts, Epicures, Paynims, Iewes, Mahumetiſts, &c. By Philip of Mornay Lord of Pleſſie Marlie. Begunne to be tranſlated by Sir Philip Sidney Knight, and at his requeſt finiſhed by Arth. Golding." Device the Loſt Sheep, as p. 902. "Printed by him, for I. B.—at the great North doore of S. Pauls Church, at the ſigne of the Bible, 1592." Dedicated "To—Robert Earle of Leyceſter, &c.—13 Maie,

Maie, 1587."—To the Right High & mightie Prince, Henrie king of Nauarre, &c.—Du Plessie.---The Preface." 552. pages. W.H. Quarto.
" Marcelli Palengenii Zodiacus vitæ." &c. as p. 862. 1592.
" A Spirituall, and most precious Pearle," &c. See p. 744. Licensed. 1593.
" An Enemie to Idlenesse." &c. as p. 968. Licensed in 1588. For 1593.
the Company.

" TETRASTYLON PAPISTICVM, That is, The foure principal pil- 1593.
lers of Papistrie, 1 conteyning their raylings, slanders, forgeries, vntruthes. 2 their blasphemies, flat contradictions to scripture, heresies,
absurdities: 3 their loose arguments, weak solutions, subtill distictions:
4 the repugnant opinions of New Papistes with the old: of the new with
one another; of the same writers with themselves: yea, of Popish religion
with and in itselfe. Compiled as a necessarie supplement—to the Author's
former worke, intituled Synopsis Papismi: To the glorie of God for
diswading of light-minded men from trusting to the sandie foundation
of poperie, and to exhort good Christians stedfastlie to hold the rockie
foundation of the Gospell.—Printed—for Tho. Man,—1593." Dedicated
" To—Sir Iohn Puckering Knight, Lord Keeper, &c.—Anth. Willet.—
The prefaae." 176 pages. W.H. Quarto.

" Terentianus Christianus, siue Comoediæ duæ Terentiano stylo con- 1595.
scripta: ad vsum Scholarum seorsum excusa. Tobæus, Juditha. His
accessit Peudostratiotes, Fabula iocosa ac ludicra. Authore Corn. Scho-
næo. Excudebat—impensis R. D. 1595. Cum priuilegio Regiæ Maiestatis." Octavo.

" A Comparison betweene the auncient fayth of the Romans, and the 1595.
new Romish Religion. Set foorth by Fr. Bunny, sometime fellowe of
Magdalen College in Oxforde. Matth. 15: 13.—." The lost sheep.
" Printed—for Raph Jackeson, 1595." Dedicated " To—Katheren Coun-
tise of Huntingdon." 78 pages. W.H. Quarto.

" A learned and excellent treatise, containing all the principall grounds 1595.
of Christian religion,—by way of conference. Written in French by
Mat. Virell, after translated into Latin, and now translated into English;"
probably by Steph. Egerton, who wrote the preface, an admonition
touching reading. For R. Dexter. Again 1597. Octavo.

" The Couenant betweene God and Man, plainly declared in laying 1596.
open the first & smallest pointes of Christian Religion. 2 Cor. 11; 9.—
Printed—for Raph Jackson, 1596." Dedicated " To—M. Maior, and
the rest of the magistrates of the towne' & libertie of Feuersham in Kent,
and also to all others there & thereabout that feare God vnfaignedly, &c.
—Octo. 20, 1595. J. F." 360 pages. W.H. Octavo.

" A profitable Exposition of the Lords Prayer," &c. as p. 1195. 1596.
Babington's Exposition of the x Commandements. See p. 1062. Quarto. 1596.
" SALOMON, or A treatise declaring the state of the kingdome of Israel 1596.
as it was in the daies of Salomon. Whereunto is annexed another trea-

* For the copy-right of this, and The pen- || ton's bond for 5l. to the company. See p.
sive man's practice, which were Singleton's || 740.
copies, he undertook to discharge Single-

tise,

tife, of the Church: or more particularly, Of the right conſtitution of a Church. Pſal. 45; 14. Adducetur Regi." The queen's arms. " Iere. 29; 7.—for Rob. Dexter, 1596." On the back, " Regni Angloiſraelitici typus." The queen's arms encompaſſed with laurel and crowned, ſupported on the dexter ſide by a queen, over whoſe head, on a ſcroll, is " Flos de Jeſſe, Reg. Pacis;" On the ſiniſter, by a king, over whoſe head is " Leo de Iuda, Rex pacis." Both theſe royal perſonages ſet one of their feet on a lion couchant. At the bottom, " Bellum de pace." Beneath this cut, is " Iacobs prophecie of the kingdome of Iſrael. Gen. 49; 8," (9, 10.) Dedicated " To the moſt Mightie and Maieſticall, peaceable & happie Monarche Elizabeth, &c.—Tho. Morton."—Latin verſes, applicable to the ſubject.—To the Reader.—The Arguments of theſe treatiſes. That of the kingdom of Iſrael ends on p. 71, with Dexter's rebus, &c. The other treatiſe of the church has freſh ſignatures, and paging to p. 144. W.H. Quarto.

1596. " The Diſcouerie of the large, rich and bevvtiful empyre of Guiana, with a relation of the great & golden Citie of Manoa (which the Spanyards call El Dorado) And of the Prouinces of Emeria, Arromaia, Amapaia, &c. Performed in the yeare 1595, by Sir W. Ralegh Knight, &c. Imprinted—by him, 1596." Dedicated " To—my ſinguler good Lord & kinſman, Charles Howard, knight of the Garter, Barron, and Councellor, and of the Admirals of England the moſt renowmed: And to—Sr. Robert Cecyll Knight, Counceller in her Highnes priuie Councels.—W. R.—To the Reader." 112 pages. W.H. Quarto.

1596. " A commentarie vpon the whole booke of the Prouerbs of Salomon. The ſecond time peruſed, much enlarged, &c. Whereunto is newly added an Expoſition of a few choiſe—Prouerbs—here & there in the Scriptures. Prov. 22; 17.—." Dexter's rebus. " Printed—for R. Dexter." Dedicated " To—Edward Earle of Bedford.—P. M. (Pet. Muffet) The proverbs of Solomon on 483 pages, and a table; the ſcripture proverbs on 18 pages; then " The Author to the Reader," in verſe. W.H. Octavo.

1597. " Certaine worthye manuſcript Poems of great Antiquitie Reſerued long in the Studie of a Northfolke Gentleman. And now firſt publiſhed By J. S. 1 The ſtatly tragedy of Guiſtard & Siſmond. 2 The Northern Mothers Bleſſing. 3 The way to Thrifte." Dexter's rebus. " Imprinted at London for R. D. 1597." On the back, " To the worthieſt Poet Maiſter Ed. Spenſer." The two latter poems have a ſeparate title-page. " The Northern Mothers Bleſſing. The way of Thrift. Written nine yeares before the death of G. Chaucer.—Printed—for R. Dexter, 1597." The whole F 6, in eights. W.H. Sixteens.

—— " EVERARD DIGBIE his Diſſuaſiue From taking away the lyuings and goods of the Church, &c. Hereunto is annexed Celſus of Verona his Diſſuaſiue tranſlated into Engliſh. Beatius eſt dare quā recipere. Act. 20. Printed by him, and Tho. Newman." In the elegant compartment as p. 926. Dedicated " To—Sir Chriſtopher Hatton, Lord High Chaunceller," whoſe arms are on the back of the title-page. 242 pages. W.H. Quarto.

He

ROBERT ROBINSON.

He had alſo licenſes to print the following, viz. In 1586, " A true report of vnknown Fowles." In 1587, " A French Grämer." In 1588, " On condition y̆ he agree with Mrs. Middleton, The Godly Garden; and Chriſtian Prayers." In 1589, " A ſermon' preached at Paules Croſſe, 17 Nov. 1589, in remembrāce & thankeſgiving for her Maieſties reign, now 32 yeres, by Tho. White profeſſor of divynyty." In 1590, " The tragicall murder of y̆ lord Burgh, with the ſorrowfull ſighes of a ſadd foule for his vntymely Loſſe: provided y̆ yf it be hurtfull to y̆ copie entered the laſt day for Edw. White touching Coſbyes condemnation, &c. then this entrance to be voyd."

GEORGE ROBINSON,

WAS admitted and ſworn a freeman of the Stationers' company, being tranſlated from the Grocers'. His widow was married to Tho. Orwin.

" Tho. Wilſon's Art of Rhetoric. See p. 837.	Quarto.	1585.
Warner's Poetry.	Octavo.	1586.
" Albion's England, or a Hiſtorical Map of the ſame Iſland, by Wm. Warner."—for Tho. Cadman.	Quarto.	1586.
Sir Philip Sidney's epitaph. Licenſed.		1586.
" The Catechiſme,—by John Caluin," &c. See p. 840.	Sixteens.	1586.
" Juſtini et Aurelii Victoris hiſtoria."	Octavo.	1586.
" A ſhort Diſcourſe—of all the ſaте—Treaſons," as p. 1199.		(1586.)
" A Woorke concerning the trewneſſe of Chriſtian Religion, &c. By Philip of Mornay," &c. as p. 1234. "—for Tho. Cadman,—at the great North-doore of S. Paules Church, at the—Bible." W.H.	Quarto.	1587.

" THE Worthines of Wales: Wherein are more than a thouſand ſeueˉ rall things rehearſed: ſome ſet out in proſe to the pleaſure of the Reader, and with ſuch variety of verſe for the beautifying of the Book, as no doubt ſhal delight thouſands to vnderſtand. Which worke is interlarded with many wonders & right ſtrange matter to conſider of. All the which labour & deuice is drawen forth & ſet out by Tho. Church-yard, to the glorie of God and honour of his Prince & Countrey.—for Tho. Cadman, 1587." Dedicated " To—the Queenes moſt excellent Maieſtie, Elizabeth," &c. At the end is the author's coat of arms. George Maſon Eſq; Quarto. 1587.

" THE ENGLISHEMANS Treaſure: With the true Anatomie of Mans 1587. bodie: Compiled by—Maiſter Tho. Vicary Eſq; Sergeant Chirurgion to

---

[1] Mr. Ames not improperly attributed this book to one Richard Robinſon, after Maunſell: but as there is no mention of that name as a printer or bookſeller throughout the Stationers' regiſter, but on the contrary it is there licenſed to Robert, there appears ſufficient reaſon to apprehend Maunſell or his printer made a miſtake; further, as a corroborating circumſtance, it is ſaid to be printed with Tho. Newman, with whom Robert was concerned in printing Digby's Diſſuaſive.

K. Hen.

K. Hen. 8. K. Edw. 6. Q. Mary. And to our Soueraigne Lady Q. Elizabeth: And also chiefe Chirurgion to S. Bartholmewes Hospitall. Whereunto are annexed many secretes appertayning to Chirurgerie, &c. Also the rare treasure of the English Bathes:—by Wm. Turner, Doctor in Phisicke. Gathered &c. by Wm. Bremer Practitioner in Phisicke & Chirurgerie. Imprinted—for Iohn Perin, dwelling in Paules Churchyard at the—Angell, and are there to be solde, 1587." Dedicated "To —Sir Rouland Hayward Knight, President of little St. Batholmewes in West Smithfield, Sir Ambrose Nicolas Knight, Sir Thomas Ramesay Knight, with the rest of the worshipfull Maisters & Gouernours of the same: Wm. Clowes, Wm. Beton, Ric. Story & Edw. Bayly, Chirurgions the same Hospital.---To the Reader.---Tho. Vicary to his Brethren, practising Chirurgerie." 110 pages. W.H.   Quarto.

EDWARD ALLDE, or ALDEE, stationer,

SON of John Allde, was made free by patrimony, 18 Feb. 1583-4; and resided at first with his father, at the long shop adjoining to St. Mildred's church in the Poultry. In 1590, he appears to have dwelt near the conduit, without Cripple-gate, at the sign of the gilded cup. In 1560, he was fined vs. for printing disorderly a ballad of The wife of Bath. See p. 1197, note e.

1584.   Beware the Cat: in three parts. My copy wants the title page. It has prefixed, some verses by "T. K. to the Reader." From which, to convey some idea of the book, the first four stanzas are inserted below."---The dedication "To the right worshipful Esquire John Yung—G. B.---The "argument." F 4, in eights. "Imprinted—at the long Shop adioyning vnto St. Mildreds Church in the Pultrie by Edw. Allde, 1584." Octavo.

* "This little book Bevvare the Cat
moste pleasantly compil'd:
In time obscured was and so,
since that hath been exilde.
Exilde, becaufe perchaunce at first,
it shewed the toyes and drifts:
Of such as then by wiles and willes,
maintained Popish shifts.
Shifts, such as those in such a time,
delighted for to vse:
Wherby ful many simple foules,
they did ful sore abuse.
Abuse? yea sure and that with spight,
when as the Cat gan tel:
Of many pranks of popish preests,
bothe foolish, mad and fel."

† Wherein we learn that the publisher, G. B. had a controversy with one Mr. Streamer, "whether Birds & beasts had reason; the occasion therof was this. I had heard that the Kings Players were learning a Play of Esops Crowe, wherin the moste part of the actors were birds, the deuice wherof I discommended.—Wel quoth maister Streamer I knowe, what I knowe & I speak not onely what by hearsay—but what I my self haue prooued. Why?—haue you proofe of beasts & foweles reason? Yea quoth he I hau herd them and vnderstand them bothe speak and reason aswel as I hear & vndersta ᵈ you.—and at last said, If that I thought ᵒ you could be content to hear me, and with out any interruption,—I would tel you such a story of one peece of myne owne experimenting, as should bothe make you wonder and put you out of dout concerning this matter." &c. And then begins his oration, the subject of this book.

"The

"The Mirrour of mans miseries, or a brief summarie of the first parte 1584. of the Resolution, in verse." Licensed, 1594. Octavo.

"A Briefe and pleasaunt treatise, Intituled, Naturall & Artificiall 1586. conclusions:" See it in p. 840. Licensed. Sixteens.

"The safegard of sailers, or great rutter: containing the courses, dis- 1587. tances, depthes, foundings, flouds and ebbes; with the markes for the entringes of sundry harboroughs bothe of England, France, Spaine, Ireland, Flaunders, and the sounds of Denmark; with other necessarye rules of common nauigation. Translated out of Dutch into English, by Robert Norman, hydrographer. Imprinted—in the Pultrie." Licensed. Again 1590. Quarto.

"The Good huswiues treasurie, being a very necessarie booke, instruct- 1588. ing to the dressing of meates." Octavo.

"Tho. Hood his speach made vnto the worshipfull Companie, pre- 1588. sent at the house of the Worshipfull Maister Smith, 4 Sept. 1588." Quarto.

"THE Quintesence of Wit, being A corrant comfort of conceites, 1590. Maximies, & poleticke deuices, selected & gathered together by Francisco Sansouino.—Translated out of the Jtalian tung, &c. Printed by him, dwelling without Cripple-gate, at—the gilded Cuppe. Octobris 28, 1590." Dedicated "To—Maister Robert Cicell Esquire, one of the Sonnes of the —Lord high treasurer.—Rob. Hichcock.---Franc. Sansouino to the Emperour Rodolph the second.—Venice, 24 Feb. 1578.---The Author to the Reader." &c. 803 conceits. "This is the end of the first Booke,— 1590." Hitchcocks armes; then a table signed "Captaine Hichcock. This saide Capt. Hichcoc'k seruing in the Lowe Cuntries, Anno 2586 (1586) with two hundreth Souldiours: brought from thence with this Booke, the second booke of Sansouinos politick Conceites, which shall be put to Printing so soon as it is translated out of the Italian into English." W.H. Quarto.

"A treatise against Traitors: taken out of Ieremie 40; 13—16, and 1591. 41; 1—4. Meete for all faithful subiects in these dangerous dayes. By Sam. Cottesford. Printed for William Holme." Octavo.

"The golden chaine," &c. as p. 891. 1591.

"A Book of Cookerye.—Gathered by A. W. And now newlye enlarged 1591. with the seruing in of the Table. With the proper Sauces to each of them conuenient." E, in eights. W.H. Octavo.

"Hill's profitable art of gardening," &c. See p. 857. 1594.

"The battell of Barbarie, betweene Sabastian king of Portugall, and 1594. Abdelmelec king of Marocco; with the death of captaine Stukeley. As it was sundrie times plaid by the lord high admirall his seruants." Quarto.

"A Table for Gauging, Or Speedy measuring of all manner of Vessels; 1594. eyther VVine, Oyle, Hony or Beere. Exactly calculated by Iohn Goodwyn, Teacher & Practicioner of Arithmetique & Geometry in the Cittie of London. VVith most easie rules," &c. A tun lying on its bouge. "Printed—for Hugh Astley, and are to be sold at S. Magnus corner, 1594." Dedicated "To—Cuthbert Buckle, Lord Maior of—London: and the—Aldrmen hys Bretheren." 16 leaues. W,H. Sixteens.

The

1240   EDWARD ALLDE.

1595.   The book of Homilies. Both tomes. See p. 720, &c.   Quarto.
1595.   " The French Alphabeth, teaching in a very fhort time by a moſt eaſie way to pronounce French naturally, to reade it perfectly, to write it truely, and to fpeake it accordingly. Together with the Treafure of the French tung, containing the rareſt Sentences, Prouerbes, &c. The one diligently compiled, and the other painfully gathered—after the Alphabeticall maner, &c. By G. D. L. M. N.—Printed by E. Allde, and are to be fold by H. Iackfon,—in Fleetſtreet,—1595." Dedicated " A tres-illuſtre, et tres-heroique le fieur Henry Walloppe Cheualier, & Threforier General de la Sereniſſime Maieſté en Irlande.—De Londres ce 11. d'Aouſt, 1592.—G. Delamothe N.---Anagramme du dict Sieur.---Quatrain.---Sonnet Acroſtiche.---To the Reader, warning him of the methode.---A table.---An aduertifement:" that the author was to be heard of at the fign of St. John the Evangeliſt in Fleetſtreet, or at the Helmet in St. Paul's Church-yard. The French Alphabet on 161 pages. The Treafure of the French tongue has a feparate title-page, and is paged anew, but the fignatures are continued. Dedicated " A tres-noble et tres-vertueufe damoifelle Madamoifelle Tafburh.—De Lōdres ce 10 d'Aouſt, 1596.—G. de la Mothe. N." 58 pages. W.H.   Octavo.
1596.   " The Regiment of Life." &c. See p. 549. Y, in fours. W.H. Quarto.
1596.   " Ioyfvll Newes Out of the New-found VVorlde." &c. as p. 879. " Newly corrected.—Wherunto are added three other bookes, treating of the Bezaar ſtone, & the herb Efcuerconera; the properties of Jron & Stéele in Medicine, and the benefit of Snow.—Printed by him, by the afsigne of Bonham Norton, 1596." Thefe treatifes have feparate title-pages, but the fignatures and pages are continued to Fol. 187. W.H.
                                                                                       Quarto.
1596.   The Book of Hawking, Hunting & Fiſhing corrected.   Quarto.
1596.   The Paradice of Dainty Deviſes. See p. 685.   Quarto.
1596.   " The arte of Navigation. Containing a breife defcription of the Spheare, vvith the partes & Circles of the fame: as alfo the making & vfe of certaine Inſtruments.—Firſt written in Spaniſh by Martin Curtis, and tranſlated—by Ric. Eden: and laſtly corrected & augmented, with a Regiment or Table of declination, and diuers other neceſſary tables & rules of common Nauigation. Calculated (this yeare 1596, being leap yeare) by J. T." A fhip under fail. " Imprinted—by him, for Hugh Aſtley, by the afsignes of Ric. Watkins, and are to be folde at St. Magnus corner, 1596." Infcribed " To the induſtrious Seamen & Mariners of England.—Iohn Tap." See p. 719. W.H+   Quarto.
1599.   " The Tragedie of Solimon and Perfeda. Wherein is laide open loues conſtancie, fortunes inconſtancie, deaths triumphs." Printed for Edw. White.   Quarto.
——   " A lamentable tragedie," &c. as p. 892.
——   " The Spaniſh Tragedy, containing the lamentable end of Don Horatio, and Bel-imperia, with the pitiful death of old Hieronimo." See thefe three tragedies in the origin of the Eng. Drama.   Quarto.
——   " Graces to be faid before and after meat."   Octavo.

                                                                                    " A worthy

EDWARD ALLDE

"A worthy work, profitable to this whole kingdom. Concerning the mending of all high wayes; as alſo, for waters and iron works. By Thomas Proctor eſquier, and to be ſold at his houſe on Lambard hill, near old Fiſh ſtreet. Quarto.

He had alſo licenſes for the following, viz. In 1586, "A godly exhortac'on, whereby England may knowe what ſinfull abhominac'on doth flowe. A merie & pleaſant prognoſticac'on. The old book of Reignold the Foxe: to be printed for y̆ cūpanie." In 1587, "Entred vnto him for his copie vpon condyc'on y̆ he get yt orderly aucthoriſed & allowed to y̆ print, when yt is tranſlated into Engliſh Hiſtoire Palladinne &c. per Claude Collet." In 1588, "The firſt 4 books of Amadis de Gaule: to be tranſlated—& orderly allowed &c. A coniecturall diſcourſe vpon y̆ hieroglipicall letters & carecters found vpon iiij fiſhes taken neere Maſtrane in Denmarke, 28 Nov. 1587." In 1589, "A Frenche mans ſonge, made vpon y̆ deathe of y̆ French king, who was murdered in his owne Courte by a traiterouſe Fryer of St. Jacobs order, 1 Aug. 1589. A diſcourſe of vij murders cōmitted by a merchant of Brabant." In 1590, "The Engliſh Scholemaſter, ſet forth by Rob. Edwards." In 1591, "A ſermon called y̆ Trumpet of the ſoule—by Hen. Smythe." In 1594, "A glaſſe of foly or abuſes." Alſo at times for ſundry ballads. He, or another Edw. Allde appears to have printed in 1626.

THOMAS ORWIN, ſtationer,

WAS made free by Tho. Purfoot, 5 May, 1581. "March 4, 1587-8. At a court holden this day yt is ordeyned & decreed—That Tho. Orwyn ſhall from henceforth leave off from further dealinge with printinge—whatſoever, till ſuch time," &c. as in the order for Mrs. Middleton. See p. 1055, note c. However on the 3 June following, "Yt is agreed y̆ Orwins admiſſion* to be a printer ſhalbe entred on the booke." He came on the livery 1 July, 1592. He dwelt in Pater-noſter row, over againſt the ſign of the checker. Sometimes he uſed the device of an urn, marked with T. O. reſting on the ſhoulders of two ſatyrs ſitting on the

---

* "Vpon the letter of the Abp. of Cant. the Bp. of London, Dr. Coſen, & Dr. Walker, directed to this Court, and dated 7 M'cij, 1587. T. Orwin, according to the Starchamber decree for theleẽtion of Printers, was elected a printer, and preſented 14 May 1588, to the ſaid Abp. the Lieut. of the tower, Mr. Recorder, The deane of Weſtm. Mr. Dr. Awbrey. Mr. Dr. Cozen, Mr. Dr. Lewyn, by Mr. Judſon maſter, Mr. Coldock & Mr. Conneway wardens, Mr. Watkins, Mr. Cooke, Mr. Denham, & Mr. Howe; and thereupon, 20 May 1588, The ſaid Orwin was by the ſaid Abp. the deane of Weſtm. & Mr. Dr. Awbrey admitted to be a printer according to the ſaid decree." On this account, Marprelate upbraids the archbiſhop with "Did not your grace of late erecte a new printer contrary to the—decree? one Thomas Orwine (who ſometimes wrought popiſh bookes in corners: namely Jeſus Pſalter, our Ladies Pſalter, &c.) with condition he ſhould print no ſuch ſeditious bookes as Walde-graue hath done?" Marprelate's Epitome, Part 1. p. 25.

ground,

ground, each of them holding a rose; at the top of the urn is a phœnix issuing out of flame, with this motto, SEMPER EADEM. Also on a scroll flying over all, DELAMIA MORTE ETERNA VITA VIVO. Sometimes that of two hands clasping each other, and holding two cornucopias with a caduce upright between them and T. O. beneath: about it, BY WISDOME PEACE, BY PEACE PLENTY. Sometimes that of Mars standing, with sword and shield. He lived till 1593, when his widow carried on the business, at the sign of the Bible, for a few years.

1587. "The Politicke and Militarie Discourses," &c. as p. 1167, &c. This
1588. was placed to Tho. Chard by mistake.
1588. "The Education of Children in learning: declared by the dignitie, vtilitie, & methode thereof, by W. K.." (Wm. Kempe, who seems to have been a schoolmaster at Plymouth.) Dedicated "To MaisterWm. Hawkins Esq; maior of---Plymouth, &c. Imprinted by him, for Iohn Porter & Tho. Gubbin." The Rev. Dr. Lort. Quarto.
1588. "A discourse of the great & furious batraile," &c. as p. 1168.
1588. "A profitable Exposition of the Lords Prayer," &c. as p. 1196.
1588. "An Oration Militarie to all naturall Englishmen, whether Protestants, or otherwise in Religion affected to moue resolution in these dangerous times. VVherein is expressed the delight of libertie, & the tyrannie of the enemie. With a Praier both pithie & necessarie. Written by a zealous affected Subiect. Non nobis solùm nascimur.—Printed by him, & Tho. Cadman, 1588." Eight leaves. W.H. Octavo.
1588. "A godly and profitable Sermon—shewing the true fruites of peace & warre, as also the rare vertues of godly wise captains, &c. Preached at Paules crosse 25 August last past, by Adam Hill, preacher of the word of God. Pro. 25; 11.—Prou. 24; 26.—Printed—for John Hill dwelling in Pater-noster row, at the signe of the three Pigeons." Dedicated "To—the Ladie Walsingham."—Westburie, 14 Octob. 1588." Text; 2 Chron. 20; 1. F, in eights. W.H. Sixteens.
1588. "ORDERS, Set downe by the Duke of Medina, Lord generall of the Kings Fleet, to be obserued in the voyage toward England. Translated out of Spanish—by T. P." A Ship. "Imprinted – for Tho. Gilbert dwelling in Fleetstreete neere to the signe of the Castle 1588." 8 leaves., W.H. Quarto.
1588. "Abrahami Fransi, Insignium, Armorum, Emblematum, Hieroglyphicorum, et Symbolorum, quæ in Italis Imprese nominantur, explicatio: Quæ Symbolicæ philosophiæ postrema pars est. Excudebat—Impensis Thomæ Gubbin & Tho. Newman." Dedicated "Illustris. Domino D. Roberto Sydneio," in 2 distichs. R 2, in fours. W.H. Quarto.
1588. "THE Coronation of Dauid : Wherein out of that part of the Historie of David, that sheweth how he came to the Kingdome, wee have set forth unto us what is like to be the end of these troubles that daylie arise for the Gospels sake. By Edm. Bunny." King David with his harp. "Psalm 89; 20.—Imprinted—for Tho. Gubbin & Iohn Perin." Dedicated "Vnto—Henry Earle of Huntingdon, Knight of the—Garter, L. President of her Ma. Counsell in the North parts established; and Lieut. Generall of

of her Ma. people & forces there.—At York, 28 Aug. 1588." An analytical table, and another of the contents. 108 pages. W.H. Quarto.

"Elizabetha Trivmphans. Conteyning The Damned practizes, 1588. that the diuelifh Popes of Rome haue vſed euer ſithence her Highneſſe firſt comming to the Crowne, by mouing her wicked & traiterous ſubiects to Rebellion & conſpiracies, &c. VVith a declaration of the manner how her excellency was entertained by her Souldyers into her Camps Royall at Tilbery in Eſſex: and of the ouerthrow had againſt the Spaniſh Fleete:—by I. A. Poſt-victoriam gloria.—Printed—for Tho. Gubbin & Tho. Newman." Dedicated "To—Iulius Cæſar, Doctor of the Ciuill Law, &c.—Ia. Aſke.—-To the gentle Reader.---An acroſtic, Elizabetha Triumphans." 35 pages; in blank verſe. W.H. Quarto.

"A prooued practice for all young Chirurgians concerning burnings 1588. with Gunpowder, and wounds made with Gunſhot, &c. Publiſhed—by Wm. Clowes, Mayſter in Chirurgery. Printed—for Tho. Cadman." Again 1591, for the Widow Broome. Quarto.

"Remember Lots wife. Two godly & fruitfull Sermons very conve- 1588. nient for this our time lately preached on a Sunday in the Cathedral Church of St. Peters in Exceſter. By Iohn C." (Carpenter.) Licenſed. John Fenn, Eſq; Octavo.

"The Chriſtians Sacrifice. Seene, & allowed.—Printed—for T. Man. 1589. 1589." The author Hen. Smith. See p. 1164. W.H. 4 Sixteens.

"P. Rami profeſſoris regii Grammatica, ab eo demum recognita: 1589. Et Ex varijs ipſius ſcholis ac prælectionibus breviter explicata. Editio poſtrema, á ſuperioribus longé diverſa." Device, the phœnix, &c. "—Ex ædibus T. Orwini, Impenſis T. Man & T. Gubbin." 190 pages. Then, on a ſeparate title-page, "Rudimenta grammaticæ, ex P. Rami —poſtrema Grammatica, breviter collecta." &c. The paging continued to p. 135. W.H. Octavo.

"Inſtructions for the warres, Amply, learnedly, & politiquely, diſ- 1589. courſing of the method of Militarie Diſcipline. Originally written in French by that rare & worthy Generall, Monſieur William de Bellay, Lord of Langey, Knight of the order of Fraunce, and the Kings Lieutenant in Thurin. Tranſlated by Paule Iue, Gent." The phœnix, &c. "—Printed for Tho. Man, & Tobie Cooke." Dedicated "To—Wm. Daviſon Eſq; one of her Maieſties principall Secretaries: &c.---To all Gentlemen Souldiers, &c. In verſe.—Tho. Newton, Ceſtreſh.—The preface.---The Contents." 312 pages. W.H. Quarto.

"The Practiſe of Fortification:—in all ſorts of ſcituations, with 1589. the conſiderations to be vſed in delining & making of royal Frontiers, Skonces, and renforcing of ould walled Townes. Compiled—by Paule Iue, Gent." The Phœnix, &c. "Printed—for Tho. Man & Toby Cooke." Dedicated "To—Sir William Brooke,—of the Garter Knight, &c. And vnto—Sir Fraucis Walſingham, Knight," &c. 40 pages, with plans. W.H. Quarto.

"An Antidotarie Chyrurgieall, containing great varietie & choice 1589. of all ſorts of medicines that commonly fal into the Chyrurgians vſe:
partlie

partlie taken out of Authors, olde and new, printed or written: partlie obtained by free gifte of fundrie worthie men of this Profeffion within this Land. By Iohn Banefter Mafter in Chirurgerie. Imprinted—for Tho. Man." Dedicated " To—Ambrofe Earle of Warwicke, &c.— 10 Aug. 1589.---To his verie louing frend, & worthie Brother in the Art—John Banefter:—Wm. Goodrus.----Salutem in Chrifto Gul. Clowes." The letter in Englifh. 359 pages, and a table. W.H. Octavo.

1589. " A SHORT YET found Commentarie; written on that woorthie worke called The Prouerbs of Salomon: and now publifhed for the profite of Gods people. Pfalm 78; 1, 2.—Printed—for T. Man." Dedicated " To—the Ladie Bacon,—25 Sept. 1589.—T. W." *(Tho. Wilcoxe)* 105 leaves. W.H. Quarto,

1589. " The Remedie of Reafon: Not fo comfortable for matter as compendious for memorie.—.By Charles Gibbon.—1 Peter 3; 15, 16.— 2 Cor. 8; 2. Imprinted at London by *him*, 1589. Dedicated " To— Sir Robert Iarmin Knight: And—Mafter Henrie Blagge Efq;---To the Chriftian & curteous Render.—Vale. Bury S. Edmond."---An acroftic, *Anglia*. K, in fours. Licenfed. W.H. Quarto.

1589. " THE French Hiftorie. That is; A lamentable Difcourfe of three of the chiefe & moft famous bloodie broiles that haue happened in France for the Gofpell of Iefus Chrift. Namelie; The outrage called The winning of S. Iames his Street, 1557. The conftant Martirdome of Annas Burgæus one of the K. Councell, 1559. The bloodie Marriage of Margaret, Sifter to Charles the 9. Anno 1572. Publifhed by A. D.—1 Tim. 3; 2." Truth crowned, ftanding naked, with a fcourge at her back: about it, VIRESCIT VULNERE VERITAS. " Imprinted—for T. Man," The arms of Edgecombe, on the back of the title-page. Dedicated " To—her louing Bro-Mafter Pearfe Edgecombe, of Mount Edgecombe in Deuon, Efq;—Honiton, 25 Julie, 1589. Your louing Sifter Anne Dowriche."---An acroftic, PEARS EDGCOMB, in Alexandrines, two verfes to each letter, introduced with this anagram:

" The fharpeft EDGE will fooneft PEARSE and COME vnto AN end: Yet DOWT not, but be RICHE in hope, and take that I doo fend. A. D. To the Reader." The whole hiftory in Alexandrine verfe. 38 leaves. The cut of Truth is exhibited again on the laft leaf, with defcriptive verfes under it. W.H. Quarto.

1589. " The Bucoliks of Virgilius Maro," &c. as p. 1110.

1589. " The firft and fecond parts of Albions England. With hiftoricall intermixtures, invention, and varietie, profitably, briefly and pleafantly performed, in verfe and profe, by William Warner." 167 pages. Quarto.

1589. " Principum, ac illuftrium aliquot, & eruditorum in Anglia virorum, Encomia, Trophæa, Genethliaca, & Epithalamia. A Joanne Lelando Antiquario confcripta, nunc primum in lucem edita. Quibus etiam adiuncta funt," with a feparate title-page, . " Illuftriffimorum aliquot Heroum, hodie viventium, aliorumque hinc inde Anglorum, Encomia quædam: à Thoma Newtono, Ceftrefhyrio, fuccifivis horulis exarata. Londini, Apud Tho. Orwinum Typographum, 1589." Lelandi Collectanea, Vol. 5, p. 79.

" The

THOMAS ORWIN. 1245

"The Haven of Health:" &c. as p. 880, &c. 1589.
"Sophroniftes. A dialogue perfwading the people to reuerence and 1589.
attend the ordinance of God, in themi niftrie of their own paftors."
Printed for Thomas Man. Quarto.
The hiftory of the valerous fquire Aleƈtor, by John Hammon. Quarto.
"A Verie fruitefull Expofition of the Commandements," &c. as p. 1195. 1590.
"Not fo newe as true, being a caueat for all Chriftians to confider of, 1590.
wherein is truely defcribed the Iniquitie of this prefent time, by occafion
of our confufed liuing, And iuftly approued, the world to be never worfe
by reafon of our contagious lewdnes: by Char. Gibbon." Quarto.
"A Diall of Dreames, iudicially poynting to the fucceffe that followes 1590.
euerie fancie appearing in fleepe." Oƈtavo.
"The Writing Schoolemafter; Conteining three Bookes in one; The 1590.
firft teaching Swift writing: The fecond, True writing; The third, Faire
writing. The firft Booke Entituled, The arte of Brachygraphie; that is,
to write as faft as a man fpeaketh, treatably writing but one letter for a
word, &c. The knowledge whereof may eafilie be attained by one moneths
praƈtife, The proof alreadie made by diuers fchollers therein. The fecond
booke named, The order of orthographie; fhewing the perfeƈt Method to
write true Orthographie in our Englifh tongue, as it is now generally
printed, &c. to be attained by the right vfe of this booke without a
Schoolemafter, in fhort time, &c. The third booke is, The key of Caly-
graphie; opening the readie waie to write faire in verie fhort time.—Inuented by Peter Bales, 1 Janu. 1590.—Imprinted—by him: and are to be
folde at the Authors houfe, in the vpper ende of the Old Bayly, where he
teacheth the faid Artes." Dedicated "To—Sir Chr. Hatton, Knight,"
&c. whofe arms are on the back of the title-page. R 3, in fours.
Licenfed. W.H. Quarto.
"A Briefe difcourfe of Warre. Written by Sir Roger Williams Knight; 1590.
VVith his opinion concerning fome parts of the Martiall Difcipline." Device, the hand in hand, &c. "Jmprinted—by him, dwelling in Paternofter Row, ouer againft—the Checker, 1590." Dedicated "To—Robert
Earle of Effex, &c.---To all men of warre in generall." 62 pages. Two
editions. W.H. Quarto.
"A Plaine Declaration that our Brownifts be full Donatifts, by com- 1590.
paring them together—out of the writings of Auguftine. Alfo a replie
to Mafter Greenwood touching read prayer, wherein his groffe ignorance
is deteƈted, &c. By Geo. Gyffard Minifter of Gods word in Maldon."
Hand in hand, &c. "Printed for Toby Cooke,—1590." Dedicated "To
—Sir William Cecill, Knight, &c.---To the Reader." 126 pages. W.H.
Quarto.
"Miles Moffe his Catechifme." Maunfell, p. 30. Oƈtavo. 1590.
Tho. Newton's Staff to lean on: paraphraftically expounding Pfalm 1590.
22; 9, 10. Oƈtavo.
"A Short view of the Perfin Monarchie, and of Daniels weekes: 1590.
Beeing a peece of Beroaldus workes: with a cenfure in fomepoints." Hand
in hand, &c. "—Imprinted by him, 1590." By Hugh Broughton. 46 pages.
W.H. Quarto.
Complaints.

7 G

1591. Complaints. Containing sundry small poems of the Worlds Vanitie, by Edm. Spenser. These are, The Ruines of Time; The Teares of the Muses, with Virgils Gnat; Prosopopoia, or Mother Hubberds Tale, with The Ruines of Rome, by Bellay; Muiopotmos, or The Fate of the Butterflie; with Visions of the worlds vanitie, The visions of Bellay, and The visions of Petrarch. My copy wants the general title-page, and the last poem. The Tears of the Muses, Mother Hubberd's tale, and The fate of the Butterfly, have separate title-pages, all in compartments with Moses and David on the sides; this had been used by John Day, and after him by Tho. Marsh, but now commonly by Orwin, to books in quarto.; we may therefore conclude he printed them " for Wm. Ponsonby." The fate of the Butterfly is dated 1590, probably by mistake, as the signatures are regularly continued to Z, in fours. W.H.+ Quarto.

1591. " A very godly and learned Exposition, vpon the whole Booke of Psalmes.—Heretofore penned—and now faithfully reuieued by the Author—and newly published at the no smal cost of the printer, for the glory of God, &c. 2 Tim. 3; 16, 17.—Printed—for Tho. Man." In a compartment with the queen's arms at top, Mars and Pallas on the sides, and Sir Chr. Hatton's arms at the bottom. Dedicated " To—Roger Herlackinden Esq; and Maister Wm. Herlackinden.—27 Apr. 1591.— T.W. (*Tho. Wilcoxe*.)---The printer to all godly readers." 600 pages. This had been printed in 1586 for T. Man and Wm. Brome. W.H. Quarto.

1591. " A Work worth the Reading: Wherein is contayned, Five profitable and pithy Questions, very expedient as well for Parents to perceive howe to bestowe their Children in Mariage, and to dispose of their Goods, at their Death, as for all other Persons to receive great Profit by the rest of the Matters herein expressed. Newly published by Charles Gibbon. Imprinted by *him*, and sold by Hen. Kyrkham," &c. 34 leaves. Harl. pamphlets, No. 521. Quarto.

1591. " SERMONS, Vpon the 101 Psalme, conteyning profitable instruction for all, especially for such as haue any gouernement ouer others. By O. P. Iosua 24; 15.—Pet. 2; 13, 14.—Imprinted—for T. Man." Dedicated " To—Edward Earle of Bedford, and to—Ladie Brydgit the Countisse of Bedford.—London, 12 May, 1591.—Oliuer Pigg." 141 pages. W.H. Sixteens.

1591. " Campaspe." Also, " Sapho and Phao." &c. as p. 1203.

1591. " The Compound of Alchymy; or the ancient hidden Art of Archemie; Conteining the right & perfectest meanes to make the Philosophers Stone, Aurum potabile, with other excellent experiments. Diuided into 12 Gates. First written by the rare & excellent Philosopher of our Nation George Ripley, sometime Chanon of Bridlington in Yorkeshire, and dedicated to K. Edward the 4. Whereunto is adioyned his Epistle to the King, his Vision, his Wheele, and other his workes neuer before published: with certaine briefe Additions of other notable Writers concerning the same. Set foorth by Raph Rabbards Gent. studious & expert in Alchemicall Artes. Pulchrum pro Patria pati. Imprinted—1591." Dedicated " To the moste high & mightie Princesse Elizabeth," &c. whose picture is portrayed in the first letter E, as p. 1132.---"To the—industrious Students

Students in the fecrets of Philofophie." Commendatory verfes. &c. In verfe. M. in fours. Licenfed. W.H. Quarto.

" A fhort treatife of hunting, compyled for the delight of noblemen, 1591. by Thomas Cockaine, knight." Quarto.

" The examination of uferie, in two fermons, Taken by chara&terie 1591. and after examined."* Sixteens.

" A treatife of the Lord's Supper, in two fermons."* Sixteens. 1591.

" The True Hiftory of the Ciuil VVarres of France," &c. as p. 1110. 1591.

" Elizabeth Brewer her laft will & teftament." Quarto. 1591.

" Richard Coxe his Catechifme." O&tavo. 1591.

" The Counteffe of Pembrokes Emanuel. Conteining the Natiuity, 1591. Paffion, Buriall & Refurre&tion of Chrift: togeather with certaine Pfalmes of Dauid. All in Englifh Hexameters. By Abr. Fraunce. Imprinted—for Wm. Ponfonby,—1591." In the compartment with Mofes & Dauid. Dedicated " To—the Lady Mary, Counteffe of Pembroke." E 3, in fours. W.H. Quarto.

" THE Counteffe of Pembrokes Yuychurch. Conteining the affe&tionate 1591. life, and vnfortunate death of Phillis & Amyntas: That in a Paftorall; This in a Funerall: Both in Englifh Hexameters. By Abr. Fraunce.— Printed—for Wm. Ponfonby,—1592." Dedicated "To—the Ladie Marie, Counteffe of Pembroke." The firft part is divided into five a&ts, with fcenes; the fecond part into twelve days, to which are annexed, The Lamentation of Corydon for the loue of Alexis, verfe for verfe out of Latine; and the beginning of Heliodorus his Aethiopian Hiftory." M, in fours. W.H. Quarto.

" THE Third part of the Counteffe of Pembrokes Yuychurch:" &c. 1592. as p. 1111.

" The Sermons of Mafter Henrie Smith gathered into one volume. 1592. Printed according to his corre&ted Copies in his life time." Hand in hand, &c. For T. Man. O&tavo.

" A Prognoftication euerlafting." &c. as p. 867. O, in fours. Quarto. 1592.

" A Cafe of Confcience, the greateft that euer was; How a man may 1592. knowe whether he be the Child of God or no. Refolued by the worde of God. Whereunto is added a briefe difcourfe, taken out of Hier. Zanchius. 2 Peter 1; 10.—Imprinted—for Iohn. Porter & T. Man." By Wm. Perkins. 83 pages. W.H. O&tavo.

" ALBIONS England: The Third time corre&ted & Augmented. Con- 1592. tinuing an Hiftory of the fame Countrey &.Kingdome, from the originals of the firft Inhabitants of the fame: With the chiefe Alterations & Accidents there inhappening, vntill her nowe Maiefties mofte bleffed Raigne. VVithin termixture of Hiftories & Jnuention, performed in verfe, by Wm. Warner." Device, Mars. " Imprinted—for I. B. dwelling at the great North Doore of S. Pauls church, at the figne of the great Bible, 1592." Dedicated " To—Henrie Carey, Baron of Hunfdon: Knight" &c.---- To the Reader." 208 pages, and a table. At p. 144 is a cut of the lineage of the two noble houfes of Lancafter and York: the fame that had been ufed for the title-page of Hall's chronicle. W.H. Quarto.

" A Booke named Te&tonicon," &c. See p. 860. Quarto. 1592.

" Cambrobrytannicæ Cymbraecæve Linguae Inftitutiones & Rudimenta 1592. accuraté accuratè & (quantum fieri potuit) fuccinctè & compendiofè confcripta à Joanne Dauide Rhæfo Monenfi Lanuaethlæo Cambrobrytanno, Medico Senenfi: Ad Jlluft. virum Edouardum Stradlingum Equeftris ordinis Cambrobrytannum: Ad intelligend. Biblia facra nuper in Cambrobrytahnicum fermonem & caftè & eleganter verfa,—: Cum exacta carmina Cymracca condendi Ratione, & Cambrobrytannicorum Poeinatum generibus," &c. Device, Mars. " Excudebat—1592." Dedicated " Illuftr viro—Edouardo Stradlingo—Equiti aureo.—Brychanij prid. Non. Julias, 1590." Licenfed. Folio..

1592. Calvin's Catechifm. See p. 840. Octavo

1592. " THE Solace for the Souldier and Saylour: Contayning a Difcourfe and Apologie out of the heauenly word of God, how we are to allow, and what we are to efteeme of the valiant attempts of thofe Noblemen, and Gentlemen of England, which incurre fo many daungers on the feas, to cut off or abridge the proude and haughtie power of Spayne, by Simon Harward. For T. Wight,—at the great North-doore of S. Paules Church." Dedicated " Ad Reuerendifs.—D. Johannem Cantuarienfem Achiep. totius Angliæ primatem, &c. Et Honoratifs. D. Georgium Comitem Cumbriæ. Dom. Weftmerlandiæ, &c." In 15 Latin diftichs.--- The preface to the Chriftian Reader.—Jdib. Oct. 1592." G, in fours. At the end, " A Prayer—In time of feruice:" Licenfed. W.H. Quarto.

1592. " The nine VVorthies of London; explaining the honourable exercife of Armes, the vertues of the valiant, and the memorable attempts of magnanimous minds. Pleafaunt for Gentlemen, not vnfeemely for Magiftrates, and moft profitable for Prentifes. Compiled by Ric. Iohnfon." Mars. " Imprinted—for Humf. Lownes,—at the weft doore of Paules." F, in fours. W.H. Quarto.

1592. " The moft excellent, Profitable and pleafant book of the famous Doctor, and expert Aftrologian Arcandam, or Aleandrin, to find out the fatal deftiny, conftellation & natural inclination of euery man & child by his birth. With an addition of Phifiognomy, very pleafant to reade. Now newly turned out of French,—by Wm. Ward. Imprinted—by *him*, 1592." Q 4, in eights. W.H. Sixteens.

It appears by a minute in the Stationers' regifter, 18 Dec. 1592, that he had begun an impreffion of Juftin in Latin, which when finifhed he he was ordered to bring into the Hall; and then the right of Mr. Norton, or any other pretending intereft to yt to be difcuffed:" &c.

1593. " Tho. Fale his Art of Dialling, teaching an eafie & perfect way to make

---

7 The names of thefe worthy men, &c. 1. Sir Wm. Walworth, Fifhmonger, in the time of K. Richard II. 2. Sir Henry Pitchard, Vintner, in the time of K. Edw. III. 3. Sir Wm. Sevenoake, Grocer; in the time of K. Hen. V. 4. Sir Tho. White, Merchant-taylor, in the time of Q. Mary. 5. Sir John Bonham, Mercer; in the time of K. Edw. I. 6. Sir Chr. Croker Vintner; in the time of K. Edw. III. 7. Sir John Haukwood, Merchant-taylor; in the time of K. Edw. III. 8. Sir Hugh Caverly, Silk-weaver, in the time of K. Edw. III. 9. Sir Henry Maleveret, Grocer; in the time of K. Hen. IV.

This is written in the manner of The mirror for magiftrates. Fame flies to Parnaffus, and takes Clio with her to the Elyfian fields, where thefe champions are required to declare their own fortunes, which Clio records with her golden pen.

make all kinde of Dialls vpon any plaine platte, howfoever placed, with the drawing of the 12 fignes & houres vnequall in them all; whereunto is annexed, the making and ufe of other Dialls, & inftruments, whereby the hower of the day & night is known." Licenfed. Quarto.

"The order of keeping a courte leete, and a court baron; with the charges appertayning to the fame; truely & plainly deliuered in the Englifh tongue, &c. By Jonas Adames. Quicquid agas, prudenter agas, & refpice finem." Hand in hand, &c. "Imprinted—by *him* and Wm. Kirkham, and are to be fold at the little north doore of S. Paules Church, at the—black boy." 20 leaves. Licenfed. Quarto. 1593.

"A Treatife concerning—Bees." &c. as p. 1111.

"A Dictionarie French and Englifh:" &c. as ib. 1593.

"ARCADIAN RHETORIKE, or the precepts of Rhetoricke, made plaine by examples Greeke, Latyne, Englifhe, Italyan, Frenche & Spanifhe, By Abr. Fraunce." Licenfed to T. Gubbyn & T. Newman, 11 June, A. D. 1588. Octavo.

"A right godly rule, how all faithfull Chriftians ought to occupie & exercife themfelues in their dayly praiers." Sixteens.

"He had alfo licenfes for the following, viz. In 1588, "The complaint of Tyme. A miraculous.—difcourfe," &c. as p. 1201. "The arte of Eng. Poefye in 3 bookes. The ftraight waie ȳ leadeth to eternall life. A Joyefull fonet of the royall receaving of the queenes maieftie into ȳ cyttye of London on Sonday 24 Nov. all along Fleteftreete to the Cathedrall Church of St. Paule, &c. Broke his furfeyt in love, with a farewel to ȳ folies of his owne phantafie. Secunda pars Elizabethe. The monarchie of the Greeks." In 1589, The type or figure of frendfhip. A mofte extellent & Mytheologicall hiftorye of the valerous knighte Alector fone of Macrobius francgal & of quene Prifcaraxe. A manuel of godly praiers. The perfect pathway to Salvac'on. Godly devout meditac'ons &c. by John Sydley. A tablet for gentlewomen." In 1590, "The amorous paffions of twoo gentlemen; a flatterer & a true lover. A confort of ȳ creatures with the Creator, & with themfelues." Sundry copies "granted vnto him by ȳ confent of Edw. Marfhe." See p. 847, &c. In 1591, "The Schole of fayre wrytinge." 3 Maye, A. D. 1593, "Entred for his copies, by affent of a Court *of Affiftants*, thefe books, which were firft Kingftons, and after Geo. Robinfons, whofe widow Orwin hath married, viz. The Whetftone of wytt. Wilfons Retorik & Logik. Calvins Catechifme, Virgil, in Latin. Sturmis epiftles. Tullies offices, Latin. Sufembrots figures. Acolaftus. Pueriles confabulatiücule. Pueriles fentencie."

JOAN ORWIN, the Widow

Of Tho. Orwin, who appears to have lived but a fhort time after the Court's allowance of the laft copies entered to him 3 May, 1593; for on the 25 June following fhe binds an apprentice by the name of Jone "Orwin widow, late wife of Tho. Orwin deceafed." How long they had been married is not quite apparent, but it feems probable that he would have a confirmation of her former hufband's copies to him as foon as

might

JOAN ORWIN.

might be. She frequently used her late husband's device of the Hand in hand, &c. but always without the T. O. And so it was used afterwards by Felix Kingston. Dec. 1, 1595, she was fined xx s. for printing books disorderly; and her son committed to ward for the same offence. Paid x s. the 29 Jan. following; the rest remitted.

1593. "A Commentarie vpon the Lamentations of Ieremie; wherein are contained first the method & order of euery Chapter laide open in seuerall tables; then a litterall interpretation of the text out of the Hebrew, with a paraphrasticall exposition of the sense thereof; afterward a collection of diuers doctrines; lastly the particular vses that are to bee made of them. Printed for Tho. Man." Maunsell ascribes this to John Udall as the author. Printed by F. Kingston in. 1599. Quarto.

1593. "The most strange & admirable discouerie of the three witches* of Warbois arraigned, conuicted & executed at the last assises at Huntington, for the bewitching of the fiue daughters of Rob. Throckmorton Esq; and diuers other persons with sundry diuelish & grieuous torments; And also for the bewitching to death of the Lady Crumwell, the like hath not bene heard of in this age. Printed—for T. Man & Iohn Winnington, —1593." See Brit. Topogr. Vol. I. p. 439. Quarto.

1594. "SINOPSIS PAPISMI, that is a general view of Papistrie, wherein the whole misterie of iniquitie, and sūme of Antichristiā doctrine is set down, which is maintained this day by the Sinagogue of Rome against the Church of Christ, together with an Antithesis of the true Christian faith, and an Antidotum, or counterpoyson out of the Scriptures, &c. Devided into 4 centuries—of Popish heresies & errors. The second edition. Printed—for T. Man." Quarto.

1594. "An exposition of the Lords Prayer, made in diuers Lectures, and now drawne into Questions & Answers,—: Whereunto is prefixed a briefe treatise of prayer for all men.—By W. B. Minister of the Word at Reading in Barkshire. Printed—for T. Man." Dedicated "To Robert Earle of Essex, &c.—Wm. Burton." 215 pages. W.H. Sixteens.

1594. Wm. Blackwall's "Fruitfull & Pithie Treatise of Death." Sixteens.
1594. "The Chatechisme—by the Excellent doctor Ihon Calvin," &c. Octavo.
1595. "The Tragedie of Dido," &c. as p. 1112.
1595. "The Doctrine of the Sabbath,—Declaring first from what things God would haue vs straightly to rest vpon the Lords day, and then by what meanes we ought publikely & priuatly to sanctifie the same. Diuided into two Bookes by Nicolas Bownde, D. D.—Printed—for Iohn Porter and T. Man." On the back are the arms of Robert Earl of Essex, &c. to whom the book is dedicated, from "Norton in Suff. June 1595.— To the Christian Reader." 286 pages, and tables to each book. W.H.
Quarto.
"The

---

* Their names were John and Alice Samuel, and Agnes their daughter. On occasion of this pretended discovery, a sermon is preached annually at Huntingdon on Lady-Day, being their Fair day, by one of the Fellows of Queen's Coll. Camb. For which the preacher receives 40 sh. of that corporation; so much being settled on the town by sir Hen. Cromwell, out of the estate or Monies belonging to the supposed witches; forfeited probably at their death. See Hutchinson on Witchcraft, p. 130, &c.

JOAN ORWIN.

"The Caftell of Health, Corrected, and in fome places augmented by 1595. the firft Author thereof, &c. 1595" Hand in hand, &c. without T. O. "Printed—and are to be fold by Matthew Lownes." 140 pages. W.H. See p. 435. Quarto.

"A fhort, yet true and faithful narration of the fearfull fire, that fell 1595. in the town of Wooburne, in the county of Bedford, the 13th of September. By Tho, Wilcocks. Printed—for Thomas Man." Octavo.

"The arraignment and conviction of vfurie,—in fixe fermons, preached 1595. at St. Edmunds Burie, in Suffolke, vpon Pro. 28 : 8. By Miles Mofſe. Seene & allowed.—Reade all, or cenfure none.—" David ftanding with his harp. "Printed—for Iohn Porter." Dedicated "To—Iohn archbifhop of Canturburie, Primate, &c.—Bury S. Edmunds, 1 Jan. 1595.---To the Reader.—6 Feb. 1594.---The names of Authors" &c. 171 pages. W.H. Maunfell has the fame for T. Man. Quarto.

"A treatife tending vnto a declaration, whether a man be in the eftate 1595. of damnation or—grace: &c. Reuiewed & corrected by the Author.— 2 Pet. 1; 10.—" Hand in hand, &c. "Printed—for Iohn Porter, and Iohn Legate, 1595." Dedicated "To—Mafter Valentine Knightlie Efq; Cambridge, 24 Nouember, 1589.—Wm. Perkins.---To the Chriftian Reader.—1595." Pages 182. W.H. Again 1597, for John Porter. Quarto.

"Ariftotle, or the Philofophers and Phyfitians problems wherein are 1595: contained diuers queftions with their anfwers, touching the eftate of Mans body." Octavo.

"A Pleafaunt Satyre or Poefie: Wherein is difcouered the Catholicon 1595. of Spayne, and the chiefe leaders of the League Finelie fetched over, & laide open in their colours. Newly turned out of French into Englifh. Pro. 19; 25—." Hand in hand, &c. "At London by—for T. Man." Quarto.

"The Pathway to Perfection." &c. as p. 1034. "Printed—for Andr. 1596. Wife, dwelling in Paules Church-yeard, at the—Angel, 1596." Sixteens.

"The Meane in Mourning. A Sermon preached at St. Maryes Spittle 1596. in London on Tuefday in Eafter weeke, 1595. By Tho. Playfere D. D. Printed—for Andr. Wife," &c. Sixteens.

"The Order of Houfhold inftruction: By which euery mafter of a Fa- 1596. milie, may eafily & in fhort fpace, make his whole houfhold to vnderftand the—chiefe points of Chriftian religion: &c. Deut. 6; 6, 7—2 Tim. 3; 15.—Printed—for T. Man." Dedicated "To—Robert—Earle of Effex, &c.—At Eaftwell in Kent, 26 Feb. 1596.—Iofias Nichols.---To all gouernours of Families, &c.---The Booke to the Houfeholder;" in verfe. H 4, in eights. W.H. Octavo.

"A treatife fetting forth the worthie praife of true Wifdom : collected 1596. out of the facred fcriptures, by Tho. Gibfon. Printed—for Wm. Blackwall." Octavo.

"Certaine Sermons vpon diuers textes of Holie Scripture.—By M. 1597. Geo. Giffard, Preacher of the worde of God at Mauldon in Effex. Printed—for Tho. Man." Dedicated "To—M. Iohn Hutton;" 288 pages. W.H. Octavo.

"Albions

JOAN ORWIN.

1597. " Albions England; a continued historie of the same kingdome, from the original of the first inhabitants thereof; and most of the chiefe alterations and accidents there happening vnto, and in the happie reigne of our now gracious soueraigne, queene Elizabeth. By W. Warner. Printed—for I. B.—1597." See p. 1247. Quarto.

1597. " The Mirror of Honor: Wherein euerie professor of armes, from the Generall—to the priuate officer & inferiour souldier, may see the necessitie of the feare & seruice of God, &c. Exod. 14; 14—" Hand in hand, &c. " Printed—for T. Man." On the back, " The Contents," &c. Dedicated " To—Robert Earle of Essex, &c.—Iohn Norden.---To the Reader." 93 pages. W.H. Quarto.

1597. " The foundation of Christian Religion: gathered into sixe principles.— to be learned of ignorant people that they may be fit to heare Sermons with profit, and to receiue the Lords Supper with comfort. Psalm 119; 30.—Printed—for Iohn Porter." In the compartment as p. 926. Prefixed is an epistle, " To all ignorant people that desire to be instructed.— Wm. Perkins." 16 pages. Hereunto is annexed his " treatise—whether a man be in the estate of damnation," &c. as in 1595; the paging continued to p. 142. Then, " A case of conscience" &c. as p. 1247. with the pages still continued to p. 175. To these are also annexed the author's " Exposition of the Lord's Prayer:" &c With the same device of the Hand in hand as she used, but " Printed by Felix Kingston, for Iohn Porter & Ralph Iackson, 1597." The paging further continued on to p. 236. W.H. Quarto.

RICHARD FIELD, stationer,

SON of Henry Field of Stratford on Avon, Tanner, was put apprentice to Mr. Geo. Bishop for 7 years from Michaelmas 1579, made free 6 Feb. 1586-7, and came on the livery, 1 July 1598. In 1588, he married Jakin the daughter of Tho. Vautrollier, whom he succeeded in his house and business, using the same devices, and printing occasionally the same copies. May 12, 1589, he was fined for printing a book contrary to order, by his own consent x s. Also 3 Nov. 1589 for keeping an apprentice unpresented, ij s. vj d. March 26, he was chosen renter, and paid 10 l. for his dispensation.

1588. " The copie of a letter sent out of England" &c. as p. 1075.
1589. " THE ARTE OF ENGLISH POESIE. Contriued into three Bookes: The first of Poets and Poesie, the second of proportion, the third of ornament." The anchor, &c. " Printed by Richard Field, dwelling in the black-Friers neere Ludgate." Webster Puttenham is allowed to have been the author. Dedicated " To—Sir William Cecil Knight, &c.— Black-friers, 28 May, 1589.—R. F." Then a portrait of Queen Elizabeth,

bĕth; over it, " A colei;" under it, " Che fe. ſteſſa raſsomiglia & non altrui." 258 pages, and a table. Licenſed by T. Orwin's conſent. George Maſon Eſq; Quarto.

" P. Ovidii Naſonis Metamorphoſeon Libri XV. Ab. Andrea Naugerio caſtigati, & Vict. Giſelini ſcholijs illuſtrati.—Excudebat—impenſis Iohannis Harriſoni, 1589." Neat Brevier Italic. E e, in eights. W.H. Sixteens. 1589.

" The Reformed Polliticke. That is, An Apologie for the generall cauſe of Reformation, written againſt the ſclaunders of the Pope and the League.—.Whereto is adioyned a diſcourſe vpon the death of the Duke of Guiſe, &c. Imprinted—1589." Dedicated " To the King.—From London, 12 December, 1588.—Iohn Fregeuille of Gaut." 90 pages. Licenſed both in French and Engliſh. W.H. Quarto. 1589.

" The Reſtorer of the French Eſtate, Diſcouering the true cauſes of theſe vvarres of France & other countries, and deliuering the right courſe of reſtoring peace & quiet to all Chriſtendome. Wherein are handled (6) principall queſtions touching Religion, Policie, and Juſtice: &c. Tranſlated out of French, Eccleſiæ & Reipub, D." This is introduced with a ſhort addreſs " To the Reader." 172 pages. Licenſed. W.H. 1589.

A letter from a Gentleman employed in diſcovery on the coaſt of Spain. Quarto. 1589.

Sir Francis Drake's Weſt Indian Voyage. See it in 1596.--Quarto. 1589.

" A briefe diſcourſe of the Spaniſh ſtate, with a Dialogue annexed intituled Philobaſilis." Dedicated " To the moſt—vertuous Princeſſe Elizabeth—queene of England, &c.—Edw. Daunce.---To the noble & vertuous Reader." 52 pages. Licenſed. W.H. Quarto. 1590.

" The Methode of Phiſicke, contayning the cauſes, ſignes & cures of inward diſeaſes in mans bodie,—:Wherevnto is added the forme of making Medicines, &c. By Phillip Barrough." Licenſed. Again 1596. Quarto. 1590.

" CHRONOGRAPHIA. A Deſcription of time" &c. as p. 1228. " The ſecond edition corrected & augmented.—Printed—for Rob. Dexter." W.H. Octavo. 1590.

" A Treatiſe named Lucarſolace" &c. as p. 1157. 1590.

" A Comfortable Treatiſe for the reliefe of ſuch as are afflicted in conſcience. By R. Linaker." Octavo. 1590.

" A ſoueraigne ſalue for a ſick ſoule.—teaching the right vſe of patient bearing the croſſe,—and the euils that come of impacience.—Engliſhed by W. F." An introduction " To the Reader," wherein we learn that " the work was firſt penned by D. Chytræus." D. 4, in eights. See p. 1069. Licenſed. W.H. Sixteens. 1590.

" An Arithmetical vvarlike treatiſe, named Stratioticos." See p. 983. " Lately reuiewed & corrected by the author himſelfe, and alſo augmented with ſundry additions, Aſwell concerning the Science or Art of great Artillerie, as the offices of the Sergeant Maior Generall," &c. 280 pages, but printed 380: after p. 124 follows 225, &c. W.H. Quarto. 1590.

" Of the markes of the children of God," &c. See p. 1228. W.H. 1591.

" The Popes Parliament, containing a pleaſant & delightful hiſtorie, wherin are throughly deliuered, and brightly blazed out the paltry traſh 1591.

7 H and

and trumperies of him and his pelting Prelats, their mutinies, difcord anfl diffentions, their ftomack and malice at Pope Joane, their fhifting and foifting of matters for defence of her, and their Antichriftian practifes for maintenance of their pompe & auarice. Whereunto is annexed an Anatomie of Pope Joane, more apparently opening her whole life, &c. Written by John Mayo." Dedicated " To—Sir George Trenchard Knight, &c.—To the Chriftian Reader." 28 pages. The "Anatomie of pope Joan" has a feparate title-page, but the paging is continued to p. 42. W.H. Again 1594. Quarto.

1591. Knox's Anfwer to the cavillations of an Anabaptift, &c. p. 1196.

1591. " Orlando Furiofo in Englifh verfe, by John Haringtō. Principibus placuifse viris non vltima laus eft. Horace." This title is in an engraved compartment, with a buft of Ariofto at top, Mars on one fide, Venus and Cupid on the other, Sir John Harrington's portrait, ætatis fuæ 30, at bottom. " Tho. Coxonus fculp." Dedicated " To the moft excellent, vertuous and noble Princeffe Elizabeth—queene of England, &c.----A preface, A briefe apologie of poetrie, and of the author and tranflator of this Poem---An aduertifement to the reader before he read this poeme," &c. To each of the 46 Cantos is prefixed a neat folio copper-plate cut; every canto alfo is illuftrated with notes, declaring the moral, hiftory, allegory and allufion thereof. At the end are annexed, " A brife & fummarie Allegorie of Orlando Furiofo, not vnpleafant nor vnprofitable for thofe that haue read the former poeme.---The life of Ariofto," &c. An alphabetical table. The (24) principal tales—that may be read by themfelves. O o 4, in fixes—On the laft leaf, under his device, " Imprinted—in the Black-friers by Ludgate, 1591." Licenfed. W.H. Folio.

1591. The preparative to marriage.---A treatife on the Lord's fupper.----The examination of ufury.----Seven fermons; three prayers &c. All newly perufed and corrected by the author H. Smith. Printed for T. Man. Sixteens.

1592. " Sixe fermons preached by Maifter Henry Smith at Clement Danes church without Temple barre. VVith two Prayers of the fame Author hereunto annexed." Dexter's rebus. " Imprinted—by R. F. for. Rob. Dexter,—1592." Pages 176. W.H. Again 1599, corrected. Octavo.

1592. " A Commentarie upon the—Proverbes of Salomon.—Ecclefiaftes 12; 13." Dexter's rebus. " Printed—for R. Dexter." Dedicated " To—Edward Earle of Bedford.—P. M." (Peter Muffet) 321 pages. Octavo.

1592. " Apocalypsis. A briefe and learned Commentarie vpon the Reuelation of St. John the Apoftle & Euangelift, applied vnto the hiftorie of the Catholike and Chriftian Church. Written in Latine by Mr. Fran. Junius, D.D. &c. And tranflated into Englifh—. Deut. 29; 29.—Imprinted—for R. Dexter,—1592." There are prefixed, " An admonition vnto the reader.—5 Feb. 1591.----The authors epiftle.----The order of time," &c. An analytical table. W.H+ Again in 4to, 1594. Octavo.

1592. " The Spaniards Monarchie," &c. as p. 1157.

1592. " Simonis Verepaei de epiftolis Latine confcribendis libri v. denuo exactiore

exactiore methodo schematismis & scholiis illustrati, et accessione nova postremum aucti." Quarto.

" Psalmorum Davidis paraphrasis poetica Georgii Buchanani, Scoti." 1592.
Sixteens.

" Thomas Masterson his first booke of arithmeticke. Shewing the 1592. ingenious inuentions, and figuratiue operations, by which to calculate the true solution, or answeres of arithmeticall questions; after a more perfect, plaine, briefe, well ordered Arithmeticall way, then any other heretofore published; verie necessarie for all men. Nothing without labour. All things with reason." Dedicated " To— Robert Earle of Essex, &c.—London, 20 Aug. 1592.----To the Reader.----The contents." 108 pages. W.H. Quarto.

"—his second booke," &c. Dedicated " To—Sir William Webbe 1592. knight Lorde Maior of the famous Citie of London,—the Aldermen, and Commons of the same.----To the Reader." 142 pages. W.H. Quarto.

" Tabulæ Analyticæ," &c. as p. 881. Printed also for John Oxen- 1592. bridge, with his rebus of an Ox, with the letter N on its back, going over a bridge. Quarto.

" Spirituall Preseruatiues against the pestilence:—Chiefly collected 1593 out of the 91 Psalme,—By H. H. Leuit. 26; 25.—Printed—for T. Man,— 1593." Dedicated ". To—the Lord Maior of the most renowmed Cittie of London, and to—the Sheriffes; and also to all the—Aldermen their brethren, and to—my verie good friend M. Tho. Aldersey Esq; and Citizen of the said Citie.—Hen. Holland.---The epistle to the Reader.— Febr. 18, 1593." Leaves 87. W.H. Sixteens.

" The French Littelton: a most easy perfect and absolute way to 1593. learne the French tongue: set forth by Claudius Holyband, Gentilhomme Bourbonnois.— Dum spiro, spero." Dedicated " To—Lord Edward Souche.---Geo. Gascoine squire in commendation, &c.----Sonnet. In French.—Pax in Bello." 223 pages. Licensed. W.H. Again 1597.
Sixteens.

" A caveat for suerties. Two sermons of suertiship, on Prov. 6 ;1— 1593. 5: made in Bristol; by Wm. Burton. Printed—for Toby Cooke." Octavo.

" A Christian and godly view of Death and Life, as also of human actions, by Phil. Morney. Translated by A. W. Printed by R. F. for R. B." Octavo.

" A briefe discourse of mans transgression, and of his redemption by 1593. Christ; with a particular surueigh of the Romishe religion, and Rome itselfe:—by F. Clement." Licensed. Octavo.

" A discouerie of the vnnatural and Traiterous conspiracie of Scottish (1593.) Papists against God his Church, their natiue Country, the Kings Maiesties Person and Estate: set down as it was confessed and subscribed by Maister Geo. Ker, yet remaining in Prison; and Dauid Grahame of Fentrie, iustly executed for his treason in Edenburgh, 15 Feb. 1592. Whereunto are anexed certain intercepted Letters, written by some of that faction to the same purpose. First printed and published in Scotland.

land, at the speciall commandment of the Kings Maiestie. London, Printed by R. F. for John Norton." Quarto.

1594. " Jesu Christi D. N. Novum Testamentum, Theo. Beza interprete. Impensis J. Harisoni." Sixteens.

1594. " Sixe bookes of politickes, or ciuil doctrine, written in Latine by Iustus Lipsius; which doe especially concerne principalitie. Done into English by Wm. Iones, gent.—Printed—for Wm. Ponsonby." Dedicated " To—his singular good lord & maister, Sir Iohn Puckering, Lord keeper &c.—At Newington Buts, 1 Jan. 1594.---To the courteous Reader.---The Author his Epistle.---An Alphabet of the authors," &c. 207 pages. W.H. Quarto.

1594. " A learned—Treatise, containing all the principall grounds of Christian Religion.—By maister Math. Virell." &c. See p. 1235. For R. Dexter. 290 pages. W.H. Octavo.

1594. " The History of Christ his natiuitie, life, actes, miracles, doctrine, death, passion, resurrection & ascention, in meeter. By Hen. Holland." Licensed. Octavo.

1594. " The Pearle of Practise, or Practisers Pearle for Phisicke & Chirurgerie, found out by I. H. (*John Hester*) a Spagericke or Distiller, amongst the learned obseruations & proued practises of many expert men in both faculties. Published & drawn into methode by Iames Forestier." Licensed. Quarto.

1594. " The Rape of Lucrece," &c. as p. 1157.

1594. " P. Ovidii Nasonis heroidum epistolæ. Amorum, libri III. De arte amandi, libri III. De remedio amoris, libri II. Aliaq; huius generis, quæ sequens pagella indicabit. Omnia ex accuratiss. Andreæ Nauigerij castigatione. Guidonis Morilloni in Epistolas.—Excudebat R. F. impensis Iohannis Harisoni, sub signo Canis Leporarij, 1594." W.H. Octavo.

1594. " Questiones of profitable & pleasant concernings, talked of by two olde Seniors, the one an ancient retired gentleman, the other a midling or new vpstart frankeling, vnder an Oake in Kenelworth Parke, where they were met by an accident to defend the partching heate of a hoate day, in grasse or Buck-hunting time, called by the reporter the Display of vaine life, together with a Panacea or suppling plaister to cure if it were possible the principall diseases wherewith this present time is especially vexed." Dedicated " To—Robert Earle of Essex, &c.—O. B.---The Epistle to the Reader." 34 leaves. Then, " The Argument—according to the Authors meaning," 4 leaves more. W.H. Quarto.

1594. " Analysis logica Evangelii secundum Matthæum; una cum scholiis & obseruationibus locorum doctrinæ, &c. Authore M. Johan. Piscatore. Impensis B. Norton." Licensed to John Norton; as also, Octavo.

1595. " ——— secundum Marcum;—Authore M. Johan. Piscatore, professore sacrarum literarum in illustri schola Nassovica Sigenensi." Sixteens.

1595. " Thomas Masterton his thirde booke of Arithmeticke." &c. See his first and second books in 1592. Dedicated " To—Sir Iohn Puckering,—Lord Keeper" &c. whose arms are on the back of the title-page.---" To the Reader.---The contents;" 68 pages. W.H. Quarto.

" Methodus

RICHARD FIELD. 1257

"Methodus de conscribendis epiftolis, a Georgio Macropedio Secun- 1595.
dum veram artis rationem tradita: Eiufdem Epitome præceptionum de
paranda copia verborum & rerum per quæftiones: item de nouem fpe-
ciebus argumentationum Rhetoricarum, rem omnem breuiter explicans.
Acceffit Chr. Hegendorphini Epiftolas confcribendi Methodus. Hac
editione longè quàm antea emendatior.—Ex officina typographica Richardi Field." Infcribed, "Ad fcholæ Traiectinæ ftudiofum iuuentutem."
121 leaves, and an index. Licenfed. W.H. Sixteens.

"Deffynniad ffydd Eglvvys Loegr: Lley ceir gweled, a gwybod, 1595.
dofparth gwir Grefydd Crift, ag anghywirdeb Crefydd Eglwys Rufain:
—Wedi ei gyfieuthu o Ladin, yn Gymraeg, drwy waith M. Kyffin.—
Richard Field a'i printiodd yn Llunden, 1595." Befides the prefixes, 107
leaves; on the back of the laft, are the author's arms. W.H. Octavo.

"Phrafes linguæ Latinæ, ab Aldo Manutio P. F. confcriptæ; 1595.
nunc primum in ordinem abecedarium adductæ, et in Anglicum fermonem converfæ." See p. 1157. Licenfed to John Harrifon and him. 16°.

"The Haven of Health:" &c. See p. 880, &c. 1596.

"Daniel his Chaldie vifions, and his Ebrevv: Both tranflated after 1596.
the original; and expounded, both by reduction of heathen moft famous
ftories vnto the exact proprietie of his wordes (which is the fureft cer-
taintie, what he muft meane) and by ioyning all the Bible, and learned
tongues to the frame of his worke. Let him that readeth (Daniel) vnderftand. Mat. 24. The vvife vvill vnderftand. Dan. 12. Printed—for
Wm. Young, dwelling neare the great North doore of Paules, where
the other workes of the fame author are to be fold, 1596." Dedicated
"To—the LL. of her M. moft honorable priuie counfel.—Hugh Broughton.---To the Chriftian Reader:" &c. P, in fours, with copper-plate
cuts, and much Hebrew type. Licenfed. W.H. Quarto.

Orationum M. T. Ciceronis, a Joan. Michaele Bruto emendatum, &c. 1596.
In three volumes. Licenfed. For the company. Tully's Offices, alfo.
Sixteens.

Sermons upon the whole book of the Revelation by George Giffard, 1596.
preacher of the word at Maulden in Effex. Again 1599. Quarto.

"Catechifmo que fignifica, forma de inftrucion; que contiene los prin- 1596.
cipios de la religion de Dios, vtil y neceffario para todo fiel Chriftiano;
compuefto en maner a de dialogo, donde pregunta el maeftro, y refponde 1596.
el difcipulo." Sixteens.

"El Teftamento Nueuo de nueftro fenor. Jefu Chrifto, Luc. ii. 10. 1596.
Heaqui os doy nueuas de grangozo, que ferà á todo el pueblo. En cafa
de Ricardo del Campo." 742 pages, befides the preface. Parallel places
in the margin.* Octavo.

"Aphorifmes of Chriftian Religion: or a verie compendious abridge- 1596.
ment of M. I. Caluins Inftitutions, fet forth in fhort fentences methodically by M. I. Pifcator: And now Englifhed according to the Authors
third & laft edition, by H. Holland.—Heb. 13. 9." The Anchor. "Imprinted by him and Rob. Dexter,—1596." Dedicated "To the reuerend
father, the right worfhipfull Mr. Dr. Goodman, Deane of Weftminfter.
—18 Maij, 1596.---To the Reader.---The authors preface.---A Table
of

of the common places." 197 pages; on the back of the laſt is Dexter's rebus. Licenſed to him, and R. Dexter. W.H. Octavo.

1596. "The elements of arithmeticke, moſt methodically deliuered. Written in Latine by C. Vrſtitius, profeſſor of the mathematickes in the vniuerſitie of Baſill, and tranſlated by Thomas Hood, doctor in phyſicke and well-willer of them, which delight in the mathematicall ſciences." 216 pages. Licenſed. Octavo.

1596. "A ſummarie and true diſcourſe of Sir Francis Drakes Weſt Indian Voyage. Wherein are taken the Townes of Saint Iago, Sancto Domingo, Cartagena, & Saint Auguſtine: With Geographical Mappes exactly deſcribing each of the townes with their ſcituations, and the manner of the Armies approaching to the winning of them." The anchor. "Imprinted—for Wm. Ponſonby." Dedicated "To—Robert Earle of Eſſex, &c.—Tho. Cates." 52 pages. Licenſed. Alexander Dalrymple, Eſq; Quarto.

1596. "A nevv diſcourſe of a ſtale ſubiect, called the Metamorphoſis of Aiax. Written by Miſacmos, to his friend & coſin Philoſtilpnos." In MS. Seen and diſallowed." Alſo, on the back of the title-page, "An Epigram of the booke hanging in cheynes, to the Ladayes." Lifewiſe, a dedication "To the right worſhipfull Thomas Markham Eſquyer this bee dd.—iij of Auguſt, 1596. By the Author." The prefixes to the book are, "A letter written by a gentleman of good worth to the Author of this booke.---The anſwer.---The prologue to the reader of the Metamorphoſis of Aiax.---A ſhort aduertiſement." K, in eights. To which is annexed "An anatomie of the metamorphoſed A iax. Wherein by a tripartite method is plainly, openly & demonſtratiuely declared, explained & eliquidated, by pen, plot & precept, how vnſauerie places may be made ſweet, noyſome places made wholeſome, filthie places made cleanly. Publiſhed for the common benefite of builders, houſekeepers, & houſe-owners. By T. C. Traueller, Aprentice in Poetrie, Practiſer in Muſicke, profeſſor of Painting; the mother, daughter & handmayd of all Muſes, artes & ſciences." Continued by ſignatures to Q 1, in eights. W.H. To this may be added as a third part, Octavo.

1596. "Vlyſſes vpon Aiax. Written by Miſodiaboles to his friend Philaretes." Device a griffin. "Printed at London for Tho. Gubbins, 1596." F 3, in eights. Mr. Reed. Octavo.

1597. "Inſtitucion de la religion chriſtiana;—Por Juan Caluino. Y ahora nuevamente traduzida en Romance Caſtellano, Por Cypriano de Valera. En caſa de Ricardo del Campo, 1597." Pages 1032, beſides the prefixes and affixes. The ſignatures in eights, the ſize of a modern royal octavo. W.H.

1597. "A briefe diſcourſe of certaine points of the religion which is among the common ſort of chriſtians," &c. See p. 1123. Sixteens.

1597. "Hen. Locke gent. his tranſlation of Eccleſiaſtes into Engliſh poeſy, and paraphraſtically dilated: whereunto are added ſundry ſonnets of Chriſtian Paſſions." See Ath. Oxon. 1; 289. Warton, 111; 445. Quarto.

1597. "Theorique and practiſe of warre. Written to Don Philip Prince of Caſtil, by Don Bernardino de Mendoza. Tranſlated out of the Caſtalian tonge

RICHARD FIELD.

tonge into Englifhe, by Sir Edwarde Hoby Knight. Dedicated to Sr. George Carew Knight. 1597." After the dedication, which is in Spanifh, and dated, Queenborowe, 31 March, 1596. " The Autors epiftle. —Madrid, the laft of Auguft, 1594." Pages 165. At the end is the vignette of a female head between two cornucopias, ufed by Field, the only indication of the printer. On the laft page is the " Cenfure," or approbation of Don Francifco Arias de Bobadilla, Capt. of Light horfe, &c. W.H.
Quarto.

" The Theorike and Practike of moderne warres, Difcourfed in Dia- 1598. loguevvife. VVherein is declared the neglect of Martiall difcipline: the inconuenience thereof:—a redreffe by due regard had:—the imbattailing of men in formes now moft in vfe: with figures & Tables to the fame: &c. VVritten by Rob. Barret.—in fixe Bookes.—." The anchor. "—Printed for VVilliam Ponfonby, 1598." Dedicated " To—Henrie Earle of Pembroke," &c. whofe arms are on the back of the title-page.-- " To the right noble young lord, William Lord Harbert of Cardiffe, fonne & Heyre apparant to the—Earle of Pembroke.---To all men of warre in generall.-- To all gallant minded gentlemen," &c.---Three fix-lined ftanzas " in praife of the Author & his worke.—Wm. Sa. Gent." 247 pages, and a table explaining foreign terms. On the laft page are the author's arms. W.H. Folio.

" Synonymorum Sylva olim a Simone Pelegromio collecta, & alpha- 1598. beto Flandrico ab eodem authore illuftrata; nunc autem e Belgarum fermone in Anglicanum transfufa, & in alphabeticum ordinem redacta, per H. F. Accefferunt huic editioni fynonyma quædam poetica, in poefi verfantibus perquam neceffaria." Sixteens.

" A Difcourfe of the Felicitie of Man: or his Summum bonum. 1598. Written by Sir Richard Barkley knight." The anchor. "—Printed for VVilliam Ponfonby, 1598." Dedicated " To the moft renowmed & vertuous, learned & prudent Prince, the Queenes moft excellent Maiefty, —Neftors yeares, with the felicitie of both worlds.---To the Reader." 618 pages. W.H. Quarto.

" Nofce teipfum. This Oracle expounded in two Elegies. 1 Of Hu- 1599. mane Knowledge. 2 Of the Soule of Man, and the immortalitie thereof. —Printed by him, for John Standich, 1599." By Sir John Davies. Dedicated to the queen. Quarto.

" The hiftorie of—the warres of Italie," &c. See p. 1070. Pages 943, 1599. befides prefixes, and index. Licenfed. W.H. Folio.

An appendage to Geminie's Anatomy. " Imprinted—in the Blacke 1599. Friers-1599." Folio.

" A difcourfe of Life and Death. Written in French by Phil. Mornay. 1600. Done in Englifh by the Counteffe of Pembroke.—Printed for Wm. Ponfonby, 1600." G 2, in eights, but F has only 4 leaves. At the end, " The 13 of May, 1590. At Wilton." W.H. Sixteens.

" An Expofition vpon the prophet Ionah: Contained in certaine Ser- 1600. mons preached in S. Maries Church in Oxford, By George Abbot, Profeffor of Diuinitie, and Maifter of Vniuerfitie Colledge—Imprinted by him, and are to be fold by Rd. Garbrand, 1600." Dedicated " To—
Thomas

RICHARD FIELD.

Thomas Baron of Buckhurſt, Lord Treaſurer." 638 pages. Lamb. Libr. Quarto.

1600. " The treaſurie of Catechiſme, or chriſtian inſtruction. The firſt part, which is concerning the morall Law or ten Commandements of Almightie God : with certaine Queſtions & Aunſwers preparatory to the ſame.— Printed by him, for Tho. Man." Dedicated " To—my ſingular good patron Sir Nicholas Bacon Knight, And to the lady Anne Bacon his verie worthie & vertuous wife—Rob. Allen,---To the reuerend & learned examiners & readers," &c. V, in eights. W.H. The ſize, and ſhape of a modern octavo.

He had alſo licenſes for the following, viz. In 1588, " The declarations of ý Fr. king, & ý king of Nauarra vpon ý truce concluded betwene their maieſties : together with ý king of Navarras declaration at his paſſag ouer ý Ryver of Lorre. Davids Faith & Repentance. Vray diſcours ſur la deffaicte du Duc D'aumalle & Sieur de Ballagny auec leurs troupes par le Duc de Longue-ville & autres ſeigneurs, &c. Lettre D'un gentil-hôme de Beauſſe a vn ſien amy Bourgeois de Paris. Letter du Roy de Nauarre, a meſſieurs D'orleans, du 22 Maij, 1589, a Bangerey. Diſcours brief mais tres ſolide monſtrant clairement quil eſt loiſible, honeſte, vtile & neceſſarie au Roy de Nauarre, &c." In 1589, " Le vray Agnus Dei pour deſarmer le peuple Francois, Eſcrit pour le roy tres chreſtien Hen. iij roy de Fraunce & de Pologne ſur le point de ſon maſſacre. Dedié au roy tres chreſtien, Hen. iiij, roy de Fraunce & de Navarre. The furious: tranſlated by James the ſixte king of Scotland; with the Lepanto of the ſame king. A brief diſcourſe dialogue-wiſe, ſhewing how falſe & dangerous their reports are which affirme the Spanyards intended invaſion is not for reſtabliſhment of ý Romiſhe Religion, &c. The treatiſe of Chriſtian Righteouſnes." See p. 1070. In 1591, " Parte prima della breui diuini ſtationi & precette vtiliſſimi di diverſi propoſiti morale politici & iconomici, da Petruccio Ubaldino cittadin Fiorentine. The French alphabet, with ý treaſure of ý French tongue." In 1592, " A brief apologie of certen newe invenc'ons cōpiled by H. Plot. Venus & Adonis. The firſt part of chriſtiān paſſions, cōteyninge 100 ſonets of meditac'on, humiliac'on & prayer." See H. Locke, &c. in 1597. " The theater of fine devices, cōteyning 100 morall emblems tranſlated out of Fr. by Tho. Combe. A deſcription of ý properties of—myneralls." In 1594, " Tho. Campiani poema." He printed after 1600.

THOMAS LUST.

1585. " The treaſury of health, containing many profitable medicines, gathered out of Hippocrates, Galen, and Avicen, by one Petrus Hyſpanius, and tranſlated into Engliſh by Humphry Lloyd," &c. as p. 899. Octavo.

TOBY

TOBY COOKE, stationer,

SON of James Cooke late of London, yeoman, was put apprentice to Mr. John Harrison senior, for 12 years from Christmas 1564, and was made free 14 Jan. 1567-8. He was a very orderly member, being only once fined vjd. for keeping open shop on St. Luke's day. He was rather a bookseller than printer; yet Mr. Ames entered Calvin's two sermons translated by M. Rob. Horne, late Bp. of Winchester, under his name, as if he had printed them for Hen. Car; but as no printer's name or device, &c. appears to the book, and no other book has been found with Cooke's name to it, as the printer, it cannot claim a place here; but will be inserted in our General History. He dwelt at the Tiger's Head in St. Paul's church-yard. As this was the residence also of Chr. Barker, Mr. Ames supposed him and John Cook, Hugh Corne, and Hen. Car, to have been servants to Mr. Barker, and seems to intimate as if they all dwelt there. I have not met with any thing printed either by or for John Cook, or Hugh Corne, and so have nothing to say of them. Car's directions to the book above-mentioned run thus, " Printed for Henry Car, and are to be sold in Paules Churchyard, ouer against the signe of the blasing Starre." In January 1597-8 he was chosen beadle of the company, on the resignation of John Wolf. Octob. 8, 1599, he appears to have been too infirm to do his duty; the Court of Assistants therefore " ordered & agreed ỹ John Hardie, a freeman of this company, exercising the office of ỹ Bedleship for Toby Cooke, Bedell, shall exercise ỹ same during ỹ life of ỹ said Toby, & the companies lyking, reserving to ỹ said Toby ỹ fee of xl. by the yere, for his life time." Which appears to have been but short, for "3 Jan. 1599-600, John Hardy took his oath for the Bedelship," &c.

"Of the happines of this our age, and the ingratitude of men to God for his benefits, by Io. Riuius; translated by Wm. Watkinson. Printed for Tobie Cooke." Licensed. Quarto. 1578.

"Thirteene Sermons of Maister Iohn Caluine," as p. 1117. Licensed. 1579.

"Meditations on the 32 psalme, by Anth. Sadeell.—translated by Wm. Watkinson. Printed for *him* & T. Man." Licensed. Octavo. 1579.

Rodolph Gualter's sermon on Zephaniah, translated by Moses Wilton. Licensed. Octavo. 1580.

"A sermon on the parable of the Sower," &c. as p. 1145, & 1172. 1581.

"Foure Sermons vppon the—effects of faith," &c. as p. 1121, & 1224. 1581.

St. Augustin's Ladder to paradise; &c. as p. 1121. 1581.

"G. Gifford's Country Diuinity. Again 1582. See p. 1123. 1581.

"A Dialogue betweene a Papist & a Protestant. applied to the capacitie of the vnlearned." Printed for him. Licensed. Octavo. 1583.

"A Godlie, Zealous & profitable Sermon vpon the second Chapter of S. James. (14—26) Preached at London, by M. Geo. Gifford, published at the request" &c. D 3, in fours. For him. W.H.4 Sixteens. 1583.

" Against

TOBY COOKE.

1584. " Againſt the Prieſthood & ſacrifice of the Church of Rome wherein you may perceaue their impietie in vſurping that office & aćtion, which euer appertained to Chriſt onely. By Geo. Gifford." Octavo.
1586. " The Blazon of Gentrie:" as p. 1226. Licenſed.
1586. Z. Urſinus's Catechiſm abridged by Iohn Moorecroft. Octavo.
1586. " Geo. Gifford his catechiſme, giuing a moſt excellent light to thoſe that ſeeke to enter the pathway to Saluation." Octavo.
1586. A ſermon on James 2; 14—26. By Geo. Gifford. Octavo.
1587. Chardon's ſecond ſermon at St. Mary's, Oxford. See it, p. 1226.
1587. " A Diſcourſe of the ſubtill Practiſes of Deuilles by VVitches and Sorcerers. By which men are & haue bin greatly deluded : the anti-quitie of them: their diuers ſorts & Names. With an Aunſwer vnto diuers friuolous Reaſons which ſome doe make to prooue that the De-uils did not make thoſe Alterations in any bodily ſhape. By G. Gyf-ford. Imprinted—for *him*, 1587." Dedicated " To—Maiſter Ric. Mar-tin, Alderman, and Warden of her Maieſties Mint." J, in fours. W.H.
Quarto.

1589. " A Skeltonicall Salutation,   That in a bravado
    Or condigne gratulation,   Spent many a Cruſado,
    And iuſt vexation   In ſetting forth an Armado
    Of the Spaniſh Nation,   England to invado.
Imprinted—1589." B, in fours. W.H. Quarto.
1589. Eight Sermons on the firſt 4 Chap. & part of the 5th of Eccleſiaſtes.
Octavo.
1589. " Inſtructions for the warres." &c. as p. 1243.
1589. " The Practiſe of Fortification:" &c. as ibid.—Alſo, in 1597, F. Kingſton for him. 64 pages.*
1590. " The vſe of the Celeſtial Globe in plano," &c. as p. 1228. Licenſed.
1590. " The vſe of the Iacobs ſtaffe." Dexter's rebus. " Jmprinted—for *lim* and Rob. Dexter, 1590." Dedicated " To—L. Iohn Lumley, Baron Lumley, &c.—Th. Hood." To this is annexed " The vſe of the Croſſe Staffe." 20 leaves, and a copper-plate cut of projections, T. Hood. Licenſed. W.H. Quarto.
1590. " A Plaine Declaration &c. by Geo. Gvffard," &c. as p. 1245.
1590. " A ſhort treatiſe againſt the Donatiſts of England," &c. as p. 1229.
1591. " A ſhort reply vnto the laſt printed bookes of Hen. Barrow & John Greenwood the chief ringleaders of the Donatiſts in England, by Geo. Gif-ford." Licenſed. Quarto.
1591. A ſermon at Paules croſſe 30 Maie, 1591, on Pſalm 133. by G. Gif-ford. Licenſed. Octavo.
1591. " Wm. Burton's catechiſme—concerning the knowledge of God, and the right vſe of the law." Octavo.
1592. " Wm Burton his 7 ſermons on Pſalm 41; 11,—13." Octavo.
" Vpon the letters of Mr. Wilbraham, yt is ordered ẏ Toby Cook (and none other) ſhall have ẏ printinge of ẏ truthe of ẏ murther of Rob. Hayton, as yt ſhalbe found & deliv'ed to ẏ ſaid Toby by ẏ ſaid Mr. Wilbraham; and ẏ yf any ſhall preſume to meddle therewith he ſhalbe ſtaied." Stat. Reg. 5 Feb. 1592-3.

" A caveat

"A caveat for fuerties.—by Wm. Burton." &c. as p. 1255. 1593.
"A Dialogue concerning Witches and Witchcraftes." &c. as p. 1230. 1596.
"Prælections vpon the—Reuelation of S. John," &c. See p. 996. 1596.
"Four fermons vpon feuerall partes of Scripture, preached by George 1598.
Gyffard—at Mauldin in Effex. Printed by Tho. Judfon for *him* & Rob.
Walker." Sixteens.

He had alfo licenfes for the following, viz. In 1580, " Knoxes anfwere to the cavillations of an Anabaptift." In 1582, " Fourdes Catechifme." In 1589, " Memoires de ce qui eft advenu en L'armee du roy, depuis la prife de fauxbourge de Paris, iufques a celle de la ville de Fallaize : to be tranflated into Eng." In 1591, " The principles of Geometry, Aftronomie, & Geographie."

※※※※※※※※※※※※※※※※※※※※※※※※※※※※※※※

GABRIEL SIMSON and WILLIAM WHITE, ftationers,

WERE both made free 10 Apr. 1583, by Mrs. Jugge, began trade together, and continued in actual partnerfhip from 1590 to 1597, at the White Horfe in Fleet lane, over againft Seacoal-lane. They were fined jointly, 18 Aug. 1595 " for printing part of a book of Mr. Broughtons without authority, x f. and ordered to bring the leaf printed into ỹ Hall, and not to proceed—till they have authoritie for it: Alfo their imprifonment referred over till further order be taken. They agreed to pay ỹ fine next Court daie. Paid vj f. viij d. ỹ reft remitted." They frequently ufed Jugge's compartments.

" A letter to a friende, touching Mardochai his age, which helpeth 1590. much to holde the trueth, for that chiefe prophecie of our faluation, in Gabriels feuenties, which fhew that moft exactly 490 yeares after the Angels fpeech Chrift the moft holy fhould be killed to giue life. Dan. 9; 23.—Imprinted for G. S. and W. W. 1590." By Hugh Broughton. " The Printer to the Reader." The letter is infcribed " To his good friende A. T." (*Alex. Top*). B 2, in fours. W.H. Quarto.

" A Direction to the waters of lyfe. Come and beholde, How Chrift 1590. fhineth before the Law, in the Law, and in the Prophetes : and withall the iudgementes of God vpon all Nations for the neglect of his holy worde, wherein they might haue feene the fame :—by R. C. Imprinted at London, for Gabr. Simfon & Wm. White, and—folde by Wm. Barley." Dedicated " To the godlie & learned Maifter Hugh Broughton, teacher of Diuinitie.—Roger Cotton.[a] ---To the Reader." 51 leaves. At the end are the Drapers' arms. Licenfed. Again 1592. W.H. Quarto.

" A treatife

---

[a] " Mr. Roger Cotton, a Draper in Canning ftreet, was a true Scholar of fuch a Mafter, and fo conftantly plied the Scriptures, according to the admonitions he had received

GABRIEL SIMSON and WILLIAM WHITE.

1591. " A treatife of Melchifedek prouing him to be Sem, The father of all the fonnes of Heber, the fyrft king, and all kinges glory: by the generall confent of his owne fonnes, by the continual liudgement of ages, and by plentifull arguments of fcripture. Heb. 7; 4. Now confider how great HE is. Imprinted at London for *them*, 1591." In a compartment ufed by Jugge. See p. 725. Dedicated " To—Syr William Cecill,—Lord Treafurer &c.—Hugh Broughton." J, in fours. W.H. 4°.

1591. " TEXTES of Scripture, Chayning the holy Chronicle vntyll the Sunne loft his lyght, and the Sonne brake the Serpentes head: dying, rifing, and afcending. Search the Scriptures:—John 5; 39. Imprinted—for *them*, and are to be folde at their houfe in Fleete lane, 1591." It is introduced with an epiftle " To the Chriftian Reader.—H. B." (*Hu. Broughton*) C 2, in fours. Licenfed. W.H. Quarto.

1594. " A SEDER OLAM, that is: Order of the worlde: or yeeres from the fall to the reftoring. A feconde Apologie for the Angel Gabriels proprietie of trueth in his holy and healthy meffage, or the cleerenes and certainty for our redemption: &c. With a long Preface touching the humanity of the Gentry of Cambridge, and higher, in fauour of ancient Learning. Job 24; 25. Yf it be not fo now, &c. 1594." To—Henrie Earle of Huntingdon, &c.—H. Broughton.---To the Reader." 32 pages. This book is without printer's name, or place; but is placed here as having the title page enclofed in a border of metal flowers, like the head-piece to the preface of the laft mentioned article, and as they printed moft of this authors pieces; and is perhaps the book for which they were afterwards fined. W.H. Quarto.

1595. " Two learned and godly fermons, preached by that reuerende and zelous man M. Richard Greenham: on—Pro. 22; 1.—1 Theffa. 5; 19. London, Printed by *them*, for Wm. Jones, dwelling neare Holborne Condite, at the figne of the Gunne: where they are to be folde, 1595." F 5, in eights. W.H. Sixteens.

1596. " An Armor of Proofe, brought from the Tower of Dauid, to fight againft the Spannyardes, and all enemies of the trueth, By R. C—Pro. 18; 10. Imprinted—by *them*, 1596." Dedicated " To—Gilbert Talbot, Earle of Shrewesburie, &c.—R. Cotton.---To the chriftian Reader." D 3, in fours. In fix lined ftanzas. W.H. Quarto.

1596. " A Spirituall Song: conteining an Hiftoricall Difcourfe from the infancie of the world, vntill this prefent time:—Drawen out of the holy Scriptures, By Roger Cotton.—At London, Printed by *them*. 1596." Dedicated " To—Sir Francis Drake Knight:---To the Reader." Commendatory verfes by P. K. G. W. A. W. and R. I. This Song is in five lined ftanzas, and divided into 6 parts: Thereunto is annexed, " A defcription of olde Rome, or miftical Babylon:" &c. 18 pages. W.H. Quarto.
" A Con-

received from him, that he read over the | cimen of in that little Book he publifhed, Bible twelve times in one year; and what | called Directions to the Waters of Life." proficiency he made therein, he gave a fpe- | Preface to Mr. Broughton's works.

GABRIEL SIMSON and WILLIAM WHITE. 1265

" A Concent of Scripture by H. Broughton." In the compartment used by R. Jugge to the Bible, 1577. Over the title, in a tablet formed by metal flowers, " Come and fee." Dedicated " To the moſt high and mightie prince Elizabeth,—queene of Englande, &c.—-The Preface, ſhewyng the ſumme of the booke,—and the olde reading of the Law and prophetes." G 3, in fours, with a map of the world, both hemiſpheres in one globular projection, engraved on copper plate. W.H. I have another copy with an engraved title-page. Quarto.

" Sundry workes, defending the certayntie of the holy Chronicle: dedicated togeather vnto her Maieſtie : with requeſt, that authoritie might ſtablyſh the trueth vnto publique agreement. Job 12; 11. and 34; 3, 4." This title page and dedication, without printer's name, or date, but with the ſame compartment as to Melchiſedek, are ſet before the following pieces, which had been before printed at different times, viz. Textes of Scripture; Seder Olam; An apologie—that our Lord died in the time foretold to Daniel; A treatiſe of Melchiſedek· This dedication is reprinted in his works Tom. 1. p. (153) but without the leaſt intimation of this purport of it. W.H. Quarto.

They had licenſed to them, In 1588, " A ſignee of Sight."

GABRIEL SIMSON, ſtationer,

Continued printing alone at the ſame houſe in Fleet lane. He was fined 3 Mar. 1599-600, " for printing the Table of good Counſell, contrary to order, and never to meddle with the printing it again vpon the peril that belongeth thereto, iij ſ." He uſed the device of Daniel on his knees praying, the angel Gabriel appearing to him in the air; ſometimes a pink with the letters G. S.

" Daniel his Chaldie viſions," &c. See p. 1257. Device Daniel, &c. 1597. With, O Lord let thine anger be turned away, &c. Dan. 9; 23. " Printed by *him*—in Fleet-lane, and are there to be ſold; as alſo the reſt of the ſame Authors workes are. 1597." P, in fours. W.H. Quarto.

" A Spirituall Grammer, or The eight partes of ſpeech Moralized, 1597. which by analogicall alluſion, may put Schollers in minde of many good leſſons : Aptly accommodated to their Studies. Viuere diſce Deo. Imprinted—ouer againſt Seaco-lane, 1597." Dedicated " To—ladie Marie Ramſey, ſometime wife to—Sir Tho. Ramſey Knight, deceaſed late Lord Maior of—London.—G. A." D 6, in eights. W.H. Sixteens.

" A diſcoverie of the Knights of the poſte, by E. S. Printed by G. S." 1597. Quarto.

" An introduction into the bookes of the prophetes and apoſtles. 1598. Written by Peter Palladius, D. D. and biſhop of Rochil. Faithfully tranſlated out of Latin—by Edw. Vaughan." Octavo.

" The Arte of vulgar arithmeticke, both in Integers and Fractions, &c. 1600. Newly collected, digeſted," &c. Device, a pink. " Imprinted—by *him* in Fleete-lane, 1600." Dedicated " To—ſir Thomas Sackville knight. Baron of Buckhurſt, Lord Treaſurer &c.—Tho. Hylles." A preface, and commendatory verſes. 266 leaves, and a table. W.H. Quarto.

He

GABRIEL SIMSON.

He had a licenſe ſo early as 4 May, A. D. 1584 to print the following ballad, "A dyſputac'on of twoo faythfull Louers, In prayſe of Taylors and cōmendac'on of Glovers."

WILLIAM WHITE, ſtationer.

After his ſeparation from Gabr. Simſon, dwelt in Cow lane. Mr. Ames ſeems to have been miſtaken in aſcribing to him " A breeſe and true reporte of the execution of certaiue Traytors at Tiborne the 28, and 30 dayes of May, 1582." &c. as he was not made free till 10 Apr.

1583. 1583. I have the ſame " Imprinted at London for Wm. VVright, and are to ſolde at his ſhop adionying vnto S. Mildreds Church in the Poultrie, the middle ſhop in the rowe, 1582." Quarto. He was fined in June 1560 for printing with Edw. Allde a ballad of the Wife of Bath. &c. as p. 1205, note k.

1598. " A hedgerow of buſhes, brambles and briars; or a field full of tares thiſſels and time; of the vanities and vain delights of the world, &c. Now newly compiled by I. D." Printed for John Brown. Quarto.

1598. " A Health to the Gentlemanly profeſſion of Seruing men: or the Seruing mans Comfort: With other thinges not impertinent to the premiſſes, &c. Imprinted at London by W.W. 1598." W.H4 Quarto.

1598. Loves labour loſt, by Wm. Shakeſpeare. W.W. for Cuthbert Burbey. Quarto.

1599. " A new Booke of good Huſbandry, very pleaſaunt, and of great profite both for Gentlemen and Yeomen: Conteining The order and maner of making of Fiſh-pondes, with the breeding, preſeruing and multiplying of the Carpe, Tench, Pike and Troute, and diuers kindes of other Freſh fiſh. Written in Latine by Janus Dubrauius, and tranſlated into Engliſh at the ſpeciall requeſt of George Churchey, fellow of Lions Inne, the 9 of Februarie, 1559." The author's coat of arms. " Imprinted—by him, dwelling in Cow-lane. 1599." Dedicated " To—Sir Edmonde Anderſon Knight, Lord chiefe Juſtice of the common Plees.— Geo. Churchey.—To the Reader." 36 leaves, and a table. Sir John Hawkins. Quarto.

1600. " The letting of Humours blood in the Head-vaine; with a new Moriſſco, daunced by ſeauen ſatyrs upon the bottom of Diogenes tubbe." Octavo.

" Octob. 29, 1600. Yt is ordered that the next Court-day/ two bookes ately printed; thone called the letting of humors blood in the head vayne, thother A mery metiuge, or tis mery when knaues mete, ſhalbe publiquely burnt; the whole impreſſion of them, for that they conteyne matters vnſytt to be publiſhed/ Then to be burnt in the Hall kytchen, with other popiſhe bookes and thinges that were lately taken. And alſo Mr. Darrell's book lately printed, concerning the caſting out of the Devil."

This ſeems to be the reprinted edition after the book had been forbidden and burnt. Mar. 4, 1600-1. Several of the trade were fined 2-6 a piece a piece for buying them; but i do not find that either he, or any other, was fined for printing it.

King Henry vi. By Wm. Shakefpeare. For Tho. Millington. Quarto. 1600.
He continued printing after 1600.

ROBERT DEXTER, ftationer,

SON of Rob. Dexter of Ipfwich, Suffolk, failor, was bound apprentice to Mr. Fr. Coldock, for 9 years from Michaelmas 1580; and made free 25 June, 1589. In 1591, he had a controverfy with Rd. Jones, the iffue of which may be feen in p. 1047, note z. July 1, 1598, he was taken on the livery. He dwelt, or kept fhop, at the brazen ferpent in St. Paul's church-yard; had for his device, or rebus, a right-hand pointing with the fore-finger to a ftar; about it DEVS IMPERAT ASTRIS. He employed others to print for him; and was a benefactor to his Company.

" The vfe of the Iacobs ftaffe," &c. By Thomas Hood. p. 1262. Licenfed. 1590.
" Chronographia." &c. p. 1228. Again 1590, as p. 1253. Licenfed. 1590.
" Guilielmi Saluftii Bartaffii Hebdomas. A Gabriele Lermæo latinitate donata. Ad ferenifsimam, atq; illuftrifs. Elizabetam, Angliæ—Reginam. Opus argumento facrum, ftylo perelegans, doctis gratum, ftudiofæ iuuentuti perutile." The queen's arms. " Londini. Apud Ro. Dexter, in cœmeterio D. Pauli, fub infigni Serpentis ænei. 1591." K 5, in twelves. On the laft leaf Dexter's rebus. Licenfed. W.H. 24°: 1591.
" The Chriftians combat, wherein is fet downe that daungerous fight wherevnto all the elect children of God are called: with a moft fure hope of victorie ouer all their enemies in Chrift their Captaine, &c. tranflated out of French by Geo. Capelin, gent." For him. Licenfed. Octavo. 1591.
" Giles Whiting his fhort queftions & aunfweres to be learned of the ignorant before they bee admitted to the Lords Supper." For him. Octavo. 1591.
" The Triall of truth, or a treatife wherein is declared who fhall be Iudge betweene the reformed Church & the Romifh: written by a Hungarian, tranflated out of Latin by Rd. Smith." For him. Licenfed. Quarto. 1591.
" Sixe fermons—by Maifter Henry Smith," &c. as p. 1254. Licenfed. 1592.
" Apocalypfis.—by Mr. Fran. Iunius," &c. as p. 1254. Licenfed. 1592.
" Ten fermons on 2 Sam. 24; 11—25, concerning God's late vifitation, by Wil. Cupper." For him. Licenfed. Octavo. 1592.
" A Commentarie vpon the—Proverbs" &c. as p. 1254. See p. 1236. 1592.
" The—Science of Surueying Lands," &c. as p. 1230. Licenfed. 1592.
Peter Ramus's arithmetic. For him. Licenfed. Octavo. 1592.
Ste. Egerton's brief method of Catechifing. For him. Octavo. 1594.
" M. Virell's Grounds of Chriftian Religion." See p. 1256. Again 1597. 1594.
" Terentianus

1595.. "Terentianus Christianus." &c. as p. 1235.
1596.. "Prioris Corinthiacæ Epist. expositio quædam." By Tho. Morton. Octavo.
1596. "The Alphabet of the holy Proverbs of king Solomon, specially from the beginning of the 10 Chap. to the end, by R. A." For him.
1596.. "Salomon, or—the state of the kingdome of Israel," &c. as p. 1235.
1596. "Aphorismes of Christian Religion," &c. p. 1257.
1596. "A Treatise of the threefolde state of man—, 1 His Created holinesse in his innocencie. 2 His Sinfulnesse since the fall of Adam. 3 His Renewed holinesse in his regeneration Eph. 4; 22—24.—" His rebus. "Printed—for *him*, & Raph Iackeson." Dedicated "To—the Ladie Elizabeth Cary, wife to—Sir Rob. Cary Knight.—Cambridge, 30 Mar. 1596.—Tho. Morton.----To the reader, &c. The argument." A table. 426 pages. Licensed to him and Raiph Jackson. W.H. Octavo.
(1597.) "A sermon—at Paules Crosse, 6 Feb. 1596. By Iohn Doue D. D. Dedicated "To—Sir Thomas Egerton—Lord Keeper," &c. 79 pages. W.H. Sixteens.
1597.. "Certaine worthie manuscript Poems" &c. as p. 1236..
1597. "Two treatises concerning Regeneration. 1 Of Repentance. 2 Of the Diet of the Soule. &c. Printed by Tho. Creede for *him*, & Raph Iackson." The first tract on 104 pages, the last on 119. Each of them has an argument prefixed. W.H. Octavo..
1597. "VIRGIDEMIARVM, Sixe Bookes. First three Bookes, of Tooth-lesse Satyrs. 1 Poeticall. 2 Academicall. 3 Morall.—Printed by T. Creede for *him*, 1597." By Jos. Hall. F 2, in eights. W.H. Again. 1598. 16°. Virgidemiarum. The three last Bookes of byting Satyrs." Rd.
1598.. " Virgidemiarum. The three last Bookes of byting Satyrs." Rd. Bradock for him. 106 pages.* Both parts reprinted 1599. Sixteens.
1598.. "Baptistæ Mantuani Carmelitæ theologi Adolescentia, seu Bucolica, breuibus Jodoci Badii commentariis illustrata. His accesserunt Joan. Murmelii in singulas eclogas argumenta, cum annotatiunculis eiusdem in loca aliquot obscuriora," &c. Octavo..
1599. "Dialectica Joannis Setoni Cantabrigiensis," &c. See p. 1205.
1599. "The Workes of the Reuerend & faithfull seruant of Iesus Christ, M. Richard Greenham, Minister &c. The second edition, reuised, corrected & published by H. H. (*Hen. Holland.*) Printed by Felix Kingston for *him*—1599." Quarto.
1599.. A treatise of the nature of God. Anonymous. T. Creed for him. Octavo. He had also licenses for the following, viz. In 1590, "An Allphabet & playne pathwaie to—Reading,—ỹ spelling A.B.C." See p. 1047." "The end of the world, &c. The old faithe done by M. Coverdale. The guide vnto Godlines; set forthe by Ryvius. A familiar xpian instruc'on; translated from the French. That part of P. Martirs Cōmon places, which belonged to Andr. Maunsell. The English Creede cōpiled by Tho. Rogers; both parts. Prayers & xpian cōsolac'ons, translated out of Fr. by G. Capelin. A sermon of sing'ler cōfort for soe manye as see their sundrye syns, & are inwardlie afflicted with a conscience & feeling thereof. The seconde sounde, by Martyn. A blazon of Gentrye,

trye, by Ferne. Cato conftrued." In 1591, " Propria que maribus, & As in prefenti, cõftrued: Provided it do not hurt Mr. Fr. Flowers priviledge." In 1593, " A brief methode of cathechifinge, with certen brief exercifes of religion thereunto annexed."
He printed after 1660.

WILLIAM KEARNEY.

" A NEW Booke containing all fortes of Hands vfually written at this daye in Chriftendomε,—with Examples of each of them in their proper tongue and Lettter. Alfo an Example of the true and iuft proportion of the Roman Capitals: Collected by the beft approued writers in thefe Languages, now firft publifhed." See p. 1066. Broad octavo. 1590.

" The magiftrates fcripture, which treateth of their election, excellency, qualities, dutie and end; with two godly prayers annexed thereunto. By Henry Smith, from Pfalm lxxxii. 6, 7." Sixteens. 1591.

" The preachers proclamation. Difcourfing the vanity of all earthly things, and prouoing, that there is no contentation to a chriftian minde, but only in the fear of God, Eccl. i. 2." By Henry Smith. Sixteens. 1591.

" The Heroicall Deuifes of M. Claudius Paradin Canon of Beauieu. Whereunto are added the Lord Gabriel Symeons and others. Tranflated out of Latin into Englifh by P. S. London, Imprinted by Will. Kearney dwelling in Adlingftreete. 1591." Dedicated " To—the renowmed Capteine Chr. Carleill Efq; chief Commander of her Maiefties forces in the prouince of Vlfter—Ireland, and Senefhall there of the counties of Clandeboy, the Rowte, the Glins, the Duffre, and Kylultaugh."—London, 3 Jan. 1591.—Wm. Kearney.---To the moft worthie knight Theodot of Marze, Lord of Belleroche, Laffenaz, &c. Claudius Paradin fendeth greeting." 374 pages; the devices are neatly cut on wood. W.H. 24°. 1591.

" An Apologie in briefe affertions defending that our Lord died in the time properly foretold to Daniel. For fatisfaction of fome ftudentes in both Vniuerfities. H. Broughton.—Imprinted by *him* dwelling within Creeple-gate, 1592." Dedicated " To—Sir Peregrine Bertye Knight, Lord of Willoughby and Erefby." L 3, in fours, with 2 leaves of a fragment of Phlegon in Greek, with the Englifh tranflation, interpolated, which was printed by " a Printer dwelling farre off." W.H. 4°. 1592.

He had alfo licenfe to print in 1590, " Of ỹ treuneffe of Chriftes humane nature: Of ỹ fpirituall eating of xpiftes body: Of ỹ facramentall eating thereof."

WILLIAM SAUNDERSON.

Printed a book intitled, " The globes coeleftiall, and tereftriall, fet forth in plaine, by Emery Molineux." Octavo. 1592.

ROBERT BOURNE, and JOHN PORTER.

"AN Expofition on the Lordes praier, by way of catechifing, by William Perkins." Maunfell p. 80. Licenfed. Octavo.

JOHN DANTER, ftationer,

SON of John Danter of Enfham, Oxon, weaver, was bound apprentice to John Day for eight years from Michaelmas, 1582; and put over 15 Apr. 1588 to Rob. Robinfon, to ferve him unto Michaelmas 1589; and whereas there was one year more then to come of his apprenticefhip, that year was freely remitted by confent of Mrs. Day, alias Stone. He was accordingly made free 30 Sept. 1589. Before he had ferved half his apprenticefhip he was involved in a fevere decree with Rob. Bourne, &c. See note b, below. But which was alleviated by an order of the Court of Affiftants, 3 Aug. 1591. See p. 1113, note t. And by a fubfequent order, 7 Aug. 1592, he was appointed to print "The inftruction of chriftian woman, and Ovid's metamorphofis," for the company. allowing 6d in the pound for the ufe of their poor. He dwelt in Hofier lane, near Holbourn conduit; and feems vpon the whole to have been but a diforderly member, according to another order of Court, 10 Apr. 1597,ᵈ which
overfet

ᵇ I find one Rob. Burne made free of the Stationers' company, 24 May, 1581, by Mr. Marfhe; very probably the fame perfon with this Rob. Bourne. However, doubtlefs he is the perfon indicated in the following minute in that Company's regifter, Nov. 3, 1586. "Whereas the wardens on Monday the laft daye of October 1586, did by vertue of the late decree of Starrechamber, exemplyfied under her maiefties greate feale of England, fieze one preffe belonging to Rob. Bourne, Hen. Jefferfon and Laurence Tuck, or to fome of them, for that they contrary to the faid decree had printed therewith the grammer belonginge to thaffignes of Mr. Fr. Flower by vertue of her Ma. l'res patents to him in that behalf graunted; which beinge broughte to the Stac'oners Hall according to the faid decree, Yt is now ordered and decreed, by force of the fame, that the faid preffe, letters and printinge ftuffe fhall be made vnferviceable, defaced and v(ed in all points according to the faid decree, And alfo it is ordered and decreed that the faid R. Bourne, H. Jefferfon and John Danter, Gilbert Lee and Tho. Dunne, and all others that wrought upon thimpreffion of the faid booke fhall from henceforth be dyfabled to prynte otherwife then as Journemen in pryntinge, and fhall never hereafter keepe any printinge houfe to their or any of their owne behoof, but be vtterly barred therefrom, according to the faid decrees."

ᶜ He does not appear to have been a printer, but had feveral books printed by different printers for him, either alone or with other connections, which may be found by the index.

ᵈ " Whereas there was, in Lent laft, found in the houfe of John Danter twoo printinge preffes and certen letters – in fourmes and cafes, which were employed in printinge a book called Jefus pfalter, and other thinges without authority; which preffes &c. were by vertue of the decrees of the Starre chamber feifed and brought to the Stac'oners Hall, with certen leaves of the faid booke, Yt is nowe, accordinge to the faid decrees, ordered in full court that the faid preffes and l'res fhalbe defaced and made vnferviceable for pryntinge, as the faid decrees in fuch cafes appointe."

JOHN DANTER.

overset him quite, for we do not find any thing printed by him after that time. Yet in the Return from Parnassus, a play acted by the students of St. John's Coll. Cambridge in 1606, Danter is introduced as living in St. Paul's church-yard, and chaffering for a pamphlet entitled a Chronicle of Cambridge cuckolds.

Hen. Smith's sermon, on 1 Thes. 5; 19—22. See p. 1113.   1591:
" The Affinitie of the faithfull :" &c. as ib.   1591.

" The lamentable ruines of the towne of Shuffnall, alias Idsall, in Shropshire, by fire; with the most rare, and wonderfull burning of the parish church, standing on the other side of a water; and the miraculous preservation of certaine houses, which stoode close by the saide church. Set forth by Edward Mullard, parson of Idsall, alias vicar." Quarto.

" The ground worke of conny catching, their manner of the pedlers   1592. French, and the means to vnderstand the same, with the cunning flights of the counterfeit cranke. Therein are hancled the practises of the visitor, the fetches of the shifter and rufflar, the deceipts of their doxes, the deuises of priggers, the names of the base lovtering lofels, and the means of euery Backe-art mans shifts; with the reproofe of all their diuellish practises. Done by a iustice of peace of great authoritie, who hath had the examining of diuers of them." Written in 1567. Printed for William Barley. Quarto.

" The blacke bookes messenger. Laying open the life and death of   1592. Ned Browne, one of the most notable cutpurses, crosbiters, and connycatchers, that euer liued in England. Herein he telleth verie pleassently in his own person strange pranks, and monstrous villanies, by him and his consorte performed, as the like was neuer heard of in any of the former bookes of connycatching. Read and be warned, laugh as you like, iudge as you find : Nascimur pro patria. By R. G." Printed for W. Barley. Quarto.

" Jurisprudentiæ Medicinæ & Theologiæ Dialogus dulcis. Authore   1592: H. Smith Theologe." Device, Fortune. " Londinum, Excudebat I. Danter, Impensis Tho. Man, 1592." In hexameters and pentameters. 15 leaves. To this is annexed,

" Vita Supplicium : siue de misera Hominis conditione querela.   1592. Heu vitæ legem, cui annexum est angi & dolore in omni sorte. Authore H. S." Device, Fortune, Londini, Impensis Tho. Man." 8 leaves; in Sapphics. The rev. Dr. Lort. Octavo.

" The English Phlebotomy :" &c. as p. 1114.   1592.
"—The instruction of a christian woman." See p. 937. Octavo. 1593.
Ovid's Metamorphosis translated by Ar. Golding. Quarto. 1593.
" Strange news of the intercepting certaine letters, and a convoy of   1593. verses as they were going priuilie to victuall the Low Countrie. Vnda impellitur vnda. By Thomas Nashe, gent."

" A Free schoole for God's children, &c. by I. R."—Licensed. 8°.   1593.
" The affectionate shepherd, Containing the complaint of Daphnis   1594. for the loue of Gangueda." In six-lined stanzas. Quarto.
" PIERS PLAINNES seauen yeares Prentiship, by H. C. (Hen. Chettle)   1595.

7 K 2   Nuda

JOHN DANTER.

Nuda veritas. Printed—by *him* for Tho. Goſſon, and are to be ſold at his ſhop by London-bridge gate. 1595." Quarto.

1595. "The Old wiues Tale, a pleaſant conceited Comedie, plaied by the Queenes Maieſties players. Written by G. P. (*Geo. Peele*). Printed—by *him*, and are to be ſold by Ralph Hancocke & John Hardie, 1595." 4°.

1595. "Eglvryn Phraethineb. Sebh, doſparth ar retoreg, uno'r ſaith gel- bydhyd, yn dy dyſculhuniaith ymadrodh, a'i pherthynaſſau. Printiedig gann Joan Danter yn Lhundain." Ends, DYBEN. 104 pages. Quarto.

1595. "Fulfordo & Fulfordae. A ſermon preached at Exeter, in the cathe- drall church, the ſixth day of Auguſt, commonly called, Jeſus day 1594, in memoriall of the cities deliuerance in the daies of king Edward the ſixt. Wherein is intreated of the goodnes of God toward man, and of the ingratitude of man toward God. By John Charldon, doctor of divi- nity. In which alſo ſome ſewe thinges are added, then omitted through want of time." Another edition printed the year before.* Twelves.

1595. "The true Copie of a Lamentable Petition exhibited in the names of the afflicted Chriſtians in the Eaſt parts to the Chriſtian Kingdomes in the Weſt." Head title, on ſignature C. 3 leaves. Dated, "From Prage the 13 of Januarie. Your poore afflicted Chriſtian Brethren of Boheme, Hungarie, Auſtria, Polonia, and Heluetia. Printed—by *him*, for Tho. Goſſon, 1595." W.H.+ Quarto.

1595. "MARIE Magdalens Loue." A practical diſcourſe on John 20; 1— 18. Device, Opportunity. E.7, in eights. "Printed by *him*, and are to bee ſold by Wm. Barley—1595." To which is annexed

1595. "A Solemne Paſsion of the Soules Loue. Printed—by *him* and are to be ſold by Wm. Barley,—1595." In ſix-lined ſtanzas. The running title and catch words between borders of metal flowers throughout. The ſig- natures continued to G 8. At the end, "Nicholas Britten." W.H. 16°.

1596. "A merrie, pleaſant, and delectable hiſtorie betweene K. Edward the Fourth, and a Tanner of Tamworth." Bodl. Libr. Quarto.

1596. "Haue with you to Saffron Walden; or, Gabriell Harueys hunt is vp. Containing a full anſwer to the eldeſt ſonne of the halter maker; or, Naſhe his confutation of the ſinfull doctor." Quarto.

1597. Romeo and Juliet by Wm. Shakeſpear. Quarto.

He had alſo licenſes for the following, viz. In 1592, "The repentance of a Conycatcher: with y̌ life & death of—Mourton & Ned Browne, twoo notable conycatchers; the one latelie executed at Tyborne, y̌ other at Aix in Fraunce. The Repentaūce of Rob. Greene, M. A. The Apo- logie of Pierce Pennyleſſe, or ſtrange newes of y̌ interceptinge certen letters, & a convoy of verſes, as they were goinge to victuall y̌ Lowe cūtries. The pleaſant hiſtory of Edward lord of Lancaſter, knight of y̌ Holy Croſſe, with his adventures, &c. Gods Arrow againſt Atheiſme & irreligion. A manuarye, or hande dyall,—to knowe y̌ houre of y̌ clock when y̌ ſūne ſhineth by y̌ hand, without other inſtrument. The tyrror of y̌ nyght, or a diſcourſe of apparifions." In 1593, "The teares of Fanſie, or Love diſdained. A playne declarac'on or deſcrp'on of ſīne, death, y̌ devil & hell. An interlude of y̌ life & death of Jack Strawe. Juſtruction whereby

JOHN DANTER.

whereby a man may learne—to playe on y̓ Cylterne, without y̓ helpe of any teacher. Greene his funerall. A noble Roman Hiſtorye of Titus Andronicus. A newe booke of newe cōceites. The nomber of novelties. The woūndes of Civill warre timely ſett forth in y̓ true tragedies of Marius & Scilla. An enterlude; Godfrey of Bulloigne, with y̓ conqueſt of Jeruſalem. An other—of y̓ lyfe & death of Heliogabilus." In 1594. The cruell handlinge of one Nich. Burton M'chanttailor of London, by y̓ blody Spaniards in y̓ cittye of Cyvill, who was there burned for y̓ teſtimony of Jeſus Chriſt. Strange ſightes ſeene in y̓ ayre, &c. about y̓ cittie Roſenbergh, 19 Jan. laſt. The terror of the night, or an apparition of Dreames. Cornucopia or divers ſecrets in man, beaſts, &c. The wonderfull ſincking of certen grounde in—Worley, Somerſetſh. &c. The hiſtorie of Gargantua." Likewiſe for ſeveral ballads.

WILLIAM PONSONBY, ſtationer,

APPRENTICE to Mr. Wm. Norton for 10 years from Chriſtmas 1560, was made free 11 Jan. 1570-1, choſen on the livery 6 May 1588, and ſerved Renter in 1595 and 1596. He married Joan the eldeſt daughter of Francis Coldock, and dwelt at the Biſhop's head in St. Paul's Church-yard.

"Of the woorthyneſſe—of the holy Scripture:" &c. as p. 1117. 1579.
"Praiſe and diſpraiſe of Women. Written in the French tongue, and 1579. brought into our vulgar by John Allday." See p. 1038. Licenſed. Octavo.
"A diſcouerie—of unwritten Traditions:" as p. 998. 1582.
"The Summe of the 4 Euangeliſtes, comprehending both the courſe 1582. of the hiſtorie, and aiſo the ſeuerall pointes of doctrine ſet forth in the ſame, by Hen. Bullinger; tranſlated by Io. Tomkis." Printed for him. Licenſed. Octavo.
"Rodolph Gualter his ſermons on the prophet Ioel, tranſlated by Iohn 1582. Ludham." Printed for him. Licenſed. Octavo.
"An examination of the Councell of Trent, touching the decree of 1582. traditions, by Mart. Kemnicius, tranſlated by R. V. Licenſed. Quarto.
"Briefe inſtructions for all families by S. S." For him. Octavo. 1583.
"M. Tullii Ciceronis Conſolatio." &c. as p. 1061. Licenſed. 1583.
"A Perſwaſion to Godlie purpoſes:" &c. as p. 1123. Licenſed. 1583.
"Iohn Tomkis on the Lords prayer." For him. Licenſed. Sixteens. 1585.
"THE FAERIE QUEENE. Diſpoſed into twelue* books, Faſhioning XII. 1590. Morall vertues." John Wolf's device, the flower de luce feeding. London

---

* Though the title mentioned twelve books, this volume contains only the firſt three. Three more were publiſhed in 1596, and, as it is ſaid, a ſecond edition of the former three, with alterations and additions; but whether the remaining 6 books were finiſhed by the author, and loſt by his ſervant, as related by Sir James Ware, is not for me to determine. See Biogr. Brit. 3813, note U, However, an edition was publiſhed in 1609, entitled, "The Faerie Queen, containing two new Cantoes, the only remains of a loſt book, intituled The Legend of Conſtancie."

don. Printed for *him*, 1590." On the back, " To the moſt mightie &
magnificent empreſſe Elizabeth—queene of England &c.—Ed. Spenſer.
The third book ends on p 589. Then, " A Letter of the Authors expounding his whole intention in the courſe of this worke: &c. To—Sir
Walter Raleigh knight," &c. with ſeveral copies of verſes to the author,
and verſes by the author to moſt of the nobility in Q. Elizabeth's court.
W.H. Licenſed. Quarto.

1590. Sir Philip Sidney's Arcadia. See it in 1593. Licenſed. Quarto.
1590. " A theological diſcourſe of the Lambe of God & his enemies. Containing a briefe Commentarie of Chriſtian faith: together with a detection
of old and new barbariſme, by R. H." (*Richard Harvey*.) Dedicated to
—Robert Earle of Eſſex," &c. John Windet for him. Licenſed. Quarto.
1591. " Complaints: &c. by Edm. Spenſer." See p. 1246. Licenſed.
1591. " The Counteſſe of Pembrokes Yuychurch." &c. as p. 1247. Licenſed.
1591. " The Counteſſe of Pembrokes Emanuel." &c. as ib. Licenſed.
1591. " A report of the truth of the fight about the iſles of Açores, this laſt
ſommer, betwixt the Reuenge, one of her maieſties ſhippes, and an armada of the king of Spaine." 14 leaves. For him. See Hackluyt's
Voyages, 11; p. 169. Licenſed. Quarto.
1592. " The defence of the Article: Chriſt deſcended into Hell. With arguments objected againſt the truth of the ſame doctrine: of one Alex.
Humes. All which reaſons are confuted, &c. By Adam Hyll, D. D.
Magna eſt veritas & praeualet." A cut of the reſurrection, as p. 668.
For him. Dedicated " To—Iohn—Archbiſhop of Canterburye." 70
leaves. Licenſed. W.H. Quarto.
1592. " Amintae Gaudia. Authore Thoma Watſono, Londinenſi, Juris ſtu-
dioſo," In Hexameters. Licenſed. Quarto.
1592. " A Diſcourſe of Life and Death. Written in French by Ph. Mornay.
Antonius, A Tragoedie written alſo in French by Ro. Garnier. Both done
in Engliſh by the Counteſſe of Pembroke." Windet's device of Peace
& Plenty, as p. 1229. O 2, in fours. See p. 1208, & 1259. Licenſed.
W.H. Quarto.
1593. " The Counteſſe of Pembrokes Arcadia. Written by Sir Philip Sidney
Knight. Now ſince the firſt edition augmented and ended. London
Printed for *him*,—1593." In a compartment, ſeemingly done on purpoſe
for this book, having at top a boar paſſant, on a wreath; on the ſides,
an Arcadian ſhepherd, and an Amazon; at bottom, a tablet with a boar
making towards ſome flowers growing about which is twiſted a ſcroll
with this motto, NON TIBI SPIRO." Dedicated " To my deare lady and
ſiſter, the counteſſe of Pembroke.—Philip Sidney.---To the Reader."
243 leaves, ." Printed for *him* dwelling in Paules Church yard, neere
vnto the great north doore of Paules. A. D. 1593." W.H. Folio.
1594. " Sixe bookes of politickes, &c. as p. 1251. Licenſed.
1594. The Florentine Hiſtory. T. Creed for him. Again, Folio.
1595. " The Florentine Hiſtorie. Written in the Italian tongue, By Nicholo
Macchiavelli, citizen and ſecretarie of Florence. And tranſlated into
Engliſh by T. B. Eſq; London Printed by T. C. for W. P. 1595." In
the

the fame compartment as the Arcadia. Dedicated "To—Syr Chriftopher Hatton,—Lord Chancellour.—Tho. Bedingfeld.---The Proeme of the Authour.----To the Reader.---The Contents." 222 pages. " Printed by Tho. Creede for *him*, 1595." W.H. Folio.

" Two Difcourfes of Mafter Frances Guicciardin, vvhich are wanting 1595. in the thirde & fourth Bookes of his Hiftorie, in all the Italian, Latin, & French Coppies heretofore imprinted ; Which for the worthineffe of the matter they containe, were publifhed in thofe three Languages at Bafile 1561. And are now for the fame caufe doone into Englifh." Device the brazen ferpent, ufed by P. Short. " Printed at London for Wm. Ponfonbie, 1595." On the back are three fonnets of Petrarch's in Englifh. Then, four epiftles to the different readers, in Englifh, Latin, French, and Italian. Every leaf contains a column of one of thefe four languages. 67 pages. Licenfed. W.H. Quarto.

" The defence of Poefie, by Sir Philip Sidney." T. Creed for him. 4°. 1595.

" COLIN CLOVTS Come home againe." T. Creed's device of Truth. 1595. " Printed for *him*, 1595." Dedicated "To—Sir VValter Raleigh, Captaine of her Maiefties Guard, &c.—From my houfe of Kilcolman, 27 December, 1591.—Ed. Sp." Annexed are " Aftrophell : A paftorall Elegie vpon the death of—Sir Philip Sidney ;" near the conclufion of which are 16 ftanzas, that appear to have been written by his fifter, the amiable Counteffe of Pembroke, under the name of Clarinda. " The mourning Mufe of Theftylis ;" and " A paftorall Aeglogue vpon the death of Sir Phillip Sidney Knight, &c.—L. B." H, in fours. W.H. Quarto.

" Amoretti, or Sonnets : And Epithalamion. By Edm. Spenfer." Li- 1595. cenfed. Octavo.

" The fecond part of the Faerie Queene. Containing the fourth, fifth, 1596. and fixth bookes. By Ed. Spenfer." Field's deuice. " Imprinted—for *him*, 1596." W.H. See the firft three books in 1590. Quarto.

Four hymnes. Daphnaida. Prothalamion. By Edm. Spenfer. Quarto. 1596. "—Sir Francis Drakes Weft Indian Voyage." &c. as p. 1258. Licenfed.

" The Hiftorie of George Caftriot, furnamed Scanderberg, king of 1596. Albanie. Containing his famous actes, his noble deedes of Armes, and memorable victories againft the Turkes, for the Faith of Chrift. Comprifed in twelue Bookes : By Iaques de Lauardin, Lord of Pleffis Bourrot, a Nobleman of France. Newly tranflated out of French—by Z. I. Gent." Field's device. " Imprinted for *him*, 1596." Dedicated " To —Sir George Carey, Knight Marfhal of her Maiefties Houfe," &c.---- To the reader.----The preface of the author to the nobilitie of France.---- An advertifement.----A catalogue of authors." Commendatory verfes by Ed. Spenfer, R. C. Gent. and C. C. Gent, 498 pages, and an index. Licenfed. W.H. Folio.

" The Hiftorie of the great emperour Tamerlan. VVherein are ex- 1597. preffed, encounters, &c. with diuerfe Stratagems of warre, &c. which fhould not be vnknowen of them that would attaine vnto the knowledge of armes. Drawen from the auncient Monuments of the Arabians, by. Mefsire Iean du Bec, Abbot of Mortimer. Newly tranflated out of French into

into Englifh,—by H. M." Field's device. "Printed for *him*, 1597." Recommended "To the friendly reader," for his correction, rather than unto any man's particular protection. Dated, 15 Octob. 1597. Pages 265. W.H. Again 1599. Quarto.

1598. "—The felicitie of man :" &c. as p. 1259.

1598. "A treatife parænetical, That is to fay: An exhortation. Wherein is shewed by good & euident reafons, infallible arguments, moft true & certaine hiftories, and notable examples; the right way & true meanes to refift the violence of the Caftilian king: to breake the courfe of his deffeignes: to beat down his pride, and to ruinate his puiffance. Dedicated to the Kings, Princes, Potentates & Common-weales of Chriftendome: and particularly to the moft Chriftian King: By a Pilgrim[f] Spaniard, beaten by time, & perfecuted by fortune. Tranflated out of the Caftilian tongue into the French, by I. D. Dralymont, Lord of Yarleme. And now Englifhed.—Printed for *him*, 1598." Dedicated "To—Maifter Fulke Grevil, Gentleman of her Maiefties moft Hon. Priuie chamber.— W. P. (*the printer*)----The author vnto the moft Chriftian King, to the Princes, Potentates, &c.----The epiftle of the French tranflator, To the moft Chriftian King.—From your towne of Pau, 1 Octob. 1597.---- The French Tranflator to the Reader.----A Table—of the principall thinges:" at the end of which is one of Field's Vignettes. 160 pages. W.H. Quarto.

1598. "The Theorike and Practike of Modern Warres,"' &c. as p. 1259.

(1598.) "The Tragicomoedi of the vertuous Octauia. Done by Samuel Brandon, 1598.—Printed for *him*, and are to be foulde at his fhop in S. Paules Churchyarde." Dedicated "To—the Ladie Lucia Audelay."--- All'autore.—Mia.---- Profopopeia al libro.—S. B.----The Argument." There are annexed an epiftle from Octavia to M. Antony, and his anfwer. Dedicated "To—Miftrefse Mary Thinne." H, in fours. W.H. Octavo.

1599. "The Manfion of Magnanimitie. Wherein is fhewed the moft high & honorable acts of fundrie Englifh Kings, Princes, Dukes, Earles, Lords, Knights & Gentlemen from time to time performed, in defence of their Princes & Countrie: Set forth as an encouragement to all faithfull fubiects, by their example, refolutely to addrefs themfelues againft all foreign Enemies. Publifhed by Ric. Crompton, an apprentice of the Common Law, 1599. Whereunto alfo is adioyned a collection of diuers lawes and ftatutes meete to be known of all men: With a briefe table, fhewing what Munition ought to be kept by all fortes of her Maiefties fubiects for the defence of her Highneffe Realmes & dominions. Printed for *him*, 1599." See Oldys's Catal. of pamphlets in the Harl. Libr. No. 395. 4°.

1600. "Certaine experiments concerning Fifh and Fruit: Practifed by John Taverner Gent. and publifhed by him for the benefit of others. Printed for *him*, 1600." See Oldys's Catal. No. 225. Quarto.

He had alfo licenfes for the following, viz. In 1579, "A declaration of the Pr. of Orange." In 1581, "The ioyfull Entry of Monf. the Frenche

[f] My Copy has in M.S. of the time, "Vz. to Philip the 2d, who came hitherinto Don Antonio de Peroz, Secretarie of ftate England."

French kings brother." In 1582, "Remedies to take out spots, &c." In 1583, "Mamilia: The 2d part of y tryumphe of Pallas, wherein with perpetual Fame ⁊ conſtancie of gentlewomen is canonized." In 1589, "Jobus Theod. Bezæ; to be printed in Eng." In 1590, "The politike & martiall difcourfes of y lord De la noüe: turned over to him from Edw. Aggas." In 1593, "Scianuctos, or the ſhadowe of nighte." He did not print long after 1600.

WILLIAM BARLEY,

ASSIGNEE of Tho. Morley, whoſe patent may be ſeen in our General Hiſtory. He dwelt in Gracechurch-ſtreet, at the upper end, near Leaden-hall gate; and appears to have been more of a bookſeller, or publiſher, than a printer; yet he was fined xlſ. for printing three ballads, and a book, diſorderly. He uſed for his device a death-head, hour-glaſs, &c.

"A Direction to the waters of lyfe." &c. as p. 1263. Sold by him. 4°. 1590.
"Preſent remedies againſt the plague," &c. For him. Quarto. 1592.
"The ground worke of conny catching," &c. as p. 12ẏi. 1592.
"The blacke bookes meſſenger" &c. as ib. 1592.
God's Arrow againſt Atheiſts, by Hen. Smith. For him. Quarto. 1593.
"A new booke of Citterne Leſſons, with a plaine & eaſie inſtruction 1593. for to learne the Tableture, to conduct & diſpoſe thy hand; ſette forth to the Tunes of many Pſalmes, as they be ſung in Churches: alſo Pauins, Galliards, and diuers other ſweet & eaſy Leſſons." For him. Quarto.
"A new boke of tabliture. Containing inſtructions to guide and diſ- 1593. poſe the hand to play on ſundry inſtruments, as the lute, orpharion, and bandora. With new leſſons to each of theſe inſtruments. Whereunto is added, the pathway to muſick. Containing ſundry eaſy rules for the vnderſtanding the ſcale, or gamma, ut. Alſo, a treatiſe of diſcant, and certain tables, which doth teach how to moue any ſong higher or lower, from one key to another." For him. Quarto.
"Tho. Ratliffe his Catechiſme." For him. Octavo. 1594.
"A treatiſe deſcribing the nature of Tabacco." For him. Octavo. 1595.
"Marie Magdalens Loue." as p. 1272. 1595.
"Menæcmi; a pleaſant—comœdie, taken cut of—Plautus, by W. W." 1595. Printed by T. Creed, and ſold by him. See it in Six Old Plays, 1779. 4°.
"A world of wonders. A maſſe of murthers. A covie of Coſmages." 1595. For him. Quarto.
"The glaſs of Mans folly, and means to amendment of life." For 1595. him. Quarto.
"John Charldon, D. D. his ſermon at Exeter, on Iſa. 1; 2, 3." 8°. 1595.
"John Hawkins his ſermon on Prov. 1; 4, entitled A ſallade for the 1595. ſimple." Octavo.
"Strange & wonderful things happened to Rd. Haſleton, borne at 1595. Braintree in Eſſex, in his ten yeares trauailes in many forraine countries.

Penned

WILLIAM BARLEY.

Penned as he delivered it from his owne mouth." His device. E 3, in fours. Cuts. A. I. for him. Quarto.

1595. "The noblenes of the Affe, by Attabalibu of Peru." T. Creed, for him. Quarto.

1596. A Cornucopia of divers fecrets. For him. Quarto.

1596. The pathway to knowledge. A book of Arithmetic, tranflated out of Dutch by W. P. With tables of weights and meafures, &c. Quarto.

1598. " Epulario, or the Italian banquet. A. I. for *him*." Quarto.

1598. " The difcouery of the vnnatural Spaniolized Scotifh papifts." With John Norton, Efq; Folio.

1598. " A Looking Glaffe for London & Englande. Made by Tho. Lodge Gent. and Rob. Greene. In Artibus Magifter." Creed's device of Truth. " Printed by Tho. Creede, and are to be folde by Wm. Barley, at his fhop in *Gratious* ftreete, 1598." This hiftorical play reprefents the ftate of the Ninevites, their abominations, and their reformation. In it are introduced the prophets Hofea and Jonah. It is not divided into acts, but has the ftage directions; one of which is, " Ionas the Prophet caft out of the Whales belly on the ftage." I, in fours. W.H. Quarto.

1599. " The pfalmes of Dauid in meter, the plaine fong beeing the common tunne to be fung and plaide vpon the lute, orpharyon, citterne, or bafe violl, feuerally or altogether, the finging part to be either tenor or treble to the inftrument, according to the nature of the voyce, or for fowre voyces: With tenne fhort tunnes in the end, to which for the moft part of all the pfalmes may be vfually fung, for the vfe of fuch, as are of mean fkill, and whofe leyfure leaft ferueth to practize. By Richard Allifon, gent. practitioner in the art of muficke; and are to be folde at his houfe in the Dukes-place, neere Aldegate." Dedicated to the countefs of Warwick; with Morley's patent.* Folio.

***

THOMAS SALISBURY, or SALBERYE, ftationer,

SON of Pierce Salberye of Clokanok in Denbighfhire, was apprentice to Oliver Wilkes for feven years from St. Simon & Jude, 1581. His family feem to have been great promoters of the Britifh tongue among us, as may be feen in this effay of Englifh printing.

1593. " Grammatica Britannica in vfum eius linguæ ftudioforum fuccincta methodo et perfpicuitate facili confcripta; & nunc primum in lucem edita: Henrico Salefburio Denbighienfi autore." K, in eights. Octavo.

He had alfo licenfe to print, in 1594, " The perpetuity of Faythe: written in Latine by Mr. Dr. Some, now tranflated into Englifh, & foe to be printed."

RICHARD BOYLE, or BOILE, stationer,

SON of Tho. Boile of the city of Hereford, was bound apprentice to Tho. Woodcock for eight years from St. Bartholomew's 1576, and was made free, 15 Sept. 1584. In July 1589 he was fined ij s. vj d. for keeping an apprentice unprefented, contrary to the ordinances. He dwelt at the Rose in St. Paul's Church-yard; was a bookseller, and a puritan, as says bishop Tanner in his MSS.

"The Razing the foundations of Brownisme, wherein are diuers con-   1588.
clusions or propositions maintained against that sect, by S. B." (*S. Bred-*
well.) For him. See p. 1227.   Quarto.
"Tho. Michelthwait his Catechisme for howsholders."—For him. 8°.   1589.
"A Christian and godly view of death & life, as also of humane   1593.
actions: By Phil. Morney; translated by A. W. Printed by R. F. for R. B." See it translated by the countess of Pembroke, and printed for Wm. Ponsonby, this year and 1600.   Octavo.
"A dialogue concerning the vnlawfullness of playing at cards." Octavo.   1593.
"A Comfortable Treatise for the reliefe of such as are afflicted in   1595.
conscience. Perused the second time and enlarged in diuerse places, but especiallie with manie profitable & comfortable notes in the margent. By R. Linaker. Luk. 6; 21,—Iam. 1; 2, 3.—.Imprinted by Val. Simmes, for *him*, 1595." On the back is "A short view of those things which bee handled in the Treatise following," &c. Dedicated "To—Robert Deuoreux, Earle of Essex, *&c.*—Rob. Linaker." 71 pages. W. H. 16°.

❦❦❦❦❦❦❦❦❦❦❦❦❦❦❦❦❦❦❦

THOMAS CREED, or CREEDE, stationer,

WAS made free of that company, 7 Octob. 1578, by Tho. East. In 1595, he appears to have given bond to the company in the penalty of £40, not to teach John Wilkinson the art of printing. At the same time he was fined, "for having kept a prentize vnpresented above the tyme lymitted by the ordinances, and having bound & enrolled him contrary to the same, ij s. vj d. for the dutie of his presentment, and ij s. vi d. for breach of the ordinances, and soe he to enioye the service of the apprentice,—Hen. Vawse." He dwelt at the sign of the Catharine Wheel, in Thames-street, near the Old Swan, and frequently used for his device an emblem of Truth, with a hand, issuing from the clouds, striking on her back with a rod; T. C. at bottom, and this motto round it, VERITAS VIRESCIT VVLNERE. Sometimes a griffin sitting on a stone, to which is linked a round ball, with a pair of wings.

Bart. Traheron on Rev. iv. See p. 935. For him.   Octavo.   1583.
"The shephards Calender." &c. as p. 1159. Again 1597.   1586.
"A Mirrour of Popish Subtilties:" &c. as p. 1111.   1594.

" ARISBAS

THOMAS CREED.

1594. "ARISBAS, Euphues amidſt his ſlumbers, or Cupids Iourney to hell,. &c. By J. D." (*J. Dickenſon.*) For him. Licenſed. See Hiſt. of Eng. Poetry, III; 417, note a.   Quarto.

1594. " The firſt part of the tragicall raigne of Selimus, ſometime emperour of the Turkes, and grandfather to him that now reigneth. Wherein is ſhewne, how he moſt vnnaturally raiſeth warres againſt his own father Baiazet, and preuailing therein, in the end cauſed him to be poyſoned. Alſo, with the murthering his two brethren, Corcut and Acomat. As it was played by the queenes maieſties players."   Quarto.

1594. The Florentine Hiſtory: &c. Again 1595. See p. 1274.
1595. "The defence of Poeſie, by Sir Philip Sidney." For Wm. Ponſonby. 4°.
1595. "Colin Clovts Come home again." &c. as p. 1275.
1595. "Menæcmi —taken out of—Plautus," &c. as p. 1277. Licenſed.
1595. " The noblenes of the Aſſe," &c. as p. 1278. Licenſed.
1595. "Antipapa, vel Antitheſis Xti & Papæ. A Compariſon between Chriſt & the pope of Rome; together with a deſcription of our Sauiour Chriſt, as he was man incarnate vpon the Earth; with his laſt will & teſtament, called Magna Charta de libertatibus mundi; as it was found in an ancient abbey in Kent."   Octavo.

1595. "The lamentable tragedie of Locrine, the eldeſt ſonne of king Brutus, diſcourſing of the warres of the Britaines, and Hunnes; with their diſcomfiture. The Britanes victorie, with their accidents, and the death of Albanact. Newly ſet foorth, ouerſeene & corrected by W. S." Licenſed. See Mr. Steevens's Remarks, prefixed to Shakeſpear's plays, p. 240, &c. Alſo, Hiſt. of Eng. Poetry, III; 402, note b.   Quarto.

1595. "The moſt wonderfull & ſtrange finding of a Chayre of Gold, neare the Iſle of Iarſie, with the true diſcourſe of the death of eight ſeuerall men: And other moſt rare Accidents thereby proceeding." This title over a cut ſeemingly of the tranſaction, but torn off from my copy. 14 pages, including not only the title-page, but a blank leaf before it, as was frequent about this time. Licenſed. W.H.   Quarto.

1595. "Bromleion. A Diſcourſe of the moſt ſubſtantial points of Diuinitie, handled by diuers Common places: With great ſtudie, ſinceritie, & perſpicuitie." &c. Dedicated "To—Sir Henry Bromley Knight.—S. I. ----The Epiſtle to the Reader." 565 pages. "Printed by him, dwelling in Thames ſtreete, at the ſigne of the Katheren-wheele, neare the olde Swanne." Mr. Voet.   Quarto.

1596. " The thirteene bookes of Aeneidos." &c. See p. 777. Licenſed.
1596. "Dele table demaundes, and pleaſant Queſtions, with their ſeuerall Anſwers, in matters of Loue: Naturall cauſes, with Morall & politicke deuiſes. Tranſlated out of French" &c. His device of Truth. "London Printed by *lim*, 1596." Inſcribed "To the ſtudious & well diſpoſed youthes of England." 232 pages. W.H.   Quarto.

1596. " The Engliſh Mans Treaſure:" See p. 1237. Licenſed. Again 1598.
1596. "Of Ghoſtes & Spirites, Walking by Night." &c. See p. 971.
1596. "Three ſermons on Famine, and dearth of victualls, &c. By Lod. Lavater, on 2 Chron. 6; 26—31. Tranſlated by Wm. Barlow, B. D." 8°.

" The

" The second part of the Faerie Queene." &c. as p. 1275. 1596.
" The—conqueſt of the Weſt India." &c. See p. 983. Licenſed. 1596.
" THE MOST PLEAſaunt and delectable Hiſtorie of Lazarillo de Tormes, 1596.
a Spanyard: And of his maruellous Fortunes & Aduerſities: The ſecond
part. Tranſlated out of Spaniſh—by W. P." Device, an ox, &c. as
p. 1255. " Printed—by T. C. for Iohn Oxenbridge, dwelling in Paules
Church-yard at the ſigne of the Parrot, 1596." J, in fours. W.H. 4°.

" An expoſition on Acts 11; 27—30: The prophecy of Agabus, con- 1597.
cerning famine, by Peter Baker." Octavo.

" The wil of wit, wits will, or wils wit, chuſe you whether. Con- 1597.
taining five diſcourſes, the affects whereof follow. Reade and judge.
Compiled by Nicholas Breton, gentleman." Licenſed. Quarto.

" THE Mirror of Alchimy, Compoſed by the thrice-famous & learned 1597.
Fryer, Roger Bachon, ſometime fellow of Martin Colledge: and afterwards of Brazen-noſe Colledge in Oxenforde. Alſo a moſt excellent &
learned diſcourſe of the admirable force & efficacie of Art and Nature,
written by the ſame Author. With certaine other worthie Treatiſes of the
like Argument. Vino vendibili non opus eſt hedera." Truth. " Printed
for Rd. Oliue, 1597." A ſhort preface. 84 pages. W.H. Quarto.

" THE Mutable and wauering eſtate of France, from the yeare of our 1597.
Lord 1460, vntill the yeare 1595. The great Battailes of the French Na-
tion, as well abroad with their forraigne enemies, as at home among
themſelues, in their ciuill & inteſtine warres: With an ample declaration
of the ſeditious & treacherous practiſes of that viperous brood of Hiſpa-
niolized Leagueis. Collected out of ſundry, both Latine, Italian, &
French Hiſtoriographers. London Printed by him, 1597." Dedicated
" To—Maiſter Iulius Cæſar Eſq; Doctor of the Ciuill Lawe, &c.—
Though nameleſſe, yet alwayes by your worſhip to be commaunded.----
To the Reader." 148 pages. Licenſed. W.H. Folio.

" PROVISION FOR the poore, now in penurie. Out of the ſtore-houſe 1597.
of Gods plentie: Which they ſhall be ſure to find in all places, that are
indued with his graces, to thinke ſeriouſly on this ſentence following.
Pſal. 41; 1. Bleſſed is he that conſidereth the poore, the Lord will
deliuer him in the time of trouble. Explaned by H. A." His device of
Truth. " Printed—1597." Inſcribed " To the Chriſtian Reader, in
what calling ſoeuer.—Hen. Arth." An analytical table. E, in fours.
W.H. Quarto.

" Two treatiſes concerning Regeneration." &c. as p. 1268. 1597.

" The Cognizance of a true Chriſtian, or the Outward Markes whereby 1597.
he may be the better knowne: Conſiſting eſpecially in theſe two duties:
Faſting and giuing Almes, &c. By Sam. Gardiner B. D.—Printed by him,
and—ſold by Nic. Ling, 1597." Pages 213. Licenſed. Octavo.

" ——Vtopia—by Sir Thomas More," &c. See p. 773. Quarto. 1597.
" VIRGIDEMIARVM, Sixe Bookes." &c. as p. 1268. Licenſed. 1597.
" Regimen ſanitatis Salerni:" &c. See p. 426. Licenſed. 1597.

" A preparation to the moſt Holie Miniſtrie:—: Alſo a liuely exhor- 1598.
tation to all youth to giue themſelues to the ſtudie thereof: and a confutation

tation of the obiections which may be brought in any fort to touch the fame: &c. Written in French by Peter Gerard, and tranflated into Englifh by N. B.—Imprinted—by *him*, for Tho. Man,—1598." Dedicated " To—Sir Wm. Periam Knight, Lord chiefe Baron, &c.—Nich. Becket.----To his louing Brethren, the Paftors & Minifters of Deuon, & Cornwall, &c.—N. B." 328 pages. W.H. Octavo.

(1598.) " A Method for Trauell. Shewed by taking the view of France. As it ftoode in the yeare of our Lord 1598." Truth. " London Printed by Tho. Creede." Without date. Infcribed " To all gentlemen that haue Trauelled.—Rob. Dallington."----An analytical table. Y 2, in fours. W.H. Quarto.

1598. " An hiftorical collection of the moft memorable accidents, and Tragicall Maffacres of France, vnder the Raignes of Henry 2, Francis 2, Charles 9, Henry 3, Henry 4, now liuing. Conteining all the troubles &c. vntill this prefent yeare, 1598.—Tranflated out of French into Englifh." The device, an arm, clad in armour, holding a flaming fword, with this motto on a fcroll twirled about it, CONTRAHIT AVARITIA BELLVM; and an unknown cypher, or mark, at bottom. " Imprinted—by *him*, 1598." Dedicated " To—George earle of Cumberland, baron of Clifford, &c.---To the—Reader.---The contents" &c. It is divided into two parts; the former, concluding with the reign of Charles the 9th, to which is annexed A difcourfe on the reduction of the city of Lyons. 310 pages. The latter part, 292 pages, and A difcourfe of the delivery of Brittaine. Licenfed. W.H. Folio.

1598. " MOTHER BOMBIE. As it was fundrie times plaied by the Children of Powles." Truth. " Printed by *him*, for Cuthbert Burby, 1598." H, in fours. W.H. One of the 6 Court comedies by John Lylie. 4°.

1598. Richard III. by Wm. Shakefpear. For Andr. Wife. Quarto.

1598. " A LOOKING Glaffe for London & Englande." &c. as p. 1278. Licenfed.

1598. ' PARISMVS, the renowned prince of Bohemia, his moft famous, delectable, and pleafant hiftory: Conteining his noble battailes fought againft the Perfians; his loue to Laurana, the kings daughter of Theffaly; and his ftrange aduentures in the defolate ifland," &c. Licenfed. 4°.

1599. " PARISMENOS. The fecond part of the moft famous delectable hiftory," &c. Licenfed. Quarto.

1599. " Micro-cynicon. Sixe fnarling Satyres." Octavo.

1599. A treatife of the nature of God. For Rob. Dexter. 239 pages. Octavo.

1599. " DYETS DRY DINNER: Confifting of eight feuerall Courfes: 1 Fruites. 2 Herbes. 3 Flefh. 4 Fifh. 5 whitmeats. 6 Spice. 7 Sauce. 8 Tabacco. All ferued in after the order of Time vniuerfall. By Henry Buttes, Maifter of Artes, & Fellowe of C. C. C. in C.—: Printed—by *him*, for Wm. Wood, and are to be fold at the Weft end of Powles, at the figne of Tyme, 1599." Dedicated " Partem Parentes To—the Lady Anne Bacon, fole heire to the Worfhip. Edw. Buttes Efq; her Father, as alfo to her Vncles,—Syr Wm. Buttes Knight, and Tho. Buttes Efq; deceafed. ---Partem Amici To—Ric. Thekefton Efq; and Elynor his wife.---Partem Patria. To my Country-men Readers.----The Authors Method

thod comprifed in Verfe, by Sam. VVallfall.---The fame man in prayfe
of this learnedly witty Booke.---Eiufdem ad Libri nomen allufio.---Grace
before Diets dry Dinner ferued in by Time." On the front page of fignature B 1 is a device of Time bringing Truth to light, as p. 978. P 7,
in eights. Before the diet of Tabacco is "A Satyricall Epigram, vpon
the wanton, & exceffive vfe of Tabacco." At the end, "Epipofion. Grace
after Diets dry Dinner, wherein Diets Drinking is promifed.---Ioa.
VVeeueri Epicrifis ad Henricum Butfum.---Eiufdem ad eundem de eodem
Palinodoia." W.H. Sixteens.

Romeo and Juliet, by Wm. Shakefpeare. For Cuthbert Burby. Quarto. 1599.

"Of MARIAGE and Wiuing. An excellent, pleafant and Philofophicall 1599.
Controuerfie, betweene the two famous Tafsi now liuing, the one Hercules
the Philofopher, the other, Tarquato the Poet. Done into Englifh by
R. T. Gentleman." Truth. " Printed by him, and—fold by Iohn Smy-
thicke, at his fhop in Fleetftreete, neare the Temple Gate, 1599." Dedicated " To—the Lord Giouan Battifta Licino.—Bergamo, 1 May,
1598.—Hercole Taffo." L.2, in fours. Licenfed. W.H. Quarto.

"The hiftory of the two valiant knights, fyr Clyomer, knight of the 1599.
golden fheeld, fonne to the king of Denmarke; and Clamydes the white
knight, fonne to the king of Suauia. As it hath bene fundry times acted
by her maiefties players." Quarto.

"A treafurie or ftore-houfe of Similies: Both pleafaunt, delightfull 1600.
& profitable, for all eftates of men in generall. Newly collected into
Heades & Common places: By Rob. Cawdray." Truth. " Printed by
him, dwelling in the Old Chaunge, at—the Eagle & Childe, neare Old
Fifh-ftreete, 1600." Dedicated " To—Sir Iohn Harington Knight,
as alfo to—Iames Harington Efq; his brother.---To the Religious &
Chriftian Reader." 860 pages, and a table. Licenfed. W.H. Quarto.

Henry V. by Wm. Shakefpear. For T. Millington and J. Bufby. 4°. 1600.

"Ouidius Nafo His Remedie of Loue. Tranflated & Intituled to the 1600.
Youth of England,—". Truth. " Printed by T. C. for Iohn Browne,
and—fold at his fhop in Fleet ftreet, at—the Bible, 1600." Quarto.

"Greens Groatfworth of wit." With John Danter; for Rd. Oliff. 4°.

He had alfo licenfes for the following, viz. In 1593, " An epitaph
vpon the death of ye righte hon. Henrie erle of Derbie. A memorial, or
epitaphe, of Sir Wm. Rowe knight, late Lord maior of—London. A
memoriall of the life & deathe of ye right hon. & renownd warrior ye
valiant lorde Graye of Wilton. The famous victories of Henrye the ffyft,
conteyning the hon. battell of Agincourt. The Scottifhe ftory of James
ye ffourthe flayne at ffodden, intermixed with a pleafant comedie prefented by Obocom kinge of ffayns. The Pedlers prophecie. Letters
patentes concerninge the reducing of the townes of Roan, Newe haven,
Harfleur, &c. The troublefome & hard adventures in Love, with many
fyne conceyted fonnetts, & pretty poemes,—, written in Turkey by R. C.
An enterlude, The tragedie of Richard ye third: wherein is fhewed ye
death of Edward ye ffourthe, with ye fmotheringe of ye two princes in ye
tower, with a lamentable end of Shores wife, and ye coniunction of ye twoo
houfes

houses of Lancaster & York." In 1594, " Mother redde cappe her last will & testament conteyning sundry conceipted & pleasant tales, &c. Pheander the mayden knight." In 1595, " The second parte of y͏̈ moste delectable & plesant historye of prynce Edward knight of y͏̈ Holye Crosse of Jerusalem. A goodly Gallery." In 1596, " Twoo partes of Palmerin of England. Twoo partes of Palmerin D'ulius. The second parte of The Anatomye of abuses, called The Displaye of Corrupc'ons. These 5 copies were assigned from Wm. Wright to Tho. Scarlet, and from T. S. to T. C. The hystorie of Euryalus & Lucretia. Willes witt, with the miseries of Mabilla; printed by Tho. Scarlet. Kind hartes dreame. The true report of certen ground to y͏̈ quantatie of 9 acres & more, lyenge together in one trenche which lately was moved, & stirred, & carried from his former place, with the trees there vppon verye straunglie in y͏̈ parish of Westram in Kent, about 16 myles from London, &c. A preparatif to contentation. Hitiphon & Loutippe translated from the Greeke by W. B." In 1597, " The historye of y͏̈ last troubles of Frauce vnder Hen. III. of Frauce & Poland, and Hen. IV. of Frauce & Navarre: leaving out y͏̈ discourse of y͏̈ quene of Scotts. Illustriū poetarū Flores," &c. In 1598, " Of mans Justification before God, twoo bookes opposed to y͏̈ sophismes of Rob. Bellermine Jesuit, by Jo. Piscator, translated into Englishe." In 1599, " The first tome or tablet of y͏̈ differences of Religion, &c. Gathered by Phil. de Mornay, lord of Mount St. Aldegond. Quintus Curtius: for the companye: to pay vj d. in the li. to y͏̈ poore: one impression only, and no more." Also for several ballads. He printed a considerable time after 1600.

****

ADAM ISLIP, stationer,

APPRENTICE to Hugh Jackson, but turned over to Tho. Purfoot, 7 Octob. 1578, was made free 8 June, 1585. Febr. 5, 1598-9, he was fined ij s. vj d. for printing The xv joys of marriage, disorderly; and all the leaves confiscated to the House. Again, 2 Apr. 1599, for printing The fountain of fiction, without entering. ij s. He used the device of a flourishing palm tree, &c. as p. 1180. Also: a pillar, the capital of which represents a bird's head, crowned with a ducal coronet, having a long bill with a ring on it, and holding a string in its mouth, seemingly coming from the middle of the beam of a pair of scales, behind the pillar; in one scale is a serpent which preponderates the other with a cat on her hind legs, as going to jump out. About the pillar is twisted a cornucopia. The base represents a bird's claw, steadying itself on a helmet laying on a shield. The whole in a compartment, with this motto, QVIBVS RES PVBLICA CONSERVETVR.

1594. " The Triall of Bastardie: that part of the second part of Policie, or maner of Gouernement of the Realme of England: so termed, Spirituall, or Ecclesiasticall. Annexed at the end of this Treatise, touching the

the prohibition of marriage, a Table of the Leuitical, Englifh, and Pofitiue Canon Catalogues, their concordance and difference, By Wm. Clerke. Iohn Chrifoft. Nemo verè indè, aut obfcurus, aut clarus eft." The palm tree.'"—Printed by Adam Iflip, 1594." Dedicated "To—Robert Redmayn, doctor of law, Iudge Delegate, and Commiffarie (for the vacancie of the See) within the Citie and dioces of Norwich.----The Preface.---Gregor. Nazian. de bono Coniugij.---A Table of the contents." The triall of Baftardie, with three folding tables, ends on p. 82, on fignature M 1. The table of the Levitical—catalogue is not paged, but the fignatures are continued to O 4.* Licenfed. W.H. Quarto.

" Ten Introductions: how to reade, and in reading, how to vnderftand: and in vnderftanding, how to beare in minde all the bookes, chapters, and verfes contained in the holie Bible. With an anfwer for lawers, phyficians, minifters. By Edward Vaughan." Octavo. 1594.

The gardiners labyrinth. See p. 979. Quarto. 1594.

" The doctrin of Prayer for all men, proued againft the pofition of thofe that fay, and preach, that all men are not to be prayed for,. by John Smith," Fellow of St. John's Coll. Oxon. and " minifter at Reading." 4°. 1595.

" The Mount of Caluarie." &c. as p. 1200. The fecond part, 1597. 1595.

" Politicke, Moral, and Martial Difcourfes. Written in French by M. Iaques Hurault, lord of Vieul & of Marais, and one of the French kings priuie Councell.—tranflated into Englifh by Arth. Golding." The palm tree, &c. " London, Printed—1595." Dedicated " To—William Lord Cobham, L. warden of the Cinque ports, &c.—27 Jan. 1595.---To the *French* King.—Paris, 28 Octo. 1588.—Iames Hurault, &c.---The Contents." 495 pages. W.H. Quarto. 1595.

" Strange—things happened to Rd. Hafleton," &c. as p. 1277. 1595.

" The Contents of Scripture: Containing the fum of euery Booke & chapter of the old & new Teftament. Gathered from Tremelius, Iunius, Beza, Pifcator, and others.—Printed—for Ra. Iackfon, 1596." In a compartment with the Refurrection at top, as p. 976. Dedicated " To —Sir William Fitz-william Knight, with the vertuous Lady his wife.— From your houfe, Park Hall in Effex, Octo. 12. 1596. Your Worfhips bounden in the Lord, Rob. Hill." At the end is " A moft neceffarie Table to know the rifing & fetting of the Sunne, and how the dayes in euery Moneth encreafe & decreafe by the fame." 526 pages. To which is annexed, " The Content of the foure Euangelifts: Or, The life of Chrift: Collected by C. L. and placed before his Harmony. Englifhed for an Appendix to the Contents of Scripture. To this are added an hundred Aphorifmes,—containing the matter & method of M. Caluins Inftitutions, in a far other order then that fet out by Pifcator." &c. Dedicated " To—M. Roger Ofield, citizen & Merchant of London.—18 Octob. 1596.—R. H." 63 pages, and the Aphorifmes 46. W.H. 24°. 1596.

" The Orator: Handling a hundred feuerall Difcourfes, in forme of Declamations: Some of the Arguments being drawne from Titus Liuius & other ancient Writers, the reft of the authors owne inuention: Part of which are of matters happened in our Age. Written in French by Alex. Siluayn, & Englifhed by L. P." The fmall flower-de-luce with motto 1596.

motto as p. 1183. "London Printed—1596." Dedicated "To—Iohn, Lord St. Iohn, Baron of Bletſho.—Lazarus Piot.---To the Reader." 436 pages. Licenſed. W.H. Quarto.

1596. "A BOOKE OF SECRETS: ſhewing diuers waies to make & prepare all ſortes of Jnke, & Colours: as Blacke, White, Blew, Greene, Red, Yellow, and other Colours. Alſo to write Gold & Siluer, or any kind of Mettall out of a Pen: with many other profitable ſecrets, as to colour Quills & Parchment of any colour: and to graue with ſtrong Water in Steele & Jron.—.Tranſlated out of Dutch—by W. P. Hereunto is annexed a little Treatiſe, intituled Jnſtructions for ordering of Wines: Shewing how to make Wine, &c. Written firſt in Italian, & now tranſlated into Engliſh, by W. P.—Printed by him for Edw. White,—1596." E, in fours. Alex. Dalrymple, Eſq; Quarto.

1596. "Examen de Jngenios.[f] THE EXAMINATION of mens Wits. In which, by diſcouering the varietie of natures, is ſhewed for what profeſſion each one is apt, and how far he ſhall profit therein. By John Huarte. Tranſlated out of the Spaniſh tongue by M. Camillo Camilli. Engliſhed out of his Italian, by R. C. Eſq;" The palm-tree, &c. "London Printed—1596." Dedicated "To—Sir Francis Godolphin Knight, one of the deputie Lieutenants of Cornwaile.—R. C.---To the maieſtie of Don Philip, our Soueraigne.--The ſecond Proeme to the Reader." 333 pages, and A Table of all the chapters, &c. W. H. Mr. Ames had an edition, printed for C. Hunt, at Exeter. Quarto.

1596. "Hawking, Hunting, Fouling & Fiſhing, with the true meaſures of blowing, &c. Whereunto is annexed, the maner and order in keeping of Hawkes; their diſeaſes, and cures, &c. Now newly collected by W. G. (Wm. Gryndall) faulkener." Mr. Haworth. Quarto.

1596. "THE DIVEL coniured. London Printed by him for Wm. Mats,—in Fleetſtreet at the ſign of the Hand & Plough, Anno 1596." Dedicated "To—ſir Iohn Forteſcue, Knight, Chancellor of the Exchecker, &c.—15 Apr. 1596.—T. L. (T. Lodge)----To the Reader" &c. M 2, in fours. W.H. Quarto.

1596. "Certaine Sermons, taken out of ſeuerall places of ſcripture. Rom. 10; 16.—London, Printed by him, for Tho. Man, 1596.---A Summarie."---Dedicated "To—Francis Earle of Bedford,—Iohn Vdall.---To the Reader." V u 2, in eights. W.H. Octavo.

1596. "Wits miſery & the worlds Madneſs diſcovering the Devills Jncarnate of this age. By T. Lodge." The Rev. Dr. Lort. Quarto.

1597. "Progymnaſma ſcholaſticum. Hoc eſt, Epigrammatum Græcorum ex Anthologia ſelectorum ab He. Stephano, &c. Opera & induſtria, Iohannis Stockvvoodi, Scholae Tunbridgienſis olim Ludimagiſtri. Græca præterea ſunt omnia per lineas interlineares Latinis expreſſa typis, ad faciliorem eorundem lectionem, in ſtudioſæ juventutis gratiam." Flower de luce and motto as p. 1183. "Londini, Ex Typographia Adami Iſlip, M.D.XCVII." Dedicated "Herôi illuſtriſsimo & inclytiſſimo, Domino Roberto

---

[f] Thus printed, as alſo in the edition 1604; but ſeems to have been corrected in Mr. Ames's copy.

berto Deuorax, Eſſexiæ Comiti &c.—Ioan. Stockwoodus.—Ad candidum Lectorem.—Othoni Comiti Solmenſi, Domino Minzebergæ & Sonnevaldi, Hen. Stephanus. S. D.—Lectori, S. D." H h, in eights. W.H. Octavo.

"Mount Caluarie. The ſecond part:" &c. as p. 1200. 1597.

"THE THEATRE of Gods Iudgements: Or, A collection of hiſtories 1597. out of Sacred, Eccleſiaſticall and prophane Authours, concerning the admirable Iudgements of God vpon the tranſgreſſours of his commandements. Tranſlated out of French, and augmented by more than three hundred Examples, by Th. Beard." Device, a palm-tree, &c. "—Printed by him, 1597." Dedicated "To—Sir Edward Wingfield Knight.—Th. Beard.—The Preface.—The names of the Authours from whom the moſt part of the examples contained in this Booke are collected." G g, in eights. Licenſed. W.H. Quarto.

"Ariſtotles Politiques, or Diſcourſes on Gouernment. Tranſlated 1597. out of Greeke into French, with Expoſitions taken out of the beſt Authours, &c. Concerning the beginning, proceeding, and excellencie of Ciuile Gouernment. By Loys Le Roy, called Regius. Tranſlated out of French into Engliſh." Device, a pillar, &c. "At London printed by him,—1597." Dedicated "To—Sir Robert Sidney Knight, Lord Gouernour of the Cautionarie towne of Vliſſing and Caſtle of Ramakins.— I. D.—To the courteous Reader.—To—Henrie king of Fraunce and Poleland, the third of that name.—At Paris in the month of Iuly,—1596. —Loys Le Roy.—Interpres ad Lectorem." In ten Latin hexameters. 393 pages, and a table. Again 1598. Licenſed. W.H. Folio.

"HERO AND LEANDER. By Chriſtopher Marloe." A flower de luce. 1598. "—Printed by him, for Edward Blunt, 1598." Dedicated "To—Sir Thomas Walſingham, knight.—Edward Blunt." At the end, "Deſunt nonnulla." It has no diviſions or arguments, as afterwards when printed in octavo "for Iohn Flaſket, 1600." Alexander Dalrymple, Eſq; 4°.

"The betraying of Chriſt. Iudas in deſpaire. The ſeuen words of 1598. our Sauiour on the croſſe. With other poems on the paſſion." Dedicated, by Samuel Rowland, to Nicholas Walſh, knight. Quarto.

"THE MIRROVR OF POLICIE. A Worke no leſſe profitable than neceſ- 1598. ſarie, for all Magiſtrates, and Gouernours of Eſtates and Commonweales." Device, a pillar, &c. "—Printed by him, 1598.—The Printer to the Reader." L l, in fours, with ſeveral genealogical ſchemes, or projections. Licenſed. Again, 1599. W.H. Quarto.

"THE Workes of—Geffrey Chaucer," &c. See p. 1152. This edi- 1598. tion has a copper-plate portrait of Chaucer, at length; with his pedigree and arms, as deſcribed by Thomas Occleve, his ſcholar. Licenſed.

The neceſſary, fit and convenient education of a young gentlewoman. 1598. Italian, French and Engliſh. Octavo.

"The Key to vnknowne Knowledge," &c. as p. 1200. 1599.

"The fountaine of ancient fiction. Wherein is liuely depictured the 1599. Images and Statues of the gods of the Ancients, with their proper and perticular expoſitions. Done out of Italian into Engliſh, by Richard Linche, Gent. Tempo è figliuola di verita." The palm tree, &c. "— Printed

Printed by *him*, 1599." Dedicated " To—M. Peter Davison Esquire.— Richard Linche.---To the Reader." C c, in fours. Licensed. W.H. 4°.
1600. Philemon Holland's English Livy: or, Roman history. Dedicated to the queen. Licensed. Folio.
1600. " A toile for two-legged foxes.—By I. B. Preacher of the word of God. Canticles 2; 15.—2 Chron. 15; 8.—.A Maxima. As Poperie and treacherie goe hand in hand, whileſt Poperie is kept vnder; ſo Poperie and crueltie are companions vnſeparable, if once Poperie get the vpper hand.—Imprinted by *him* for Thomas Man, 1600." Inſcribed " To all firme & faithfull louers of true religion & loyaltie.—I. B.---The ſumme of the chapters." 220 pages. W.H. Sixteens.

VALENTINE SIMS, or SIMMES, ſtationer,

SON of Richard Simmes of Adderbury, Oxon, was bound apprentice 11 Feb. 1576-7, to Hen. Sutton for 8 years, and made free, 8 March 1584, by Joan, the late wife of Hen. Sutton deceaſed: he was afterwards ſervant to Hen. Bynneman, as appears by a minute in the Stationers' regiſter[b]; and dwelt at the ſign of the White Swan, in Addle or Adling-ſtreet, or hill, near Baynard's caſtle.

1594. " The common-wealth of England," &c. See p. 1228. "—Printed by Valentine Simmes, for Gregorie Seton, and are to be ſolde at his ſhoppe vnder Alderſgate, 1594." An epiſtle " To the Reader." 148 pages. W.H. Quarto.
1595. " A Comfortable Treatiſe for—ſuch as are afflicted" &c. as p. 1279.
1595. " Mæoniæ. Or, certaine excellent Poems and ſpirituall Hymnes: Omitted in the laſt Impreſſion of Peters Complaint; being needefull thereunto to be annexed, as being both Diuine & Wittie. All compoſed by R. S." (*Robert Southwell.*) " Printed by *him*, for John Busbie 1595." Inſcribed by " The Printer to the Gentlemen Readers.—I. B." 32 pages, W.H. Again 1596. Quarto.
1595. " Truth and Falſhood: Or, A Compariſon betweene the Truth now taught in England, and the Doctrine of the Romiſh Church: with a briefe confutation of that Popiſh doctrine.—By Francis Bunny, ſometime fellow of Magdalen College in Oxford. Gal. 1; 9.—Printed by *him*, for
Rafe

---

[b] 1 March 1595-6, " At the motion of Mr. Sled, who married the wydowe of H. Bynneman deceaſed, yt is ordered that Valentine Syms who was ſervant to the ſaid H. Bynneman ſhall haue entred for his copies all ſuche copies as did apperteine to the ſaid H. Bynneman which are not otherwiſe entred & diſpoſed. And that the ſaid Val. Syms ſhall haue the printinge of all ſuch of them as are already diſpoſed for the behoof of ſuche perſons as they belonge vnto. And that from tyme to tyme as any of them by death, or otherwiſe ſhall fall to be newly diſpoſed, the ſaid Val. ſhall haue the ſame entred for his copies, provided that for ſo many of them as are already granted to any meere printer exerciſinge printinge, the ſaid Val. ſhall not clayme the printinge thereof from any ſuche printer."

Rafe Iacſon——." Dedicated " To the right worſhipfull Companie of Iron-mongers in London,—From my houſe at Ryton in—Durham—1595.—— A neceſſarie Table, of all the principall matters." &c. 160 leaves. Annexed is

" A Short Anſwer to the Reaſons, which commonly the Popiſh Recu- 1595. ſants in theſe North parts alleadge, why they will not come to our Churches. By Fr. Bunny.—Eccleſ. 4; 13.—Printed by V. S. for Rafe Iacſon, dwel-ling in Paules Church-yard, at the—white Swanne 1595." Dedicated " To—the Lord Ogle: To—ſir Iohn Foſter Knight, Lord Warden of the middle marches fore-aneinſt Scotland: and to all the reſt—of Nor-thumberland,—that haue a deſire to know & obey the trueth &c. - From Ryton vpon Tine, this ſeuenth of February. 1592." Not paged, but con-tinued from the former article by Signatures to E e, in eights. W.H. 4°.

" The Gentlemans Academie. Or, The Booke of S. Albans: Con- 1595. taining three moſt exact & excellent Bookes: the firſt of Hawking, the ſecond of all the proper termes of Hunting, and the laſt of Armorie: all compiled by Iuliana Barnes, in the yere—1486. And now reduced into a better method, by G. M. (Gervaiſe Markham). London Printed for Humfrey Lownes, and are to be ſold at his ſhop in Paules church-yard, 1595." Inſcribed " To the Gentlemen of England: and all the good fellowſhip of Huntſmen & Falconers.—G. M." The treatiſes of Hunt-ing and Armory haue ſeparate title-pages. The whole, D d in fours. W.H. Quarto.

" A BRIEFE TREATISE containing many proper Tables," &c. See 1595. p. 1101. " The Contentes whereof, the leafe that next followeth doeth expreſſe.—printed by V. S. for Tho. Adams dwelling in Pauls Church-yard at the ſigne of the white Lion, 1595." 127 pages. W.H. 16°.

" A Survey of the Popes ſupremacie: VVherein is a triall of his title, 1595. and a proofe of his practiſes: and in it are examined the chiefe arguments that M. Bellarmine hath, for defence of the ſaid ſupremacie, in his bookes of the biſhop of Rome. By Francis Bunny.—Hoſ. 8, 4,—At London, Printed by him for Ralfe Iacſon—1595." Dedicated " To—Henry Earle of Huntingdon, &c.—To the Reader." E e, in fours. W.H. Quarto.

" The Gouernment of Health: A Treatiſe written by William Bul- 1595. leia.—Both pleaſant and profitable to the induſtrious Reader.—Printed by him, dwelling in Adling ſtreet, at the—white Swan, neare Bainards caſtel, 1595." Dedicated " To—ſir Thomas Hilton knight, Baron of Hilton, &c.—Wm. Bullein." 87 leaves and " The Epilogue." Prefixed are ſome verſes and a table of Contents. Licenſed. W.H. Sixteens.

" THE Triumphs ouer Death: or A Conſolatorie Epiſtle, for afflicted 1595. minds, in the affects of dying friends. Firſt written—by R. S. the Authour of S. Peters Complaint, and Mœoniæ his other Hymnes.—Printed by him for Iohn Busbie and are to be ſolde at Nicholas Lings ſhop at the Weſt end of Paules Church, 1595." A dedication in verſe " To—M. Richard Sackuile, Edward Sackuile, Cicilie Sackuile and Anne Sackuile, the hope-full iſſues of—Robert Sackuile, Eſquire.—Iohn Truſſell." An acroſtic on " Robert Southewell.—-To the Reader," and a letter from the authors E, in fours. Again 1596. W.H. Quarto.

The

1596. The triall of true Friendſhip. By M. B. Quarto.
1596. "FRVITEFVLL SERMONS preached by—Hugh Latimer, to the edifying of all which will diſpoſe themſelues to the reading of the ſame. Seene and allowed," &c. See p. 671. Device, a ſwan on a wreath. "—Reprinted by Valentine Sims,—1596." 331 leaves, and a table of the ſermons. Licenſed. W.H. Quarto.
1596. "The caſtle of knowledge. To Knowledge is this Caſtle ſet," &c. See p. 603. "—Printed by him, aſſigned by Bonham Norton, 1596." 232 pages. W.H. Quarto.
1596. "God wooing his Church: Set fporth in three godly Sermons. By William Burton, preacher at Reading." The ſwan, &c.—"Printed by V.S. for Iohn Hardie, dwelling in Paules Church-yard at—the Tygers head, 1596." 148 pages. W.H. Sixteens.
1596. "A hundred and fourteene Experiments and Cures of the famous Phyſitian Philippus Aureolus Theophraſtus Paracelſus; Tranſlated out of the Germane tongue into the Latin. Whereunto is added certaine— workes by B. G. a Portu Aquitano. Alſo certaine Secrets of Iſacke Hollandus concerning the Vegetall and Animall worke. Alſo the Spa- gericke Antidotarie for Gunne-ſhot of Joſephus Quirfiranus. Collected by John Heſter." The ſwan, &c. "—Printed by him dwelling on Adling hill at the—white Swanne, 1596." Licenſed. Quarto.
1596. Thomas Bell's ſurvey of Popery, &c. Z 2 ſheets. Quarto.
1596. The key of Philoſophy. 1 and 2 parts. Bagford's papers.
1597. "Laura. The toyes of a traueller; or, the feaſt of fancie, diuided in- to 3 parts, by R. T. gent. of London." Dedicated "To the ladie Lucie, ſiſter to Henry, earle of Northumberland." Quarto.
1597. Richard 11. By Wm. Shakeſpeare. Againe 1598. For Andr. Wiſe, and licenſed to him. Quarto.
1597. Richard 111. By W. S. For Andrew Wiſe, & licenſed to him. 4°
1598. "T. Tyros roaring Megge planted againſt the Walles of Melancholy." Bibl. Bodl. Quarto.
1599. "The ſilk wormes, and their flies; liuely deſcribed in verſe, by T.M. a countrie farmer, and apprentice in phyſicke." Quarto.
1599. "Certaine Errors in Navigation, Ariſing either of the ordinarie erroneous making or vſing of the ſea Chart, Compaſſe, Croſſe ſtaffe, and Tables of declination of the Sunne, and fixed Starres detected and corrected. By E. W. Printed—by him, 1599." Dedicated "To— George Earle of Cumberland.——Edw. Wright.---The Præface to the Reader.---The ſumme of this treatiſe." After ſignature O, is A a, and ſo on regularly to Q q. Then, (the catch-word continued,) begins, "The Voyage of—George Earle of Cumberl. to the Azores, &c." A to D. in fours. On the laſt leaf, "Faultes eſcaped in the E. of Cumb. voiage." W.H. Quarto.
1600. "Cato Chriſtianus. In quem coniiciuntur ea omnia, quae in ſacris literis ad parentum, puerorumque pietatem videntur maximè pertinere." For Matt. Law. Sixteens.

"The

VALENTINE SIMMES.

" The lawes and ſtatutes of the ſtannarie of Deuon." It begins, with 1600.
" The charter granted to the tinners of the countie of Deuon, by King
Edward the third, and ſince confirmed by diuers kings of this realme, as
followeth, Henry, by the grace of God, &c." At the end, " Actes made
in the 42 year of Queen Elizabeth, and a coat of arms with this motto,
AMORE ET VIRTVTE.   Folio.

Paſquil's Mad-cappe's meſſage.   Quarto.   1600.
" Much Ado about Nothing. By Wm. Shakeſpeare." For Andrew 1600.
Wiſe and Wm. Aſpley. Licenſed to them.   Quarto.   1600.
Henry VI. See p. 1267.   1600.
Henry IV. Second part. By Wm. Shakeſpeare. For Andrew Wiſe 1600.
and William Aſpley, and licenſed to them.   Quarto.

" A PEARLE of Price or, The beſt Purchaſe: For which the Spirituall 1600.
Marchant Ieweller ſelleth all his Temporalls. By Samuel Gardiner, B. D.
——Job. 28; 16." Buſhell's rebus: Juſtice ſtriking a buſhel of corn,
in a compartment with this motto, SVCH AS I MAKE, SVCH WILL I TAKE.
" Printed by V. S. for Tho. Buſhell, and are to be ſold at his ſhop at the
North Doore of Paules, 1600." On the back is the creſt of Sir Thomas
Egerton, to whom the book is dedicated.---An epiſtle to the reader, and a
table of contents. 207 pages. W.H.   Sixteens.

" THE Courtiers Academie: Comprehending ſeuen ſeuerall dayes
diſcourſes: wherein be diſcuſſed, ſeuen noble and important arguments,
worthy by all Gentlemen to be peruſed. 1 Of Beautie. 2 Of Humane
Loue. 3 Of Honour. 4 Of Combate and ſingle fight. 5 Of Nobilitie.
6 Of Riches. 7 Of precedence of Letters or Armes. Originally written
in Italian by Count Haniball Romei, a Gentleman of Ferrara, and tranſlated into Engliſh by I. K. L'occhio Linceo, ha l'intendimento cieco."
A roſe and crown. " Printed by Valentine Sims." Dedicated " To
—Sir Charles Blunt, Lord Mountioy, &c.---The Preface," 295 pages.
W.H.   Quarto.

He had alſo licenſes for the following books, viz. In 1595, " The hiſtory
of Palladine of England:" conditionally, " and to pay vj d in the li.
to thuſe of ẏ poore;" therefore printed for the company. " Lucius
Apuleus de Aſino aureo, in Engliſh:" conditionally, &c. as above.
" The diſcovery of Popery." In 1596, " Colloquia Eraſmi: For the
company, for this time, paying vj d. in the li. according to thorder."
In 1598, " vij ſeu'all diſcourſes of Counte Haniball Rinei gent. of Ferrara. A table of good Counſell. A table concernynge the rates of ẏ
intereſt of money after x per Cent. Seuen ſermons, or the exerciſe of 7
Sabbaths, together with a ſhort treatiſe vpon ẏ cōmandements by Lewes
Thomas: vpon Condition that none of theſe—be printed alredy." In
1600, " A diſcourſe of ẏ vnnatural & vile cōſpiraſie attempted againſt
ẏ kinges maieſties perſon at St. Johns Towne vppon Tweſday the 1 Aug.
1600. The deſtruction of Troye: one impreſſion only; for the company, and to pay vj d. in the li. The paſſions of the mynde." He
printed after 1600.

RICHARD

RICHARD HUDSON, stationer.

WAS apprentice to Mr. Tho. Berthelet, before the company had their charter, but was prefented to them (together with Hen. Wykes) 15 Oct. 1556; and was made free, 4 Oct. 1557.

"Certaine Englifh verfes penned by Dauid Gwyn, who for the fpace of eleuen yeeres and two mounths, was in moft greuious feruitude in the gallies vnder the King of Spaine, and nowe lately by the wonderfull prouidence of God, deliuered from captiuitie, to the ouerthrow of many of the Spaniards, and the great reioycing of all true hearted Englifhmen. Prefented to the Queen's moft excellent Maieftie in the parke at Saint James, by Dauid Gwyn, as followeth. Imprinted at London by Richard Hudfon, dwelling in Hofier Lane, at the Signe of the Wool-Packe."

HENRY BALLARD, stationer.

WAS made free, 15 Aug. 1586, by Mr. Tottell, and dwelt at the fign of the bear, without Temple-bar, over againft St. Clement's church.

1597. "The Accedence of armorie." By Gerard Leigh. See p. 813.
1597. "Workes of Armorie." &c. See p. 820. "At London, Printed by Henrie Ballard, dwelling without Temple-barre, ouer againft Saint Clements Church at the figne of the Beare, An. Do. 1597." Dedicated "To—Sir William Cecill Baron of Burghleigh, &c.—Cyllenius cenfure of the Author, in his high Court of Herhaultrie.—Nicholas Rofcarrocke. (In verfe) The names of the Authors" &c. 136 pages; the third books begins on a new fet of pages; but my copy being imperfect at the end, cannot tell how far they continue. W. H. Quarto.

FELIX KINGSTON,

WAS tranflated from the Grocers' Company, 25 June, 1597, for which he paid a fine of xviijd. He frequently ufed T. Orwin's device of the hand in hand, &c. but without the T. O. See p. 1242.

1597. "A true Chronologie of the Times of the Perfian Manarchie, & after to the deftruction of Ierufalem by the Romanes. Wherein—is handled the day of Chrift his birth: with a declaration of the Angel Gabriels meffage to Daniel—againft the friuolous conceits of Matthew Beroald

Beroald. Written by Edw. Liuelie, Reader of the holie tongue in Cambridge.—Printed by Felix Kingſton for Tho. Man, John Porter, and Rafe Iacſon, 1597." Dedicated " To—the Archbiſhop of Canterburie his Grace.—24 Nouemb. in the 1597 yere of Chriſt our Lord—Edw. Liuelie." 258 pages, and a chronological table, the whole C c, in eights. W.H. Octavo.

" Analyſis Logica libri S. Lucae, qui inſcribitur, Acta apoſtolorum, 1597. vna cum ſcholiis et obſeruationibus locorum doctrinae. Authore M. Joannæ Piſcatore." Octavo.

" The Practiſe of Fortification," &c. See p. 1262. 1597.

" An Expoſition of the Lords Prayer:—Hereunto are adioyned the 1597. Prayers of Paul, taken out of his Epiſtles. By Wm. Perkins;" as p. 1252. Walter Cary's farewell to phiſic. See p. 957. 1598.

" Ad lectorem. Gemma Fabri: Qua ſacri Biblij margaritæ,—fere omnes 1598. continentur, methodo analytica, & verſu Alphabetico: &c. Bonus Textualis bonus Theologus. Adiecimus etiam in nouum Teſtamentum Schepreui Oxonienſis carmen à Doctore Humphredo olim editum. Et Chronologicam temporum obſeruationem.—Londini, Ex Typographia Felicis Kingſtoni, impenſis Iohannis Porteri, 1598." Dedicated " Ad ſereniſſ. & chriſtianiſſ. Elizabetham, Angliæ,—Reginam, &c.---Vere nobili." I, in eights. W.H. Octavo.

" The workes of Iohn Heiwood, newlie imprinted. Namelie, A dia- 1598. logue, wherein are pleaſantlie contriued the number of all the effectuall Prouerbs in our Engliſh tongue: Compact in a matter concerning two maner of Mariages. Together with 300 Epigrammes vpon 300 Prouerbes. Alſo, a fourth, fifth, and ſixth hundreth of other very pleaſant, pithie, and ingenious Epigrammes." Device, the hand in hand, &c. " At London, Imprinted by *him*, 1598." The 4, 5, and 6 hundred epigrams have ſeparate titles without the device. At the end is " An epilogue or concluſion of this worke:—1598. Thomas Newtonus Ceſtreſhyrius." C c, in fours. W.H. Quarto.

" Alba, the mounths minde of a melancholy louer. By R. T. at 1598. London." Octavo.

" A watch word to all religious, and true hearted Engliſhmen. By ſir 1598. Francis Haſtings knight." Octavo.

" The moral Philoſophie of the Stoicks. Written in French, and 1598. engliſhed for the benefit of them which are ignorant of that tongue, by T. I. (*Thomas James*) Fellow of New Colledge in Oxford. Non quæro quod mihi vtile eſt, ſed quod multis.—Printed by *him* for Tho. Man, 1598." Dedicated "To—Sir Charles Blunt, Lord Mountioy, Knight of the Garter &c. Th. Iames.---To the French Reader." 205 pages. W.H. 16°.

" St. Peters Pathe to the Joys of heaven, wherein is deſcribed the frailtie 1598. of the fleſhe & the power of the Spirit, the labouring of this life, Satans ſubtilltye & the Soules ſaluation; and alſo the elexion, lyfe & martyrdome of the xij Apoſtles." Licenſed. Quarto.

" Tetraſtylon papiſticum," &c. See p. 1235. 1598.

" Of the Redemption of Mankind three buokes: Wherein the contro- 1598. uerſie of the vniuerſalitie of Redemption & grace by Chriſt, and of his death for all men, is largely handled. Hereunto is annexed a treatiſe of Gods

Gods Predeſtination in one booke. Written in Latin by Iacob Kimedoncius D. & profeſſor of Diuinitie at Heidelberge, and tranſlated—by Hugh Ince, Preacher of the word of God." Hand in hand, &c. "—Imprinted by *him*, 1598." Dedicated "To—Sir Thomas Egerton Knight, &c. Greenſteed in Eſſex, Octob. 31, 1598—Hugh Ince.---To—Frederike the fourth, Count Palatine on the Rhene, &c.—At Heidelberge, 12 March, 1592. Iacobus Kimedoncius D.---A table of certaine places of Scripture expounded in theſe Bookes." 406 pages, and a table of the principal matters. W.H. Quarto.

1598. "A godlie Forme of Houſeholde Gouernment: for the ordering of priuate Families,—Whereunto is adioyned—The ſeuerall duties of the Huſband towards his Wife: & the Wiues duty towards her Huſband, &c. Gathered by R. C.—Printed by *him* for T. Man, 1598." Dedicated "To—Robert Burgaine, of Roxall, &c.—R. C." 384 pages. Again 1600. W.H. Octavo.

1599. "A Commentarie vpon the Lamentations of Ieremy:" &c. p. 1250. " Ierem. 13; 22.—" Hand in hand. &c. "—Imprinted by *him*, for Tho. Man, 1599." Inſcribed "To the Chriſtian Reader, &c." 195 pages. W.H. Quarto.

1599. "Of the Marks of the Children of God," &c. See p. 1253. "—Imprinted by *him*, for T. Man,—1599." W.H.

1599. "The Sermons of Maſter Henry Smith, Gathered into one volume," &c. as p. 1247. "—Imprinted by *him*, for T. Man, 1599." On the back, "Nobiliſsimo viro, Gul. Cecilio.—Cantabrigienſis Academiæ Cancellario: H. Smithys hæc pignora in grati animi teſtimonium conſecravit." 595 pages. W.H. Quarto.

1599. "The Works of—Richard Greenham, Miniſter & Preacher of the word of God: Examined, corrected, and publiſhed, for the further building of all ſuch as loue the trueth, & deſire to know the power f. godlines: By H. H. Eccles. 12; 11.—Imprinted by *him*, for Ralph Iacſon, & are to be ſold—in Paules Churchyard at—the Swan, 1599." Dedicated "To—the Ladie Margaret Counteſſe of Cumberland, and the Ladie Katherine Counteſſe Dowager of Huntington:—Hen. Holland. ---The Preface to the Reader;" and commendatory verſes. There are different titles to the ſeveral diviſions of the work, but all printed by him, excepting the laſt, which is "Printed—by Richard Bradocke for Robert Dexter, 1599." The whole 476 pages and a table. W.H. Again the ſame year. See p. 1268. Quarto.

1599. "Times iourney to ſeeke his daughter Truth, and Truths letter to Fame of Englands excellencie." In verſe. Quarto.

1599. "The Art or Skil well and fruitfullie to heare the holy Sermons of the Church. Written firſt in Latin, by—Gul. Zepperus, & now newly Tranſlated into Engliſh by T. W. &c. Imprinted—1599." Licenſed.

1599. "A Diſcourſe of the preſeruation of the Sight: of Melancholike diſeaſes; of Rheumes, & of Old age. Compoſed by M. Andreas Laurentius, ordinarie Phiſition to the King, & publike profeſſor of Phiſicke in—Mompelier. Tranſlated out of French—according to the laſt Edition, by Richard Surphlet, Practitioner in Phiſicke." Hand and hand, &c. "—Imprinted by *him*, for Ralph Iacſon,—1599." Dedicated "To—Sir

'Sir Tho. Weſt Knight, Lord la Ware, & — Ladie Anne his Wife. Ric. Surphlet.---To the Reader.---To the noble Ladie Madame, Dutcheſſe of Vzez, and Counteſſe of Tonnera —Andr. Laurentius.--The Author to the Reader." Several commendatory verſes----The Contents. D d 1, in fours. W.H. Quarto.

An anſwer to Wm. Alabaſter's motives, by Rog. Fenton. Quarto. 1599.

" The Hiſtorie of the troubles of Hungarie : Containing the pitifull 1600. loſſe & ruine of that Kingdome, & the warres happened there, in that time, betweene the Chriſtians and Turkes. By Mart. Funce Lord of Genillé, Knight of the Kings Order. Newly tranſlated out of French— by R. C. Gent. Tout pour l'Egliſe." Hand in hand, &c. "——Imprinted by *him*, 1600." Dedicated " To—Sir Robert Cecil knight.— R. C.---To the Reader.---The Authors Epiſtle to the people of France." At the end is a copious table. K k, in ſixes. Licenſed. W.H. Folio.

" An Apologie or Defence of the Watch-vvord, againſt the—Ward- 1600. vvord publiſhed by an Engliſh-Spaniard, lurking vnder the title of N. D. Deuided into eight—Reſiſtances—by Sir Francis Haſtings Knight. Pſ. 122; (6)—Pſ. 109; (28)—" Hand in hand, &c. "—Imprinted by *him*, for Ralph Iacſon, 1600." Inſcribed " To the Chriſtian Reader." E e, in fours. W.H. Quarto.

He had licenſes alſo for the following. viz. In 1597, for the uſe of the poor, " The firſte parte of the Reſolutions, or Chriſtian exerciſe : and to allow vjd. in the li." In 1598, with T. Man, and R. Field, viz. T. Man one half; and R. Field, & he the other; " The Revelation of St. John :" See p. 1257. With R. Field; aſſigned to them from Toby Cooke: " Mr. Giffordes xij ſermons vppon Eccleſiaſtes. His Catechiſme. Twoo ſermons vppon part of the Epiſtle of St. James. iiij ſermons vppon part of the Epiſtle of St. Peter. His parable of the Sower. His ſermons at Pauls croſs. His Dialogue betweene the Papiſts & Proteſtants. His countrie Divinitye. His booke of worthies. His booke againſt the Donatiſtes'." In 1599, " Parkins vppon the Lordes prayer, in welſhe." In 1600, " A Dictionarie of the Bible." He printed after 1600.

JOHN de BEAUCHESNE.

WE have had information only of " A book, containing the true 1597. portraiture of the countenances and attires of the kings of England, from William the Conqueror vnto our ſoueraigne lady queene Elizabeth, now raigning : Together with a briefe reporte of ſome of the principall acts of the ſame kings, eſpecially ſuch as haue beene leaſt mentioned in our late writers. Diligently collected by T. T. [*Thomas Timmes.*] London : Printed by John de Beauchaſne, dwelling in Black Fryers." 4°.

JOHN

---

[1] He ſeems rather to have been an ingenious ſchoolmaſter from a book printed in 1602, by Richard Field, his neighbor, with this title, " A booke containing Diuers Sortes of hands." &c. See p. 1066. It opens long ways, & is very curious.

JOHN NORTON, Efq; ftationer,

SON of Richard Norton of Billingefley, Salop, Yeoman, deceafed, was bound apprentice to Mr. Wm. Norton for eight years, 8 Jan. 1577-8; made free 18 July, 1586; came on the livery 1 July, 1598. and in 1602, paid a fine of x li. to be excufed ferving the Renterfhip. Mr. Ames ftyles him queen's printer, in Latin, Greek, and Hebrew, but Fran. Flower had a patent for the fame, dated 15 Dec. 1573, for his life. I do not find when Flower died; but the company feem to have taken the privilege of printing the Accidence, and the Grammar, "whiche, by publique aucthoritie are taught in the Scholes" into their own hands, granting licenfe for printing them to Mr. Stirrup & Mr. Dawfon, the wardens for the time being, 18 January 1596-7." I have not found any book printed by John Norton, as a Royal printer, before 1604. So that it appears doubtful, whether Mr. Norton fucceeded to that office in the queen's life. He died in 1612, aged 55 years; and was a benefactor[k] to his company.

(1593.) "A difcouerie of the—confpiracie of Scottifh Papifts" &c. as p. 1255.
1594. "Analyfis logica Evangelii fecundum Matthæum—Mr. Johan Pifcatore."[1] &c. as p. 1256.
1595. "———fecundum Marcum;" &c. as ib. Licenfed.
1596. "The Hiftorie of P. de Comines" &c. p. 1213. Licenfed.
1597. "The Herball," &c. by J. Gerarde. See p. 1216. Licenfed. Again 1598.
1598. "Balfami, Opobalfami, Carpobalfami, & Xylobalfami," &c. as p. 1213.
1598. "The Annales of Cornelius Tacitus," &c. as p. 1213.
1598. "The Ende of Nero and Beginning of Galba," &c. as p. 1217.
1598. "The difcouery of the vnnattural—Scotifh papifts.| See p. 1278.
1600. "Ecloga Oxonio-Cantabrigienfis," &c. as p. 1153. Licenfed.

He had licenfe alfo for the following, viz. In 1589, "Tractatus pius & moderatus de vera excomunicatione chriftiano prefbiterio, Theod. Beza aucthore:" by Wm. Norton's confent. In 1594, "De juftificatione hominis coram Deo, Lib. 2. oppofiti fophifmatis Rob. Bellarmini Jefuite. Per. Joh. Pifcatorem." In 1597, Centum fabulæ Faernij." In 1598, "The Riddles of Heraclites & Democritus."

GEORGE

---

[k] By his will, dated 21 May, 1612 (and proved 10 Jan. following) he bequeathed to them 1000l. to purchafe lands to the value of 50l. per an. and part to be lent to poor young men of the faid company; alfo, he gave 150l. to the parifh of St. Faith, under St. Paul's, to purchafe 7 l. 10f. yearly for ever, to be given to the poor.

[l] 17 May, 1596. "Whereas Mr. John Norton hath before this printed Pifcator vpon Mathew: Jtem vpon Marke: And nowe is in hand with the fame Author vpon Luke. At a court—it is ordered for remedy of fuche preventions as he hath ufed therein, yt it fhall not be Lawfull for him or any other hereafter to prevent others from any copie cominge from beyond the Seas; but that the fame fhall be in comon to fuche of the Company as will haue partes therein. Jtem that he fhall enioye the Jmpreffion of the faid Author vpon Luke as he hath enioied the firft Jmpreffions of the faid Author vpon Mathew & Marke, payinge for this vpon Luke vjd. in the L. to thufe of the poore of the Company. Jtem that hereafter he fhall not Reprint any of the fame copies without makinge the Mafter & Wardens of the Company privy therevnto & acceptinge fuche of the companies as will to haue partes therein for paper & printinge."

GEORGE SHAW, ſtationer,

SON of Thomas Shaw, citizen and cordwainer of London, was firſt bound apprentice, 25 Nov. 1577, to Roger Ward for 8 years, from St. Andrews next following, and 1 Dec. 1579, was turned over to Hen. Denham, and 3 Feb. 1583-4 was put over to Henry Middleton, by whom he was made free, 31 Jan. 1585-6.

"Encomium illuſtriſs. Herois, D. Rob. Comitis Eſſexii, et alia poe- 1598. mata, autore Guil. Vaughanno Maridunenſi." Octavo.

He had licenſe alſo for the following, viz. In 1595, with Wm. Blackwall, "A true reporte of 3 Murders lately cōmitted. The firſt in Eſſex, the 2 in Oxfordſhire, and the 3 in Middleſex."

THOMAS JUDSON, ſtationer,

WAS made free 16 Jan. 1580-1, by patrimony; and concerned at firſt in partnerſhip with John Windet. See p. 1223. He quitted buſineſs about the year 1600, as appears by a minute* in the Stationers' regiſter.

"Peters Fall," &c. as p. 1224.. 1584.
Ovid's Metamorphoſes. See p. 1224. 1584.
Geo. Gifford's four ſermons, on the 7 effects, of faith, &c. See p 1224. 1584.
——four ſermons on ſeveral parts of ſcripture, &c. See p. 1263. 1598.
"THE SCHOOLE OF SKIL: Containing two Bookes: The firſt, of the 1599. Sphere, of heauen, of the Starres, of their Orbes, and of the Earth, &c. The ſecond, of the Sphericall Elements, of the celeſtiall Circles, and of their vſes, &c. Orderly ſet forth—by Tho. Hill.---At London Printed by T. Judſon, for W. Jaggard, 1599." Prefixed is an epiſtle to the reader. 267 pages, and a table with many projections. W.H. Quarto.

"A BRIEFE DESCRIPTION OF THE whole worlde. Wherein is particu- 1599. larly deſcribed, all the Monarchies, Empires, and kingdomes of the ſame: with their ſeuerall titles and ſcituations thereunto adioyning.—Printed by him, for Iohn Browne, and are to be ſould at—the Bible in Fleeteſtreete, 1599." E 2, in eights. W.H. Quarto.

RICHARD

---

* 4 Feb. 1599—600; "This day came Tho. Judſon before a whole Court of Aſſiſtants & acknowledges he hath ſold all his preſſes, letters, & printinge ſtuffe to John Haryſon the yonger, ſon of Mr. John Haryſon thelder: and doth now before the Court, relinquiſh, Reſigne & gyve vp his place of printinge as a maſter printer, and agreeth never to erect a printinge preſſe agayne, nor vſe or kepe a printing houſe as a Maſter printer hereafter. Agreynge alſo, that the ſaid John Haryſon the yonger having thus bought his letter, preſſes, & printinge ſtuffe ſhalbe preferred therewith to vſe printinge, and kepe a printing houſe."

RICHARD BRADOCK, stationer,

WAS made free, 14 Octob. 1567, by Mr. Middleton; and came on the livery, 1 July, 1598, for which he paid ij li. About this time he appears to have had a controversy[a] with the company, concerning certain books, entitled The godly gardens; and Chriftian prayers. In September following, he was ordered[o] to bring into the hall, the 6 leaves of The penfive mans practife, which he had printed contrary to a promife made to Math. Lownes, to whom he had fold the laft impreffion; and that he fhould defift from printing any more till the 25 March next. And 9 Nov. Mr. Rd. Watkins entered into an engagement for his performing the fame. The 7 Octob. 1598, He, John Windet, John Standifh & Tho. Adams, were fined xijd each for not attending the Mafter, Wardens, &c. to the burial of Mr. Conway. He dwelt in Aldermanbury a little above the conduit.

(1570) Stows Chronicle. See p. 853, notes. Again 1598. W.H. Sixteens.
1581. " An excellent new Commedie, Intituled: The Conflict of Confcience. Contayninge, A moft lamentable example, of the dolefull defperation of a miferable wordlinge, termed, by the name of Philogolus, who forfooke the trueth of Gods Gofpel, for feare of the loffe of lyfe, and worldly goods. Compiled, by Nathaniell Woodes, Minifter, in Norwich, &c.—Printed, by Richarde Bradocke dwelling in Aldermanburie, a little aboue the Conduict. Anno 1581.----The Prologue." I, in fours. W.H4 Quarto.
1584. Thomas Baftard's Chreftoleros. Seven books of epigrams. Octavo.
1589. " The Enemie of Idleneffe :" &c. See p. 908. Again, "Imprinted at
1598. London by him—1598." Dedicated in verfe " To —, Mafter Anthonie Radcliffe, Mafter of the worfhipfull cqmpanie of the Merchant Tailors of London, and other the VVardens & Commonalty of the fame: &c.—W. Fulwood.7--To the reafonable Reader.---The booke to the lookers on," in verfe. 257 Pages and a table. W.H. Sixteens.
1598. " A Difcourfe vpon the catalogue of doctors of Gods Church, to witt, afwell of thofe that haue beene from the beginning of the world, mentioned in the holy fcriptures; as of manie which haue fithens by order fucceeded, Together with the continuall fucceffion of the true Church of God, vntill the yeare 1565. Written in French by Simon de Voyon, a worthie member, and Minifter of the fame Church, and novv faithfully tranflated

---

[a] Aug. 7, 1598. " Ric. Braddock hath promifed to vtter none of the godly gardens, nor xpian praiers till an ende be made betwene him and the company for the fame. And he hath a fort'night refpit gyven him to delin' his anfwere, Whether he will bring them into the hall accordinge to his former promife or not; within which tyme he hath promifed to gyve his anfwere. The number be thefe, viz. 500 of the godly garden, & 12 C. of the other. Richarde Bradocke." Stat. Reg.

[o] Sept. 19. " Ordered, &c. that Math. Lownes fhall have tyme onely till the 25 of m'che to fell all fuche numbers as nowe reft in his hands vnfold of an imp'ffion which he bought of Rd. Bradock of a praier book called The penfive mans practife, And that the faid Bradocke by Satterdaie next fhall bringe into the hall, there to remaine till the faid 25 m'che next thofe vj leaves that he hath now printed of the faid booke contrary to his former promife made to the faid Lownes, And fhall defift & forbeare vntill the faid xxvth of m'che next from printinge any other leaf or leaves of the faid book."

RICHARD BRADOCK. 1299.

translated into Englifh By Iohn Golburne.—Pfal. 68; 13. Imprinted at London by *him* for Iohn Browne, and are to be folde at his fhop in Fleet-ftreete, ourr againft the white Friars, at the figne of the fugarloafe, 1598." Dedicated to Sir Tho. Egerton. 207 pages. W. H. Sixteens,
"Greene in conceipt new raifed from his grave to write the tragicke Hif- 1598. tory of the fair Valeria of London. By Iohn Dickenfon." The rev. Dr. Lort. 4°.
"VIRGIDEMIARVM." Firft three books. By Jof. Hall. See p. 1268. 1598.
"Virgidemiarum, the three laft Bookes," &c. as p. 1268. 1598.
A fhort form of cathechifing, &c. By Ric. Greenham. See p. 1294. 1599.
A difplaying the wilful devifes of worldlings, by S. Harward. See p.1051. 1599.
Simon Harward's fermon on the horror of difobedience. See p. 1051. 1599.
"A briefe defcription of the whole worlde." &c. See p. 1297. "At 1600. London Printed by R. B. for Iohn Browne, and are to be fould at the figne of the Bible in Fleete-ftreete, 1600." H, in fours. W.H. Quarto.

He had alfo licenfes for the following, viz. In 1598, "The hunting of ý Romifh Fox. The deputation of thaffignes of Sir John Parkington, concernynge Starche." In 1599, "Natural & artificial directions for Health."

SIMON STAFFORD

DWELT on Adling-hill, near Carter-lane; and had for his device, Opportunity with her forelock, holding a knife in her left hand, and ftanding on a wheel floating on the fea. In the offing, a fhip failing, on one fide; and a wreck, on the other: the whole inclofed in a compartment, with this motto "AVT NVNC, AVT NVNQVAM." The fame was ufed by John Danter: See p. 1272. He and Wm. Barley appear to have encroached upon the privilege for Grammars & Accidences; whereupon in Jan. 1597-8, the parties interefted applied to the Stationers' company, who refolved to join them in a fuit againft the faid offenders, and to execute the decrees of the Star-Chamber: the refult of this does not appear. Stafford however had his preffes, &c. feized in March 1597-8, apparently for erecting a prefs, not being free of the Stationers' company: whereupon he appears to have appealed to the Lords of the Council, who in Sept. 1598, made an order concerning him; a copy of which may be feen below'. But this order does not appear to have quite fatisfied him,

' "At the Court at Greenwiche, 10 Sept. Prefent. Lord Keeper, Lord Admiral, Lord Chamberlain, Lord North, Lord Buckhurft, Mr. Comptroller, Mr. Secretary, Sir John ffortefcue.

It is ordered by their Lordfhips vpon the hearinge of the caufe betweene Symon Stafford & the Company of Stationers. That forafmuche, as the faid Stationers haue offered before their honors to receaue the faid Stafford to be of their Commonalty & Corporac'on yf he will chaunge hymfelf from the Company of Drapers whereof he nowe is, and to procure him to be admitted a maifter printer accordinge to the decree of the honorable Court of Star-chamber of the 23 of June 1586. Therefore the faid 8tafford fhall defifte from erecting any printe, and fhall not vfe the fame as a Maifter printer vntill he fhall be admitted & made a freeman & member of the faid Corporac'on of Stationers & be elected & admitted a prynter to printe according to the faid decree. And that the faid Stationers for their partes fhall readily & frely admitt him fo to printe, yieldinge himfelf to be of their Corporac'on, as aforefaid.

Sic fignator originalis Copia. Concordat cum originali. Ex. per Th. Smithe.

SIMON STAFFORD.

him, as he did not accept of the Company's freedom, but seems to have made further application to the Council for the redelivery of his printing materials. Accordingly, 5 March 1598-9, at a court of affistants, an order was made for that purpofe, as below.[q] On the 7 May following, to prevent further litigation, he was tranflated from the Drapers', and made free of the Stationers' company; and then unmolefted he printed the following books:

1599. "The raigne of king Edward the third. As it hath bene fundry times played about the citie of London." Printed for Cuthbert Burby. Quarto.

1599. "A pleafant conceyted comedie of George a Greene, the Pinner of VVakefield. As it was fundry times acted by the feruants of the right Honourable the Earle of Suffex." Device, Opportunity, &c. "Imprinted at London by Simon Stafford, for Cuthbert Burby: And are to be fold at his fhop neere the Royall Exchange, 1599." W.H.+ Quarto.

1599. "The hiftorie of Henrie the fourth; with the battell at Shrewfburie, betweene the king and lord Henry Percy, furnamed Hotfpur of the North. With the humourous conceits of fir John Falftaffe. Newly corrected by W. Shakefpeare." Printed by him for Andrew Wife. Quarto.

1600. "THE Golden-groue, moralized in three books: a worke very neceffary for all fuch, as would know how to gouerne themfelues, their houfes, or their countrey. Made by W. Vaughan, M. A. and ftudent in the Ciuill law." Device, Opportunity, &c. "Printed—by him, dwelling on Adling hill, 1600." Dedicated "To—Sir Iohn Vaughan of Goldengroue, Knight.—Frō Iefus Colledge in Oxford. Your louing brother, William Vanghan.---To the Reader."---Sundry commendatory verfes, in Latin & Englifh, by "Iohn Williams, D. D. Wm. Ofborn, Henry Pricius, B. D. Griffinus Powel, Iohn Budden, Nich. Lengford, M.A. Tho. Came, M. A. Gabriel Powel, Tho. Storer, M. A. Sam. Powel, M. A. Iohn Raulinfon, M. A. Charles Fitz-Geffrey, Tho. Michelborne."---A table of contents. At the end are other commendatory verfes, by "Mat. Gwin M. D. Iames Perrot, T. Fl. M. A. Tho. Iames, M. A. I. Pr." Contains C c, in eights. W.H. Sixteens.

1600. "THE Picture of a perfit Common wealth, defcribing afwell the offices of Princes and inferiour Magiftrates ouer their fubiects, as alfo the duties of fubiects towards their Gouernours. Gathered forth of many Authors, afwel humane, as diuine, by Thomas Floyd, M. A. Printed—by him, dwelling on Adling hill, 1600." Dedicated, in Latin, to Sir Thomas Egerton, "Oxonii, e collegio Iefu, Anno a partu virginis, 1600. ---To the Reader.---A Table." O, in eights. W.H. Sixteens.
"A Sermon

---

[q] "It was concluded that Symon Stafford, according to my L. Keepers order, fhall haue his printinge ftuffe redelivered vnto him, which was feifed in or about the beginning of Marche 1597, whiche Redeliuery is made in refpect of the faid order, & that he hathe accordinge to the fame entred into bond to her Maiefties vfe in the office of the Starre chamber, not to print contrary to the decrees of that Court.
Whiche printinge ftuffe was redeliuered vnto him accordingly the xij day of March 1598. Viz. All the letters & partes of a printing preffe, and all the reft of the goods & ftuffe whiche were feifed & brought from the faid Symon Staffords by vertue of the decrees of the Starre-chamber or otherwife.
Simon Stafford."

SIMON STAFFORD.

"A Sermon preached at great Yarmouth, vpon VVednesday, the 12 1600. of September, 1599. By W. Y. The argument whereof was chosen to minister instructions vnto the people, vpon occasion of those present troubles, which then were feared by the Spaniards. Ecclef. 9: 18.—Imprinted—by *him:* and are to be sold by Thomas Man. 1600." Dedicated "To—Master Iohn Felton the elder, and Master Thomas Manfield, Bayliffes of the Towne of great Yarmouth,—From the Priory in Yarmouth, Oct. 24. 1599.—William Yonger". The text, Jeremiah 4; 14. F 4, in eights. W.H. Sixteens.

Ovid's epistles, &c. See p. 943. "Printed at London by *him,* dwel- 1600. ling on Adling hill, neere Carter lane, 1600." X, in eights. Licensed. 16°.

"Vicissitudo rerum. An Elegicall Poeme of the interchangeable 1600. course & varietie of things in this world. By Jo. Norden." Device, Opportunity, &c. "Imprinted—by *him*—1600." Quarto.

"The Tears of the Beloued, or the Lamentation of St. John, con- 1600. taining the Death & Passion of Christ. By I. M." Device, Opportunity, &c. "Imprinted—by *him:* And are to be sold by John Browne at—the Bible in Fleete-streete, 1600." Quarto.

"The care of a Christian Conscience. Ten sermons on the 25 Psalm." 1600. Dedicated "To the Lady Dorothee Stafford, Widow, one of the Ladys of her Maiesties Bedchamber—London, 20 April, 1600.—Simon Stafford." Octavo.

He had also licenses for the following, viz. In 1599, "Bradfords 2 sermons, on Repentance & the Lords supper. Rams lyttle Doddeon. Preces in usum Scholæ Paulinæ." He printed after 1600.

JOHN BUSBY, stationer,

SON of Wm. Busby, citizen and cordwainer of London, was bound apprentice, 10 Dec. 1576, to Oliver Wilkes, for 9 years, and made free, 8 Nov. 1585. He appears to have been rather a bookseller than a printer, having his books printed for him.

"The affinitie of the faithfull:" &c. as p. 1113. 1591.

"T. Roberts Catechisme in meeter, for the easier learning, & better 1591. remembring of those principles of our faith which wee ought familiarly to be acquainted withal," Printed for him. Sixteens.

Absalom's fall, &c. by W. F. Printed for him. Octavo. 1591.

"Euphues shadow," &c. as p. 1163. Licensed to him & Nicholas 1592. Linge.

"Mæoniæ," &c. as p. 1288. Licensed. 1595.

The triumphs over death. &c. as p. 1289. Licensed. 1595.

"Sermons of the churches loue to Christ, on Cant. 3; 1—4. By Wm. 1595. Burton." Printed for him. Octavo.

"A Margarite of America. By T. Lodge." For him. See Warton's 1596. Engl. Poetry, 111; 481, note x. The rev. Dr. Lort. Quarto.

He

1600. Henry V. by Wm. Shakefpear. See p. 1283.
He had licenfes alfo for the following, viz. In 1590, with N. Ling;
"Euphues golden legacye, found after his deathe at his Cell at Selepidra.'
In 1592, with T. Gubbin; "The defence of Conye Catchinge, or a confuta⸗
c'on of thofe ij Jniurious pamphlets publifhed by R. G. againfte ÿ practi⸗
fioners of many nymble wytted & mifticall fciences." In 1593, with N.
Ling; "Pierce Gavifton Erle of Çornewell, his life, deathe, and fortune."
In 1594, with N. Ling; "Cornelia, Tho. Kydd beinge the Author." In
1595, "Endimion and Phebe. By affignment of Eliz. widow of John
Winnington Stationer, deceafed, 1. The returninge againe of hym ÿ is fallen
in Religion. 2. The fynners falve: 3. Tullies love. Provided alwaies ÿ yf
ÿ faid Eliz. marrie againe to any of the Companie, then fhe fhall have thefe
copies again as in her former ftate." In 1599, "The Aegles flight."
He printed after 1600.

BONHAM NORTON, ftationer,

ONLY fon of Wm. Norton, (See p. 877) was made free by patri⸗
mony, 4 Feb. 1593-4, & came on the livery, 1 July, 1594, and
ferved Renter in 1595. He was fined, 12 June, 1597, for binding Wm.
Barret his apprentice without firft prefenting him according to order,
xviij d. And, 29 Dec. 1592 for not ferving the Underwardenfhip xx li.
He with Tho. Wight had a patent for printing Common Law books, dated
10 March, 1598-9; which was notified to the company 2 Apr. 1599.

1596. The caftle of knowledge, &c. as p. 1290.
1596. "Joyfull Newes Out of the Newfound VVorlde," &c. as p. 1240.
1597. "The Herball, &c. by John Gerarde." See p. 1216.
1598. "The Ende of Nero and Beginning of Galba." &c. as p. 1217.
1598. "The Annales of Cornelius Tacitus," &c. as p. 1213.
1598. "The Dialogue in Englifh, betweene a Doctor of Diuinitie, and a
Student in the Lawes of England. Newly corrected and Imprinted, with
new Additions. At London, Printed by Thomas Wight, and Bonham
Norton, 1508. Cum Priuilegio Regiæ Maieftatis." 176 pages, and a
table. W.H. Sixteens.
1598. "A treatife and Difcourfe of the Lawes of the Forreft.—Alfo a Trea⸗
tife of the Puralleε,—Collected &c. by Iohn Manwood.—Printed by
Thomas Wight and Bonham Norton, 1598. Cum Priuilegio." Dedi⸗
cated "To - Charles, Lord Howard, Earle of Nottingham, &c.---
To the Reader.---The Contents" &c. 167 pages, and a table. W.H.
Again 1599. Quarto.
1598. "The firft part Of Symboleography." See p. 1132. "Lately peru⸗
fed and amended by William Weft of the Inner Temple Efquire, firft
Author thereof.—Printed by Thomas Wight and Bonham Norton, Anno
1598. Cum priuilegio Regiæ Maieftatis." Dedicated "—Edmundo
Anderfon Militi, Regiæ Maieftatis Iudici primario ciuilium—W. Weft.---
The Table." Pp, in eights. W.H. Quarto.
"LA

BONHAM NORTON.

"La Novel Natura Breuium" &c. See p. 808. "Londini, In Edibus Thomæ Wight, & Bonhami Norton,—1598. Cum gratia & Priuilegio Regiæ Maieſtatis." 271 leaves, and a table prefixed. W.H. 8°. 1598.

"A Learned Commendation of the politique Lawes of England:" &c. See p. 816. "Written in Latine by the learned & Right Honorable maſter Fortefcue Knight, &c. in the time of King Henrie the ſixt. And tranſlated in Engliſh by Robert Mulcaſter.—Printed by T. Wight, and him, 1598. Cum priuilegio." On the back of the title is an addreſs "Pio Lectori." Dedicated "To—John Walſhe, Eſquire, one of the—Court of common plees." 132 leaves and a table. Latin and Engliſh. Again 1599. W.H. Sixteens. 1598.

"L'Abridgment Des Cafes Concernants les Titles plus materiall pur les Eſtudiens & practiones des Leyes du Royalme digeſtes &c. Londini. In Ædibus Thomæ Wight, & Bonhami Norton—1599. Cum Gratia & Privilegio Regiæ Maieſtatis." It contains 370 leaves; and Garrantie on 35 more. At the end "Floreat Legum noſtrarum Hercules." It generally goes by the name of Gregories Moot Book. Quarto. 1599.

"Eirenarcha:" &c. See p. 823. With Thomas Wight. 1599.

"Les Tenures de Monſieur Littleton," &c. See p. 819. "Reuieu & Corrige en diuers lieux queux vous troueres ſignes oueſque ceux Signes. ** Londini. In ædibus Thomae Wight, & Bonhami Norton. Cum Priuilegio, 1599." Q, in twelves. Twenty-fours. 1599.

"Les Commentaries, ou Reportes de Edmund Plowden" &c. as p. 822. With Thomas Wight; Cum privilegio. W.H. Folio. 1599.

"The Inſtitution of Chriſtian Religion," &c. as p. 1213. 1599.

"Maiſon Ruſtique, or the Countrie Farme." &c. as p. 1217. Licenſed. 1600.

He had alſo licenſes for the following, viz. In 1594, "Aphoriſmi doctrinæ chriſtianæ maximam partem ex inſtitutione Calvine extracti." In 1596 "Piſcator in Evangelium Secundum Lucam." In 1597, "Analyſis logica libri ſcripturæ qui inſcribitur Acta Apoſtolorum. Cum Scholiis," &c. He printed after 1600.

THOMAS WIGHT,

PROBABLY was related to John Wight, many of whoſe copies he printed. See p. 779. He was concerned with Bonham Norton in the patent for law books. See p. 1302. In the year 1600, and afterwards he printed the patent books in his own name only.

"A booke of the arte and maner how to Plant" &c. as p. 1018. Again 1592. 1590.

Wm. Borne's Regiment for the Sea, with the Mariners guide. Printed for him. Again 1596, by Tho. Eaſt for him. See p. 1020. W.H. 4°. 1592.

"The Solace for the ſouldier and Saylour;" &c. as p. 1248. 1592.

Edmund Bunny's Chriſtian Exerciſe. See p. 1216. 1594.

"A briefe

THOMAS WIGHT.

1595. " ABriefe Defcription of Hierufalem" &c. p. 1208.
1595. " The Secrets of—Maifter Alexis" &c. as p. 1208.
1596. " Foure Bookes of Hufbandrie," &c. as p. 1090. Again 1600.
1596. Chaucer's works. See p. 1152.
1598. " The Dialoge in Englifh, betweene a Doctor of Diuinitie," &c, as p. 1302.
1598. " A treatife and Difcourfe of the lawes of the Forreft :" &c. as p. 1302.
1598. " The firft part Of Symboleography." &c. as p. 1302.
1598. " LA NOVEL Natura Breuium" &c. as p. 1303.
1598. Fortefcue in Commendation of the lawes of England. See p. 1303. Again 1599.
1599. " L'Abridgment Des Cafes Concernants les Titles" &c. as p. 1303.
1599. " EIRENARCHA." &c. See p. 823. With Bonham Norton.
1599. " Les Tenures de Monfieur Littleton," &c. as p. 1303.
1599. " Les Commentaires d'Edmund Plowden" &c. See p. 822. With Bonham Norton.
1600. " LITTLETONS TENURES in Englifh. Lately perufed and amended. Imprinted at London by Thomas Wight, 1600. Cum priuilegio Regiæ Maieftatis." 142 leaves, and a table. W.H. Sixteens.
1600. " A DIRECTION or Preparatiue to the ftudy of the Lawe: Wherein is fhewed, what things ought to be obferued & ufed of them that are addicted to the ftudy of the Law, and what on the contrary part ought to be efchued & auoyded. At London. Printed by *him*.—1600.---The Contents of this Booke.---Lectifsimis & generofifsimis iuuenibus in hofpitijs curialibus connutrijs, & Juris Anglicani ftudio operam nauantibus afsiduam.——: Ex hofpitio Graiano: pridie Nonas Septemb. An. falutis humanæ. 1599.—Guilielmus Fulbeckus." 95 leaves. W.H. Sixteens.
1600. " Les Reports De Edvvard Coke L'attorney generall le Roigne, de diuers Refolutions, & Iudgements donnes auec graund deliberation per les trefreuerendes Iudges, & fages de la ley, de cafes & matters en ley queux ne fueront vnques refolue, on aiuges par deuant, & les raifons, & caufes des dits refolutions & Iudgements, durant-les trefheureux regiment de trefilluftre & renomes Roigne Elizabeth, le fountaine de tout Juftice, & la vie de la ley.—Londini in ædibus Thomæ Wight. Cum Priuilegio Primo Ianuarij 1600." It is introduced with "Præfatio ad Lectorem.---The preface to the Reader." 177 leaves. At the end, " Cafuum iftius libri feries :" &c. W.H. Folio.

JOHN BOWEN, ftationer,

SON of William Bowen, capper of Hereford, was bound apprentice 20 Sept. 1578 to Tho. Dawfon for 7 years, and made free 31 Jan. 1585-6
1588. " Patr. Galloway his Catechifme." &c. as p. 1168. Licenfed.

He had licenfes alfo for the following, viz. In 1590, " The oration & declaration of Henrie iiij of ý name ý French King, by ý grace of God, & King of Navarre ý viij of Aug. 1590." With Iohn Morris, " A Sermon preached at Hitchin. 1587, 17 daie of Nouemb. by Edw. Harris, Maifter of artes."

.LPH

RALPH BLOWER, ftationer,

SON of Wm. Blower, hufbandman of Worthing, Salop, was bound apprentice to Richard Tottle for 7 years, from 1 Oct. 1587, but not prefented to the Company before 1 July 1588; and made fr e 3 Oct. 1594. He, and Wm. Jagger, were fined " for printing without Licence & contrary to order, A little booke of Sir Anthony Shelleies voiage. Alfo that he fhall prefently, according to thordinances in that behalf, forfayt & bringe into the hall all the faid Books fo printed. vjf. viijd. And their Jmprifonment for this offence is referred out to another time."

R. Bird's catechifm, entitled "A Communication dialogue wife, to bee. 1595. learned of the ignorant." Printed for him. Licenfed. Octavo.

Fifton's eftate and defcription of Germany. Printed for him in Fleet 1595. ftreet, near the Middle Temple. Licenfed. Quarto.

"A Rich Store-houfe, or Treafury for the Difeafed," &c. as p. 1001, 1595. Licenfed.

"Remedies Againft Difcontentmēt, drawen into feuerall Difcourfes, 1596, from the writinges of auncient Philofophers. By Anonymus. Rebus aduerfis conftans. London Printed for Rafe Blower An. Do. 1596." Dedicated "To—Edward Cooke Efquier, her Maiefties Attorney generall.---- Anonymus to his Friend.----A briefe Table of all the Difcourfes conteyned in this Booke." H 2, in eights. Licenfed. W.H. Sixteens.

"The Rare & moft wonderful thinges whiche Edward Webbe an Englifhman borne, hath feene & paffed in his troublefome trauailes, in the Citties of Jerufalem, Dammafko, Bethlem & Galely: and in the Landes of Jewrie, Egipt, Grecia, Ruffia, and in the Land of Prefter Iohn. Wherein is fet foorth his extreame flauerie fuftained many yeres togither, in the gallies and wars of the great Turk againft the Landes of Perfia, Tararia, Spaine & Portugall, with the manner of his releafement, & comming into England in May laft. London, Printed by him, for Thomas Pauier, & are to be folde at his fhop in Corn-hill, at the figne of the Cat & Parrats, ouer againft Popes head alley, nere the Royal Exchange." The epiftle to the Reader is dated at "Black-wall, this 19th of May 1590." C 2 in fours. . Quarto.

He had licenfes alfo in 1597 for, "A gentle crafte intreatinge of Shoomakers." In 1599 by the confent of Abell Jeffes, "A book of prettye romances, taken out of Latyne, Frenche and Dutche." He printed after 1600.

❀❀❀❀❀❀❀❀❀❀❀❀❀❀❀❀❀❀❀❀❀❀❀❀❀❀❀

THOMAS ADAMS, ftationer,

SON of Thomas Adams, yeoman of Nyenfavage, Salop, was firft bound apprentice to Oliver Wilkes, 4 Dec. 1582, and turned over 14 Oct. 1583 to George Bifhop for feven years; was made free, 15 Oct. 1592; and

1306   THOMAS ADAMS.

and came on the livery, 1 July 1598. He dwelt at the White Lion,. St. Paul's Church-yard; and had, by affignment, all the copies of Robert Walley. See p. 734.
1591.   " A brief Treatife contayning many proper tables," &c. as p. 1101. Again 1595. See p. 1289.
1595.   " Francifci Junii de peccato primo 'Adami," &c. See p. 882. Licenfed.
1597.   " A Sermon preached at Pauls Croffe, 4 Dec. 1597." &c. as p. 1223;.
1598.   " A Second Sermon,—21 May 1598," &c. as p. 1223.
  He had licenfes alfo for the following, viz. In 1592," with ¶. Oxenbridge, "Tabulæ analiticæ." &c. See p. 881. With I. Oxenbridge, " The Adventures of Brufanius prince of Hungaria." In 1593, with. I. Oxenbridge, " Greenes newes bothe from Heaven & Hell." In 1597, " The ffift booke of ỹ laſt troubles of ffrance cōteininge ỹ hiſtorie of ỹ moſt memorable thinges happened Synce ỹ death of Henry the third of fraunce in ỹ moneth of Auguſt 1589, vnto the fiege of La fere." In 1598, with Simon Waterſon, & Peter Short; " The woorkes of fflave Joſeph, fon of Matthias, viz, xx bookes of ỹ ancient hiſt. of Judea, vij bookes of ỹ Warres of the Jues, ij bookes againſt Appian of ỹ ancientnes of ỹ Jues : a booke touching ỹ Machabees : The life of Jofephus, written by himſelf. Treafur des Remedies Secretes pour les maladies des femmes." to be tranflated into Englifh. The birth of Mankind, otherwife, The Womans booke." by affignment from Mr. Watkins. In 1599, " Of the manner of Inthroniſinge Byſſhops in ỹ former times." He printed after 1600.

r This book was placed by Mr. Ames, to Richard Adams, in 1595. I was doubtful of the date, and therefore omitted it, but have fince been favoured with intelligence ‖ from the Rev. Mr. J. Price of the Bodl; Libr. that the book was printed by Tho. Adams.

※※※※※※※※※※※※※※※※※※※※※※※※※※※※※

TOBY SMITH, ſtationer,

APPEARS to have been bound apprentice to Luke Harriſon in the time of the chaſm in the Stationers' regiſter, and was turned over, to Richard Sergier 24 Nov. 1578, with the Conſent of Robert Farmer, who married the widow of the ſaid L. Harriſon, and made free by her, 26 March 1580. He dwelt in St. Pauls church-yard, at the ſign of the crane.
1581.   " A Checke, or reproofe of M. Howlets," &c. as p. 1120. Licenfed.
1581.   A conference by Dr. Fulke with the papiſts at Wiſbich caſtle, 4 Oct. 1580. Printed for him.
1581.   " AN EXPOSITION of the Symbole of the Apoſtles," &c. as p. 1060: Licenfed.
1582.   " An Epiſtle to the Faithfull," &c. as p. 1123.

" A

"A Tragicall Hiftorie of the troubles & Ciuile warres of the lowe (1583.) Countries, otherwife called Flanders." &c. as p. 841. Licenfed.

" AN ANSWEARE for the time," &c. as p. 1123. 1583.

" An Aunfwere to Howlets Epiftle to the Queenes Maieftie; alfo to 1584. the reafons aledged in a difcourfe thereunto annexed, why Catholickes (as they are called) refufe to goe to Church, by P. Wiborne." Printed for him. Quarto.

" An anfwere to a fupplicatorie Epiftle, of G. T." &c. See p. 1073.

" Rom. 13; 4.—Apoc. 20, 21—At London; Printed for Tobie Smith, dwelling in Paules Church-yard at the figne of the Crane." 400 pages, and the errata. Licenfed in 1582. WH. Sixteens.

" A caueat for Parfons Howlet," &c. as p. 1145.

He had alfo licenfes for the following, viz. In 1580, " Mounfhier Fountayns catechifm." In 1581, De la verité de la religion Chreftienne contre Les Athées, Payens, Epicuriens, Juifs, Mahumediftes, & autres infideles. Par Phillippes de Morney."

SOME book-binders in old time in most countries impressed upon the outsides, or covers of books, many curious pretty devices, coats of arms, and mottos; but now very much left off. However, i shall exhibit only a singular one used by John Reynes in King Henry the VIIIth's time, mentioned in page 413.

INDEX of the PRINTERS contained in this VOLUME.

A.
	Page
ADAMS, Richard	882
Adams, Thomas	1305
Aggas, Edward	1167
Allde, Edward	1238
Allde, John	889
Awdeley, or Sampſon, John	884

B.
Ballard, Henry	1292
Bamford, Henry	1138
Barker, Chriſtopher	1075
Barker, Robert	1090
Barley, William	1277
Beaucheſne, John	1295
Beddell, or Riddell, William	799
Berthelet, William	1202
Biſhop, George	1146
Blower, Ralph	1305
Bollifant, Edmund	1215
Bourne, Robert	1270
Bowen, John	1304
Boyle, Richard	1279
Braddock, Richard	1298
Broome, William	1202
———, his Widow	1203
Burrel, James	875
Buſby, John	1301
Bynneman, Henry	965

C.
Caly, Robert	828
Car, Roger	707
Carter, William	1204
Caſe, John	771
Cawood, John	785
Chard, Thomas	1194
Charlewood, John	1093
———, his Widow	1105
Charlton, Richard	771
Coldock, Francis	918
Colwell, Thomas	930
Cooke, Toby	1261
Creed, Thomas	1279
Crowley, Robert	757

D.
Danter, John	1270
Dawſon, Thomas	1115
Denham, Henry	942
Dewes, Gerard	940
Dexter, Robert	1267
Diſley, Henry	685

E.
Eaſt, Thomas	1006

F.
Field, Richard	1252

G.
Geminie, Thomas	872
Griffith, William	922
Gualtier, Thomas	765

H.
Hacket, Thomas	894
Hall, Rowland	800
Harriſon, John, Senior	1155
Harriſon, John, Junior	1158
Harriſon, Luke	924
Harriſon, Richard	883
Hatfield, Arnold	1211
Hill, Nicholas	708
Hill, William	755
Hoſkins, William	1113
Howe, William	1036
Hudſon, Richard	1292

I, and J.
Jackſon, John	1218
Jackſon, Hugh	1133
Jeffes, Abel	1160
Jones, or Johns, Richard	1039
Iſlip, Adam	1284
Judſon, Thomas	1297
Jugge, John	728
Jugge, Richard	713
———, his Widow	728

INDEX, &c.

K.
Kearney, William - - - - 1269
Kele, Richard - - - - 746
Kingſton, John - - - - 832
Kingſton, Felix - - - - 1292
Kytſon, Anthony - - - 873
Kynge, John - - - - - 762

L.
Lacy, Alexander - - - - 1005
Luſt, Thomas - - - - 1260
Lynne, Gualter - - - - 752

M.
Madeley, Roger - - - - 827
Marſhe, Henry - - - - 1205
Marſhe, Thomas - - - - 846
Mather, John - - - - 884
Maunſell, Andrew - - - 1134
Mierdman, Stephen - - - 770
Middleton, Henry - - - 1055
Moptid, David - - - - 884

N.
Newbery, Ralph - - - 900
Newton, Ninian - - - 1214
Norton, Bonham - - - 1302
Norton, John - - - - 1296
Norton, William - - - 877

O.
Orwin, Thomas - - - - 1241
——, his Widow - - - 1249

P.
Ponſonby, William - - - 1273
Porter, John - - - - 1270
Powell, Humphrey - - - 749
Powell, Thomas - - - 874
Powell, William - - - 735
Purfoot, Thomas - - - 993

R.
Redborne, Robert - - - 686
Riddell, or Beddell, William - 799
Roberts, James - - - - 1030
Robinſon, George - - - 1237
Robinſon, Robert - - - 1233
Rogers, Owen - - - - 876

S.
Saliſbury, Thomas - - - 1278
Sampſon, or Awdeley, John - 884
Saunderſon, William - - 1269
Searlet, Thomas - - - 1164

Scoloker, Anthony - - - 748
Seres, William - - - - 686
Serll, Richard - - - - 964
Shaw, George - - - - 1297
Shepperde, John - - - 1114
Short, Peter - - - - 1207
Simmes, Valentine - - - 1288
Simpſon, Gabriel - - - 1263
Singleton, Hugh - - - 740
Smith, Toby - - - - 1306
Smyth, Henry - - - - 706
Stafford, Simon - - - - 1299
Stoughton, Robert - - - 750
Sutton, Henry - - - - 843

T.
Tilly, William - - - - 764
Tiſdale, John - - - - 766
Tottel, Richard - - - 806
Toy, Humphrey - - - 933
Turk, John - - - - 771

V.
Vautrollier, Thomas - - 1065
Veale, Abraham - - - 772
Venge, Edward - - - 1222
Venge, Walter - - - 1221

W.
Waldegrave, Robert - - 1139
Walley, John - - - - 729
Walley, Robert - - - 734
Ward, Roger - - - - 1189
Waterſon, Simon - - - 1222
Watkins, Richard - - - 1023
Webſter, Richard - - - 1138
White, Edward - - - 1197
White, William - - - 1263
Wight, John - - - - 779
Wight, Thomas - - - 1303
Williamſon, William - - 1063
Windet, John - - - - 1223
Wolfe, John - - - - 1170
Woodcock, Thomas - - 1106
Wyer, John - - - - 712
Wyer, Richard - - - 711
Wykes, or Weekes, Henry - 937

Y.
Yardeley, Richard - - - 1205
Yetſweirt, Charles - - - 1130
——, his Widow - - - 1131

www.ingramcontent.com/pod-product-compliance
Lightning Source LLC
Chambersburg PA
CBHW021226300426
44111CB00007B/435